International Scientific Conference on Radiation-Thermal Effects and Processes in Inorganic Materials 2015 (RTEP 2015)

IOP Conference Series: Materials Science and Engineering
Volume 110

Tomsk, Russia
31 August - 10 September 2015

ISBN: 978-1-5108-2086-9
ISSN: 1757-8981

Printed from e-media with permission by:

Curran Associates, Inc.
57 Morehouse Lane
Red Hook, NY 12571

Printed by Curran Associates, Inc. (2016)

For permission requests, please contact the Institute of Physics
at the address below.

Institute of Physics
Dirac House, Temple Back
Bristol BS1 6BE UK

Phone: 44 1 17 929 7481
Fax: 44 1 17 920 0979

techtracking@iop.org

Additional copies of this publication are available from:

Curran Associates, Inc.
57 Morehouse Lane
Red Hook, NY 12571 USA
Phone: 845-758-0400
Fax: 845-758-2634
Email: curran@proceedings.com
Web: www.proceedings.com

Table of contents

Volume 110

International Scientific Conference on Radiation-Thermal Effects and Processes in Inorganic Materials 2015 (RTEP2015)
31 August to 10 September 2015, Tomsk, Russia

Accepted papers received: 23 December 2015
Published online: 23 February 2016

Papers

The **International Scientific Conference "Radiation-Thermal Effects and Processes in Inorganic Materials"** is a traditional representative forum devoted to the discussion of fundamental problems of radiation physics and its technical applications. The first nine conferences were held fourfold in Tomsk, Ulan-Ude (Russia), Bishkek (Kyrgyzstan), Tashkent (Uzbekistan), Sharm El Sheikh (Egypt), the island of Cyprus. The XI conference was held in Tomsk, Russia.

The program of the Conference covers a wide range of technical areas and modern aspects of radiation physics, its applications and related matters. Topics of interest include, but are not limited to:

- Physical and chemical phenomena in inorganic materials in radiation, electrical and thermal fields;
- Research methods and equipment modification states and properties of materials;
- Technologies and equipment for their implementation;
- The use of radiation-thermal processes in nanotechnology;
- Adjacent to the main theme of the conference issues

The conference was attended by leading scientists from countries near and far abroad who work in the field of radiation physics of solid state and of radiation material science.

The School-Conference of Young Scientists was also held during the conference.

The event was held with the financial support of the Russian Foundation for Basic Research, projects № 15-02-20616.

IOP Conf. Series: Materials Science and Engineering **110** (2016) 011002 doi:10.1088/1757-899X/110/1/011002

THE ORGANIZING COMMITTEE OF THE CONFERENCE RTEP-2015:

A. Surzhikov, Chairman of Scientific Program Committee, Deputy Director on Research, Professor, Tomsk, Russia
G. Remnev, co-chair of Scientific Program Committee, TPU, Professor, Tomsk, Russia
V. Krivobokov, co-chair of Scientific Program Committee, TPU, Professor, Tomsk, Russia

S. Gyngazov, Academic Secretary of the Conference, TPU, Tomsk, Russia

E. Lysenko, Secretary of the Conference, TPU, Tomsk, Russia

V. Lisitcin, Section Chairman, Professor, TPU, Tomsk, Russia

N. Koval, Section Chairman, Professor, TPU, Tomsk, Russia

Yu. Yakovlev, Section Chairman, Professor, TPU, Tomsk, Russia

Yu. Ivanov, Section Chairman, Professor, TSU, Tomsk, Russia

Yu. Tyurin, Section Chairman, Professor, TPU, Tomsk, Russia

A. Potekaev, Section Chairman, Professor, TSU, Tomsk, Russia

E. Naiden, Section Chairman, Professor, TSU, Tomsk, Russia

B. Aduev, Section Chairman, Professor, KSU, Kemerovo Tomsk, Russia

PROGRAM ORGANIZING COMMITTEE

K. Baktybekov, Professor, ENU, Astana, Kazakhstan
R. Bayazitov, Professor, KPhTI, Kazan, Russia
G. Bondarenko, Professor, MIEM, Moscow, Russia
V. Brudnyi, Professor, TSU, Tomsk, Russia
B. Goshchitskii, Professor, corresponding member of RAS, IPhM, Ekaterinburg, Russia
E. Gorkunov, Professor, academician of RAS, Institute of Engineering, Ekaterinburg, Russia
B. Kalin, Professor, MEPHI, Moscow, Russia
M. Kidibaev, Professor, Corresponding Member of the Kyrgyz Academy of Sciences, IPhTPMS, Bishkek, Kyrgyzstan
N. Kolin, Professor, IAE, Obninsk Russia
V. Kortov, Professor, UFU, Ekaterinburg, Russia
A. Kupchishin, Professor, KNU, Almaty, Kazakhstan
Yu. Gulyaev, Professor, academician of RAS, KIREE, Moscow, Russia
E. Martynovych, Professor, ISU, Irkutsk, Russia
V. Ovchinnikov, Professor, IEP, Ekaterinburg, Russia
N. Ratakhin, Professor, academician of RAS, IHCE, Tomsk, Russia
A. Sigov, Professor, academician of RAS, MSUREA, Moscow, Russia
P. Selishchev, Professor, University of Pretoria, South Africa

Tributes to colleagues

Editorial board of the collection (journal) thanks the organizers of the XI **International Scientific Conference "Radiation-Thermal Effects and Processes in Inorganic Materials"**:

- A. Surzhikov, 1th Section Chairman, Professor, TSU, Tomsk, Russia
- B. Aduev, 2th Section co-Chairman, KSU, Professor, Kemerovo, Russia
- E. Gorkunov, 3th Section Chairman, academician of RAS Institute of Engineering, Ekaterinburg, Russia
- V. Ovchinnikov, 4th Section Chairman, Professor, IEP, Ekaterinburg, Russia
- A. Potekaev, 5th Section Chairman, Professor, TSU, Tomsk, Russia
- Yu. Tyurin, 6th Section Chairman, Professor, TPU, Tomsk, Russia
- S. Gyngazov, Academic Secretary of the Conference, TPU, Tomsk, Russia
- E. Lysenko, Secretary of the Conference, TPU, Tomsk, Russia
- I. Sharkov, Department for development of publication activities, TPU, Tomsk, Russia

IOP Conf. Series: Materials Science and Engineering **110** (2016) 011004 doi:10.1088/1757-899X/110/1/011004

Peer review statement

All papers published in this volume of *IOP Conference Series: Materials Science and Engineering* have been peer reviewed through processes administered by the proceedings Editors. Reviews were conducted by expert referees to the professional and scientific standards expected of a proceedings journal published by IOP Publishing.

Radiation annealing of radiation embrittlement of the reactor pressure vessel steel

E Krasikov[1] and V Nikolaenko[2]

[1] Laboratory Head, National Research Centre «Kurchatov Institute»,
1, Kurchatov Sq., Moscow, 123182, Russia

E-mail: ekrasikov@mail.ru

[2] Group Head, National Research Centre «Kurchatov Institute»,
1, Kurchatov Sq., Moscow, 123182, Russia

E-mail: vadim1927@bk.ru

Abstract. Influence of neutron irradiation on RPV steel degradation are examined with reference to the possible reasons of the substantial experimental data scatter and furthermore – nonstandard (non-monotonous) and oscillatory embrittlement behavior. In our glance this phenomenon may be explained by presence of the wavelike component in the embrittlement kinetics. We suppose that the main factor affecting steel anomalous embrittlement is fast neutron intensity (dose rate or flux), flux effect manifestation depends on state-of-the-art fluence level. At low fluencies radiation degradation has to exceed normative value, then approaches to normative meaning and finally became sub normative. Data on radiation damage change including through the ex-service RPVs taking into account chemical factor, fast neutron fluence and neutron flux were obtained and analyzed. In our opinion controversy in the estimation on neutron flux on radiation degradation impact may be explained by presence of the wavelike component in the embrittlement kinetics. Therefore flux effect manifestation depends on fluence level. At low fluencies radiation degradation has to exceed normative value, then approaches to normative meaning and finally became sub normative. Moreover as a hypothesis we suppose that at some stages of irradiation damaged metal have to be partially restored by irradiation i.e. neutron bombardment. Nascent during irradiation structure undergo occurring once or periodically transformation in a direction both degradation and recovery of the initial properties. According to our hypothesis at some stage(s) of metal structure degradation neutron bombardment became recovering factor. As a result oscillation arise that in tern lead to enhanced data scatter.

1. Introduction

As a main barrier against radioactivity outlet reactor pressure vessel (RPV) is a key and irreplaceable component in terms of safety and extended LWR plant life. For this reason, it is important for RPV materials degradation forecasting to be as accurate as possible.

The surveillance programmes call upon to predict ahead RPV materials characteristics conservatively to guarantee RPV structural integrity without any compromise. But for any length of time results of the RPV steels behavior under neutron irradiation study are characterized by high data scatter [1].

Despite the great amount of experimental results coming from surveillance programmes and research fields, the comprehension of this phenomenon still remain unanswered.

Applicability of dose-damage relations to operating reactors thoroughly analyzed [2]. The overall effect of the all the input uncertainties (in the measurement of TTS, composition and exposure conditions) on the development of the dose-damage relations and their ability to predict data is illustrated in Figure 1.

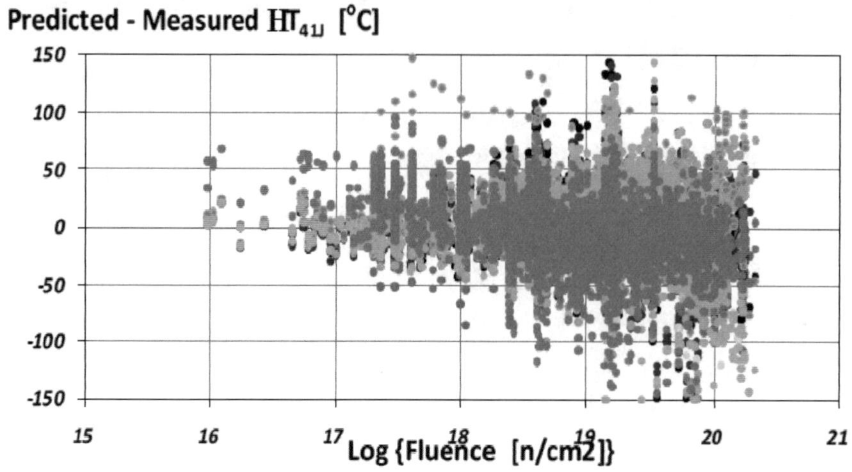

Figure 1. Dependence of the range of the residuals for different dose-damage relations on the fast neutron fluence.

2. Prerequisites to the substantiation of the possible (potential) RPV steel embrittlement data scatter sources

Among unaccounted factors dose rate (neutron flux) is the most enigmatic. For example, it was established that irradiation at low flux enhanced embrittlement of the steel with high or medium copper contamination [3]. Moreover flux effect can lead to the alteration of the standard monotonous dose-damage dependence to the non-monotonous one [4].

2.1. Background.

The changing condition of the RPV steel during service is predicted by dose-damage relationship also known as embrittlement curve. The general dependence of the temperature transition shift (TTS) on radiation embrittlement coefficient A_F is as follows:

$$TTS=A_F \times F^n \quad (1),$$

where A_F is the radiation embrittlement coefficient and F is the fast (E≥0,5MeV) neutron fluence in units of $10^{18}sm^{-2}$(fluence factor), n – coefficient (~1/3).

For Russian RPV Cr-Mo-V base steel $A_F=A \times (P+0,07Cu)$, where A_F – coefficient (800 at 270°C), P and Cu – are the weight concentrations of these elements.

Along with routine investigations in Russia systematic research on actual radiation embrittlement and flux effect manifestation of the decommissioned PWR pressure vessels via through samples (trepans) has been carried out.

The earliest commercial PWR prototype unit Novovoronezh-1 (NV1) RPV after 20 years of operation was trepanned in 1987. Then Novovoronezh-2 (NV2), the oldest PWR type experimental reactor-prototype ERP and, finely, nuclear icebreaker NIB) «Lenin» RPVs also were trepanned. The most interesting and unexpected data were discovered during trepans of the first nuclear ship – icebreaker «Lenin» investigation.

Chemical analyses of the icebreaker RPV material were carried out with FSQ «Baird» optical emission spectrometer (Table 1).

Table 1. Chemical composition of the icebreaker RPV materials under study (wt. %%).

Material	C	Mn	Si	P	Cu	Mo	Ni	Cr	V
Weld	0.05	1.03	0.41	0.035	0.15	0.49	0.17	1.39	0.15
Base	0.17	0.45	0.28	0.018	0.09	0.67	0.35	2.75	0.09

Fast (E≥0,5MeV) neutron fluence evaluation was based on the specific Mn-54 and Nb-93m activities of the vessels steel and on the Nb-93m activity of the RPV cladding.

The RPV materials radiation degradation (embrittlement) was determined by finding the ductile-to-brittle transition temperature shift (TTS). RKP-300 impact pendulum test machine for standard Charpy specimens testing was used.

2.2. Test results

The unexpected results of the icebreaker RPV weld and base metal studying are given in Figure 2, where one can recognize that the actual (measured) radiation embrittlement coefficients of the trepan materials for the periphery (remote) zone of the vessel are significantly higher that for the inner part. One can see also that hardness measurement and Charpy impact testing results agree.

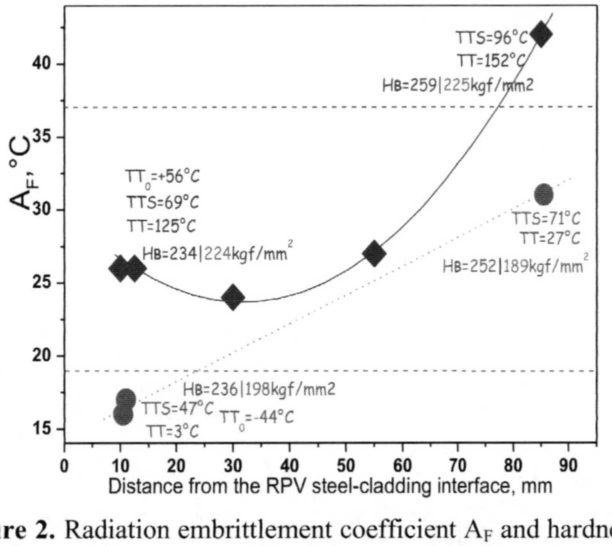

Figure 2. Radiation embrittlement coefficient A_F and hardness value HB through the icebreaker RPV wall distribution. TT – transition temperature, TT_0 - transition temperature in initial state.
Note: H_B=252|189kgf/mm^2 - hardness values at as-received condition|after annealing 650°C/2h.

3. Data analysis and discussion

First impression from foregoing decommissioned PWR pressure vessel material properties study – enhanced degradation rate at low neutron fluxes. Examples for icebreaker base metal and weld metal are demonstrated in Figures 3,4. Registered facts denote that known as «flux effect» factor was in action. Unexpected circumstance however is reduced embrittlement zone appearance that follows after of previous part of enhanced embrittlement. It is seen by means of TTS on neutron flux and fluence dependencies observation.

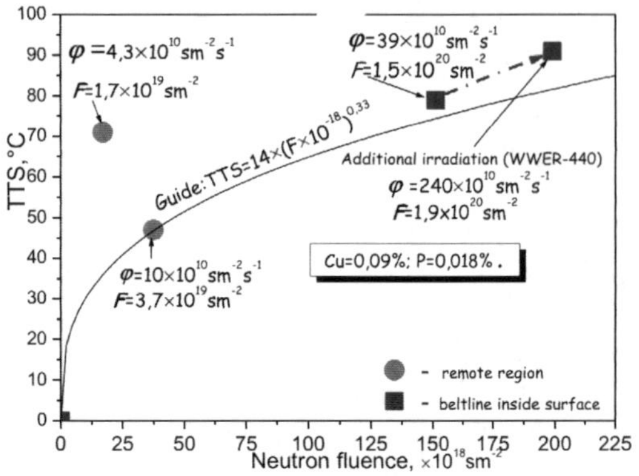

Figure 3. Comparison of the TTSs between «remote» and «inside» icebreaker RPV base metal specimens.

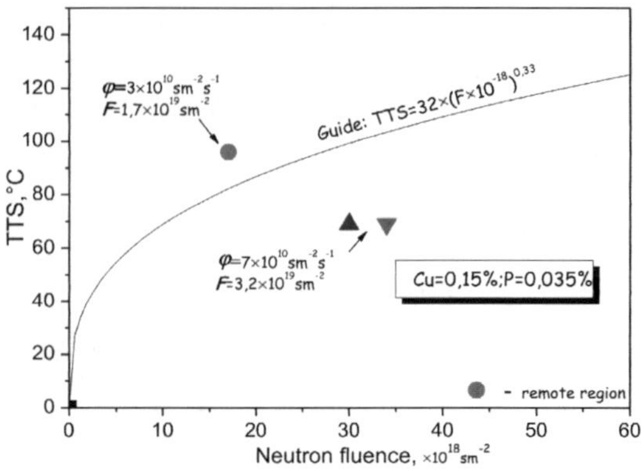

Figure 4. Comparison of the TTSs between «remote» and «inside» icebreaker RPV weld metal specimens.

In support of the unusual data gained, we started «flux effect» purposeful study using surveillance specimens from WWER-440/213 RPV. Taking into account the flux level that is the irradiation position of the selected standard Charpy-type surveillance specimens additional sub size specimens were manufactured and tested. Analysis shows that 3 fold difference in flux level may lead to evident distinction in terms of TTS:~60°C at neutron fluence of $4\times10^{19}cm^{-2}$ and copper content 0,1% mass.[5].

Flux effect (FE) depends on Cu content in the steel: the lower Cu concentration the lower FE However super cleaning of the metal up to less than 0,05 wt.% Cu diminishes FE but does not reduce data scatter [6].

It is seen from Figures 3,4 that alteration of the standard monotonous dose-damage dependence to the non-monotonous one is possible. Search of similar far from trivial effect brings to example where it characterized by authors as «quite atypical» [7]. For the sake of correctness, it is necessary to underline that the first mention concerning distinction between test reactors and low-lead-factor (surveillance) data had appeared as early as 1980 [8].

Careful surveillance programs database analysis shows that damage-dose relationship can contain oscillating component. Figure 5 shows TTS-fluence dependence for RPV materials with A_F=15-24°C:

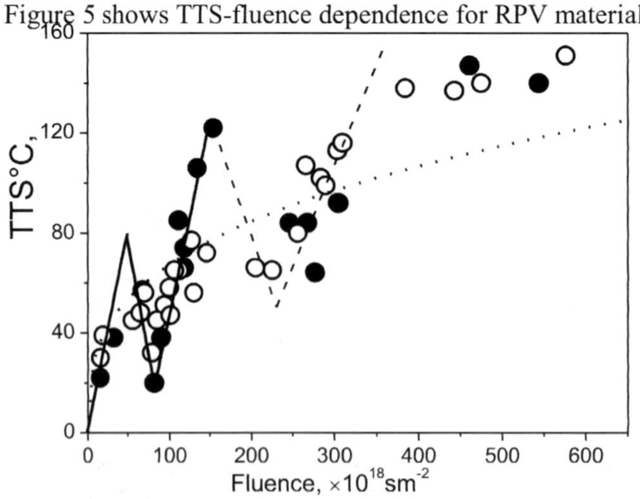

Figure 5. TTS-fluence dependence for RPV steels with A_F = 15-24°C:
• – weld metal (flux $4\times10^{11}sm^{-2}s^{-1}$);
○ – base metal (flux $30\times10^{11}sm^{-2}s^{-1}$);
- - - - Regulatory guide.

With surprising we have seen the serrated-type curve and independence on neutron flux and material origin (base or weld metal). It is necessary to remember that there is an example when contrary to the famous radiation embrittlement in metals neutron irradiation at some range of fast neutron doses improves both the strength and ductility of steel [9].

The very wonderful fact is the effect of the temporary weakening of the embrittlement appearance. Possible comprehensible explanation is as follows: the radiation-induced copper-rich precipitates nature (dimensions and concentration) alteration. Evidently, we have fixed phenomenon when neutron irradiation in some range of low doses improves the mechanical properties of the steel. In this case we have to consider irradiation temporarily as a recovery factor.

Conclusion

Influence of neutron irradiation on RPV steel degradation are examined with reference to the possible reasons of the substantial experimental data scatter and furthermore – nonstandard (non-monotonous) and oscillatory embrittlement behavior. In our glance this phenomenon may be explained by presence of the wavelike component in the embrittlement kinetics. As a result of dose rate effect manifestation peripheral RPV's zones in some range of fluencies have to be damaged to a large extent than situated closely to core. This finding recently was confirmed [10].

As a hypothesis we suppose that at some stages of irradiation damaged metal have to be partially restored by irradiation i.e. neutron bombardment. In this case we have to consider that provisionally irradiation became a recovering factor. As a result oscillation arises that in turn lead to enhanced data scatter.

References

[1] Fabry A 1996 Research to understand the embrittlement behavior of Yankee/BR3 surveillance plate. Effects of Radiation on Materials. 17-th International Symposium. *ASTM STP 1270* 138-87. Ed. by Gelles D and Nanstad R West Conshohocken, USA

[2] Ortner S and English C 2013 Applicability of dose-damage relations to operating reactors. IAEA Technical Meeting "Degradation of Primary Components of Pressurized Water-Cooled Nuclear Reactors. Vienna, Austria 5-8 November

[3] Soneda N 2007 Multiscale Computer Simulations and Predictive Modeling of RPV Embrittlement. Materials Issues for Generation IV Systems. MATGEN-IV, Cargese, Corsica, September 29

[4] Krasikov E and Nikolaenko V 2008 Neutron Flux Influence on the WWER-440/213 Reactor Pressure Vessel Materials Embrittlement. *Atomic energy* **105** 201-5 (in Russian)

[5] Nikolaenko V and Krasikov E 2004 Neutron Flux Influence on the WWER-440/213 Reactor Pressure Vessel Materials Embrittlement. *Atomic energy* **97** 177-82 (in Russian)

[6] English C, Ortner R, Carter A 2002 A review of dose rate effects. EPRI-CRIEPI Workshop on Dose Rate Effects in RPV Materials. *Proceedings*

[7] Gerard R and Fabry A 1996 In-service embrittlement of the pressure vessel welds at the Doel I and II NPPs, in D. Gelles, R. Nanstad (Eds.), Effects of radiation on materials, *ASTM STP 1270* 294-319. Ed. by Gelles D and Nanstad R West Conshohocken, PA, Ann Arbor

[8] Marston T 1980 Radiation embrittlement significance of its effects on integrity and operation of PWR pressure vessels. *Nuclear Safety* **21** 724-30

[9] Murty K 1984 Is neutron radiation exposure always detrimental to metals (steels)? *Nature* **308** 51-2

[10] Viehrig H 2012 Fracture Mechanics Characterization of the Beltline Welding Seam of the Decommissioned WWER-440 RPV of NPP Greifswald Unit 4 *International Journal of Pressure Vessels and Piping* **89** 129-35

Processing line for industrial radiation-thermal synthesis of doped lithium ferrite powders

A P Surzhikov[1], O V Galtseva[1], E A Vasendina[1], V A Vlasov[1], E V Nikolaev[1]

[1]National Research Tomsk Polytechnic University, Tomsk, Russia

E-mail: vlvitan75@mail.ru

Abstract. The paper considers the issues of industrial production of doped lithium ferrite powders by radiation-thermal method. A technological scheme of the processing line is suggested. The radiation-thermal technological scheme enables production of powders with technical characteristics close to the required ones under relatively low temperature annealing conditions without intermediate mixing. The optimal conditions of the radiation-thermal synthesis are achieved isothermally under irradiation by the electron beam with energy of 2.5 MeV in the temperature range of 700–750 ^{0}C within~ 120 min.

Keywords: lithium ferrites, solid-state synthesis, electron beam, radiation-thermal annealing

1. Introduction

Modern technological electron accelerators (for example, ILU and ELV developed at BINP SB RAS, Novosibirsk [5–6]) are sufficiently small to be easily integrated in the production cycle. The design of these accelerators provides their continuous round-the-clock operation in the industrial production. In addition, electron beams provide broad options to control the radiation mode and do not induce radioactivity. They are cost-beneficial compared to other ionizing radiation sources, and provide high efficiency of conversion of the electric energy into the electron beam energy.

The studies which have been carried out in Russia showed that RT effects intensify a number of solid-phase reactions such as synthesis and sintering of some oxide compounds, Portland cement clinker, opening and processing of minerals, etc. [7]. RT proved to be effective for synthesis and sintering of ferrite ceramics [8–10]. In the latter case, the RT effect was characterized by increased speed of compact compression, particularly in the early stages of sintering. RT sintering reduced the processing time tenfold and enabled production of polycrystalline ferrites with a required combination of properties. The radiation was found to intensify the formation of the magnetic parameters for ferrite both in terms of time and maximum achievable values.

This suggests the existence of a number of physical effects which include the phenomenon of multiple acceleration of the diffusion mass transfer in crystalline and ceramic materials under excitation by power beams of charged particles and homogenization of heterogeneous structures in radiation fields. The phenomenon of radiation intensification of sintering of ferrite powder systems under the combined action of high temperatures and intense fluxes of accelerated electrons was unambiguously identified.

This paper addresses the problems to be solved for industrial production of doped lithium ferrite powders via radiation-thermal method and suggests a technological scheme of the production line.

IOP Conf. Series: Materials Science and Engineering **110** (2016) 012002 doi:10.1088/1757-899X/110/1/012002

2. Experimental techniques

Initial technical requirements claimed in the technological scheme design are as followers:

1) The heater for RT-annealing of reaction mixtures is pulsed electron accelerator ILU-6.

2) The homogeneity of the temperature field on the irradiated substrate with samples is not less than 3%.

3) The temperature-time mode of irradiation provides the content of the basic phase in the synthesized product not less than 70%.

The electron accelerator ILU-6 with sufficiently high parameters and relatively small dimensions has been chosen as a heat source. The electron beam is released into the air through the foil, and it can provide the power density of up to 400–500 W/cm^2 to allow heating of the process material up to ~2000 ^{0}C. A protected area with internal dimensions of 3×4×5 m^3 is sufficient to install the accelerator.

The maximum energy of accelerated electrons was E = 2.5 MeV, pulse current was 400 mA, pulse duration was 500 μs, the pulse repetition rate varied from 4.5 to 11.5 Hz, and the rate of compact heating was 130 ^{0}C/min. The durations of the nonisothermal annealing stages (heating and cooling) in the radiation-thermal synthesis were ~ 3 min.

Irradiation was carried out in air in the heat insulating cell [11] of lightweight chamotte with a mass thickness of the horizontal plates of ~ 0.16 g/cm^2. The electron energy loss in the plates of that thickness did not exceed 8% and could be neglected. The annealed samples inside the cell were placed on a thin plate of lightweight chamotte 3 mm thick located on ceramic tubes. To reduce the temperature gradient, the cell was covered with a radiation transparent cover 0.1 g/cm^2 thick. The temperature was monitored by the platinum/platinum-rhodium (90% of platinum, 10% of rhodium) thermocouple with the measuring junction in the test sample located in the immediate vicinity of the annealed compacts. To suppress the interference generated by the injected charge, a platinum earthing electrode was welded to the measuring junction. The platinum/ platinum-rhodium thermocouple was chosen due to its chemical resistance to most substances in the corrosive ionized medium at high temperatures, ease of producing sufficiently pure measuring junction, flexibility and a wide range of operating temperatures [12]. The thermocouple ceramic tube was located in the shadow of the metal radiation shield to improve the accuracy of the measured temperature. Mixture heating and maintenance of the given temperature were carried out without external heat sources due to the energy of decelerated electrons. The relation of the electron extrapolated paths to the sample mass thickness ensured the minimum temperature gradients on the pellet.

3. Result and discussion

The experimental data obtained in [11, 13] was used as the basis for development of the technological scheme of radiation-thermal synthesis of lithium ferrites (Fig. 1).

The working chamber is a container of lightweight chamotte with a compacted reaction mixture inside. The containers are placed on the remotely operated conveyor system "loading-unloading" opposite the accelerator outlet port. The measurements obtained with a system of thermocouple sensors showed that the required temperature uniformity in the range of 600–800 ^{0}C can be reached if the irradiated surface is located at a distance of 16 cm from the accelerator outlet port and the accelerator operates in the nominal mode with the electron beam scanning. In this case, the size of the irradiated area is 6×30 cm^2. In the area of the reaction mixture preparation, mixing and briquetting of the initial reagents is carried out under the pressure of 200 MPa. The compact thickness is to be about two-thirds of the extrapolated electron path for uniform depth heating of the compact.

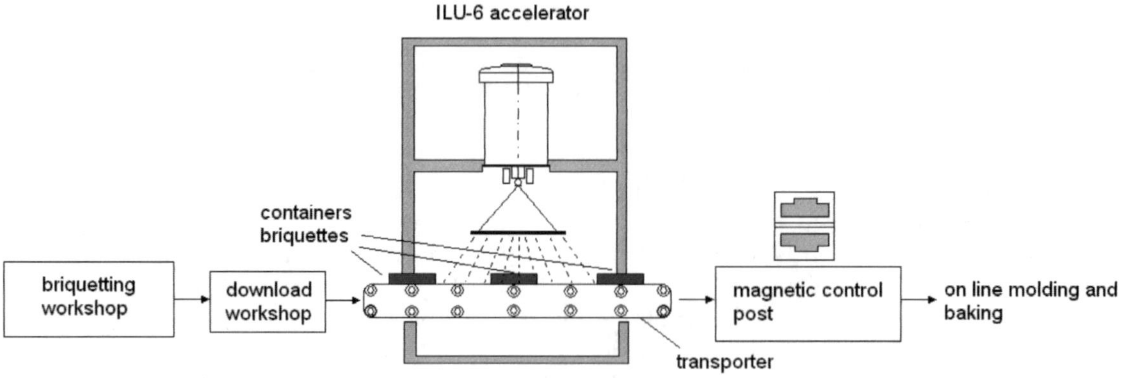

Fig. 1. Technological scheme of radiation-thermal synthesis of ferrite powders

The calculation shows that if the electron energy is 2.4–2.5 MeV, the optimum thickness of the compacted mixture is 3 mm. Under these parameters, a single container loading is ~55 cm^3. The mixture weight is ~170 g for the compact density of ~ 3 g/cm^3.

The proposed scheme was used to perform synthesis of a batch of lithium substituted samples of all compositions in a single container within 120 minutes at isothermal excerpt temperatures of 600 ^{0}C, 700 ^{0}C and 750 ^{0}C [14].

The comparison of sample batches shows that the developed scheme of radiation-thermal technology provides the production of powders with specifications close to the required ones at relatively low temperature conditions of annealing without intermediate mixing.

4. Conclusions

A technological scheme for radiation-thermal synthesis of substituted lithium ferrites is proposed. It includes heating of the compacted reaction mixtures by intense electron beam with energy of 2.5 MeV without external heat sources. The optimum mode of radiation-thermal synthesis is achieved through isothermal excerpt under irradiation in the temperature range of 700–750 ^{0}C within ~ 120 min.

5. Acknowledgements

The research was supported by the Ministry of Education and Science of the Russian Federation in part of the "Science" program.

References

[1] Neronov V A, Voronin A P, Tatarintseva M I et al 1986 *Journal of The Less-Common Metals* V **117** Issue 1-2 391–394
[2] Auslender V L, Lazarev V N and Tsetlin V V 1983 *Soviet physics. Technical physics* V **28** Issue 3 321–322
[3] Boldyrev V V, Voronin A P, Gribkov O S et al 1989 *Solid State Ionics* V **36** Issue 1-2 1–6
[4] Lyakhov N Z, Boldyrev V V, Voronin A P et al 1995 *Journal of Thermal Analysis* V **43** Issue 1 21–31
[5] Auslender V L 1994 *Nuclear Instruments and Methods in Physical Research* No **B89** 46–48
[6] Salimov R A, Cherepkov V G, Golubenko J I et al 2000 *Radiation Physics and Chemistry* No. **57** 661–665
[7] Auslender V L, Korchagin A I, Salimov R A et al 2004 *Special issue of journal "Science for Manufacture"* № **7** 23–27
[8] Surzhikov A P, Pritulov A M, Lysenko E N et al 2012 *Journal of Thermal Analysis and Calorimetry* V **110** Issue 2 733–738

[9] Surzhikov A P, Pritulov A M, Lysenko E N et al 2012 *Journal of Thermal Analysis and Calorimetry* V **109** 63-67

[10] Surzhikov A, Lysenko E, Vlasov V et al *Proceedings - 2012 7th International Forum on Strategic Technology, IFOST 2012* Article number 6357503

[11] Galtseva O V 2009 Solid-phase synthesis of lithium ferrites in the beam of accelerated electrons [in Russian] (Tomsk) PhD theses in Engineering Science p 21

[12] Hudson R P 1980 *Pribory dlja nauchnyh issledovanij* [in Russian] No **7** 4–6

[13] Vasendina E A 2011 Radiation-thermal synthesis of doped lithium ferrite in the beam of accelerated electrons [in Russian] (Tomsk) PhD theses in Engineering Science p 21

[14] Vlasov V A, Vasendina E A, Galtseva O V et al 2013 *Modern problems of science and education* (in Russian) № **5** URL: http://www.science-education.ru/111-10075

IOP Conf. Series: Materials Science and Engineering **110** (2016) 012003 doi:10.1088/1757-899X/110/1/012003

Atomic structure of cascades of atomic displacements in metals and alloys after different types of radiation

V A Ivchenko

Yeltsin Ural Federal University, ul. Mira 19, Yekaterinburg, 620002 Russia
Institute of Electrophysics, Ural Branch, Russian Academy of Sciences, ul. Amundsena 106, Yekaterinburg, 620016 Russia

E-mail: ivchenko2008@mail.ru

Abstract. Using the methods of field ion microscopy, we studied radiation induced defects on an atomically clean surface and within a subsurface volume of platinum initiated by the interaction of neutron (E >0.1MeV) and Ar^+ beams (E = 30 keV). It is shown that the interaction of fast neutrons (E > 0.1 MeV) F = $6.7 \cdot 10^{21}$ m^{-2}, F = $3.5 \cdot 10^{22}$ m^{-2} with matter leads to the formation in the amount of platinum such as radiation damage which occur after ion irradiation by beams of charged Ar^+ ions with E = 30 keV, F = 10^{20} ion/m^2. They are observed at a depth of about 1.5–2 nm irradiated under the surface of Pt by ions Ar^+. Thus, we have carried out modeling of neutron impact with matter when replacing the neutron beam by an ion beam that causes the same radiation damage in the bulk of the material. Experimental results on atomic–spatial investigation of radiative defect formation in surface layers of materials, initiated by neutron bombardment (of Pt, E > 0.1 MeV) and ion implantation (in Cu_3Au: E = 40 keV, F = 10^{20} ion/m^2, j = 10^{-3} A/cm^2), are considered. Quantitative estimates obtained for the size, shape, and volume fraction of cascades of atomic displacements formed under various types of irradiation in the surface layers of the materials. It is showing that the average size of radiation clusters after irradiation of platinum to a fast neutron fluence of $6.7 \cdot 10^{22}$ m^{-2} (E > 0.1 MeV) is about 3.2 nm. The experimentally established average size of a radiation cluster (disordered zone) in the alloy after ion bombardment is 4×4×1.5 nm.

1. Introduction

Investigating interaction mechanisms of accelerated particles with matter and studying the atomic rearrangement and, therefore, formation of crystal lattice defects and changing the phase state of the material are important tasks in radiation physics of solids. Radiation clusters formed under irradiation during the evolution of cascades of atomic displacements are the regions of strong elastic distortions, which affect the motion of dislocations under straining, induce radiation hardening, reduce plasticity, and change the characteristics of elasticity. Information on the characteristics of radiation clusters such as concentration, size, internal structure, and the number of point defects contained in them is essential for quantitative analysis of the effect of irradiation on the structure and physico-mechanical properties of alloys. This study is devoted to experimental investigation of fundamental physical processes in solids, which are initiating by the interaction of flows of charged gas ions (Ar^+) and neutron beams with the substance.

The main goal was analysis of radiation defects on an atomically pure surface and in the bulk of materials. Materials were induced by neutron bombardment of Pt (99.99) with E > 0.1 MeV and ion implantation in Pt and Cu_3Au alloy in the ordered state (E = 30-40 keV) by the methods of field ion microscopy (FIM).

One of the tasks of this work was to establish the adequacy of the effects of different forms of radiation affecting on the same material (Pt) when analyzing radiation damage of the same type. For this purpose, using the methods of FIM, we have studied radiation defects on an atomically clean surface and within a subsurface volume of platinum that are created because of neutron and ion beam bombardment

(E > 0.1 MeV and E = 30 keV, respectively). FIM allows precise studying to be carried out of the changes in the real crystal lattice structure of metals and alloys occurring as a result of irradiation on the atomic scale. At the same time, this method allows one to analyze the structure in the bulk of the sample by means of consecutive removal of surface atoms by the electric field.

2. Experimental techniques

The experimental samples were prepared in the form of needle shaped point emitters with a radius of curvature of 30–50 nm at the tip from wires and fillets by electrochemical polishing. The field emitters certified for ion implantation have an atomically smooth surface of the vertex prepared by field evaporating surface atoms in situ. Radiation regimes of beams of Ar^+ charged particles: with E = 30 keV, F = 10^{20} ion/m^2 for platinum, Cu_3Au irradiated perpendicular to an edge sample axis with E = 40 keV; at the density of ion current of j = 10^{-3} A/cm^2 and pulse duration $\tau \sim 10^{-3}$ s, the radiation dose was F ~ $6 \cdot 10^{20}$ ion/m^2. Point samples implanted after preliminary certification were again placed into a field ion microscope for analyzing the bulk of the alloy. In contrast, after neutron bombardment, specimens were manufactured from previously irradiated samples. Blank samples had the form of platinum wire segments with a diameter of 0.2 mm and a length of 20 mm. They were irradiated in an IBB 2M reactor at T = 310 K for an hour at neutron energies E >0.1 MeV, and fluencies $F = 6.7 \cdot 10^{21}$ m^{-2} and $F = 3.5 \cdot 10^{22}$ m^{-2}. Sharpened specimens with a curvature radius of 30–50 nm near the apex were prepared by electrochemical polishing and were then were placed into the FIM chamber. The structural state of the alloy in the bulk was analyzed using ion images of the surface obtained by a recording device (photo or video camera) upon controllable removal of the atomic layers. This allows one to analyze the structural state of the alloy in the bulk. The field ion microscope was equipped with a microchannel ion electron converter that allows one to increase the brightness of ion image by 10^4 times. As a refrigerant, we used liquid nitrogen (T = 78 K), and as the imaging gas we used spectrally pure neon.

3. Results and Discussion

As a result of studying structural state of pure Pt when irradiating by neutrons up to fluence of $6.7 \cdot 10^{21}$ m^{-2} (E > 0.1 MeV), a great amount of crystal lattice damage was found [1]. Among them, we observed isolated point defects, vacancies, displaced interstitial atoms, and vacancy clusters.

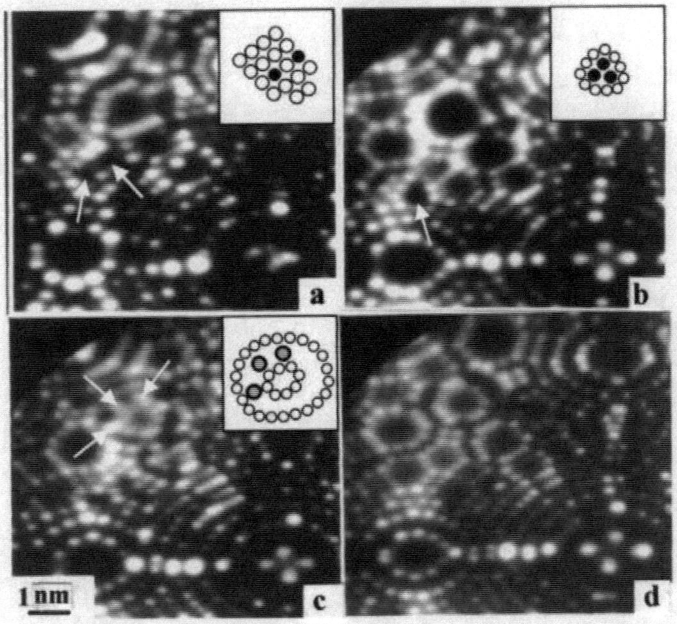

Figure 1. Micrographs of platinum surface areas irradiated by neutrons with $F = 6.7 \cdot 10^{21}$ m^{-2} obtained using Ne as the imaging gas. The arrows show: (a) isolated vacancies, (b) a tetrahedral vacancy cluster, (c) interstitial atoms, and (d) the same surface region without defects. The corresponding schemes of defects are shown in the right upper corner of the figure: ○ – atom, ● - vacancy, and O - interstitial atom.

Figure 1 shows the ion contrast of typical spatial radiation damage distribution in the crystal lattice of Pt upon irradiation by the beam of neutrons. Micro image of the surface of platinum that was subjected to irradiation by fast neutrons is nearly similar to that of the unirradiated surface. However, in some parts of the irradiated platinum surface, violations in the ion contrast and regularity in the annular picture were observed. It is the violation of the ion contrast of the annular picture determines the imperfection of crystal structure. One or another type of defect appearing in material after a certain external irradiation is identified according to the known contrast [2]. In this case, variation in the ion contrast of the irradiated platinum as compared to the initial Pt is caused by radiation damages, which appear as a result of interaction of neutrons with crystal lattice atoms. The defect structure was analyzed in the volume during the process of controlled removal of platinum atoms by the electric field. As a rule, the observed radiation damage was represented by individual point defects (vacancies and interstitial atoms), or small vacancy complexes with the dimensions comparable to the interatomic distances, figure 1.

The depleted zones with the locally higher concentration of vacancies and annular zones of interstitial atoms were found with increasing neutron fluence up to $3.5 \cdot 10^{22}$ m^{-2}, figure 2.

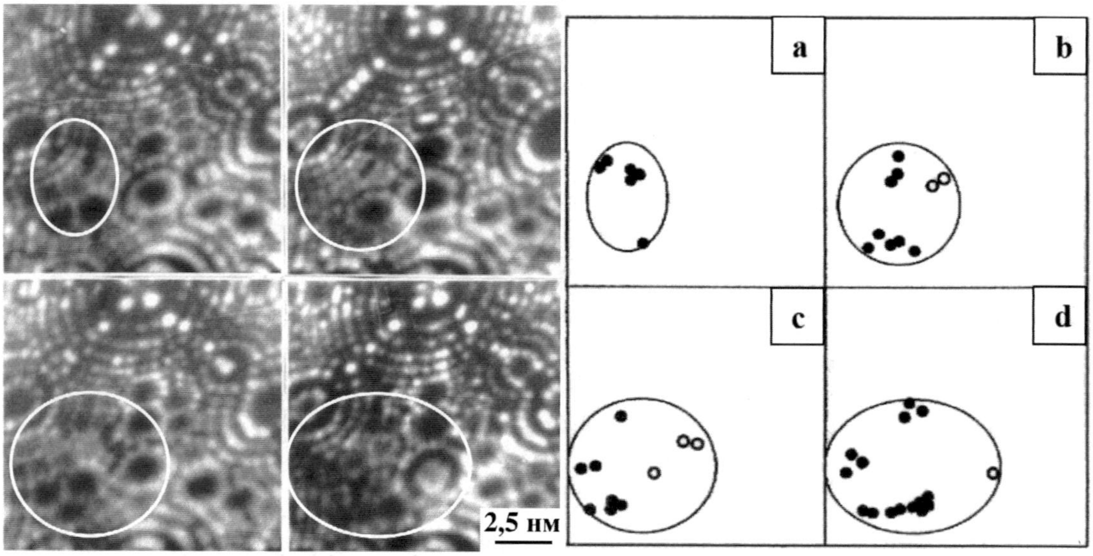

Figure 2. Regions of sequential field ion images of the platinum surface bombarded by neutrons and corresponding scheme of spatial distribution of defects. The depleted zone formed as a result of sequential controllable removal of atomic layers (a. l.): (a) initial distribution; (b) differs from (a) by one a. l.; (c) differs from (b) by two a. l.; (d) differs from (c) by two a. l.: • - vacancy; ○ - interstitial atom.

This observation supports the hypothesis of [3], according to which the development of the cascade in metal occurs in such way that a great number of atoms are removed from the central part of cascade (the most disturbed area) by means of substitution chains. According to our estimates, the average concentrations of vacancies and interstitial atoms in the depleted areas proved to be equal to 9 and 1.5%, respectively. An attempt to clarify spatial arrangement of the depleted zones in platinum when irradiated by neutrons of high and intermediate energies ($E > 0.1$ MeV) with a fluence of $3.5 \cdot 10^{22}$ m^{-2} was made in [1]. The shape of these depleted zones was analyzed in the standard regime by controlled electric field evaporation of atomic layers to determine the characteristic anisotropy. As a result of analyzing the contrast in defect areas, we did not find anisotropy of shape in the depleted areas. As follows from the data, the configuration of the zones does not correspond to any simple geometrical figure, since the vacancies are arranged extremely irregularly.

Studying a great number of micro images of the irradiated platinum allowed us to measure longitudinal and transverse dimensions of separate depleted areas. The estimated average radiative cluster diameter was equal to 3.2 nm, figure 3. The preliminary ion images of the field emitters obtained before the ion irradiation showed an almost perfect annular contrast for single crystals of pure metal, indicating the absence of structural defects [4]. Figure 4 shows a neon image of an atomically clean platinum surface irradiated by Ar^+ ions with $E = 30$ keV and $F = 10^{20}$ ion/m^2.

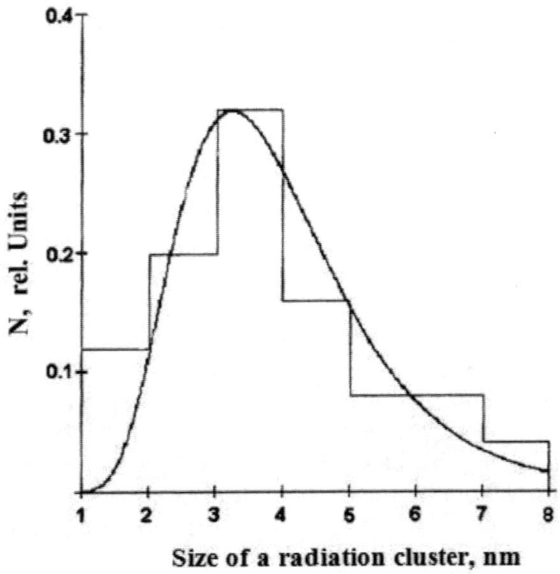

Figure 3. Size distribution of radiation clusters in platinum bombarded by neutrons (F=6.7·10^{22} m^{-2}).

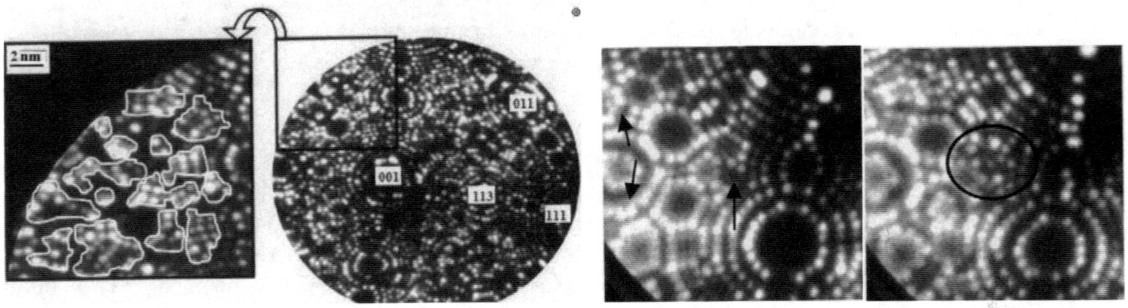

a b c d

Figure 4. Neon micrographs of the Pt surface region after irradiation by Ar^+ ions with $F = 10^{20}$ ions/m^2 (T = 343 K) (a-b). Nano blocks are indicated (a). (c) - at a depth of 2 nm from the irradiated surface (arrows indicate isolated vacancies and interstitial atoms); (d) - at a depth of 1.5 nm from the surface (the depleted zone is shown).

Violations in the annular contrast on the ion micrographs of the crystal faces were found. Precisely these violations in the annular ion contrast point to the imperfectness of an ideal crystal structure and determine the contrast from various defects arising in the material after external irradiation. In this case, the changes in the ion contrast of irradiated platinum as compared to the contrast of the initially certified Pt are observed within a surface layer of 1.5 nm thickness and indicate the presence of a block like structure in the subsurface layer of the material [4].

As we can see, the ion contrast of radiation damage after the ion irradiation (figure 4 a, b) differs from the contrast after the neutron interaction with material (figure 1). Upon further study of the surface of the ion irradiated Pt by the consecutive field evaporation of atomic layers, radiation damage was found at a depth of 2 nm, (figure 4 c) and was identical to that shown in figure 1. The ion contrast of such radiation damage is shown in figure 1 (F= 6.7·10^{21} m^{-2}) leads to the formation of the same amount of radiation damages which is observed at a depth of 2 nm after the Ar^+ irradiation (E = 30 keV, F = 10^{20} ions/m^2). The radiation damage (depleted zone) at a depth of 1.5 nm, (figure 4 d) was identical to that shown in figure 2. From this, we can conclude that all the types of radiation defects arising in the bulk of the material after neutron irradiation depend only on the energy and fluence of neutrons. In contrast to this, upon the interaction of charged ion beams with matter, the type of radiation damage, apart from the parameters of radiation, depends on the distance from the irradiated surface. At a different depth of the material, one can observe transition from one type of the radiation damage to the other. Therefore, one can correctly simulate the impact of fast neutrons when replacing the neutron beam by a beam of ions only at a certain depth of a subsurface layer of material irradiated by the charged particles, given the corresponding fluence. For studying the atomic structure of defects formed in the region of single displacement cascades, atomically ordered alloy Cu_3Au was irradiated perpendicularly to the needle axis with E = 40 keV; for ion current density $j = 10^{-3}$ A/cm^2 and pulse duration $\tau = 10^{-3}$ s, the radiation dose was F~ $6\cdot10^{20}$ ion/m^2. This ensured (on the average) the incidence of one ion on an area element 4×4 nm in size. Analysis of resultant ion microphotographs of the surface revealed radiation defects such as disordered zones and segregations of copper atoms [5]. The average size of disordered zones determined from analysis of four samples was 4×4 ×1.5 nm. Disordered zones were identified as violations of the annular structure of the image of the surface of the atomically ordered alloy, whose contrast is analogous to the ion contrast of the surface of the pure metal. The ion contrast of the surface of Cu_3Au alloy in the ordered state formed only by gold atoms. The contrast of radiation defects was preserved upon field evaporation of surface atoms to the depth of distorted zones. The experimentally established average size of the distorted zones coincided in order of magnitude with the calculated diameter of a displacement cascade (5–11 nm), which was estimated as the mean free path of the primary knocked-on atom and in the approximation of the spherical shape of the cascade region. In the course of investigation, we succeeded in detecting only the segregations of copper atoms. Since copper atoms do not form a visible image on ion microphotographs of the surface, the contrast from a copper segregation is observed as a dark region. Aggregates or segregations of copper atoms are three dimensional and contain 200–500 atoms as a rule. The presence of vacancies in these aggregations can not be determined experimentally, although analysis of the ion contrast of the boundaries of the dark regions shows that their possible number is insignificant.

4. Conclusion

We have obtained the results of direct analysis of radiation induced defect formation in the subsurface bulk of the Pt, Cu_3Au by the FIM method. The sizes, shapes, and volume fractions of clusters formed under various types of irradiation in the surface layers of the materials have been estimated. On atomic scale, we have analyzed various types of defects formed in ordered solid solutions as a result of implantation of gas ions as well as upon the interaction of metals with the neutron beam and appearing as a result of evolution of single cascades of atomic displacements and radiation stimulated diffusion and segregation processes. It is shown that the average size of radiation clusters after irradiation of platinum to a fast neutron fluence of $6.7\cdot10^{22}$ m^{-2} (E > 0.1 MeV) is about 3.2 nm. The experimentally established average size of a radiation cluster (disordered zone) in the alloy after ion bombardment is 4×4×1.5 nm. Using the methods of field ion microscopy, we studied radiation induced defects on an atomically clean surface and within a subsurface volume of platinum initiated by the interaction of neutron (E >0.1MeV) and Ar^+ beams (E = 30 keV). It is shown that the interaction of fast neutrons (E > 0.1 MeV) F = $6.7\cdot10^{21}$ m^{-2}, $F = 3.5\cdot10^{22}$ m^{-2} with matter leads to the formation in the volume of platinum such as radiation damage which arises after ion irradiation by beams of charged Ar^+ ions with E = 30 keV, F =$6.7\cdot10^{20}$ m^{-2}. They are observed at a depth of about 1.5–2 nm the irradiated surface of Pt by ions Ar^+.

Acknowledgments
The research was carried out within the state assignment of FASO of Russia (No. 0389-2014-0002), supported in part by RFBR (project No. 15-08-06744-A).

References

[1] Ivchenko V A, Medvedeva E V and Ovchinnikov V V 2009 *Poverkhnost'. Rentgen., Sinkhrotr. Neitron. Issled.* **8** 26
[2] Boowkett K M and Smith D A 1970 *Field Ion Microscopy Defects in Crystalline Solids* (Amsterdam, London: North Holland Publishing Company)
[3] Seeger A K 1958 *Proceeding of The Second Intern. Conf. on Peaceful Uses of Atomic Energy* (Geneva: United Nations)
[4] Ivchenko V A and Medvedeva E V 2009 *Izv. Vysch. Uchebn. Zaved. Fiz.* **8** 410
[5] Bunkin A Yu Ivchenko V A, Kuznetsova L Yu, et al 1990 *Fiz. Met. Metalloved.* **7** 111

Studying of nanocomposite films' structure and properties obtained by magnetron sputtering

Y O Tleukenov[1], **S V Plotnikov**[1], **N K Erdybaeva**[1], **A D Pogrebnyak**[2]

[1] *Ust-Kamenogorsk, 070003, Kazakhstan,* D. Serikbayev East Kazakhstan State Technical University, 23/2-54 Serikbayev Street
[2] *Sumy, 40000, Ukraine,* Sumy State University, Rimsky-Korsakov Street, 2

E-mail: erick_090@mail.ru

Abstract.

Various approaches to creating multicomponent na-nocomposite coatings of high and superhigh hardness (from 30 to 100 ± 120 GPa) are reviewed with particular emphasis placed on mechanisms underlying the increase in hardness in thin coatings. Nanocomposite Nb-Al-N films fabricated by magnetron sputtering were researched in this work. Two stable crystalline structural states were found in the films: NbN_{ch} and solid solution B1-$Nb_xAl_{1-x}N_yO_{1-y}$, and also an amorphous component associated with aluminum oxynitride with reactive magnetron sputtering. A relationship of substructural characteristics sensitivity with the current and nanohardness and Knoop hardness characteristic was determined in this paper. Recent changes in the range of 29-33.5 GPa and 46-48 GPa, respectively. Initial principle calculations of NbN and Nb_2AlN phases and NbN/AlN heterostructures were carried out for the interpretation of the results. Deposited nanocomposite films with the given mechanical properties may be used as wear resistant or protective coatings. On the basis of these results, it can be assumed that two stable crystalline structural states were found in the films: B1-NbN_x and solid solution with a composition close to the B1-$Nb_{0.67}Al_{0.33}N$. The films also contain an amorphous component associated with aluminum nitride.

1. Introduction

Creation of new materials, including multicomponent nano-composite coatings with a characteristic grain size below 100 nm, is a priority area of modern science and technology. Nanotechnologies are expected to become the principal driving force in scientific and technical developments by the mid-21st century. Nanothechnology is a field focusing on the control of matter on the atomic, molecular, and supramole-cular scales (in a range of sizes from 1 to 50 nm) for the creation, processing, and application of materials, devices, and systems having novel properties and functional capabil-ities due to the small size of their structural elements. Nanomaterials may be classified by the size and geometric shape of their structural elements. The main types of nanomaterials in terms of grain size are cluster and fiber materials, films and coatings, multilayer (multi-component) structures, and bulk nanocrystalline materials whose grains have nanometer dimensions in all three directions. Nanocomposite coatings exemplify materials of a new generation, composed of at least two phases with a nanocrys-talline and/or amorphous structure. Because of the very small size (4 10 nm) of their grains and more important role of boundary zones surrounding single grains, nanocomposite materials behave unlike traditional materials with a grain size over 100 nm and display quite different properties. The novel unique physical and functional properties of nanocomposites constitute a driving force for the rapid development of nanocomposite materials [20 ± 30]. Films with the hardness H 4 40 GPa and H > 40 GPa are currently described as hard and superhard, respectively.

Films based on NbN and (Ti-Zr-Hf-V)N show many interesting properties such as high hardness and electrical conductivity, thermal stability and chemical inertness [1]. NbN and (Ti-Zr-Hf-V)N films are used as the cathode material for field electron emission in vacuum of microelectronic devices [2]. It was shown that the introduction of Al atoms into the lattice led to the formation of solid solution $Nb_{1-x}Al_xN$. For solid solutions $Nb_{1-x}Al_xN$, B1 (type - NaCl) structure is more preferable for x below 0.45. In the range of x = 0.45-0.71, the mixture of structures B1 and B4 was observed, while when x> 0.71, B4 structure (type - wurtzite with a hexagonal structure) was formed [6,7]. Films Nb-Al-N composed of solid solution $Nb_{1-x}Al_xN$ with B1 (rare B_K) and B4 structures or mixtures thereof [3-7]. On the other hand, hitherto, nanocomposite NbN/AlN films were not investigated. Thus, in the present study, we set out to research the Nb-Al-N film and multi-element (Ti-Zr-Hf-V)N coatings, arguing that under certain conditions can be formed nanocomposite structure of the films, which may have improved mechanical properties.

2. Experimental details

Nb-Al-N and (Ti-Zr-Hf-V)N coatings deposited on the mirror-polished wafer Si (100) using a DC magnetron sputtering Nb (99.9%, Ø72 x 4 mm) and Al (99.999%, Ø72 x 4 mm) under argon and nitrogen with the following parameters of deposition: the substrate temperature $T_S = 350°C$; substrate bias voltage $U_B = -50$ V; The flow rate (F) $F_{Ar} = 40$ sccm; $F_{N2} = 13$ sccm; operating pressure $P_C = 0.17$ Pa. The current supplied to the target Al (I_{Al}) was 100, 150, 200, 250 and 300 mA, which corresponds to the discharge power density P_{Al}= 5.7, 8.6, 11.4, 13.7 and 17.1 W/cm^2, respectively. Current supplied to the target Nb (I_{Nb}) was 300 mA ($P_{Nb} = 17.1$ W/cm^2). The base pressure in the vacuum chamber was better than 10^{-4} Pa. The distance between the target and the substrate holder was 8 cm. The dihedral angle between the targets was ~ 45^0. Substrates were ultrasonically cleaned before they were placed in a vacuum chamber. Furthermore, prior to deposition, the substrates were etched in a vacuum chamber in the hydrogen plasma for 5 min.

3. Results and discussion

Structural and mechanical properties were analyzed depending on I_{Al}.values. The structure of the coatings was researched by X-ray diffraction (XRD, diffractometer DRON-3M) in CuK_α radiation. The program of authoring profiles sharing was used with the application of complex diffraction profiles. Substructural characteristics (crystallite and microstrain size) were determined by approximation using the approximation function as - Cauchy function. The spectrum of Fourier spectroscopy (FTIR) was measured at room temperature in the range of 400–4000 cm^{-1} by spectrometer "TSM 1202" LTD «Infraspek». Knoop hardness (HK) was assessed using microhardnessmeter Microhardness Tester Micromet 2103 BUEHLER LTD under a load of 100 mN, and by nanoindentation using nanohardnessmeter G-200, equipped with a Berkovich indenter. The loads were chosen in such conditions that the indenter penetration wouldn't exceed 10-20% of the film hickness. The film thickness was determined by optical profilometer "Micron-gamma". The thickness values of the Nb-Al-N coatings depends weakly on I_{Al}. D and were in the range of 0.7-0.9 μm, for (Ti-Zr-Hf-V)N the thickness values were in the range of 30-90 μm.

IOP Conf. Series: Materials Science and Engineering **110** (2016) 012004 doi:10.1088/1757-899X/110/1/012004

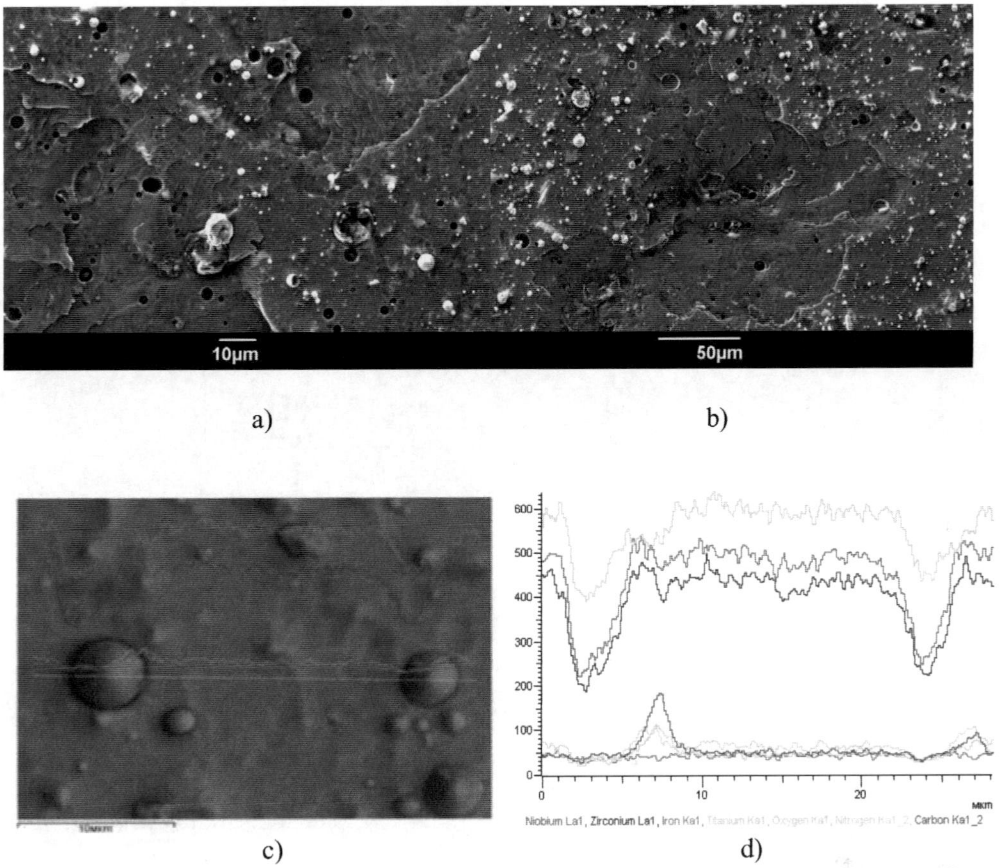

Figure 1. SEM images showing the surface (a-c) of (Ti-Zr-Hf-V)N coatings and elements contribution deposited by magnetron sputtering.

Figure 1 shows as an example the composition morphology of (Ti-Zr-Hf-V)N coatings deposited by reactive magnetron sputtering under applied steel 45 substrate. Where, the coatings deposited under magnetron sputtering show low concentrations of oxygen which originated from the water absorbed during exposure of the coatings to air after their deposition. These examples confirm that care must be taken when depositing hard and superhard coatings for application on tools. Elements concentration is shown in the table 1.

Table 1. Elemental analysis, in%.

Spectrum	O	Ti	V	Zr	Hf	Au	Hg	Total
Total spectrum	8.86	19.03	6.05	12.54	34.10	12.16	7.27	100.00
Average	8.86	19.03	6.05	12.54	34.10	12.16	7.27	100.00
Std. deviation	0.00	0.00	0.00	0.00	0.00	0.00	0.00	
Max	8.86	19.03	6.05	12.54	34.10	12.16	7.27	
Min	8.86	19.03	6.05	12.54	34.10	12.16	7.27	

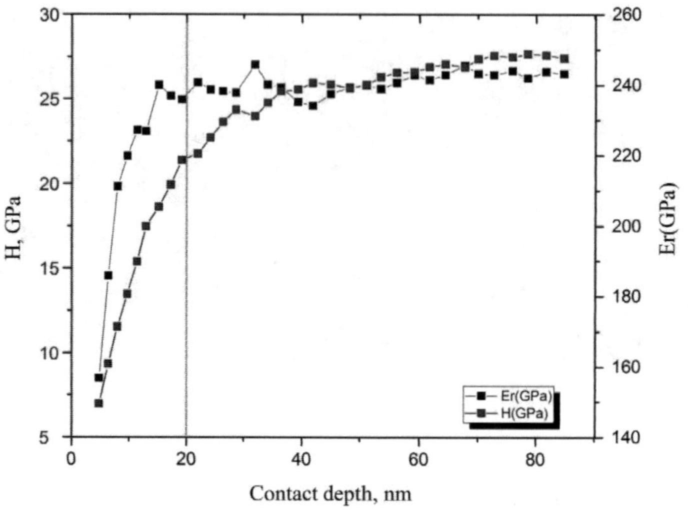

Figure 2. The results of the mechanical studies of (Ti-Zr-Hf-V)N coatings.

Figure 2 shows the results of dynamic indentation pyramid Berkovich (triboindentor). We can estimate the depth of the indentation to determine the value of the reduced modulus of elasticity. The measurements were made when the load on the indenter 500 10 000 μN (Ti-Zr-Hf-V)N coatings. The corresponding dependence of nanohardness and reduced modulus of elasticity Er on the penetration depth for the mode was also shown. Nanohardness increased to almost 27 GPa where the layer thickness was 30 nm with gradual access to the horizontal line to the penetration depth of 85 nm.

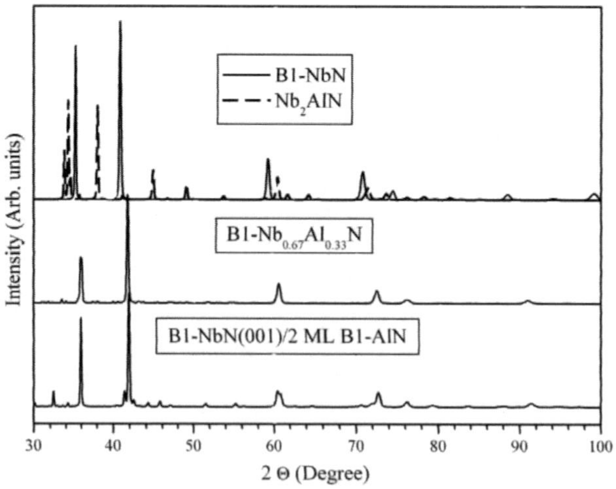

Figure 6. The calculated X-ray diffraction patterns.

Using atomic configurations resulting from first-principles calculations, we calculated diffraction patterns for the B1−NbN$_ч$, B1−$Nb_xAl_{1-x}N$, x ~ 0.67 и Nb_2AlN. The calculated XRD spectra are shown in Figure 4. Comparison of the calculated and experimental spectra (Figure 6) shows that reflexes about 2Θ ~ 32^0 associated with the heterostructure, and about 2Θ ~ 38^0, resulted by phase Nb_2AlN and does not occur at the experimental spectra. Therefore, we can assume that our films contain neither

Nb_2AlN, nor any epitaxial layers B1-AlN, but rather consist of crystallites $B1-NbN_{ch}$ and $B1-Nb_xAl_{1-x}N$, $x \sim 0.67$. As the matter of fact, the difference between the peak positions $\Delta 2\Theta = 2\Theta \cdot (B1-NbN_{q})-2\Theta \cdot (B1-Nb_xAl_{1-x}N)$ for each diffraction peak (200) and (400) on experimental and theoretical diffraction patterns is almost identical. We should also note that films of niobium nitride are prone to accumulate small amounts of oxygen [5]. Oxygen can replace part of the nitrogen in solid solution and in the amorphous matrix [8-17]. Therefore, structure $Nb_xAl_{1-x}N_yO_{1-y}$, $x \sim 0.67$, $1-y << 1$will be more realistic for the solid solutions, as for the amorphous matrix - a-AlNO, which was shown as a result of elemental analysis, obtained by SIMS, RBS and EDS in these films.

4. Conclusion

Films Nb-Al-N were deposited on silicon substrates by magnetron sputtering targets of Nb and Al at different discharge powers at the target of aluminum. Experimental and theoretical studies show that the films obtained at selected deposition parameters have the nanocomposite structure that represents the nanocrystals $B1-NbN_x$ and $B1-Nb_xAl_{1-x}N_yO_{1-y}$, embedded in a-AlNO matrix ($nc-B1-NbN_x/nc-B1-Nb_xAl_{1-x}N_eO_{1-y}/a-AlNO$). Nanocomposite coating with high microstrain action due to the difference in atomic radius of the crystal lattices metal components shows high hardness values (up to 32 GPa). Deposited nanocomposite films may be recommended with the given mechanical properties as wear resistant or protective coatings.

References

[1] Barnett S A, Madan A, Kom I, Martin K 2003 *Mrs Bulletin*. 28, pp. 169.
[2] Gotoh Y, Nagao M, Ura T, Tsuji H, Ishikawa J 1999 *Nucl. Instr. Methods Phys. Res.* Vol. **148** pp. 925.
[3] Selinder T I, Miller D J, Gray K E 1995 *Vacuu* vol. **46** pp. 1401.
[4] Makino Y, Saito K, Murakami Y, Asami K 2007 *Solid State phenomena* vol. **127** pp. 195.
[5] Barshilla H C, Deepthi B, Rajam K S. Mater J 2008 *Res.* Vol. **23** pp. 1258.
[6] Franz R, Lechthaler M, Polzer C, Mitterer C 2010 *Surf. Coat. Technol.* Vol. **204**, 2447.
[7] Holec D, Franz R, Mayrhofer P H, Mitterer C 2010 *J. Phys.* Vol. **41**, 145403.
[8] X-ray powder diffraction file [038-1155].
[9] Umanski Y S, Skakov Y A 1978 Physica metallov. Atomnoe stroenie metallov i splavov *Atomizdat M.* pp. 352.
[10] Jadannadham K, Sharma A K, Wei Q, Kalyanraman R, Narayan 1998 *Vac. Sci. Technol* vol. **16**, 2804.
[11] Ivashchenko V I, Veprek S, Scrynskyy P L, Lytvyn O, Butenko O O, Sinelnichenko O K, Gorb L, Hill F, Leszczynski J, Kozak A O 2014 *Superhard Materials* vol. **36**, pp. 1.
[12] Ivashchenko V I, Veprek S, Turchi P E A, Shevchenko V I 2012 *Phys. Rev.* Vol. **85,** 195403-1.
[13] Giannozzi P, Baroni S, Bonini N, Calandra M, Car R, Cavazzoni C, Ceresoli D, Chiarotti G L, CococcioniI M. Dabo, Dal Corso A, de Gironcoli S, Fabris S, Fratesi G, Gebauer R, Gerstmann U, Gougoussis C, Kokalj A, Lazzeri M, Martin-Samos L, Marzari N, Mauri F, Mazzarello R, Paolini S, Pasquarello A, Paulatto L, Sbraccia C, Scandolo S, Sclauzero G, Seitsonen A P, Smogunov A, Umari P, Wentzcovitch R M 2009 *Phys.: Cond. Matter* vol. **21,** 395502-19.
[14] Perdew J P, Burke K, Ernzerhof M 1996 *Phys. Rev. Lett.* **77,** 3865.
[15] Kraus W, Nolze G 2000 PowderCell for Windows, version 2.4.
[16] Ivashchenko V, Veprek S, Pogrebnjak A,Postolnyi B 2014 *Sci. Tech. Adv. Mater.* Vol **15,** 025007-11.
[17] Pogrebnjak A D, Bagdasaryan A A, Yakushchenko I V, Beresnev V M 2014 *Rus. Chem. Rev.* Vol **83**, 1027.

IOP Conf. Series: Materials Science and Engineering **110** (2016) 012005 doi:10.1088/1757-899X/110/1/012005

Conversion of the electrostatic energy analyzer of charged particles into double spectrograph regime in energy and angle

A O Saulebekov[1], Zh T Kambarova[2] and S N Assylbekova[3]

[1] Lomonosov Moscow State University, Kazakhstan branch, Kazhimukan str.11, Astana 010010, Kazakhstan
[2] Ye.A. Buketov Karaganda State University, Universitetskaya str. 28, Karaganda 100028, Kazakhstan
[3] Nazarbayev Intellectual School of Physics and Mathematics in Astana, 31th str. 37, Astana 010000, Kazakhstan

E-mail: saulebekov@mail.ru

Abstract. The possibility to convert the combined electrostatic energy analyzer of charged particles into double spectrograph regime in energy and angle has been studied. The analyzer consists of the hyperbolic and cylindrical mirrors. The correlation, expressing condition of the first-order focal line straightening of the particles beams of various energies, has been obtained. The possibility of simultaneous registration of the finite energy interval of charged particle distribution by means of a position-sensitive detector has been substantiated.

1. Introduction

In modern research rapidity of analysis and its informative value play an important role. Spectrograph regime allows to the conduct simultaneously analysis of charged particle beams in a certain energy interval, and angular analysis allows to get information about the electronic structure of the surface region of the solid. In space research for analysis of the energy spectrum of particles electrostatic analyzers are used for a long time. Using the spectrograph regime is also actually.

The conversion of electrostatic energy analyzer of charged particles into the spectrograph regime involves the solving of the problem on focal line straightening. The analyzer focal line is a focus of foci of beams of various energies. Usually its shape is complicated and differs from straight. Position-sensitive detectors have a simple configuration as a plane strip, part of a cylindrical surface or a right circular cone. Consequently it is necessary to solve of the problem on the focal line straightening and to identify the optimal regime of its alignment with the detector surface. Only on the basis of these data it is possible to predict the effectiveness of the energy analyzer conversion into spectrograph regime.

As a criterion of focal line straightening the equality of the slope angle of the tangent to the focal line to zero was accepted, which leads to the condition $d\Delta/d\varepsilon = 0$, where $\varepsilon = (W - W_0)/W_0$ is the kinetic energy spread in a beam. This condition should be considered together with condition of the angular focusing of charged particles beam.

In work [1] criteria of the second-order focal line straightening on divergence angle in the symmetry plane for systems, composed of electrostatic mirrors with two-dimensional fields having a

IOP Conf. Series: Materials Science and Engineering **110** (2016) 012005 doi:10.1088/1757-899X/110/1/012005

symmetry plane, was derived. The system composed of successively arranged two coaxial cylindrical mirror analyzers, allowing focal line straightening in terms of maintaining second-order angular focusing in a wide energies range, has been calculated. The results of calculations of the characteristics of the energy analyzer composed of two successively arranged electrostatic hyperbolic mirror analyzers in regime of the focal line straightening and the second-order angular focusing were presented. It was shown that in these two cascade systems the high quality of focal line straightening and focusing of beams of various energies along it is carried out.

In work [2] the possibility to convert the electrostatic energy analyzer composed of a spherical mirror (SM) and a cylindrical mirror (CM) to double spectrograph regime (in energy and azimuthal angle) has been theoretically substantiated. The second-order focusing of charged particles beams of various energies along the straightened focal surface is conserved. This surface shaped as a belt of a right circular cone. The concrete scheme of energy-angle resolved spectrograph has been calculated. The advantage of the proposed scheme of spectrograph is the property to analyze charged particles beams emitted by the source at angles close to a right angle (90^0).

Earlier, in work [3] we proposed scheme of combined energy analyzer composed of successively arranged hyperbolic mirror (HM) and CM. The scheme for an energy- and angle-resolved spectrometer which provides angular second-order focusing, high luminosity and exit angle 90^0 from the source has been theoretically substantiated. This focusing angle allows building an effective diagram of angular measurements. The proposed electron-optical scheme can be oriented in the double spectrograph regime in energy and angle.

The aim of this work is study the possibility of conversion of HM and CM system into spectrograph regime.

2. Solving of the problem on focal line straightening in combined system composed of hyperbolic and cylindrical mirrors

We shall use the formalism proposed in work [2] for study the possibility to convert an electrostatic energy analyzer into the double spectrograph regime in energy and angle. It is necessary to maintaining angular focusing along a straightened focal line.

The scheme of energy analyzer is shown in Fig.1 [3]. The energy analyzer consists of successively arranged HM and CM. A charged particle fan-shaped beam is injected into the system, the beam mid-plane being perpendicular to the symmetry axis z ($\alpha \leq 90^0$). A point-like source is placed at the z axis.

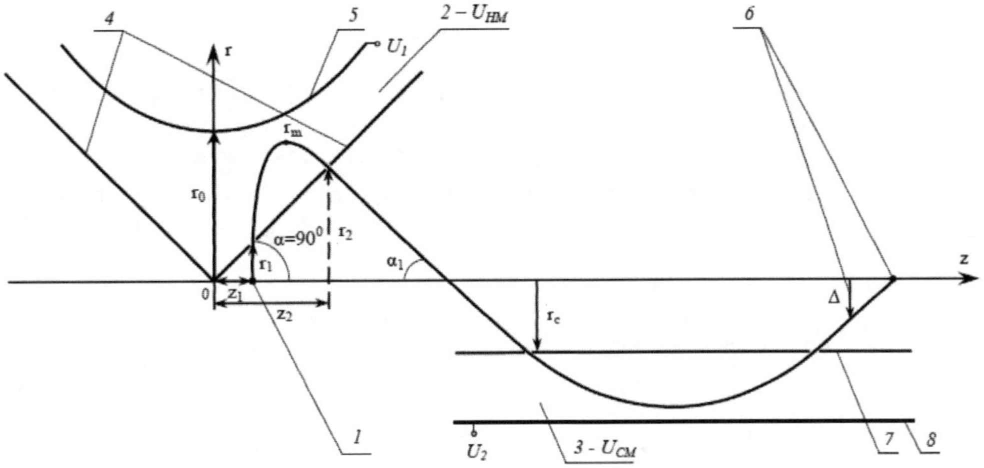

Figure 1. The energy analyzer scheme. 1 – the source of a charged particle beam, 2 – HM, 3 – CM, 4 – the transparent conical electrodes, 5 – the hyperbolic electrode, 6 – the source image, 7 and 8 – the CM block electrodes.

The slope angle of a tangent to the focal line is determined by the relation

$$tg\gamma = \left.\frac{d\Delta}{d\varepsilon}\right|_{\varepsilon=0} \Big/ \left.\frac{d\ell}{d\varepsilon}\right|_{\varepsilon=0} \tag{1}$$

Focal line straightening requirement implies that $\frac{d}{d\varepsilon}(tg\gamma)=0$, it results in the following equation:

$$\left.\frac{d^2\Delta}{d\varepsilon^2}\right|_{\varepsilon=0} - tg\gamma \cdot \left.\frac{d^2\ell}{d\varepsilon^2}\right|_{\varepsilon=0} = 0 \tag{2}$$

The derivatives $d\Delta/d\varepsilon$ and $d^2\Delta/d\varepsilon^2$ are defined by the relations derived as a result of differentiation of the angular first-order focusing condition $dl/d\alpha|_{\alpha=\alpha_0}=0$ with respect to ε, which involves $\Delta(\varepsilon)$ as a parameter. "Strengthening" of focal line straightening criterion (2) carried out by the imposition of extra requirement. This requirement is the angular second-order focusing condition along the straightened focal line:

$$\left.\frac{dl}{d\alpha}\right|_{\alpha=\alpha_0} = \left.\frac{d^2l}{d\alpha^2}\right|_{\alpha=\alpha_0} = 0 \tag{3}$$

The set of Eqs. (2) and (3) is the criterions of the second-order focal line straightening.

All parameters with length dimension are expressed in units of central radius r_0 of HM deflecting electrode. The length of the projection of a charged particle trajectory to the symmetry axis *z* in the section from the point-like source to its image [3]:

$$l = l_{HM} + l_{CM}, \qquad l_{HM} = z_2(1+\cot\alpha_1) - z_1,$$

$$l_{CM} = \frac{l_{CM}}{r_0} = \mu\cot\alpha_1(2+4P_1\theta_1-\Delta) \tag{4}$$

$$P_1^2 = \frac{W}{qU_2}\sin^2\alpha_1 \tag{5}$$

- reflection parameters of CM, $\theta_1 = e^{P_1^2}\int_0^{P_1} e^{-x_1^2}$, Δ is a quantity defining the distance of an image from the energy analyzer's z axis, $\mu = r_c/r_0$ is a coefficient that determines the quantity of inner cylindrical electrode radius r_c in units of r_0, the radius of the trajectory vertex in the CM field is calculated from equation $r_m = \mu\exp(P_1^2)$.

As a criterion of focal line straightening of system, we accepted the equality of slope angle of a tangent to the focal line at the point of its cross with the axial trajectory of the beam.

The first requirement imposed on the parameters of a mirrors system is the angular first-order focusing condition $dl/d\alpha|_{\alpha=90^0}=0$. From it for the mirrors system follows:

$$\Delta = 2+4P_1\theta_1+\frac{z_2}{\mu}-\frac{\sin^2\alpha_1\left(4\mu\omega\, z_1\cot\alpha_1 - S^2\sin\alpha_1\right)}{\mu\left(1+\cot^2\alpha_1\right)\left(z_1+\dfrac{z_2\sin\alpha_1}{(1+\cot\alpha_1)}\right)} \tag{6}$$

$$S = \sqrt{2\frac{W}{qU_1}}\sin\alpha \qquad (7)$$

- reflection parameters of HM, connecting its geometrical and energy characteristics, $\omega = P_1\theta_1 + P_1^2(1+2P_1\theta_1)$.

The second requirement is the focal line straightening of the charged particle beams of various kinetic energy. We shall differentiate of the expression (6) with respect to ε and equate the result to zero. We obtain the equation expressing the first-order focal line straightening condition of α in a direction parallel to the symmetry axis of the system $dl/d\alpha|_{\alpha=90^0} = d\Delta/d\varepsilon|_{\alpha=90^0} = 0$.

Independent variables μ, S, α_1, P_1 included in the equation of first-order focal line straightening can vary over a wide range. However, there are limits on the ratio of the radii $\mu = 0{,}5-1$. This is justified, because it reached a reasonable compromise in the selection values of the electron-optical characteristics of the combined energy analyzer. In addition requirements for construction of real scheme of spectrograph will also be simplified.

The base detail of spectrograph is an inner cylindrical electrode, as well as in commercial devices with widely known cylindrical mirror analyzer [4, 5]. In the body of this electrode must be made two aperture windows. Aperture windows covered one-dimensional grids of nichrome wire with thickness of 0.15 mm, placed along the generators of cylinder by 1 mm spacing.

When selecting the radius size of the inner cylinder through the reflection parameters of the HM and CM (Eqs.(5) and (7)), the others geometric and energy parameters of the combined analyzer are calculated.

3. Conclusions

The double spectrograph of charged particle beams (in energy and angle) composed of electrostatic hyperbolic and cylindrical mirrors has been studied. It is shown that in case of simultaneous registration of the distribution of charged particles of the finite interval in both energy and angle the focal surface has shape of truncate cone. As a criterion of focal line straightening the equality of slope angle of a tangent to the focal line at the point of its cross with the axial trajectory of the basic beam was accepted. The relation for the condition of first-order focal line straightening of particles of various energies in the combined system of mirrors has been established.

Acknowledgement

This work was supported by the grant GF4-0815 of the Ministry of Education and Science of the Republic of Kazakhstan.

References

[1] Zashkvara V V and Ashimbaeva B U 1991 The focal line straightening in the mirrors with two-dimensional electrostatic fields *Zhurnal Tekhnicheskoi fiziki.* **61(10)** 8-18 [in Russian]

[2] Zashkvara V V and Ashimbaeva B U 1994 Theoretical basis for the scheme of an elecrtrostatic energy and angle-resolved spectrograph *Nucl. Instr. and Meth.* A **340** 514-523

[3] Ashimbaeva B U, Chokin K Sh, Saulebekov A O, Kambarova Zh T 2012 The combined energy analyzer composed of electrostatic mirror fields *J. Electron Spectrosc.and Relat.Phenom.* **185** 518-522

[4] Zashkvara V V, Ashimbaeva B U 1998 Cylindrical mirror analyzer with adjustable angular focusing *J. Electron Spectrosc. and Relat. Phenom.* **94 (1–2)** 89-96

[5] Trubitsyn A A 1995 A cylindrical mirror analyser with high energy resolution *J. Electron Spectrosc. and Relat.Phenom.* **73 (3)** 305–310

IOP Conf. Series: Materials Science and Engineering **110** (2016) 012006 doi:10.1088/1757-899X/110/1/012006

The Influence of High-Power Ion Beams and High-Intensity Short-Pulse Implantation of Ions on the Properties of Ceramic Silicon Carbide

A V Kabyshev[1], F V Konusov[2], S K Pavlov[3] and G E Remnev[4]

[1] Professor, National Research Tomsk Polytechnic University, 30, Lenin Avenue, 634050, Tomsk, Russia
[2] Senior researcher, National Research Tomsk Polytechnic University, 30, Lenin Avenue, 634050, Tomsk, Russia
[3] Engineer, National Research Tomsk Polytechnic University, 30, Lenin Avenue, 634050, Tomsk, Russia
[4] Head of laboratory, National Research Tomsk Polytechnic University, 30, Lenin Avenue, 634050, Tomsk, Russia

E-mail: konusov@hvd.tpu.ru, kabyshev@tpu.ru

Abstract. The paper is focused on the study of the structural, electrical and optical characteristics of the ceramic silicon carbide before and after irradiation in the regimes of the high-power ion beams (HPIB) and high-intensity short-pulse implantation (HISPI) of carbon ions. The dominant mechanism of transport of charge carriers, their type and the energy spectrum of localized states (LS) of defects determining the properties of SiC were established. Electrical and optical characteristics of ceramic before and after irradiation are determined by the biographical and radiation defects whose band gap (BG) energy levels have a continuous energetic distribution. A dominant p-type activation component of conduction with participation of shallow acceptor levels 0.05–0.16 eV is complemented by hopping mechanism of conduction involving the defects LS with a density of $1.2 \cdot 10^{17}$–$2.4 \cdot 10^{18}$ $eV^{-1} \cdot cm^{-3}$ distributed near the Fermi level.The effect of radiation defects with deep levels in the BG on properties change dominates after HISPI. A new material with the changed electronic structure and properties is formed in the near surface layer of SiC after the impact of the HPIB.

1. Introduction

Silicon carbide SiC is a wide-band semiconductor material promising for high-power, high-temperature and radiation-resistant electronic devices. That stimulates study of the influence of the various kinds of radiation on the characteristics of SiC and of devices based on it [1–3]. SiC has a high chemical and mechanical resistance [3–5]. The band gap (BG) of SiC depending on the polytype varies in range 2.83–3.23 eV, and the threshold energy of defect formation it is 25–35 eV, that defines the high radiation resistance of the material. The characteristics of localized states in the BG of radiation defects (RD) induced by high-energy particles and their effect on properties of material depend on the type of particle and on the mode of irradiation [1–3]. The study of the effect on the properties of RD in the ceramic SiC is difficult and leads to ambiguous results owing to its complex structural hierarchy and high content of impurities and biographical defects (BD). Improvement of electrical characteristics of the epitaxial layers of SiC in devices is achieved by the ion implantation and subsequent thermal annealing owing to an annihilation and redistribution of RD and modify of the structure of the material surface [4–7]. High-intensity short-pulse implantation (HISPI) of ions followed by heating of the surface layer provides simultaneous annealing of RD [8, 9].

The aim of the work is to study the structural, electrical and optical characteristics of ceramic SiC before and after irradiation in the regimes of high-power ion beams (HPIB) and HISPI of carbon ions C^+ and to establish the reasons of the properties change.

2. Experimental

The investigations of morphology of the ceramic surface before and after irradiation by methods described in [8] were performed by scanning electron microscopy. The elemental composition was studied by energy dispersive microanalysis. Raman spectra were studied by using spectrometer Nanofinder (λ=532 nm). The dark surface conduction σ and photoconduction $\Delta\sigma_{ph}=\sigma_{ph}-\sigma$ (σ_{ph} is conduction under lighting), photosensitive K(U, T, hν)=$\Delta\sigma_{ph}/\sigma$ were measure data constant voltage on the electrodes U=0.01–300 V, at temperature T=300–700 K, and photon energy hν=1.5–4.0 eV. The temperature dependences of σ, $\Delta\sigma_{ph}$(T) were approximated by the equation for activation mechanism

$$\sigma_a(T) = \sigma_0 \times e^{(-\varepsilon_\sigma / k \cdot T)}, \qquad (1)$$

where σ_0 is a pre–exponential factor, ε_σ is an activation energy, k is the Boltzmann constant and by the equation for the hopping mechanism of transfer between the localized states (LS) near the Fermi level E_F in BG in the Mott model

$$\sigma_p(T) = \sigma_0' \times e^{(-(T_0/T)^{0.25})}, \qquad (2)$$

where σ'_0 is a pre–exponential factor, T_0 is an activation energy [10]. The LS density N(E_F) near the Fermi level was calculated from T_0 according to [10]. The sign of the dominant charge carriers was determined by the photo- and thermostimulated current amplitude I_{PhTSC}(I_{TSC})(T, hν, U=0) [10]. The spectral dependence of the absorption coefficient α(hν) in the intervalhν=1.4–3.6 eV was calculated from the spectra of the diffuse reflection [11].

3. Result and discussion

Ceramics after irradiation by HPIB (samples SiC1,2) at the energy density J=2.4 J/cm^{2}andatthe number of pulses n=3, 50is characterized by the presence of cracks and pores on the surface due to the high-speed cooling of the melt and peculiarities of structure of SiC. Melting occurs more rapidly and in a higher volume of the near surface layer of SiC with growing of n. Melting not fixed after HISPI at J=0.2–0.3 J/cm^{2}and n=300 (samples SiC 3–5). The inclusions of metals are identified: Fe (0.03–0.31 at.%), Cr and Ni ($\leq$0.03 at.%). The obvious differences between the unirradiated SiC and SiC, irradiated by HPIB and HISPI, is not found according to X-ray analysis. Both modifications of the silicon carbide such as cubic SiC-3C and hexagonal presented by the polytypes 2H, 4H and 6H were revealed. Raman spectra indicate the presence of impurity of amorphous a–C, existing prior to irradiation, or introduced after implantation. The line of silicon at ν=520 cm^{-1}after the HPIB increases with n growth and after HISPI this line is suppressed. The lines at ν=500–1000 cm^{-1}fixed after the HPIB correspond to a carbide phases. The lines atν=795 and 972 cm^{-1}correspond to optical transverse and longitudinal oscillations of the 3C–SiC and line at ν=778 cm^{-1}to optical transverse oscillations of the 4H (6H)–SiC. Both polytypes of 4H(6H)–SiC and 3C–SiC were detected after HISPI. Here with the ratio of the hexagonal phase to a cubic was higher than after HPIB.

Current voltage characteristic (CVC) I(U) of the SiC before the irradiation are nonlinear and obeys the power $I \propto U^s$ (s=1.5–2) or the polynomial law $I=a \cdot U^2+b \cdot U+c$ (a=(4–20)$\cdot 10^{-7}$A$\cdot$V^{-2}, b=(2–9)$\cdot 10^{-7}$ S, c=(2–40)$\cdot 10^{-8}$ A, a/b=2.5–3.3) (figure 1). It is typical for epitaxial SiC [1, 2]. CVC allows us to conclude that the concentration of electrical active defects N', their population by charge carriers n' and their degree of population n'/N' increase with depth of level ε in BG. The exponential distribution of the LS on energy N'(ε) is realized in analogy with [1, 10, 11]. The coincidence of the CVC at U>0 and U<0 indicates a weak influence of the space charge on the dependence $\sigma(U)$. Annealing until T=700–800 K increases the value of σ in 2–3 times due to the redistribution of charge carriers between shallow (ε_σ<0.1 eV) or more deep traps ($\varepsilon_\sigma \geq$0.1 eV). Ceramics has a weak photosensitivity K≤0.01 due to the influence of the BD with a high concentration of N>10^{18} cm^{-3}distributed on the grain boundaries. Then–type of σ_{ph} and σdominates, as photo and thermostimulated currents $I_{PhTSC}(I_{TSC})(T)$ show. Influence of the donor BD having the impurity or vacancy nature on the transport of charge carriers prevails. Temperature dependences ofσ, $\sigma_{ph}(T)$ within the intervalT=300–700 K are determined by the thermally stimulated electron exchange between shallow donor levels with the activationenergies$\varepsilon_{\sigma1}$=0.06–0.061 eV, $\varepsilon_{\sigma2}$=0.2–0.21 eV,$\varepsilon_{\sigma3}$=0.4–0.43 eV and conduction band (CB) (figure 2). Population of traps by charge carriers is redistributed between the single levels of energy $\varepsilon_{\sigma1-3}$aftersequential heating owing to low values of their degree of population n'/N'=10^{-10}–10^{-6}. The factor σ_0 indicates that $(n'/N')_3 \gg (n'/N')_2 \gg (n'/N')_1$. The values $\varepsilon_\sigma$below their values for intrinsic defects of SiC [1–3] due to the interaction between BD and a continuous distribution of LS in the BG, by analogy with [10, 11]. The shallow donor levels with ε_σ<(0.1–0.3) eV may be due to impurity atoms N and Ti or vacancy V_{Si}^{2-} [1, 2]. Oxygen atoms (content<5.5 at. %) in complexes with BD as well as Vatoms have a deep levels (ε=0.65–1.59 eV [2]), which impact on reduce of values of σ.

The electrical parameters ε_σ, σ_0, K, n'/N' indicate the effect on transport of the hopping conduction that is verified by approximation $\sigma(T)$ by the Eq. (2) within the interval T=300–700K. LS density with the participation of which carried out $\sigma_p(T)$ is $N(E_F)=(2–9)\cdot10^{17}eV^{-1}\cdot cm^{-3}$. The value of the most probable jump distance calculated according to [10] with the values $N(E_F)$ is R=7–11 nm. The electronic structure of materials such as ceramics can be described in frame of the model with the function $N(\varepsilon)$, which varies little with depth of LS in BG [10, 11]. The sharp edges of the valence band (VB) and CB as in crystals do not exist. Fermi level in SiC is pinned near the middle of the BG at ε=1.5 eV due to the high concentration of defects and impurity atoms O, V, Ti. This is confirmed by low values of $\sigma \leq 2\cdot10^{-6}$S (figures 1 and 2) and K≤0.01 and by localization of a strong bands at energies ε=1.4 and 1.7 eV in the absorption spectra $\alpha(h\nu)$. Despite the dominance of n–type of σ and σ_{ph} we cannot exclude the influence of acceptor defects on the properties.

Irradiation of SiC significantly alters the morphology and structure of the surface and thin near surface layer (until 200 nm), which is reflected in CVC, $\sigma(T)$, in change of the type of σ (n→p) and in parameters ε_σ, σ_0, T'_0, σ'_0, $N(E_F)$, R and in spectra $\alpha(h\nu)$ (figures 1–3). Influence of HPIB and HISPI on properties change is different (figures 1–4). Changes of the elemental composition of SiC after HPIB and HISPI showed that increase of the content of silicon from 20 to 40 at. % has the greatest effect on the growth of conduction (figure3). At the same time the content of carbon atoms in layers of SiC decreases from 80 to 40 at. % and oxygen decreases from 6 to 1 at. % (figure 4). A similar correlation between the content of C, Si and concentration of the donors was fixed in 6H–SiC (n–type ofσ) [2]. HPIB significantly increases σ and changes its characteristics (figures 1 and 2). CVC becomes almost linear (s=0.99–1.05, a/b=0.01–0.02) (figure 1). The depth of the shallow centers decreases from $\varepsilon_{\sigma1}$=0.06 to 0.034 eV and from $\varepsilon_{\sigma2}$=0.21 to 0.123 and 0.155 eV and values of $(n'/N')_{1,2}$ increases in 10^2–10^3 times as shown a factor σ_0. The deep centers with $\varepsilon_{\sigma3}$ do not appear. Currents $I_{TSC}(T)$ and their $\varepsilon_\sigma$indicate the predominance of exchange between holes and VB and LS with $\varepsilon=\varepsilon_v+\varepsilon_{\sigma1,2}$ (ε_v is the top of VB).The manifestation of stable defects $V_C–C_{Si}$ is more probable in SiC with p–type of conduction while the defects $V_{Si}^{(2-)}$ dominate in case of n–type of σ[1, 2]. Shallow acceptor levels are associated with impurity of Al (≤0.09 at.%) whose energetic levels at $\varepsilon=\varepsilon_v$+(0.1–0.3) eV activated by interaction with RD[2]. $I_{PhTSC}(T)$<0 show that deep donor levels withε>1.5 eV have the effect on parameters also.LS density for hopping conduction increases from $N(E_F)=(2–9)\cdot10^{17}$to $(1.2–2.4)\cdot10^{18}eV^{-1}\cdot cm^{-3}$,and jump distance decreases from R=7–11 to 5–7 nm, which means increasing the role ofσ_p. The surface morphology change occurs due to its melting and the impact of nanoparticles of Si, C and metals. Partial annealing of RD and their association in a complex with BD occur owing to heating of surface.

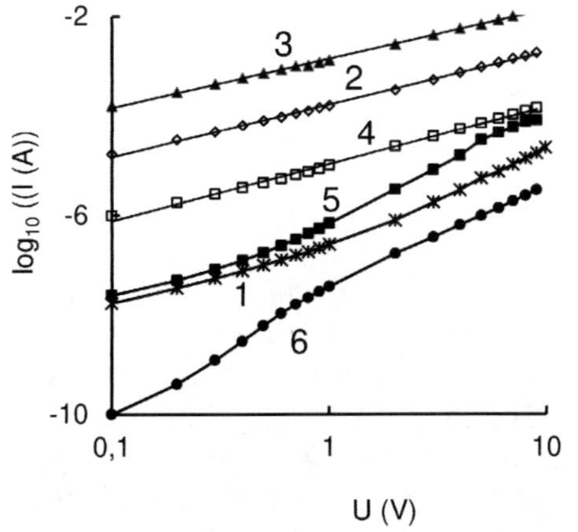

Figure 1. CVC of SiC before (curve 1) and after HPIB (curves 2 and 3) and HISPI (curves 4–6): J=2.4 J/cm^2, n=3 (curve 2), 50 (curve 3); J=0.2–0.3 J/cm^2, n=300 (curves 4–6).

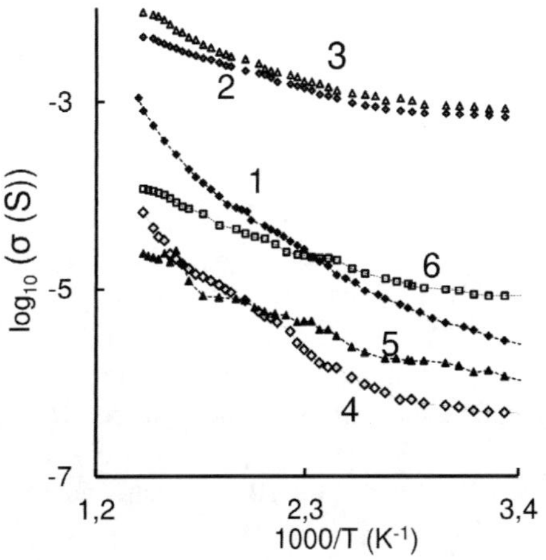

Figure 2. Temperature dependencies $\sigma(T)$ of SiC before (curve 1) and after HPIB (curves 2 and 3) and HISPI (curves 4–6): J=2.4 J/cm^2, n=3 (curve 2), 50 (curve 3); J=0.2–0.3 J/cm^2, n=300 (curves 4–6).

After HISPI σ decreases or not changes and nonlinearity of CVC increases (s=1.2–2.3, a/b=2–5) (figure1).This is caused by the accumulation of RD with deep LS that are centers of trapping of charge carriers. Parameters of RD with$\varepsilon_{\sigma1}$are close to their values prior to irradiation and of centers with $\varepsilon_{\sigma2}$– to parameters levels induced by HPIB. Deep centers with $\varepsilon_{\sigma3}$ appear also. Effect of donors increases, as shown $I_{TSC}(T)$.LS density for realization of σ_p is $N(E_F)=(1.2–11)\cdot10^{17}eV^{-1}\cdot cm^{-3}$ at R=6–12 nm. LS density decreases with σ increasing. Comparison of parameters of σ_p and σ_a shows that the influence of BD after HISPI is more greatly than after HPIB. Interrelation between electrical parameters allows to clarify mechanism of transport charge (figure 4).

Value of α decreases in 2–5 times after HPIB while α not changes or increases in 1.2–1.5 times after HISPI. Shape of $\alpha(h\nu)$ after HPIB changes similarly as in SiC after a powerful laser irradiation[6]. Absorption band at 1.4–2.8 eV induced by BD (concentration $N=(2–20)\cdot10^{18}cm^{-3}$) separates from the band of interband absorption at 2.8–3.3 eV characteristic for amorphous and highly

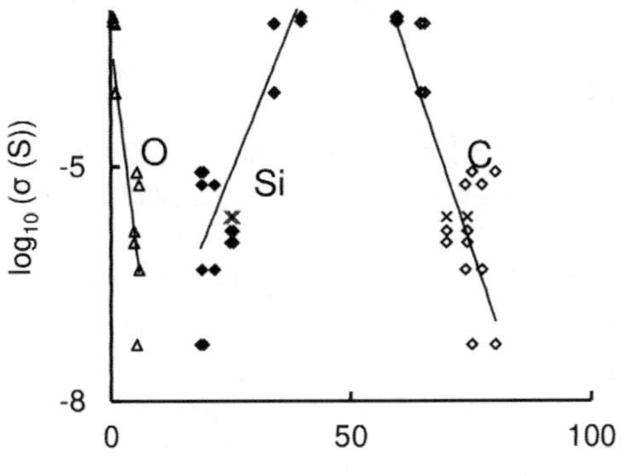

Figure 3. Effect of the elemental composition of SiC on its conduction σ before and after irradiation in the regimes of HPIB and HISPI.

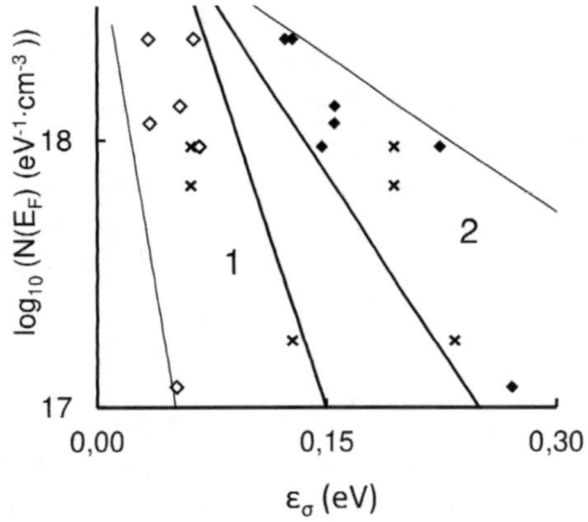

Figure 4. $N(E_F)$ vs. $\varepsilon_\sigma$in SiC before (x) and after irradiation (♦, ◊).

defective materials. Average value of BGE_{g0}=3 eV is in limits of the values E_g=2.83–3.23 eV peculiar to different polytypes of SiC [6].Conservation of this band after HISPI indicates a negligible change of the absorption edge.Accumulation of the RD levels after HISPI strengthens the pinning of E_F near the middle of BG. There are three groups of bands with centers at ε_1=1.4 eV; ε_2=1.72 eV, ε_3=2.4 eV. The concentration of these centers before irradiation is: $N_1=(2–2.3)\cdot10^{18}cm^{-3}$, $N_2=(5–7)\cdot10^{19}cm^{-3}$, $N_3=(6–7.3)\cdot10^{18}$ cm^{-3}. After HPIB $N_1=(1–1.2)\cdot10^{18}cm^{-3}$,$N_2=(2–2.2)\cdot10^{19}cm^{-3}$, $N_3=(2.5–2.7)\cdot10^{18}cm^{-3}$. Taking into account [1, 2] bands with $\varepsilon_{1–3}$were identified with vacancy RD $V_{C,Si}$. Single RD clearly manifest after HISPI. Transitions between the LS with a continuous distribution dominate after HPIB. Formation after HPIB of a new material with changing electronic structure and properties has an effect on dependencies $\alpha(\sigma,\varepsilon_\sigma)$ also. Dependency $N(E_F)(\varepsilon_\sigma)$ in SiC before and after irradiation is typical for materials containing a high concentration of defects (figure4) [10]. Parameters $N(E_F)$ and ε_σ are distributed on two arrays for shallow and more deep centers (figure4). Decrease of LS density from $N(E_F)=2.4\cdot10^{18}$to $1.2\cdot10^{17}eV^{-1}\cdot cm^{-3}$correlates with the increasing of absorption coefficient from $\alpha=3\cdot10^3$to $10^4cm^{-1}$and decreasing of $\sigma$by analogy with [10]. The impact of σ_p on transport charge for values $N(E_F)$ and ε_σ in band 1 is higher than in band 2 (figure4).

4. Conclusion

Electrical and optical properties of SiC ceramic before and after irradiation are conditioned by BD and RD whose levels are continuously distributed on energy in BG. Defects and impurities with high concentration $(2–20)\cdot10^{18}cm^{-3}$are heterogeneously distributed along the boundaries between the structural fragments of ceramic. The activation component of the p–type conduction after irradiation is realized in the exchange of holes between the VB and shallow acceptor levels with the activation energy of ε_σ<0.3 eV and is complemented by the hopping mechanism of transport on LS near the Fermi level which localized in the middle of BG. Density of LS for realization of hopping conduction is changed after irradiation from $(2–9)\cdot10^{17}$ to $(1.2–24)\cdot10^{17}eV^{-1}\cdot cm^{-3}$ and impact of hopping component on electrical parameters is amplified. Influence of RD with deep levels on the optical and electrical properties of SiC dominates after HISPI. A new material with the changed electronic structure and properties is formed in the near surface layer of SiC after the impact of the HPIB.

Acknowledgments

The research was particularly supported by the grant from Russian funds for basic research RFBR No. 13-08-98061 and megaproject of National Research Tomsk Polytechnic University (VIU_IHTPh_85_2014) «Materials for Extreme Conditions».

References

[1] Kalinina E V 2007 *Semiconductors (in Russian)* **41** 769
[2] Lebedev A A 1999 *Semiconductors (in Russian)* **33** 2129
[3] Castaldini A, Cavallini A, Rigutti L, Nava F and Giorgis F 2005 *J. Appl. Phys.* **98** 053706
[4] Leclerc S, Beaufort M F, Declémy A and Barbot J F 2008 *Appl. Phys. Lett.* **93** 122101
[5] Cavallini A, Castaldini A and Nava F 2008 *Appl. Phys. Lett.* **93** 153502
[6] Zhao Q Z, Ciobanu F, Malzer S and Wang L J 2007*Appl. Phys. Lett.* **91** 121107
[7] Lemaitre M G, Tongay S, Wang X, Venkatachalam D K, Fridmann J, Gila B P, Hebard A F, Ren F, Elliman R G and Appleton B R 2012 *Appl. Phys. Lett.* **100** 193105
[8] Konusov F V, Kabyshev A V, Pavlov S K and Remnev G E 2014 *J. of Surface Invest.* **8** 1168
[9] Remnev G E, Ivanov Yu F, Naiden E P, Saltymakov M S, Stepanov A V and Shtanko V F 2009 *Technical Physics (in Russian)* **79** 156
[10] Konusov F V, Kabyshev A V, Linnik S A, Gaydaychuk A V and Remnev G E 2014 *Journal of Physics: Conf. ser.* **552** 012046
[11] Konusov F V, Kabyshev A V and Remnev G E 2011 *J. of Surface Invest.* **5** 228

IOP Conf. Series: Materials Science and Engineering **110** (2016) 012007 doi:10.1088/1757-899X/110/1/012007

Research of metallic materials irradiation with high energy pulsed laser impact

A I Blesman[1,3], D V Postnikov[2,3], G M Seropyan[4], E A Tkachenko[5], A A Teplouhov[2] and D A Polonyankin[2,3]

[1]Head of the Physics Department, *Physics Department*,
Omsk State Technical University (OmSTU), 11 Mira Avenue, Omsk, 644050, Russia

[2]Associate Professor, *Physics Department*,
Omsk State Technical University (OmSTU), 11 Mira Avenue, Omsk, 644050, Russia

[3]Research scientist, *Scientific-educational center of nanotechnology*,
Omsk State Technical University (OmSTU), 11 Mira Avenue, Omsk, 644050, Russia

[4]Associate Professor, *General Physics Department*,
Omsk State University n. a. F.M. Dostoevsky, 55A Mira Avenue, Omsk, 644077, Russia

[5]Postgraduate, *Physics Department*,
Omsk State Technical University (OmSTU), 11 Mira Avenue, Omsk, 644050, Russia

E-mail: nano@omgtu.ru, polonjan@mail.ru

Abstract. In the process of metallic materials treatment by pulsed laser beams with nanosecond duration occurs extremely rapid and intensive heating of their surface. In this case a thin surface layer of material is heated to the boiling point and rapidly evaporates. This leads to arising substantial forces of reactive nature which significantly influence on the shape of the solidified melt and in some cases may cause deformation of the underlying layers. The considered question is relevant in the research of precision treatment of miniature products by laser beams. A metallic powder with microfine material structure was selected as the object of research and was exposed to laser irradiation with nanosecond duration. At the core of reactive forces calculation used the approach similar for laser rocket engines. The paper also presents the model and the results of the forces and the reactive recoil impulse calculation occurring during laser impact to the microfine metallic powder.

1. Introduction

High energy pulsed laser impact is the actual problem in different applications of laser interaction with materials in various areas of science. Some studies are dedicated to the questions of laser pulse spreading inside a bulk of transparent media (sapphire, glass and polymer) [1]. In some works laser ablation of dielectrics [2, 3] and semiconductors [4] is investigated. The physical principles of generation of laser pulses with high power and short duration and their interaction with matter are considered in fundamental works, for example in [5] and [6].

The papers [7] and [8] present the idea of an engine based on thrust caused by the laser evaporation. The important characteristics of the laser engines are: reactive recoil impulse, which is defined as the ratio of magnitude thrust to power of the laser radiation; specific impulse; thrust referred to fuel consumption [9, 10]. The most interesting from the standpoint of energy efficiency are the engines based on a solid materials ablation. In particular there are several possible options:

- direct laser ablation when thrust is generated only by the vapor pressure of the ejected material from the surface [11];
- mixed ablation, when the shock wave is added to pressure of the evaporating material [12];
- laser ablation of structured materials, which include micro and nano powders.

The present paper is devoted to the study of laser interaction with the microfine metallic powder with approach which is similar to laser rocket engines.

2. Research objects and methods

Nickel-based alloy metallic powder with microfine material structure was selected as the object of research. Metallic powder SEM microphotography presented on the figure 1 with different magnification. The elemental composition of the microfine metallic powder will be considered in more detail in next paragraph.

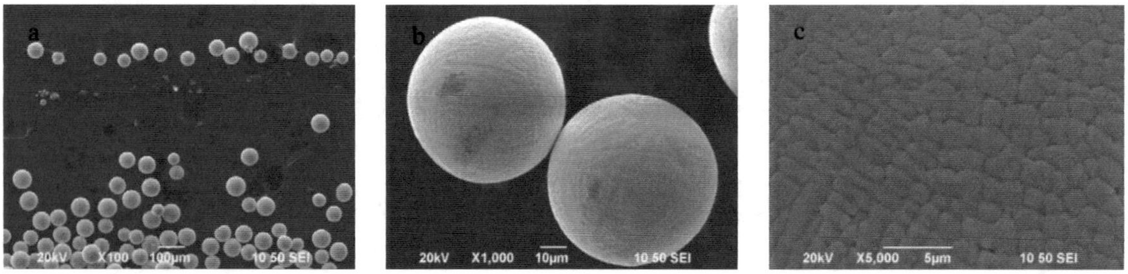

Figure 1. Metallic powder SEM microphotography, x100 (a), x1000 (b), x5000 (c) magnification.

For metallic powder irradiation was applied solid state pulsed aluminium-yttrium garnet doped with neodymium ions (Nd: YAG) laser. The basic principles of operation of such lasers can be found in the study [13]. YAG laser with the neodymium rod generates radiation with a wavelength of 532 nm, the pulse duration is 16 ns in free generation mode, the pulse repetition rate is 10 Hz, pulse energy up to 0.5 J, the laser beam diameter up to 6 mm. The appearance of the laser system is shown in figure 2. Laser equipment is located at the area with ISO-5 clean room class.

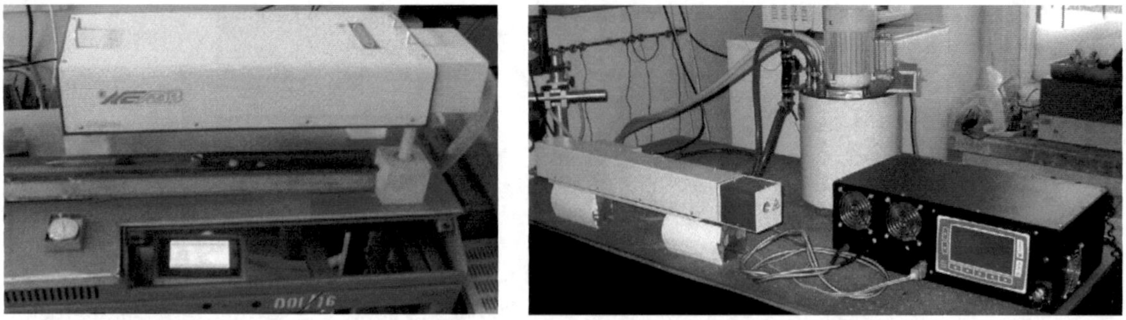

Figure 2. The appearance of the Nd: YAG laser systems.

3. The quantitative elemental composition of the microfine metallic powder

The quantitative elemental composition of the microfine metallic powder was obtained with EDS method using a scanning electron microscope JEOL JCM–5700 [14, 15]. The results of analysis are presented in table 1. As the analysis shows the greatest content of the powder occupy such metals as nickel, cobalt and chromium.

The particle size of the microfine metallic powder ranges from 50 micron and to 60 micron (figure 3). Also was performed EDX analysis of the powder, the results of which are presented in figure 4.

Table 1. Quantitative elemental composition of the microfine metallic powder.

Element	Mass%	Atom%
C	3.41	14.40
Al	2.88	5.43
Ti	1.78	1.89
Cr	10.67	10.42
Co	15.84	13.65
Ni	57.78	50.04
Sr	3.67	2.13
Mo	3.89	2.06
Total	100.00	100.00

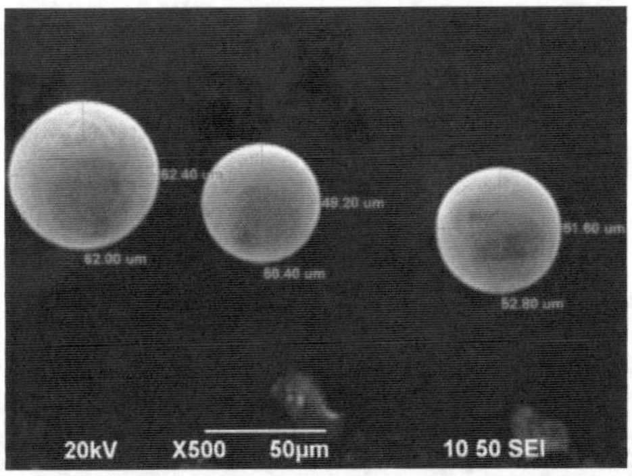

Figure 3. The metallic powder particle size.

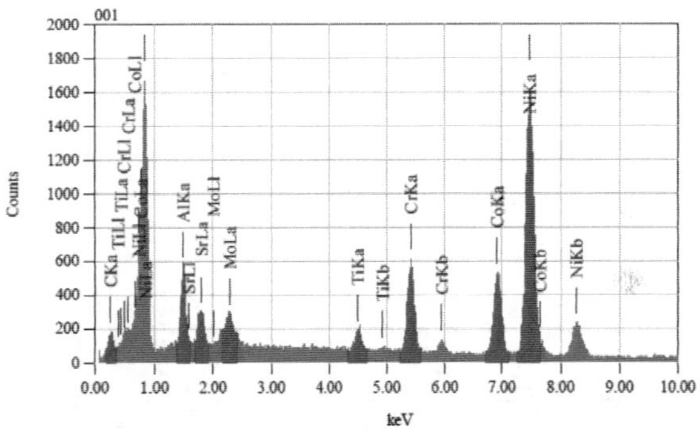

Figure 4. The metallic powder spectrogram.

4. The reactive recoil impulse calculation

As a result of nanosecond laser irradiation the material dispersion occurs due to intense evaporation with a laser beam from the part of heating. To calculate the reactive recoil impulse using a formula borrowed from [7]:

$$C_m = \left(\frac{A}{L_{s.h.b.}}\right)\left(\frac{\gamma T_{b.p.}}{m}\right)^{1/2} \tag{1}$$

Here A is the absorption coefficient of laser radiation at the stage of material evaporation, $L_{s.h.b.}$ is the specific heat of the boiling, γ is the effective adiabatic index of vaporized substance, $T_{b.p.}$ is the boiling point and m is the mass of evaporated substance.

Consider ablation at the final stage when the remaining portion of material warmed up to a temperature close to boiling point. In this case, the transition velocity of the phase boundary tends to a constant value determined by the specific heat of the boiling. Using the approaches outlined in [16] it is possible to derive a formula of the intensive evaporation front velocity:

$$\upsilon = \left(\frac{4q_0}{21\rho L_{s.h.b.}}\right)\left(1 - \frac{L_{s.h.b.}}{cT_{b.p.}}\right)\left(1 + \frac{1}{1 + (3L_{s.h.b.})(cT_{b.p.})^{-1}}\right) \quad (2)$$

Here c is the heat capacity, ρ is the material density, q_0 is the intensity of the laser radiation, r is the radius of the laser radiation and t is the exposure time. Then the reactive recoil impulse for the powder material can be found using the following formula:

$$C_m = \left(\frac{A}{L_{s.h.b.}}\right)\left(\frac{\gamma T_{b.p.}}{\upsilon\rho\pi r^2 t}\right)^{1/2} \quad (3)$$

The impulse shape at energy E = 0.19 J is presented on the figure 5. The distribution of reactive recoil impulse normalized to the beam power depending on the exposure time for nickel and aluminium is presented on the figure 6.

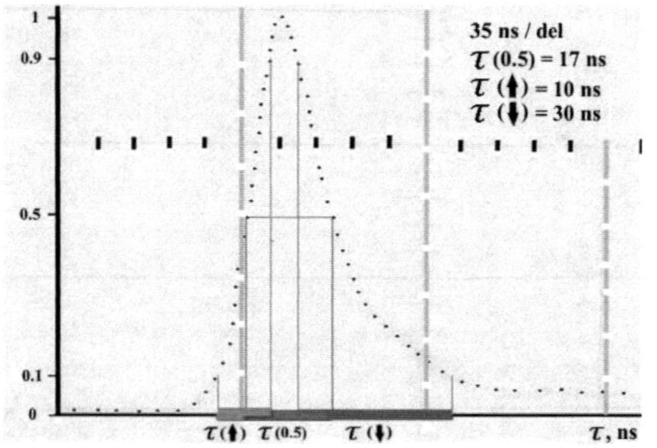

Figure 5. The laser impulse shape at energy E = 0.19 J.

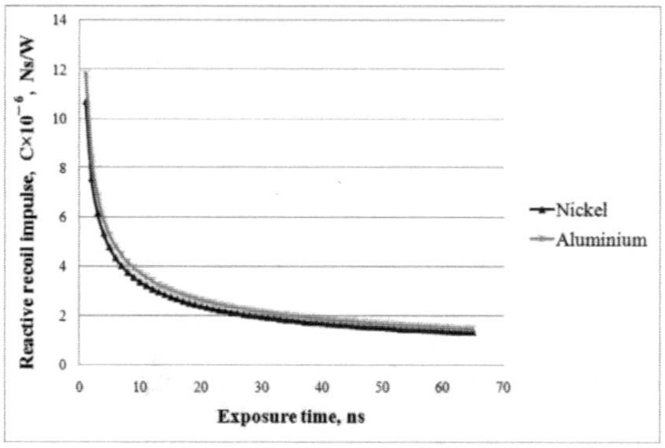

Figure 6. Reactive recoil impulse for nickel and aluminium.

Comparing formulas (1) and (3) it is possible to conclude that the mass of evaporated substance could be estimated according to the formula (4).

$$m = \upsilon\rho\pi r^2 t \quad (4)$$

The dependence of the evaporated substance mass from the intensive evaporation front velocity has a linear character.

5. Results and discussion

As a result of pulsed laser exposure of nanosecond duration nickel-based alloy metallic powder is melted, heated to the boiling point and then evaporated (see figure 7a and 7b). On the surface of the powder is formed spherical droplets of different sizes, which were absent before irradiation.

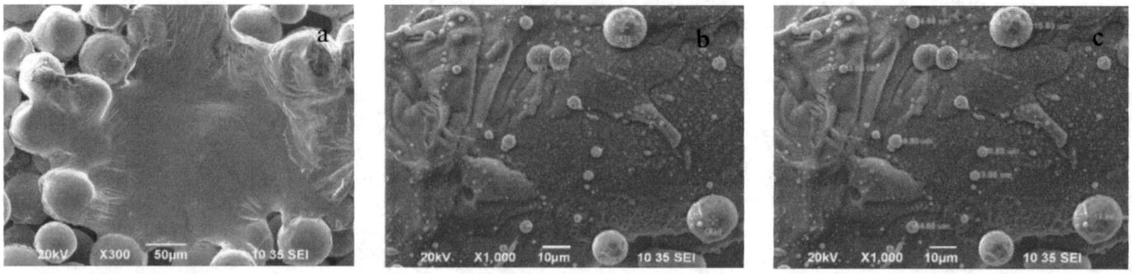

Figure 7. Metallic powder SEM microphotography after laser irradiation (a). Formation of droplet phase on the surface of irradiated powder (b). The size of the droplet phase (c).

Most clearly these changes are observed when comparing figures 1c and 7a. The droplet size on the surface of the evaporated substance ranges from 3 to 18 microns (figure 7c) in particle size of the powder distribution from 50 to 60 microns. The elemental composition of the resulting droplets is significantly different from the composition of the initial powder, occurs the elemental redistribution. The greatest change occurs at the oxygen atomic concentration, its composition increases to 34–44% (data from marked droplets, figure 7b) with 14.4% in the initial powder. Also it appears oxygen with a concentration up to 12% in the droplets whereas the initial powder composition doesn't include it.

6. Conclusion

The developed approach permits to calculate the reactive recoil force and impulse in dependence of the material density and heat capacity, the intensity, the absorption coefficient and the radius of the laser radiation. To important parameters included in the calculation formula are also belong the specific heat of the boiling and the boiling point of evaporated material, the effective adiabatic index of vaporized substance. It is proposed a formula for the evaporated substances mass calculation depending on the intensive evaporation front velocity. Using the developed approach obtained the distribution of reactive recoil impulse normalized to the beam power depending on the exposure time for nickel and aluminium. Reactive recoil impulse decreases inversely proportional to the square root of time with a rise of exposure duration.

The appearance of the droplet phase on the surface of the evaporated substance can be attributed to the substantial results of the study, wherein the droplet size ranges from 3 to 18 microns in particle size of the powder from 50 to 60 microns. Thanks to high-energy laser impact a redistribution of elements in the droplet phase occurs; its composition is significantly different from the initial.

References

[1] Gamaly E G, Juodkazis S, Misawa H, Luther-Davies B, Hallo L, Nicolai P and Tikhonchuk V T 2006 *Phys. Rev. B* **73** 214101
[2] Balling P and Schou J 2013 *Rep. Prog. Phys.* **76** 036502
[3] Stuart B C, Feit M D, Herman S, Rubenchik A M, Shore B W and Perry M D 1996 *J. Opt. Soc. Am. B* **13** 459
[4] Gamaly E G, Rapp L, Roppo V, Juodkazis S and Rode A V 2013 *New J. Phys.* **15** 025018
[5] Luther-Davies B, Gamaly E, Wang Y, Rode A and Tikhonchuk V 1991 *Laser Phys.* **1** 325
[6] Kruer W L 1988 *The Physics of Laser Plasma Interactions* (New York: Addison-Wesley)
[7] Rezunkov Yu A 2011 *Izvestiya vysshikh uchebnykh zavedeniy. Priborostroenie* **54** 7
[8] Rezunkov Yu A 2007 *Opt. Zh.* **74** 20
[9] Ignat'ev A B 2008 *Integration of high-power laser sources with the means of formation, orientation and precise beam pointing* (Moscow: Mirea)
[10] Ageev V P, Barchukov A I, Bunkin F V, Konov V I, Prokhorov A M, Silenok A S and Chapliev N I 1977 *Sov. J. Quantum Electron.* **7** 1430
[11] Pakhomov A V, Thompson M S and Don A G 2003 Ablative laser propulsion: a study of specific impulse, thrust and efficiency *Beamed energy propulsion: First International Symposium on Beamed Energy Propulsion (Huntsville, Alabama, USA, 5–7 November 2002) (AIP Conf. Proc.* vol 664*)* ed. A V Pakhomov pp 194–205
[12] Phipps C R, Seibert D B, Rouse R W, King G, Campbell J W 2000 Very high coupling coefficient at low laser fluence with a structured target *High-Power Laser Ablation III (Santa Fe, NM, USA, 24 April 2000)* (*Proc. SPIE.* vol 4065*)* ed. Phipps C R pp 931–938
[13] Fibrich M, Hambálek T, Němec M, Šulc J and Jelínková H 2014 *Laser Phys.* **24** 035803
[14] Blesman A I, Postnikov D V and Polonyankin D A 2015 *IOP Conf. Ser.: Mater. Sci. Eng.* **81** 012031
[15] Blesman A I, Postnikov D V and Polonyankin D A 2015 *Procedia Eng.* **113** 413
[16] Rykalin N N, Uglov A A, Zuev I V and Kokora A N 1985 *Laser and electron beam processing of materials* (Moscow: Mashinostroenie)

IOP Conf. Series: Materials Science and Engineering **110** (2016) 012008 doi:10.1088/1757-899X/110/1/012008

Onset of Ultrasonic Oscillation at Impulse Processing of Metal Melts

N A Shaburova
Ph.D., assistant professor of «Physical metallurgy and physics of solids» department, South Ural State University (NRU), Chelyabinsk, Lenin Avenue, 76, Russia

E-mail: nata8855@gmail.com

Abstract. Processing of metal melts powerful electromagnetic impulses - a new and effective way to change the structure and properties of metal. Numerous experiments on ferrous and non-ferrous metals have shown that such processing usually allows both increase in strength and ductility. A theoretical explanation is given to the impact mechanism of powerful electromagnetic impulses on ferrous and non-ferrous metal melts. The possibility to convert electromagnetic impulses to sound waves is shown.

1. Introduction

The results of the numerous experiments on the impact of powerful electromagnetic impulses (EMI) applied to metal melts are shown in the previously published papers [1-3]. The generator used in the experiments has the following characteristics: pulse height – 10KV, duration – 1nsec, repetition rate – 1000Hz. The study of properties and structure of ferrous and nonferrous metals in the cast state after melts treatment supports the practicability of such processing [4-5]. Metal castability rises, porosity reduces, and there is a simultaneous increase in strength and ductility. There are several theories explaining the mechanism of impulse effect on metals. The most promising is the acoustic transformation model, when a powerful electromagnetic impulse is converted into a sound wave in the melt. The same effects as at the ultrasonic treatment are observed. The similarity of effects, produced by the electromagnetic pulse and ultrasonic treatment was shown [6-7].

2. Theoretical study

Let's consider the process of electromagnetic acoustic transformation (EMAT) in the form of the energy transition from the electromagnetic into the acoustic field [8]. Figure 1 schematically shows the direct EMAT. As it can be seen from figure 1, the nanosecond electromagnetic pulses from the generator output 1 are fed to the radiating inductor 2, which can take any shape and size: removable, pass-through, flat, volumetric, axial or mirror-symmetrical, in the form of a meander or a phased array, etc. The impulses, emitted by the pulses inductor 2, reach the surface of the metal sample 3 and in its surface layer (skin layer) are converted into acoustic-pulses of approximately the same time form and the same duration. The last propagate in the sample in the direction of the radiation, reach its opposite surface, where they are recorded by the receiver, for example, piezoelectric transducer, the electrical signals from which are fed to the input of the nanosecond acoustic impulse registrator (NAI). If the metal sample 3 is placed in the magnetic field, the effectiveness of the direct EMAT increases significantly.

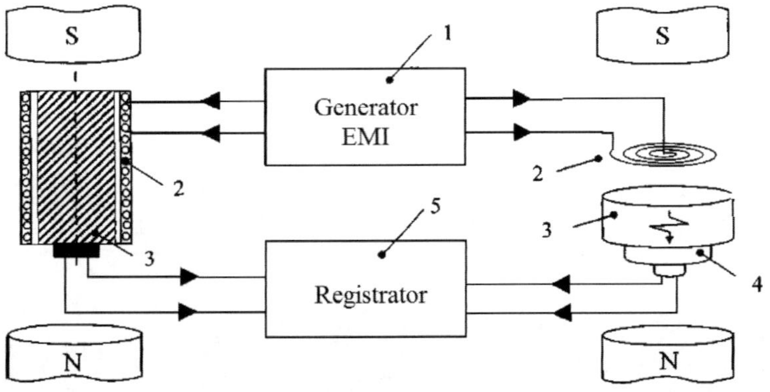

Figure 1. Schematic layout of an arrangement to study the impact mechanism of nanosecond electromagnetic impulses on metals: 1 – EMI generator; 2 – radiating inductors; 3 – metal samples; 4 – acoustic inducer; 5 – NAI registrator.

In the field of practical applications as well as in the scientific research, qualitatively different approaches to the study of EMAT in regular (well conductive weak-magnetic) and magnetically ordered metals has been evolved. The transformation of electromagnetic impulses in the elastic ones due to the Lorentz force is effective under the conditions where the thickness of the skin layer is much less than the spatial length of the elastic impulse as the "smeared" in the space by force is not possible to initiate short-wave elastic impulses. This condition is satisfied in pure metals at frequencies not less than 10^{10} Hz and determines the frequency range of the EMAT study. The transformation of electromagnetic impulses in elastic ones due to magnetostriction, in contrast, is effective when this process involves the entire volume of the metal.

If a metal is placed in the constant magnetic field $\mathbf{H}_0$, then, the density exerted by the Lorentz force acts on a metal in places, where the density of the current $\mathbf{j}$ differs from zero:

$$\mathbf{f}^L = \left[\mathbf{j},\mathbf{H}_0\right]/c \quad . \tag{1}$$

Besides the Lorentz force there is, however, still the power of a completely different nature. The electromagnetic wave disturbs the equilibrium between electrons and ions. The energy, acquired from the electromagnetic wave, is transferred from electrons to ions as a result of the collision. This results in the emergence of a kind of the force dipole. The importance of this transformation mechanism is the greater the larger is the "leverage" of the dipole – the length of the electron free path.

We assume that the conduction electrons are strongly degenerate gas of charged particles with an arbitrary dispersion law, i.e. the energy of the electron ε and its velocity $\mathbf{v}$, generally speaking, are complex periodic functions of the quasi-momentum $\mathbf{p}$ ($\mathbf{v} = \partial\varepsilon/\partial\mathbf{p}$). Because of the degeneracy of the electron gas in the EMAT, as in many other "purely metallic-like" phenomena (electrical conductivity, skin-effect, etc.), the decisive role is played by the Fermi-electrons, the electrons at the Fermi surface:

$$\varepsilon(\mathbf{p})=\varepsilon_F \tag{2}$$

The basis of the theoretical study of linear EMAT is a coupled system of linear equations, consisting of elasticity equations and Maxwell's equations. The equations of the theory of the metal elasticity differ from the usual ones by the presence of the right-hand side, containing a force density **D**, which describes the effect of either directly the electric field of an electromagnetic wave, or the conduction electrons on the grid [8].

The analysis of the forces, acting on the metal lattice, shows that the force density **D** can be represented as a sum of three terms:

IOP Conf. Series: Materials Science and Engineering **110** (2016) 012008 doi:10.1088/1757-899X/110/1/012008

$$\mathbf{D}=\mathbf{f}^{L}+\mathbf{f}^{ST}+\mathbf{f}^{D} \tag{3}$$

where, $\mathbf{f}^L$ is the density of the Lorentz force (3.2); $\mathbf{f}^{ST}$ – the Stewart-Tolman force

$$\mathbf{f}^{ST}=-(m/e)\partial\mathbf{j}/\partial t, \tag{4}$$

emerging because the electrons under the influence of electromagnetic waves move relative to the oscillating crystal lattice, which is a non-inertial system. In (4) m is a regular electron mass. It occurs only in equations, bound to the inertia force, as the latter does not depend on the periodic field of the crystal lattice [9].
For monochromatic oscillations with angular frequency $\omega=2\pi f$:

$$\mathbf{f}^{ST}=(im\omega/e)\mathbf{j}. \tag{5}$$

The values of $\mathbf{f}^L$ and $\mathbf{f}^{ST}$ allow a simple comparison: according to (1) and (5) the Stewart-Tolman force exceeds the Lorentz force if $\omega>\omega_c=e\mathbf{H}_0/mc$ is the free electron cyclotron frequency. The last term $\mathbf{f}^D$ is the density of the deformation force.
Maxwell's equations after the elimination of the magnetic field tension of the electromagnetic wave associate the current density $\mathbf{j}$ and the intension of the transverse part of the electric field $\mathbf{E}$:

$$\text{rot rot }\mathbf{E}=-(4\pi/c^2)\partial j/\partial t,\ \text{div }\mathbf{E}=0. \tag{6}$$

The longitudinal component of the electric field $\mathbf{E}$, if it is not identically equal to zero, is excluded by the condition

$$\text{div}\mathbf{j}=0 \tag{7}$$

In general, the current density is determined by the of non-equilibrium correction $\delta f=(\partial f_F/\partial\varepsilon)\chi$ to the Fermi electron distribution function f_F

$$\mathbf{j}=-e\langle\mathbf{v}\chi\rangle. \tag{8}$$

Here the angle brackets denote averaging over the Fermi surface. Function χ is the solution of the kinetic equation:

$$\frac{\partial\chi}{\partial t}+\mathbf{v}\frac{\partial\chi}{\partial\mathbf{r}}+\frac{e}{c}\left[\mathbf{v},\mathbf{H}_0\right]+\frac{\partial\chi}{\partial\mathbf{p}}+\hat{\nu}\chi=-e\mathbf{v}\left(\mathbf{E}+\frac{1}{c}\left[\dot{\mathbf{U}},\mathbf{H}_0\right]-\frac{m\ddot{\mathbf{U}}}{e}\right)-\Lambda_{ik}(\mathbf{p})\dot{U}_{ik} \tag{9}$$

where $\hat{\nu}$ is the operator of electron collisions. Almost always we will use τ – approximation, i.e. replace the integral in the $\mathbf{p}$–space operator by the $\hat{\nu}$ relaxation frequency $\nu=1/\tau$, where τ is the time of the electron free path.
The right side of the kinetic equation (9) shows that the electronic system is taken out of balance not only by the electric field, but also by the movement of the (ion) metal crystal lattice. The movement of the crystal lattice is manifested in the presence of terms (in the right side of equation (9)), depending on the time derivatives of the vector $\mathbf{U}$ and tensor U_{ik}. The terms, containing time derivatives $\dot{\mathbf{U}}$ and $\ddot{\mathbf{U}}$, are stipulated by the electrons moving along the vibrating lattice: $e[\dot{\mathbf{U}},\mathbf{H}_0]/c$ is the Lorentz force acting on the metal ion, $m\ddot{U}$ is the Stewart-Tolman force. The last term $\Lambda_{ik}(\mathbf{p})\dot{U}_{ik}$ is the effect of the

direct bond between the electrons and the lattice. $\Lambda_{ik}(\mathbf{p})$ tensor is called the renormalized tensor of the deformation potential:

$$\Lambda_{ik}(\mathbf{p})=\lambda_{ik}(\mathbf{p})-\langle\lambda_{ik}(\mathbf{p})\rangle,\quad \lambda_{ik}=\delta\varepsilon(\mathbf{p})/\delta U_{ik}. \tag{10}$$

Renormalization is the effect of the requirement of the metal quasi-neutrality. The tensor of the deformation potential $\lambda_{ik}(\mathbf{p})$, describing the bond between the electron and the long-wavelength phonons, that serves as a "phonon charge" of the electrons. This is an analogy between the expressions $\delta\varepsilon=-e\varphi$ and $\delta\varepsilon=\lambda_{ik}(\mathbf{p})U_{ik}$, φ is the potential of the electric field, U_{ik} is the phonon potential. It is necessary to note how the sound wave disturbs the system of electrons from equilibrium: first, creating an electric field **E** – the electrochemical potential gradient; secondly, due to the **U** and U_{ik} time dependence. Spatial heterogeneity of these characteristics could lead to non-uniform distribution of electrons in a conductor, but cannot disturb the electron state of equilibrium. The task of the metal spectroscopy is to define tensor $\lambda_{ik}(\mathbf{p})$ components for various cavities of the Fermi surface and at various points in these cavities. At present the structure (topology, the size of the cavities) of the Fermi surfaces of almost all metals is known with great precision, but the study of the deformation potential tensor is just at its beginning.

The definition applied to $\lambda_{ik}(\mathbf{p})$ is the deformation potential tensor and it's not apt enough. The reason is that it describes two different effects: a true change of the electron (its energy) dispersion law stipulated by the elastic deformation of the crystal, and the electron impulse transfer from one point of the crystal to another on account of velocity different from zero. Thus, even at the free electron approximation, when the electron energy does not depend on the capacity of the lattice ions, the tensor $\lambda_{ik}(\mathbf{p})$ differs from zero and is equal to $p_i v_k = m v_i v_k$.

For the free-electron approximation the equation for the "deformation" force is:

$$\mathbf{f}^D=-ne(\mathbf{E}-\mathbf{j}/\sigma_0), \tag{11}$$

and allows clear understanding that the deformation force differs from zero only due to the spatial dispersion of conductivity ($\sigma_0=ne^2l/p_F$ is the static conductivity of metal). Therefore, the EMAT occurring due to the deformation force, is favored by the extremely anomalous skin effect ($l>>\delta$, where δ is the depth of the skin-layer), when the difference $\mathbf{E}-\mathbf{j}/\sigma_0$ is at the maximum.

The above equations constitute a system of equations that allows studying the propagation of bound electromagnetic-acoustic oscillations in metals. To receive self-contained, applicable for analysis equations, it is necessary to specify the relation between the parameters (there are a lot of them under such general problem statement). In a strong magnetic field ($\omega_c\tau>>1$), when weak decaying waves are spread in the electron subsystem of the metal [10], their interaction with sound waves alters significantly the wave properties of the metal.

To consider EMAT from the theoretical standpoint, as stated above, it is necessary to cover inhomogeneous problems. The simplest of them is: an incident electromagnetic wave falls on the metal surface, filling the half-space. This problem statement requires the formulation of boundary conditions. Firstly, these are natural electromagnetic conditions of the equation (6): the continuity of the tangential components of the electric and magnetic fields at the boundaries of the half-space occupied by metal; secondly, the conditions corresponding to a particular character of boundary "securing": if the boundary is free, then elastic stresses are equal to zero, and if it is fixed, then there should be no disposition; thirdly, it is necessary to set the boundary conditions to the equation (9), describing the interaction between conducting electrons and the surface; and finally, fourthly, it is necessary to set the conditions at infinity – deep inside the metal. For EMAT the latter conditions are formulated as the conditions of the electromagnetic field and the displacement field reducing to zero at infinity as well as the non-equilibrium increment to the electron distribution function.

Before proceeding to the solution of specific problems, we should discuss the ways to significantly simplify the problem statement. Only choosing a very specific boundary condition for the function χ, we can achieve significant simplification: either at diffuse scattering of electrons the non-differential part of the nucleus is reduced to zero - you can use the Wiener-Hopf method [11]. Or at specular reflection, expanding in a certain way all the required functions on a metal-free half-space, it is possible to artificially "create" a differential nucleus and with the integration limit from $-\infty$ to $+\infty$, that allows you to apply the Fourier method for the solution of the problem. On the other hand, the study of limiting cases indicated that the change in the boundary conditions for the distribution function of electrons has little effect on the value of the linear response of the metal.
Since the bond between electromagnetic and acoustic waves is small, then the problem of EMAT can be divided into two stages. The first stage is the solution of electromagnetic problems: the calculation of the electromagnetic field in the metal. The second stage is the calculation of the force density and the solution of the acoustic problem with the corresponding boundary conditions.
For the half-space the division may not be feasible in cases where weak decaying electromagnetic waves propagate in the metal, placed in a strong magnetic field ($\omega_c\tau >> 1$). If the frequency of the incident electromagnetic wave, falling on the metal, corresponds with the possibility of the resonance between the weak decaying waves and sound, then, it usually becomes necessary to solve electromagnetic and acoustic problems jointly.
During EMAT a geometric resonance is possible on the metal plate of the finite thickness (not on the half-space): an integer of sound half-waves is placed on the metal plate of finite thickness. In this case the calculation of the resonance of the electromagnetic impedance of the plate is more convenient to carry out self-consistently without dividing the task into two stages.
The proposed approach allows to solve the problem of the origin of the acoustic oscillations in the metal due to single-frequency sinusoidal electromagnetic oscillations. Transition to impulse exposure may be performed using Fourier transformation.
EMAT studies in liquid metals were performed by different researchers considering the fact that the transverse elastic waves do not propagate in fluids, and the generation of longitudinal elastic waves is satisfactorily described by the theory of induction interaction. In this case, however, the important role is played by the boundary conditions of the problem. For a free metal boundary the condition for the absence of stress is applied. For the liquid, drenching the walls of the crucible, the assumption of boundary conditions for a fixed surface-mount $u|_{z=2}=0$ is justified
In this case, the transformation effectiveness

$$W_1 \approx \frac{1}{\rho_M S} \cdot \frac{\beta}{1+\beta^2} \approx \frac{s\sigma}{\rho_M \omega}. \tag{12}$$

This expression can be compared to EMAT efficiency in the metal free surface, the expression for which can be written as

$$W_2 \approx \frac{1}{\rho_M S} \cdot \frac{1}{1+\beta^2} \approx \frac{S^3\sigma^2}{\rho_M \omega^2}. \tag{13}$$

We see that in limit $\beta << 1$ the transformation effectiveness in the case of the fixed surface β times exceeds W at the free surface. Due to this reason, the amplitude of longitudinal elastic waves may be significantly increased in the conductive liquid metals with an appropriate choice of frequency.
The calculation of EMAT effectiveness in certain liquid metals, normalized to the effects of EMAT in aluminum at the frequency f = 10 MHz at room temperature, is shown in the table 1 (transformation effectiveness in aluminum at room temperature is taken as a unit).
The experiment, carried out on the melts of cesium and mercury, corroborated these calculations. The measurements were carried out using a pulse-echo method with two helical coils for sound generation and recording, or with a coil encompassing the glass crucible with the metal inside.

IOP Conf. Series: Materials Science and Engineering **110** (2016) 012008 doi:10.1088/1757-899X/110/1/012008

Table 1. The calculation of EMAT effectiveness.

Metal	β	W_2	W_1
Aluminum (T=300 K)	0,033	1	0,033
Sodium	0,75	4,6	3,5
Potassium	1,81	2,4	4,3
Aluminum	0,88	1,1	0,97
Copper	0,88	0,3	0,26
Indium	3,1	0,1	0,31
Cesium	11,8	0,07	0,83
Lead	15,1	0,004	0,06
Bismuth	23,5	0,002	0,05
Quicksilver	21	0,0019	0,04

3. Conclusions

The method of excitation of longitudinal elastic waves in liquid metals as a result of exposure to EMI can be particularly useful at high temperatures, when the problems, associated with acoustic contact and emergence of metal oxides on the metal surface, create considerable difficulties. The same method, apparently, can be used to study phase transitions of liquid-solid body.

In experimental studies on processing EMI aluminum, copper, lead and bismuth were treated. The effects of changes in the properties are seen in all metals. It means that even at low coefficients of transformation (lead and bismuth) energy impulse action is enough to obtain new properties of cast metals.

4. References

[1] Belkin V S, Bukharin V A and Dubrovin V K 2001 *Nanosecond electromagnetic impulses and their use* (Chelyabinsk: Publishing House «Tatiana Lurie») p 119

[2] Shaburova N A 2006 Nanosecond electromagnetic impulses in the non-ferrous metal melts *Vestnik SUSU, Series «Math, physics, chemistry»* **152–156**

[3] Shaburova N A 2007 *Formation of structure and properties of cast aluminum alloys under the impact of deformation and electro impulse exposure*: diss. candidate of Sciences (Chelyabinsk) p 177

[4] Shaburova N A 2011 *Processing of metal melts and alloys by nanosecond electromagnetic impulses: Monograph ed. by V.V. Krymsky* (Chelyabinsk: Publishing HouseChelTsNTI) p 94

[5] Balakirev V F, Krymsky V V and Shaburova N A 2012 *Nano-impulse technologies* (Chelyabinsk) p 124

[6] Balakirev V F, Krymsky V V, Ree E H, Ree Hosen and Shaburova N A 2014 *Electro-impulse processing of metal melts* (Chelyabinsk) p 144

[7] Shaburova N A Changes in metal properties after thermal and electric impulse processing 2015 *IOP Conf. Ser.: Mater. Sci. Eng.* **81** 012016

[8] Vasiliev A N, Buchelnikov V D, Gurevich S Y 2001 *Electromagnetic excitation of sound in metals* (Chelyabinsk: SUSU Publishing House) p 339

[9] Landau L D 1982 *The electrodynamics of continuous media* (Moscow: Nauka) p

[10] Abrikosov A A 1987 *Fundamentals of the theory of metals* (Moscow: Nauka) p

[11] Mattews D 1972 *Mathematical methods of physics* (Moscow: Mir) p

The model of calculation the adhesion force and energy for coatings deposited by magnetron sputtering

E A Tkachenko[1], D V Postnikov[2,4], A I Blesman[3,4] and D A Polonyankin[2,4]

[1]Postgraduate, *Physics Department*,
Omsk State Technical University (OmSTU), 11 Mira Avenue, Omsk, 644050, Russia

[2]Associate Professor, *Physics Department*,
Omsk State Technical University (OmSTU), 11 Mira Avenue, Omsk, 644050, Russia

[3]Head of the Physics Department, *Physics Department*,
Omsk State Technical University (OmSTU), 11 Mira Avenue, Omsk, 644050, Russia

[4]Research scientist, *Scientific-educational center of nanotechnology*,
Omsk State Technical University (OmSTU), 11 Mira Avenue, Omsk, 644050, Russia

E-mail: tkeduard@yandex.ru

Abstract. The paper justifies the usefulness of preliminary ion implantation before forming the protective coating by magnetron sputtering in order to improve its adhesion and hence the coating durability. The important characteristics of coatings include the adhesion force and energy. To select the optimal modes of coatings formation, materials and equipment it is proposed the theoretical method of the adhesion force calculation in binary metallic systems. The adhesion force and energy depend on the elemental distribution in the depth of the coating and on the single bond force as in the substrate and in the coating. In addition the adhesion force is also determined by the coefficient taking into account the reduction of the possible bond number and depending on the surface purity and the structural defects presence. The developed model includes all of the above factors. The elements distribution over the depth of the coating was estimated using a kinetic model of mass transfer by vacancy mechanism. The paper presents the results of the adhesion force calculation for the chromium coating on the surface of A21382 steel.

1. Introduction

Restoration and strengthening of machine parts and tools operating under high dynamic loads is an important practical problem due to its rapid wear. The surface layer of the working part of the tool requires special protection. For this purpose are applied wear-resistant coatings deposited by magnetron sputtering. Before the protective coating formation the surface is subjected to chemical and ionic cleaning [1–3]. In addition in order to improve the adhesion and increase the products durability is proposed to produce the ion implantation deep into the substrate. In this case the implanted layer will be a continuation of the coating part embedded in the surface of the product. The important characteristics of coating technology are the adhesion force and energy [4–6]. In the course of the study was developed a theoretical method of the adhesion force calculation for binary metallic systems obtained by ion implantation and subsequent magnetron sputtering.

2. A theoretical model

In the process of ion implantation and the subsequent magnetron sputtering occurs a transition layer between the substrate surface and the coating under the influence of the temperature gradient and the internal stresses (see figure 1) [7, 8].

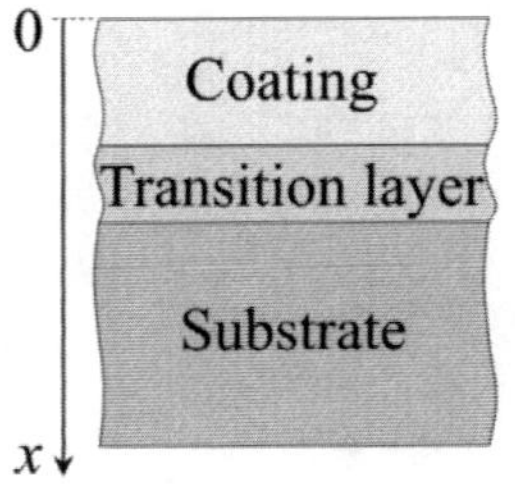

Figure 1. The position of the coating, the transition layer and the substrate surface at the depth. The x axis is directed into the sample perpendicular to the surface of the binary metallic system. The point of the depth reference starts from the coating surface.

In the process of coating by magnetron sputtering to calculate the adhesion force applied the approach described in the source [3]. The adhesion force (A) is proposed to calculate by the formula:

$$A = kNf \tag{1}$$

Here k is the dimensionless coefficient taking into account the reduction in the number of possible bonds depending on the availability of structural defects, f is the average force value of a single metallic bond of the atom near the boundary of transition layer substrate–coating ($[f] = N$) and N is the distribution of the atoms number in the layer with thickness about the crystalline lattice constant near the boundary ($[N] = m^{-2}$).

As a result of ion implantation the depth of the transition layer is significantly greater than for the magnetron sputtering. The transition layer represents a substitutional solid solution and in this case the contribution to the adhesion force from each material can be described by formulas:

$$A_1 = kN_1f_1;\ A_2 = kN_2f_2 \tag{2}$$

Here f_1 and f_2 is the force value of a single metallic bond of the atom of the each material and N_1, N_2 is the atom's quantity distribution in depth of each material in the layer with thickness about the crystalline lattice constant. Putting that N_1, f_1 is the characteristics of the substrate surface and N_2, f_2 is the characteristics of the coating. The adhesion force in the transition layer can be calculated in view of formulas (1) and (2):

$$A = A_1 + A_2 = kN_1f_1 + kN_2f_2 = k(N_1f_1 + N_2f_2) \tag{3}$$

The distributions N_1, N_2 depending on the depth of the coating formed in view of spraying the surface material due to the high radiation dose were calculated by the method proposed in [9, 10]. Denote the atom's concentration of type 1 and 2 as C_1 and C_2 respectively, moreover $C_1 = 1 - C_2$ in the transition layer. The concentration dependence of the implanted impurity from the depth x was calculated by the formula:

$$C_2(x) = \frac{D}{\sqrt{2\pi}\Delta R_p n_0}\left[exp\left(-\frac{(x-h-R_p)^2}{2\Delta R_p}\right) + exp\left(-\frac{(x-h-R_p+s_f)^2}{2\Delta R_p}\right)\right] \tag{4}$$

Here D is the integral dose per unit area of the target (ion/m^2) set in the installation options, n_0 is the concentration of target atoms (substrate surface), R_p is the projective ions range along the x axis,

ΔR_p is the longitudinal dispersion of runs (longitudinal straggling), h is the coating thickness and s_f is the sprayed layer thickness as a result of ion bombardment. Projective ions range R_p and ΔR_p was calculated by the Lindhard–Scharff–Schiott (LSS) model [9] on the basis of mass and atomic number of the target and the coating:

$$\Delta R_p = 0{,}4R_p\left[\frac{4m_1m_2}{(m_1 + m_2)^2}\right]^{1/2} \tag{5}$$

Here m_1 and m_2 is the mass of the target and the coating atoms.

With regard to the number of atoms in the crystalline lattice of the target layer N_0 and according to (4) it was found the depth distribution $N_1(x)$ and $N_2(x)$ of the transition layer.

$$N_1 = C_1N_0;\ N_2 = C_2N_0 \tag{6}$$

The single bond force of the atom was calculated using the melting point (T) of the material. The average kinetic energy of random atom's motion at such temperature goes into the overcome work of the interatomic forces of attraction thus this process can be represented by equation:

$$\frac{3}{2}k_BT = nfd \tag{7}$$

Here k_B is the Boltzmann constant, n is the number of the overcoming atomic bonds in the crystal lattice and d is the lattice constant. Thus for the calculation of the single bond force of the atom can be used the following expression:

$$f = \frac{3k_BT}{2nd} \tag{8}$$

The adhesion force A can also be estimated through the adhesion energy [11, 12].

$$W_a = \frac{W_{p1}W_{p2}(K_1 + K_2)^2}{W_{p1}K_1^2 + W_{p2}K_2^2} \tag{9}$$

Here W_{p1} and W_{p2} is the surface energy of the contacting materials (the reference values) that can be also calculated by the method described in [12] and K_1, K_2 are the coefficients calculated by the formulas:

$$K_1 = v_1(1 - v_1)^{-1};\ K_2 = v_2(1 - v_2)^{-1} \tag{10}$$

Here v_1 and v_2 are the the Poisson's coefficients of the contacting materials. Thus the estimated formula for the adhesion force has the form [12]:

$$A = (\pi h)^{-1/2}\sqrt{W_aE} \tag{11}$$

Here E is the elasticity modulus of the material surface. The described model for calculation the adhesion force and energy was tested for the binary metallic system «steel–coating».

3. Results and discussion

To verify the model it was calculated the adhesion force and energy for chromium coating formed on the surface of the A21382 steel by magnetron sputtering.

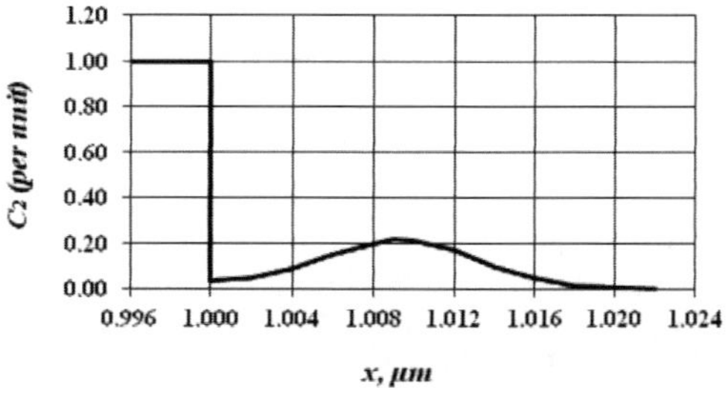

Figure 2. The distribution of the implanted impurity concentration in the A21382 steel after irradiation with subsequent forming the $1\,\mu m$ coating by magnetron sputtering (the coating is not fully shown).

Mass content of iron in constructional steels is greater than 95 percent. Dependence of the impurities content in the transition layer chromium–steel was calculated from the mass transfer model [9], using the formulas (4–6). The values required for the calculations were mostly taken from handbook of physical quantities [13] as well as other handbooks. Figure 2 shows the distribution of the implanted impurity concentration in the A21382 steel after irradiation with the energy $32\,keV$ and the dose $2.7\times\ 10^{17}\,ion/cm^2$ with followed forming the $1\,\mu m$ coating by magnetron sputtering. Figure 2 shows that the transition layer on the x axis situated in the range of 1 to 1.02 μm. In the atomic layer of the A21382 steel fits $N_0 = 1.93 \times 10^{19} m^{-2}$ atoms. The distributions N_1 and N_2 by depth obtained by the formula (6) presented on the figure 3.

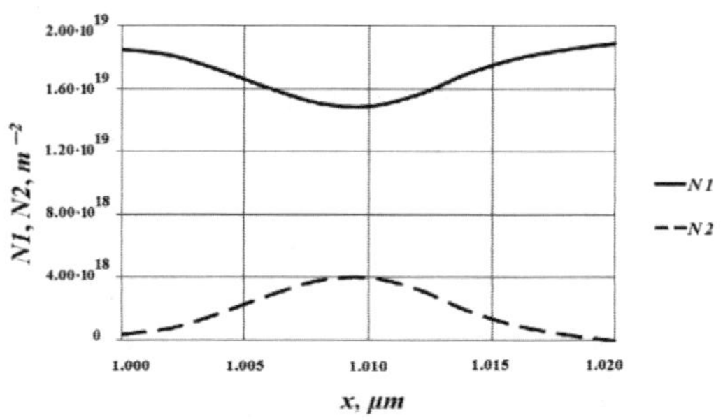

Figure 3. The distributions N_1 and N_2 by depth in the transition layer «chromium–iron» after irradiation with the energy $32\,keV$ and the dose $2.7\times\ 10^{17}\,ion/cm^2$ with subsequent forming the $1\,\mu m$ coating by magnetron sputtering.

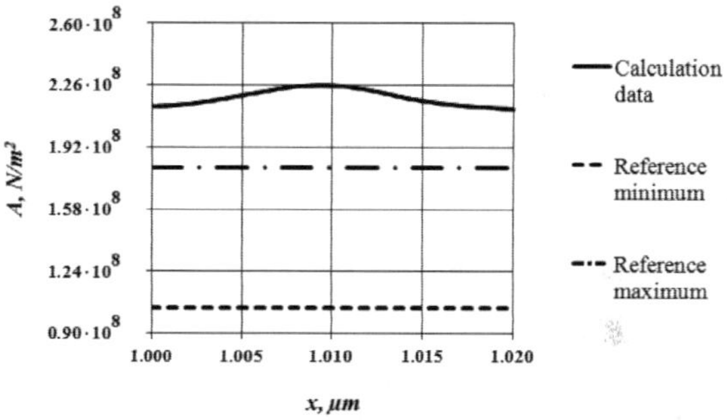

Figure 4. The adhesion force of the coating in dependence from the thickness of transition layer «chromium–iron» after irradiation with followed forming the $1\,\mu m$ coating (solid line). Maximum (dash-dotted line) and minimum (dotted line) the value of the force of adhesion for coatings obtained by ion-plasma methods.

During the single bond force calculation by the formula (8) was taken into account that the iron and chromium crystal lattice have a similar body-centered cubic structure with nine atoms (eight atoms are located at the vertices of a cube and one atom located at its center). Due to the presence of short range order in the crystal structure to remove the atom's layer from the metal matrix it is necessary to break the four strong bonds. For the iron and chromium forces take the following values: $f_1 = 3.76 \times 10^{-11}\,H$ and $f_2 = 4.81 \times 10^{-11}\,H$.

The adhesion force is also influenced by the surface purity of the substrate the presence of impurities and defects and can be influenced by various other factors. Due to these factors experiment that measure adhesion force shows that as a result of magnetron sputtering the contact area of the coating by 45%÷50% smaller than the area of the modified surface [14]. For this reason the model calculations used the adhesion force coefficient $k = 0.5$. Figure 4 shows the adhesion force of the coating (A) in dependence from the transition layer thickness calculated by formula (3) based on the distributions of N_1 and N_2 (solid line on figure 4). According to the literature sources [12, 15] the adhesion force of metals (A) to steel surfaces for a coatings with a thickness of 1 to 10 μm accepts values in the range of 100 to 250 N/m^2. The adhesion force of the chromium coatings formed by the ion-plasma methods on the steel substrate ranges from 104 to 181 N/m^2. Figure 4 shows a slight discrepancy between the calculated values of A compared to the reference 213÷226 N/m^2 so we can assume that the model results and literature data are in satisfactory agreement. The maximum value of the graph $A = A(x)$corresponds to the largest concentration of implanted impurity on figure 2 so by tearing coating off separation most likely occurs or near the surface or at a depth of 1.02 μm which ends with a transition layer. Another important characteristic in the thin–film coatings calculation also stands the adhesion energy which is calculated by the formula (9). The calculated value of the adhesion energy ($W_\alpha = 2.8 \div 4.9$ J/m^2) for the chromium coatings corresponds to the data provided by the sources [11, 12].

4. Conclusion

The results of the model verification are in good agreement with the reference data so it can be used for the modes selection of coatings formation. Calculations indicate the feasibility of the preliminary ion implantation in the protective coating formation that allows for enhanced adhesion and increase the coating durability. The calculated values of the adhesion force slightly exceeded the range of the reference data but were close to the maximum value of 181 N/m^2. It serves a theoretical basis of the ion implantation effectiveness before the formation of metal coatings by magnetron sputtering. Also it should be noted that the estimated value of the adhesion force calculated by the formula (11) substantially higher than the experimental data. The adhesion force on the steel surface for the coatings with thickness of one micrometer order can reach a value of 600 N/m^2 and the adhesion of chromium coating on the surface of pure iron can exceed 820 N/m^2 [12]. This overestimation of the results is obviously due to the fact that the valuation model considered perfectly smooth clean surface without the presence of defects. Also these estimates give an error of more than 100% due to the fact that it is not possible to accurately determine the interaction potential between the atoms that's why it is necessary to use approximate potentials – Morse and others. If in the proposed model will be used coefficient $k = 1$ (ideal case) the adhesive force will be closed to the estimated value.

References

[1] Roikh I L, Koltunova L N and Fedosov S N 1976 *Deposition of protecting coatings in vacuum* (Moscow: Mashinostroenie)
[2] Chen J and Bull S J 2011 *J. Phys. D: Appl. Phys.* **44** 034001
[3] Kim Y C, Lee D S and Lee W J *Jpn. J. Appl. Phys.* **49** 076501
[4] Gerberich W W and Cordill M J 2006 *Rep. Prog. Phys.* **69** 2157
[5] K W Allen 1988 *Physics in Technology* **19** 234
[6] Klapetek P and Nečas D 2014 *Meas. Sci. Technol.* **25** 044009
[7] Mousavi A K, Kashamolla M R and Leseman Z C 2013 *J. Micromech. Microeng.* **23** 115011
[8] Blesman A I, Postnikov D V and Polonyankin D A 2015 *IOP Conf. Ser.: Mater. Sci. Eng.* **81** 012031
[9] Abroyan I A, Andronov A N and Titov A I 1984 *Physical bases of electron and ion technology* (Moscow: Vysshaya Shkola)
[10] Shalaev A M 1988 *Radiation-induced processes in metals* (Moscow: Atomizdat)
[11] Vitkovsky I V, Konev A N, Shorkin V S, Kraev N D, Rusanov A E, Khoroshikh V M and Leonov S A 2003 *Plasma Devices and Operations* **11** 81

[12] Kostyk G I and Melkozerova O M 2011 *Aerospace technic and technology* **3** 16
[13] Grigoriev I S, Meilikhov E Z and Radzig A A 1996 *Handbook of Physical Quantities* (FL: CRC Press)
[14] Kalashnikov M P, Sergeev V P, Fedorishcheva M V and Neyfeld V V 2012 *Russian Physics Journal* **55** 133
[15] Lunev V M and Nemashkalo O V 2010 *Physical Surface Engineering* **8** 64

IOP Conf. Series: Materials Science and Engineering **110** (2016) 012010 doi:10.1088/1757-899X/110/1/012010

The regulation of secondary explosives sensitivity to laser influence

B P Aduev[1], D R Nurmukhametov, A A Zvekov and I Yu Liskov

Institution of Coal Chemistry and Materials Science, 650000, Sovetsky av., 18, Kemerovo, Russia.

E-mail: lesinko-iuxm@yandex.ru

Abstract. The explosive decomposition of pentaerythritol tetranitrate (PETN) pressed pellets containing metal nanoparticles initiated by laser pulses was studied. The critical energy density values, which are the ones that make the sample to explode with 50% probability, were experimentally determined for PETN pellets containing nickel nanoparticles with three average radii initiated by the radiation of 1st harmonics of neodymium laser. It was shown that the critical energy density for PETN -Ni samples anti-correlates with the linear absorption coefficient determined with the optoacoustic approach. The dependence of the critical energy density of PET[1]N samples containing aluminum nanoparticles on their weight percent concentration was measured using the influence of 1st and 2nd harmonics of the laser. It was determined that the minimum critical energy density for both wavelengths coincide while the minimum abscissae, providing almost equal linear absorption coefficient, values are different. The variation of dopants' nanoparticles' radius and concentration makes it possible to regulate the sensitivity of PETN to laser influence.

1. Introduction

There have been many works recently concerned on the regulation of energetic materials sensitivity to laser pulses influence. The practical aspect of the problem is linked to the development of new detonator types and explosives initiated by laser irradiation showing fewer drawbacks than contemporary detonators of other types. Such kind of detonators is essential, for instance, for application in military and special techniques [1]. In order to design the composition of new materials it is necessary to do a research into the mechanism of explosives initiation by laser pulses as it is studied incompletely. It is evident that the explosive's initiation is coupled with the laser pulse energy consumption by the material. In our paper [2] the effectiveness of light absorption by metal nanoparticles of different radii in pentaerythritol tetranitrate (PETN) were calculated at different wavelength. This calculation results needs to be verified experimentally.

The aims of the present work are: (i) experimental measurement of linear absorption coefficient k_{eff} of the laser irradiation for different radii of nickel nanoparticles and the link between k_{eff} and the laser initiation threshold determination; (ii) Experimental verification of the explosion initiation threshold dependence on the concentration of aluminum nanoparticles with fixed size under the influence of first and second harmonics of neodymium laser.

[1] To whom any correspondence should be addressed.

IOP Conf. Series: Materials Science and Engineering **110** (2016) 012010 doi:10.1088/1757-899X/110/1/012010

2. Experimental section

The samples were prepared from PETN powder with narrow size distribution, the average size was 1-2 μm. The three type of nickel powder were used as the additives, the particles' size at the size distribution maximum was 280 nm ($Ni^{(1)}$), 160 nm ($Ni^{(2)}$), and 130 nm ($Ni^{(3)}$). We also used the aluminum powder with the size in the distribution maximum 100-120 nm. The samples were pressed in a copper plate with thickness 1 mm having a round hole with diameter 3 mm at its centre. The samples' density was ρ=1.77±0.03 g/cm^3.

A YAG:Nd^{3+} laser operated in Q-switch mode was used as the initiation radiation source. The main (1064 nm) and the second (532 nm) harmonics of the laser were applied in the present study. The pulse duration fwhm was $\tau_i \approx 14$ ns for the first and 12 ns for the second harmonics of the laser. The intensity distribution over the beam's radius was nearly rectangular. The radiation was focused unto the sample surface with a lens. The beam diameter on the sample was 2.5 mm. The pulse energy was varied with a set of neutral light filters that attenuated the light intensity.

The dependencies of explosion probability on the pulse energy density were measured as follows. The samples were set on an aluminum plate. The samples' surface was protected with an optical glass plate 1 mm thick that partly resisted the gas-dynamic response of the high-pressure zone forming after the irradiation. The explosion event was fixed if the laser influence was accompanied by a sharp sound, the experimental cell destruction and leaving of an imprint on the aluminum plate with the diameter equal to the sample's one. The experimental explosion probability was plotted against the initiation energy density then. The initiation threshold was determined as the energy density at the explosion probability level 50%.

In another experimental series the similar samples were used in the absorption coefficient measurements with optoacoustic approach [3]. In that case the samples were set in the acoustic contact with a piezo detector made of CTS-19 ceramics. The irradiated surface was not protected with a glass plate. The pulse energy was 5 mJ which was enough for the measurable optoacoustic signals stimulation and was significantly less than samples' destruction threshold. The signal of piezo detector was fixed with an oscilloscope LeCroy WJ332A.

3. Results and discussion

The influence of nickel nanoparticles' size on the initiation threshold of PETN – Ni composite in the conditions of 1st harmonics of the neodymium laser irradiation was studied. The nanoparticles' weight percent concentration was 0.1%. The critical energy densities of the laser pulse were measured for three samples containing nanoparticles with different sizes $Ni^{(1)}$, $Ni^{(2)}$, and $Ni^{(3)}$. The corresponding dependence is plotted on the fig. 1 (curve 1). One is able to notice that the optimal size of the nanoparticle exists which leads to the maximum sensitivity to the laser irradiation.

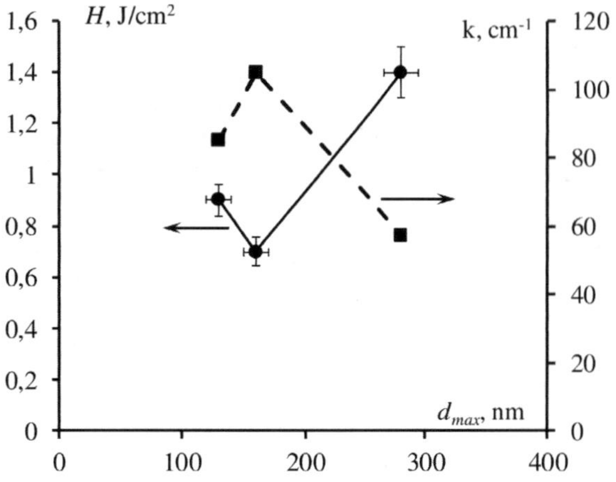

Fig. 1. The experimental dependencies of the critical laser pulse energy of PETN-Ni composite initiation H_{cr} (1) and the effective absorption coefficient k_{eff} (2) on the nanoparticles' size at the size distribution maximum d_{max}.

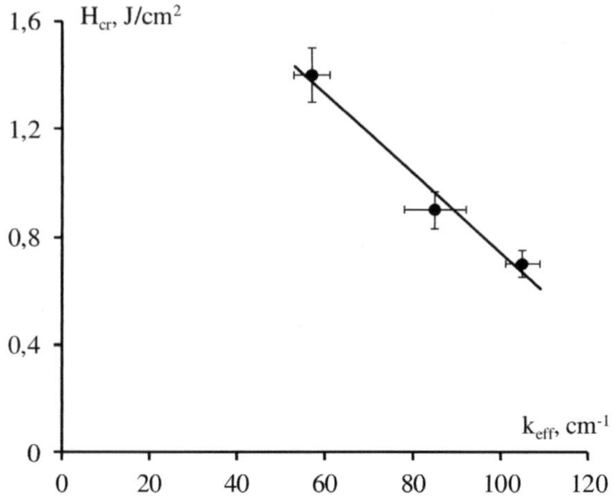

Fig. 2. The critical laser pulse energy density of PETN-Ni composite initiation H_{cr} dependence on the effective absorption coefficient k_{eff} measured with optoacoustic method.

The absorption coefficient was measured with the optoacoustic approach using the same samples. The temporary acoustic signal curve registered by a piezo detector corresponds to the distribution of the thermal sources in the medium generated by the laser pulse, which is the foundation of the optoacoustic method [4]. The ascending exponential part of the experimental signal was used for the time-constant determination:

$$\tau_c \approx (k_{eff} c_0)^{-1} \quad (1)$$

where τ_c is the character time of the signal rise corresponding its e times augmentation, k_{eff} is linear absorption coefficient of light concerned on its absorption by nanoparticles, and c_0 is sound velocity in the sample. The measured c_0 value was 2500±200 m/s in the experimental samples. The determined values of τ_c and c_0 were utilized to calculated k_{eff} using equation (1). The resulting effective absorption coefficients are shown on the fig. 1 (curve 2). It is seen that the dependence $k_{eff}(d)$ is a non-monotonic function. The effective absorption coefficient one is able to write as

$$k_{eff} = \sigma n = Q_{abs} \sigma_g n$$

where n is concentration, σ is absorption cross section, σ_g is geometric cross section, and $Q_{abs}(d)$ is efficiency of light absorption by metal nanoparticles in the PETN matrix that is a function of nanoparticles' material, size and light wave length according to the calculation results [2] performed in terms of Mie theory. The dependence $Q_{abs}(d)$ is non-monotonic that correlates with the results depicted on the fig. 1.

Fig. 2 presents the same results as a H_{cr} dependence on k_{eff}. The augmentation of linear absorption coefficient makes the critical energy density decrease providing that the nanoparticles' concentration is constant.

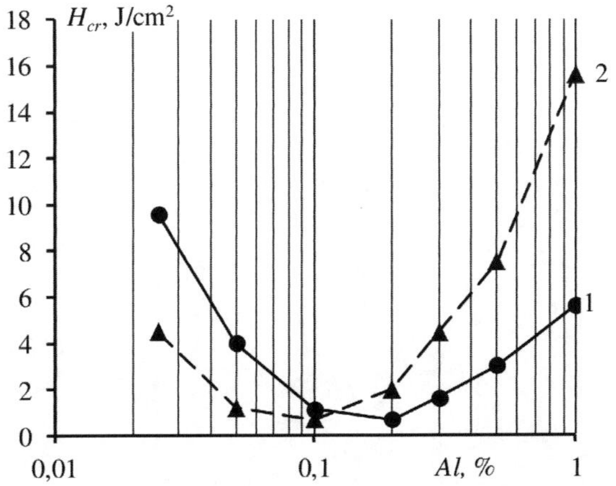

Fig. 3. The critical pulse energy density of PETN-Al composite initiation H_{cr} dependence on the nanoparticles' weight percent at the laser wavelengths 1064 (1) and 532 nm (2).

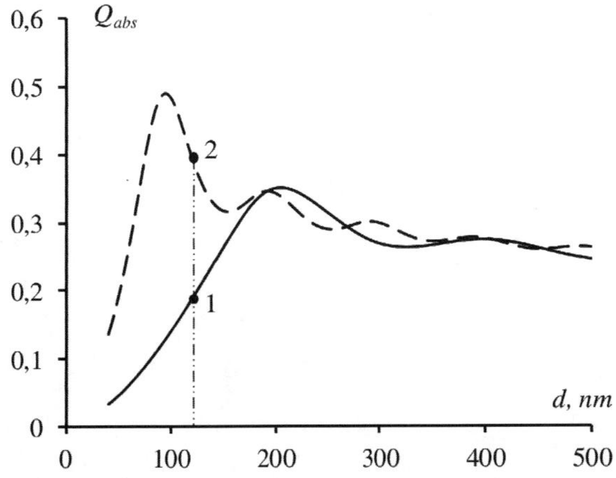

Fig. 4. The calculated dependencies of aluminum nanoparticles' absorption efficiencies (Q_{abs}) on their diameter at wave lengths 1064 (1) and 532 nm (2).

A research into the influence of the nanoparticles' concentration on the critical energy density of PETN –Al composites was carried out. The composites' explosion was initiated with 1st and 2nd harmonics of the neodymium laser. As it was done in the previous cases the explosion probability dependence on the initiation energy density were measured. The nanoparticles' weight percent was in the range 0.025 – 1 %. The results are presented on the fig. 3. The both curves obtained at the wave lengths 1064 and 532 nm shows the minimum critical energy density values H_{cr} =0.7 J/cm^2 coinciding inside the experimental precision. However, the minimum point for 1st harmonics is observed at 0.2 weight % of nanoparticles while for 2nd – at 0.1 %. For that reason we can assume that the same value of H_{cr} means that

$$k_{eff.}^{1}= k_{eff.}^{2} \tag{3}$$

The absorption effectiveness of the aluminum nanoparticles in PETN calculated at the wave lengths 1064 and 532 nm in [2] are shown on the fig. 4. The minimum points positions' give:

$$n_1=2n_2.$$

So, taking into account equations (2) and (3), an equation is expected

$$Q_{abs}^{1}/\ Q_{abs}^{2}=2. \tag{4}$$

According to the fig. 4, the equality in formula (4) is provided for the nanoparticles' diameter 120 nm, which matches the maximum position of the size distribution of the aluminum nanoparticles.

The following experiment was carried out in order to understand the dependence of H_{cr} on the nanoparticles' weight percent concentration. A dependence of the optoacoustic signal amplitude, which is proportional to the pressure amplitude, was measured at constant pulse energy density using 1st harmonics of the neodymium laser. The resulting curve is plotted on the fig. 5. The pressure increases with the weight percent concentration of nanoparticles increase until the optimum value 0.2%. The consequent weight percent concentration increasing makes the pressure amplitude decrease.

The effect could be qualitatively explained as follows. According to [4], if the condition

$$k_{eff}c_0\tau_i<<1 \tag{5}$$

where τ_i is laser pulse duration, is fulfilled, the instantaneous mode of laser excitation arises. The radiation energy consumed by the nanoparticles is converted into heat. The consequent temperature rise generates thermal stress and pressure increasing. The pressure jump appearing in the radiation absorption layer propagates inside the sample as an acoustic pulse and is measured by a piezo detector. As long as the condition (5) is fulfilled, the pressure measured increases with the nanoparticles' concentration increasing. The critical initiation energy density decreases in the same range (figs. 3 and 5). When an opposite condition $k_{eff}c_0\ \tau_i>1$ is met, the quasi-stationary mode of excitation evolves [4]. The surface layer having width $x=k_{eff}^{-1}$, where the pulse energy is absorbed, has enough time to expand during the laser pulse in the case. As a consequence, the pressure amplitude diminishes and the critical energy density H_{cr} increases (figs. 3 and 5).

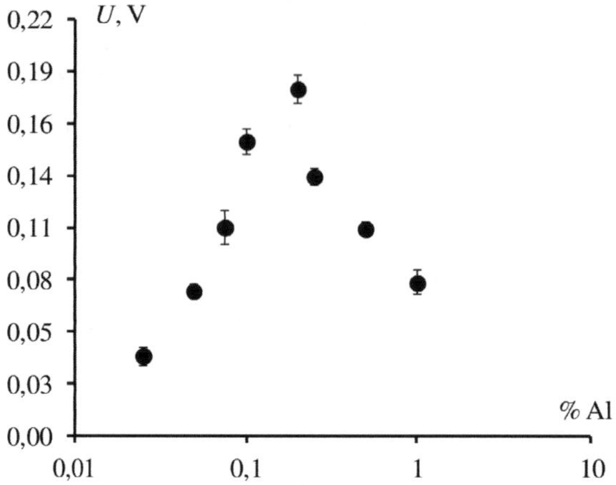

Fig. 5. The optoacoustic signal amplitude as a function of Al nanoparticles weight percent in the sample.

In the direction of condition (5) meeting range the measurement of the effective absorption coefficient k_{eff} for optimal weight percent 0.2% of aluminum nanoparticles in the case of 1st harmonics laser excitation was done with the optoacoustic approach. The value obtained was $k_{eff}=240\ cm^{-1}$. It means that $k_{eff}c_0\ \tau_i=0.7$. Thus the optimal concentration of nanoparticles (and corresponding k_{eff} value) matches the approximate boundary between the instantaneous and quasi-stationary modes.

Acknowledgement
The authors thanks Kemerovo Scientific Center for the equipment used. The work was partially supported by the grant of Russian Foundation for Basic Research (13-03-98032_r_sibir_a) and the grant of the President of Russian Federation (MK-4331.2015.2).

References
[1] Barbarisi M J and Kessler E G 1969 *Initiation of Secondary Explosives by Means of Laser Radiation* (*Technical Report 3861*) (NewJersey: Picatinny Arsenal, Dover).
[2] Kriger V G, Kalenskii A V, Zvekov A A, Zykov I Yu and Aduev B P 2012 Effect of laser radiation absorption efficiency on the heating temperature of inclusions in transparent media *Combustion, Explosion, and Shock Waves* **48** 705.
[3] Magunov A N 2009 Spectral pyrometry *Instruments and Experimental Techniques* **52** 451.
[4] Gusev V E and Karabutov A A 1993 *Laser Optoacoustics* (New York: AIP)

Radiation Hardening of Ni-Ti Alloy Under Implantation of Inert Gases Heavy Ions

V Poltavtseva[1], A Larionov[1], D Satpaev[1] and M Gyngazova[2]

[1]Institute of Nuclear Physics, Almaty, Republic of Kazakhstan
[2]National Research Tomsk Polytechnic University, Tomsk, Russia

E-mail: poltavtseva@inp.rk

Abstract. The consistent patterns of changes in nano- and micro-hardness of Ni-Ti alloy with the shape memory effect after implantation of ${}_{40}Ar^{8+}$ and ${}_{84}Kr^{15+}$ ions depending on phase composition and implantation parameters have been experimentally studied. It has been shown that softening by 4 and 14% near the surface of the two-phase Ni-Ti alloy after implantation of ${}_{40}Ar^{8+}$ and ${}_{84}Kr^{15+}$ ions is connected with the differences in the nanostructure. Hardening of the near-surface layer of this alloy maximum by 118% at h = ~3 μm and single-phase alloy in the entire region of the ${}_{40}Ar^{8+}$ and ${}_{84}Kr^{15+}$ ions range and in the out-range ($h > R_p$) area have been detected. The role of the current intensity of the ions beam in the change of nanohardness for the two-phase Ni-Ti alloy has been established.

1. Introduction

The problem of applying ion implantation and/or various beam technologies to produce the hardened coatings with the improved corrosion properties, biochemical compatibility with living tissues of the body and low toxicity for the titanium nickelide based alloys for medical application has not yet been solved. This is associated with the behavior of the dispersion processes, phase transformations, and the features of defect structure and the effect of the phase composition of the titanium nickelide based alloys and implantation parameters. In [1, 2] it was shown that as a result of implantation of MeV-energy gas elements heavy ions at relatively the same parameters the dispersion degree of the two-phase alloy surface decreases with the increase of the ion mass accompanied by the radiation-induced phase transformation B19'→B2, formation of the fine-dispersed or track structure, the two-layered radiation-hardened near-subsurface structure. The effect of E_{ion}, M, Z, J parameters on tracks evolution in two-phase and single-phase Ni-Ti alloy [2] and the effect of ion charge on the hardening of the surface layer [3] were established. The provided results suggest new possibilities of modification by implantation of inert gases heavy ions of the titanium nickelide based alloys. However, it is required to perform a comprehensive and thorough study of their structure, physical and mechanical properties, the characteristics of phase transformations, hardening/softening and etc.

The aim of the present investigations is to study the characteristics of changes in nano - and micro-hardness in the modified layer of Ni-Ti alloy after implantation of argon and krypton heavy ions of MeV energy depending on phase composition and implantation parameters.

2. Experimental methods and material

We studied the Ni-Ti commercial alloy with Ni of 53.46 wt.% and Ti of 46.54 wt.%, preferably consisting of NiTi with the B2 structure (austenite), NiTi with B19 structure (martensite) and a minor content of Ti, excess Ni in the form of solid solution and process particles similar in composition to $Ti_2Ni(C)$ [1]. Before implantation, a proven technology was used to prepare the surface of the samples: cutting out by the method of spark cutting followed by wet mechanical polishing and buffing with GOI paste. The sample quality was evaluated by the methods of optical metallography and scanning electron microscopy. Based on certification by the method of X-ray structural analysis we have selected the samples of Ni-Ti alloy with the titanium nikelid phase ratio B19'/B2 equal to 0.6-0.7 (two-phase alloy) for investigations. The single-phase alloy with the B2 structure (austenite) was produced by annealing of two-phase samples in the vacuum (~10^{-4} Pa) oven for 1 h at 510 K.

Implantation of $_{40}Ar^{8+}$ ions (70 MeV, 100 and 80 nA/cm^2) and $_{84}Kr^{15+}$ (147 MeV, 94, 70 and 30 nA/cm^2) was performed with DC-60 cyclotron of the Eurasian University (Astana) under comparable parameters A/Z = ~5, E_{ion} = 1.75 MeV/nucl, Φt = 1×10^{19} ion/m^2. The area of radiation treatment was ~1×10^{-4} m^2. The samples temperature in the process of irradiation was below 370 K.

Measurements of nano – and microhardness of the Ni-Ti alloy samples were performed by the method of sclerometry [2, 3] and by Vickers' method depending on the applied load in the range of 1 ± 200 mN and 0.098 ± 4.9 N using the nanohardness gage «NanoScan-Compact» and microhardness gage «PMT-3M» (Russia) respectively. The accuracy was 1% for nanohardness measurement and 3-4% for microhardness measurement. The thickness of surface and near-surface layers at each applied load was determined by the depth of scratches and impressions. The surface roughness was evaluated according to the scanned images.

3. Experimental results and discussion

The results of nanohardness measurement depending on surface layer thickness for two-phase and single-phase Ni-Ti alloys before and after implantation of $_{40}Ar^{8+}$ and $_{84}Kr^{15+}$ ions are represented in Fig. 1. It is shown that the values of surface nanohardness for these alloys differ in 1.8 times, caused mainly by the difference in hardness of martensite and austenite.

In the case of two-phase alloy the character H(h) of the curves is similar, with the maximum near the surface (Fig. 1a). The highest value of nanohardness is observed for un-implanted two-phase Ni-Ti alloy (Fig. 1a, curve 1). The paper [1] shows that it is connected with formation of Ni_3Ti_3O compound in the deformation-hardened layer in the process of Ni-Ti alloy samples preparation.

The implantation of $_{40}Ar^{8+}$ and $_{84}Kr^{15}$ ions results in softening of the surface layer by 4 and 14% near the surface respectively and slightly in the layer of >120 nm thickness (curves 2 and 3), compared to un-implanted two-phase alloy (Fig. 1a, curve 1). Softening of the modified surface layer of two-phase alloy, regardless of the ion mass, can be caused by sputtering of the deformation-hardened layer, as indicated by the shift of maxima on H(h) - curves to the modified surface, by decreasing of martensite due to radiation-stimulated phase transformation B19'→B2 [1, 4] and the change in the surface microstructure [1, 5].

According to X-ray diffraction analysis data [1] the process of $_{40}Ar^{8+}$ and $_{84}Kr^{15}$ ions implantation is characterized by the radiation stimulated transformation B19'→ B2 occurring in the surface layer. The phase composition of the surface consists essentially of the phase with the structure B2 (austenite). However, the degree of dispersion and change of the surface structure, as established in [1, 5], depends on the mass of heavy ions under comparable parameters of implantation. Indeed, in [1] based on the studies data obtained by scanning electron microscopy and X-ray spectral analysis it was shown that for the two-phase alloy, firstly, the degree of dispersion decreases and the role of surface ion polishing increases with the ion mass growth. This conclusion is also supported by the difference in the values of maxima shift (Figure 1a, curves 2 and 3).

Secondly, the interaction of $_{40}Ar^{8+}$ and $_{84}Kr^{15+}$ ions with the surface of two-phase alloy is accompanied by formation of uniform fine-dispersed (up to 1 nm) structure and inhomogeneous larger (up to 25 nm) track structure respectively [5]. From comparison of the above mentioned causes of the surface layer softening of the two-phase alloy, it can be concluded that the difference in the nanostructure is one of the main reasons for the differences in nanohardness after implantation of $_{40}Ar^{8+}$ and $_{84}Kr^{15+}$ ions.

In contrast to the two-phase alloy (Figure 1 a), the single-phase alloy is characterized by hardening near the surface and in the layer ≥50 nm thick by 8.6 and 64%, respectively as the result of implantation of $_{84}Kr^{15}$ ions (figure 1b, curve 3). However, in case of $_{40}Ar^{8+}$ ions implantation hardening by 66% in average is observed only in the layer ≥75 nm thick, and there is no maximum on H(h) - curve (Figure 1b, curve 2). The data obtained for the single-phase alloy also support the conclusion about the influence of ion mass on the degree of surface dispersion.

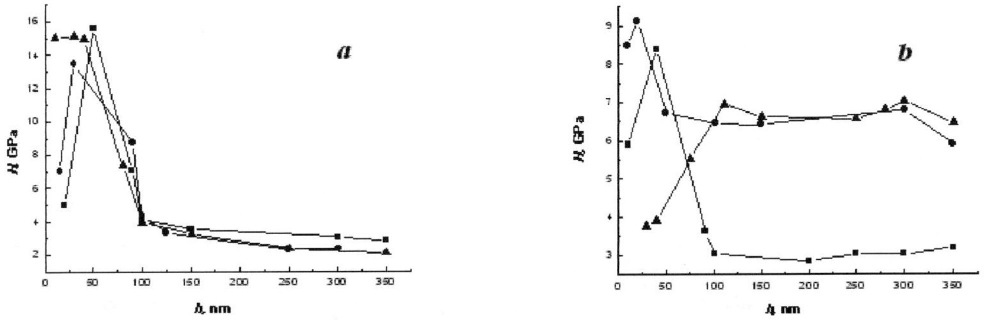

Figure 1. Dependence of nanohardness for two-phase (a) and single-phase Ni-Ti alloy (b) before (curve 1) and after implantation of $_{40}Ar^{8+}$ (curve 2) and $_{84}Kr^{15+}$ (curve 3) ions on thickness of surface layer

It follows from the analysis of the arithmetic mean values of roughness height obtained by measuring of the roughness curves, that the surface quality is practically independent from the phase composition during implantation of $_{84}Kr^{15}$ ions. However, as the result of implantation of $_{40}Ar^{8+}$ ions, the surface roughness of the two-phase alloy, on the contrary, is 1.6 times higher than for the single-phase alloy, whereas this ratio is 1.9 times for the unimplanted alloys. Thus, the marked roughness data also indicate the predominant role of the ion polishing process during implantation of heavier krypton ions, and dispersion during interaction of lighter argon ions with the surface of Ni-Ti alloy.

The patterns of microhardness changes depending on thickness of the near-surface layer for two-phase and single-phase Ni-Ti alloys before and after implantation of $_{40}Ar^{8+}$ и $_{84}Kr^{15+}$ ions are shown in Figure 2. Similar to nanohardness (Figure 1), firstly, the microhardness values of the near-surface layer of these alloys vary, but in a less (1.3-time) degree.

Secondly, $H_\mu(h)$ - curves are similar for two-phase alloy (Figure 2a) and different for single-phase alloy (Figure 2b) after implantation of heavy ions of inert gases. This measured $H_\mu(h)$ - curve of single-phase alloy after implantation of $_{40}Ar^{8+}$ ions (Figure 2b, curve 2) is similar to the corresponding curves for two-phase alloy (Figure 2a, curves 2 and 3), and after implantation of $_{84}Kr^{15+}$ ions it is more flat (Figure 2b, curve 3). Thirdly, the unimplanted two-phase alloy softens with increasing thickness of the near-surface layer (Figure 2a, curve 1), whereas in case of single-phase alloy, on the contrary, strengthening is occurring (Figure 2b, curve 1).

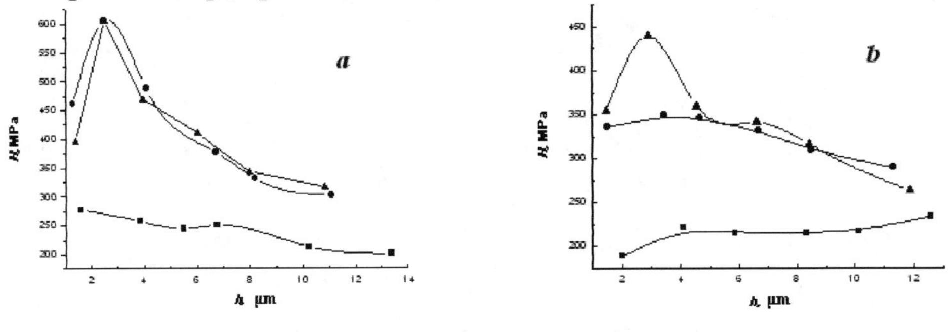

Figure 2. Dependence of microhardness for two-phase (a) and single-phase Ni-Ti alloy (b) before (curve 1) and after implantation of $_{40}Ar^{8+}$ (curve 2) and $_{84}Kr^{15+}$ (curve 3) ions on thickness of near-surface layer

We shall note that, in contrast to the surface layer (Figure 1) the radiation hardening of Ti-Ni alloy (Figure 2) is occurring for the near-surface layer, irrespective of phase composition and ion mass. The maximum degree of hardening is observed ~3 μm deep and it equals to 118% for two-phase alloy ($_{40}Ar^{8+}$, $_{84}Kr^{15+}$) and 99 ($_{40}Ar^{8+}$) and 58% ($_{84}Kr^{15+}$) for single-phase alloy. These data confirm the formation of two-layered structure of the near-surface layer with various degree of hardening [1]. Besides, the hardening of the near-surface layer is observed for both alloys in thicknesses greater than the projected range of R_p ions in Ni-Ti alloy (R_p = 8.6 and 10.1 μm respectively for $_{40}Ar^{8+}$ and $_{84}Kr^{15+}$ ions), i.e. in the out-range area.

The comparison of the implantation parameters shows that the current intensity for $_{40}Ar^{8+}$ (80 nA/cm^2) and $_{84}Kr^{15+}$ (30 nA/cm^2) ions beam in the case of single-phase alloy is lower than for $_{40}Ar^{8+}$ (100nA/cm^2) and $_{84}Kr^{15+}$ (94 nA/cm^2) ions during implantation of two-phase alloy. In [5] it was found that the current intensity of $_{84}Kr^{15+}$ ions beam has a significant impact on the microstructure of two-phase Ni-Ti alloy surface. In this regard, the measurements of nano - and micro-hardness of two-phase alloy after implantation of $_{84}Kr^{15+}$ ions were performed with varying of the beam current intensity. The obtained results are shown in Figure 3.

As it can be seen in Figure 3a, the character of H(h) - curves in the layers (up to 100 nm thick) is the same for all values of intensity. The maximum value of nanohardness is observed in case of J = 70 nA/cm^2 (curve 2). A decrease in nanohardness by 24% during implantation of ions with the maximum (94 nA/cm^2) intensity (curve 3) may be associated with a high degree of dispersion processes [1, 4] and with the presence of tracks formations in the surface microstructure [4] compared to the intensity of 70 nA/cm^2.

The minimum value of nanohardness in the layer up to 50 nm thick (curve 1) is observed at minimum (23 nA/cm^2) intensity and caused mainly by predominance of smaller tracks in the near-surface layer microstructure [4]. However, in the > 120 nm thick layer (Figure 3), H(h) curve after ions implantation with the minimum intensity, in contrary, is higher than for the case with the maximum intensity.

In case of microhardness the character of $H_\mu(h)$ - curves in the layer of < 5 microns, is the same for all values of the beam current intensity (Figure 3b). However, the maximum value of microhardness corresponds to the maximum intensity (curve 3), and its value is practically the same (curves 1 and 2) for lower intensities. And the intensity value has no evident impact on the microhardness value in the out-range area where radiation hardening of two-phase and single-phase alloys (Figure 2a) are detected.

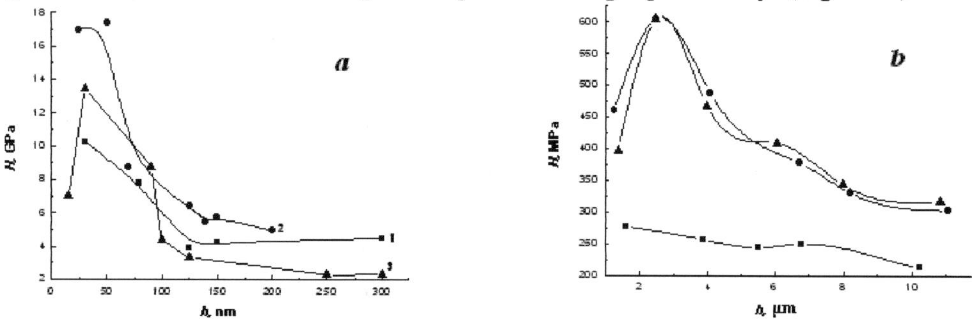

Figure 3. Change of nano – and microhardness of two-phase Ni-Ti alloy from intensity of $_{84}Kr^{15+}$ ions beam current: 1 – 23; 2 – 70; 3 – 94 nA/cm^2

The paper [3] also specified the role of $_{84}$Kr ions charge in the nanohardness change depending on thickness of the surface layer of two-phase Ni-Ti alloy. It was found that near the surface the nanohardness value increases in 2.6 times with the decrease of the ions charge from 15+ to 11+ and equalizes in 800 nm thick layers. The increase of $_{84}$Kr ions charge, as shown in [5], occurs with the slight increase in the number of individual tracks.

4. Conclusion

Thus, the study shows that softening by 4 and 14% near the surface of two-phase Ni-Ti alloy after implantation $_{40}Ar^{8+}$ и $_{84}Kr^{15+}$ ions at relatively same parameters of A/Z, E/nucleon, Φt, J is mainly caused by formation of fine-dispersed and track structure, respectively. The near-surface layer of this alloy, on the contrary, hardens regardless of the ion type, maximum by 118% at the depth ~3 μm. It is found that in the case of single phase Ni-Ti alloy hardening is observed in the entire region of heavy ion range, except the area near the surface due dispersion of the deformation-hardened layer by $_{40}Ar^{8+}$ ions. The hardening of both alloys in the out-range area (h > R_p) is alsp observed. It is shown that the effect of the current intensity of the $_{84}Kr^{15+}$ ions beam on the nanohardness value in the case of two-phase Ni-Ti alloy is associated with the change of the track structure.

Acknowledgements
This work was partially supported by The Ministry of Energy of the Republic of Kazakhstan within the framework of the budget program "Applied scientific research of technological character" and by The Ministry of Education and Science of the Russian Federation in part of the science activity program.

References

[1] Poltavtseva V, Kislitsin S and Satpaev D 2015 *IOP Conf. Series: Materials Science and Engineering.* **81** 01234
[2] Poltavtseva V 2015 *Proc. XI Int. Conf. on Interaction of Radiation with Solids (Minsk)* (Publishing House Center BSU) p 141
[3] Larionov A, Poltavtseva V and Kislitsin S 2015 *Proc. XII Int. Conf. on* Advanced *technologies, equipment and analytical systems for materials and nano-materials (Yst-Kamenogorsk)* (Publishing House Center VKGTU) p 283
[4] Gogolinskiy K, Lvova N and Useinov A 2007 *Plant laboratory* **73** p 28
[5] Poltavtseva V, Kislitsin S, Koval N and Oskomov K 2012 *Izv. Vyssh. Uchebn. Zaved. Fiz.* **12/3** p 41

Fatigue variation of surface properties of silumin subjected to electron-beam treatment

S V Konovalov[1,4], K V Aksenova[1], V E Gromov[1], Yu F Ivanov[2,3] and O A Semina[1]

[1] Siberian State Industrial University, Novokuznetsk, Russia
[2] Institute of High Current Electronics Siberian Branch of Russian Academy of Science, Tomsk, Russia
[3] National Research Tomsk State University, Tomsk, Russia
E-mail: konovalov@physics.sibsiu.ru

Abstract. The analysis of structure-phase states modification of silumin subjected to electron beam treatment with the following fatigue loading up to the failure is carried out by methods of transmission electron diffraction microscopy. The tribology and strength properties of silumin surface after electron beam treatment and fatigue tests are studied and hardness decrease, wear coefficient and friction coefficient increase with the growth of cycles number are revealed. The possible reasons of the tribology and strength properties of silumin surface layers decrease are discussed.

1. Introduction

Concentrated flows of energy are used currently for increasing the mechanical and service properties of steels and alloys. An exposure to their effect the ultrahigh speed heating, melting, evaporation, crystallization and subsequent cooling of the material are realized. In total, these processes result in the formation of wide spectra of amorphous, nano-, submicro- and microcrystalline structures being accompanied by multiple increase in physico-chemical and mechanical properties of material which are often beyond the reach of the conventional methods of surface processing [1-3].

Intensive pulsed periodic electron beams are the exceptionally flexible surface source of dynamic thermal effect on materials. The main advantage of the method in question consists in combination of practically total absorption of energy of electrons and three-dimensional character of its liberation with the possibility of a rather wide variation of depth of electron's penetration in material and correspondingly, dynamics of thermal fields and parameters of stress wave. Nowadays due to the wide variation limits of beam energy density, duration of pulse and electron energy, they are the unique and highly effective tool for both the investigations of physical nature of formation of gradient structure-phase states in a solid body and purposeful modification of structure and properties of silumins in order to improve the service characteristics of products [4-6].

In a number of papers [1, 7] it was shown that surface processing of stainless steels with high intensity electron beam was accompanied by multiple (by more than 3.5 times) increase in fatigue life of these materials. It is apparent that deformation effect taking place in fatigue tests will be favorable to transformation of structure and properties of materials.

[4] Department of physics, Siberian state industrial university, Kirov str., 42, Novokuznetsk, 654007, Russia. Tel. +7 3843 462277, Fax. +7 3843 465792.

Nowadays silumins find wide application in industry therefore the issues of increasing the mechanical and strength properties of silumins are especially urgent. The goal of the research was to study the change properties of silumin surface subjected to electron beam treatment (EBT).

2. Materials and methods of research

Al-Si alloy (silumin of eutectic composition) was used as test material. Fatigue tests were performed according to the scheme of asymmetric cyclic cantilever bending [1, 7]. Samples had the shape of parallelepiped with sizes of 8x14x145 mm. The imitation of a crack was done by the semicircle notch with 10-mm radius. Test temperature was 300 K, frequency of bend loading of samples was 15 Hz under load of 10 MPa.

The irradiation of samples surface prepared for fatigue tests were done on the device "SOLO" [8] with the following parameters: electron energy = 16 keV; pulse repetition frequency = 0.3 sec^{-1}; pulse duration of the electron beam τ = 150 μs; number of pulse of effect N = 1 and 5; energy density of electron beam of E_S = 20 $J \cdot cm^{-2}$. The face surface of samples was irradiated i.e. sample's surface located above the notch imitating a crack. For each mode of irradiation not less than five samples were tested. Strength properties of the material were studied by determining the nanohardness (Nano Hardness Tester NXT-S-AX-000X, load on indenter 5-300 mN). Investigations of wear rate of film-substrate system were done in disk-pin geometry by tribometer (CSEM, Switzerland) at room temperature and humidity. Volume of material's wear was determined after performance of profilometry of the formed track with the help of laser optic profilometer MicroMeasure 3D Station (Stil, France).

3. Results and discussion

In EBT of eutectic silumin in optimal regime a more than triple increase in fatigue life is attained [9]. The main reason of this effect is dispersion and quasihomogeneous distribution of silicon in the layer modified with EBT. EBT increases significantly the wear resistance of eutectic and hypereutectic silumins [10, 11]. The results of tribological tests of silumin are presented in table 1 and figure 1. The samples undergone the fatigue tests during 132000 cycles have the minimum wear coefficient. After 547000 cycles of fatigue tests the wear coefficient of the material increases, however, it remains lower in reference to coefficients registered for the initial material (unirradiated material before fatigue tests). The value of friction coefficient is similarly changed (table 1).

Table 1. Change of tribological parameters of silumin after irradiation and subsequent fatigue loading up to failure.

Regime of electron beam treatment (EBT)	Wear coefficient, 10^{-6} $mm^3/N \cdot m$	Friction coefficient
Initial state without EBT	8927	0.527
132000 cycles up to failure without EBT	13520	0.520
466000 cycles up to failure without EBT	13920	0.444
20 $J \cdot cm^{-2}$; 150 μs; 1 pulse. 132000 cycles up to failure	6466	0.457
20 $J \cdot cm^{-2}$; 150 μs; 5 pulses. 574000 cycles up to failure	8135	0.480

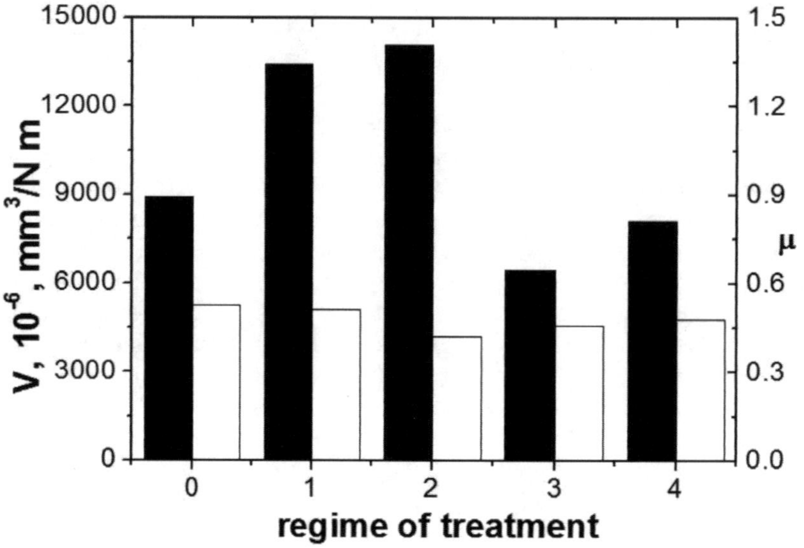

Figure 1. Values of wear coefficient V (dark columns) and values of friction coefficient μ (light columns) having place in different regimes of effect on silumin: 0 – silumin without irradiation (initial state); 1 – initial silumin after fatigue tests (130000 cycles); 2 – initial silumin after fatigue tests (466000 cycles); 3 – irradiation (20 J · cm^{-2}; 150 μs; 1 pulse) and subsequent fatigue tests (130000 cycles); 4 – irradiation (20 J · cm^{-2}; 150 μs; 5 pulses) and subsequent fatigue tests (466000 cycles).

The results of nanohardness tests are shown in figure 2. It is seen clearly that increase in cycles' number of fatigue loading by ~4.4 times results in reducing the hardness of surface layer by ~4.5 times. The value of Young modulus is changed similarly (figure 3): increase in cycles' number of fatigue tests results in multiple (by 3-4 times) decrease in Young modulus of surface layer of the material.

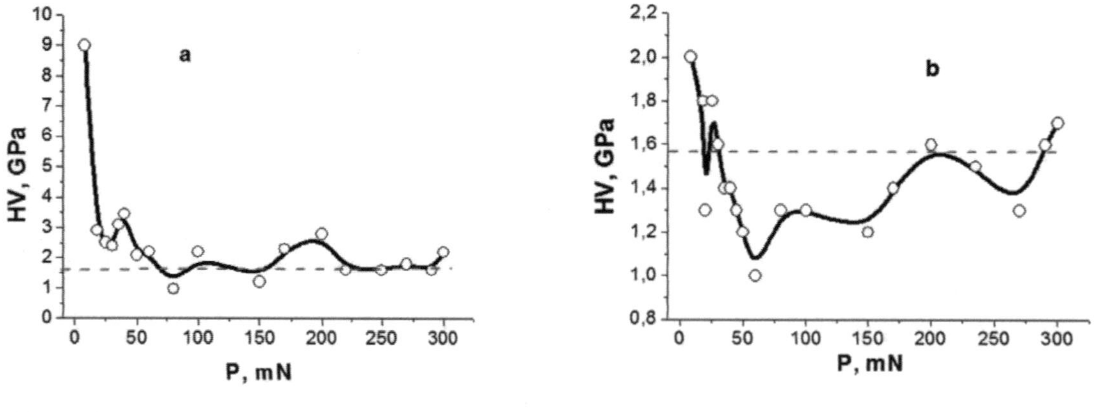

Figure 2. Dependence of hardness of silumin surface layer treated with high intensity pulsed electron beam and subjected to fatigue tests up to failure on the load to indenter; a – 132000 cycles; b – 574000 cycles.

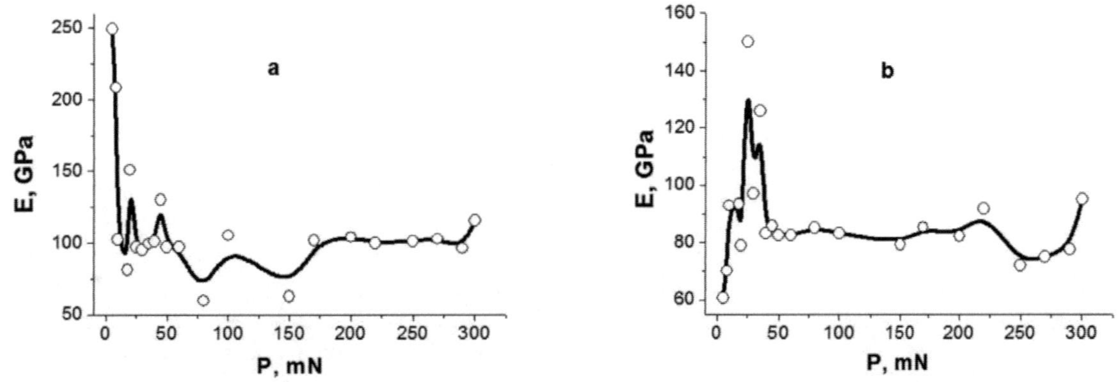

Figure 3. Dependence of Young modules of silumin surface layer treated with high intensity pulsed electron beam and subjected to fatigue tests up to failure on the load to indenter; a – 132000 cycles; b – 574000 cycles.

It is evident that the similar changes of tribological and strength characteristics are caused by change in surface layer of silumin in the process of fatigue tests. The dependences of crystal lattice parameter of Al-based (a) and Si-based (b) solid solution on the regime of sample treatment (initial state and fatigue failed) determined by methods of X-ray and phase analysis are shown in figure 4.

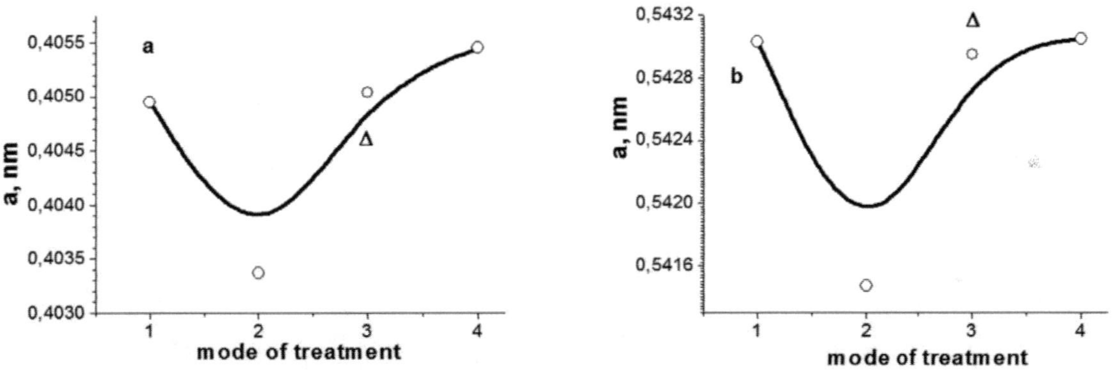

Figure 4. Dependence of lattice parameter of Al- based (a) and Si-based (b) solid solution on density of absorbed energy of electron beam. Regime of effect: 1 – silumin without irradiation; 2 – silumin without irradiation after fatigue tests (130000 cycles); 3 – after irradiation (20 J · cm^{-2}; 150 μs; 1 pulse) and fatigue tests (132000 cycles); 4 – after irradiation (20 J · cm^{-2}; 150 μs; 5 pulses) and fatigue tests (574000 cycles). Parameters of aluminium and silicon crystal lattice are designated by mark Δ, respectively for silumin irradiated according to the regime (20 J · cm^{-2}; 150 μs; 1 pulse).

When analyzing the results shown in figure 4 it can be noted that maximum change of parameter of aluminum and silicon crystal lattice after fatigue tests was revealed in silumin untreated with electron beam. The decrease in parameter of aluminium crystal lattice in fatigue tests of initial samples (figure 4, a, regime 2) is connected, apparently, with the dissolution of silicon inclusions and penetration of silicon atoms into aluminium lattice. The radius of silicon atom is less than radius of aluminium atom by 0.0054 nm [12], therefore the concentration of Al-based solid solution with silicon atoms will be accompanied by the reduction of parameter of its lattice. The irradiation of silumin with electron beam at energy density 20 J · cm^{-2} results a relatively little decrease of parameter of aluminum crystal lattice and it is connected with dissolution of silicon particles and intermetallides phases, and concentration of solid solution with silicon atoms decreasing the parameter of aluminium crystal lattice and with magnesium atoms having a larger, as compared to aluminium, atomic radius and, consequently, increasing the parameter of aluminium crystal lattice. Thus, electron beam treatment results in dissolution of Al- and Si-based phases and concentration of Al-based solid solution with these elements. The subsequent fatigue tests of silumin are accompanied by the additional increase of parameter of aluminium crystal lattice and it may be connected with the removal of silicon atoms from aluminium crystal lattice (effect of strain ageing of the material).

4. Conclusion

The fatigue tests of silumin subjected to irradiation with high intensity pulsed electron beam resulting in the increase in fatigue life by 3.5 time were carried out. Investigations of tribological and strength properties of silumin surface subjected to fatigue tests were performed and the decrease in hardness, the increase in wear coefficient and coefficient of friction with the increase in number of cycles of fatigue loading were revealed.

Acknowledgments

This work has been supported by a grant of the President of the Russian Federation for state support of young Russian scientists – Doctors of Sciences (project MD-2920.2015.8) and state task № 3.1496.2014/K.

References

[1] Gromov V E, Ivanov Yu F, Vorobiev S V and Konovalov S V 2015 *Fatigue of steels modified by high intensity electron beams*, (Cambridge, Cambridge International Science Publishing Ltd)
[2] Hao S Z, Qin Y, Mei X X, Gao B, Zuo J X, Guan Q F, Dong C and Zhang Q Y 2007 *Surf. and Coat. Technol.* **201** 8588–8595
[3] Hu J J, Zhang G B, Xu H B and Chen Y F 2012 *Materials Science and Technology* **27** (4) 300-303
[4] Hao Y, Gao B, Tu G F, Li S W, Hao S Z and Dong C 2011 *Applied Surface Science* **257** 3913–3919
[5] Hao Y, Gao B, Tu G F, Cao H, Hao S Z and Dong C 2012 *Applied Surface Science* **258** 2052–2056
[6] Hao Y., Gao B., Tu G. F., Wang Z. and Hao C. Z. 2011 *Materials Science Forum* **675-677** 693-696
[7] Gromov V E, Ivanov Yu F, Sizov V V, Vorob'ev S V and Konovalov S V 2013 *J. of Surf. Investigation. X-ray, Synchrotron and Neutron Techniques* **7** 94-98
[8] Proskurovsky D I, Rotshtein V P, Ozur G E, Ivanov Yu F and Markov A B 2000 *Surf. and Coat. Technol.* **125** (1-3) 49-56.
[9] Ivanov Yu F, Alsaraeva K V, Gromov V E, Popova N A and Konovalov S V 2015 *Journal of Surface Investigation. X-ray, Synchrotron and Neutron Techniques* **9** (5) 1056–1059
[10] Hao Y, Gao B, Tu G F, Li S W, Dong C and Zhang Z G 2011 *Nuclear Instruments and Methods in Physics Research Section B: Beam Interactions with Materials and Atoms* **269** 1499-1505
[11] An J, Shen X X, Lu Y and Liu Y B 2006 *Wear* **261** 208–215
[12] Babichev A P, Babushkina N A and Bratkovskij A M 1991 *Physical quantities: Directory* (Moscow, Jenergoatomizdat)

IOP Conf. Series: Materials Science and Engineering **110** (2016) 012013 doi:10.1088/1757-899X/110/1/012013

Tunable synthesis of copper nanotubes

E Kaniukov*[1], A Kozlovsky[2,3], D Shlimas[2,3], D Yakimchuk[1], M Zdorovets[2,3] and K Kadyrzhanov[2]

[1]Scientific-Practical Materials Research Centre NAS of Belarus, P. Brovki street, 19, Minsk 220072, Republic of Belarus

[2]L.N. Gumilyov Eurasian National University, Mirzoyan street, 2, Astana 010008, Kazakhstan

[3]Institute of Nuclear Physics, Ibragimov street, 1, Almaty 050032, Kazakhstan

*E-mail: Ka.Egor@mail.ru

Abstract. Simple method of tunable synthesis of copper nanotubes based on template synthesis was developed. A comprehensive study of the structural, morphological and electrical characteristics of the obtained nanostructures was carried out. Characterization of structural features was made by methods of scanning electron microscopy, energy dispersive spectroscopy and X-ray diffractometry analysis. Evaluation of wall thickness is made by methods of gas permeability. Electrical conductivity of nanotubes was define in the study of their current-voltage characteristics. The possibility to control of copper nanotubes physical properties by variation of the deposition parameters was shown.

1. Introduction

Over the past few years, much effort was directed to the synthesis of metallic micro- and nanostructures. Special attention is paid to copper nanostructures due to the low price, easy synthesis and high performance. Nowadays, a large number of techniques described in [1-5] allow to synthesize Cu nanocrystals with different morphologies (cubes [6], nanorods [7], nanodiscs [8], nanowires [9], and etc.). However, due to the specific structure and unique properties, more attractive for the practical application is hollow nanostructures. Due to a low density and large surface area, hollow nanostructures are very promising for use as elements of nanoelectronic devices, catalysts, targeted drug carriers, biomedical agents and chemical reactors [10-16]. The main problem to prepare Cu hollow micro- and nanostructures is the lack of a reliable method for their synthesis with the given parameters, all the above determine the relevance of the current search for an appropriate approach. To solve this problem, we propose method of template synthesis of copper nanotubes with the predetermined parameters, also a comprehensive characterization of morphological and electrical properties were carried out.

2. Experimental

Track-etched membranes based on poly(ethylene terephthalate) (PET) were used as templates with a thickness of 12 microns nominal pore diameter of 380 nm and a density of 4×10^7 cm^{-2}. Deposition of Cu was carried out by electrochemical method in potentiostatic mode with a potential difference U in the range of 1.0 to 1.5 and steps of 0.25 V. Control of the deposition was carried out by chronoamperometry using a multimeter Agilent 34410A.

Characterization of structural features was carried out by scanning electron microscopy (SEM, Hitachi TM3030), energy dispersive analysis (EDA, Bruker XFlash MIN SVE) and X-ray diffraction (XRD, D8 ADVANCE) using Cu Kα-irradiation. Determination of internal diameters of Cu nanotubes (NTs) was carried out by manometric method by determination of gas permeability (Sartocheck® 3 Plus 16290). Evaluation of electrical conductivity of nanotubes was carried out in the study of their current-voltage characteristics using a current source HP 66312A and amperemeter 34401A Agilent.

3. Results and Discussion

For the electrochemical pore filling of metal, the surface of PET templates was covered by gold layer with thickness of 10 nm using magnetron sputtering in vacuum deposition, this gold layer served as the working electrode (cathode) during deposition. The selected thickness of the deposited metal provides a partial penetration of Au on the pore walls, allowing formation of nucleation center of Cu on the cathode and initiating growth of the NTs. Electrolyte for the synthesis of NTs was: $CuSO_4{\cdot}5H_2O$ (238g/l), H_2SO_4 (21g/l). The deposition time t was determined based on chronoamperogramms. It corresponded to complete filling of the pores and was t=120 s for U=1.0 V, 100 s for 1.25 V and 30 s for 1.5 V. It is obvious, that we can tune the length of NTs by changing of deposition time. SEM pictures of the prepared nanostructures after chemical dissolution of the polymer matrix are presented in figure 1.

Figure 1. SEM microphotographs of Cu nanutubes.

SEM studies show that Cu nanostructures have a hollow NTs shape with a length corresponding to the original template thickness of PET (11.8 ± 0.2 mm) and an outer diameter D corresponding to the pore diameter of PET – 380±20 nm. Taking into account the lack of resolution of SEM, definition of inner diameters d of nanostructures in the pores of PET templates was carried out by manometric method using gas permeability at pressures in the range of 0.008 to 0.020 MPa at a pitch of 0,004 MPa. The inner diameter of synthesized nanotubes were calculated according to the equation [17]:

$$d^3 = \frac{Q \cdot 6l}{\sqrt{\frac{2\pi}{R \cdot T \cdot M}} \cdot \Delta p \cdot n}, \qquad (1)$$

where, Q – gas flow rate, l – thickness of the film, Δp – applied pressure, R – gas constant, M – the molar mass of air, n – pore density, T – temperature. Obtained results are presented in figure 2.

The results showed that calculating pore diameter of PET template is 380±5 nm that is corresponding to the data obtained from SEM pictures. Analysis of the data obtained for Cu NTs at different U showed an interesting relationships: increasing the deposition potential from 1 V to 1.5 V led to increasing of inner diameter of NTs from 170 and 200 nm that corresponds to a reduction of the wall thickness from 100 nm to 85 nm.

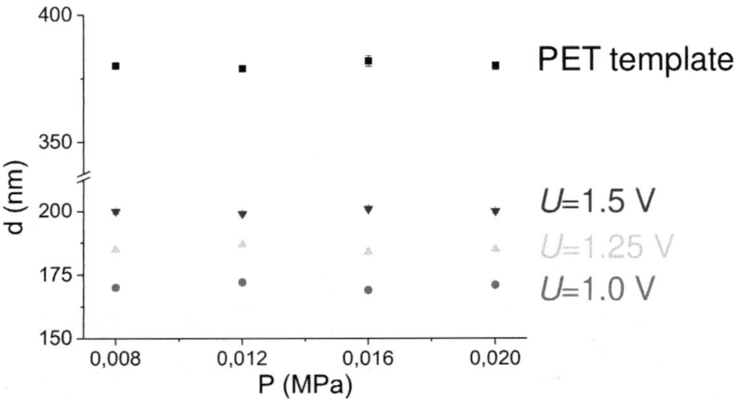

Figure 2. Pore diameter of PET template and the inner diameters of Cu NTs at various applied pressure using manometric method of gas permeability determination.

Determination of elemental composition by EDA showed that all the samples are contained 100% copper without any impurities. Analysis of the crystal structure of Cu NTs was carried out by XRD. Figure 3 shows obtained radiographs.

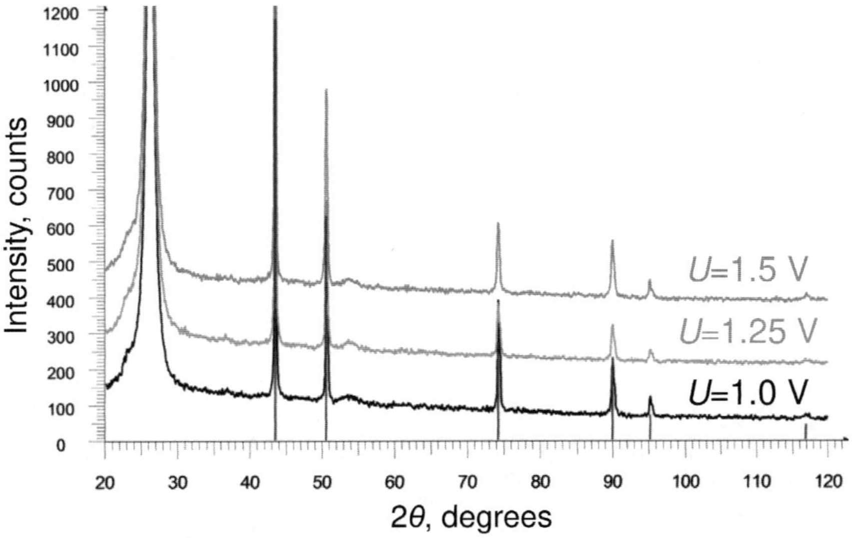

Figure 3. XRD spectra of Cu NTs in PET template at different deposition potentials.

According to the analysis of XRD, all samples of Cu NTs have the *FCC* structure with a cell parameter *a* differ from the standard (3.6130 Å). The peak with the angular position of 2θ=20–33° belongs to PET template [18]. Table 1 shows the lattice parameters and the average size of crystallites calculated by Scherrer equation [19].

Table 1. Parameters of Cu NTs obtained at different deposition potentials.

U, V	*a*, Å	The average size of crystallites, nm
1.0	3,6124±0,0009	32,23
1,25	3,6132±0,0009	28,82
1,5	3,6135±0,0007	43,85

Orientation of polycrystalline nanotubes was investigated by calculation of texture coefficient *TC*(*hkl*), which was calculated using Harris equation [20]:

$$TC(hkl) = \frac{I(hkl)}{I_0(hkl)} \Big/ \frac{1}{n}\sum \frac{I(hkl)}{I_0(hkl)}, \quad (2)$$

where *I(hkl)* – experimentally obtained relative intensity, $I_0(hkl)$ – the standard relative intensity according to the database JCPDS, *n* – number of planes. The results are presented in table 2.

Table 2. Data on the texture coefficients.

2*θ* °	(*hkl*)	TC(*hkl*)		
		U=1.0 V	*U*=1.25 V	*U*=1.5 V
43.39	(111)	1,584974	1,775872	1,780566
50.53	(200)	0,826349	1,030679	1,121368
74.18	(220)	1,145673	0,825306	0,902701
90.00	(311)	0,687315	0,661072	0,848477
95.20	(222)	0,755689	0,707071	0,897925
116.99	(400)	-	-	0,448962

The values of texture coefficients larger than one indicate a preferred orientation along the array of nanotubes in the planes, that indicate increasing the number of grains along the plane. The values of about 1 indicates that this structure has a polycrystalline nature with a preferred direction [111]. The increasing of the deposition potential leads to increasing of texture coefficients values for the plane [111] within 10%.

To determine the influence of the synthesis conditions on the electrical properties, conductivity σ of Cu NTs array in PET template were studied. The electrical conductivity was calculated by the formula:

$$\sigma = \frac{dI}{dU}\frac{l}{A}, \quad (4)$$

where l – length of nanotube, A – conducting surface area, dI/dU – tangent of the angle depending I–U. Graph of dependence σ from deposition potential is shown in figure 4.

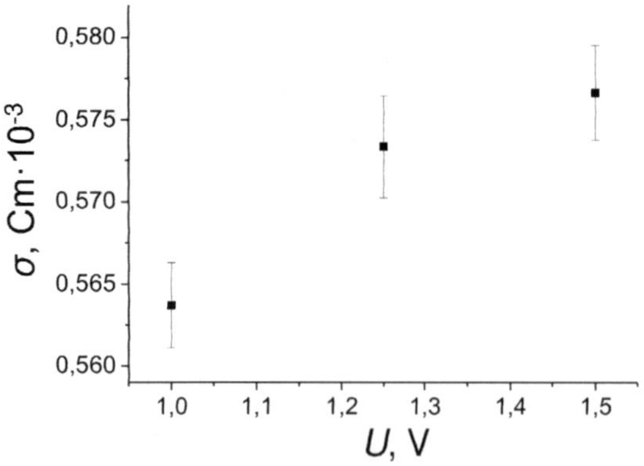

Figure 4. Dependence of Cu NTs conductivity on deposition potential.

As seen from the presented dependence, increasing deposition potential on the 0.25 V leads to increasing conductivity of Cu NTs of about 2%. The results are in good agreement with the data of XRD analysis: with increasing of the deposition potential, crystallite size of copper is increased, of which the NTs contains, that is accompanied by decreasing in the scattering of conduction electrons on these boundaries.

4. Conclusions

Using PET track-etched membranes with thickness of 12 μm as template, polycrystalline copper nanotubes were synthesized by electrochemical method with outer diameter of 380 nm and wall thickness in the range from 85 to 100 nm. A comprehensive study of the structural, morphological and electrical characteristics was carried out, the ability to control parameters of Cu NTs such as length and thickness of the wall by variation of deposition time and potential. It was found that the change in the potential of electrochemical deposition leads not only to decreasing the wall thickness of Cu NTs, but also to increasing the size of copper crystallite.

Changes in the degree of crystallinity is accompanied by decreasing in conduction electron scattering at the grain boundaries, that causes increasing conductivity of NTs. Relative availability and reliability of the proposed method, along with the ability to control physical properties of the nanostructures defines the prospects for their further application as catalysts or components for modern electrical circuits, including on flexible substrates.

References

[1] Lisiecki I and Pileni M P 1993 *J. Am. Chem. Soc.* **115** 3887–96
[2] Prunier H, Ricolleau C, Nelayah J, Wang G and Alloyeau D 2014 *Cryst. Growth Des.* **14** 6350–6
[3] Yen M-Y, Chiu C-W, Hsia C-H, Chen F-R, Kai J-J, Lee C-Y and Chiu H-T 2003 *Adv. Mater.* **15** 235–7
[4] Huang Y-F, Shih H-S, Lin C-W, Xu P, Williams D J, Ramos K J, Hooks D E and Wang H-L 2012 *Cryst. Growth Des.* **12** 1778–84
[5] Lee K B, Seo J H, Ahn J P, Kwon H and Yang H R 2012 *Met. Mater. Int.* **18** 727–30
[6] Filankembo A, Giorgio S, Lisiecki I and Pileni M P 2003 *J. Phys. Chem. B* **107** 7492–500
[7] Lisiecki I, Filankembo A, Sack-Kongehl H, Weiss K, Pileni M-P and Urban J 2000 *Phys. Rev. B* **61** 4968–74
[8] Salzemann C, Urban J, Lisiecki I and Pileni M-P 2005 *Adv. Funct. Mater.* **15** 1277–84
[9] Roussel J-M and Gailhanou M 2015 Stability of a Screw Dislocation in a (011) Copper Nanowire *Phys. Rev. Lett.* **115** 075503
[10] Yin Y, Rioux R M, Erdonmez C K, Hughes S, Somorjai G A and Alivisatos A P 2004 *Science* **304** 711–4
[11] Caruso F, Caruso R A and Möhwald H 1998 *Science* **282** 1111–4
[12] Jiang S, Sreethawong T, Lee S S C, Low M B J, Win K Y, Brzozowska A M, Teo S L-M, Vancso G J, Jańczewski D and Han M-Y 2015 *Adv. Mater. Interfaces* **2** n/a – n/a
[13] Bhanushali S, Ghosh P, Ganesh A and Cheng W 2015 1D Copper Nanostructures: Progress, Challenges and Opportunities *Small* **11** 1232–52
[14] Yin Z, Song S K, You D-J, Ko Y, Cho S, Yoo J, Park S Y, Piao Y, Chang S T and Kim Y S 2015 *Small* **11** 4576–83
[15] He G, Hu W and Li C M 2015 *Colloids Surfaces B Biointerfaces* **135** 613–8
[16] Zhou X, Zhao G, Niu H, Liu Y and Chen C 2011 *Micro Nano Lett.* **6** 269
[17] Shao P, Ji G and Chen P 2005 *J. Memb. Sci.* **255** 1–11
[18] Faraj M G and Ibrahim K 2011 *Int. J. Polym. Sci.* **2011** 1–4
[19] Langford J I and Wilson A J C 1978 *J. Appl. Crystallogr.* **11** 102–13
[20] Barrett C S and Massalski T B 1980 *Structure of metals: crystallographic methods, principles and data* (Oxford; New York: Pergamon)

IOP Conf. Series: Materials Science and Engineering **110** (2016) 012014 doi:10.1088/1757-899X/110/1/012014

Using and assessing energy efficiency of electrical ovens with unit-type releasing intended for thermal energization of sungulite-vermiculite conglomerates

A I Nizhegorodov

Head of the Department of Construction, Roadmaking Machinery and Hydraulic Systems, Irkutsk National Research Technical University, 83, Lermontov Street, 664074, Irkutsk, Russia

E-mail: nastromo_irkutsk@mail.ru

Abstract. The paper deals with assessing energy efficiency of electric ovens supplied with unit-type releasing to be used in the thermal activation technology of vermiculite – phlogopite conglomerates. The analysis of a heat absorption process is given on the basis of the conglomerate particles moving in the conditions of heat radiation induced by an outer source and heat absorption taking place inside a special nonelectric unit on the account of inner energy accumulated by the particles.

1. Introduction

Technogeneous wastes of Kovdorski phlogopit-vermiculite deposit in Murmansk region hosts hundreds of thousands tons of vermiculite and sungulite [1] that are the raw material to produce thermal vermiculite and magnesian-silicate chemicals. Expanded vermiculite participates in soil-forming processes, whereas sungulite turns into an effective chemical after thermal energization that is able to reduce the mobility of heavy metals [2] and can be used in detoxification of technogeneously polluted soils and water objects.

The basic minerals in the parent raw material are (by mass) sungulite ~ 10%, vermiculite ~ 13%, pyroxene and olivine ~ 50 %, the remaining ones are antigorite, iron oxides and other minerals [1]. The technology of raw material dividing includes fragmentation, concentration in a hydro separator with the drain turnoff, drying followed by a magnetic separation and roasting [3]. The waste is composed of particles that are 0.1 to 0.2 mm in size and contains about 20% of vermiculite and ~ 14% of sungulite. After roasting it can be used as an ameliorant. The magnetic crop yield makes up more than 53 %. The nonmagnetic crop is the bearer of target products: vermiculite is represented by leaves up to 5 mm thick in the amount of ~ 37%, the amount of sungulite is ~ 25 %, and that of olivine-pyroxene rock makes up ~ 40 %. The yield of nonmagnetic crop is 13 % [3].

2. Experimental

The sungulite bulk density is ~ 1026 kg/m^3. It is 21 % lighter than that of olivine-pyroxene phase. So, it is not difficult to release vermiculite and sungulite by means of vibratory segregation.

It is evident that the faster is the heat rise, the higher is the degree of sungulite activation [2]. A thermal shock and holding at the temperature of 600 ^{0}C for 20 minutes are the optimum conditions.

Energy-saving electrical ovens with unit-type releasing supplied with heat energy recuperators [4] can be used to realize a thermal shock (200…240 ^{0}C/c).

The oven contains a raw material dosimeter 1 with a bin2, a roller 3, and a trough 4 (Figure1). Electrical modules 5 supplied with recuperators 6 are joined to the heat chambers of an electrical unit 8 which has a Γ-shaped form by a piping system 7. Modules 5 are covered with lids 9 and electric heaters attached to conducting heads 10 are placed under the lids.

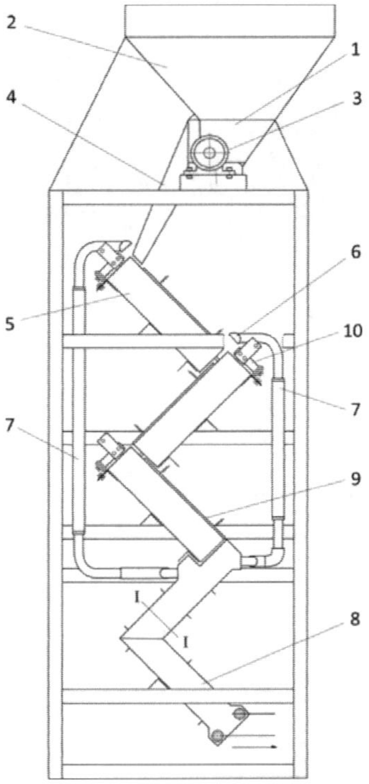

Figure. 1. The lay out of an oven with a Γ-shaped module

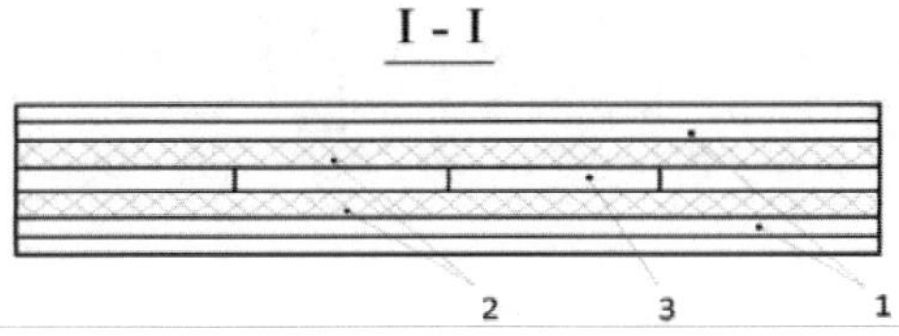

Figure 2. A I-I cross section of a Γ- shaped module

There are thermally insulated pockets 2 between the heat chambers 1 (figure 2) and they are separated by a roasting space 3. Hot air is vented into the chambers 1 through a piping system 7 (figure1), whereas it is vented out with a vent exhauster through their lower holes. In their movement through the modules the conglomerate particles find themselves in the conditions of heat radiation. Some part of the radiant energy heats the air which is being released from the modules 5 (figure1), whereas the other part of it together with its reflected flows is released through the cross-cut ends. The recuperators "collect" it from branch pipes and collectors to send it to a nonelectrical unit 8 as hot air ~380° C.

3. Result and discussion

The analysis of a vermiculite roasting process is given consideration to in paper [5]: the energy of its heat absorption, Q_{Σ}, makes up 1202 kJ/kg.

Redistribution of the energy takes place while a conglomerate is being roasted:

$$Q'_{\Sigma} = k_v\left(Q_1 + Q_2 + Q_3 + Q_4 + Q_5\right) + \left(1 - k_v\right) \times \left(Q_6 + Q_7\right) \quad (1)$$

where Q'_{Σ} is a complete conglomerate heat absorption energy, k_v is the index of a vermiculite portion by weight, Q_1 is the energy of hydrated water dehydratation, Q_2 is the heat energy of a "dry" mineral, Q_3 and Q_4 are the heat energy of change in phase and the superheating of water evaporations, Q_5 is the heat energy of adsorbed gases heating, Q_6 is sungulite heating and Q_7 is the heat energy produced by evaporation of physical water adsorbed by sungulite:

$$Q_6 = 0.955\ c_1 mT \ and\ Q_7 = 0.045\ c_2 m$$

where 0.955 and 0.045 are the portions of the dry part and moisture by weight in 1kg of sungulite, c_1 is an average sungulite specific heat capacity in the range of 20 … 700^{0}C (an average value of c_1 is ~ 943 J/kg ^{0}C [6]), m is sungulite mass (1 kg), T is the temperature of heating, c_2 is the specific heat of steam production ($c_2 = 2250 \times 10^3$ J/kg) [7].

The amount of heat energy received by "dry" sungulite while being roasted is:

$$Q_6 = 0.955 \times 943 \times 1 \times 650 = 585.4\ kJ$$

where 650 ^{0}C is the difference of roasting (750 ^{0}C) and conglomerate preheating (100 ^{0}C) temperatures before the roasting.

The heat energy of adsorbed water in the condition of change in phase is:

$$Q_7 = 0.045 \times 2250 \times 10^3 \times 1 = 101.3\ kJ$$

The data obtained being taken into consideration, equation 1 is transformed into:

$$Q_\Sigma = 1202 \times k_v + 686.7 \times (1 - k_v)$$

The index of k_v changes within the limits of 0.35 …0.4, so the available energies relation of Q_Σ and Q'_Σ is dependent on vermiculite contents by weight. In equation (1), two members are used to show the accumulated heat: Q_2 in the expanded vermiculite and Q_6 in sungulite. The heat energy of Q_2 in 1 kg concentrate makes up ~560 kJ [4], whereas Q_6 makes up ~585 kJ. k_v being equal, for example, to 0.4, the heat energy of Q_8 accumulated by them is calculated as follows: $0.4 \times 560 + 0.6 \times 585 = 224 + 351 = 575$ kJ/kg.

This is the value of heat energy accumulated in the condition of a fully completed roasting process.

An intensive heat absorption is performed by conglomerate components in electrical modules due to radiant energy transference on vermiculite and sungulite particles. While expanding, vermiculite grains form a heating "jacket" which makes radiant energy penetration into deeper layers more difficult.

If we give consideration to the indices of sungulite thermometric conductivity, χ_1, and expanded vermiculith one, χ_2, [8] it is evident that

$$\chi_1 / \chi_2 = \lambda_1 p_2 c_2 / \lambda_2 p_1 c_1 \qquad (2)$$

where λ_1 and c_1 are the thermal conductivity (~0.35 W/m ^{0}C) and specific heat capacity (~800 J/kg ^{0}C) of sungulite, λ_2 and c_2 are the thermal conductivity (~0.056 W/m ^{0}C) and specific heat capacity (~1625 J/kg ^{0}C) of expanded vermiculite, p_2 is its density (~ 200 kg/m^3), p_1 is sungulite density (~ 1000 kg/m^3) [4]. The calculation performed in accordance with formula (3) results in the ratio of χ_1 to χ_2 as 2. 4.

Sungulite gets heated faster, so the expansion of vermiculite is continued in a nonelectrical unit, the process being not caused by an outer source, as the hot air of heat chambers makes the heat leak from the roasting space smaller. In this case, we deal with the transport of radiant energy from sungulite to vermiculite grains and the conductive transport inside the grains from their outer layers to deeper ones. Besides, there is some heat exchange between vermiculite grains, too. Small grains that have already expanded are unable to absorb any energy but radiate and reflect it on greater neighboring grains instead.

Such is the model of heat absorption observed in vermiculite and sungulite conglomerates roasting.

$$Q'_\Sigma = k_v\ k_1 (Q_1 + Q_3 + Q_4 + Q_5 + Q_2) + (1 - k_v) Q_6 \qquad (3)$$

$$k_v k_1 Q_2 + (1 - k_v) Q_6 \approx k_v (1 - k_1)(Q_1 + Q_3 + Q_4 + Q_5) \quad (4)$$

In this case, Q'_Σ is adopted heat, the process being not completed; k_1 is an index registering the portion of heat adopted by vermiculite before it enters the non-electric unit which is dependent on the temperature and time t of the conglomerate moving while being heated from an outer source and also on the components relation by weight. The index of k_1 is absent before the member of equation (4) which is $(1 - k_v) Q_6$ as sungulite had completely adopted the heat before it entered the non-electrical unit which is evident from equation (2). This fact accounts for the absence of Q_7 member in equations (3) and (4) as the water adsorbed by sungulite evaporated at the starting stage. In equation (4) the approximated equality shows that the thermal energy accumulated in the conglomerate can be lacking to complete the dehydration process This depends on k_v and is expressed by the member of $k_v k_1 Q_2 + (1 - k_v) Q_6$ of equation (4). Equation (4) can be balanced by changing t (the time of roasting) and the temperature in the oven, increasing the value of index k_1 and the time of a conglomerate movement through the non-electrical unit t_1 as well as its particle concentration in the working space. All these changes will make the losses resulting from heat transport from sungulite particles to vermiculite grains smaller.

The time of heat adoption grows by ~27 % on the account of the non-electric unit. This is why the temperature of heaters in the electric modules can be lowered. (Figure3) demonstrates the temperature and chrono interrelation for an oven with the regime point of *a* characterized by the parameters of $T = 750^0$C and $t_2 = 2.88$ sec.

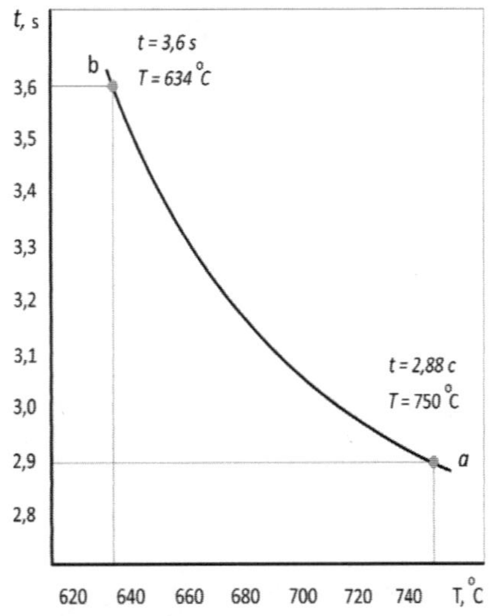

Figure 3. Chrono-thermal relationship

The oven is shifted into the temperature regime of $T = 627^0$C, point *b*, by increasing the roasting time in these limits and attaching the non-electrical unit. Using the interrelation between the parameters of a three-phase network, the area of heaters radiant surfaces, f, and the absolute temperature of T_4 (^{0}K) [4] we obtain the formula:

$$3 \times 1.27\, I\, U \,/\, f = \delta T^4 \quad (5)$$

This way it is possible to determine the ratio of the oven unit current loading, *I* (A), in the regime points *b* and *a* basing on equation (5) and taking into consideration the area f quasi-increase by 27 %:

$$I_в \,/\, I_a = 0.76$$

where σ is a constant value by Stefan-Boltzmann (σ = 5.67 x 10^{-8} W/m^2 ^{0}K) whereas U is a network line voltage (~220 W).

4. Conclusion

Thus, a non-electrical unit installing is to cause the reduction of energy consumption by 24 %. This is an extreme prediction, but it allows to recommend the ovens under consideration as a means of realizing the technologies of minerals thermal activation.

References

[1] Gershenkop A Sh 2011 Vermiculite concentration disengagement from vermiculite ores mill tailing and its use in heat resistant concrete *Mining Journal* **11** 57-9
[2] Kremenetskaya I P 2008 Reagent used to immobilize heavy metals of overburden rocks containing serpentine *Vodo'ochistka.Vodopodgotovka.Vodosnabzheniye* **4** 33
[3] Tereshenko I P 2014 *Directions of Multipurpose Utilization of Phlogopite Extraction Wastes* vol 1All-Russian Techn.Res.Conf ed Kola Res. C. Mining Inst (Apatity:Renome) pp 272-79
[4] Nizhegorodov A I 2015 Energotechnological Equipment for Vermiculite Concentrate Processing (Irkutsk:INRTU) p 250
[5] Nizhegorodov A I 2011 Improvement of vermiculite roasting technology in electrical ovens with module releasing. *J. Stroiteln. Materialy* **35** 62-4
[6] *Handbook on Making Calculations of Heating and Thermical Ovens 1983* ed V M Tymchak and V L Gusyevski (Moscow: Metallurgiya) pp 480
[7] *Handbook on Elementary Physical Science* 1972 ed N I Koshkin and M G Shirkevich (Moscow: Fizmatgiz) pp 256
[8] Teleg'in A C, Shvydkiy B C and Yaroshenko Yu G 2002 *Heat Transportation* (Moscow: "Akademkniga") pp 455

IOP Conf. Series: Materials Science and Engineering **110** (2016) 012015 doi:10.1088/1757-899X/110/1/012015

Formation of structure in Au, Cu and Ni nanoclusters: MD simulations

Yu Ya Gafner*, S L Gafner, Zh V Golonenko, L V Redel and V I Khrustalev

Katanov State University of Khakassia, pr. Lenina 90, Abakan, 655000 Republic of Khakassia, Russia

*E-mail: ygafner@khsu.ru

Abstract. The molecular dynamics method with the modified tight-binding (TB-SMA) potential has been used to study structure formation in gold nanoparticles 1.6–5.0 nm in diameter. The formation of the internal structure of gold nanoclusters is studied in terms of canonical ensembles. The stability boundaries of various crystalline isomers are analyzed. The obtained dependences are compared with the corresponding data obtained for copper and nickel nanoparticles. The structure formation during solidification is found to be characterized by a clear effect of the particle size on the stability of a crystalline modification.

1. Introduction

The last two decades were marked by heightened interest of researchers in the development and study of different nanostructures. Nanostructures are of practical and scientific significance from the viewpoint of understanding fundamental electrical, magnetic, optical, catalytic, thermal, and mechanical properties of nanoscale materials, as well as from the viewpoint of fabrication of new technical devices on their basis. It is already clear that the quantum-mechanical effects implemented in nanostructures, such as the conductance quantization, band gap renormalization, Coulomb blockade, and others, can significantly improve functional characteristics of different electronic devices.

The determination of physical and chemical properties of individual elements of nanomaterials, i.e., nanoclusters, is of particular interest due to the possibility of studying the transition from free atoms and molecules to bulk condensed systems. Clusters, being the main unit of such evolution, are unique objects of study and can give a clue to understand the nature of different processes, including nucleation, dissolution, catalysis, adsorption, phase transitions, and others.

Furthermore, already in the early 1990s, the first experiments were performed on the development of electronic elements consisting of metal units of in fact nanometer size. Further experiments showed that the circuitry of such electronics can be developed based on metal nanoclusters exhibiting quantum conductance properties at room temperatures.

Thus, the main objective of this work is to study the possible effect of electronic and structural features on stability of free small gold, nickel and copper clusters, since clusters of exactly these metals are currently efficiently synthesized by different physicochemical methods. In this study, particular attention was paid to the internal nanoparticle structure stability, since a change in the atomic configuration of the cluster can have an effect on many other properties, in particular on its electronic structure. It is well known that clusters with different sizes and internal structures have different distances between energy levels, which cannot but attract interest from the technical point of view.

IOP Conf. Series: Materials Science and Engineering **110** (2016) 012015 doi:10.1088/1757-899X/110/1/012015

2. The computer model

In solid state physics, instead of expensive experiments, a model replacing a real object can be studied using modern computers. To date, many methods have been developed for simulating the nanocluster behavior, first of all, the molecular dynamics (MD) method. The MD method is based on the calculation of classical (Newtonian) trajectories of object motion in the phase space of coordinates and momenta of its atoms. In the simplest version of this method, classical trajectories of atomic motion in the force field of the empirical atom–atom potential, i.e., a detailed microscopic pattern of the internal thermal mobility is simulated in nanosecond time intervals.

MD simulation of nanostructures is completely based on a detailed description of particles composing them. In MD, the classical viewpoint is most often used, according to which atoms or molecules are represented as point masses interacting via forces depending on the distance between these objects. In the numerical analysis, such forces are calculated using different interatomic interaction potentials, and the confidence level of the results obtained is directly dictated by the used potential choice. Therefore, after analyzing different representations of the potential energy, metallic nanoclusters were simulated using well established tight-binding (TB-SMA) potentials [1] with fixed cutoff radius corresponding to the fifth coordination shell inclusive.

These potentials based on the approximation of the second moments of the tight-binding Hamiltonian were developed by Cleri and Rosato [1]. The modified tight-binding method proposed by them was successfully used in a number of cluster studies. The method is based on the fact that a large group of properties of transition metals can be fully defined using the density of states of outer *d*-electrons. The second moment of the density of state is μ_2; it was experimentally found that the cohesion energy of metals is proportional to the width of the density of states, which, in the approximation of the second moments, is simply $\sqrt{\mu_2}$.

If only *dd*σ-, *dd*π-, and *dd*δ- orbitals are taken into consideration, then, according to this model, the band energy of atom *i* can be written as

$$E_B^i = -\left(\sum_j \xi_{\alpha\beta}^2 e^{-2q_{\alpha\beta}\left(r_{ij}/r_0^{\alpha\beta}-1\right)}\right)^{1/2} . \tag{1}$$

It depends only on the distance between atoms i and j, i.e., r_{ij}, and is formally the same as the embedding function in the embedded atom method. Furthermore, the model includes a term accounting for the repulsive interaction in the form of the sum of Born–Meyer ion pair potentials,

$$E_R^i = \sum_j A_{\alpha\beta} e^{-p_{\alpha\beta}\left(r_{ij}/r_0^{\alpha\beta}-1\right)}, \tag{2}$$

where α and β are different types of atoms. Thus, the total energy of the system is written as

$$E_c = \sum_i \left(E_R^i + E_B^i\right). \tag{3}$$

The quantities $\xi_{\alpha\beta}, p_{\alpha\beta}, A_{\alpha\beta}, q_{\alpha\beta}, r_0^{\alpha\beta}$ and the potential cutoff radius r_c define the parameters of system elements. The values of $\xi_{\alpha\beta}, p_{\alpha\beta}, A_{\alpha\beta}$, and $q_{\alpha\beta}$, are determined by fitting the cohesion energy, lattice parameter, bulk modulus B, and elastic constants C_{44} and C ' to experimental values. Despite the simple functional form, the tight-binding model quite adequately describes elastic properties, defect characteristics, and melting for a wide range of fcc and hcp metals. In our opinion, it is one of the most appropriate schemes for numerical analysis of small metal particles consisting of a few hundred atoms.

Clusters were simulated within the canonical ensemble using a Nose thermostat [2]. In this case, the number of particles N, volume V, and temperature T remain unchanged (NVT ensemble), and the total momentum p of the system is zero. The temperature during simulation was determined from the average kinetic energy of atoms, which was calculated based on the velocity Verlet algorithm [2] with time step h = 1 fs. The simulation was performed with the MDNTP program developed by Dr. Ralf Meyer (Universität Duisburg, Germany).

3. Results and discussion

To find the specific features of structure formation, we studied the role of the gold nanoparticle size on the isomer formation. A cluster structure was formed by melting of metallic nanoparticles of a certain size followed by their cooling into a crystalline phase. At the first stage, we used the Nose thermostat to perform a steplike change in the temperature: when simulating cooling, we decreased the temperature by 50 K and held clusters at every fixed temperature for 0.5 ns. To remove the side effects related to thermal noise, we held clusters in the solidification range at a fixed temperature for 2 ns but the temperature step was decreased to 5 and even 1 K, which corresponded to a total cooling time of 5–10 ns. The final temperature was taken to be 300 K, since most nanodevices should operate under these temperature conditions in practice.

When performing simulation, we found that all basic possible crystalline modifications (fcc, hcp, Ih, Dh) formed during such steplike cooling of nanoclusters from the liquid phase, and they began to form once the solidification point was passed. It is interesting to study the dependence of the fraction of a certain structure on the cluster size, and particular emphasis was placed on the possibility of formation of pentagonal, namely, icosahedral and decahedral, structures.

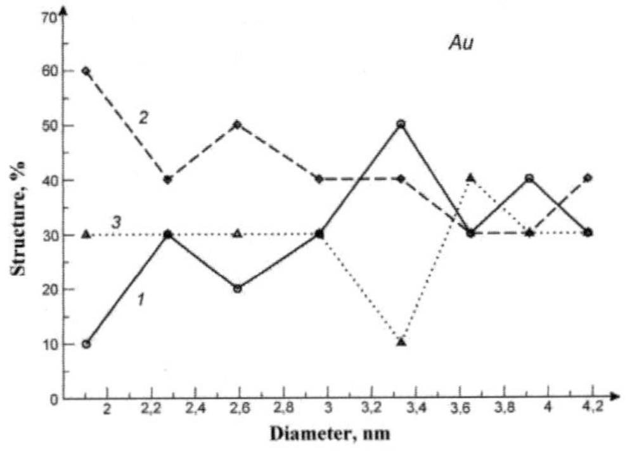

Figure 1. Structural configurations obtained for gold: (1) Ih, (2) Dh, and (3) fcc.

To process the results of computer simulation, we applied not only visual inspection using graphics editors and radial distribution functions but also statistical analysis of the internal structures of gold clusters. When performing experiments with clusters of each size, we can draw a conclusion regarding the formation of various crystalline modifications at a finite temperature. The percentages of the structures forming during the solidification of Au nanoparticles of various sizes are shown in Fig. 1. The percentage probability of the appearance of a certain structure means that M clusters have this structure among the ensemble of N clusters of the same size.

It is clearly visible that the percentage of the appearance of icosahedra increases from 10 to 50% as the particle diameter increases from 1.6 to 3.33 nm. As the particle size increases further, the percentage passes through its maximum, and the probability of formation of gold clusters with this crystallographic type of structure decreases due to size effects. For example, this percentage fluctuates at a level of 30–40% for clusters 4.2 nm in diameter; at a cluster size of 5.0 nm, the percentage of appearance of the icosahedral modification is zero in all experiments. Since fivefold symmetry is prohibited for macrocrystals, the formation of large particles is inevitably related to the appearance of internal voids or elastic strains. The elastic strain energy, which is initially very low, increases in proportion to volume and exceeds the decrease in the surface energy for large icosahedral particles; as a result, this behavior destabilizes this crystallographic type.

The percentage of appearance of the decahedral phase, which can be actually called an intermediate phase between fcc and Ih, has a more complex character. The number of decahedra is 60% (D = 1.6 nm) at the very beginning and then smoothly decreases to about 40% as the cluster size increases to D = 4.2 nm. The number of decahedra then increases sharply because of the absence of the Ih modification. The percentage of appearance of an fcc (hcp) structure is almost independent of the cluster size in the range under study and is 30% on average, which points to the stability of formation of this structural modification in sufficiently small gold clusters. However, we should note an increase in this parameter to 40% in the largest clusters (D = 5.0 nm). Thus, our experiments suggest that the process under study is characterized by a distinguishable effect of size N on the stability of a certain structural modification.

For comparison, we present the data of other authors regarding the determination of structures in gold nanoclusters in order to search for any common features. We can state that an amorphous state is most stable in clusters smaller than 1.5 nm at very low temperatures: this is supported by both an MD simulation and experimental data [3]. This conclusion is also true of clusters with a "magic" icosahedral size of 13 or 55 atoms. In the case of nickel and copper, an icosahedron is most stable at N = 13 or 55 for any simulation, and the situation with gold is more complex. For example, at a cluster size N = 13, the authors of [4] found an amorphous structure using an ab initio method, and only some multi-particle potentials resulted in an icosahedral structure [4], which again points to the importance of choosing an interaction potential.

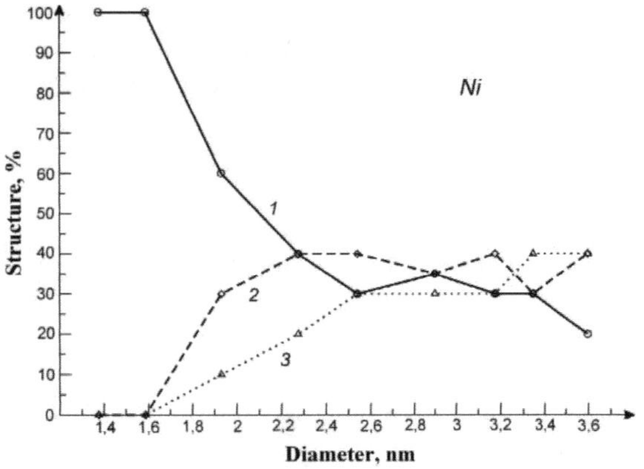

Figure 2. Structural configurations obtained for nickel: (1) Ih, (2) Dh, and (3) fcc.

With a computer simulation, the authors of [3] studied a huge ensemble of various random atomic configurations in clusters (N = 55, 75) at a temperature of 10^{-9} K and found that amorphous structures had the maximum density of states. At N = 55, approximately 80% of the structures with the minimum energies corresponded to an amorphous state and only 20% corresponded to an icosahedral state. At N = 75, a decahedron was detected among the ground states and the majority of clusters had a disordered state. As the temperature increases to room temperature, crystalline modifications, such as fcc, Ih, and Dh, begin to appear at noticeable fractions.

To understand the cause of the intense appearance of amorphous structures in small gold clusters, we performed an additional MD simulation of nickel and copper nanoclusters. In contrast to gold, the potential parameters in them demonstrated a high interaction; as a result, only ordered structures were found in the ground state. Other isomers were separated from them by a significant energy gap. The short-range interatomic interaction between surface atoms in gold clusters leads to uncoordinated formation of disordered regions, which form an amorphous structure.

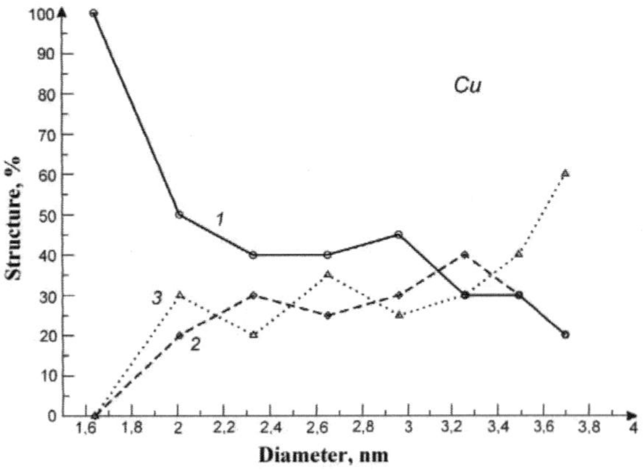

Figure 3. Structural configurations obtained for copperd: (1) Ih, (2) Dh, and (3) fcc.

When the cluster size increases to 2.0 nm, the decahedral modification becomes preferable, which follows from the results of our simulation and the data in [5]. The so-called equivalent theory of crystals developed in [6] demonstrates that a decahedron begins to appear in gold at zero temperature when a cluster has more than 150 atoms and that fcc clusters appear at N > 180, which agrees with our data even at T = 300 K.

At a cluster diameter of 3.0–3.5 nm, an icosahedron is considered to be most stable among all possible structures, which was experimentally supported in [7, 8]. In [7], clusters were formed when gold was cooled to room temperature in helium vapors and then placed onto an amorphous carbon film. Most initial particles had an Ih morphology, and some of the particles corresponded to a decahedral structure. An ensemble of gold particles of close sizes (3.5 nm in diameter) was experimentally formed in [5], and structural analysis performed there shows that an icosahedron is stable.

In addition, the authors of [7] supported the operation of the size effect during the formation of a fixed structure. They found that the number of fcc clusters in particles 3–6 nm in diameter remained almost the same. As the cluster size increased, the number of icosahedra decreased, which led to an increase in the number of decahedra [7]. We made a similar conclusion when simulating the structure formation in gold clusters of diameter D = 4–5 nm. At sufficiently large sizes (D > 5–6 nm), most researchers believe that an fcc structure is predominant [9].

It is interesting to compare our results with the data of a similar investigation of nickel and copper (Figs. 2, 3). The percentage of the icosahedral modification for small clusters of these metals is 100%. For the largest simulated cluster (N = 2243), an icosahedral structure appeared in 20% of the experiments and 60% clusters had the fcc structure characteristic of bulk samples. As the nickel cluster size increased, the number of icosahedra also decreased with the corresponding increase in the fraction of the fcc phase. The fraction of decahedra was almost independent of the cluster size and fluctuated about a certain value specific for each metal.

Thus, our experiments suggest that the structure formation during solidification is obviously characterized by the effect of the cluster size on the stability of a certain structure. Nickel and copper clusters exhibit certain common features in the formation of structural properties of nanoparticles, and gold clusters demonstrate much more complex behavior.

4. Conclusions

Using an MD simulation, we studied the formation of the internal structure in gold clusters up to 5 nm in size. To analyze the formation processes, we used the structureless clusters formed upon melting of primary fcc Au nanoparticles and cooled them to room temperature. To create fixed cluster structures during solidification, we used various techniques that can be applied in experiments. Various structural modifications were shown to form in cooling from a liquid phase, and some criteria of their stability were determined. In simulating, we revealed the role of the size factor and the heat removal rate in the structure formation in Au clusters. A comparative analysis of our results and the data on structure formation in nickel and copper clusters shows that they exhibit common features in the formation of structural properties, whereas gold clusters demonstrate much more complex behavior, which is often in conflict with the laws characteristic of nickel and copper particles of the same size. This discrepancy in the behavior is thought to be caused by the physicochemical nature of gold, which is a rather soft and ductile material; as a result, the formation of an ideal (from a crystallography standpoint) structure is hindered.

ACKNOWLEDGMENTS

This study was performed within the framework of the State Task of the Ministry of Education and Science of the Russian Federation.

References

[1] Cleri F and Rosato V 1993 Phys. Rev. B: Condens. Matter **48** 22
[2] Tao Pang An introduction to computational physics 2006 University Press, Cambridge
[3] Garzón I L, Michaelian K, Beltrán M R, Posada-Amarillas A, Ordejón P, Artacho E, Sánchez-Portal D and Soler J M 1998 Phys. Rev. Lett. **81** 1600
[4] Grigoryan V G, Alamanova D and Springborg M 2005 Eur. Phys. J. D **34** 187
[5] Barnett R N, Cleveland C L, Häkkinen H, Luedtke W D, Yannouleas C and Landman U 1999 Eur. Phys. J. D **9** 95
[6] Negreiros F R, Soares E A and de Carvalho V E 2007 Phys. Rev.B **76** 205429
[7] Kenji Koga, Tamio Ikeshoji and Ko-ichi Sugawara 2004 Phys. Rev. Letters **92** 115507
[8] Chun-Ming Wu, Chi-Yen Li, Yen-Ting Kuo, Chin-Wei Wang, Sheng-Yun Wu, Wen-Hsien Li 2010 J. Nanopart. Res. **12** 177
[9] Zhang J, Zhao B, Meng L, Wu H, Wang X and Li C 2007 J. Nanopart. Res. **9** 1167

IOP Conf. Series: Materials Science and Engineering **110** (2016) 012016 doi:10.1088/1757-899X/110/1/012016

Optimal orientations of the magnetizing field and crystal lattice of a ferrite film in spin-wave devices

V V Shagaev and Tun Tun Lin

Bauman Moscow State Technical University, Kaluga Branch, Russia

E-mail: shagaev_vv@rambler.ru

Abstract. A method of stabilization of the uniform ferromagnetic resonance frequency in a film is considered. From a condition of equality to zero of frequency derivatives with respect to magnetization and anisotropy field the angles defining crystallographic orientation of a film and its orientation in a magnetic field are calculated. The choice of such angles allows to minimize external factors influence on frequency.

1. Introduction

The spectrum of the own magnetization oscillations of a single-crystal ferrite film depends on the ferrite saturation magnetization ($4\pi M_0$) and crystallographic magnetic anisotropy field (H_c). For small deviations of parameters from the initial values the corresponding change of frequency can be calculated on a formula:

$$\delta f = a \cdot \delta(4\pi M_0) + b \cdot \delta H_c, \tag{1}$$

where $a = df/d(4\pi M_0)$ and $b = df/dH_c$. It is possible to represent the condition of weak influence of the parameters on frequency as equation $\delta f = 0$. It is possible to consider two versions of execution of a condition. In the first, the values of all values from the right part of formula (1) are taken into account, and then the condition of frequency constancy will receive a view:

$$a \cdot \delta(4\pi M_0) + b \cdot \delta H_c = 0. \tag{2}$$

In the second version it is possible to require execution of equalities $a = 0$ and $b = 0$. The first version was in detail investigated in works [1 – 3]. In these works temperature coefficients of magnetostatic waves frequencies are calculated and conditions of frequencies thermostabilization are considered. The second version in scientific literature isn't described. This article is devoted to it.

From a practical viewpoint equalities $a = 0$ and $b = 0$ are interesting the fact that it is allowed to stabilize frequency independently of change nature of the ferrite magnetic parameters. Such changes can be connected with temperature, with a chemical composition, with structure of a crystal lattice. A possibility to increase not only temperature but for example radiation resistibility of the spin-wave devices arises as a result.

2. The development of a mathematical model

The offered way was tested at the solution of a task of a uniform ferromagnetic resonance (FMR) frequency stabilization in a cubic ferrite film. The studied model is presented in figure 1.

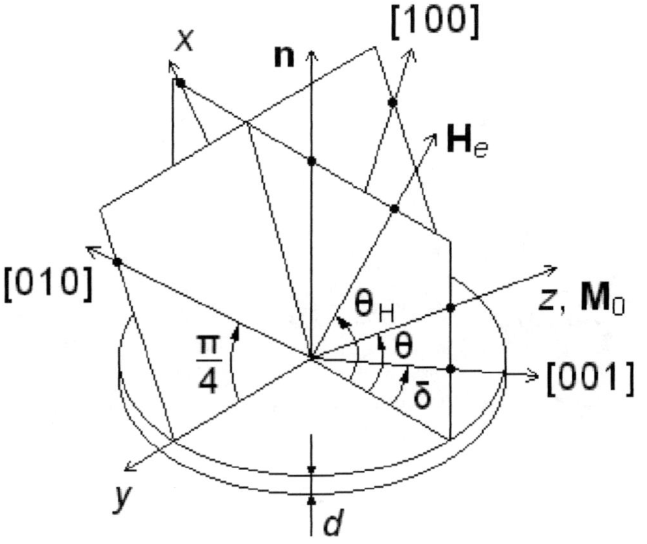

Figure 1. Film model with the plane of the symmetry of the {110} type located perpendicularly to a film and magnetized in this plane.

The used designations: $\boldsymbol{M}_0$ is the vector of ferrite magnetization; $\boldsymbol{H}_e$ is the external magnetizing field strength; $\boldsymbol{n}$ is the normal to a film; *x*, *y*, *z* are coordinate axes and $z \parallel \boldsymbol{M}_0$, $y \perp \boldsymbol{n}$; θ, θ_H, δ are the angles of inclination to a film of vectors $\boldsymbol{M}_0$, $\boldsymbol{H}_e$ and crystallographic axis [001] respectively. It was assumed that the vectors were also situated in a plane of crystallographic symmetry passing across an [001] axis. This assumption allows to reduce quantity of the angular variables used in model.

The energy density of crystallographic magnetic anisotropy in the most of the ferrites with cubic symmetry of the crystal lattice is described by the expression [4]:

$$W_a^c = -\frac{K_{c1}}{2M_0^4}\sum_p M_p^4,$$

where K_{c1} is the first constant of cubic magnetic anisotropy and summation is made on magnetization vector projections to the coordinates axes formed by fourth order symmetry axes of a crystal. This expression was used at a deducing of the FMR frequency formula and also at a deducing of a connection equation between the angles θ and θ_H.

FMR frequency in the considered model is given by a formula [5]:

$$(f_0/g)^2 = (H_{ez} + H_c n_x + 4\pi M_0 \cos 2\theta)(H_{ez} + H_c n_y - 4\pi M_0 \sin^2\theta), \quad (3)$$

where $g = 2{,}8\,\text{MHz/Oe}$ is the gyromagnetic ratio; H_{ez} is projection of $\boldsymbol{H}_e$ to a vector $\boldsymbol{M}_0$; n_x n_y are taken into account influence on the frequency of magnetic anisotropy and defined by formulas:

$$n_x = \frac{1}{2}[\cos 2(\theta - \delta) + 3\cos 4(\theta - \delta)], \quad (4)$$

$$n_y = \frac{1}{8}[-3 + 16\cos 2(\theta - \delta) + 3\cos 4(\theta - \delta)]. \quad (5)$$

It is possible to deduce the ratio connecting among themselves small changes of a frequency square with changes of ferrite magnetic parameters out of a formula (3):

$$\delta\left(\frac{f_0}{g}\right)^2 = 4\pi M_0\left[A\cdot\delta(4\pi M_0)+B\cdot\delta H_c\right], \tag{6}$$

where A and B are dimensionless sizes. At calculation have to be considered both variations $\delta(4\pi M_0)$ and δH_c and connected with them variations of δH_z, δn_x, δn_y, $\delta\theta$. Expressions for $\delta\theta$ and δH_z follow from the equations defining the magnetizing field projections to coordinate axes x and z:

$$H_{ez} = H_e\cos(\theta_H-\theta), \tag{7}$$

$$H_{ex} = H_e\sin(\theta_H-\theta) = M_0N_{xz}^c + \left(\frac{\sin 2\theta}{2}\right)4\pi M_0. \tag{8}$$

Here N_{xz}^c is the component of the anisotropy effective demagnetizing factor. It is defined by the formula [5]:

$$M_0N_{xz}^c = \frac{1}{8}H_c\left[2\sin 2(\theta-\delta)+3\sin 4(\theta-\delta)\right]. \tag{9}$$

From the equations (7) – (9) we will obtain $\delta\theta = C\cdot\delta(4\pi M_0)+D\cdot\delta H_c$ and $\delta H_z = K\cdot\delta\theta$, where:

$$C = -\frac{\sin 2\theta}{2H_{ez}+H_c\left[\cos 2(\theta-\delta)+3\cos 4(\theta-\delta)\right]+2\cdot 4\pi M_0\cos 2\theta},$$

$$D = -\frac{2\sin 2(\theta-\delta)+3\sin 4(\theta-\delta)}{4\left\{2H_{ez}+H_c\left[\cos 2(\theta-\delta)+3\cos 4(\theta-\delta)\right]+2\cdot 4\pi M_0\cos 2\theta\right\}},$$

$$K = \frac{H_c}{8}\left[2\sin 2(\theta-\delta)+3\sin 4(\theta-\delta)\right]+\frac{\sin 2\theta}{2}4\pi M_0.$$

δn_x, δn_y can also be expressed through $\delta\theta$. From (4), (5) follows: $\delta n_x = (\partial n_x/\partial\theta)\cdot\delta\theta$ and $\delta n_y = (\partial n_y/\partial\theta)\cdot\delta\theta$.

Thus, formulas for calculation of A and B will have an appearance:

$$4\pi M_0\cdot A = \frac{\partial(f_0/g)^2}{\partial(4\pi M_0)}+\left[\frac{\partial(f_0/g)^2}{\partial H_z}\cdot K+\frac{\partial(f_0/g)^2}{\partial\theta}+\frac{\partial(f_0/g)^2}{\partial n_x}\frac{\partial n_x}{\partial\theta}+\frac{\partial(f_0/g)^2}{\partial n_y}\frac{\partial n_y}{\partial\theta}\right]\cdot C,$$

$$4\pi M_0\cdot B = \frac{\partial(f_0/g)^2}{\partial H_c}+\left[\frac{\partial(f_0/g)^2}{\partial H_z}\cdot K+\frac{\partial(f_0/g)^2}{\partial\theta}+\frac{\partial(f_0/g)^2}{\partial n_x}\frac{\partial n_x}{\partial\theta}+\frac{\partial(f_0/g)^2}{\partial n_y}\frac{\partial n_y}{\partial\theta}\right]\cdot D.$$

3. Results of calculations

The purpose of calculations consisted in that from system of the equations $A=0$, $B=0$ and (3) to define values δ, θ and H_{ez}. Values of parameters $4\pi M_0$, H_c and f_0/g in the equations are given. Further the equations (7) – (8) allow to pass from parameters θ and H_{ez} to experimentally controlled parameters θ_H and H_e. Thus, the choice of parameters of the magnetizing field (H_e and θ_H) and a choice of crystallographic orientation of a film (δ) will be a result of the solution of a task. Conservation of the FMR frequency value in deflection of the ferrite magnetic parameters from the initial values will be criterion for optimality of such a choice.

As an example, results of the calculations executed for value $f_0/g = 4\pi M_0$ are given in figure 2. It is necessary to note that the built model doesn't introduce the restrictions on the value and a sign of H_c. Therefore the deduced ratios are applicable not only to ferrite, at which $|H_c/4\pi M_0| << 1$ and which are widely used in spin-wave electronics, but also to perspective materials with big anisotropy.

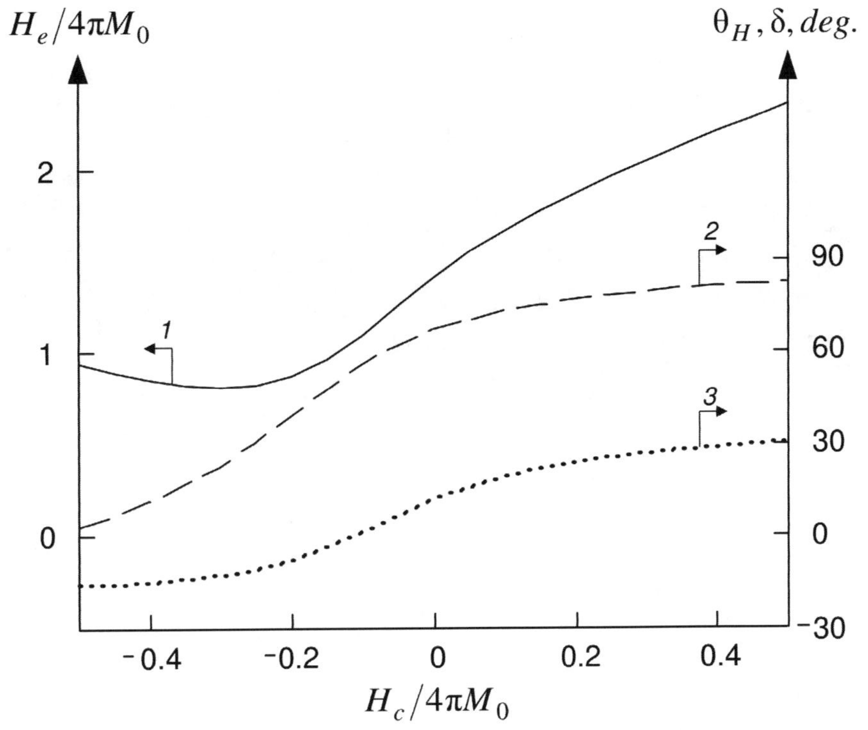

Figure 3. The calculated dependences of the optimal values $H_e/4\pi M_0$ (*1*), θ_H (*2*) and δ (*3*) on $H_c/4\pi M_0$.

4. Conclusions

Results of work can be used in the development of the spin-wave electronics devices. In particular, change of characteristics of these devices at radiative and thermal effects will be defined by changes of magnetic parameters of ferrite – magnetizations of saturation and a field of magnetic crystallographic anisotropy. The obtained analytical dependences allow both to predict these changes and to minimize them promoting development of devices with thermostable and resistant to radiation characteristics.

5. References

[1] Shagaev V V 1998 *Technical Physics* **43** 1226
[2] Shagaev V V 2003 *Physics of the Solid State* **45** 2320
[3] Shagaev V V 2006 *Technical Physics Letters* **32** 779
[4] Gurevich A G and Melkov G A 1994 *Magnetic oscillations and waves* (Moscow: Nauka)
[5] Bondarenko G G and Shagaev V V 2013 *Izv. Vyssh. Uchebn. Zaved. Ser.: Fizika* **56** (N1/2) 14 [in Russia]

Effect of ion irradiation on the properties multi-element plasma coatings

E N Eremin[1], A S Syzdykova[1], S A Guchenko[2], V M Yurov[2] and M S Gyngazova[3]

[1]Omsk State Technical University

[2]E.A. Buketov Karaganda State University

[3]National Research Tomsk Polytechnic University, Tomsk, Russia

E-mail: exciton@list.ru

Abstract. The paper presents the results of the study of ion irradiation on the properties of multi-element plasma coatings. The coatings were bombarded by argon ions using heavy current ion source with a hollow cathode. After ion irradiation, the structure and physical properties of the coatings change, however, the nature of the changes is different for different coatings. To predict the behavior of the coating exposed to irradiation is virtually impossible. Therefore, structural studies and investigation of physical properties of the coatings to determine their functional characteristics are to be conducted.

1. Introduction

The main effects of ion irradiation on the properties of the coatings occur at the stage of their emergence due to stress relaxation in the region of ion impact and restructuring of the crystal structure [1-3]. Point defects and active adsorption centers are formed on the surface. The mobility of adatoms on the surface (surface diffusion), which is enhanced by low-energy bombardment of the growing film with ions of inert gas, is also considered to be essential. The number of point defects can be increased either through an increase in the ion flux energy, or an increase in the density of the ion flux. Simultaneously with the defect formation, the reverse process of their recombination, "annealing", occurs, which reduces the concentration of defects. As a result of these two processes, the amount of nucleation centers becomes equilibrium, and it may be changed through varying parameters of ion irradiation. The effect of ion irradiation on films and coatings has been investigated in a number of studies [3–11]. Ion irradiation includes low-energy and high-energy ion bombardment. In [4] it is shown that the level of radiation impact on materials can be divided into three groups: a) "most resistant", substantial changes occur only at the substructure level (for example, Ti-Zr-V-Hf-Nb-Ta-N multiple system), b) the "madium resistance", significant changes occur in a macro stress-strain state (for example, TiN system), c) "structural variable", substantial changes occur at the macro-level and in phase composition (for example, MoN system).

This paper discusses the experimental and theoretical aspects of the effects of ion irradiation on the structure and properties of multi-element coatings.

2. Objects and experiment techniques

In the experiments, we used composite cathodes Cr-Mn-Si-Cu-Fe-Al, Zn-Al, Zn-Cu-Al, Mn-Fe-Cu-Al and Al-Fe produced through induction melting. These cathodes were used to deposit coatings on a

IOP Conf. Series: Materials Science and Engineering **110** (2016) 012017 doi:10.1088/1757-899X/110/1/012017

steel substrate under different process conditions with ion-plasma installation HHB-6.6I1. The cathode made of steel 12X18H10T and Ti, Zr, Cu, Zn-Al, Zn-Cu-Al and Al-Fe cathodes were used for simultaneous sputtering. The coatings were exposed to argon ion bombardment using the high current ion source with a hollow cathode by N.N. Koval (IHCE SB RAS, Tomsk). The arc current was 1 A, and the potential on the substrate was maintained equal to 200 V. The quantitative analysis of the elemental composition of the coating was carried out with the electron microscope JEOL JSM-5910. The microhardness of the coatings study was measured using the microhardness tester ISOSCAN OD. The atomic force microscope NT-206 was used to perform a nanoscale study of the surface coatings.

3. Results and discussion

As an example in the Figure 1 shows the AFM image images of the surface of the Fe-Al surface coating a) (before irradiation) and b) (after irradiation). Table 1 presents roughness of the multi-element coatings and Table 2 provides their microhardness before and after irradiation.

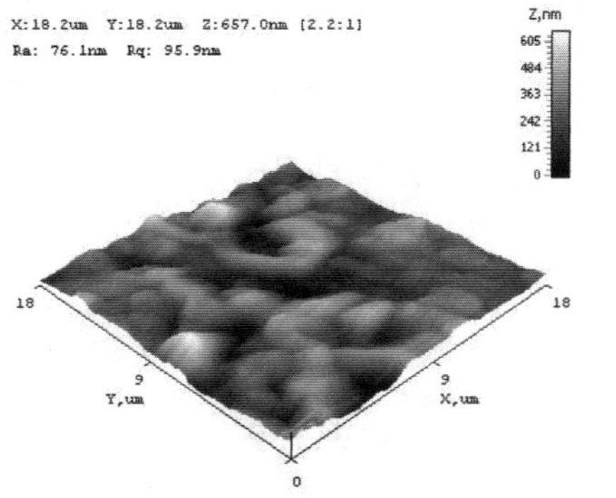

a)

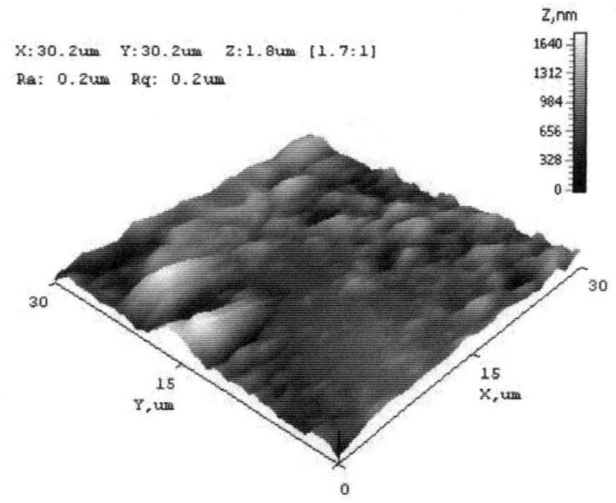

b)

Figure 1. AFM image of the Fe-Al surface coating a) (before irradiation) and b) (after irradiation)

Table 1. Roughness of the coating

Coating	after irradiation R_a (nm)	before irradiation R_a (nm)
Cr-Mn-Si-Cu-Fe-Al	22.26	13.34
Zn-Al	76.39	78.0
Mn-Fe-Cu-Al	23.08	29.89
Fe-Al	147.06	56.14

As can be seen in Figures 1, ion irradiation affects the coating structure. An exception is the Zn-Al coating, which turned out to be radiation-resistant. The roughness value measured before and after irradiation correlates with the obtained result (Table 2). We attribute this behavior of the Zn-Al coating to its pronounced globular structure. The presence of the "ball" system leads to elastic scattering of argon ions; as a result, the local deformation is negligible.

Table 2. Microhardness of the coating before and after irradiation

Coating	Microhardness (MPa)	
	before irradiation	after irradiation
12X18H10T+Zr	477.2	514.8
12X18H10T+(Zn-Cu-Al)	406.3	485.2
12X18H10T+(Zn-Al)	474.4	476.1
12X18H10T+(Fe-Al)	401.9	475.6
12X18H10T+Cu	543.1	585.7

The structure of the Fe-Al coating is significantly different from the structure of the Zn-Al coating. In this case, non-closed nanostructures can be observed (Fig. 1). The roughness of this coating after irradiation increases nearly three-fold. The same roughness pattern can be observed for the Mn-Fe-Cu-Al coating.

3.1. Surface tension of the deposited coatings

The structure and properties of solid surfaces and coatings are mainly determined by their surface energy or surface tension. We measured the surface tension through determining the dependence of the microhardness on the thickness of the deposited coating. The dependence is given by the formula [11]:

$$\mu = \mu_0 \cdot \left(1 - \frac{d}{h}\right), \tag{1}$$

where μ is the microhardness of the deposited coating; μ_0 is for "thick" model; h is the deposited coating thickness.

The parameter d is related to the surface tension σ through the formula [11]:

$$d = \frac{2\sigma\upsilon}{RT}, \tag{2}$$

where σ is the surface tension of the bulk sample; υ is one mole volume; R is gas constant; T is temperature.

In the $\mu \sim 1/h$ coordinates ($1/h$ is the reverse thickness of the deposited coating) we obtain a straight line, the slope determined by d, and formula (2) is used to calculate the surface tension of the deposited coating (σ).

Table 3. Surface tension of the multi-element coatings produced in argon

Coating	σ (J/m^2)	Coating	σ (J/m^2)
12X18H10T+Zr	0.970	12X18H10T+Zn-Al	1.098
12X18H10T+Zn-Cu-Al	1.093	12X18H10T+Al	1.144
12X18H10T+Fe-Al	1.292	12X18H10T+Cu	1445

Table 4. Surface tension of the multi-element coatings produced in nitrogen

Coating	σ (J/m^2)	Coating	σ (J/m^2)
12X18H10T+Zr	0.185	12X18H10T+Zn-Al	0.784
12X18H10T+Zn-Cu-Al	0.744	12X18H10T+Al	0.801
12X18H10T+Fe-Al	1.034	12X18H10T+Cu	1.032
Zn–Cu–Al	0.243	Cr–Mn–Si–Cu–Fe–Al	0.711
Mn–Fe–Cu–Al	0.367		

The data in Tables 3 and 4 shows that the surface tension of the coatings produced in nitrogen atmosphere is less than that of the coatings produced in argon. Hence, formation of the nitride phase causes decrease in the surface energy of the coatings.

Table 5 shows the results of the effect of ion bombardment on some of the surfaces.

As can be seen in Table 5, all the three tested coatings behave differently under ion bombardment: Zn-Cu-Al coating is radiation-resistant, and the value of the surface tension remains practically unchanged; the surface tension of the Cr-Mn-Si-Cu-Fe-Al coating increases twofold, and the Mn-Fe-Cu-Al coating decreases threefold.

Table 5. Surface tension of the coatings after ion irradiation

Coating	Before irradiation σ (J/m^2)	after irradiation σ (J/m^2)
Zn – Cu – Al	0.243	0.241
Cr – Mn – Si – Cu – Fe – A 1	0.711	1.422
Mn – Fe – Cu – Al	0.367	0.122

3.2. Evaluation of the melting point of the deposited coatings

The temperatures of steel melting and solidification depend on its composition. Typically, when calculating Ts, allowance is made for the additive effect of dopants on these values.

It should be emphasized that the T_s value is of no practical interest, since the elements between the liquid and the solid phases in crystallization are considerably redistributed and as a result, the liquid is doped with segregates, primarily carbon, sulfur and phosphorus (which determine the ability of the elements to segregation).

The above results of the experimental evaluation of the surface tension of multi-element coatings, the results of its calculation by the elemental analysis and the data on the surface tension of pure metals reported in [13] can be used to show that the averaged value of the surface tension is an additive value. In this case, the melting point of the coating can be estimated by the formula:

$$T_m = 1.4 \cdot 10^3 \cdot \sigma \ (K). \qquad (3)$$

The data from Table 3 were used to provide the evaluation in Table 6.

IOP Conf. Series: Materials Science and Engineering **110** (2016) 012017 doi:10.1088/1757-899X/110/1/012017

Table 6. Melting point of the multi-element coatings produced in argon

Coating	T (K)	Coating	T (K)
12X18H10T+Zr	1358	12X18H10T+Zn-Al	1537
12X18H10T+Zn-Cu-Al	1530	12X18H10T+Al	1602
12X18H10T+Fe-Al	1809	12X18H10T+Cu	2023

The melting point of steels depends on their chemical composition, but it keeps in the range of (1450–1520) K. As can be seen in Table 6, the melting point of 12X18H10T + Al, 12X18H10T + Fe-Al and Cu + 12X18H10T coatings produced in argon is significantly higher than that of steels. The melting temperature of Zn-Cu-Al, Cr-Mn-Si-Cu-Fe-Al and Mn-Fe-Cu-Al coatings is low. Table 7 shows the melting point of the investigated coatings after ion bombardment.

Table 7. Melting point of the coatings produced in nitrogen atmosphere, before and after irradiation in the medium

Coating	Melting point (K)	
	Before irradiation	After irradiation
12X18H10T+(Zn-Cu-Al)	1042	1250
12X18H10T+(Fe-Al)	1448	1454
12X18H10T+(Zn-Al)	1098	1296
12X18H10T+Cu	1445	1561

As can be seen in Table 7, the melting point slightly increases. For 12X18H10T+(Fe-Al) coating, this change is negligible.

4. Conclusion

The results of the studies provided in this paper show that ion irradiation largely affect all the properties of the coatings. However, the properties of radiation-resistant coatings change insignificantly. It should be noted that the behavior of the coating under irradiation can hardly be predicted so far. Therefore, structural studies and research into physical properties of the coatings to determine their functional characteristics are to be conducted.

Acknowledgements
This work was supported by The Ministry of Education and Science of the Russian Federation in part of the "Science" program.

References

[1] Pranevicius L and Dudonis Y 1980 Modification of properties of solids by ion beams. Vilnius: Mokslas 242
[2] Korotaev A D, Tyumentsev A N, Pinzhin Yu P and Remnev G E 2004 Surface and Coatings Technology **185** 38–49
[3] Ghyngazov S A, Vasil'ev I P, Surzhikov A P, Frangulyan T S and Chernyavskii A V 2015 J. Tech. Phys. **60** 128-132
[4] Andreev A A, Voevodin V N, Sobol O V et. al. 2013 *Proceedings of the International Conference Nanomaterials: Applications and Properties* **2** 03PISERE08 (4PP)
[5] Perinskaya I V, Lyasnikov V N, Perinskii V V and Muktarov O D 2014 *Beam Processing Inorganic Materials: Applied Research* **5** 159–163
[6] Novakovic M, Popovic M and Bibic N 2011 Processing and Application of Ceramics **5** 25–29
[7] Guglya A G, Neklyudov I M, Shkuropatenko V A et. al. 2005 *Problems of Atomic Science and Technology* (in Russian) **3** 171–175
[8] Belous V A, Leonov S A, Nosov G I et al. 2009 Modification of the E110 alloy surface through deposition of multilayer Zr/ZrN coating and ion bombardment FP FIP PSE **7** 76–81
[9] Krivobokov V P, Umnov S P 2009 *Proceedings of the universities. Physics* **11/2** 223–226
[10] Ostapchuk V N, Movshovich A I, Gorelik B V 2008 *High technologies in mechanical engineering Kharkov: NTU "KHPI"* **1** 211–216
[11] Pacaud J. et al. 1999 J. Appl. Phys. **86** 4848–62
[12] Jurov V M, Laurynas V C, Guchenko S A and Zavatskaya O N 2014 *Hardening technologies and coatings* **1** 33–36
[13] Jurov V M 2011 Eurasian Physical Technical journal **8** 10–14

Determination of glass transition temperature for polymers by methods of thermoactivation spectroscopy

V M Yurov[1], V C Laurinas[1] and E N Eremin[2] and M S Gyngazova[3]

[1]E A Buketov Karaganda State University

[2]Omsk State Technical University

[3]National Research Tomsk Polytechnic University, Tomsk, Russia

E-mail: exciton@list.ru

Abstract. For rapid determination of glass transition temperature for polymers, we propose a method of thermally stimulated luminescence. The experiments were carried for epoxy polymers dyed and undyed with organic dyes. It is shown that glass transition temperature depends on curing temperature and concentration of the dye. The comparison with the thermogravimetric analysis showed coincidence of the results obtained.

1. Introduction

The glass transition temperature of polymers is an important characteristic defining their practical application in a particular area [1–4].

The glass transition temperature of polymers is an important characteristic defining their practical application in a particular area [1–4].

A large number of methods have been currently developed to determine the glass transition temperature for polymers:

- Differential Thermal Analysis (DTA). In this method, when glass transition temperature is attained, the slope of the specific volume of the sample temperature changes [5, 6];

- Single Differential Thermal Analysis (Single DTA), which combines both, DTA and thermogravimetric method [7]

- Differential Scanning Calorimetry (DSC). This method measures the heat flow to/or from a small sample, when the sample is subjected to a programmed temperature variation [2, 5, 8];

-Thermogravimetric Analysis (TGA). This method measures the change in mass of the sample in a linear temperature change [5, 6, 8, 9];

- Evolved Gas Analysis (EGA). EGA is a family of methods to analyze the nature and amount of volatile gaseous product evolving from the sample which is subjected to programmed temperature variation. The most important methods are mass spectrometry and infrared spectroscopy. EGA is often used in combination with TGA. [10];

- Thermomechanical Analysis (TMA). TMA measures deformation (typically one-dimensional deformation) of the sample as a function of temperature. Different varieties of the method use the mode of constant, increasing or modulated power. [11];

Dynamic Mechanical Analysis (DMA). In this method, the sample is subjected to sinusoidal mechanical impact. The amplitude, phase shift, and other characteristics are the functions of temperature [12–14];

- and others.

Most of the methods described above demand for the realization of complex and expensive equipment.

In this paper for the rapid determination of the glass transition temperature polymers, we propose a method of thermoactivation spectroscopy.

2. Methods of thermoactivation spectroscopy (TAS)

The physical nature underlying TAS methods is implies that when the temperature of the sample changes under a certain law, the transition from the nonequilibrium state to a new state approaching the thermal equilibrium is thermally stimulated. This transition may be foolowed by thermally-stimulated luminescence, TSL) [15]), by emission electron (thermally-stimulated emission of electrons, TSE [16]), by the change in conductivity (thermally-stimulated conductivity [17]), etc. The analysis of the temperature dependence for the changing physical property of an object allows exploration of the parameters of electrically or optically active defects, mechanisms of relaxation processes occurring in the substance.

3. Results and discussion

The experimental scheme in the TSL method is simple: the polymer was irradiated with the ultraviolet deuterium lamp 400 W (any other source of ultraviolet radiation can be used) for 5–10 minutes at T = 80 K through the quartz window and a combination of light filters that produce the emission band lamp with 300 nm. The temperature of the sample (1 mm thick and 10 mm in diameter) was measured by the copper-constantan thermocouple one junction of which was mounted on the sample surface, and the other one was placed in a Dewar vessel with melting ice. The sample was heated at a constant rate of 0.15 K/s through a copper chip carrier via an electric heater. The integral emission of the sample was detected by a photomultiplier tube and then the signal entered the input of the DC amplifier and the display.

Figure 1 shows the integral curve of the thermally-stimulated luminescence for epoxy polymers obtained under various curing temperatures: 100 ° C (1), 140 ° C (2), 200 ° C (3).

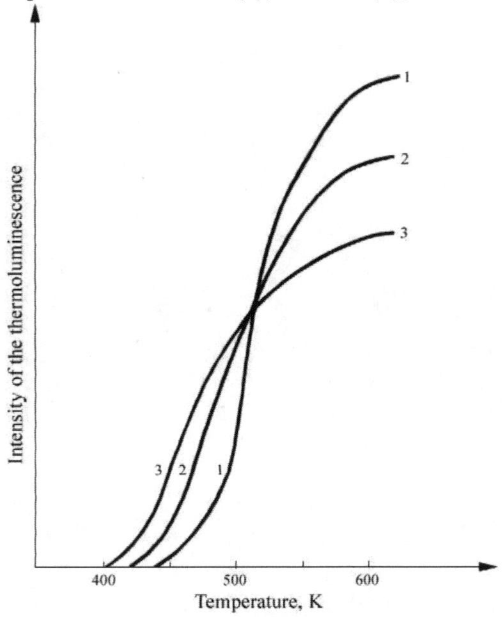

Figure 1. Thermally-stimulated luminescence of the undyed polymer produced under different curing temperatures

The corresponding curves in Arrhenius coordinates (lnI ~ $10^3/T$) intersect the horizontal axis at points with temperatures T_{C1}, T_{C2}, T_{C3}. Table 1 shows these values along with the results of the thermogravimetric analysis.

Fig. 2 shows the integral curve of the thermally-stimulated luminescence of the epoxy polymer dyed in yellow with rhodamine of different concentration.

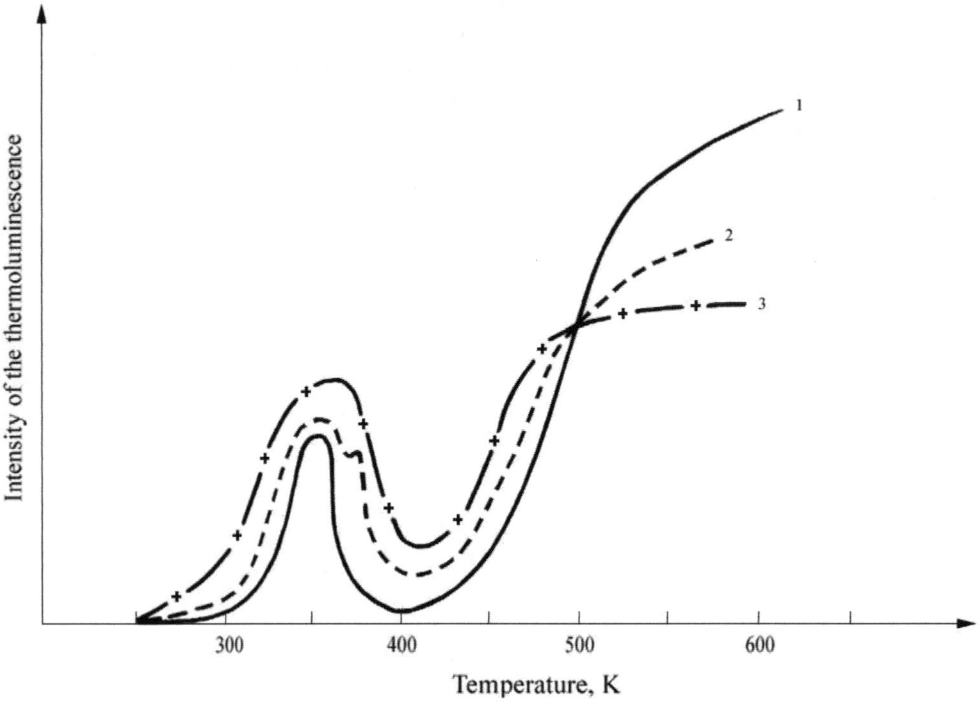

Figure 2 Thermally-stimulated luminescence of the dyed epoxy polymer

Table 1. Glass transition temperature for epoxy polymer

Sample	Method	T_{C1} (K)	T_{C2} (K)	T_{C3} (K)
Undyed epoxy polymer	TSL	471.6	446.4	423.7
	TGA	479	450	426
Epoxy polymer dyed with rhodamine 6Y	TSL	400	383	379
	TGA	406	392	384

The data provided in the Table 1 show:

1) the values of the glass transition temperature T_c obtained by the method of thermally stimulated luminescence are in good agreement with the thermogravimetric measurements, which allows its application to determine T_c along with conventional methods;

2) the glass transition temperature T_c of the polymer produced under different curing conditions varies considerably (in this case, almost by 50 K), which allows the use of the method of thermally stimulated luminescence as a rapid method when developing the polymer production technology;

3) the glass transition temperature T_c of the polymer decreases when its base is doped with dye molecules (up to 70 K in this case), and as the concentration of the latter grows, it continues to decrease.

4. Conclusion

With no regard to the issues related to the mechanism of excitation of thermally stimulated luminescence in undyed and dyed epoxy polymers after exposure to UV radiation, it should be noted that the method of thermally stimulated luminescence in its sensitivity is significantly superior to optical absorption techniques, electron paramagnetic resonance and other spectroscopic methods. It is sufficiently simple, reliable and does not require expensive equipment.

In our opinion, further investigation of the method with other types of polymers will provide engineers with a simple method to control the production technology of polymer materials with desired properties.

Acknowledgements
This work was financially supported by The Ministry of Education and Science of the Russian Federation in part of the science activity program.

References

[1] Pascault J-P. 2002 Thermosetting Polymers. - Marcel Dekker 477
[2] Reading M, Hourston J H (Eds.) 2006 Theoretical and practical applications in polymer characterisation. – Springer 330
[3] Belfiore L A 2010 Physical Properties of Macromolecules 803
[4] Campbell F C 2010 Structural Composite Materials. - ASM International 629
[5] Menczel J D e.a. (ed.) 2009 Thermal Analysis of Polymers, Fundamentals and Applications. - John Wiley & Sons, Inc., Hoboken, New Jersey 698
[6] Crompton T R 2013 Thermal Methods of Polymer Analysis. - Smithers Rapra Technology 256
[7] Qi Zhou, Nian-Sheng Cheng 2009 Chemical Engineering Journal **149** 301-310
[8] Gabbott P.(ed.) 2008 Principles and Applications of Thermal Analysis. - Singapore: Wiley-Blackwell 480
[9] Akay M 2012 Introduction to Polymer Science and Technology. - Ventus Publishing ApS 269
[10] Brown M E 2001 Introduction to Thermal Analysis. Techniques and Applications. – Kluwer 310
[11] Nicholson D W 2003 Finite Element Analysis: Thermomechanics of Solids. - CRC Press 296
[12] Siebenmorgen T J, Yang W and Sun Z 2004 American Society of Agricultural Engineers **47** 835-839
[13] Stading M 1998 Annual Transactions of the Nordic Rheology Society **6** 147-150
[14] Mahlin D J, Wood N, Hawkins J et. al. 2009 Int. J. Pharm. **371** 120-125
[15] Martini M, Meinard F 1997 Rivista del Nuovo Cimento **20** 71
[16] Ankjærgaard C, Murray A S, Denby P M, Botter-Jensen L 2006 Radiation Measurements **41** 780–786
[17] Algaer E 2010 Thermal Conductivity of Polymer Materials. – Darmstadt 74

Effect of an electric field on nucleation and growth of crystals

V M Yurov[1], S A Guchenko[1] and M S Gyngazova[2]

[1]E A Buketov Karaganda State University
[2]National Research Tomsk Polytechnic University, Tomsk, Russia

E-mail: exciton@list.ru

Abstract. The effect of the electric field strength on nucleation and growth of the crystals of ammonium halides and alkali metal sulfates has been studied. The optimal electric field strength for NH_4Cl and NH_4Br crystals was found to be 15 kV/cm, and for NH_4I, it equaled 10 kV/cm. No effect of the electric field strength on the crystal growth was found for alkali metal sulfates. This difference is analyzed in terms of the crystal growth thermodynamics. In case, when the electric field is small and the Gibbs energy is of a significant value, the influence of the electric field at the crystal growth is negligible. A method to estimate the critical radius of homogeneous nucleation of the crystal is suggested.

1. Introduction

The effect of electric, magnetic and other fields on nucleation and growth of various crystals has been studied since 50-ies of the last century. Some qualitative patterns were found and empirical models were suggested. However, the quantitative theory of these effects has not been worked out. This is primarily due to the lack of a rigorous theory of homogeneous nucleation. The main ideas of the theory of new phase nucleation were put forward by Gibbs and then developed by Volmer, Becker and Doring, Stransky and Kaishev, Frenkel and Zeldovich. According to this theory, the formation of small particles of a new phase in the metastable phase is considered as a fluctuation process. Despite this progress, there is no quantitative agreement between the theory and experimental data.

This is due to the fact that the size effects in the formation of crystal nuclei which have dimensions of the order of (1–5) nm are ignored. If we consider the size dependence of the surface tension at the interface only, classical theories of crystallization nucleation provide satisfactory quantitative agreement between theoretical and experimental data [1]. In [2], an additional force generated when applying the electric field was introduced. The results obtained in this research show that an electric field can either stimulate or inhibit the nucleation process depending on the ratio of the dielectric constant and orientation of the applied electric field relative to the surface nucleation. This approach is demonstrated in [3], which provides extensive bibliography on the effect of the electric field on crystal nucleation and growth. This issue is addressed in the reviews in [4–6].

2. Effect of the electric field on nucleation and growth of crystals of ammonium halides

A V Shubnikov [7] observed nucleation under the action of the electric field in the ammonium chloride solution. He attributed the observed phenomenon to the presence of the smallest dust particles in air which enter the solution and become the crystallization sites under the action of the electric field.

In [8], Kozlowsky experimentally investigated the effect of the electric field on the solution of halides and sulfates. The field strength was varied from 1 to 30 kV/cm. The experiments showed that the electric field does not affect the sulphate solutions of nickel and copper and weakly impacts the nucleation process in the cadmium iodide solution. However, a significant impact of the field was observed on solutions of sodium and potassium chloride and NH_4I.

We have conducted a detail study of the effect of the electric field on nucleation and growth of ammonium halide crystals. Note that in [7, 8], as in many other studies, ammonium halides were produced in the form of dendrites. Nobody succeeded to produce single crystals of ammonium halides.

A small amount of urea was added to 90 % aqueous solution of , and . The temperature in the thermostat was maintained at 35 ° C. When the intensity of the electric field increased from 0 to 3 kV/cm, the number of centers increased as well. In the range from 3 to 15 kV/cm, the number of nuclei decreased, however, larger and more transparent crystals emerged. The optimal electric field strength for and was equal to 15kV/cm, and for it was 10 kV/cm.

3. Theoretical aspects of the effect of the electric field on crystal nucleation and growth

Thermodynamic methods used in [8] showed under the impact of the electric field, the crystallization pressure changes

$$P - P_0 = \frac{\beta E^2}{V_1 - V_2} \tag{1}$$

where $\beta = \text{const}, V_1, V_2$ are the specific volumes of the first and second phases (per molecule), respectively.

The critical radius of the nucleus in the solid phase in the electric field:

$$r^* = \frac{2\sigma V_2}{\Delta\mu^0 - \beta E^2} \tag{2}$$

where σ is the coefficient of the surface tension at the solution-crystal interface; $\Delta\mu^0$ is the chemical potential in the absence of the field.

For a one-component composition , where ΔG is the Gibbs energy. According to [1] the N value is of the order of 100. Table 1 shows the values of the Gibbs energy (in modulus) for some of the substances taken from [9].

Table 1. Gibbs energy for some halides and sulfates [9]

Substance/Agent	ΔG (kJ/mol)	Substance/Agent	ΔG (kJ/mol)
KCl	408.490	$CuSO_4$	661.070
NaCl	384.120	$CuSO_4 \cdot 5H_2O$	1377.780
NH_4Cl	202.770	$NiSO_4 \cdot 6H_2O$	2680.050
AgCl	109.700	$NiSO_4 \cdot 7H_2O$	2459.640
AgBr	97.250	K_2SO_4	1318.790

Taking into account the remarks above equation (2) will take the form:

$$r_к = \frac{2\sigma\vartheta_2}{G/N - \beta E^2} \tag{3}$$

In case, when the electric field strength is low and the Gibbs energy is of a significant value, i.e. G/N >> βE^2, the effect of the electric field on the crystal growth will be negligible. As can be seen in Table 1, the sulfate solution having a larger value of the Gibbs energy will be less affected by the electric field. This is consistent with the results reported in [2, 8].

4. Evaluation of the critical radius of homogeneous nucleation of crystallization sites

The evaluation of the critical radius of homogeneous nucleation of pure metals can be carried out in the parameter d, which is determined by the size dependence of some physical property A (r) of metal (magnetic, electric, etc.) according to the formula [1]:

$$A(r) = A_0\left(1 - \frac{d}{r}\right), \tag{4}$$

where A_0 is the physical property of the bulk sample.

Equation (4) is valid for both internal and external size effects. Therefore, the developed approach is valid not only for crystallization of pure metals, but for any other material. In this case, the parameter d can be determined by one of the proposed methods [10], and the formula $r = r_c \geq d/3$ can be used to evaluate r_c.

We defined the parameter d of the intensity I (r) of the luminescence of crystalline powders of size. The parameter d = $2\sigma\upsilon$ / RT, where σ - the surface tension of the crystal, υ - molar volume, R - the gas constant, T - temperature. The coordinates I (r) ~ 1 / r a straight line is obtained according to formula (4). The slope of this straight line gives the value of d.

Table 2. Surface tension and critical radius of the alkali metal halides.

Chemical formula	Surface tension σ (J/m^2)	The critical radius r_k (nm)
NaCl	0.158	1.095
NaBr	0.138	1.181
NaI	0.118	1.283
KCl	0.141	1.414
KBr	0.123	1.420
KI	0.113	1.597
RbCl	0.130	1.497
RbBr	0.114	1.502
RbI	0.104	1.654

Tables 2 and 3 show values of the surface tension and the critical radius halides and sulfates of alkali metals. From these tables it follows that the critical radius at sulfates of alkali metals slightly larger than alkali metal halides. However, the difference in the critical radius is not as great as one would expect. This means that the electric field does not act on the nucleation process, and on their growth in accordance with the formula (3).

Table 3. Surface tension and critical radius sulphates of alkali and alkaline earth metals.

Chemical formula	Surface tension σ (J / m^2)	The critical radius r_k (nm)
Li_2SO_4	0.141	1.862
Na_2SO_4	0.124	1.753
K_2SO_4	0.117	2.044
Rb_2SO_4	0.112	2.209
Cs_2SO_4	0.110	2.504
$BeSO_4$	0.240	2.690
$MgSO_4$	0.166	2.003
$CaSO_4$	0.189	2.958
$SrSO_4$	0.180	2.226
$BaSO_4$	0.175	2.421

5. A simple model

Here we want to show a simple model, which gives some idea of the physical process of homogeneous nucleation of a new phase. In a homogeneous medium with the particle number density n probability W(r) is the first that the nearest neighbor is at a distance r_k from the selected particles, is easily obtained from the classical statistical physics, and it is equal to:

$$W(r_\kappa) = 4\pi n r_\kappa^2 \exp\left[-4\pi n^2 r_\kappa^3/3\right] \tag{5}$$

The probability of finding N particles in the embryo radius r_k is obviously $W_N(r_k) = \prod_{i=1}^{N} W_i(r_k)$.

The probability (5), we define the other hand, as the ratio of energy Bohr atom $\left(E = z^2/2a,\ z - \text{charge},\ a = \text{const}\right)$ to the total energy (ideal gas) $\left(3/2nkT_0\right)$. This is done, something to qualitatively assess whether the different critical radius r_k of the embryo, for example, for different metals. Thus, we have:

$$\frac{z^2/2a \cdot N}{3/2nkT_0} = (4\pi n)^N r_\kappa^N \exp\left[-4\pi N n^2 r_\kappa^3/3\right] \tag{6}$$

Taking the logarithm of both sides of (6), and performing simple transformations, we obtain:

$$-9\ln 10+\ln\left(\lambda\frac{Z^2}{T_0}\right)+\frac{3}{2}n\frac{T_0}{\Delta T}=\frac{4\pi}{3}Nn^2r_к^3 \tag{7}$$

The first term arises by substituting constants $\alpha = 1$ - provides a single dimension of values, T_0 - melting point, $\Delta T = T_0 - T_1$ - hypothermia. From (7) we see that the first two terms on the left are negligible compared to the rest. This immediately removes the dependence of r_k on Z, i.e, the type of atoms crystallizing substance. Finally, from (7) we have:

$$N\cdot r_к^3=\frac{1}{\pi n}\cdot\frac{T_0}{\Delta T} \tag{8}$$

If you make an estimate of the average values: r_k ~ 1 nm, T_0 about 1000 (for me-metals), ΔT ~ 100, n = N_L - number Loschmidt ~ 3 10^{19} cm^{-3}, we obtain for the number of particles in the bud N ~ 100. Below we provide data taken from [11]. We see that our estimates of the formula (8) in qualitative agreement with the experimental data. Indeed, the radius of the embryo practical ski does not depend on the type of metal atoms, the average number of particles N in the embryo has the same order.

Table 4. The parameters of nucleation kinetics at crystallization of supercooled metal [11].

Metal	T_0 (K)	ΔT (K)	r_k (nm)	N
Hg	234	90	0,8	100
Sn	505	180	0,8	130
Bi	544	200	0,8	30
Pb	600	170	1,1	180
Ge	1210	295	0,9	130
Ag	1234	253	1,2	430

From (8) it follows that the right side are the parameters, the magnitude of which has not changed over a wide range, i.e.:

$$N\cdot r_к^3 \approx \text{const} \tag{9}$$

Relation (9) is approximately true for many substances. The deep causes of these correlations are not yet clear.

5. Conclusion

From the above analysis it follows that the crystal growth in the electric field does not always lead to positive results. Thermodynamic characteristics are to be considered, and it is primarily the magnitude of the Gibbs energy.

Acknowledgements

This work was financially supported by The Ministry of Education and Science of the Russian Federation in part of the science activity program.

References

[1] Yurov V M, Guchenko S A et al. 2012 Scientific Review **5** 205–209
[2] Dhanasekaran R, Ramasamy P 1986 Nuovo Cimento D **7** 506–512
[3] Hammadi Z, Veesler S 2009 Progress in Biophysics and Molecular Biology **101** 38–44
[4] Mullin J 2001 Crystallization. Butterworth-Heinemann 600
[5] Sunagawa I 2005 Crystals: Growth, Morphology and Perfection. Cambridge University Press 295
[6] Duffar Th (Ed.) 2010 Crystal Growth Processes Based on Capillarity: Czochralski, Floating Zone, Shaping and Crucible Techniques. Wiley 533
[7] Shubnikov A V, Porvov V F 1961 Crystallography **6** 443–450
[8] Kozlovsky N M 1962 Crystallography **7** 157–159
[9] Bullakh A G, Bullakh K G 1978 Physicochemical properties of minerals and components of hydrothermal solutions. L.: Nedra 167
[10] Yurov V M, Portnov V S, Ibraev N X and Guchenko C A 2011 Advances of modern natural science **11** 55–59

IOP Conf. Series: Materials Science and Engineering **110** (2016) 012020 doi:10.1088/1757-899X/110/1/012020

Synthesis of Cu nanopowders by condensation from the gas phase

I V Chepkasov[1], Yu Ya Gafner[1], K V Zobov[2], S B Batoroev[3] and S P Bardakhanov[2,3,4]

[1] Katanov Khakassia State University, Lenin pr., 90, Abakan, 655017, Russia

[2] Khristianovich Institute of Theoretical and Applied Mechanics of SB RAS, Institutskaya str., 4/1, Novosibirsk, 630090, Russia

[3] Buryat State University, Smolina str., 24a, Ulan-Ude, 670000, Russia

[4] Novosibirsk State Technical University, Karl Marks pr., 20, Novosibirsk, 630073, Russia

E-mail: ilya_chepkasov@mail.ru

Abstract. In order to determine the most efficient regimes of copper nanoparticles synthesis, a series of experiments were conducted by evaporation and subsequent condensation of the raw material in an argon atmosphere. During the tests it was found that an increase of evaporation rate increases significantly the average size of the synthesized particles. However, the study of the dependence of dimensional parameters of the produced clusters on the intensity of the buffer gas flow rate has encountered significant difficulties associated because the results significantly divergent from the previously conducted experiments on the synthesis of transition metal oxides. In order to solve this contradiction the computer simulation was held of copper atoms condensation from the gas phase for the three different cooling rates and for the two final temperatures T = 373 K and T = 77 K. It was found after analysis that the rate of cooling of the gas mixture and the final temperature directly influences the number and the size of particles produced. For instance, with the 10 times of cooling rate decreases the average size of the particles obtained had increased by 2.7 times at a final temperature of 77 K and by 3.1 times at T_f= 373 K.

1. Introduction

Preparation of stable nanoscale copper particles having unique physical and chemical properties [1], is complicated task nowadays because the metal nanoparticles are rapidly oxidized in air or aqueous environment [2]. Currently there are many ways of nanoparticles producing of transition metals, such as mechanochemical methods, plasmachemical methods, some embodiments of a chemical, photochemical and radiation reduction, and the method of condensation from the gas phase [3]. A large volume of the nanooxides is synthesized using the method high temperature pyrolysis [4], and in much less quantity by the material's evaporation by the laser pulses in a very short femtosecond range [5].

[1] Assistant professor, PhD, Chepkasov Ilya Vasilyavich

However, not all of the above mentioned approaches are capable to provide the required purity of the nanomaterial. Among the variety of methods, in our opinion, the most promising method for the synthesis of fine particles, including the copper particles [6], is the method of evaporation of metals by powerful electron beam and subsequent condensation of vapor in the atmosphere of inert gas. Simple scalability to industrial volumes, any material allowed to evaporate and the high purity of the resulting product with the required technical application properties (electrical conductivity, strength, ductility) distinguishes particles synthesized by a gas phase method from the particles obtained by other methods.

The huge interest, shown by researchers to study nanoscale particles of transition metals, in particular to copper clusters, is due to the fact, that a decrease in the size of up to several tens of nanometers changes significantly the fundamental properties of the material, e.g. thermal and electrical conductivity, heat capacity and reactivity. This leads to the fact that these particles begin to be widely used as a feedstock in the production of variety of ceramic and composite compounds, superconductors, electronic components, filters, catalysts [7, 8] and nanoelectronic devices [9]. In particular, the nanosized copper clusters, in its electrically conductive properties can compete with silver. A huge range of applications of copper nanoparticles is available in steel production and the chemical industry [1]. However, the preparation of copper nanoparticles and other metals in fairly large quantities with the necessary properties of nanoscale clusters, depending on the application, claims very high demands to the production method.

In this paper the details are described of the copper nanoparticles synthesis through metal evaporation by the high energy electron beam in inert atmosphere to determine the most efficient modes of obtaining the material with the required size distribution at the suitable productivity of installation. The significant difficulty of the experimental study is impossibility to observe the processes of particles growth and their structuring. The materials obtained by this method are studied only after the synthesis process, without a complete understanding of the physical picture of the formation of nanoparticles. The computer simulation using a physically-based interatomic interaction potentials allows to identify the main patterns of nucleation, nanoclusters growth, consistent and zero-defect formation of coordination shells, structural isomeric transitions and morphological features of cluster surfaces. Therefore, in this paper, experimental study of copper nanoparticles condensation from the gas phase was complemented by molecular dynamics simulation of the cooling of 85000 of copper atoms system. This computer model was created in such a way to reflect maximally the processes which occur in a real experiment on the evaporation and condensation of nanoparticles.

In this paper, the condensation of copper nanoparticles was carried out on experimental-industrial facility jointly developed by the Khristianovich Institute of Theoretical and Applied Mechanics and Budker Institute of Nuclear Physics of SB RAS (Novosibirsk). The facility based on the linear electron accelerator ELV-6 with the release into the atmosphere of the focused relativistic beam with an energy of 1.4 MeV and power up to 100 kW. Such energy level allows to work at atmospheric pressure of gas flow because scattering losses in the gas are small.

On the basis of the processes occurring at the experimental-industrial installation the computer model had been built of copper nanoparticles condensation after high-temperature evaporation of raw material. A model system was a cube volume of 42600 nm^3 containing 85000 copper atoms uniformly distributed in space, with an initial temperature T_i = 1000 K. Due to the large number of atoms in the system, it was decided to model clusters condensation not in inert gas atmosphere, but for its model, in which the gas environment consists of the virtual particles having a constant temperature and generated by Andersen thermostat [10]. This thermostat allows the cooling with a fixed speed by gradual removing of excess energy from the system. To find a solution to the equations of motion of the particles in the model block the Verlet scheme was used in its "speedy" form [11] with a constant time step (h = 2 fs).

2. Results and discussion

In the present study we examined the effect of the main parameters such as the electron beam power, flow rate of the buffer gas and the temperature of coolant on the final shape and size of copper nanoparticles. To date, there are several methods for determining the size of the synthesized nanoparticles. For example, direct, but at the same time a very laborious method (even if the modern image processing software is available) is the averaging data of transmission electron microscopy (TEM). It allows also to get the estimation of the nanoparticle size distribution. However, the simpler and faster way to determine the average particle size is the specific surface measurement, which is calculated for spherical particles from the expression $S_{sp} = 6 / (d \cdot \rho)$, where d - the average particle diameter, ρ - density of the particle material. The specific surface area is readily determined by nitrogen adsorption with the BET method. Then it is possible to calculate d, as an integral parameter characterizing the size of the primary nanoparticles of powder and promptly judge the optimal conditions for obtaining particles with the specified parameters. For non spherical complex particle the average size can be obtained using by special approaches approximating experimental measurements [12]. So, in our research the size parameters (in particular during debug series) were investigated by measuring the specific surface of nanopowders, and the TEM was also used for mutual verification.

Table 1. The structure and the shape of Cu nanoclusters synthesized from the gas phase

Cooling rate (U, ps^{-1})	Tempe-ture (T_f, K)	Number of clusters	Average cluster size	Structure				Form	
				FCC (HCP), %	Ih, %	Dh, %	Amorphous,%	Worm like, %	Spherical, %
0.05	77	46	1847	22.02	45.83	20.87	11.28	76.08	23.92
0.025	77	35	2428	22.51	41.93	30.02	5.54	74.28	25.72
0.005	77	17	5000	23.59	26.82	49.59	0	58.83	41.17
0.05	373	19	4473	22.57	31.38	46.05	0	89.47	10.53
0.025	373	13	6538	29.43	30.07	40.5	0	84.61	15.39
0.005	373	6	14166	38.33	26.67	35	0	83.34	16.66

In our experimental research the possibility had been proved of particle size distribution control by varying the electron beam power and by flow rate of the argon. With the irradiation power the value of specific surface of the nanoparticles decreases (i.e., the average size increases). This is due to the fact that in the process of particles formation the concentration of vapor significantly increases. And, the higher gas flow rate leads to faster cooling rate, that eventually leads to a decrease in their average size. The experimental results were compared with calculations. The molecular dynamics simulation of copper atoms condensation from the gas phase was made for three different velocities $U = 0.005$ ps^{-1}, $U = 0.025$ ps^{-1} and $U = 0.05$ ps^{-1} (corresponding to a cooling rate of about 10^8 -10^9 K/s) and two end temperature T = 373 K and T = 77 K.

During the simulation of the atoms condensation process the temperature in the system is determined by two independent factors, such as vapor cooling rate and the thermal energy coming from atoms bonds energy. The condensation of the copper atoms in this model takes place very rapidly. In first few picoseconds of condensation the atoms combine into small cluster fragments, what leads to a significant increase of the kinetic energy in the system, which will undoubtedly entail a sharp jump in temperature. This effect is difficult to observe in the direct experimental studies, but the it makes a significant contribution to the further dynamic growth of the particles.

Figure 1. Examples of copper nanoclusters synthesized from the gas phase at a cooling rate $U = 0.005\ ps^{-1}$ and a temperature $T_f = 373$ K, N = 37140 atom.

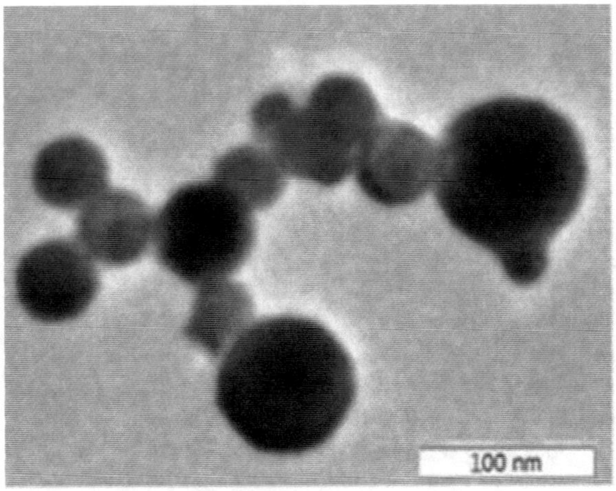

Figure 2. TEM of copper nanoparticles condensed from the gas phase.

The cooling rate of the atomic vapor is determined by the frequency of collisions and by the difference between the atomic masses of coolant and of metal atoms. In our case, as mentioned above, as the cooling buffer gas the virtual particles generated by Andersen thermostat are used. The use of this thermostat allows very rapidly to withdraw the excess kinetic energy from the system by means of elastic collisions of virtual particles and metal atoms.

As a result of the six options of modeling of copper atoms condensation from the vapor phase (with three different values of the cooling rate and the two end temperatures), a detailed analysis of the simulated systems in the last stage of the synthesis was conducted. In particular, we studied the effect of variable parameters on the amount, structure and shape of the particles obtained (Table 1). It is clearly seen that the rate of cooling of the gas mixture directly affects the number and average size of the clusters in the system provided the same number of atoms (85000).

By reducing of the cooling rate in 10 times the average size of the particles increased in 2.7 times at a final temperature of 77 K and 3.1 times at T_f = 373 K. In addition, in the system cooled to liquid nitrogen temperatures, approximately two and a half times more particles are originated than in the system cooled to the boiling point of water. This dependence is quite natural, and, as previously mentioned, is associated with an excess of kinetic energy in the system using cooling water. The size of the particles in the computer experiments varied from a few to tens of nanometers, and the particles synthesized under such a condition could be formed by combining each other of more than 20 small primary clusters (Figure 1).

As a result, the process of experimental synthesis of copper particles the clusters are formed of similar chain shape, but with a much larger size. The formation of large particles in the experiment is associated with technological features of the experimental setup, in particular with the the method of collecting of the synthesized particles on the fabric filter, what leads to a significant increase in the size of the synthesized particles. As an example, Figure 2 shows the experimentally synthesized copper clusters having a pronounced worm-like shape, what is similar obtained by the simulation (Figure 1).

3. Conclusions

Investigation of processes for copper nanoparticles production by the evaporation of the source material by an electron beam in an inert gas atmosphere and with following condensation was conducted to determine the best mode of synthesis of the material with the desired physical properties. During the experiment it was found that with increasing of the radiation intensity the average size of the synthesized particles is significantly increased. However, due to significant difficulties in interpreting the results of studies of the effect of the buffer gas flow rate on the distribution of the nanoparticles size distribution, the molecular dynamics simulation was carried out of condensation of copper nanoparticles for the different cooling rates and final temperatures.

As a result of computer simulation it was discovered that the cooling rate of the gas mixture and the final temperature directly affects the number and size of the particles of the system, provided the same number of atoms. Since the cooling rate decreases 10 times the average size of the obtained particles increased 2.7 times at a final temperature of 77 K and 3.1 times at T_f = 373 C. The experimental investigation results show the increase in the average size of the copper particles with increase of the cooling rate system may be connected with the eddy currents under the sublimator cover causing coalescence of small droplets before leaving the crucible by prolonged residence time therein. In general, this problem can be overcome through structural improvements of sublimator.

Under the computer experiment the influence of different final temperatures on the shape of the resultant particles was investigated. As a result of the conditions identified of the predominant formation of the clusters with wormlike and spherical shapes. The production of the spherical clusters can be performed most efficiently in a system with the coolant temperature of the order of 77 K, and a vermicular clusters form at the higher cooling temperature (T = 373 K).

ACKNOWLEDGMENTS
The presented work was performed as part of the base part of the public task of the Ministry of Education of the Russian Federation, including Ministry of Education and Science of the Russian Federation contract No. 16.1930.2014/K, and supported by the Russian Foundation for Basic Research, grant number 15-42-04164-r_sibir_a.

References:

[1] Zolotukhin L V Gelchinski B R, Kishkoparov N V Oglezneva S A, Ershov D V 2006 *Nanotechnics* **8** 22.
[2] Datta K K R, Kulkarni C, Eswaramoorthy M 2010 *Chem. Commun.* **46** 616.
[3] Nasibulin A G, Shurygina L I, Kauppinen 2005 *Colloid Journal* **67** 1.
[4] Strobel R, Pratsinis S E 2007 *Journal of Materials Chemistry* **17** 4743.
[5] Osterwalder N, Capello C, Hungerbuhler K, Stark W J 2006 *Journal of Nanoparticles Research* **8** 1
[6] Lin P A, Kumar A, Sankaran R M 2012 *Plasma Processes and Polymers* **9** 1184.
[7] Weber A P, Davoodi P, Seipenbusch M, Kasper G. 2003 *Journal of Nanoparticle Research* **5** 293.
[8] Weber A P, Seipenbusch M, Thanner C, Kasper G 1999 *Journal of Nanoparticle Research* **1** 253.
[9] Fissan H, Kennedy M K, Krinke T J, Kruis F E 2003 *Journal of Nanoparticle Research* **5** 299.
[10] Honeycutt J D, Anderson H C 1987 *J. Chem. Phys.* **91** 4950.
[11] Heermann D 1990 Computer Simulation Methods in Theoretical Physics. (Moscow: Nauka) [in Russian]
[12] Tsyganov S, Kästner J, Rellinghaus B, Kauffeldt T, Westerhoff F, Wolf D 2007 *J. Phys. Rev. B* **75** 045421.

The thermal evolution of Cu nanoparticles condensed from the gas phase: MD simulations

I V Chepkasov[1], S L Gafner[1], A V Nomoev[2,3], D G Bazarova[2] and I A Semenova [3]

[1] Katanov Khakassia State University, Lenin pr. , 90, Abakan, 655017, Russia

[2] Buryat State University, Smolina str., 24a, Ulan-Ude, 670000, Russia

[3]Institute of Physical Materials Science, Siberian Branch of the Russian Academy of Sciences, Sakhyanova str. 6, Ulan-Ude, 670047, Russia

E-mail: ilya_chepkasov@mail.ru

Abstract.

The gas-phase condensation of 85000 Cu atoms is examined by molecular dynamics simulation with a tight-binding potential. A detailed study of the evolution of the system cooled at a constant rate from 1000 K to 77 K is presented. The results are used to identify four distinct stages of the evolution from a hot atomic gas to a few synthesized particles. The effect of the subsequent thermal treatment on the shape and structure of synthesized particles was studied by simulating their gradual heating in a range of 100–1200 K. It is concluded that short-term heating leads to significant ordering of the internal structure in 70% of agglomerated nanoparticles with the predominant formation of spherical shapes. In order to explain this result, the main mechanisms of cluster formation from the gas phase have been analyzed and it is found that the agglomeration temperature plays the main role in the formation of clusters with unified shape and structure.

1. Introduction

Studying individual nanoparticles and the nanostructured state is currently one of the most intensively developing fields of research in physics, chemistry, and engineering. Great scientific and practical interest in such research results from the unique properties of nanoparticles that are already or will be widely used in the immediate future for the fabrication of miniature electronic devices, and production of new materials.

Copper clusters hold a most unique position among the whole range of currently used metal nanoparticles. These particles have many unique properties and are relatively inexpensive for manufacturing. Relating to conducting properties copper nanosized particles may compete with silver. Also the range of their usage is wide in the form of catalysts [1]. However the formation of nanoparticles of copper with implementation of application area dependent properties specifies very strict requirements to their producing and processing. One of the promising techniques of nanodispersed particles synthesis process is metals evaporation and condensation in the inert gas atmosphere. Theoretically, this technique makes it possible to synthesize nanoparticles with controlled

[1] Assistant professor, PhD, Chepkasov Ilya Vasilyavich,

chemical composition [2], defect density, structure, and size distribution. Moreover, it is relatively easy to control the parameters of gas-phase synthesis under experimental conditions [3-5].

However one of the main characteristics of vapor phase synthesis method is the substantial inner structure and external shape discontinuity of resulting nanoparticles. Therefore, depending on process parameters up to 90% of particles non-spherical shape and different inner structure may be formed. Production of nanoparticles of desired shape, and structure can be another important step toward their more versatile use. However, the preparation of clusters of certain morphology is a formidable technical task. To solve this problem the method of heat treatment of obtainable metallic particles can be used as a final stage of synthesis from gaseous phase

2. Simulation technique

This section gives a description of the basic principles of the MD procedure used in this study. Any simulation method must first choose some potential of interaction between atoms in the system under consideration. The interatomic potentials used in computer simulations range from ab-initio to simple Lennard-Jones type empirical ones.

The choice of a particular potential depends on the problem formulation, the properties to be examined, the available computing resources, and the desired accuracy of results. Upon analysis of various representations of the potential energy of interatomic interaction, we decided to simulate the formation of Cu nanoclusters from the high-temperature gas phase in terms of the well-established modified tight-binding TB-SMA potentials as introduced by Cleri and Rosato [6].

The starting point of the condensation process was a spatial configuration consisting of 85000 Cu atoms, which were uniformly distributed in a region of space with a volume of V = 42600 nm3 and periodic boundary conditions. The atomic velocities were set according to the Maxwell–Boltzmann distribution at an initial temperature of T_i = 1000 K, which is quite typical of the gas phase synthesis processes [7]. In the inert-gas–metal-vapor mixture used to produce Cu nanoclusters by gas-phase condensation, the inert gas is mainly required to cool the metal vapor and control its temperature. In our approach, external influences are reduced to heat removal from the simulated system. At the start of the simulation, a supersaturated metal vapor fills the condensation region.

The main idea of the proposed gas-phase condensation model is to minimize the rate of formation of primary nuclei from a supersaturated metal vapor. Otherwise, single atoms or small clusters will be captured on the collection surface. This effect is avoided by using an inert gas in the condensation region to reduce the diffusion rate of atoms outside the supersaturated vapor source via an increase in collision frequency.

3. Results and discussion

The formation of a cluster in a metastable gas phase starts with homogeneous nucleation. After termination of this process, the subsequent growth of clusters can proceed according to various mechanisms. The most important of these are agglomeration and coalescence. In the case of agglomeration, clusters combine without changing their shape, whereas during coalescence, clusters fuse and merge together with the formation of a common particle with a shape differing from that of the initial clusters prior to their collision.

To study particles shape and structural properties depending on temperature the gradual heating of copper nanoclusters available from condensation from 100 K to 1200 K with temperature steps of 100 K was modeled. For each temperature particles were held from 3 to 4 ns depending on the cluster size. Clusters under study were chosen in such a way that to fully reflect differences in size, original shape and structure of particles available from condensation. In this simulation, we revealed two typical tendencies:

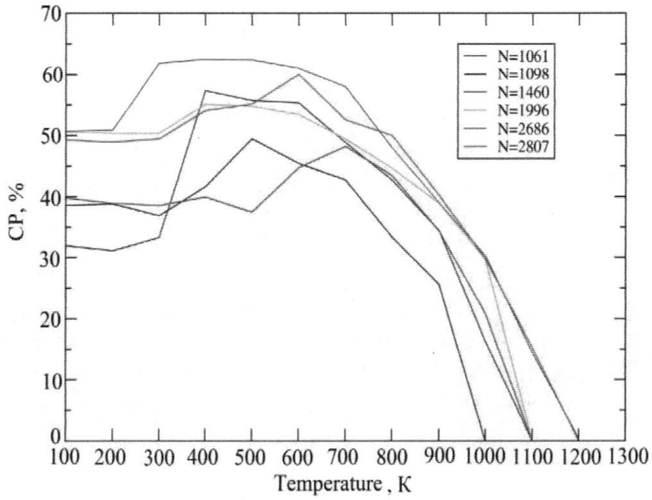

Figure. 1. The fraction of the number of atoms with coordination number $z = 12$ (in %) of total N atoms in the cluster as a function of temperature for a cluster size of 1061 to 2807 atoms.

1) if a particle has a regular internal structure, when heating a cluster to temperature 300-500 K further structure ordering takes place and the number of atoms with coordination number z = 12 (CP) increases to 55 - 60%. Long-range order destruction begins at T = 600 K, the complete particle melting occurs at T = 1100 K (Figure 1);

2) in the absence of the strongly marked closely-packed core of cluster due to intensive kinetic processes under heating, the formation of such core occurs in the temperature range from T = 300 K to T = 700 K. With further temperature rise long-range order destruction occurs with the process termination to 1100 K (Figure 1).

Relatively low maximum values of atoms with coordination number z = 12 in particles under study (not more than 65%) is attributed to the fact that surface atoms were disregarded. The atoms share in a cluster of ideal spherical form is determined from the formula $N_s/N = 4 \cdot N^{-1/3}$ (where N – number of atoms in a particle, N_s – number of surface atoms) and can reach approximately 40% for a cluster consisting of 1061 atoms and approximately 29% for a cluster, consisting of 2807 atoms. Taking it into consideration we can conclude that in condition of simulated thermal influence there is practically complete ordering of synthesized particles internal structure.

To explain such a behavior (Figure 1) in details the processes of clusters atomic restructuring in condition of thermal influence is analyzed. In Figure 2 there are "snapshots" of Cu_{2686} cluster evolution at short-term thermal influence. This cluster pertains to the first group of particles having explicit close-packed core. The colors of the atoms denote the result of a CNA analysis [8] as follows: atoms drawn in blue (green) have a local crystalline fcc (hcp) environment; yellow atoms have no identified structure, however their coordination number is z = 12. Atoms with coordination numbers z ≠ 12 are drawn in red.

We can see that in the early stages (Figure 2 a) the particle consisted of three combined with each other clusters with various internal organization. However, at T = 300 K (Figure 2 b) there are distinct processes of transforming internal structure of amorphous area dividing the biggest initial cluster and element with face-centered cubic and hexagonal close-packed structure.

As a result of heating up to T = 500 K the processes of intensive thermo-activated self-diffusion of surface atoms in the places of connection of two cluster parts begin, and by T = 700 K this leads to formation of unified oval exterior form for this particle (Figure 2 d). The further increasing of temperature just slightly changes the exterior form and in accordance with the principle of surface energy minimum the particle is aiming to take spherical shape. At T = 900 K the melting processes begin, in consequence of which the destruction of long-range order with completion of process at T = 1200 K occurs (Figure 2 f). This behavior was neither very complex nor unusual, since the driving force for the combination of small clusters and formation of a common structure of the initial agglomerated particle was the existence of a large disordered region involving almost the entire Cu_{2686} cluster.

For all the six researched particles shown in Figure 1 with temperature increasing the share of atoms with coordination number z = 12 grows and peaks by the temperature from 400 to 700 K

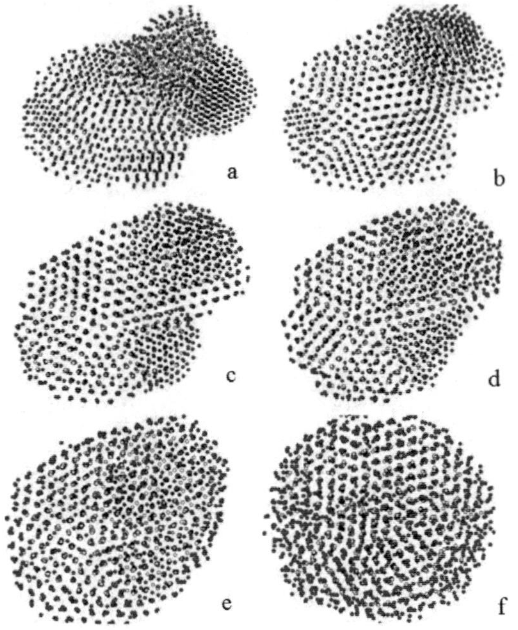

Figure 2. Evolution of a Cu_{2686} cluster with increasing temperature: a) T = 100 K, b) T = 300 K, c) T = 500 K, d) T = 700 K, e) T = 900 K, f) T = 1100 K.

depending on particle's size and ordering of internal structure. Thus, to improve the internal morphology of particle by thermal influence method there is no need in increasing temperature over T = 700 K, although the external configuration of cluster in this case does not always take perfect spherical shape and can remain at restructuring intermediate stage presented by ellipsoid.

In such a way several dozens of various clusters were thermally treated, and after analyzing the results we can conclude that the thermal processes influence the formation of the cluster external appearance and structure quite much. Although the observed particles of copper had different form and internal structure there was significant grading of such difference at higher temperatures.

However, not all investigated particles were influenced by the thermal process in cluster form and internal structure. The analysis has shown that thermal treatment in case of about 30% of clusters synthesized from gaseous environment had led neither to the formation of unified structure, nor to the formation of spherical shape. As one of the typical examples in Figure 3 there is a Cu_{1771} cluster that during the whole process of heating constituted two particles with various internal structures divided by an amorphous layer. Being heated from T = 100 K up to T = 800 K the parts of the cluster have adopted almost spherical shape with simultaneous reduction of contacting area (Figure 3 c). The particle is fully dissolved without packaging of interacting segments at further increasing of temperature (Figure 3 e and f).

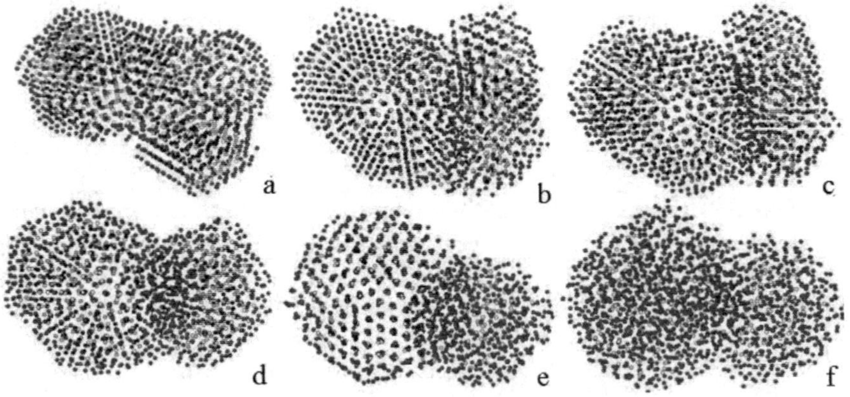

Figure 3. Evolution of a Cu_{1771} cluster with increasing temperature: a) T = 77 K, b) T = 100 K, c) T = 500 K, d) T = 800 K, e) T = 1000 K, f) T = 1200 K.

It was suggested that such a behavior of particles at thermal treatment can be the result of sintering process at rather low temperatures of perfect spherical clusters with icosahedral structures. Because of the low temperature of kinetic energies environment the collision of particles was insufficient for the activation of surface atoms interdiffusion, and the clusters were just "sticking" to each other due to their interacting forces. As the icosahedral structure corresponds to minimal value of energy, at further heating the initial lightly connected icosahedral clusters saved such a construction of atoms and simultaneously turned into maximal possible sphere. We should also note that clusters gained through direct experiments on condensation from gaseous phase often possess segments with icosahedral structure. In order to verify this assumption, we have studied other nonunified clusters. It was found that, in the overwhelming majority of cases, these clusters comprised combinations of several icosahedral clusters that agglomerated at very low temperatures (below 100 K). The competition of these structures in a cluster leads to further formation of an amorphous-like layer between the particle parts.

Thus, the results of computer simulations of the thermal treatment of copper clusters synthesized from the gas phase suggest that a rather high temperature of agglomeration is the key factor. If the agglomeration took place at the initial stages of development, a high ambient temperature favored the agglomeration of primary clusters into almost unified particles. If the agglomeration proceeded at low temperatures, the kinetic energy of collision was insufficient to fuse the primary particles together. As a result, the parts of the cluster merely stuck together, predominantly due to interatomic interaction forces. In this case, the initial shape of the cluster changed during its thermal treatment and acquired a chainlike or wormlike structure.

4. Conclusions

In this paper the processes of copper atoms condensation with formation of nanoparticles of various form and structure were modeled with the method of molecular dynamics with using of tight-binding modified potential TB-SMA. To define the mechanism of external form and internal structure evolution, the copper nanoclusters condensed from gaseous phase were influenced by thermal treatment.

As a result of thermal treatment modeling we have found that in case of 70% of observed clusters there was full ordering of internal structure and external form. The remaining 30% has failed to form unified shape and structure in the process of thermal-activated relaxation and was presented as competing parts of one particle with comparable energies of surface and chemical potentials. The analysis of the evolution of clusters synthesized from a gaseous medium suggests, that it is necessary to restrict the temperature of cooling of the working gas mixture to a level 0f 400 - 500 K to obtain a large fraction of sperical metal nanoparticles with a unified crystalline structure. Cooling the mixture by liquid nitrogen according to the commonly accepted practice leads to the formation of a considerable fraction of weakly bound (because of the small area of contact) icosahedral clusters. Upon subsequent heat treatment, these clusters form wormlike nanoparticles consisting of segments with different crystalline structures.

ACKNOWLEDGMENTS
The presented work was performed as part of the base part of the public task of the Ministry of Education of the Russian Federation, including Ministry of Education and Science of the Russian Federation contract No. 3824, and supported by the Russian Foundation for Basic Research, grant number 15-42-04164-r_sibir_a.

References:

[1] Dubey R, Srivastava P, Kapoor I P S, Singh G 2012 Thermochimica Acta. **549** 102.
[2] Starsich F H L, Hirt A M, Stark W J, Grass R N 2014 Nanotechnology **25** 505602.
[3] Kauffeldt E, Kauffeldt T 2006 Journal of Nanoparticle Research **8** 477.
[4] Weber A P, Davoodi P, Seipenbusch M, Kasper G 2006 Journal of Nanoparticle Research **8** 445.
[5] Seipenbusch M, Toneva P, Peukert W, Weber A P 2007 Part. Part. Syst. Charact. **24** 193.
[6] Cleri F, Rosato V 1993 J. Phys. Rev. B **48** 22.
[7] Backman U, Jokiniemi J K, Auvinen A, Lehtinen K E J 2002 Journal of Nanoparticle Research **4** 325.
[8] Meyer R, Gafner J J, Gafner S L, Stappert S, Rellinghaus B, Entel P 2005 Phase Transitions **78** 35.

Bio resorbability of the modified hydroxyapatite in Tris-HCL buffer

O A Golovanova[1], R R Izmailov[1] and S A Ghyngazov[2]

[1]Department of Inorganic Chemistry Omsk F. M. Dostoevsky State University
[2]National Research Tomsk Polytechnic University, Tomsk, Russia

E-mail: Golovanova2000@mail.ru

Abstract. The solubility of carbonated hydroxyapatite powders and granulated carbonated hydroxyapatite produced from the synovial biofluid model solution has been studied. The kinetic characteristics of dissolution were determined. It was found that the solubility of carbonated hydroxyapatite is higher as compared to that of hydroxyapatite. The impact of the organic matrix on the rate of sample dissolution was revealed. For HA-gelatin composites, as the gelatin concentration grows, the dissolution rate becomes greater, and a sample of 6.0 g / L concentration has higher resorbability. The results of the research can be used to study the kinetics of dissolution and the biocompatibility of ceramic materials for medicine, namely for reconstructive surgery, dentistry, and development of drug delivery systems.

1. Introduction

Currently, much attention is paid to studying and improving the bioactive properties of the materials produced based on calcium phosphates. Interest in these compounds is due to the fact that the human bone tissue is a biocomposite based on hydroxyapatite (HA) with an admixture of amorphous calcium phosphates, brushite, octacalcium phosphate [1,2] and high-protein collagen. Therefore, calcium phosphate-based materials are widely used in medicine for bone defect replacement [3].

However, widespread use of these compounds is hindered by some problems. The most significant of the problems is low dissolution (i.e. bioresorption) rate of calcium phosphate-based materials. A significant disadvantage of the HA-based material most frequently used for implantation is its low bioresorbability rate. This problem can be solved through chemical modification of calcium phosphate, since it is known that isomorphic impurities introduced into calcium phosphate may increase both its bioactive properties and solubility.

The resorption process may be enhanced through ionic substitution of phosphate and hydroxide anions in the apatite structure for carbonate, silicate and other groups [4]. The carbonate-substituted hydroxyapatites are similar to the mineral components of the natural bone in their composition and structure, and silicate ions in the HA composition play a bridging role in the physiological processes of growth and restoration of bone and cartilaginous tissues, etc. One of the most frequently used forms of phosphate calcium HA-based ceramics are granules, coatings and other forms, in which collagen, gelatin and chitosan are used as matrix-carriers. As is well-known, the human body contains buffer solutions, and the study of the resorption processes of the bioactive samples based on HA is one of the urgent issues to promote practical application of these products.

The most common buffer simulating the physiological medium of the human body is Tris-HCl buffer with, pH equal to 7.4. Since pH of this buffer is similar to the physiological values of the organism, this buffer was used as the solvent to simulate the medium of a living organism.

The study aims to estimate the kinetic parameters of the dissolution for samples based on modified HA in Tris-HCl buffer.

2. Materials and methods

Modified hydroxyapatite was synthesized from the model medium similar to the human synovial fluid in its ionic and electrolyte composition, pH and ionic strength [5]. To produce materials with properties similar to those of the bone apatite, carbonate ions (CO_3^{2-}) were introduced in the synovial fluid model solutions, their concentration being varied in the range from 0 to 32 mmol/l. These concentration ranges correspond to the content of ions in the human intracellular, interstitial and synovial fluid [5]. Crystallization of the solid phase was carried out for 30 days.

To make granules of carbonated hydroxyapatite (CHA) the resulting powder with a weight of 0.2–1 g was mixed with 1.5–7.5 ml of the 5, 10 and 15 mass. % aqueous gelatin solution and stirred to obtain a homogeneous mass. The mixture was dripped through a capillary with a diameter of 1–2 mm in oil, cooled to –1– 0 °C using a eutectic mixture of fine ice mixed with $CaCl_2$. The resulting granules were washed successively with ethanol and acetone and dried at a temperature of 25–27 ° C in air for 6– 7 hours.

Dissolution of the produced samples was performed in Tris-HCl buffer at physiological pH equal to 7.4 under continuous stirring. Several experiments we performed with different initial sample mass of calcium phosphate (m = 0.100 and 0.200 g) and under different temperatures.

At regular intervals, the values of acidity and calcium ion concentration were recorded via direct potentiometry. The experimental data obtained were used to plot the graphic dependences in $pCa = f(\tau)$ coordinates to evaluate the solubility of the synthesized samples. To determine the dissolution rate for each of the samples, the kinetic curves were analyzed using the regression analysis [7]. After that, these curves were mathematically processed according to the algorithm suggested in [8]. The morphology of the samples was studied with the JEOL JSM 6610LV Scanning Electron Microscope.

3. Results and Discussion

As a result of the synthesis from the synovial fluid model solution under varying concentration of carbonate ions, carbonated hydroxyapatite was produced. The above techniques and algorithms were used to study the solubility of the synthesized HA and KGA in Tris-HCl buffer at physiological pH of the human synovial fluid.

The data in Fig. 1 (change in the calcium ion concentration over time) were used to obtain graphic dependences (Fig. 2) to evaluate the degree of dissolution of the synthesized calcium phosphates. The data in Fig. 2 were used to determine the dissolution orders for CHA and HA (Table 1).

After that, the solubility of HA and CHA was investigated at different temperatures to calculate the activation energy of the processes. The obtained kinetic curves are shown in Figure 3; the curves and formula 3 were used to determine the values of the activation energies.

The analysis of the data in Table 1 showed that the process of CHA dissolution is characterized by lower activation energy if compared with that for HA. When comparing HA and CHA solubility (Figure 4), carbonated hydroxyapatite is found to be more soluble than hydroxyapatite. In our opinion, this may be due to different mechanisms of dissolution which is indicated by the change in the reaction order from n=1 to n=2 for CHA. A number of authors suggest that this may be caused by the predominance of internal diffusion processes over external ones [9].

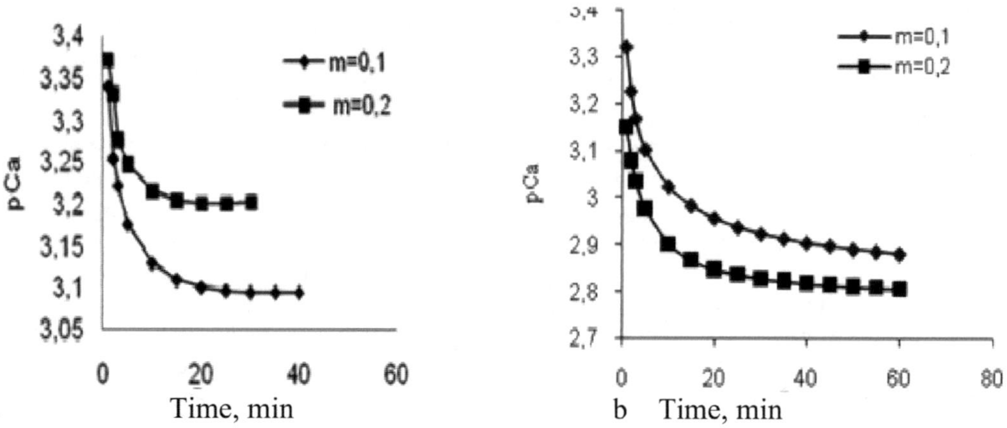

a b

Figure 1. Time dependence of HA (a) and CHA (b) dissolution

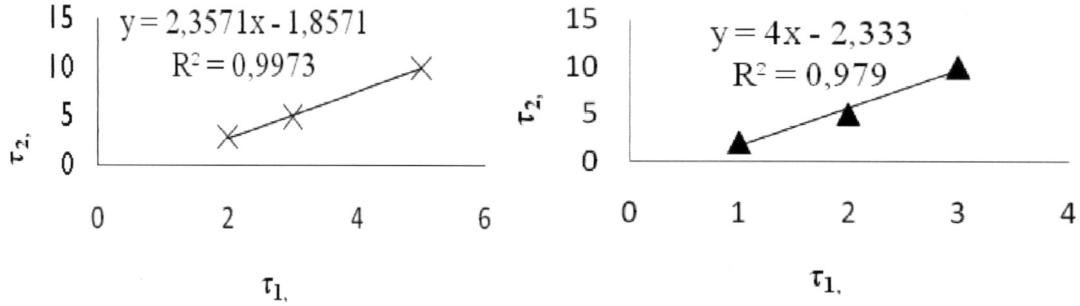

a b

Figure 2. Kinetic curves of HA (a) and CHA (b) dissolution with different masses of the initial samples, (τ- time, min).

Table 1. Kinetic parameters of HA and CHA dissolution

Phase	HA	CHA
Order of the dissolution reaction, n	1	2
Activation energy (kJ/mol)	13.12	9.92

Next, the dissolution process of CHA-gelatin granulated samples was explored. The pCa system changed significantly within the first 24 hours when the samples of granulated CHA-gelatin were dissolved in Tris buffer. Investigation of the sample solubility in the subsequent days showed insignificant change in pCa values. To determine the rate of sample dissolution, each of the kinetic curves was analyzed using the regression analysis (Fig.5). It was found that the time dependence of pCa in Tris-HCl buffer in the initial section may be approximated by a linear function.

To verify the determined patterns by the method of optical microscopy, the surface morphology of the considered samples before and after dissolution was investigated in the considered model media. Figure 6 shows the images of the granulated carbonated hydroxyapatite at 60-fold magnification.

The results of quantitative evaluations of the kinetic parameters are presented in Table. 2. The analysis of the kinetic curves showed that granule dissolution is a multi-stage process; at the initial stage (up to 1000–1200 min) the time dependence of the calcium ion concentration in the solution $C(t)=-lgCca^{2+}$ may be approximated by a linear function. The dissolution decelerates with time, and kinetics obeys the exponential dependence. It is known that exponential dependence corresponds to kinetics of the first order reaction when the rate of change in calcium ion quantity is proportional to the number of "active regions" in the dissolved material at the given moment [10].

a b

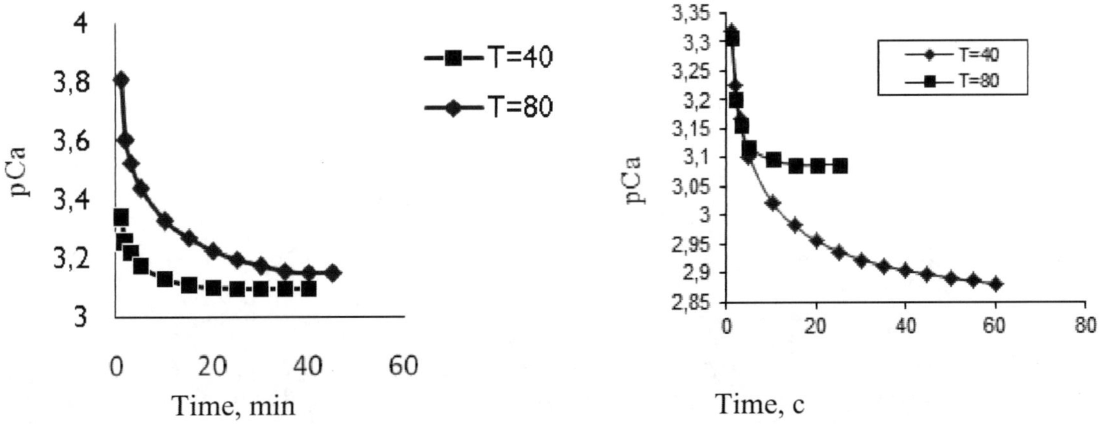

Figure 3. Solubility of HA and CHA at different temperatures

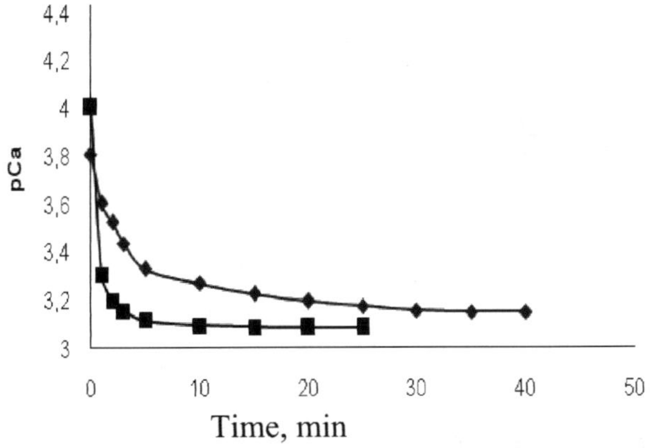

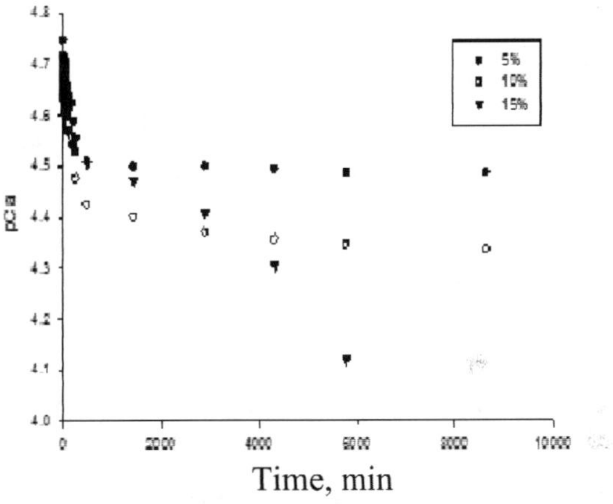

Figure 4. Comparison of HA and CHA solubility

Figure 5 Dependence of Ca^{2+} ions concentration on dissolution time of CHA-gelatin granules in Tris buffer (pH=7.4) with varying mass fraction of gelatin

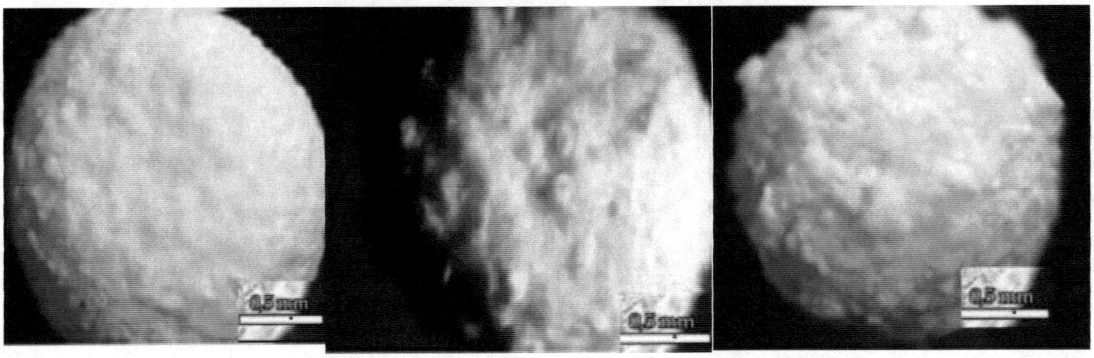

a b c

Figure 6. Images of the granulated carbonated hydroxyapatite with gelatin mass fraction of 5% (a), 10% (b) and 15% (c) (mass) after dissolution in Tris buffer (pH=7.4) within an hour

The quantitative evaluation of the rate of change in the calcium ion concentration in the solution depending on gelatin concentration in granules is of great relevance. At the "exponential" stage of dissolution, the rate of change in the concentration decreases with time, therefore, the initial dissolution rate can be considered as a quantitative measure.

Table 2. Parameters of the granule dissolution depending on gelatin concentration in the Tris buffer solution

C_{gel}, %	R^2	Equation
5	0.8976	$pCa = 4.4982 + 0.1903e^{-0.086t}$
10	0.9719	$pCa = 4.3563 + 0.2917e^{-0.0025t}$
15	0.9937	$pCa = 4.3061 + 0.3800e^{-0.0017t}$

Based on the data presented in Table 2, decrease of the initial rate of sample dissolution is seen to cause increase in the content of the granulated gelatin which may be due to hindered diffusion of the solvent into the porous internal region of granules that correlates to the data obtained with the scanning electron microscopy (Fig. 7).

The analysis of the sample micrographs showed that as the gelatin concentration in the initial suspension of samples increases, the number of open pores decreases due to formation of a thin gelatin film on the surface of the granule mineral component.

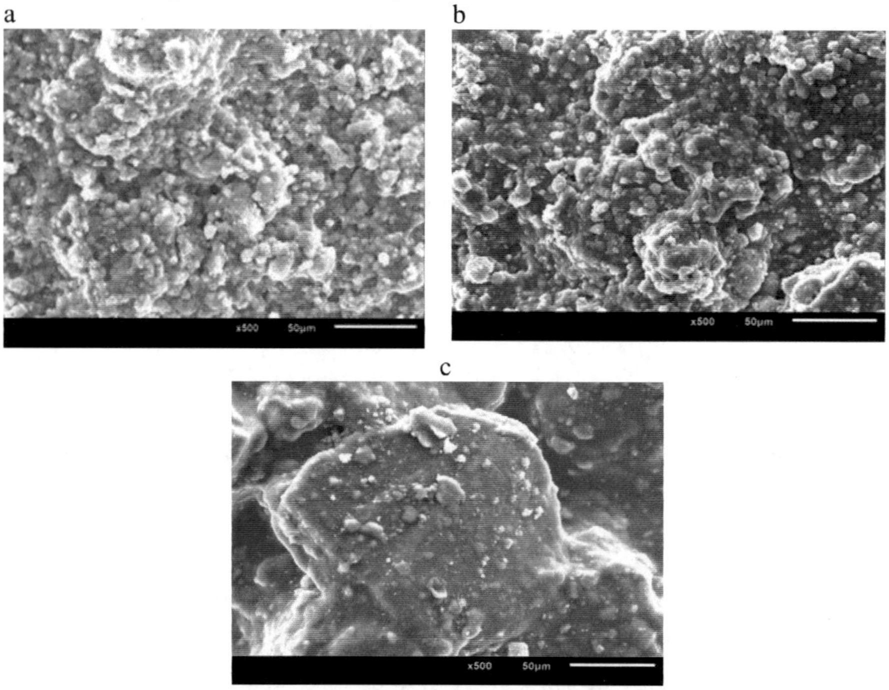

Figure 7. Surface morphology of CHA-gelatin composite: a, b, c are for $\omega_{gelatin}$= 5 mass%, 10 mass%, 15 mass%, respectively

The kinetic patterns of dissolution can be explained based on the polyampholyte properties of gelatin: minimum swelling of gelatin is observed at pH approximately equal to 4.7, and pH of the maximum swelling is at pH close to 3.2. In the alkaline medium, the –COOH group dissociates to form the $–COO^-$ group. The molecules of gelatin acquire a net negative charge. Change in the macromolecule charge leads to change in macromolecule conformations which results in the change in solution viscosity. This is due to the fact that the tendency of –COOH groups to dissociate is greater than the tendency of $–NH_2$ groups to protonate, that is proteins are stronger acids than bases. This means that to achieve the pI in the gelatin solution, the excess of acid (pH <7) is required to suppress ionization of carboxyl groups. Therefore, acid medium may be most favorable for rapid dissolution of the CHA-gelatin composite.

4. Conclusion

Thus, the study has shown that:

CHA solubility in Tris-buffer at physiological pH of the human synovial fluid is higher than that of HA.

Different patterns of the kinetic curves and impact of the organic matrix on the dissolution rate have been determined for granulated CHA samples.

For HA-gelatin composites, growth of the gelatin concentration results in the increase of the dissolution rate. The sample with concentration of 6.0 g/l in the Tris buffer solution has higher resorbability.

The results of the research can be used to study the biocompatibility and kinetics of dissolution of ceramic materials for medicine, namely for traumatology, reconstructive surgery, dentistry, and development of drug delivery systems.

5. Acknowledgements

The research was partially supported by the Russian Foundation for Basic Research (research projects No. 15-33-50314 and No. 15-29-04839).

References

[1] Korzh A A, Belous A M and Panov E Y 1972 215
[2] Luneva S N, Romanenko S A, Shikhaleva N G 2013 *VKGU* **3** 29
[3] Fomin A S, Komlev V S, Barinov S M, Fadeeva I V and Rengini K 2006 *J. Adv. Mat.* **2** 51
[4] Frank-Kamenetskaya O, Kol'tsov A 2011 *J. Of Molecular Structure* **9** 9
[5] Izmailov R R, Golovanova O A 2014 *J. Inorganic Materials* **50** (6) 641
[6] Safronova T V, Shekhirev M A, Putlyaev V I and Tretyakov Y D 2007 *J. Inorganic Materials* **43** (8) 1005
[7] Akselrud G A, Molchanov A D 1977, M. *Chemistry* 262.
[8] Krasnov K S 2001 M .: High School **1** 512
[9] Kubasov A A 2004 Moscow *MSU* 4
[10] Panchenkov G M and Lebedev V P 1985 M. *Chemistry* 276

Crystallization of modified hydroxyapatite on titanium implants

O A Golovanova[1], R R Izmailov[1], S A Ghyngazov[2] and A V Zaits[1]

[1]Department of Inorganic Chemistry Omsk F. M. Dostoevsky State University
[2]National Research Tomsk Polytechnic University, Tomsk, Russia

E-mail: Golovanoa2000@mail.ru

Abstract. Carbonated-hydroxyapatite (CHA) and Si-hydroxyapatite (Si-HA) precipitation have been synthesized from the model bioliquid solutions (synovial fluid and SBF). It is found that all the samples synthesized from the model solutions are single-phase and represent hydroxyapatite. The crystallization of the modified hydroxyapatite on alloys of different composition, roughness and subjected to different treatment techniques was investigated. Irradiation of the titanium substrates with the deposited biomimetic coating can facilitate further growth of the crystal and regeneration of the surface.

1. Introduction

Currently, much attention is paid to improving the quality of life and lifespan [1]. In medical practice, to save lives and restore important physiological functions of patients we have to resort to reconstruction of bone and tooth tissue through replacing the defect with an implant [2]. Every year, about 2 million of these operations are made throughout the world, while in Russia the number of operations such as joint prosthetics per year is ten times smaller than it is required.

To date, the research, development and production of medical supplies and proper implants are the most urgent issues for material scientists, physicists, chemists and engineers [3–8]. Modern trends in the development of high technologies and laboratory base of research centers require advances in the technology of medical products and expansion of their diversification to meet the requirements of international standards and to reduce their prime cost.

The solution of this problem is production of a bioactive coating on metal implants [9–13]. Bioactive implants can reduce the time of treatment for severe diseases and eliminate loosening and rejection (due to toxicity, inflammation and foreign body response). One of the important trends of modern material science is the development of biomaterials based on calcium phosphates [3–13]. The advantage of these materials is absolute biocompatibility due to the identity of chemical composition and crystal structure of hydroxyapatite (HA) of the natural bone. Thus, the hydroxyapatite coating on the implant surface ensures rapid and effective engraftment in the bone structure due to high level of biological activity of the surface.

In biomimetic synthesis, identification of optimal conditions for producing powders of highly dispersed phase with high resorption and biocompatibility with the human body are relevant. Therefore, the study and development of new methods for synthesis of fine crystalline hydroxyapatite, development of low-cost and technologic techniques of formation of bioactive coatings made of composite material on titanium are of critical importance.

The paper aims to investigate the possibility of crystallization of modified hydroxyapatite on the titanium alloy surface with different surface roughness and processing techniques applied.

2. Materials and methods

Synthesis of the modified hydroxyapatite was carried out in two ways:

1. From the model medium similar to the human synovial fluid in its ionic and electrolyte composition, pH and ionic strength [14]. To produce materials with properties close to those of the bone apatite, carbonate ions (CO_3^{2-}) were added to the synovial fluid model solutions. Their concentration was varied from 0 to 32 mmol/l. Crystallization of the solid phase was carried out within 30 days. The carbonate-hydroxylapatite prepared via biomimetic synthesis in the form of 5.00 mass% suspension was deposited on the sample surface subjected to different treatment techniques.

2. From the model medium similar to the human synovial fluid in its ionic and electrolyte composition, pH and ionic strength [15]. Na2SiO3 was chosen as the basic compound, the modifier of the silicate groups, for the synthesis from the intercellular fluid model solution. The synthesis time was 48 hours. The silicon-containing hydroxyapatite produced by biomimetic synthesis in the form of 1.00 mass% suspensions was deposited on the sample surface subjected to different treatment techniques. Upon completion of the solid phase crystallization in the supernatants, pH and Ca2+ concentration was measured by potentiometric method. The photometric method was used to determine the residual concentration of the phosphate ions in the solution [RD 52.24.382-2006], the residual concentration of the silicate ions (RD 52.24.382-2006) was measured with the colorimeter CPC-2. The solid phases were filtered, dried and analyzed by FTIR spectroscopy and X-ray diffraction analysis (XDA). IR spectra were recorded with the spectrometer FT-801 (samples were prepared in the form of KBr pellets). XDA was performed with the Bruker D8 Advance X-ray diffractometer (Germany) under monochromatic Cu-kα radiation. Statistical data processing was carried out using StatSoft Statistica 6.0. To deposit hydroxyapatite on the VT1-0 titanium surface, the titanium plates 15 mm*15 mm*1.2 mm in size were pretreated: polished; etched (etching mixture HNO_3 :HF 1:1 enlarging the surface of the titanium substrates was used); exposed to power ionization beam (the setup "Temp" was used for irradiation); perforated (etched, not subjected to mechanical impact and chemical attack, subjected to laser ablation) and plates which were not subjected to mechanical impact and chemical attack.

After that, the plate was immersed into the model prototypes of the synovial and intercellular fluids [14,15]. This system was soaked for 3, 6 and 12 days (the original SBF solution being changed and not changed), and after that, the plates were removed from the model solution and dried at room temperature. Optical microscopes Neophot 2 and MBS-9, and the scanning electron microscope JEOL JSM-6610LV were used to study the surface morphology.

3. Results and discussion

The results obtained by XDA indicate that throughout the range of varied carbonate ion concentration, the produced hydroxyapatites are of low crystallinity, for which the lattice parameters are evaluated [14]. The value of the parameters a and c of the elementary cell of the synthesized samples differ from that of the bone tissue and stoichiometric hydroxyapatite. The a parameter of the produced samples compared to the bone apatite is larger, and the c parameter, on the contrary, is smaller. Such characteristics of the crystal lattice are typical of the non-stoichiometric calcium-deficient hydroxyapatite, carbonate-bearing hydroxyapatite among them[14]. Thus, A-type carbonated-hydroxyapatite (CHA) was synthesized [14], which is currently widely used in surgery [16].

The carbonated-hydroxyapatite produced through biomimetic synthesis in the form of 5.00 mass% suspension was deposited on the prepared titanium sample surface subjected to different treatments. Fig. 2 (as an example) shows the surface morphology of the samples coated with the suspension (CHA). The coating formed on all the titanium samples contained cracks regardless of the pretreatment technique.

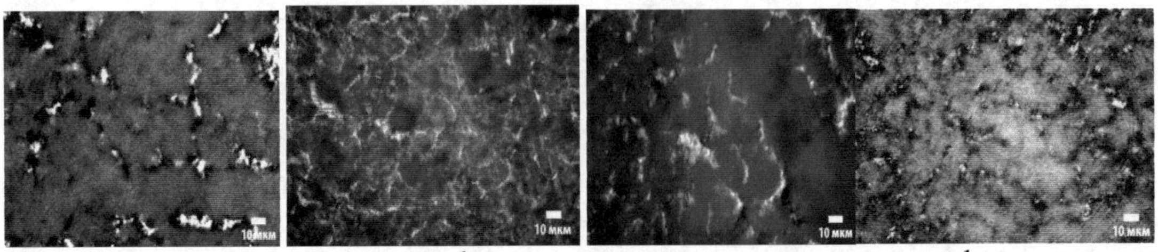

Figure 1. Morphology of the BT1-0 surface of different roughness coated with the hydroxyapatite suspension: polished surface (a), perforated surface (b), etched surface (c), ground surface (d).

Fig. 2 shows the morphology of the VT1-0 titanium sample obtained via scanning electron microscopy. It indicates CHA particles of a rounded shape clearly seen on the titanium alloy surface.

Figure 2. Surface morphology of the perforated surface of the VT1-0 titanium alloy coated with hydroxyapatite suspension.

A thickness gauge was used to measure the thickness of the deposited CHA layer. The measurement error was ± 3 μm. The thickness of the deposited layer was approximately identical for all the samples and equaled 30 μm. After producing the CHA layer, the samples were exposed to a power ion beam (PIB) with the "Temp" setup. Under PIB action, the deposited layer and the substrate interfused and partially intermixed due to high temperatures and gradients on the surface. Figure 3 shows the morphology of the CHA layer on the sample surface after irradiation obtained via scanning electron microscopy. It indicates intermixed portions of the substrate and CHA (Fig. 5) and shows the film structure mixed with the metal formed on the surface of VT1-0 titanium in some areas with cracks along the cleavage planes (Fig. 3b).

The study of crystallization of the metal samples from the synovial fluid model solution under irradiation with the varying number of pulses (n = 3 and 10) has shown that as the number of pulses increases, the CHA crystal growth activates. It is apparent that when n=10, the sample surface becomes more developed which contributes to an increase in the number of nucleation sites on its surface (Table. 1) and the area of the CHA layer coating the titanium alloy enlarges.

Table 1. Coated areas on the irradiated sample surface during crystallization

Alloy	Exposed to PIB with $j = 150$ A/cm^2 n=10	Exposed to PIB with $j = 150$ A/cm^2 n=3
Titanium VT1-0	28 %	20 %

It was found that under irradiation by PIB, the layers of the coatings based on the CHA suspension get fused and reliably fastened. This increases the biocompatibility of the titanium implant which is confirmed by further crystallization of CHA from the model solution on the surfaces of the synovial fluid samples. The surface morphology is shown to affect the coating uniformity and the rate of crystal growth. The more developed the surface morphology, the more active the crystal growth (Table. 3).

a *b*

Figure 3. Surface morphology of the perforated VT1-0 titanium alloy surface coated with hydroxyapatite suspension after exposure to PIB.

For a group of samples with different surface roughness, deposition on the surface was performed from the model solution of the human synovial fluid and the solid phase crystallization lasted for 40 days. The dependence of the CHA crystal growth and the areas of the coated surface on the method of the titanium alloy treatment were studied using optical microscopy at different time intervals (7, 11, 18 and 40 days), and each time the initial solution for crystallization was renewed. After crystallization, the coated area of the CHA was determined with respect to the sample surface (Table. 2).

Table 2. Areas of the sample surface coated with CHA

Alloy	Polished	Perforated	Etched	Ground
Titanium VT1-0	15 %	85 %	26 %	40 %

The intensive crystallization of the solid phase was found to occur on the VT1-0 titanium samples with highly perforated surface. The coated area of this sample is about 85%, whereas, on the polished surface of the titanium alloy it is smaller.

The next stage involved synthesis of the modified hydroxyapatite silicate ions from the model medium of the human intercellular fluid. XDA has revealed that all the samples synthesized in the medium of the model solution of the extracellular fluid under varying concentration of silicate ions are single-phase and represent hydroxyapatite [15]. The sizes of the produced unmodified crystallites are 6.3 nm for HA and 6.0÷7.1 nm for Si–HA, which indicates the formation of nanocrystalline compounds. The parameters of the crystal lattice of the solid phase increase as compared to the unmodified hydroxyapatite, which confirms possible substitution of the orthophosphate-ion by the silicate ion (distances Si-O = 1.66 Å, P-O = 1.55 Å).

The study of the surface and morphological characteristics of the produced phosphate coatings modified by silicate ions identified more complete deposition of Si-HA on the titanium substrate surface for etched samples (Fig. 4).

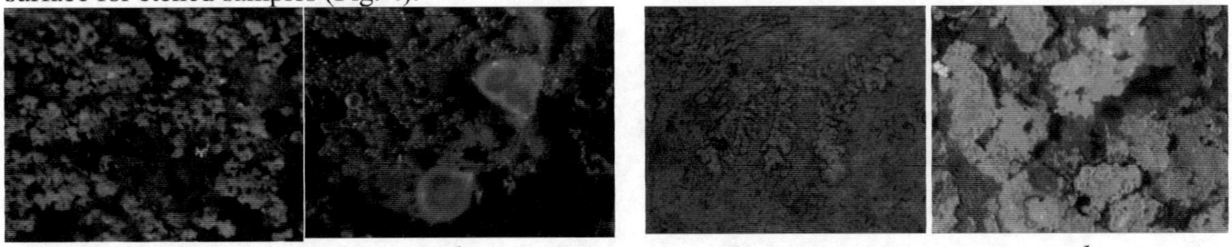

a *b* *c* *d*

Figure 4. Deposition of Si-HA on titanium substrates (etched surface) within 3 days (a); 3+3 days, the model solution being changed (b); 6 days without changing the solution (c); 3+3+6 days, the model solution being changed

This treatment technique provides the coating which is uniform, dense, highly dispersed, and the HA crystals grow in the form of dendrites. The crystals tend to grow in a more structured form. Increase in the time of titanium soaking in the model sample solution leads to non-uniform growth of columnar-shaped crystals, which indicates the start of Si-HA surface structuring. At this stage, the height of the formed aggregates is found to be different (Fig.4b). Further increase in the time of the soaking in the model solution for 6 days is characterized by the formation of non-uniform coating, and the crystal grow in the form of dendrites. The crystals are recorded to grow in the form of cylindrical columns that characterizes the start of Si-HA structuring on the substrate surface (Fig.4c). During longer soaking without changing the model solution, the crystals accumulate in the form of aggregates and crystal growth goes beyond the substrate (Fig.7d).

The analysis of the coatings produced on titanium implants under power ion beam (PIB) shows that after this treatment followed by soaking in the SBF model solution, the coating of the titanium surface is uniform, and the Si-HA residuum on the samples is dense and highly dispersed (Fig. 5 *a*). After synthesis of the Si-HA layer on the titanium substrates, the samples were subjected to PIB with j=50 A/cm^2 (Fig. 5 *c–d*). As a result, the coating was found to be uniform and dense with the edges fused. Crystals in the form of dendrites were observed on the substrates, which in our opinion, will contribute to further growth and renewal of the implant surface in physiological conditions.

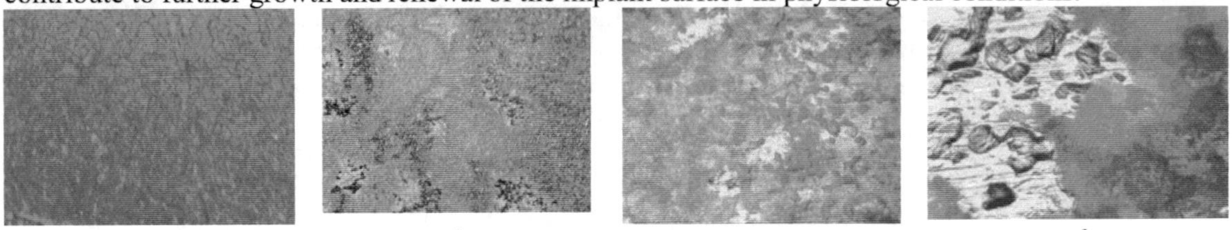

a *b* *c* *d*

Figure 5. Surface morphology of the of hydroxyapatite crystal grown on the surface of VT1-0 Ti: exposure of the surface to PIB before deposition (*a*); after exposure to PIB with j = 50 A/cm^2 (*b–d*)

While determining the coated area (Table 4), it was found that the highest percentage of deposition on the titanium substrate proceeds within the first three days. This is attributed to the fact that Si-HA dissolution in a static condition occurs mainly within three days, after which the dissolution of Si-HA is not observed. As can be seen in Table 3, to attain the best characteristics of the coatings, they are to be synthesized with the change of the suspension, which results not only in larger coated surface of the substrate, but in more uniform coating in terms of size and homogeneity of the crystals.

Table 3. The area of the coated surface depending on the technique of the titanium substrate treating

Treatment	S_c (%)				
	3 days	6 days	3+3 days	3+6 days	3+3+6 days
Etching	72	75	79	-	85
No treatment	60	67	70	-	78
Perforation etching	30	-	-	73	-
Perforation without treatment	25	-	-	79	-
PIB	-	60	-	-	-
Perf. with laser n=1	19	-	-	26	-

The analysis of the data provided in Table 3 shows satisfactory coating of the Si-HA surface for non-treated titanium samples, which confirms that the substrates made of the titanium alloy are highly biocompatible, non toxic, resistant to corrosive attack, possess characteristics are close to the mechanical properties of the bone tissue. The coatings are uniform, dense, with characteristic cracking indicating high thickness of the coating; large and small Si-HA aggregates can be found (Fig. 6 *a*, *d*). As the time of soaking in the model solution increases, the growth of columnar-shaped crystals is found to be non-uniform that characterizes the start of Si-HA surface structuring. No significant difference is observed in the height of the newly grown crystals (Fig. 6, *b*), the crystal growth goes beyond the substrate surface, and the bottom HA Si layer crystal lattice takes the form of the titanium crystal lattice. During long-term soaking of the titanium substrates in the SBF model medium, the coating is uniform, dense and highly dispersed across the whole surface, no agglomerates of individual Si-HA can be found (Fig. 6 *c*, *d*).

Figure 6. Deposition of Si-HA on titanium substrates without surface treatment: 3 days (a); 3+3 days, the model solution being changed (b); 6 days (c); 3+3+6 days, the model solution being changed (d)

4. Conclusions

The results of the study allow us to draw the following conclusions:

– varying the content of carbonate ions in the synovial fluid solution model can provide carbonate-containing materials of different crystallinity; the surface morphology affects the uniformity of the coating and the CHA crystal growth rate; the more developed the surface, the more active the crystal growth; perforated surface of the VT1-0 titanium alloy is most favorable for CHA solid phase crystallization; exposure of the sample surface coated with hydroxyapatite suspension to PIB contributes to fixation of the layer and formation of the nucleation sites for the subsequent crystal growth;

– all the samples synthesized in the model solution of the extracellular fluid under varying concentration of silicate ions are single-phase and represent hydroxyapatite; deposition of Si-HA on the titanium substrate surface preferably occurs on the etched samples; it is shown that the produced coating on the titanium is formed in several stages: growth in the form of the titanium crystal lattice; formation of crystals in the form of dendrites; crystal growth upwards in the form of cylindrical columns to form islands; it is found that after exposure of the titanium substrates to PIB with $j=50\ A/cm^2$, further growth of crystals and the regeneration of the surface are possible.

Acknowledgments

The research was partially supported by the Russian Foundation for Basic Research (grant No. 15-33-50314 and No.15-29-04839).

References

[1] Golovanova O A 2007 Omsk 395
[2] Innovative development of the health system in the Omsk region 2009
[3] Kalita V I 2000 *J. Physics and Chemistry of Material Processing* **5** 28–45
[4] Kosjakov AN 2010 *J.Orthopedics, Traumatology and Prosthetics* **4** 105
[5] Shestrikov E V 2012 *Author's abstract* Tomsk 22
[6] Kirilova I A 2004 *J.Spinal Surgery* **3** 110
[7] Ignjatovic N Delijic K 2001 *J. Zeit. Metal* **2** 145
[8] Okuno M Shikunami Y 1999 *J. Biomaterials* **2** 859
[9] Gnedenkov SV 2011 *J. Advanced Materials* **2** 49
[10] Sidel'nikov A I 2000 *J. Info-Dent* **5** 10
[11] Ferro D Barinov S M Rau J V 2005 *J. Biomaterials* **26** 805
[12] Antonov E N Bagratashvili VN Ball M 2001 *J. Key Eng. Mater* **192** 107
[13] Hamdi M Ektessabi A M 2001 *J. Thin Solid Films* **398** 385
[14] Ismailov R R Golovanova O A 2014 *J. Inorganic Materials* **6** 641
[15] Golovanova O A Berdinskaya NV Zaitv A V 2014 *J. Structural Chemistry* **5** 992
[16] Bharati S M Sinha D Basu 2005 *J.Bull. Mater Sci* **6** 128

Study of thermal effects of silicate-containing hydroxyapatites

O A Golovanova[1], A V Zaits[1], N V Berdinskaya and T S Mylnikova[2]

[1]Department of Inorganic Chemistry Omsk F. M. Dostoevsky State University
[2] National Research Tomsk Polytechnic University, Tomsk, Russia

E-mail: Golovanoa2000@mail.ru

Abstract. The possibility of modifications of hydroxyapatite silicate ions, from the extracellular fluid prototype solution under near-physiological conditions has been studied. Formation of silicon-structured hydroxyapatite with different extent of substitution of phosphate groups in the silicate group has been established through chemical and X-ray diffraction analyses, FTIR spectroscopy and optical microscopy. The results obtained are in agreement and suggest the possibility of substitution of phosphate groups for silicate groups in the hydroxyapatite structure when introducing different sources of silica, tetraethoxysilane and sodium silicate, in the reaction mixture. Growth in the amount of silicon in Si-HA results in the increase in the thermal stability of the samples. The greatest mass loss occurs at temperatures in the range of 25–400 ^{0}C that is caused by the removal of the crystallization and adsorption water and volatile impurities. It is shown that the modified apatites are of imperfect structure and crystallize in a nanocrystalline state.

1. Introduction

Hydroxyapatite (HA) is widely used for orthopedic implant manufacturing which employs an ion-plasma magnetron technique and electrochemical deposition, therefore, the temperature pattern in different processes is an important property of the material. The thermal analysis [1] indicates that pure HA is thermally less stable compared to its anion-exchanged modifications including silicon-substituted hydroxyapatite (HA-Si) of different extent of substitution.

It is known that silicon plays a bridging role in physiological processes of growth and re-construction of bone and cartilaginous tissues; therefore, the development of the techniques for silicon-substituted hydroxyapatite (Si-HA) synthesis is a relevant physical-chemical problem [2]. For complete physiological response of the organism when restoring a damaged bone, hydroxyapatite chemical modification should be carried out in the near-physiological solutions.

Model systems with the electrolyte content identical to biological fluids used as an initial solution are considered to be promising for composite biomaterials production. For this purpose, we have chosen a composition of the extracellular fluid model solution "Simulated Body Fluid" (SBF), which can be used for synthesis of calcium phosphates, as well as for evaluation of their biological activity [3]. No data on the possibility of producing silicon-substituted hydroxyapatite in this system is available.

An important characteristic of synthesized Si-HA is its thermal stability. [4] In [5] it was found that heat treatment of the biological bone at 1000 °C provides formation of silicon-magnesium-substituted oxyapatite which transforms into hydroxyapatite upon cooling, retains a balanced composition of natural nanoimpurities that support the stability of the structure, phase and chemical composition to ensure the best features of the biocoatings on titanium implants under all application techniques (electrochemical, plasma, slip processing). In [6], it was found that heat treatment of silicon-substituted hydroxyapatite at t = 500⁰S within 1 day results in Si-HA crystalline phase. The data obtained through scanning microscopy indicated bimodal (inter-aggregate and intergranular) porosity of the sample, and the grain size was 20 to 60 nm, which theoretically facilitates osteoinduction [6].

An important factor is that, regardless of the degree of the texture and stoichiometry of the Si-HA sample, the obtained nanohardness varies in the range from 11 to 12 GPa, and thermal stability and uniformity of the chemical composition are provided. However, the most important characteristic is preservation of bioactivity, which allows full use of these samples [7.8].

The paper aims to study the process of thermal transformation of Si-HA-based samples produced through deposition from the model solution of the SBF and SBF + SiO_4^{4-} extracellular fluid under varying concentration and the nature of the silicate ion supplier.

2. Materials and methods

The SBF model solution was used as a prototype of the extracellular fluid. The solid phase was prepared by deposition from aqueous solutions, wherein $CaCl_2$, $MgCl_2$, K_2HPO_4, $NaHCO_3$, Na_2SO_4 and NaCl were used as the initial system components [9].

The main compounds, suppliers of silicate groups, were Na_2SiO_3 and tetraethoxysilane (TEOS). Silicate ions were doped in the form of $Na_2SiO_3{\cdot}5H_2O$ and TEOS (tetraethoxysilane, alcohol solution). The weight content of silicon was varied in the range of 0.5 ÷ 5.0. %. The chosen ratio Ca/P = 1.67 was similar to that specified for production of stoichiometric hydroxyapatite [10]. While preparing the reaction mixture, the solution containing K_2HPO_4 and Na_2SiO_3/TEOS was the last introduced in the mixture. The synthesis was carried out at room temperature (t = 25 °C). The silicate-phosphate solution was introduced into calcium solution at a rate of 4.5–5.0 ml/min under continuous mixing of the reaction mixture. After 48 hours, the residuas were filtered under vacuum, dried at 80 ° C, weighed, and examined using a group of physical and chemical techniques.

Solid precipitates were analyzed via FTIR spectroscopy and X-ray diffraction (XRD) analysis – qualitative and quantitative phase analysis and calculation of the crystallite sizes (coherent scattering regions (CSR)). IR spectra were recorded with the FT-801 spectrometer (the samples were prepared as KBr pellets). XRD analysis of the powder samples was performed with the Bruker D8 Advance diffractometer, and the diffraction patterns were interpreted with the EVA Bruker software. After that, the thermogravimetric analysis was performed. The samples Si-HA m = 0.2000 ± 0.0005 g were placed in crucibles and then in a muffle furnace for 2 hours. The calcination temperature varied from 200 to 1000 ⁰C with 200 ⁰C space.

3. Results and Discussion

The results of the X-ray diffraction analysis of the solid phases indicate the formation of hydroxyapatite [11,12]. No other calcium phosphates were found. High intensity of the background in all the diffraction patterns can be related to the features of the SBF solution which is of more complex ion composition as compared to the systems used in conventional procedures for preparation of hydroxyapatite-based compounds [13]. When the concentration of silicate ions in the system increases, the peaks observed in the region of $2\theta = 25^o–35^o$ broaden due to low degree of sample crystallinity.

It should be noted that the diffraction patterns corresponding to solid phases with different sources of the silicate groups Na_2SiO_3/TEOS are similar. Thus, the nature of the reagent containing SiO_4^{4-} ions does not affect the hydroxyapatite structure.

The sizes of the produced modified hydroxyapatite crystallites are 6.1 ÷ 7.1 nm, which indicate the formation of compound in a nanocrystalline state.

The presence of silicon ions in the obtained powders was determined by the chemical analysis of the supernatant fluid after solid phase mineralization (Table 1). As the initial concentration of silicate ions in the system increases, the number of silicon ions significantly grows leads to the reduction of the ratio Ca/P+Si. The ratio Ca/P in the synthesized compounds exceeds the given value of 1.67 typical of stoichiometric hydroxyapatite, and corresponds to the phase of excess calcium HA. This indicates the decrease in the amount of phosphate ions in the resulting solid phases, which we believe is due to the possible replacement of phosphate groups by silicate ones.

Table 1. Concentration of silicate-ions in a solid phase, solid phase composition

C Si init., %	C Si in solid ph., %		Ca/P+Si		Ca/P		Solid phase composition
	Na_2SiO_3	TEOS	Na_2SiO_3	TEOS	Na_2SiO_3	TEOS	
0.5	0.0076	0.0024	2.44	2.32	2.48	2.34	excess calcium HA
2.5	0.11	0.11	1.92	1.88	2.31	2.23	

Investigation of the modified hydroxyapatite samples by the thermo-gravimetric analysis showed a continuous mass loss (Tables 2). Silicon-containing hydroxyapatites formed in the sodium silicate medium are of the highest thermal stability. It is found that increase in the silicon content in the solid phase results in the reduction of the total mass loss during heating up to 1000°C. The analysis of the obtained data revealed that hydroxyapatite is the prevailing phase within the entire temperature range (Fig. 1, 2). The greatest reduction in the sample mass corresponds to the temperature range of 25–400°C which is caused by the removal of the crystallization and adsorption water and volatile impurities in samples with the organic additive of silicon ions.

Table 2. Sample mass loss

T (^{0}C)	Samples SBF+Na_2SiO_3 init. (%)					Sample SBF+TЭC init. (%)				
	0.50	1.00	1.50	2.50	5.00	0.50	1.00	1.50	2.50	5.00
	m (g)					m (g)				
200	0.0110	0.0299	0.0105	0.0062	0.0113	0.0608	0.0259	0.0105	0.0104	0.0102
400	0.0137	0.0223	0.0116	0.0143	0.0155	0.0159	0.0263	0.0148	0.0174	0.0130
600	0.0252	0.0332	0.0296	0.0325	0.0383	0.0782	0.0442	0.0383	0.0311	0.0268
800	0.0356	0.0385	0.0339	0.0244	0.0355	0.0804	0.0348	0.0358	0.0354	0.0389
1000	0.0289	0.0355	0.0507	0.0357	0.0384	0.0840	0.0514	0.0400	0.0389	0.0447

Comparison of the data from Tables 2 and 3 showed that a great mass loss was observed for Si-HA samples produced in the presence of TEOS due to the removal of the organic component from the structure under temperature increase. It should be noted that the greatest mass loss is observed at low concentrations of TEOS as the HA structure is not stabilized by silicate ions, probably, due to the predominant influence of the organic component decomposition. Further removal of carbonate ions from the apatite structure and partial decomposition of the excess calcium HA followed by formation of the stoichiometric compound and phase of β-tricalcium phosphate occurs in the temperature range of 600–1000 [10].

As can be seen from the diffraction patterns (Fig. 3), the beginning of phase transition and Si-HA peak splitting are observed at 600°C, and splitting of the peak from singlet into doublet can be clearly seen at 800°C. However, the temperature increase up to 1000°C does not cause further changes in the phase composition of samples. Increase in the content of silicate ions in HA is beneficial for thermal activity of samples (Fig. 4). As can be seen from the comparison of the XRD spectra presented above, the form of doublets in the angular range (30–32) 2θ is different that indicates their different thermal stability at the given temperature. Thus, the presence of SiO_4^{4-} groups in the HA structure enhances thermal stability of HA-based materials. Fig. 5 shows IR absorption spectra for samples produced under variation of the silicate ion concentration and the product of their calcination (at 800°C) synthesized at C Si init. = 0.5%. The absorption bands corresponding to those for hydroxyapatite are observed in IR spectra of the obtained products [9]. The bands at 1020–1080 (ν_3), 960, 840 (ν_2), 600, 570 (ν_4) and 473 (ν_2) cm^{-1} correspond to the vibrations of the phosphate groups, the bands at 3570 and 630 cm^{-1} correspond to the vibrations of the hydroxyl groups, the bands at 1650, 1410–1570 (ν_3), 880 and 873 (ν_2) cm^{-1} correspond to the vibrations of the carbonate groups. The presence of the carbonate groups indicates the presence of all silicon substituted apatites in the structure, which is caused by the presence of HCO_3 ions in the initial synthesized solution. The peak of deformation vibrations of the OH^- groups at 3570 cm^{-1} is observed in the background of the high intensity band in the region of 3200–3700 cm^{-1}, which may be due to the presence of the adsorbed water in the produced apatite form [10]. As the silicate concentration in the reaction medium grows, the mode intensities of OH^- ions HA vibrations (3570 cm^{-1}) decrease.

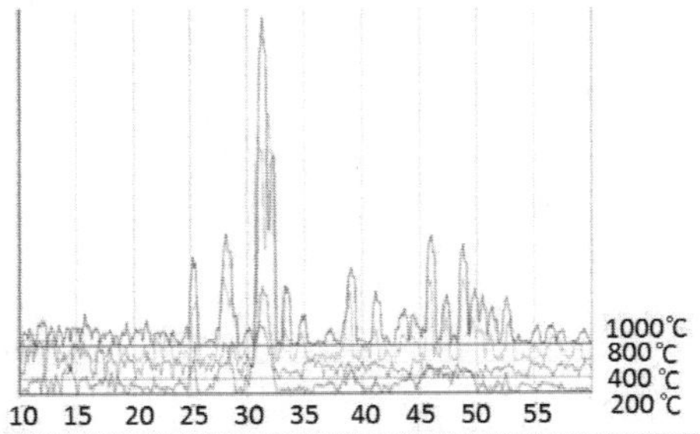

Figure 1. Diffraction patterns of Si-HA with different heat treatment (agent of silicate ions is Na_2SiO_3).

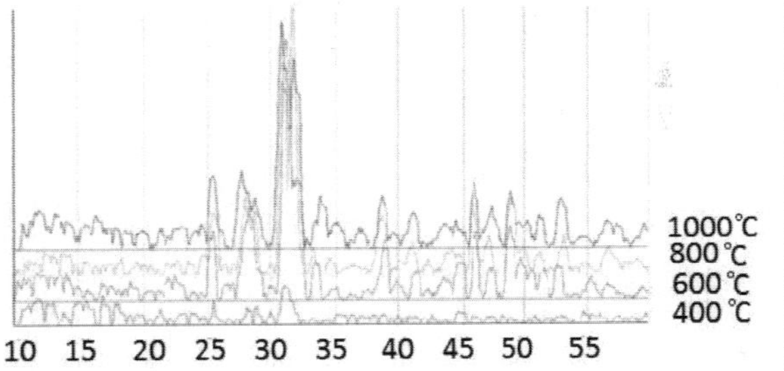

Figure 2. Diffraction patterns of Si-HA with different heat treatment (agent of silicate ions is TEOS).

The presence of SiO_4^{4-} groups in precipitates can be determined by the low intensity peaks with maxima at 504 cm^{-1} and 800 cm^{-1}, which correspond to the modes of deformation vibrations of Si-O bonds and symmetric valent vibrations of Si-O-Si bridges in SiO_4 tetrahedra. [9]. Insignificant increase in their intensity can be observed with the increase of silicon content in the initial reaction medium. Simultaneous decay of the vibrational modes of OH^- ions HA (3570 cm^{-1}) indicates the substitution of phosphates for silicates with the hydroxyl groups involved: $PO_4^{3-} + OH^- \Leftarrow SiO_4^{4-} + \square_{OH}$ that agrees with the literature data [1, 9–13].

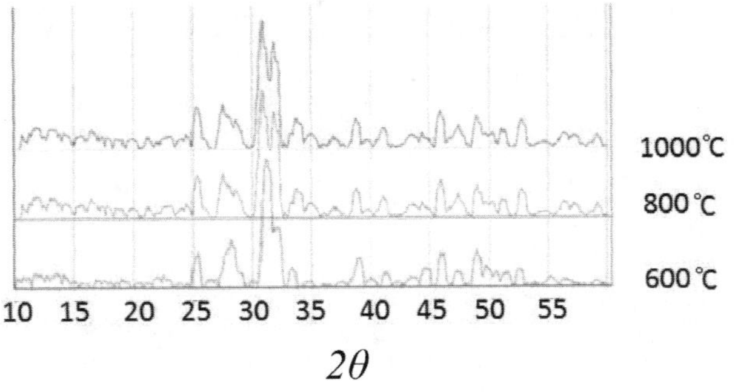

Figure 3. Diffraction patterns of Si-HA with different heat treatment (agent of silicate ions is TEOS)

Broadening of the band in the region of 3570 cm^{-1} is typical of the hydroxyapatite samples modified by silicate ions from the tetraethoxysilane structure, which we believe is due to the superposition on the absorption band at 2800–2900 cm^{-1} that corresponds to the vibrations of C-H groups. The occurrence of these bands in the spectra can be attributed to the fact that the synthesis was carried out in the TEOS alcohol solution. The IR absorption spectrum of the calcination product of silicon-substituted HA at 800°C is characterized by an intense narrow band at 3570 cm^{-1} typical of crystalline calcium hydroxyapatite. The IR spectrum contains the bands of valent and deformation vibrations 1020–1080, 960, 840, 600, 570 and 473 cm^{-1}. Thus, the thermal effect on the synthesized samples causes the removal of carbonate ions from the HA structure.

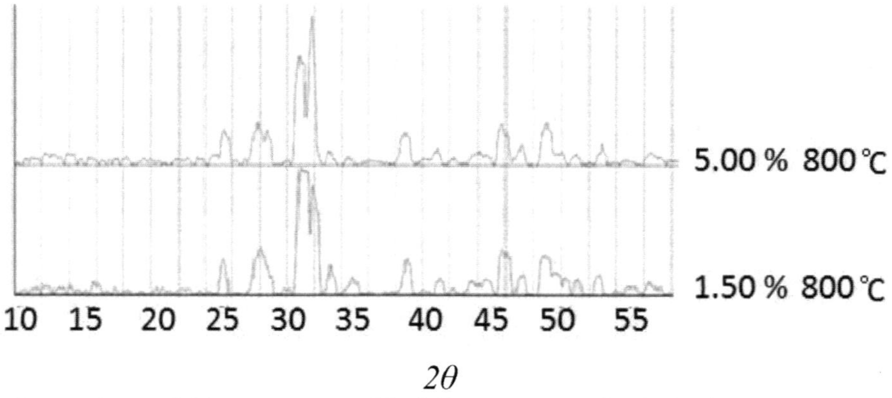

Figure 4. Comparison of Si-HA spectra with different extent of substitution (agent of silicate ions is Na_2SiO_3).

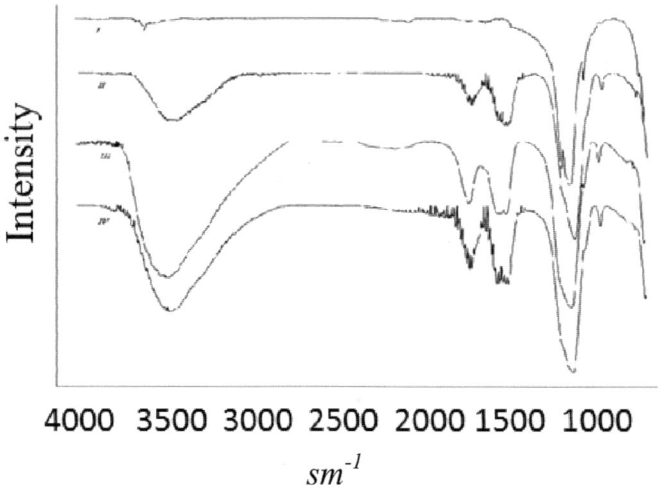

Figuire 5. IR spectra of HA samples modified (mass fraction (C%)): *(I)* Na_2SiO_3 is Si init. 0.5% at 800°C; *(II* and *III)* TEOS is Si init. 0.5 and 1.0%, respectively; *(IV)* Na_2SiO_3 is Si init. 0.5%.

4. Conclusion

The following conclusions can be drawn based on the thermogravimetric analysis of Si-HA:

1. Formation of silicon-substituted hydroxyapatite from the model solution of human extracellular fluid was found via the chemical method, and XRD and FTIR analyses. No significant difference in the structure of the formed solid phases was identified due to different types of reagents containing SiO_4^{4-}. The results of all the analyses are in agreement and indicate the possibility of substitution of

phosphate groups for silicate groups in the HA structure when introducing various sources of silicon such as tetraethoxysilane and sodium silicate into the reaction mixture.

2. As the amount of silicon in Si-HA grows, thermal stability of the samples increases. The greatest mass loss is observed at temperatures of 25–400 °C, which is apparently caused by the removal of the crystallization and adsorption water and volatile impurities.

3. The structure of the modified apatites is imperfect and they crystallize in a nanocrystalline state.

These properties together with the synthesis carried out under near-physiological conditions, potentially increase the extent of resorption, solubility and bioactivity of the material based on the obtained silicon-substituted hydroxyapatite capable of the bone tissue integration.

Acknowledgments

The research was partially supported by the Russian Foundation for Basic Research (research project No. 15-29-04839).

References

[1] Bogdanova E A 2012 *Author's abstract* 13
[2] Carlisle E M 1970 *J. Science* **167** 279
[3] Yongsheng Wang Sam Zhang Xianting 2007 *J. Materials Science and Engineering* **27** 244
[4] Soin A V Evdokimov P V Veresov A G Putlyaev VI 2007 *J. Alternative Energy and Ecology* **45** 130
[5] Petrovskaya T S 2013 *Author's abstract* 12
[6] Ivanyutina I S Mukhanova E A 2012 *Author's abstract* 27
[7] Surmeneva M A 2012 *Author's abstract* 5
[8] Safronova T V Putlyaev V I Sergeeva A I 2009 *J. Reports of the Academy of Sciences* **4** 491
[9] Solonenko A P Golovanova O A 2013 *J. Inorganic Chemistry* **12** 1420
[10] Barinov S M 2010 *J. Russian Chemical Reviews* **1** 15
[11] Golovanova O A Berdinskaya NV Zaitv A V 2014 *J. Structural Chemistry* **5** 992
[12] Komlev V S Fadeeva I V Gurin A N Kovaleva E S 2009 3 373
[13] Kokubo T Kushitani H Sakka S Kitsugi T Yamamuro T 1990 *J. Biomed Mater Res* 24

IOP Conf. Series: Materials Science and Engineering **110** (2016) 012025 doi:10.1088/1757-899X/110/1/012025

Investigation of Changing Volt-Ampere Characteristics of AlGaInP Heterostructures with Multiple Quantum Wells under Ionizing Radiation

A V Gradoboev[1] and K N Orlova[2]

[1]Open Joint Stock Company National Research of Semiconductor devices, Tomsk, Russia
[2]National research Tomsk polytechnic university, Tomsk, Russia
E-mail: gradoboev1@mail.ru

Abstract. The results of research into degradation of volt-ampere characteristics of light emitting diodes produced on the base of AlGaInP heterostructures with multiple quantum wells are presented on the example of light emitting diodes (emission wavelengths 623 nm and 590 nm) under gamma quantum and fast neutron radiation in passive powering mode. The shifts of volt-ampere characteristics into the higher voltage range have been observed in conditions of increasing neutron fluence and radiation dose. The observed increase in the resistance of ohmic contacts is caused by the rising resistance of adjacent area, which in its turn results from the changing mobility of charge carriers. The latter varies with the growth of introduced defects under irradiation. Two different areas of current generation have been identified. A mechanism of current generation depends on injected charge carriers in the range of mid-level electron injection. Moreover, the range of high electron injection is distinguished by changing resistance of light emitting diode cores alongside with current generation conditioned by charge carrier injection.

1. Introduction

A great number of semiconductor devices based on quaternary compounds AlGaInP have been developed recently and used in nuclear and atomic electronics, in the field of optical signal processing, and in various civil and special facilities [1-4]. As these light emitting diodes (LED) are applied in outer space and other conditions with higher radiation background, researchers are confronted with the issue of determining and predicting the radiation stability of AlGaInP LED. Volt-ampere characteristics are selected as a criterion parameter necessary to identify changes caused by ionizing irradiation in a semiconductor device – in LED produced on the base of AlGaInP heterostructures with multiple quantum wells. These characteristics are viewed as basic criterion parameters used by design engineers and other personnel involved in searching for LED optimal operational conditions, maximum values of supplied voltage and current intensity.

At the present the issue of degrading characteristics of gamma-and-neutron-irradiated LED produced on the base of AlGaInP heterostructures with multiple quantum wells hasn't been sufficiently covered in experimental and theoretical literature. This fact makes investigation of ionizing irradiation exposure on the characteristics of mentioned above diodes relevant both for scientific and practical purposes.

Corresponding author: Gradoboev Alexander Vasilievich, professor, Doctor of Technical Science,
+7 913 866 84 05, e-mail: gradoboev1@mail.ru,
634050, Tomsk polytechnic university, 30 Lenina av., Tomsk, Russia.

IOP Conf. Series: Materials Science and Engineering **110** (2016) 012025 doi:10.1088/1757-899X/110/1/012025

The work is aimed at research into changing volt-ampere characteristics of LED produced on the base of AlGaInP heterostructures with multiple quantum wells under gamma-quantum ^{60}Co and fast neutron radiation.

2. Research objects and methods

LED based on AlGaInP heterostructures with quantum wells were produced by various homeland and foreign manufacturers. The produced crystals were placed into a standard case TO-18. Optical compound was used to make a lens to form a directed radiation beam. According to preliminary research results the optical compound, used for lenses production, does not change its optical properties under gamma-quantum ^{60}Co radiation all over the area of neutron fluence impact, even if the dose of radiation is up to 10^7 Gy. Therefore, all the changes in optical properties of irradiated diodes can be conditioned by changing optical characteristics of LED produced on the base of AlGaInP heterostructures.

A set of diodes was divided into parts, for each of them both a radiation dose and a corresponding interval were determined sequentially, as well as all the mid-measurements were taken and a single dose was identified. The obtained results were compared and no radiation defect annealing was revealed, so, inaccuracy of measurements caused by annealing was somehow minimized.

A facility like with the dose rate (0.1 - 10) Gy per second was used as a source of gamma radiation. Isotope Co60 with the efficient energy of gamma quantums 1.25 MeV is an active element in facilities of this kind. The dose was determined according to the period of radiation, and the data on facility operation at the moment of experiments was taken into consideration. Gy (Si) was used to measure the dose of gamma radiation. The certified volume of radiation zone was 3 liters.

Static and pulse reactors with radiation pulse duration 1.5·10-3s were used as sources of fast neutrons, their energy exceed 0.1 MeV. Neutron exposure level was described by fluence of neutrons Fn, that is the number of neutrons per square centimeter (n/cm2).

Since electric field application can influence greatly on the input rate of radiation-induced defects [5-6], reference structures and devices were always irradiated in conditions of passive mode (without external electric supply). The particle flux was directed perpendicularly to the plane of the substrate in all experiments.

3. Experimental findings and discussion

Let us consider the results of measurement, which were obtained before irradiation of facilities as a part of studying LED produced on the base of AlGaInP heterostructures with multiple quantum wells.

As it has been already mentioned [7-9], volt-ampere characteristic, viewed in semilogarithmic scale, makes it possible to identify mechanism of direct current flowing unambiguously and highlight zones with different injection levels of minority carriers into the active LED layer at forward bias [10-11]. On the straight branch of LED volt-ampere characteristics, made in semilogarithmic coordinates, there are two precise zones (marked I and II in Fig. 1).

While current flows through a LED, minor charge carriers are injected into the LED core. The density of electrons in the conduction band and that one of unfilled levels in the valence band are to be simultaneously high in order to intensify recombination. These conditions can be provided if the level of electron injection is high.

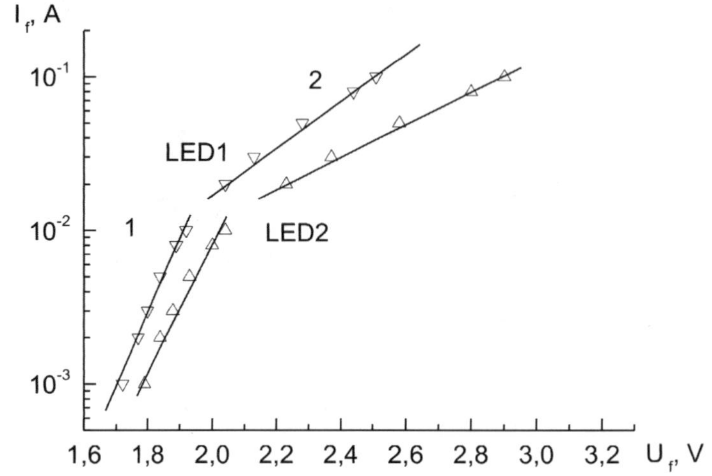

Figure 1. Volt-ampere characteristic of LED in semilogarithmic scale: 1 – mid-level injection zone; 2 – high injection zone; symbols – results of measurement; lines – identified regularities

Approaching the high injection zone the energy of electrons increases in the zone of emitter; Fermi level rises, thus, the height of a potential barrier decreases, so a high level of electron injection can be observed. After injected holes concentration in the diode base and concentration of donors in p-n junction had been approximately assessed, characteristic correlations for mid-level and high injection zones (1) and (2) were confirmed.

$$p \approx N_d \quad (1)$$

$$p \gg N_d \quad (2)$$

Therefore, zones I and II can be considered mid-level and high injection zones, respectively.

The slope of a tangent to the graph and the value of voltage on LED can help to highlight zones, where volt-ampere characteristics depend on various mechanisms of current flowing. There is a straight line on the graph where volt-ampere characteristics are an exponential function. Linear dependence can get lost while switching to high direct currents because of failure on p-n junction or ohmic contact conditioned by thermal effects, change in life span of carriers, conductance channels etc.

The first highlighted zone in Fig. 1 is distinguished by higher angular coefficient of the tangent, which is probably related to injection of charge carriers. Therefore, we can conclude that the first zone is one of electron mid-level injection with predominance of current depending on charge carrier injection.

If current is increased subsequently modulation of LED core is important and determines the second zone of volt-ampere characteristics of reference LED (Fig. 1) with a lower angular coefficient besides electron injection resistance. That is in line with the zone of high electron injection. Therefore, volt-ampere characteristics of LED in high injection zone aren't conditioned by impurity concentration in LED core. If the level of injection is high, properties of a semiconductor in the core are specified mainly by injected carrier concentration. And the resistance of an active layer is similar to the resistance of a semiconductor.

Here we consider results of measured red and yellow LED volt-ampere characteristics under ionizing irradiation, to be precise, fast neutron irradiation, and that is submitted in linear coordinates (Fig. 2). It is worth saying that the decrease in radiation power was $\frac{P_x}{P_0} = 0,1$ under irradiation by fast neutrons with fluence up to $5.06 \cdot 10^{15}$ n/cm2

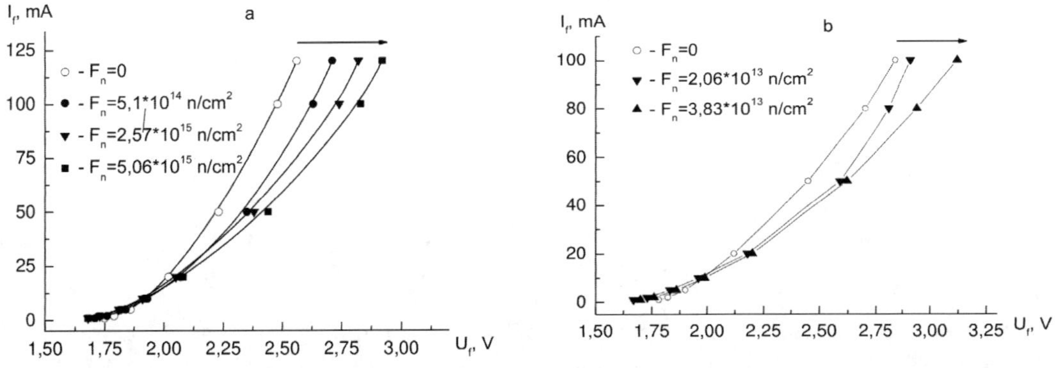

Figure 2. Volt-ampere characteristic of LED produced on the base of AlGaInP heterostructures with multiple quantum wells irradiated by fast neutrons: a) – LED (λ=623 nm); b) – LED (λ=590 nm); symbols – measurement results; lines – identified regularities; arrows indicate the direction of changing volt-ampere characteristics

All the changes in volt-ampere characteristics typical both for LED (λ=623 nm) and LED (λ=590 nm) are observed in the range of high currents. In the range of low currents changes in volt-ampere characteristics are insignificant. Moreover, volt-ampere characteristics are shifted into the higher voltage range because of the increase in neutron fluence. Since the resistance of LED core is minimal in the range of high currents the mentioned above shift can be caused by rising resistance of ohmic contacts. In compliance with information presented in the literature the resistance of ohmic contacts to GaAs materials rises proportionally to the number of radiation-induced defects, dislocations and other structural imperfections. Thus, in the range of considerably reduced irradiation power and high currents, where the shift of LED volt-ampere characteristics was registered, defects are principally radiation-induced and, as the consequence, the resistance of ohmic contacts rises. Furthermore, according to the available data [12], an ohmic contact based on Au-Ge-Ni – n-GaAs has lower irradiation stability than that of GaAs, so its degradation under irradiation it accelerated.

The identified increase in resistance in the range of high currents can't be viewed as rising resistance caused by the crossed junction metal – semiconductor, because alloyed near-to-the-contact layers, which are surfaced on both sides of the diode and reduce the potential barrier, further efficient penetration of electron flowing through material depth. Rising resistance can be probably conditioned by growing resistance of near-to-the-contact zone. The latter results from the changing charge carrier mobility, which depends on the number of radiation-induced defects and other imperfections in solid structure.

Here we consider the change in volt-ampere characteristics of gamma-quantum irradiated LED produced on the base of AlGaInP heterostructures with multiple quantum wells.

Emission power reduction under gamma-quantum 60Co radiation in the dose up to $5{\cdot}10^6$ Gy came up $\frac{P_x}{P_0}=0,4$.

All the changes in volt-ampere characteristics specific both for LED (λ=623 nm), and for LED (λ=590 nm) are identical to the changes registered for LED produced on the base of AlGaInP heterostructures with multiple quantum wells under irradiation by fast neutrons. In the range of low currents volt-ampere characteristics don't change significantly. There is a shift of LED volt-ampere characteristics into the range of higher voltages in high current area. Although the radiation mechanism of fast neutrons differs from that of gamma-quantums, emission power also reduces significantly in the area of shifted volt-ampere characteristics. It indicates exceptionally radiative nature of defects and confirms the hypothesis of degrading ohmic contacts of diodes. The mentioned structure imperfections

are the reason for changing charge carrier mobility, as the consequence, resistance grows in near-to-the-contact zone of ohmic contact.
The following conclusions can be drawn on the ground of the presented results.

4. Conclusion

1. Mid-level and high injection of charge carries can be clearly highlighted in volt-ampere characteristics of LED AlGaInP with multiple quantum wells (λ =623 nm, λ = 590 nm) before the test of ionizing irradiation influence. Therefore, we can conclude that LED produced on the base of AlGaInP heterostructures with multiple quantum wells operate in the mode of mid-level and high injection.
2. Two areas of different current generation were highlighted according to the slope of volt-ampere characteristics of non-irradiated LED in a semilogarithmic scale. The first zone of volt-ampere characteristics (Fig. 2) is distinguished by the mechanism of current generation in dependence on injected charge carriers in the range of mid-level electron injection, the change in the resistance of LED core and current generation caused by charge carrier injection are important for the second zone of volt-ampere characteristics in the range of high electron injection.
3. The slope of volt-ampere characteristics of LED produced on the base of AlGaInP heterostructures with multiple quantum wells in a semilogarithmic scale is identical in the range of mid-level electron injection. On the other hand, the slope of volt-ampere characteristics rises as the energy gap gets narrower in the range of high electron injection.
4. If LED produced on the base of AlGaInP heterostructures with multiple quantum wells is radiated by fast neutrons with fluence up to до 5.06·1015 n/cm2 its volt-ampere characteristics change insignificantly in the low current range. As the fluence of neutrons increases in the high current range, volt-ampere characteristics are shifted into the range of higher voltages. This shift is possible because the resistance of ohmic contact rises. The increase in resistance is probably conditioned by growing resistance of near-to-the-contact zone. That is the result of changing charge carrier mobility, which varies according to the augmenting number of defects conditioned by fast neutrons.
5. If LED produced on the base of AlGaInP heterostructures with multiple quantum wells are radiated by gamma-quantums 60Co in the dose up to $5{\cdot}10^6$ Gy, the results of volt-ampere characteristics are identical to those, which were obtained under radiation by fast neutrons, that is, the shift of volt-ampere characteristics in the high current area into the range of higher voltages.

References

[1] Modak P *et al.* 2000 *Journal of electronic materials,* **29** 80-85
[2] Dan Sporea 2006 *Proceedings of the Symposium on Photonics Technologies for 7th Framework*, 281-185
[3] Bei Zhang *et al.* 1999 *Proceedings of SPIE,* **3899** 232-238
[4] Chu D Y *et al.* 1993 *IEEE Photonics Technology Lett,* **5** 1353-1355
[5] Gusakov V V, Latyshev A V, Stelmakh V F 1990 *News of Higher Educational Institutions. Physics*, 15 – 30
[6] Grusha S.A. *et al.* 1990 *Electronic engineering. Series 2. Semiconducting devices,* **5** 68-72
[7] Nikiforov S G Dissertation … candidate of engineering 2006 Moscow 158 p
[8] Orlova K. N. , Gradoboev A. V. 2012 *Proceedings of IFOST - 2012*, **1** 192-195
[9] Orlova K N. , Gradoboev A V. 2014 *Proceedings of CriMiCo 2014* 874-875
[10] Gradoboev A V, Orlova K N 2015 *Phys. Stat. Sol. C* **12** No.1-2 35-38
[11] Gradoboev A V and Orlova K N 2014 *Advanced Material Research* **880** 237
[12] Gradoboev A V Surzhikov A P 2005 *Radiation stability of GaAs microwave devices Tomsk* (Izdatelstvo TPU) 277p

Deterioration of Watt and Voltage Characteristics of AlGaInP Heterostructures under Irradiation by Fast Neutrons

A V Gradoboev[1] and K N Orlova[2]

[1]Open Joint Stock Company National Research of Semiconductor devices, Tomsk, Russia
[2]National research Tomsk polytechnic university, Tomsk, Russia
E-mail: gradoboev1@mail.ru

Abstract. The paper presents the results of studying characteristic deterioration of AlGaInP heterostructures with multiple quantum wells. The research was completed for light emitting diodes (emission wavelengths 623 nm and 590 nm) under fast neutron irradiation in passive mode. It has been revealed that the change in emission power and operating current under irradiation is conditioned by band gap and level of electron injection. Here, the change of current flowing mechanism is a distinctive parameter of the boundary between the first and second stages of emission power reduction caused by fast neutron irradiation of AlGaInP heterostructures (λ=625 nm).

1. Introduction

In recent years the sphere of applying optoelectronic devices has been widen significantly due to rapid development of optoelectronics, which is grounded on the aggregate of optical and electronic properties of materials necessary to generate and transform, store and memorize information, due to the unique properties characteristic exclusively for optoelectronic devices. The following list represents only some fields of application:

- fiber-optic systems [1];
- optical signal processing [2-4];
- photoconverters and solar batteries;
- laser printers;
- traffic control devices and road signs on highways;
- movable information boards;
- car lamps;
- sea and river markers;
- and components of train and aircraft board equipment etc. [5].

The base of any optoelectronic system is a source of emission, and its properties determine characteristics of the device. The incoherent radiation source is a light emitting diode. High-performance light emitting diodes produced on the base of aluminum-gallium-indium-phosphide heterostructures (AlGaInP or InGaAlP, AlInGaP etc.) have been developed recently. They are the most effective ones in the visible region, for instance, in red, orange and yellow ranges. These four-component semiconductor heterostructures according to the order of named chemical elements are different, being a single physical entity though, and the order of designation depends.
mainly on the amount of substance of a given chemical element.

Corresponding author: Gradoboev Alexander Vasilievich, professor, Doctor of Technical Science, tel. +7 913 866 84 05, e-mail: gradoboev1@mail.ru, 634050, Tomsk polytechnic university, 30 Lenina av., Tomsk, Russia.

AlGaInP heterostructures with multiple quantum wells are produced by metalorganic chemical vapor deposition (MOCVD)-based gaseous epitaxy [6-9]. However, a number of technological problems in production of heterostructures have not been solved yet. These are as follows: hydrogen passivation, timely oxygen removal from the case, decrease of imperfection on heteroboundaries of layers and improvement of grown layer homogeneity. The presented facts are to be taken into consideration for studying radiation-induced defects in the given heterostructures and devices made of them.

Quantum wells, introduced into the active area of a heterostructure, facilitate concentration growth of minor charge carriers in it due to the state of two-dimensional electronic gas; it increases internal quantum efficiency [4]. Here, relaxation of arising on the boundary of layers resilience conditioned by discrepant lattice parameters under quantum well introduction into the core of AlGaInP heterostructures doesn't lead to dislocations or point defects, which can be centers of non-radiative recombination.

At present, the change in parameters of AlGaInP heterostructures with multiple quantum wells exposed to ionizing irradiation has not been adequately investigated [10-11]. The issue of radiation stability of light emitting diodes produced on their base is gaining its urgency because they can be used in conditions of various ionizing irradiation [12-13].

2. Research objects and methods

The objects of research were red light emitting diodes (LEDR) produced on the base of AlGaInP heterostructures with multiple quantum wells (λ = 623 nm) and yellow (LEDY) ones (λ = 590 nm). Therefore, the diodes under investigation were distinguished by the broad band gap, that is, composition of the core. Some studied heterostructures were produced as ready-made crystals, while others were heterostructures used for batch production of crystals in similar manufacturing processes.

At first, a p-layer of the structure was dielectrically SiO_2 coated (0.5 μm) and a part of the substrate was chemically and chemical-dynamically etched to obtain the total thickness of the structure (200-210 μm).

Then an ohmic contact was formed towards n-GaAs substrate all over the surface of plates via magnetron-coated AuGeNi+Au deposition (0.15-0.2 μm). An ohmic contact was burnt into in the atmosphere of hydrogen at temperature (425-430) ^{0}C for 5 min. Finally, the plates were glued onto ceramized substrate and cut by the disk in several crystals (chips). Dimensions of crystals after cutting are 1020x1020 μm.

Batch produced in this way crystals were mounted into corresponding cases via thermal-compression assembly. For this purpose metallic cases were used (KI1 and KI2, they differ in geometric dimensions).

One of the mostly applied type of ionizing irradiation to carry out radiation tests on simulating facilities is neutron flux, its effect bases on crystal lattice displacement.

The purpose of the work is to investigate deterioration of watt and volt characteristics of AlGaInP heterostructures with multiple quantum wells under fast neutron radiation.

For the purpose of research different LED sets were made, and each set included at least 10 diodes. For various LED sets exposure level was augmented via sequentially increased radiation dose. LED were irradiated in standard conditions in passive powering mode, that is without direct current flowing.

In all the experiments focused on irradiation particle flux was directed perpendicularly to the substrate plane. The degree of impact under irradiation was described in terms of neutron fluence F_n [cm^{-2}]. Watt and volt characteristic was measured before and after irradiation in a sphere. Each experimental value of the below given dependences conforms to a mean parameter value for LED set.

We applied both one-time and sequential sets of required levels of exposure. There is no difference in the results obtained this way so we can conclude that methods to measure the change in LED parameters don't cause annealing of introduced radiation defects.

3. Experimental findings and discussion

Watt and volt characteristics are ones of main LED characteristics, which determine operating features of an electronic appliance. Relying on the analysis of watt and volt characteristics optimal parameters of appliance maintenance can be selected including voltage, which provides maximum external LED quantum efficiency.

As fast neutron fluence was increased the change in watt and volt characteristics was registered. However, at first, let us consider measurement results relevant for studying non-radiated diodes. Typical watt and volt characteristics of the non-irradiated diodes under consideration are shown in Fig. 1.

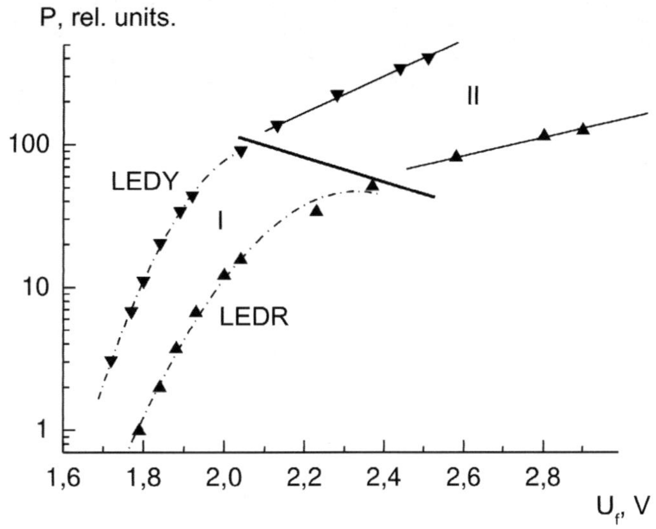

Figure 1. Watt and volt characteristic of reference light emitting diodes symbols – measurement results; lines – identified; I – mid-level injection zone; II – high injection zone

As one can see, two zones are highlighted for reference diodes, and there is a boundary between them. Each pronounced zone has its specific dependence of analyzed regularities. First, marked differences of obtained watt and volt characteristics of non-radiated red and yellow diodes are worth mentioning. These differences are conditioned by various band gaps of red and yellow heterostructures and, as the consequence, quantum efficiency (or quantum output) and LED efficiency factor are different:

$$\eta_q = \frac{N_{ext}}{N_{int}} = \eta_e \times \eta_s \quad (1)$$

where η_q – external quantum efficiency,

N_{ext}– flow of emitted photons,

N_{int} – flow of recombined photons,

η_e – efficiency factor of emission output,

η_s– efficiency factor of light emission.

Therefore, LED external quantum efficiency depends on the sort of material, technology, and rate of emitting and full recombination of charge carriers, life time of radiative and non-radiative recombination. More efficient LED (LEDY in our case) is distinguished by shorter lifetime of minor charge carriers.

Two typical zones (zone I and zone II) can be highlighted for each LED type according to obtained watt and volt characteristics. We can sum up that highlighted zones are those of mid-level and high electron injection. The frontier between mid-level and high electron injection is specified by equal concentration of injected electrons and holes in p-region. Our assessment has revealed that concentration of injected electrons is really comparable with concentration of holes in p-region in conditions of the identified frontier.

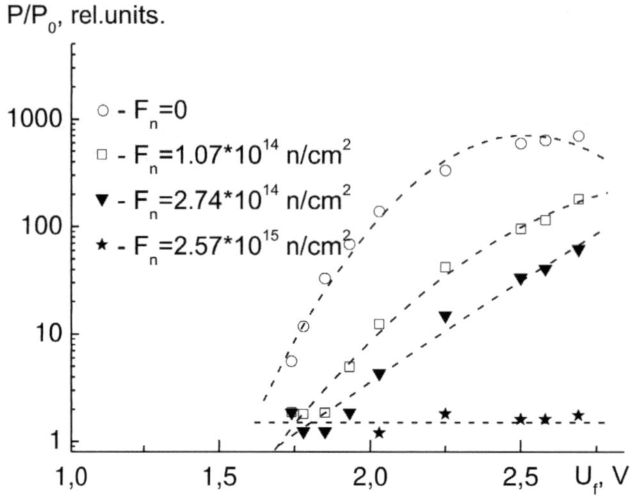

Figure 2. The changes in watt and volt characteristics of LEDR under fast neutron irradiation: symbols – experimental data; lines – identified regularities

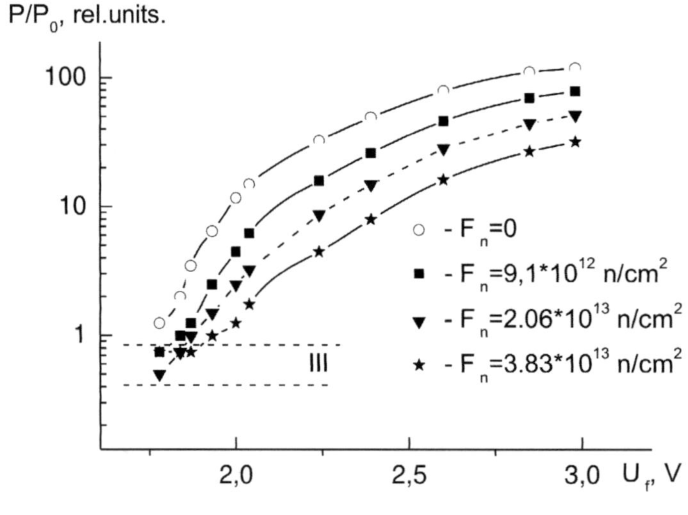

Figure 3. The change of watt and volt characteristics of LEDY under fast neutron irradiation: symbols – experimental data; lines – identified regularities.

Fig. 2 demonstrates the change in watt and volt characteristics of LEDR heterostructures under fast neutron irradiation. As one can see zones of mid-level, high and low injection of electrons into the core of a heterostructure are highly pronounced and confirm the drawn conclusions for non-radiated LED. The zone of low electron injection (III) allows deducing an inference that final phase of reducing emission power is conversion into a low electron injection mode for which dependence of emission power on operating current is not relevant.

Two typical stages of reducing emission power (bright and dark symbols in Fig. 2) differing in the mechanism of current flowing can be highlighted on the base of presented above results. All identified regularities are confirmed by accuracy of approximation more 99.9%.

Then, one can assume that the first stage of reducing LEDR emission power doesn't correlate with introduced centers of non-radiative recombination, while the second stage is conditioned by them, therefore, principles of current flowing are changed. Here, reducing at the first stage emission power of the core under fast neutron irradiation accounts for introduced absorption centers of generated irradiation, which are irrelevant for current flowing. These absorption centers can be located both in the core and adjacent layers, and in view of technological peculiarities of AlGaInP heterostructures production principles of absorption centers can be caused by mentioned above impurity compounds. The main difference of studied watt and volt characteristics of yellow light emitting diodes under fast neutron irradiation (Fig. 3) is unchanged mechanism of current flowing in conditions of first – second stage conversion of reducing emission power under fast neutron irradiation. Here reducing emission power can result from absorption of irradiation emitted by the core in adjacent layers.

4. Conclusion

In conclusion we sum up main results of investigation into deterioration of watt and volt characteristics of LED produced on the base of AlGaInP heterostructures ($\lambda 1$ = 590 nm and $\lambda 2$ = 630 nm) with multiple quantum wells under fast neutron irradiation:

- Quantum efficiency of LEDY and LEDR depends on the band gap, rate of non-radiative and complete recombination, life time of minor charge carriers.
- The change in emission power under irradiation is dependent on the band gap and level of electron injection.
- While conversion from mid-level to high level injection of electrons the mechanism of current flowing changes as the result of fast neutron irradiation of red LED produced on the base of AlGaInP heterostructures.
- The increased irradiation dose causes more pronounced differences between mid-level and high electron injection, but the mode of low electron injection is a final stage of the reducing emission power process.
- The mechanism of current flowing is not changed while conversion of mid-level injection into high injection of irradiated by fast neutrons LEDY.
- Unsoundness of reference heterostructures, that is occurrence of special impurity compounds, is relevant for arising absorption centers and influences on the current flowing mechanism through a LED. Thus, unsoundness of reference heterostructures has significant impact on efficiency of reference LED and its radiation stability. The higher efficiency of LED is the higher is its radiation stability.

References

[1] Dan Sporea 2006 *Proceedings of the Symposium on Photonics Technologies for 7th Framework*, 281-185
[2] Bei Zhang *et al.* 1999 *Proceedings of SPIE,* **3899** 232-238
[3] Chu D Y *et al.* 1993 *IEEE Photonics Technology Lett,* **5** 1353-1355
[4] Hodapp M.W 1997 *High brightness light emitting diodes New York,* Academic press 87-92
[5] Craford George M 2000 *MRS bulletin*, **1** 113-118
[6] Zeller H R 2004 *ABB Semiconductors AG Lenzburg*, 8 p
[7] Feng Z C *et al.* 1999 *Journal Of Applied Physics,* **85** 7 3824-3831
[8] Sugawara H *et al.* 1992 *Appl. Phys. Lett,* **61** 15 1775-1777
[9] Maranowski S A *et al.* 2005 *Proc. SPIE,* **3002/110** 110-118
[10] Gradoboev A V, Orlova K N 2015 *Phys. Stat. Sol. C* **12** No.1-2 35-38
[11] Gradoboev A V and Orlova K N 2014 *Advanced Material Research* **880** 237
[12] Orlova K N. , Gradoboev A V. 2012 *Proceedings of IFOST - 2012*, **1** 192-195
[13] Orlova K N. , Gradoboev A V. 2014 *Proceedings of CriMiCo 2014* 874-875

Combined ion (Ar^+, 20 keV) and light irradiation of the quenched Fe-8.25 at % Mn alloy. Separation between thermal and radiation induced long-range effects

V V Ovchinnikov[1, 2], N V Gushchina[1] and S A Bedin[1]

[1] Institute of Electrophysics, Ural Branch of Russian Academy of Sciences, Amundsena Street 106, Yekaterinburg, 620016, Russia
[2] Ural Federal Technical University named after the First President of Russia B.E. Yeltsin, Mira Street 19, Yekaterinburg, 620002, Russia

e-mail: viae05@rambler.ru

Abstract. Mössbauer and X-ray diffraction investigations of the radiation-induced $\alpha \rightarrow \gamma$ phase transformation and short-range-order formation processes in the quenched Fe-8.25 at % Mn alloy under combined exposure (simultaneous visible light and Ar^+ 20-keV ion beam irradiation) are carried out. Combined exposure made it possible to fix the target stationary temperature, and hence, the intensity of thermally-stimulated processes; an energy and ion current density could independently be varied in a wide range. As a result, an important contribution of a non-thermal constituent of ion beam exposure to the structural state of alloy was proved. Only in the presence of ion beam, an $\alpha \rightarrow \gamma$ (bcc $\rightarrow$ fcc) phase transformation and accelerated intraphase processes preparing this transformation are observed in the deep layers of the target (about 10^3 R_p). With allowance for the relatively low level of thermally and radiation-stimulated processes, radiation-dynamic effects associated with propagation of intense post-cascade solitary waves, which can rearrange metastable matters, are considered as the cause of the observed transformations.

Introduction

In the last decades, increasing attention is paid to research in the field of the impact of charged particle (electrons and ions) beams on the surface of a solid. The results of the research are extensively used for the technological purpose, in particular, for the modification of the properties of engineering materials [1, 2].

Information about the impact of accelerated ion beams with energies from a few tens to several hundreds of keV on materials is mainly limited to that about the change in the composition and structure in a very thin surface layer of the matter, the thickness of which is determined by the projected range of bombarding particles. The projected range of ions with the above energies in solids is only from a few tens to several hundreds of nanometers.

Changes in the chemical composition and structure (in the area of ion penetration and its immediate vicinity) are associated with ion implantation, selective ion sputtering of components from the target, the formation of radiation defects, the static stresses arisen from implanted ions, phase formation and dissolution due to radiation-enhanced diffusion, rapid quenching of the target material in the regions of thermal spikes (resulting from the evolution of cascades of atomic displacements), and others [3].

At the same time, a number of unusual effects of the "long-range" nature was revealed (at depth up to several hundreds micrometers and more) under ion irradiation [4-9, 11], which has not yet received a full explanation. Effects increasing the impact depth from a few tens projected ranges (R_p) to huge quantities of the order of 10^3-10^5 R_p can be divided into the following groups:

(1) effects related to the peculiarities of motion equations of accelerated ions in crystals (such as channeling and focused substitution), which provide an insignificant increase in the impact depth for heavy ions with low and medium energies to ~5-10 R_p [10, 11];

(2) effects caused by radiation or both radiation and thermally stimulated processes (radiation-enhanced diffusion and tribodiffusion) at a depth to several tens R_p [10, 11];

(3) effects in static stress fields from implanted impurities under *high-dose* irradiation: martensitic transformations at a depth of several R_p [12], generation and movement of dislocations deep into material by several tens of micrometers [4, 5];

(4) radiation-dynamic (RD) effects, which are related with the fact that the cascades of atomic displacements are nanoregions of explosive energy release, the temperature in which can achieve 5000-6000 K and above for 10^{-12} s the rapid expansion of these "thermal spikes" [13] results in the emission of post-cascade shock waves, which initiate the rearrangement of metastable media [7-9, 11, 14] at *lower fluences* (beginning from 10^{14} cm^{-2}) and at the depth of up to 10^3-10^5 R_p.

To understand the nature and to construct the reliable theory for describing RD-processes, they should be thoroughly experimentally investigated. In this case it is necessary to separate thermal from radiation induced long-range effects.

Based on the above, we investigated the regulations of the combined influence of Ar^+ *ion (E=20 keV)* and *light* beams on the *metastable* (quenched from 900°C) Fe-8.25 at % Mn alloy. In the course of the experiment, the *heating rate* and the *steady-state* temperature of the target were fixed at a given level due to the replacement of the ion beam power density P_S (W/cm^2) from 0 to 100% by the light irradiation. As a result, the temperature and the rate of thermal processes was managed to fix on the preassigned level; the energy (E), ion current density (j), and fluence (F) of argon ions could independently be varied in a wide range. Note that according to the equilibrium phase diagram, the Fe-8.25 at % Mn quenched alloy is supersaturated solid solution of manganese in iron, which is metastable.

Experimental

The Fe-8.25 at % Mn alloy under study according to chemical analysis had the following content of impurities: 0.008 C, 0.003 N, 0.007 S, and 0.010 P wt %. The alloy was melted in an induction furnace in the argon atmosphere. The ingots were forged to obtain rods 6×12 mm in cross section and rolled at 1000°C to obtain sheets 0.5 mm thick. Pretreatment consisted of normalizing (900°C), annealing (600°C), and quenching from 850°C (from γ region) in a 15% solution of NaCl. After quenching the alloy has a martensitic structure (100% bcc α'- martensite). Thin foils for Mössbauer studies in transmission geometry were mechanically and electrolytically polished in electrolyte (12 g Cr_2O_3 and 100 mL H_3PO_4). The thickness of the samples was 25 μm.

The samples were irradiated in a continuous mode in the ion implanter equipped with a PEP-18 ion source based on a low-pressure glow discharge with a hollow cold cathode [15]. The power density of the Ar^+ ion beam was 1-2 W/cm^2 at its energy in the range 10-20 keV and current densities from 15 to 75 μA/cm^2.

We used a visible light source with a comparable power (in the range 0-3 W/cm^2) in the form of an incandescent electric lamp with a reflector forming a parallel beam of light.

The sample temperature was controlled with the help of thin chromel-alumel thermocouples connected with an Advantech Adam 4000 automated system for digital signal registration.

The Mössbauer study was performed using an SM-2201 Mössbauer spectrometer in the constant acceleration mode. Isotope ^{57}Co in Pd was served as the source of γ-quanta. X-ray structural analysis was carried out using a DRON-4M diffractometer. All measurements were done at the room temperature.

Details of exposure, measurements and results

The quenched alloy samples in the form of foils 25 μm thick were irradiated with combined Ar^+ ion and light beams in different modes (Table 1). They were hanged in an implanter's vacuum chamber ($P \sim 5 \cdot 10^{-5}$ mm Hg) on thin threads with a low heat conductivity so that heat sink (removal) from the samples was carried out almost by the heat radiation only.

In this case, the kinetic curve of the target heated by energy currents has two asymptotes [16] corresponding to the energy conservation principle at $t=0$ (linear growth of temperature T in the absence of thermal radiation) and Stefan-Boltzmann law at $t \to \infty$ in a stationary mode, when $T=const$.

The study of the regularities of initiating and proceeding of RD processes is complicated by the fact that the change in energy E and ion current density j inevitably leads to a change in the target temperature. For example, a change in the ion current density from 15 to 40 μA/cm^2 at the ion energy E = 20 keV leads to a change in the stationary temperature of the foils from about 250 to 380°C[1]. It is very difficult to separate the contribution of RD effects to j and E from the accompanying thermal effects.

In this regard using an additional source of energy, namely, the light beam, the power density introduced into the sample was fixed:

$$P_S = P_S^i + P_S^l = \frac{1}{e} jE + P_S^l ,$$

at the level P_S = 1.5 W/cm^2; where indices i and l mean the ion beam and the light beam, respectively; e is the singly-charged ion charge (e = 1,6·10^{-19} C).

Table 1. Modes of sample processing (T = const = 480°C)

No.	Type of exposure	Parameters of irradiation		
		j, μA/cm^2	E, keV	Fluence F, cm^{-2}
1	Ar^+ ion irradiation	75	20	10^{16}
2	Ar^+ ion irradiation	75	20	$3 \cdot 10^{17}$
3	Light irradiation	-	-	–
4	Combined irradiation	40	20	10^{16}
5	Combined irradiation	15	20	10^{16}
6	Combined irradiation	75	15	10^{16}
7	Combined irradiation	75	10	10^{16}

The table shows that, every time, any of the parameters j, E, or fluence (F) took two different values when two other parameters were constant. When varying j and E (from maximum to lower values), the respective part of the power that was caught by the target was compensated by a visible light beam.

XRD data suggest that α (bcc) → γ (fcc) phase transformation takes place (at a projected range of penetrating ions of only two-three tens no more nanometers) only at ion current density j = 75 μA/cm^2, E = 20 keV in the entire volume of the Fe-8.25%Mn alloy specimens for 20 s and 10 min. In this case, there was no a compensating light beam.

XRD patterns taken from the irradiated and unirradiated sides of the foils include additional weak lines typical of a fcc phase in addition to the lines corresponding to bcc lattice. The most intense of which is the line (220). This shows the long-range nature of the ion bombardment effects on the structure of the Fe-8.25 at % Mn alloy, which initiates the accelerated (already within 20 s of

[1] Stationary temperatures for various metals are different from each other only due to different emissivity α of the metals (in our case $\alpha \approx$ 0.40-0.42).

exposure) γ-phase formation. An increase in the exposure time to 10 min has almost no influence on the diffraction patterns.

It was established that no α (bcc) $\rightarrow$ γ (fcc) phase transition occurred at all lower ion current densities in the presence of the compensating power introduced into the sample by the light beam.

XRD patterns of the samples subjected to the combined exposure are the same that those of the initial sample. They only exhibit the lines of the α (bcc) solid solution. However, this does not exclude the possible intraphase processes of atom redistribution preceding the γ(fcc) phase appearance.

Since the atomic (electron) scattering amplitudes of manganese and iron are very close, the efficiency of XRD methods for the analysis of short-range atomic order in the present case is extremely low. In this regard, the Mössbauer effect was used to investigate the possible redistribution of Mn and Fe atoms in the solid solution and to confirm the fact of the $\alpha \rightarrow \gamma$ structural phase transition at a maximum current density.

Figure 1 demonstrates the Mössbauer spectra of the Fe-8.25 at % Mn alloy (a) in the initial state, after exposure to (b) a light beam (in the absence of ion irradiation), and (c) ion beam irradiation.

The presence of intense paramagnetic line in the central part of the spectra of the samples irradiated with Ar^+ ions (j = 75 μA/cm^2, E = 20 keV) at 480°C for 20 s (Figure 1c) and 10 min confirms the $\alpha \rightarrow \gamma$ (bcc $\rightarrow$ fcc) phase transformation in them with the formation of approximately the same amount of the γ-phase under irradiation in both cases.

Assuming the small difference in the probabilities of the Mössbauer effect for α and γ phases, the areas of respective subspectra were used to find the amount of the precipitated γ phase ~ 8% (in both cases: t = 20 s and 10 min).

Both Mössbauer study and XRD analysis indicate that the samples *heated with a light beam* and held at a temperature of 480°C (in the absence of ion irradiation, Figure 1b) for 10 min and those *subjected to combined exposure* at the same temperature do not exhibit the $\alpha \rightarrow \gamma$ phase transition (they contain only traces of a γ-phase, i.e., retained austenite).

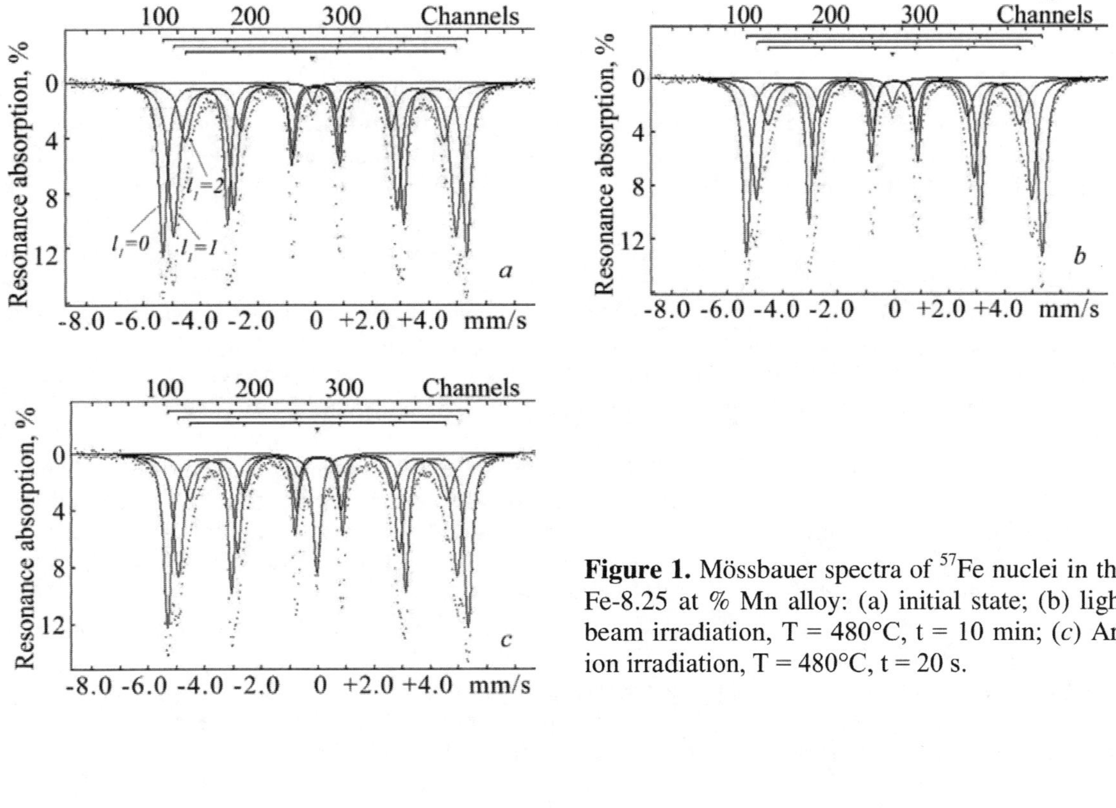

Figure 1. Mössbauer spectra of ^{57}Fe nuclei in the Fe-8.25 at % Mn alloy: (a) initial state; (b) light beam irradiation, T = 480°C, t = 10 min; (*c*) Ar^+ ion irradiation, T = 480°C, t = 20 s.

IOP Conf. Series: Materials Science and Engineering **110** (2016) 012027 doi:10.1088/1757-899X/110/1/012027

Although the intensity of the paramagnetic line does not increase after light heating of the samples (in the absence of an ion beam) as well as after different modes of the combined exposure (Table 1), nevertheless, Mössbauer subspectra intensities significantly change. Lines of the subspectra corresponding to $-1/2 \rightarrow -3/2$ nuclear transitions, as components of the outer peak of the experimental spectrum of initial sample are shown in Figure 1a.

The relative intensity of the subspectra corresponding to the Fe atoms, in the first coordination sphere of which there is no Mn atoms (i.e., the number of manganese atoms in the first coordination sphere of iron $l_1 = 0$), increases, as compared with that for the quenched state (Figures 1b, 1c).

The intensities of the components which correspond to the presence of one or two manganese atoms ($l_1 = 1, 2$) near a Fe atom, on the contrary, decrease. The observed changes in the intensities of the subspectra, which are proportional to the respective probabilities $P(l_1)$ showing the presence of l_1 Mn atoms near the Fe atom, indicate the solid solution short-range separation into regions enriched and depleted in manganese under combined irradiation (for 20 s and 10 min) and under sufficiently long (10 min), purely electromagnetic radiation.

As a result of fitting of the Mössbauer spectra with the superposition of Zeeman sextets of lines (corresponding to $l_l = 0$, 1 and 3), we obtained an average effective magnetic field $\langle H \rangle$ at ^{57}Fe nuclei for various modes of the combined irradiation. Considering the linear relation between $\langle H \rangle$ and Cowley' short-range order parameters on the nearest coordination spheres α_i [17], and taking into account that the influence of the second, third, and more remote spheres for the considered Mn concentration is insignificant [18], short-range atomic order parameters α_1 on the first coordination sphere were calculated.

Parameter α_1 is zero for the initial state and the state after a brief (20 s) light exposure. The value of α_1 becomes positive (short-range decomposition) after ion irradiation, all modes of combined irradiation, and also after prolonged (5 and 10 min) light irradiation ($\alpha_l = 0.05$ and 0.16). The maximum value of α_l for all kinds of radiation used is + 0.16 –+0.17 with an error ± 0.02. To achieve this degree of short-order separation, 20 s is enough upon argon ion irradiation and all modes of combined exposure (a further exposure up to 10 min change nothing). Upon light irradiation, in the absence of ion bombardment, this requires at least 10 min. The latter indicates that the combined irradiation (when there is a certain fraction of the power of the ion beam) leads to a substantial increase in the rate of short range order formation in the sample volume.

Note that the Mössbauer effect in transmission geometry gives information about the whole volume of the sample (foil 25 μm thick in our case), since a fraction of resonant nuclei, which are located in the area of ion path (no more than 20-30 nm) is less than 0.1% and does not provide a significant contribution to the spectrum.

Conclusions

It was established that under thermal exposure modes being the same, $\alpha \rightarrow \gamma$ (bcc→fcc) phase transformation took place over the entire depth of 25-μm-thick target of the Fe-8.25 at %Mn alloy *only at maximum* ion energy $E = 20$ keV and ion current density $j = 75$ μA/cm^2 under irradiation for 20 s and 10 min. This means that only those parameters of the Ar^+ ion beams exceed a certain threshold values (E^* and j^*).

This result is principally important. The fact that this is the *only single set* of parameters at which the transformation takes place is accidental to a certain extent, since the range of variation and the specific values E, j and F have been chosen arbitrarily. In addition, it was found that the exposure to argon ion beam repeatedly (in dozens of times) accelerates the process of short-order separation of the alloy. It can be argued based on these results that the lower threshold of the ion current density, which provides the multiple acceleration of intraphase separation j^{**}, does not exceed 15 μA/cm^2 and the lower energy threshold E^{**} does not exceed 10 keV.

According to the obtained data, the fluence threshold causing phase and intraphase processes does not exceed 10^{16} cm^{-2}. In general, the question about threshold values of the considered parameters (including their interdependence) is difficult enough and a further research is needed to clarify it.

Under the same heating modes for all samples, the obtained data are unambiguous evidence of the existence of a nonthermal component in the ion beam effect, which initiates the $\alpha \rightarrow \gamma$ phase transformation over the entire depth of the target (about 10^3 R_p) and accelerates the intraphase processes foregoing to them.

With allowance for the fact that: (1) all radiation defects form under ion bombardment in the ion path zone (i.e. in the layer with a thickness below than 0.03 μm); (2) the concentration of interstitial atoms and vacancies survived in atomic displacement cascades does not exceed 0.01 % of the total number of atoms in the target; (3) diffusion length of vacancies does not exceed 1 μm for 20 s; (4) there are a large number of traps and sinks for point defects in the initial quenched alloy; it allows us to draw the conclusion that the steep increase in the rate of structural-phase transformations under ion beam exposure can hardly be explained by only radiation-enhanced diffusion mechanism. Is excluded also the role of thermally-activated process, because the temperature was the same for the combined beams and a light beam.

With that said, the observed transformations can be considered as the additional evidence in favor of the radiation-dynamic mechanism [7-9, 11, 14, 19] of accelerated ions effects on the substances.

Acknowledgments
This work was supported by the Russian Scientific Foundation, project no. 15-19-10054.

References

[1] Proceedings of 12th International Conference on Modification of Materials with Particle Beams and Plasma Flows. 2014 (Tomsk. September 21-26)
[2] Nastasi M and Mayer J W 2006 *Ion Implantation and Synthesis of Materials* (Springer Series in Materials Science)
[3] Ovchinnikov V V 1996 *Izvestiya RAN, Metally* **6** 104
[4] Didenko A N, Kozlov E V, Sharkeev Yu N, Tailashev A S, Rajabchikov A I, Pranjavichus L and Augulis L 1993 *Surface and Coatings Technology* **56** 97
[5] Sharkeev Yu P and Kozlov E V 2002 *Surface and Coatings Technology* **158-159** 219
[6] Tetelbaum D I, Azov A Yu, Kurilchik E V et al 2003 *Vacuum* **70** 169
[7] Borodin S N, Kreindel Yu E, Mesyats G A and Ovchinnikov V V 1989 *Technical Physics Letters* **15** 13 87
[8] Kreindel Yu E and Ovchinnikov V V 1990 *Vacuum* **42** 81
[9] Ovchinnikov V V, Chernoborodov V I and Ignatenko Yu G 1995 *Nucl.Instrum. and Meth. in Phys.Res. B* **103** 313
[10] Ryssel H and Ruge I *Ionenimplantation* ((Chichester: Wiley, 1986)
[11] Goloborodsky B Yu, Ovchinnikov V V and Semenkin V A 2001 *Fusion Technology* **39** 1217
[12] Johnson E, Johansen A, Sharlot-Kristesen L. et. al. 1967 *Nucl. Instr. and Meth. in Phys. Res. B* **19-20** 171
[13] Ovchinnikov V V, Makhin'ko F F, Solomonov V I, Gushchina N V and Kaigorodova O A 2012 *Technical Physics Letters* **38** 86
[14] Ovchinnikov V V 2008 *Physics-Uspekhi* **51** 955
[15] Gavrilov N V, Mesyats G A, Nikulin S P, Radkovskii G V, Eklind A and Perry A J 1996 *J. Vac. Sci. Technol. A* **14** 1050
[16] Ovchinnikov V V, Ovchinnikov S V, Makhin'ko F F and Povzner A A 2011 *Izv. Vyssh. Uchebn. Zaved. Fiz.* **54** 1/3 101
[17] Ovchinnikov V V 2006 *Mossbauer Analysis of the Atomic and Magnetic Structure of Alloys* (Cambridge Int. Sci. Publ. Cambridge. UK)
[18] Litvinov V S, Ovchinnikov V V, Dovgopol S P and Karakishev S D 1979 *Fiz. Met. Metalloved.* **47** 1 96
[19] Ovchinnikov V V 1994 in: Proc. XVI International Symposium on Discharges and Electrical Insulation in Vacuum **2259** ed. Mesyats G A (Moscow-St. Petersburg. SPIE) pp. 605-608

Tungsten nanoparticles influence on radiation protection properties of polymers

V M Gavrish, G A Baranov, T V Chayka, N M Derbasova, A V Lvov and Y M Matsuk

Laboratory of biotechnologies and environmental monitoring, Sevastopol State University, 33 Universitetskaya str., 299053, Sevastopol, Russian Federation

E-mail: 5brachman5@gmail.com, julia.matsuk@gmail.com, lvov@sevsu.ru

Abstract. In the presented article the results of the study of metal-polymer composites based on the ultra-high molecular weight polyethylene GUR 4122 with the addition of superdispersed tungsten nanopowders with 5, 10, 20, 40, and 50 mass percent content levels are given, their thermophysical, radiation-shielding, and mechanical properties are shown, and the influence of content levels of tungsten superdispersed nanopowders on these properties is analyzed. The conducted studies have shown the increase in the listed properties depending on the content level of tungsten superdispersed and nanopowders in the ultra-high molecular weight polyethylene GUR 4122. Owing to their properties, the obtained materials may be used in various fields, such as aviation, space technologies, mechanical engineering, etc.

1. Introduction

In recent years, in almost every field of industry, there has been a tendency to substitute products made of metal for parts, structures, and coatings made of polymers [1]. As a result, stricter operational requirementsв are imposed to polymers. One way to improve the quality of polymeric materials is to use various micron and submicron-sized fillers, such as aluminum oxide ($A1_2O_3$) [2], copper (Cu) [3], alumina [4], titanium dioxide (TiO_2) [5], graphite, molybdenum disulfide (MoS_2) [6], etc. Modification of polymers is accompanied by an increase in their physical and mechanical properties which permit to largely control the structure and properties of materials due to nucleation and orientation effects, a change in conformation of macromolecules, their chemical binding with the surface of nanosized particles, and the healing of structural defects [7].

The most promising polymer binder for creating composites is ultra-high molecular weight polyethylene (UHMW PE), which is the cheapest most affordable binder with a unique set of physical and mechanical properties, such as a strong wear resistance, stability in the corrosive environment, high impact elasticity, etc. [8]. Nevertheless, for the time being, polymer composites have not been widely adopted as radiation-resistant and radio-protective materials.

Traditionally, lead and barium are used to solve radiation shielding tasks. These materials are toxic and subject to quick ageing, that is why the problem of search for alternative radio-protective materials is of topical interest. Global trends in practical solution of the tasks of radiation shielding are focused on the development of effective, light, and economically rational materials. Therefore, UHMW PE represents an attractive material intended for an effective attenuation of particle radiation [2]. Besides, its use as a filler in heavy metal composites or their compounds lends it radio-protective (RP) properties [9]. To impart radio-protective properties, papers [10, 11] suggest the use of tungsten, boron carbide, and tungsten carbide nanopowders as a UHMW PE. However, a widespread use of these nanofillers is restrained by a rather high cost of nanopowders.

2. Objects and Methods of Research

The object of research was the following: a metal-polymer composite, based on ultra-high molecular weight polyethylene with the addition of the tungsten powder obtained as a result of processing hard-alloy production wastes employing the biodegradation method, based on the destruction of the alloy matrix with specially prepared solutions using the adapted association of microorganisms [12]. During the microbiological processing, binding components of the tungsten-nickel-iron alloy (Fe, Ni) oxidize and go to a liquid phase; the alloy matrix is destroyed; as a result, mechanical properties of tungsten change, it becomes cold-short, which leads to a faster crushing of raw materials. Tungsten represents the major elemental composition of the powder, over 96.7 % (table 1); its appearance is shown in figure 1.

Table 1. The elemental composition of the tungsten powder.

Element	W	Fe	Co	Ti	Cr
Concentration, %	96.943	1.297	1.016	0.525	0.219

Figure 1. A picture of the tungsten powder obtained through processing hard-alloy wastes using the biodegradation method.

It is established that the size of particles of the powder W lie within the range of 0.19 … 2.344 microns, while 90% of them are smaller than 0.9 microns The bulk density is 19.2 g/cm^3. The specific surface is 4.2 m^2/g.

The study of the influence of the tungsten powder on thermo-physical, mechanical, and radio-protective properties of the composite was conducted based on ultra-high molecular weight polyethylene GUR 4122 (Ticona GmbH); its main characteristics are given in Table 2 [13].

Table 2. Properties of ultra-high molecular weight polyethylene.

Parameters	Density, (g/cm^3)	Thermal linear expansion coefficient at 296...353 K	Thermal conductivit y at 296 K (W/(m·K))	Specific heat capacity at 296 K, (kJ/kg·K)	Tensile yield stress, (MPa)	Tensile modulus, (MPa)
	0.93	$\approx 2\cdot10^{-4}$	0.41	1.84	≥ 17	790
Standard	ISO 1183	ISO 11359, part 1-2	-	Adiabatic calorimeter	ISO 527	ISO 3167

The uniform filler distribution was done in the rotating electromagnetic field by means of non-equilibrium ferromagnetic particles. The resulting compositions were processed into samples using the method of compression molding ($T = 433$ K, $p = 20$ MPa). As a result, six samples of the same size (d – 50 mm, h – 10 mm), containing 0%, 5%, 10%, 20%, 40%, and 50% of tungsten powder respectively, were obtained.

The heat conductivity coefficient (λ, W/m·K) was determined using devices ИТ-λ-400 in accordance with State Standard 25630.2 at temperatures 173 … 473 K.

The heat diffusivity coefficient of the materials was calculated using the equation (1):

$$\alpha = \frac{\lambda}{C_P \cdot \rho} \tag{1}$$

where α is the heat diffusivity coefficient, m^2/s; λ – heat conductivity coefficient, W/(m·K);
Cp – specific heat capacity, kJ/Kg·K; ρ – density of the sample, kg/m^3.

The compressive yield stress was determined in accordance with State Standard 4651-78 using the testing machine FP – 100. To determine the tensile modulus (E), a diagram was used to estimate the values of loads, which correspond to the values of relative deformation of 0.1% and 0.3% (State Standard 9550-81). Calculations were carried out using the equation (2):

$$E = \frac{(F_2 - F_1)}{A_0(\Delta h_2 - \Delta h_1)} \tag{2}$$

where F_1 and F_2 are loads corresponding to the relative deformation of 0.1% and 0.3 % respectively, H;
h_0 is the initial height of the sample, mm;
A_0 – area of the transverse sample, mm;
Δh_1 and Δh_2 – changes in the height, corresponding to loads F_1 and F_2.

To investigate radiation-shielding properties of the metal-polymer composite, based on UHMW PE with the addition of the tungsten powder, the resulting samples were exposed to gamma radiation; the installation flow chart is shown in figure 2.

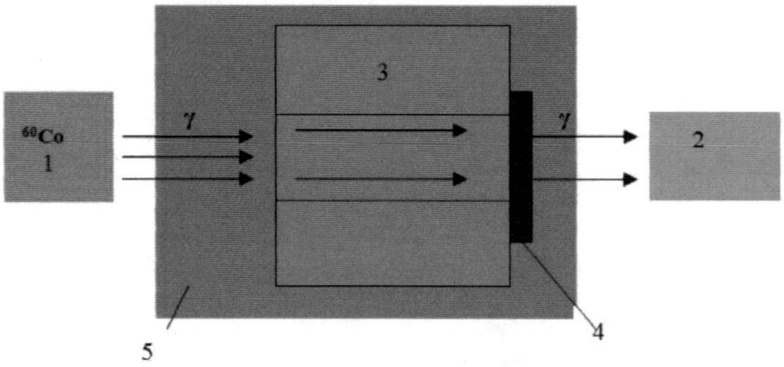

Figure 2. The installation flow chart: 1 – the source of gamma radiation ^{60}Co (E = 1252 keV); 2 – dosimeter-radiometer DKC-96; 3 – the collimator; 4 – the sample; 5 – lead housing.

Passing through the collimator *(3)*, γ–radiation from the cobalt source *(1)*^{60}Co (E = 1252 keV) forms a narrow directed beam of γ–quanta (figure 1). The samples of ultra-high molecular weight polyethylene are placed in turn after the collimator *(4)*. The dosimeter-radiometer DKC-96 was used as a detector of the γ–quanta dose rate. The γ–quanta dose rate (without the samples) equaled 701.3 mSv/h.

3. Results and Discussion

The modification of the ultra-high molecular weight polyethylene using the tungsten powder results in the change of its thermo-physical properties (the coefficient of heat diffusivity, thermal conductivity) figure 3.

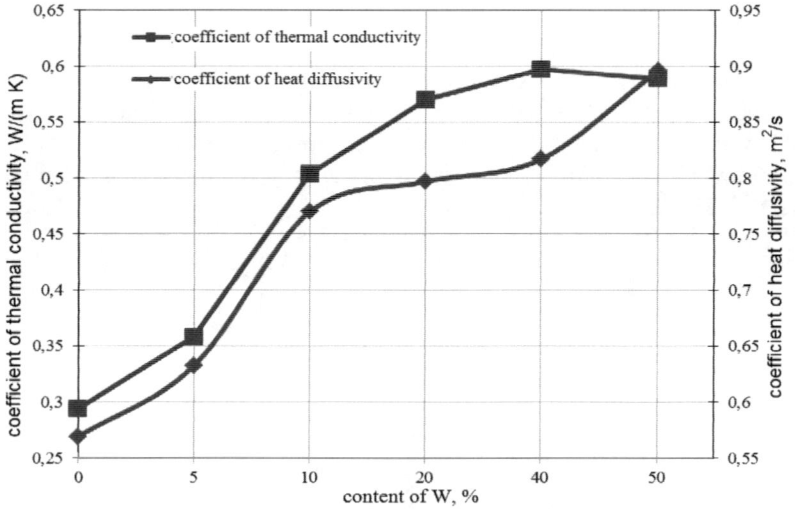

Figure 3. A change of thermo-physical properties of the metal-polymer composite depending on the tungsten powder content.

It has been established that, as a result of the modification of the polymer, as the tungsten powder content increases, so do the thermophysical properties of the polymer. Besides, a surge of rates is seen at 10% modifier content level; in particular, the coefficient of heat conductivity increases by more than 71%, heat diffusivity – by more than 35%.

The study of mechanical properties (yield stress and tensile modulus) of the investigated metal-polymer composites is shown in figure 5.

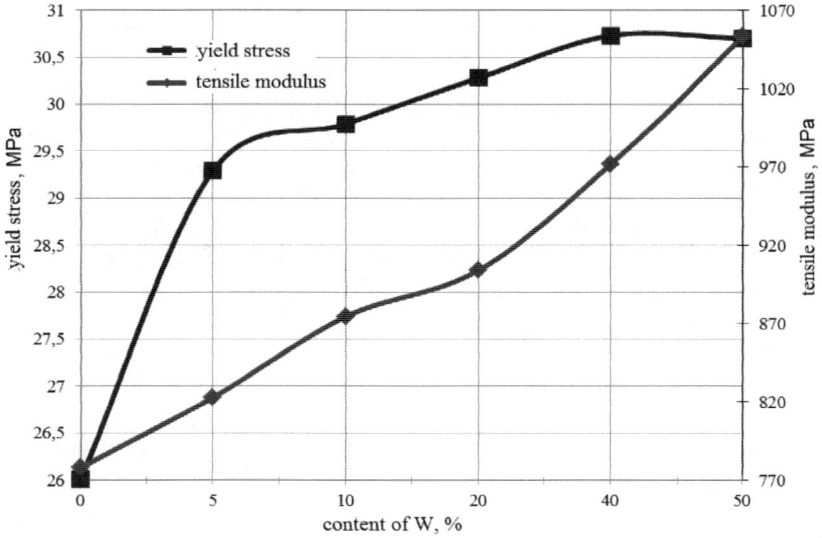

Figure 4. A change of mechanical properties of the metal-polymer composite depending on the tungsten nano-powder content.

With the addition of 5% of tungsten powder to the metal-polymer composite, a leap in the yield stress coefficient by more than 12% is observed; then, as the content level is increased two times, the coefficient is also increasing by 2% on average, with the exception of 50% concentration, where, on the contrary, a decrease in the investigated coefficient by 0.12% is observed. As to the tensile modulus, it is steadily rising over the whole investigated range, reaching its maximum value of 1053.5 MPa at 50% filler content level, which is by more than 35% higher than the value of the initial polymer.

It has been established that the composites based on the ultra-high molecular weight polyethylene, modified with the tungsten powder, excel the base polymer in thermophysical (by more than 11...100%) and mechanical (by more than 5...35%) properties respectively.

When passing gamma quanta through metal-polymer composite samples under study it is obvious that as the tungsten powder content increases up to 10%, the attenuation factor of γ-quanta remains virtually unchanged (figure 5). With further increase in the tungsten powder content, the attenuation factor rises according to the non-linear dependence. The increase in the tungsten content in the sample by 50% results in the growth of the attenuation factor by 5%.

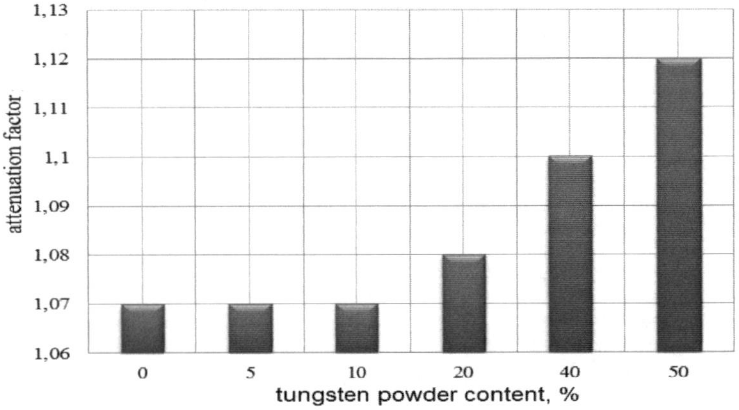

Figure 5. A change of attenuation factor of γ-quanta depending on the tungsten powder content in the metal-polymer composite samples

It is known that γ-quanta, passing through the samples under study, are exposed to three types of interaction, whose probability depends on the material of samples and γ-quanta energy: these are a photoelectric effect, Compton scattering, and the formation of electron-positron pairs.

In case of a photoelectric effect and the formation of electron-positron pairs, primary γ-quanta are absorbed by the protection matter. The formation of pairs is accompanied by the appearance of an electron and a positron; a photoelectric effect results in the emergence of a photoelectron. Thus, during these types of interaction the primary γ-quantum is dropped out of the beam. In case of a Compton effect, γ-quanta are scattered, and their energy decreases, resulting in the appearance of a recoil electron.

In the proposed installation (figure 2) the scattered γ-quanta of the beam are absorbed by the collimator material, and, having experienced the act of scattering, are dropped out of the beam. In case of a narrow beam, these three types of interaction do not result in the emergence of scattered γ-quanta in the vicinity of the detector of the γ-quanta dose rate. Therefore, in order to investigate the dependence of changes in the attenuation factor of γ-quanta on the thickness of materials under study, future work will require that additional research of γ-quanta passing through metal-polymer composite samples of different thickness be conducted. As in practice one generally encounters scattered radiation of γ-quanta ("broad-beam geometry"), it is planned to conduct a series of studies of the attenuation factor of γ-quanta in a broad beam.

4. Conclusion

Thus, as a result of the work performed, new metal-polymer composites based on the ultra-high molecular weight polyethylene and the superdispersed tungsten powder have been created. It has been determined that metal-polymer composites excel the base polymer in thermophysical and mechanical properties 1.1-2 times and by 17-35% respectively. The presence of tungsten particles in the structure of the ultra-high molecular weight polyethylene gives it a set of new properties enhancing its applicability in the parts of movable joints.

5. Acknowledgments

The research was funded by the Ministry of Education and Science of the Russian Federation according to base part of state tasks of the Sevastopol State University №2014/702 (Project 3871).

References

[1] Selyutin G E, Gavrilov Y U, Voskresenskaya E N, Zakharov V A, Nikitin V E and Poluboyarov V A 2010 Composite materials based on ultra high molecular polyethylene: properties. *Application Prospects. Chemistry for Sustainable Development* **18** 301-314

[2] Nambiar S and Yeow J T W 2012 Polymer-composite materials for radiation protection. *ACS Appl. Mater. Interfaces* **4** 5717−26

[3] Panin S V, Kornienko L A, Vannasri S et al. 2010 Comparison of the efficiency of modification of SHMPE by nanofibers (C, Al_2O_3) and nanoparticles (Cu, SiO_2) when obtaining antifriction composites. *Journal of Friction and Wear* **31 (6)** 460–468

[4] Panin S V, Kornienko L A, Sergeev V P, Sonjaitham N and Tchaikina M V 2012 Wear-resistant ultrahigh-molecular-weight polyethylene-based nano- and microcomposites for implants. *Journal of Nanotechnology.* Article ID 729756

[5] Saba N, Tahir P M and Jawaid M 2014 A review on potentiality of nano filler/natural fiber filled polymer hybrid composites. *Polymers* **6** 2247-73

[6] Panin S V, Kornienko L A, Nguen S T, Ivanova L R, Poltaranin M A and Shilko S V 2014 Wear resistance of composites based on ultrahigh molecular weight polyethylene filled with graphite and molybdenum disulfide microparticles. *Journal of Friction and Wear* **35(4)** 290-296

[7] Baranov G A, Burya A I, Gavrish M V, Maronchuk I I, Smirnov S B and Sanikovich D D 2011 The study of the influence of the superdispersed tungsten in the composites based on the ultra-high molecular polyethylene on the attenuation factor of γ-quanta *Collection of studies, SNUNEI.* **4 (40)** 140 – 146 (in Russian)

[8] Kovalevskaia O V, Gordeev Y I and Abkaryan A K 2014 Polymer composite materials based on ultra high molecular weight polyethylene matrix filled by alumina powders. *ADV MAT RES* **1040** 124-129

[9] Ivanov M, Kuznetsov S A, Volkov A E, Terekhin P N, Dmitriev S V, Tcherdyntsev V V, Gorshenkov M V and Boykov A A 2014 Photons transport through ultra-high molecular weight polyethylene based composite containing tungsten and boron carbide fillers. *J. Alloys Compd* **586 (1)** S455–S458

[10] Soylu H M, Lambrecht F Y and Ersöz O A 2015 Gamma radiation shielding efficiency of a new lead-free composite material. *J. Radioanal. Nucl. Chem.* **305 (2)** 529-534

[11] Kaloshkina S D, Tcherdyntseva V V, Gorshenkova M V, Gulbina V N and Kuznetsov S A 2012 Radiation-protective polymer-matrix nanostructured composites. *J. Alloys Compd* **536S** S522–S526

[12] Gavrish V M, Mikhailov V V, Shapovalov V A, Chaika T V and Derbasova N M 2013 Possible applications of nanopowders for the modification of lubricants, polyethylene and alloys. *Collection of studies, SNUNEI* **3 (47)** 49 – 55 (in Russian)

[13] Burya A I 2011 The study of the influence of the superdispersed tungsten on the properties of composites based on the ultra-high molecular weight polyethylene. *The Journal of East-Ukrainian National University named after Dahl. – Lugansk* **8 (162)** 56 – 61 (in Russian)

Development and investigation of silicon converter beta radiation ^{63}Ni isotope

A A Krasnov[1,3], S A Legotin[1], V N Murashev[1], S I Didenko[1], O I Rabinovich[1,3], S Yu Yurchuk[1], Yu K Omelchenko[1], E B Yakimov[1,2] and V V Starkov[2]

[1]National university of science and technology "MISiS" 119049, Leninskiy pr., bl.4, Moscow, Russian Federation
[2]Institute of Microelectronics Technology and High Purity Materials, Russian Academy of Sciences (IMT RAS), 142432 Moscow region., Noginsk, Chernogolovka, Institutskaya street, d. 6, Russian Federation

E-mail: a_krasnov_a@mail.ru, rawork2008@mail.ru

Abstract. In this paper the results of the creation and researching characteristics of, experimental betavoltaic converters (BVC), based on silicon are discussed. It was presented the features of structural and technological performance of planar 2 D- structure of BVC. To study the parameters of the converter stream the beta particles of the radioisotope was simulated by ^{63}Ni electron flux from scanning electron microscope. It was investigated the dependence of the collecting electrons efficiency from the beam energy current-voltage characteristic was measured when irradiated by an electron beam, from which the value of the short-circuit current density equal to 126 nA / cm^2 and the value of the open circuit voltage of 150 mV were obtained. The maximum power density at 70 mV is 9.5 nW / cm^2, and the conversion efficiency is 2.1%. It was presented the results of experimental studies of the current-voltage characteristics of samples by irradiating a film ^{63}Ni. The values of load voltage 111 mV and short circuit current density of 27 nA / cm^2 were obtained. Maximum power density was 1.52 nW / cm^2.

1. Introduction

Batteries that are based on using BVC with ^{63}Ni beta isotope, have a high energy density, long life with very small sizes, moreover, BVC does not require any additional maintenance during operation [1]. The creation of such batteries is extremely promising for use in devices special and household purpose [2, 3].

By the principle of operation to the BVC is similar to solar cells which convert photons into electric energy. The only difference is that the BVC work on the effect of the electron-hole pairs generation in the silicon bulk during the span not photons but beta particles. Subsequent separation of the generated current carriers in the semiconductor, as in the case of solar cells, is carried out by p-n junction or Schottky barrier.

Theoretically, the most effective material for the producing such batteries are semiconductors with a large band gap such as SiC, GaN, and C. The broadband materials have lower values of the induced current, but due to the induced voltage, these structures have a high rate of conversion into the electric

[3] To whom any correspondence should be addressed.

power. Thus in [4, 5] BVC based on SiC, GaN were investigated, the conversion efficiency of these structures did not exceed 1.1%.

From a practical point of view the values of the diffusion length in available crystals of SiC, GaN and C are much smaller than the corresponding values in silicon. In addition, leakage currents that have a significant impact on the value of the induced voltage in silicon can reach values close to the theoretical. In structures based on SiC and GaN it is not observed such that this is not seen and it connected with higher concentration of structural defects in these crystals. Today, the technology of producing silicon structures has a quite high level, which makes it possible to produce large-area elements, consistent inclusion of which will raise the output voltage of the battery.

Energy conversion efficiency of beta decay on the silicon is currently quite low and does not exceed 0.5% [6, 7]. Low efficiency may be due to several reasons, the most studied presently include: charge carrier recombination, inelastic scattering of beta-particles, the absorption of the beta-particles in the layers where there is no generation of electron-hole pairs, etc.

The aim of this work is the development of constructive and technological methods for creating BVC structures with planar structure of the 2-D, as well as the further development of experimental methods for the study of such structures. The paper used as the BVC silicon p-i-n-structure irradiated by electrons flow, which imitated a radioisotope ^{63}Ni with a specific activity of 10 Ci / g. The spectrum of electron irradiation was determined by scanning electron microscope.

Furthermore, it was also performed experimental investigation of characteristics 2D BVC structures were also investigated with the deposited ^{63}Ni on film. ^{63}N had the activity 2.7 mCi and external ionized current 3 nA. The experimental data obtained for 2D-structures can be used for creating 3D-structures that enable to considerably increase the area of the p-n junction and, hence, the efficiency of batteries based on BVC [8-10].

2. Technology for creating silicon beta radiation converter

The technology for manufacturing studied structures was developed taking into account the conducted computer simulations in [11]. As the initial substrate it was used a silicon wafer grown by floating zone melting with a resistivity of 5 Ohm · cm, thickness of 420 microns. Ohmic contact to the reverse side was carried out by diffusion of phosphorus (P), followed by speed up at a temperature (T) of 1000 ^{0}C for 45 min. A second ohmic contact was formed by ion doping of boron (B) with the energy of 60 keV and a dose of 500 μCl / cm^2. Formation of high-quality p-n junction with shallow depth was carried by ion implantation of boron out through a film of silicon oxide (SiO_2) thickness of 20 nm with a minimal energy of 10 keV and a dose of 10 μCl / cm^2. In forming the working area of the plate it was held getter annealing with slow cooling of about 1 ^{0}C / min [12]. Then, for p$^+$ regions it was produced ohmic contacts coated aluminum (Al) with thickness of 1 mm, heating the Al was at T = 475 ^{0}C, the reverse side is not metalized. Custom sample with the crystal working area of 1 cm^2 is shown in Figure 1.

Figure 1. Photo of betavoltaic element.

3. Results and discussion

For the semiconductor converter of beta energy particles parameters optimization into electrical current it is used electron beam scanning electron microscope (SEM) to simulate the beta radiation. This approach allows for a particular semiconductor structure to predict betavoltaic element parameters for any film thickness and activity of ^{63}Ni film. For the method of measuring the induced current is used (IC) [13, 14]. In the method of IC focused electron beam SEM acts as a local source of nonequilibrium electron-hole pairs. The electric field inside the space charge region (SCR) separates the electron-hole pairs, which leads to the appearance of current in the external circuit. Current collected by the collector, is a detectable signal in the method of IC at this junction, which plays the role of a collector can be positioned both parallel and perpendicular to the incident beam. The value of IC quantity is generally determined by the geometry of the sample, as well as the spatial distribution of generation rates and recombination of electron-hole pairs.

For the calculation of induced current in the p-i-n diode caused by the emission of a radioactive isotope ^{63}Ni from a film, it was used the approach proposed in [15, 16]. In these studies it was shown that induced by irradiation of the semiconductor structure current can be presented as

$$I_c = q\int_0^{\infty}\int_{-\infty}^{\infty}\int_{-\infty}^{\infty} g(x,y,z)\Psi(x,y,z)dxdydz \tag{1}$$

where q - electron charge, g (x, y, z) - function describing the electron-hole pairs generation and ψ (x, y, z) - the probability of collecting minority carriers, which is the current collected by the collector from a single charge, placed at the point (x, y, z).

Since in this case the p-n junction area is much greater than its depth, SCR width and thickness of the film BVC ^{63}Ni, the problem becomes one-dimensional, and the current can be calculated as:

$$I_c = q\int_0^{\infty} h(z)\Psi(z)dz \tag{2}$$

The function h (z) can be calculated by Monte Carlo for the electron beam, and for the film thicknesses of radioactive nickel, as well as for any converter material.

In the present study we used a program developed in [17, 18]. Specific activity of ^{63}Ni film was assumed to be 10 Ci / g (89 Ci / cm^3), and the film thickness was varied. When calculating the limiting element parameters the film thickness was 3 microns, as in further its energy radiation practically is not changed due to the absorption of energy in the film. The thickness of the protective coating of the diode is sufficiently small and energy loss in it can be ignored, and that the calculations it can be replaced by a silicon layer, in this layer the generation of charge carriers does not contribute to induced current.

The function ψ (z) depends only on the parameters of the semiconductor structure, and can be obtained [16] by solving an equation with the appropriate boundary conditions.

$$\Delta\Psi(z) - \Psi(z)/L^2(z) = 0 \tag{3}$$

For example, a Schottky barrier boundary conditions: ψ (W) = 1 and $\psi \rightarrow 0$ for $z \rightarrow \infty$. For the p-n junction ψ (z) = 1 on the borders of the SCR in p- and n-regions. It should also be noted that in the p- and n-regions in general, the value of the diffusion length L is different, and you need to add as a boundary condition the recombination velocity on the surface.

Thus, to calculate the induced current it is necessary to know the depth of the p-n junction, the width of the SCR, values L in p- and n-regions and the surface recombination velocity. In the paper, these parameters were determined by fitting the experimentally measured dependence of IC on the beam energy.

Figure 2 shows the efficiency of collecting electrons from the beam energy. The efficiency of collecting electrons was calculated by the formula:

$$f(E_b) = \frac{I_c \cdot E_i}{I_b \cdot E_b \cdot \eta}, \tag{4}$$

where I_c - measured induced current, E_i - the average energy of formation electron-hole pairs (E_i = 3.6 eV for Si), I_b and E_b - current and the energy of the electron probe, respectively, η - the proportion of the beam energy absorbed by the sample (η ~ 0.9 for Si).

As it is shown in Figure 2, it is already at an energy of 5 keV more than half of non-equilibrium carriers is collected.

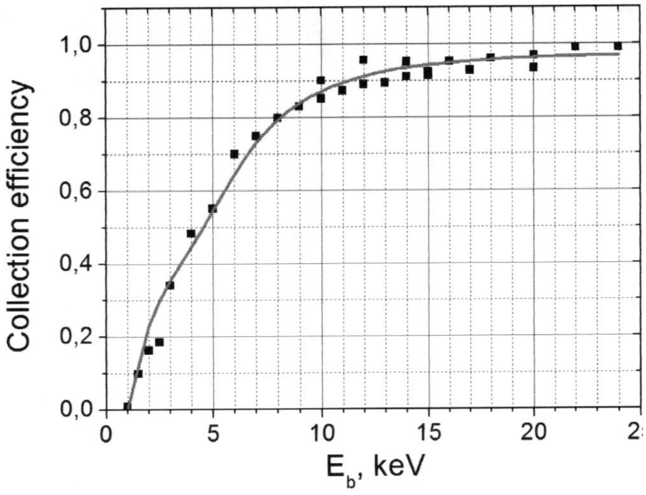

Figure 2. Dependence of the collection efficiency of the energy beam.
Red- measured, black- calculated.

Next it was calculated the probability of collecting ψ (z) and by using the function of generating h (z), the induced current was calculated. For a ^{63}Ni film with thickness of 3 microns calculated induced current in the test structure was 126 nA / cm^2.

Then the studied structure was placed in an electron microscope, and the beam current is chosen such that the induced current took the calculated value. Since the structure of the diffusion length was large enough, almost all generated non-equilibrium charge carriers reach the SCR and contribute to the current by irradiation as electron beam and beta particles. Therefore the choice of the beam energy in this case is not critical. At the same beam current it was measured by the current-voltage characteristics and structural parameters were determined for the energy converter of radiation from the ^{63}Ni film into the electric current. The measured current-voltage characteristic of the electron beam irradiation is shown in Figure 3.

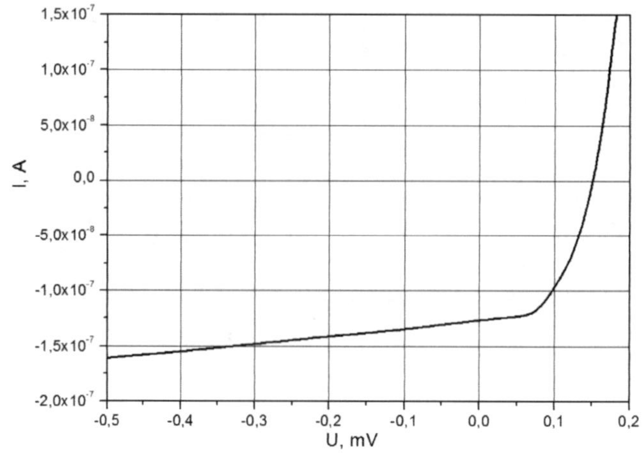

Figure 3. The current-voltage characteristic at an electron beam irradiation.

The following values were obtained after the measurements: a short-circuit current density J_{sc} = 126 nA / cm^2, the open circuit voltage V_{oc} = 0.15 V and a maximum power density P_{max} = 9.5 nW / cm^2. The value of the fill factor is 0.51, a low value of the fill factor is due to the low intensity of the beta radiation. The conversion efficiency under these conditions was 2.1 %.

Thus, if ^{63}Ni films is applied on a produced structure (film thickness 3 microns), and wherein the leakage current remains at the same level the resulting BVC will have the obtained above parameters.

Investigation of beta radiation converter for irradiation it by film ^{63}Ni radioisotope. Changes in the current-voltage characteristics of the structure was carried out on the installation of Agilent 1500B «Keysight» in a light-tight chamber. The results are shown in Figure 4.

To analyze the effectiveness of the structures obtained to use an external source of ^{63}Ni beta particles of RITVETS company. Film source is a nickel foil area of 1 x 1 cm^2 coated with a layer of ^{63}Ni, which is covered with a protective layer of Ni (thickness is about 0,2 microns). The activity of radioactive ^{63}Ni layer is 2.7 mCi / cm^2, the density of ionization current is 3 nA.

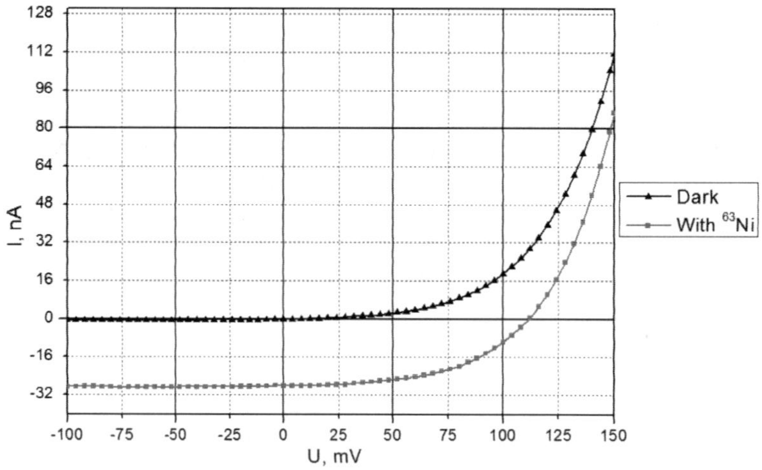

Figure 4. The current-voltage characteristic when irradiated radioisotope ^{63}Ni.

Under an actual irradiation by ^{63}Ni radioisotope following values were obtained: short circuit current density J_{sc} = 27 nA / cm^2, the open circuit voltage V_{oc} = 0.111 V. Low current density is due to the presence of the protective layer of Ni, which absorbs about 30% of energy. Maximum power

P_{max} was 1.52 nW / cm^2 at a voltage of 76 mV. The conversion efficiency under these conditions can only be determined by theoretical calculation, it is about 1.68 %.

4. Conclusions

This paper presents the technology of planar element betavoltaic converter production. For the beta radiation simulation of commercially available ^{63}Ni radioisotope with large activity it was used the method of induced current. The structure effectively collects nonequilibrium charge carriers, e.g. at an energy of 5 keV more than half of electron-hole pairs is collected.

For a ^{63}Ni film of 3 μm thickness calculated induced current density for the investigated structure was 126 nA / cm^2, the open circuit voltage V_{oc} = 0.15 V, maximum power density P_{max} = 9.5 nW / cm^2. At the same time the value of the fill factor is 0.51, a low level of fill factor is due to the low intensity of the beta radiation. The conversion efficiency under these conditions was 2.1 %.

At irradiation by ^{63}Ni radioisotope source following values were obtained: short-circuit current density J_{sc} = 27 nA / cm^2, the open circuit voltage was V_{oc} = 0.111 V, maximum power density P_{max} = 1,52 nW / cm^2 was obtained at a voltage 76 mV. BVC characteristics reduction compared with the theoretical ones and with data obtained by electron beam irradiation received by SEM, is due to losses in the film layers of non-radioactive nickel.

The experimental data can be used at producing a 3D-structure which allows high conversion efficiency.

Acknowledgements:
The current study was supported by the Federal Targeted Program "Research and development on priority directions of scientific-technological complex of Russia for 2014-2020, state contract № 14.575.21.0051 (a unique identifier for applied scientific research (project) RFMEFI57514X0051).

References

[1] Romer M., Miley G.H. and Gimlin., R.J. 2008 *IEEE Transactions on Energy Conversion*. **23**(1) 171.
[2] Reznev A.A., Pustovalov A.A., Maksimov E.M., Peredepiy A.N., and PetrenkoN.S. 2009 *Nano & microsystem techniqe* **3** 14.
[3] Murashev V.N., Mordkovich V.N., Legotin S.A., and Rabinovich O.I., et al., 2014 *J. Nano-Electron. Phys.* **6(4)** 04012.
[4] Xiao-Ying Li, Yong Ren, Xue-Jiao Chen, Da-Yong Qiao and Wei-Zheng Yuan 2011 *J. Radioanal Nucl. Chem.* **287** 173–176.
[5] CHENG Zai-Jun, SAN Hai-Sheng, CHEN Xu-Yuan, LIU Bo and FENG Zhi-Hong. 2011 *CHIN. PHYS. LETT.* **28(7)** 078401.
[6] Ulmen B., Desai P. D., Moghaddam S. and Miley G. H., 2009 J Radioanal Nucl Chem **282**. 601–604.
[7] Guo H. and Lal A., 2003 *IEEE Transducer* **6** 36.
[8] Murashev V.N., Legotin S.A., Didenko S.I., Rabinovich O.I., Krasnov A.A. and Yurchuk S.Yu. 2015 *Advanced Materials Research* **1070-1072** 585-588.
[9] Miley George H. and Lou Nie 31 July - 03 August 2011. 9th Annual International Energy Conversion Engineering Conference, San Diego, California. 356
[10] Chu Jinkui and Xianggao Piao 2009 *J. Micro/Nanolith. MEMS MOEMS* **8** **(2)** 021180.
[11] Murashev V.N., Legotin S.A., Didenko S.I., Yurchuk S.Yu, Nagornov U.S., Krasnov A.A. and Rabinovich O.I. 2015 Materials 5th internet conference. Ul'yanovsk. 188-194.
[12] Rohatgi A. and Rai-Choudhury P. 1980 *J. Electrochem. Soc.* **127 (5)** 1136.
[13] Yakimov E.B. 1992 *Izv RAS,Phys ser..* **56(3**) 31-44.
[14] Yakimov E.B.. 2002 *Zav. Lab.* **68(1)** 63-70.
[15] Possin G.E. and Kirkpatrick C.G., 1979 *J. Appl. Phys.* **50** 4033-4041.
[16] Donolato C. 1985 *Appl. Phys.Lett.* **46** 270-272.
[17] Pavlov V.N., Panchenko V.Ya., Polikarpov M.A, Svintsov E.A. and Yakimov E.B. 2013 *Surface* **9** 46.
[18] Polikarpov M.A. and Yakimov E.B. 2015 *Semiconductors* **49(6)** 763-766.

Properties of Bismuth-Based Superconductors Precursors obtained under the influence of the Radiant Flux

D E Uskenbayev[1], A S Nogay[2] and E B Aynakulov[3]

[1] Senior lecturer, Radio engineering, Electronics and Telecommunications department S.Seifullin Kazakh Agro Technical University, 010011, Astana. Kazakhstan
[2] Professor, Radio engineering, Electronics and Telecommunications department S.Seifullin Kazakh Agro Technical University, 010011, Astana. Kazakhstan
[3] Professor, Radio engineering, Electronics and Telecommunications department S.Seifullin Kazakh Agro Technical University, 010011, Astana. Kazakhstan

E-mail: usdan@mail.ru

Abstract. The results of the study on the synthesis of high-temperature superconductors in the system Bi-Pb-Sr-Ca-Cu-O based on amorphous precursors were considered in this article, and were obtained under the influence of the radiant flux. A study of the elemental composition of amorphous precursors has showed the presence of 20 at.% superstoichiometric oxygen. The increase in the rate of formation of the superconducting phase and the high texture of the particles in the ceramic samples along the plane [00I] were established. The synthesized samples were composed of superconducting phase $Bi_{1,7}Pb_{0,3}Sr_2Ca_2Cu_3O_y$. Research has established a presence in samples of HTSC phase with T_c up to 181K.

1. Introduction

HTSC are based on the system of Bi-Sr-Ca-Cu-O and recognized as one of the most promising, as are characterized by high critical parameters (T_c, J_c), a significantly lower degradation and constant composition. It was established that the HTS compounds are based on this system of homologous series with general formula $Bi_2Sr_2Ca_{n-1}Cu_nO_y$ (n = 1, 2, 3...) [1, 2]. The critical temperature of the superconducting transition (T_g) increases with the number "n" and makes 35K, 90K, 110K, respectively. Also the interest is the possible existence in the system of superconducting phase ($Bi_2Sr_2Ca_{n-1}Cu_nO_y$ (n = 4, 5, 6 ....), on the model calculations, far above the critical temperature of the phase $Bi_2Sr_2Ca_2Cu_3O_y$ (110 K) [3]. From the above HTS practical application are found the compounds having the highest T_c - $Bi_2Sr_2CaCu_2Oy$ and $Bi_2Sr_2Ca_2Cu_3O_y$.

For practical application, in addition to the critical temperature, very important are the critical current (J_c) and critical magnetic field (H_c). If the thin-film high-temperature superconductors have been reached high values of J_c, then the massive high-temperature superconductors (ceramic) to achieve high values of J_c remains as difficult task. Creating ot technology of superconducting ceramics with high and reproducible critical parameters, largely constrained by the lack of reliability of the developed synthesis process, as initial precursors, and, to a lesser extent, the process of getting the target ceramics. The problem is the need of accounting simultaneously influence of factors that lead to the flow of competing reactions, separation of the individual phases, local melting, a breach of the composition due to selective evaporation of components, formation of dendritic structures, and others.

A very common method of solid-phase reaction does not provide a high critical current density due to the formation of the granular structure, which is the main cause of weak Josephson bonds at the grain boundaries.

Melt technology (QMG, MQMG, MC, et al.) can provide texture, eliminating this deficiency. Glass-crystalline technologies, aimed to getting of amorphous precursors are positive addition to the melt methods. The advantage of this method is: the ability to control the grain size, achieving a high texture, high homogeneity, reduced segregation, lightweight molded and getting of high density products of desired shape.

With respect to the glass-crystal technology the using of concentrated light flux as a heating source can bring a number of positive factors [4, 5]. This type of heating eliminates the introduction of impurities from the crucible and technological equipment, can process a given amount of starting material in a high-gradient conditions with high speed to change the conditions for getting the melt and cooling. In addition, the concentrated light output includes UF.

UV, visible and infrared regions of the spectrum that can affect the formation of the fine structure and the critical parameters of the target HTSC ceramics.

The aim is to study of the impact of characteristics of radiant energy on the formation of the superconducting phase, morphology, and the critical temperature (T_c) superconducting ceramic based system Bi-Pb-Sr-Ca-Cu-O.

2. Methods of Experiment

For the synthesis of the original samples of nominal composition mixture $Bi_{1,7}Pb_{0,3}Sr_2Ca_{n1}CunO_\delta$ (n = 2-12) are made from pre-annealed at 873 °K powders of Bi_2O_3, PbO, $SrCO_3$, CaO, CuO, and compressed into rods with size 5x10x150 mm. Synthesis of the melt and subsequent rapid quenching was carried out using as a heating source of concentrated radiant flux (LPC) - solar furnace 3 kW and optical URAN type furnace equipped with a xenon arc lamp power of 10 kW. The hardening of the melt was carried out at a rate of up to 10^6 deg/sec by spraying.

The phase composition was monitored by X-ray diffraction on a DRON-4-M, CuK_α-radiation and Diffractometer Model Rigaku Co, Ltd., Tokyo, Japan. The microstructure was examined on the microscope Neophot and TEM, Model JEM-1200EX11, JEOL Co., Ltd., Tokyo, Japan). The critical temperature was determined by measuring the temperature dependence of the resistivity by a four and the temperature dependence of the magnetic susceptibility of a one-induction method.

3. Results and Discussion

Amorphous precursors were the plates and needles. The plates have a thickness of less than 0.3 mm and needles of up to 10mm and a diameter less than 0.4 mm. Depending on the content of Ca and Cu phase composition of needles and plates changed. If compositions $Bi_{1,7}Pb_{0,3}Sr_2Ca_{n-1}Cu_nO_\delta$ (n=2-5) and the needle plate is an amorphous phase, the compositions for $Bi_{1,7}Pb_{0,3}Sr_2Ca_{n-1}Cu_nO_\delta$ (n = 6-12) with increasing Ca content and Cu tendency of precursors to crystallization was increasing.

The microstructure analysis showed the increase of crystalline inclusions in the samples with increasing content of Ca and Cu. For example, figure 1 shows the microstructure of the samples was $Bi_{1,7}Pb_{0,3}Sr_2Ca_{n-1}Cu_nO_\delta$ (n = 5, 9).

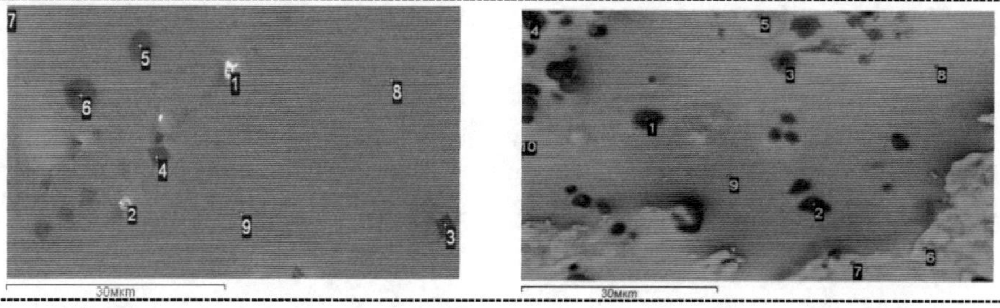

Figure 1. The microstructure of the original plate precursor compositions $Bi_{1,7}Pb_{0,3}Sr_2Ca_4Cu_5O_\delta$ (a) and $Bi_{1,7}Pb_{0,3}Sr_2Ca_8Cu_9O_\delta$ (b), produced by quenching of the melt with using a concentrated radiant flux.

It is particularly important to note the presence in the amorphous precursors to 15-20% excess oxygen. Presumably, it is due to the passage of the melting process in strongly oxidising atmosphere - atmospheric ozone. Under the influence of concentrated radiant flux is ionized oxygen to form ozone. Ozone has increased solubility in the melt than oxygen molecules. This leads to oxidation of polyvalent cations to high valent condition. In research [6] it indicates that in the smelting process is a reduction of oxygen content in the melt and stabilize Cu^+, where the monovalent copper is $R(Cu^+) = Cu^+/(Cu^+ + Cu^{2+}) = 0,7\text{-}0,8$. Perhaps a strong deficiency of oxygen in the dense amorphous precursors reduces the speed of the formation of the superconducting structure that is connected slow diffusion of oxygen into the sample during heat treatment. The oxygen content in the starting precursors that determine the valence conditions of the cations, can affect to the mechanism of superconducting compounds, the rate of their formation, defective of crystal structure and the critical parameters of the target material.

Synthesis of superconducting phase was carried out by heat treatment in the range from 973K to 1138K, with a holding time of 5 hours to 150 hours.

For all compositions superconducting phase $Bi_{1,7}Pb_{0,3}Sr_2CaCu_2O_\delta$ (2212) began to take the shape at 973-1023K. For the composition $Bi_{1,7}Pb_{0,3}Sr_2CaCu_2O_\delta$ temperature of maximum rate of formation of the 2212 phase was lying under 1123-1133K. In this temperature range the single-phase samples were formed 5-10 hours of heat treatment that is 6-7 times greater than the rate of formation of HTSC phase in compareison with solid phase method. Increasing the annealing temperature to partial melting (1133-1138K) and surface treatment with a concentrated radiant flux led to the sealing of the sample and the formation of high-texture of the particles along the plane [00I] (figure 1).

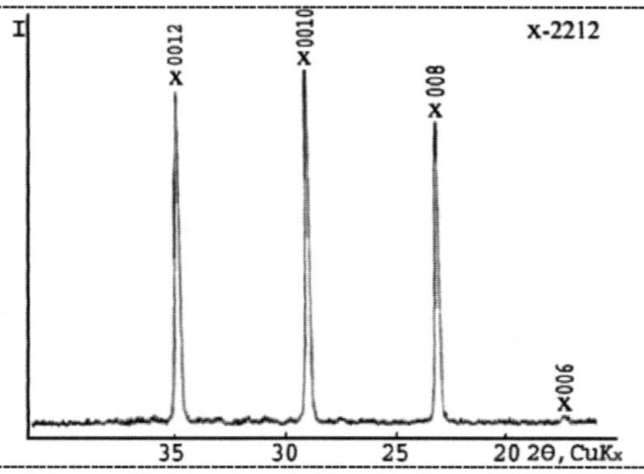

Figure 2. XRD pattern of textured HTSC ceramic samples of nominal composition $Bi_{1,7}Pb_{0,3}Sr_2CaCu_2O_\delta$ 1133-1138K synthesized at the surface and processed concentrated radiant flux.

Samples of the compositions $Bi_{1,7}Pb_{0,3}Sr_2Ca_{n-1}CunO\delta$ (n = 3-12) were synthesized in the range 1115-1125K with a holding time of 60-120 hours. X-ray diffraction study revealed that in the diffractograms of nominal compositions $Bi_{1,7}Pb_{0,3}Sr_2Ca_{n-1}Cu_nO_\delta$ (n = 3-5) were present only reflected ray reflexes related to the superconducting phase in 2223 (figure 3).

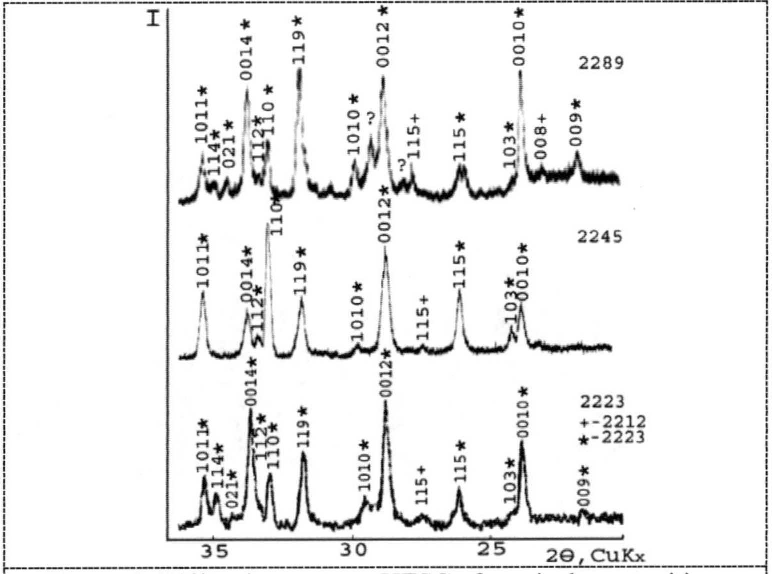

Figure 3. Diffraction patterns HTSC of nominal compositions $Bi_{1,7}Pb_{0,3}Sr_2Ca_{n-1}Cu_nO_\delta$ (n=3, 5, 8), synthesized on the basis of amorphous precursors.

The duration of a complete formation of 2223 phase was 60-80 hours. The increase in heat treatment time up to 150 hours led to the partial collapse of the superconducting 2223 phase with the formation of the 2212 phase, which may be due to tensions at the junction of the crystallites by increasing their size. In samples with a high content of Ca and Cu temperature of formation of the superconducting 2223 phase decreased. If the nominal composition $Bi_{1,7}Pb_{0,3}Sr_2Ca_2Cu_3O_\delta$ optimal synthesis temperature ranged 1123-1125K, then ended $Bi_{1,7}Pb_{0,3}Sr_2Ca_{11}Cu_{12}O\delta$ it was lying in the range 1115-1117K.

Research of the microstructure composition of the HTSC samples showed that the crystals have a lamellar structure. The samples are also set an excessive content of oxygen, as in the amorphous precursors. For example, figure 4 shows the microstructure of samples of HTSC compounds $Bi_{1,7}Pb_{0,3}Sr_2Ca_{n-1}Cu_nO_\delta$ (n = 3, 5).

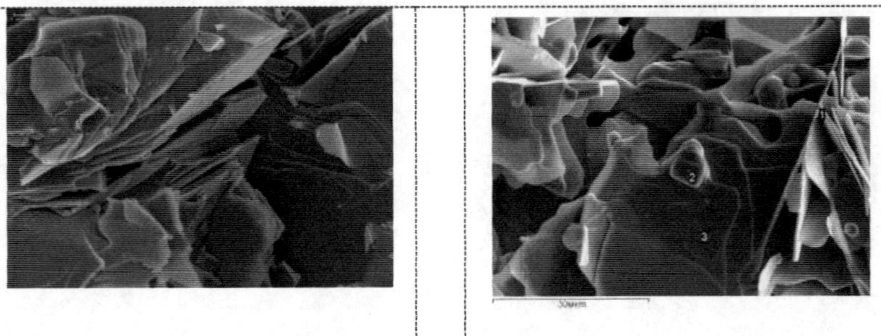

Figure 4. The microstructure of ceramic samples of HTSC compounds $Bi_{1,7}Pb_{0,3}Sr_2Ca_{n-1}Cu_nO_\delta$ (a) and $Bi_{1,7}Pb_{0,3}Sr_2Ca_{n-1}Cu_nO_\delta$ (b), derived from amorphous precursors with using of radiant flux.

For practical application in high technology the main parameter HTS is the critical current, which depends, in particular, from the texture and density of the particles. In order to obtain high texture, sample previously synthesized pattern HTS of composition $Bi_{1,7}Pb_{0,3}Sr_2Ca_{n-1}Cu_nO_\delta$ milled to grain fraction 2-3mkm layers and compressed into tablets with a diameter of 15 mm and a thickness of 1.5 - 2 mm. After the heat treatment at the temperature 1123-1125K, 24 hours in the diffraction pattern appears only intensive reflections of the 2223 phase [00<u>10</u>] [00<u>12</u>] [00<u>14</u>] that can be said to be a high-texture of the particles in the ceramic along the plane [00I] (figure. 5) .

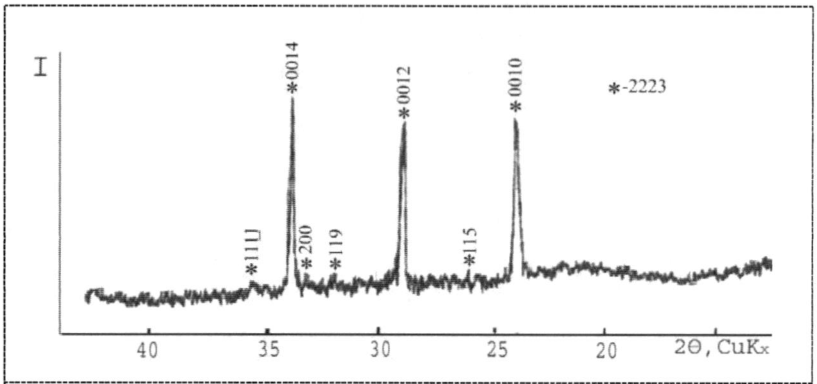

Figure 5. The XRD pattern of textured HTSC samples of nominal composition $Bi_{1.7}Pb_{0.3}Sr_2Ca_2Cu_3O_y$ (2223) obtained with stratified compression.

Studying of critical parameters showed that the samples consisted of 2223 superconducting phase with T_c = 107K. It is important to note that the samples except 2223 phase were presented the system of superconducting phases exceeding 107K (2223). For example, in a sample of nominal composition $Bi_{1,7}Pb_{0,3}Sr_2Ca_4Cu_5O_\delta$ besides the main 2223phase, the superconducting phase system is detected with T_c up to 181K (figure 6).

Formation of superconducting phases in samples of HTSC with T_c in the 107-181K can be explained as follows: firstly, in the process of melting the raw material non-equilibrium condition due to melting of the anisotropic effects of concentrated radiant flux creates a strong gradient conditions in the melt, which can form a region with microscopic structure; secondly, the concentrated radiant flux including a wavelengths of ultraviolet, visible and infrared regions of the spectrum can affect the resonant atoms and change their energy state; thirdly, the excessive dissolution of ozone into the melt under the influence of the concentrated light flux can change valence condition of cations. Together, these factors in rapid quenching of the melt can stabilize the amorphous condition, which includes a certain number of microregions have: a certain energy condition; appropriated on the composition; and prestructure (amorphous or semi-amorphous) of superconducting phases $Bi_{1,7}Pb_{0,3}Sr_2Ca_{n-1}Cu_nO_\delta$ (n = 4, 5, 6). When heat treatment, these microregions can transfer to the superconducting phase with lattice parameters corresponding to compositions $Bi_{1,7}Pb_{0,3}Sr_2Ca_{n-1}Cu_nO_\delta$ (n = 4, 5, 6) having elevated T_c in the range 107-181K.

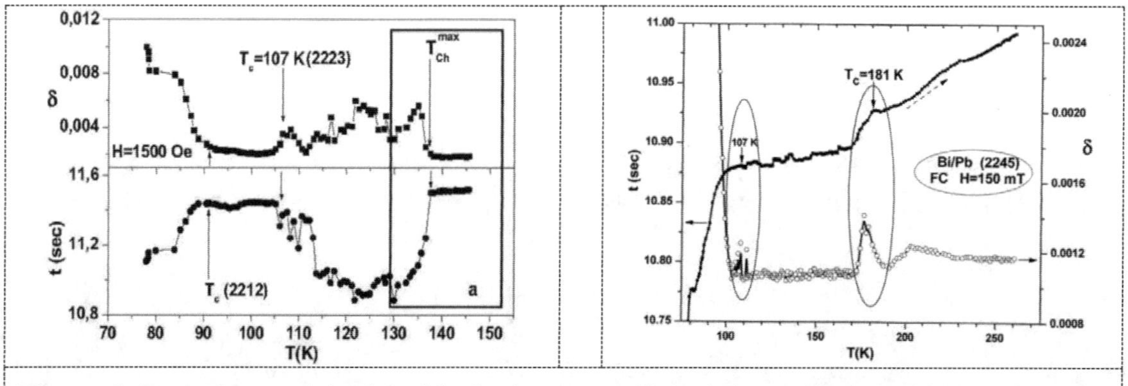

Figure 6. Dependence of the logarithmic decrement (δ) and the oscillation period (t) from the temperature of HTSC samples of nominal composition $Bi_{1.7}Pb_{0.3}Sr_2Ca_2Cu_3O_y$ (a) and $Bi_{1.7}Pb_{0.3}Sr_2Ca_2Cu_3O_y$ (b).

4. Conclusion

On the basis of amorphous and crystalline glass precursors obtained by quenching the melt under the influence of concentrated radiant flux were synthesized HTSC ceramic nominal composition $Bi_{1,7}Pb_{0,3}Sr_2Ca_{n-1}Cu_nO_\delta$ (n = 2-12). For all samples of n = 3 - 12, the main phase was the superconducting 2223 phase. The increase of the rate of formation of HTSC phases was established, which may be associated with the stabilization of highvalent variable valence condition of cations and metastable condition of amorphous precursors. In the samples, except for the main superconducting 2223 phase (110K) were found the systems of HTSC phase having a higher T_c, up to 181K. The high texture of the particles along the plane [00l] is established in the samples.

References

[1] Tretyakov U D, Os'kina T E and Putlyaev V I 1990 J. of Inorganic Chemistry B **35** 1635-1644
[2] Tarascon J M, McKinnon W R, Barboux P, Hwang D M, Bagley B G, Greene L H and Hull G W 1989 Phys. Rev. B **38** 8885-8892
[3] Eab Chai-Hok, Tang I-Ming 1989 Phys. Lett. A B **134** 253
[4] Gulamova D D and Uskenbaev D E 2006 Applied Solar Energy B **42** 40-43
[5] Gulamova D D, Uskenbaev D E, Turdiev Z S, Chigvinadze D G and Magradze O V 2009 J. Applied Solar Energy B **45** 196-200
[6] Yoshihiro Abe 1997 Superconducting Glass-Ceramics in BiSrCaCuO Nogayo 79-94

Study of nanostructured (Ti-Zr-Nb)N coatings' physical-mechanical properties obtained by vacuum arc evaporation

S V Plotnikov[1], A D Pogrebnjak[2], L N Yerokhina[1], D K Yeskermessov[1], N K Yerdybaeva[1]

[1] *Ust-Kamenogorsk, 070003, Kazakhstan,* D. Serikbayev East Kazakhstan State Technical University, 23/2-54 Serikbayev Street
[2] *Sumy, 40000, Ukraine,* Sumy State University, Rimsky-Korsakov Street, 2

E-mail: didar.eskermesov@mail.ru

Abstract.

The coatings were formed by vacuum arc deposition. Unit cast target (cathodes) was used on the basis of 30 atm. % Ti, 35 at. % Zr and 35 atm. % Nb as the vaporized materials. Molecular nitrogen was used as the working gas. The thickness of the coatings in the experiments was 4.0 microns. The surface morphology fractograph fracture, track friction were investigated in a scanning electron microscope JSM-6390 LV. The use of multicoatings based on carbides is very promising to ensure the high performance properties of the complex, nitrides and silicides of transition metals.
Findings – nanostructured coating of (Ti-Zr-Nb) N was obtained by vacuum arc evaporation cathode-cast in a nitrogen gas reaction medium. Multicomponent films have a pronounced columnar structure. Elemental composition was obtained by the vacuum arc deposition of coatings (Ti-Zr-Nb) N, depending on the physical parameters of the deposition process, in particular the pressure of the reaction gas nitrogen.

1. Introduction

During the process of work the surface layer of machine elements and mechanisms are subjected to strong mechanical, heat and chemical influence. The loss of working functionality in most cases is the result of surface wear, erosion, corrosion, etc. Significant resource to increase the working operating capacity may be referred to material of which the details having been made. Due to this reason modern machine building pays great attention to tribotechnical material technology. To apply for this reason volumetric alloyed steel is often uneconomic and sometimes technologically non-profitable. However, the necessary results may be received by many functional coatings on working surfaces. They may simultaneously unite high hardness, wear resistance and heat endurance. For ensuring the complex of high working service property the usage of multielements coatings on the bases of carbides, borids, nitrides and silicides of transitional metals is rather perspective [1 – 4]. Stability and structure of composition, also high running quality of multicomponental elements of nitride systems make possible development of surface physic-mechanical characteristics and their application in the form of safety films protecting from ingress of contamination into subsurface of items layers [5, 6].

At present the most widely used are ion-plasmous techniques by film deposition coatings, in particular vacuum-arc and magnetron sputtering [7, 8].

We investigate the peculiarities of ion-plasma coatings by multielements system sputtering on the basis of Ti+Zr+Nb and we also perform analysis of physic-mechanical properties of the received coatings.

2. Experimental details

The coatings have been made by vacuum-arc depositions. For vaporizing materials we use solid accumulation electrode (cathode) on the basis of system: 30 at. % Ti, 35 at.% Zr and 35 at.% Nb. For active gas we apply molecular nitrogen. The thickness of all coatings in our experiments was 4.0 μm. Deposition parameters are given in Table 1.

Table 1. Physic-technological deposition parameters of coatings on the basis of (Ti-Zr-Nb)N.

# series	Vaporized material	I_o, A	U_{sm}, B	P_N,Torr
1a	Ti+Zr+Nb	95	100	3×10^{-4}
1b	Ti+Zr+Nb	95	100	7×10^{-4}
1c	Ti+Zr+Nb	95	100	4×10^{-3}

Surface morphology of fractography break, friction tracks were under research on scanning electron microscope JSM-6390LV. The research of elements composition of coatings was conducted by methods of X-ray characteristic spectrum analysis generated by electron beam in scanning electron microscope. X-ray structure research of samples with coatings were conducted on diffractometer DRON-4 in Cu-k_α radiation. The coatings hardness was determined with the help of hardness testing machine mode DM 8 according to micro-Vickers method with load on indenter 0.05 H. Adhesion-cohesion solidity, firmness to tracking and mechanism of coatings damages were under research in the air with the usage of scratch-tester Revetest (CSM Instruments). Tribological tests were conducted in the air according to scheme «ball-disk». As for friction machine we use «Tribometer», CSM Instruments. The coatings thickness comprised ~ 3.5 – 4.0 μm. For rider we use a ball 6.0 mm in diameter, made of sintered certified body - Al2O3. The load was 3.0 H, slide rate 10 sm/s. Tests correspond to international standards ASTM G99-959, DIN50324 and ISO 20808. The quantitative evaluation of samples wear resistance and rider were conducted according to wear-out factor W [9], methodology of calculation was given in work [10].

3. Results and discussion

Image of surface coatings and also fractography break are shown on Fig.1.

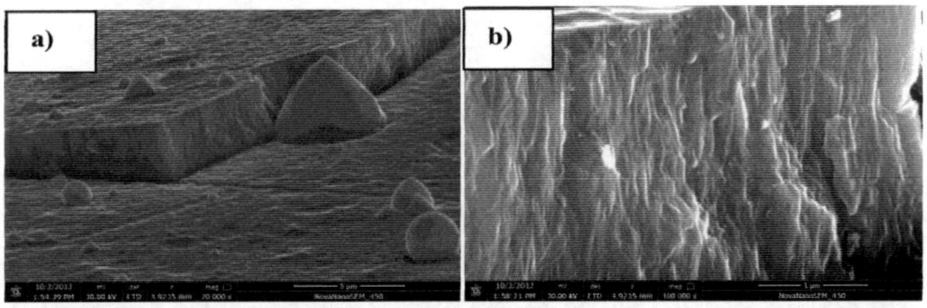

Figure 1. Image of coatings fractography break (Ti-Zr-Nb)N, received at nitrogen partial pressure: a – P = 4×10^{-3} Torr.; b – magnified image of coating part P = 4×10^{-3} Torr.

The research of surface morphology points that the pressure increase by nitric agent reaction leads to lessening of macroparticles quality and diameter.

That is particularly significant in active gases vacuum chamber forming with evaporated material refractory compounds [11]. We also observe the coating roughness lowering.

Coating element composition received by vacuum-arc deposition method was analyzed by energy dispersed method. (Table 2).

Table 2. Chemical composition of elements in coating (Ti-Zr-Nb)N.

# series	Composition of elements, at. %			
	N	Ti	Zr	Nb
a	38.72	20.91	20.38	19.99
b	40.00	22.57	18.04	19.39
c	40.86	20.52	19.36	19.26

If to compare elements composition of coatings series *b* and *c*, we can see that the first series samples the number of nitrogen atoms practically is equal. For samples *c* series, received at more higher pressure of N_2 atmosphere, the significant increase of zirconium atoms is noticeable, also several decrease of titanium atoms part. While this the niobium atoms for both series samples is actually the same. The increase of titanium atoms content in condensate of *b* series is possibly explained by more effective interaction of titanium atoms with nitrogen in subsurface layers.

The research of surface coatings fractography break (Fig. 1), received at different nitrogen partial pressure, testifies the formation of columnar structure (Fig. 1b), characteristic for coatings received by vacuum-arch precipitation method.

X-ray diffractometrical spectrum analysis presented on Fig.2 shows that the determining phase composition is phase with face-centered cubic alloy with crystal lattice. Low-intensity peak in section 2θ = 38° testifies the minor inclusion presence with BCC lattice characteristic while vacuum-arch method for drip-feed stage [12].

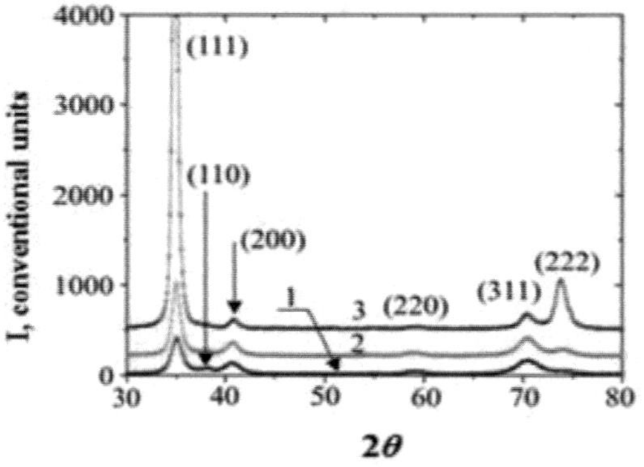

Figure 2. Sections of X-ray diffraction spectrums coatings received at different nitrogen partial pressures: curve 1 – P = 3×10^{-4} Torr; 2 – P = 7×10^{-4} Torr; 3 – P = 4×10^{-3} Torr; indicated planes of GCC lattice.

It should be marked that with pressure increase the peak intensity becomes lower (ibid. spectrum 1 and 3 on Fig. 2), it may be determined by significant decrease of drip-feed phase in coating and correlate with results of surface research. The set of planes {111} increase is a characteristic feature that is determined by perfection increase of preferred crystal growth orientation with axis [111] perpendicular to surface flatness.

The crystal size determined by approximation method with increase of pressure grows from 10 nm at lowest pressure 3×10^{-4} Torr up to 63 nm at highest nitrogen work pressure 4×10^{-3} Torr.

Adhesive-corrosion solidity research, endurance to coating tracking are given on Figures 3 – 4. Based on friction coefficient modification graph and acoustics emission from scribing load we determine the following critical loads: L_{C1} – the appearance of first chevron crack on the bottom and diagonal on the sides of tracking; L_{C2} –chevron cracks lots formation at the bottom of scratch and local chalking of coating, formation of chevron cracks at the bottom of scratch; L_{C3} – cohesive adhesive coating demolishing; L_{C4} – plastic abrasion of coating.

For criteria of adhesive firmness we adopt the critical load L_{C4}, leading to coating abrasion. Fig. 3 presents the dependence of friction coefficient and acoustic emission signal from applied load while c test samples scratch testing.

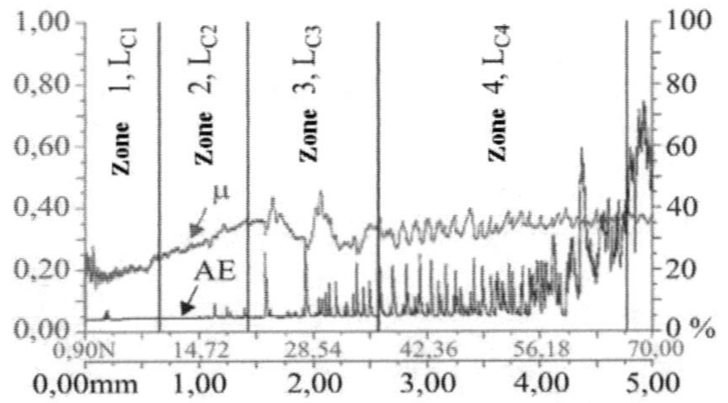

Figure 3. Dependence of friction coefficient from applied load while coating (Ti-Zr-Nb)N scratch testing received at $P = 4\times10^{-3}$ Torr.

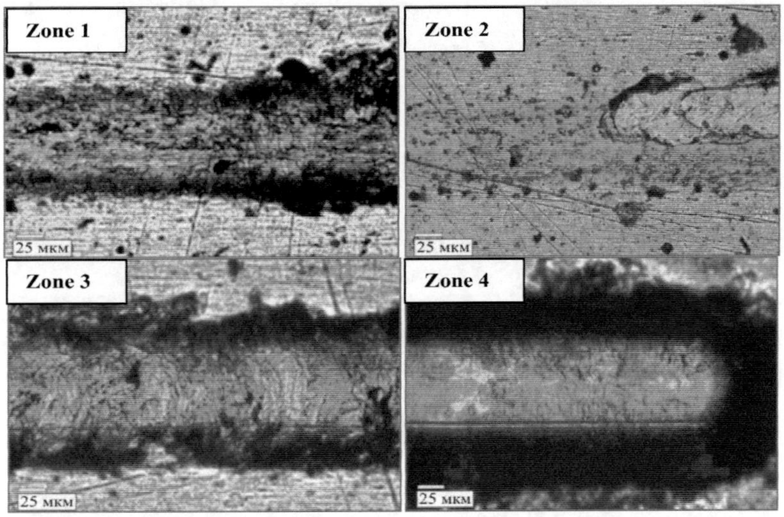

Figure 4. Diamond indenter contact zones with (Ti-Zr-Nb)N coating.

Conventionally the process of coating demolishing while indenter tracking may be divided into four stages. In the load range from F = 0,9 H up to F = 9.89 H indenter monotonous penetration comes into coating: friction coefficient slightly increases, acoustic emission signal preserved without changes At load F = 15.81 H indenter fully submerges into coating.. The sliding of diamond indenter along the coating has friction coefficient 0.35.

According to load increase (F = (20.6 – 36.4) H) runs out of stuff occurs in front of the indenter in the form of knob and increases the depth of indenter penetration.

Table 3 shows the samples test results with coatings (Zr-Ti-Nb)N comparing with coatings (Ti-Zr-Si)N and TiN received by us [13].

Table 3. Comparative results of adhesive coatings tests in systems (Zr-Ti-Nb)N and(Ti-Zr-Si)N, TiN.

Crytical loads	Coatings				
	(Zr-Ti-Nb)N Series, a	(Zr-Ti-Nb)N Series, *b*	(Zr-Ti-Nb)N Series, c	(Ti-Zr-Si)N [12]	TiN [12]
L_{C1}	2.91	0.9	9.89	3.91	21.31
L_{C2}	29.04	15.82	20.62	18.15	30.91
L_{C3}	43.18	42.37	36.43	24.29	40.28
L_{C4}	59.26	66.24	66.77	43.15	48.84

According to research [14] while testing for adhesive solidity depending upon different meanings of critical loadings several physic-chemical processes are being carried out simultaneously while abrasion, however only L_C is directly connected with adhesive demolishing.

The results of mechanical characteristics measurements, in particular hardness, received for coatings (Zr-Ti-Nb)N are given in Table 4.

Table 4. Average values of coatings hardness on the system (Zr-Ti-Nb)N base.

#series	hardness $HV_{0,05}$GPa	Endnote
a	37.21	Forward flow
b	40.21	Forward flow
c	44.57	Forward flow

The way it may be seen from Table 4 maximum meaning of hardness H = 44.57 GPa was received at pressure of reactionary gas $P = 4\times10^{-3}$ Torr may be referred according classification [15], to superhard coatings with hardness $HV_{0.05} = 40$ GPa. Before the wearing tests profile record of steel disks surfaces on which coverings were overcoated were taken off.

Overcoatings of (Zr-Ti-Nb)N system on steel disk polished surface leads to roughness increase at the account of drip-feed component of plasma flow.

The average meaning of friction coefficient (μ) in system «covering–Al_2O_3» while tests comprises 1.1.

As it follows from Table 6, coverings of system (Zr-Ti-Nb)N, received at reaction gas pressure $P = 4\times10^{-3}$ Torr, possess hard constituency.

Tribological tests of samples with coatings are given in Table 5. Under the friction conditions being held in our experiments, sample with covering is in more heavy conditions of loading comparing with rider. Rider's surface – ball is under constant loading and constantly undergoes cyclic loading connected only with elastic joining, breaking of adhesion tires in the limits of separate micro areas of real contact one and the same area

Table 5. Tribological characteristics of covering system «covering (Zr-Ti-Nb)N – Al_2O_3».

#series	Friction coefficient, μ		Wear factor, $mm^3\times H^{-1}\times m^{-1}$	
	Initial	During tests	Rider ($\times10^{-5}$)	Sample ($\times10^{-5}$)
a	0.61	1.95	0.391	9.69
b	0.45	1.19	2.84	3.1
c	0.491	1.05	3.21	2.4

The results of friction tracks elements' analysis are given in Table 6.

Table 6. Friction tracks of elements composition.

#series of samples	elements, at. %							
	N	O	Al	Zr	Nb	Ti	Fe	Mn
a	41.5	11.58	0.28	11.86	12.57	22.54	-	-
b	43.66	11.79	0.32	11.12	11.42	21.24	0.34	-
c	-	41.98	0.64	1.7	1.73	3.32	50.37	0.27

It is well known that while forecasting the working characteristic of different wares connected with friction it is necessary to treat friction characteristics of tribosystem (body - rider) not for separately taken element but in general because structural-phase coating condition and its physic-mechanical characteristics plays decisive role in the process of tribosystem operation.

4. Conclusion

1. We received nanostructural coating systems (Ti-Zr-Nb)N by method of solid cathode vacuum-arc evaporation in the medium of reacting gas nitrogen Multicomponent films have clearly expressed columnar structure.

2. X-ray analysis displays that the basic phase is the phase with face-centered cubic crystal lattice. Nanocrystals dimentions increase under pressure up to 10 nm from minor pressure of 3×10^{-4} Torr up to до 63 nm while the major pressure of running nitrogen atmosphere is 4×10^{-3} Torr. While pressure increase of reacting gas we observe set of peaks reinforcement {111}, that may be determined via the perfection of crystal growth increase with primary axial orientation [111] perpendicular to plane of growth.

3. Physic-technological parameters on solid coverings deposition were explored. Covering system (Ti-Zr-Nb)N hardness received at partial pressure P == 4×10^{-3} Torr forms $H_{0.05}$ = 44.57 GPa, and at pressure P = 3×10^{-3} Torr hardness makes $H_{0.05}$ = 37.21 GPa.

4. For covering series 1c on the system basis (Ti-Zr-Nb)N, overlayed on steel X18H10T substrate material, the starting point of cracks L_{C2} appearance occurs at load F = 20.62 GPa, adhesive destruction at load F = 66.77 GPa. In case 1c series L_{C2} occurs at F = 29.14 GPa, and destruction appears at F =59.26 GPa, that is 15 % higher.

5. Adhesion solidity of coverings on the basis (Ti-Zr-Nb)N is evidently higher comparing with coverings on the basis (Ti-Zr-Si)N and TiN, and adhesion destruction is observed at load F = 66.77 GPa for covering (Ti-Zr-Nb)N for covering on the basis (Ti-Zr-Si)N F = 48.84 GPa; and for TiN – F = 55.2 GPa.

6. Coverings wear resistance on the system basis (Zr-Ti-Nb)N series 1c is higher than wearing resistance of covering on basis (Zr-Ti-Nb)N series 1a. Significance of friction coefficient for coating (Zr-Ti-Nb)N while system testing «covering – Al_2O_3» makes $\mu \sim 1.1$.

References

[1] Pogrebnjak A D, Shpak A P , Azarenkov N A, Beresnev B M. 2009 *Struktura I svoistva tverdih I sverhtverdih nanokompositnih pokritii.UFNA*. Vol. 1 pp. 35-64.

[2] Pogrebnjak A D. 2013 Structure and Properties of Nanostructured (Ti-Hf-Zr-V-Nb)N Coatings. *J. of Nanomaterials*. Article ID 780125, pp. 12.

[3] Azarenko N A, Sobol O V, Beresnev B M, Pogrebnjak A D, Kolesnikov D A, Turbin P V, Toryanik I N. 2013 Vakuumno-plazmennie pokritii na osnove mnogoelementnih nitridov. *Metallophizika noveishie tehnologii*. Vol. 8 pp. 1001-1024.

[4] Pogrebnjak A D, Beresnev V M, Demianenko A A, Baidak V S. 2012 Adhesive strength, superhardness, and the phase and elemental compositions of nanostructured coatings based on Ti-Hf-Si-N. *Physics of the Solid State*. Vol. 54 pp. 1882-1890.

[5] Veprek S, Veprek-Hejiman M G L, Karvankova P, Prohazka J. 2005 Different approaches to superhard coatings and nanocomposites. *Thin Solid Films*. Vol. 476 pp. 1-29.

[6] Musil J, Baroch P, Zeman P. 2008 Hard nanocomposite coatings. Present Status and Trends: in Book Edit. R Wei "*Plasma Surface Engineering Research And Its Practical Applications*". Kerala Research Signpost Publ, pp. 1-34.

[7] Andreev A A, Sablev L P, Grigorev S N. 2010 *Vakuumno-dugovie pokritiya*. X.: NNC HFTI, pp. 317.

[8] Berlin E V, Seidman L A. 2010 Ionno-plazmennie processi v tonkoplenchatoi tehnologii. *M.: Tehnosphera*, pp. 528.

[9] Ibatullin I D. 2008 *Kinetika ustalostnoi povrezhdaemosti I razrusheniya poverhnostnih sloev: monographiya*. Samara: Samarskii gosudarstvenni tehnicheskii universitet, pp. 396.

[10] Vershinin D S, Smolyakova M U, Manohin S S, Druchinina O A, Ahmadeev U H. 2010 Issledovanie tribologicheskih svoistv azotirovannogo titanovogo splava BT16 s ispolzovaniem avtomatizirovannoi mashini treniya. *Zavodskaya laboratoriya. Diagnostika materialov*. Vol. 12 pp. 45-49.

[11] Aksenov I I, Andreev A A, Belous B A, Strelnickii B E, Horoshih B M. 2012 *Vakuumnaya duga: istochniki plazmi, osazhdenie pokritii, poverhnostnoe modiphicirovanie*. K.: Naukova dumka, pp. 727.

[12] Azarenkov N A, Sobol O V, Beresnev B M, Pogrebnjak A D, Litovchenko S V, Ivanov O N. 2013 *Materbalovedenie neravnovesnogo sostoyaniya poverhnosti*. Summi: SumGU, pp. 683.

[13] Pogrebnjak A D, Yakushchenko I V, Bagdasaryan A A, Bondar O V, Krause-Rehberg R, Abadias G, Chertier P, Oyoshi K, Takeda Y, Beresnev V M, Sobol O V. 2014 Mycrostructure, Physical and Chemical Properties of nanostructured (Ti-Hf-Zr-V-Nb)N coatings under different deposition conditions. *Mater. Chem. and Phys*. Vol. 147 pp. 1079-1091.

[14] Valli J. 1986 A review of adhesion test methods for thin hard coatings. *Journal of Vacuum Science and Technology*. Vol. 4 pp. 3007-3014.

[15] Azarenkov N A, Beresnev B M, Pogrebnjak A D, Kolesnikov D A. 2013 Nanostrukturnie pokritiya I nanomateriali. *Osnovi polucheniya, svoistva, oblasti primeneniya.* M.: Knijni dom «Librikom», pp. 368.

Influence of dispersion of ion irradiation D + on distribution of deuterium in titanium

G A Raspopova and V L Arbuzov

Institute of Metal Physics, Ural Branch RAS, Ekaterinburg, Russia
S.Kovalevskaya str. 18, Ekaterinburg, 620990, Russia

E-mail: raspopova@imp.uran.ru

Abstract. In single crystals of titanium was studied deuterium segregation induced by ion irradiation. Segregation created at room temperature deuterium ion implantation energy of 700 keV and analyzed by the nuclear reaction D (d, p) T. During the experiment changed the duration of continuous exposure and the duration of stops. It was found that with increasing radiation dose continuous average concentration of the implant in the irradiated volume gradually increases. It has been found that a long interruption of ion bombardment leads to an abrupt increase of the concentration. Analysis of the depth distribution of deuterium showed that during stopping the accumulation of radiation near the surface of the implant due to the redistribution of the target volume. The obtained results are discussed in terms of the evolution of the defect structure of the target in the course of continuous exposure and stop time of implantation.

1. Introduction

In recent years it was established that upon the operation of different-type nuclear reactors the rate of generation of hydrogen proved to be higher than it was expected. [1]. This experimental fact gave rise to the reinterpretation of the approach to the problem of hydrogen accumulation in reactor materials and brought about a rebirth of interest in it. The accumulation and retention of hydrogen in reactor materials can be to some extent simulated using radiation-induced segregation (RIS) of hydrogen, when high concentrations of radiation-induced defects and the accompanying high concentrations of hydrogen are formed in small volumes. Under the conditions of possible levelling-off of hydrogen concentrations, the magnitude of hydrogen trapping during RIS is an indicator of the defect state of materials. In other words, the application of RIS as a tool for studying the behaviour of hydrogen in irradiated materials allows one to investigate the evolution of the defect structure in the segregation regions from the dynamics of hydrogen trapping under irradiation.

It is evident that information on the magnitude of hydrogen trapping and evolution of a radiation-induced defect structure can be used to predict the level of radiation damage that is sufficient to nucleate critical elements of steels, such as gas pores. For identification of influence of interphase surfaces on the capture of hydrogen it is necessary to attract as objects of studying the pure model materials c with initial sink of point defects. The material modelling a role of surfaces is the sample with perfect crystal structure - single crystal titanium.

The existence of deuterium segregation induced by ion implantation in titanium at room temperature was determined for the cases of irradiation of titanium target by deuteron of low and high energies [2]. In the first case the nuclear reaction of deuterium atoms with $^{3}He^{+}$ ions was used to measure the concentration of implanted deuterium. In the second case the radiation-induced segregation (RIS) of deuterium was created and analyzed using one and the same beam of deuterons with the energy of 700 keV with the help of nuclear microanalysis method NRA. The measurements were carried out ***in situ*** using reaction $D(d, p)T$, where p is the protons, d is the deuterons, T and D are the atoms of tritium and deuterium correspondingly.

It is well known that accumulation and distribution of implanted deuterium in metals depends on the existence of effective sinks of point defects and their distribution in target and also on the existence, distribution and capacity of hydrogen traps [3-5]. The surface of sample is the only effective sink in single crystal target. To study the formation and evolution of RIS such target presents a real interest since it helps to minimize a number of uncontrolled factors during the experiment.

The aim of the present work is to study the formation and stability of segregation of deuterium in single crystal titanium implanted by deuterium ions at room temperature.

1. Experimental

2.1. Methods

The work was carried out on the nuclear-physics accelerator complex of the Institute of Metal Physics of the Ural Division of RAS using the reaction $D(d,p)T$ with deuteron energy of 700 keV. The use of deuteron beam as a probe gave the opportunity to measure the deuterium concentration during implantation and simultaneously to control the composition of the analyzed zone regarding oxygen and nitrogen (according to reaction $D(^{16}O,p)^{17}O$ и $D(^{14}N,\alpha)^{12}C$)). The choice of reaction $D(d,p)T$ ensured the coincidence in time and space of implantation processes and measurement of concentration of implanted deuterium in the irradiated volume. Besides, it ensured experimental purity of the introduction of additional component to the target (helium) and excluded the appearance of ion-induced damages beyond the irradiation zone.

The content of deuterium was determined from the energy spectrum of nuclear reaction products – protons. The sample of titanium with constant concentration of deuterium $TiD_{1.78}$ served as a standard sample. The value of RIS was estimated according to the average concentration of implanted deuterium in the irradiated region (C_D). The depth of analysis was conditioned by the length of projective range of deuterons (~ 4.7×10^{-6} m) and by depth resolution of the used nuclear reaction (0.5×10^{-6} m). The analyzable volume was 4×10^{-8} m^3.

Besides the C_D measured discretely during the continuous deuterium implantation, α trapping of implanted deuterium of RIS was calculated as a fraction of the whole implanted deuterium trapped in the irradiated volume. The total of deuterium amount trapped all kinds of traps was calculated

$$\alpha_O = C_D/C_D^{theor} \times 100\%,$$

where C_D^{theor} is the average concentration of the implanted deuterium under the condition of its complete conservation in the irradiated volume.

The stability of deuterium segregation in titanium after irradiation was also studied. For that the beam of deuterons was shut off at different stages of ion irradiation. The duration of pauses in the process of implantation was from 1.8×10^3 c to 245.4×10^3 c. After the end of shutting the process of segregation measurement was continued.

2.1. Materials

The work was carried out with single crystals α-Ti [6] made by electron-beam zone melting of titanium with the content of iron and silicon to 0.12 and 0.08 mass % correspondingly. The samples were cut of the cylinder single crystal (length 0.1 m and diameter 0.01 m) in such a way as to make the surface of sample perpendicular to the ***c*** –axis. The deformed in the process of cutting layer of crystal was removed by electro polishing. The content of oxygen in the samples was 0.05 at. %, the content of nitrogen, hydrogen and carbon was by the order lower.

Single crystals contain grains characterized 300-500 μm in size and density of dislocations less than 10^{11} m^{-2}. The neighboring blocks disorientation been not more than 0.1 degree, microstresses were absent.

3. Results and discussion

The results of measurements of dose dependence of the average content of deuterium and calculation of the corresponding trapping of deuterium in the analyzable zone during continuous irradiation and after the interruptions of deuterons beam are illustrated by figure 1 and table 1.

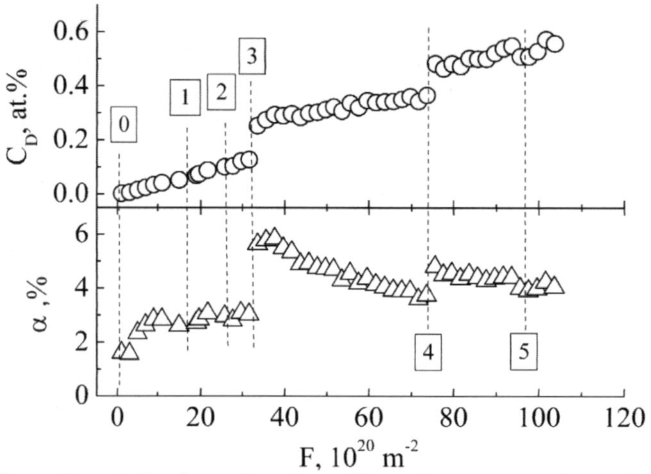

Figure 1. Effect of total implantation dose F on the average deuterium content in the analyzable zone $\boldsymbol{C_D}$ and on the fraction of implanted deuterium $\boldsymbol{\alpha}$, preserved in the analyzable zone. *Dashed line and numbers show the stops 0, 1, 2, 3, 4 and 5. Duration stops see Table 1.*

Table 1. Parameters of irradiation stops of single crystal titanium

Stops of irradiation	The dose of continuous exposure to the stop of radiation, 10^{20} m^{-2}	The total dose of radiation, 10^{20} m^{-2}	The duration of the pause, 10^3 c	Q, appm
0	0	0	0	0
1	15	15	1.8	47
2	2.6	21.6	13.2	77
3	6	31.6	245.4	189
4	40	73.6	77.4	362
5	20	95.6	2.6	382

As it was expected the $C_D(F)$ increases in the parts of continuous growth of implantation doses. Long-term stops of irradiation lead to the jumping growth of this dependence.

The first part of measurement of α(F) deuterium trapping at continuous irradiation is italicized (Figure 1) with solid line which qualitatively coincides with the dose dependence of the implanted deuterium trapping into in-cascade clusters in the alloyed austenite without initial traps [3]. It gives us a ground to suppose that in titanium single crystal there are no initial deuterium traps or their capacity is extremely small. The part of α(F) dose dependence at continuous irradiation with total doses from ~32×10^{20} m-^{2} to ~74×10^{20} m^{-2} has more complicated character. In spite of the fact that $C_D(F)$

increases with the growth of irradiation dose a remarkable decrease implanted deuterium trapping in it is observed.

It is known [3] that trapping of deuterium during irradiation occurs into initial traps IT (existed before the beginning of irradiation) and radiation induced traps RIT, i.e. $\alpha F) = \alpha_{IT}(F) + \alpha_{RIT}(F)$. Before overlapping of a scope of radiation traps $\alpha_{RIT}(F) \sim F$. After filling of initial traps $\alpha_{IT}(F) \sim 1/F$. The descending character of $\alpha(F)$ dependencies between the stops speaks about the appearance of a large amount of deuterium traps and their filling during the stops 3 and 4. For the resumed irradiation such traps play a part of the initial ones, i.e. existed before this irradiation.

Considering the renewal of irradiation as an independent experiment with preliminary created defect structure of the sample we obtained $\alpha(F^*)$ dependencies, where F^* is the dose of continuous irradiation (Figure 2). To determine the ccapacity of traps Q for each subsequent radiation, formed during the previous irradiation and stop time necessary to compare the initial portions of descending dependencies $\alpha(F^*)$. It is here $\alpha_{IT}(F^*) = Q / C_D^{teor}(F^*) \times 100\%$. Capacities of initial deuterium trap by the method described in [3] are after stops 0-5 are given in Table 1.

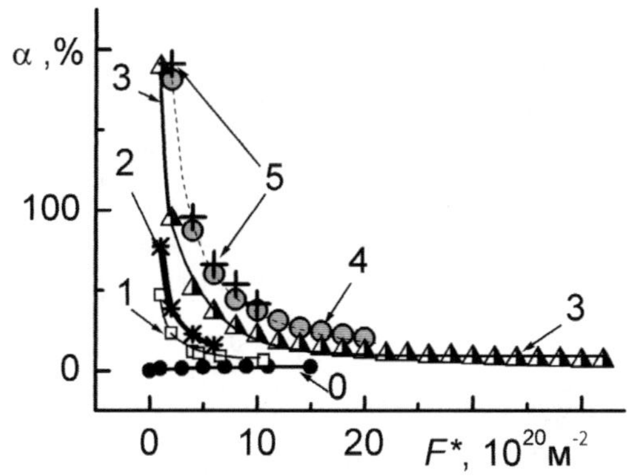

Figure 2. Dose dependencies of deuterium traps in the initial titanium (0) and after stops 1-5 (*see figure 1 and Table 1*).

Distribution of implanted deuterium without taking into account its diffusion mobility and also the mobility of radiation vacancies and SIs was calculated using SRIM program. It has a pronounced maximum near the depth 4.7×10^{-6} m which is equal to the length of projective range of deuterons.

The obtained experimentally dependencies of concentration of the implanted deuterium on the depth of analysis differ from the calculated profiles. They are more diffused and the concentration of deuterium in the irradiated volume is rather less than the calculated one.

However, in spite of the low depth resolution of the used nuclear reaction the profile analysis of deuterium in titanium during the continuous irradiation showed the existence of the maximum of deuterium content near $\sim 4.5\times10^{-6}$ m (Figure 3).The interruptions of irradiation resulted in the qualitative change of depth profiles that testifies first of all to the evolution of defect structure of the target during interruption and appearance of the "new" filled IT.

The depth dependencies of deuterium concentration in the irradiated target volume gave us the opportunity to detect the location of the "new" IT. If the continuous irradiation forms the distribution of deuterium with the maximum on the depth $\sim 4.5\times10^{-6}$ m then the long-term stop of irradiation leads to the considerable redistribution of deuterium toward the surface of sample (Figure 3b).

The increase of implantation dose before the first beam shut off resulted in the growth of amount of implant along the whole analyzable zone but mainly in the region if this maximum (curves 1 and 2 in Figure 3a). At the time of 1.8×10^3 c stop after implantation with the dose 15×10^{20} m^{-2} the character of $C_D(F)$ dependence did not change and the view of depth distribution of implant remained.

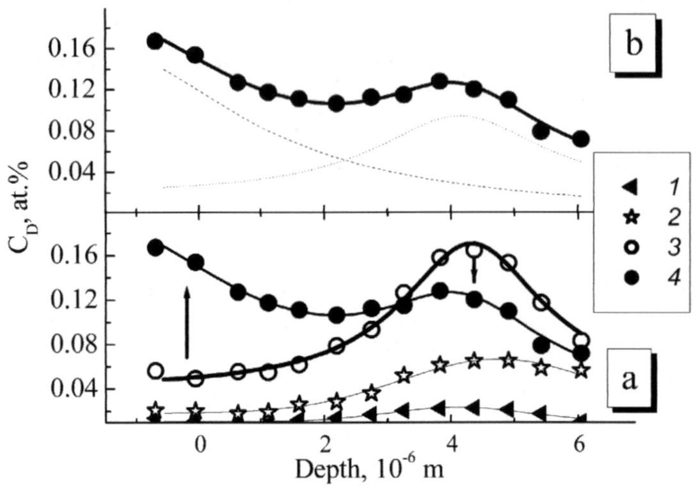

Figure.3. Distributions of deuterium on the depth of titanium sample before the first long-term stop (curves 1-3) and after interruption of irradiation for 245.4×10^3 c (curve 4). *Measurements were carried out at irradiation doses:* ***1*** *- 7 × 10^{20} m^{-2};* ***2*** *- 15 × 10^{20} m^{-2};* ***3*** *- 28 × 10^{20} m^{-2};* ***4*** *- 32 × 10^{20} m^{-2}.*

4. Conclusions

The investigation of accumulation of implanted deuterium in the irradiated volumes of titanium by nuclear reaction method made it possible to observe in situ the development of defect structure of target.

The effect of accumulation of implanted deuterium near the sample surface during the isothermal aging of irradiated titanium single crystal at room temperature was determined for the first time.

The detected peculiarity in behavior of the implanted deuterium is explained by the evolution of defect structure at room temperature and the appearance of traps of large capacity.

5. Acknowledgments

The research was carried out within the state assignment of FASO of Russia (theme "Spin" No. 01201463330), supported in part by RFBR (project № 14-02-00150).

References

[1] Garner F A, Oliver B M, Greenwood L R and others 2001 *J. Nucl. Mater* **296** pp 66-82.

[2] Arbuzov V L, Vykhodets V B, Raspopova G A 1995 *Metals* **4** pp 148-161.

[3] Raspopova G A, Arbuzov V L 2009 *Problems of atomic science and technology. Series: Physics of radiation damages and radiation material science (93)* Iss. 2 pp 14-19

[4] Raspopova G A, Arbuzov V L, Sagaradze V V and others 2007 *Physics of Metals and Metallography* **103** №2 pp 180-190.

[5] Raspopova G A, Arbuzov V L 2009 *Physics of Metals and Metallography* **107** №1 pp 63–72.

Effect of titanium doping on accumulation and annealing of radiation defects in austenitic steel 16Cr15Ni3Mo(0-1)Ti at low temperature (80 K) electron irradiation

V L Arbuzov and S E Danilov

Institute of Metal Physics, Ural Branch of Russian Academy of Sciences, S.Kovalevskaya str. 18, Ekaterinburg, 620990, Russia

E-mail:danilov@imp.uran.ru

Abstract. The effect of titanium doping on accumulation and annealing of radiation defects was investigated in austenitic stainless steel 16Cr15Ni3Mo under low temperature (80 K) electron irradiation. Steel has been taken in the quenched, aged and separation of solid solution states. The data obtained on the accumulation of radiation defects and their evolution during isochronous annealing. The types of defects and its complexes and the activation energy of the processes taking place with their participation is identified. The mechanisms of radiation-induced processes and the effect of titan doping is discussed.

1. Introduction

Currently, the austenitic stainless steel is most widely used for the reactor internals elements. It is well known that limiting phenomenon of the radiation resistance of the reactor steels has become associated with radiation-induced structural and phase changes.

Radiation defects play an important role in processes of radiation damage of metals and alloys, radiation-induced structural phase transitions [1]. Atomic segregation may be homogeneous by volume of grains, but most are mixed. In the latter case, they are observed on clusters of point defects, dislocations, grain boundaries or interphase surfaces. Freely migrating defects (vacancies and interstitials) cause the following opposite effects:

1. Accelerating the evolution of real materials often are metastable towards equilibrium;
2. The emergence of phenomena associated with the induction of non-equilibrium processes in alloys: the formation of clusters of radiation defects and segregation until separation of the phases.

It is known [2], that addition in small amounts (0.2 - 1.0 wt.%) of titanium in the austenitic steels substantially reduce the vacancy swelling. In [3.4] has been suggested that the swelling is weakened by radiation-induced formation of dispersed precipitates γ'-phase. Research on the interaction of impurity atoms of titanium with radiation defects are presented in [4,5,6].

We have previously shown, that irradiation at elevated temperatures iron-nickel alloys, leads to an ordered solid solution separation. Doping of titanium leads to the suppression of the process and the formation of precipitates γ'- phase [6]. However, knowledge of the interaction of point defects with Ti atoms is very incomplete.

Vacancy clusters (VC) is formed during the free migration of vacancies and their interaction with each other and with the dopants. In this regard, the low-temperature electron irradiation and postradiation isochronous annealing were used in this work to study the evolution of radiation defects and their interaction with each other and with the titanium atoms.

Purpose of work is to explore the effect of doping titanium on the accumulation and annealing of radiation defects and radiation-induced structural changes in the austenitic stainless steel.

2. Materials and methods

For investigation it was taken 16Cr15Ni3Mo austenitic stainless steel smelted in a vacuum induction furnace from high-purity components with titanium concentration of 0, 0.6, 1.0 wt.% Ti. Doping was carried out by adding the titanium to the melt. The concentration of carbon in the steel was 0.03 wt.%. Original state steel was obtained by annealing at 1323 K for 30 minutes, followed by quenching in water. The grain size after this treatment is 100-200 μm and a dislocation density about 10^7 cm^{-2}. According to X-ray and electron-microscopic data studied were single-phase austenitic steel.

The part of quenched samples were annealed at 923 K 5 hours to form γ'-phase precipitates. Using electron microscopy, scanning tunneling microscopy, it is shown that in aged steel γ'- phase are presented with size of 7-10 nm with concentration of $(4\div6)\cdot10^{16}$ cm^{-3} (volume ratio of ~ 3%). Another part of the sample were annealed at 773 K 3 hours to form a separation of solid solution(SSS). 5 MeV electron irradiation was carried out on a linear electron accelerator at 80 K to fluence $(1\text{-}4)\cdot10^{18}$ cm^{-2}. Measurement of residual resistivity at 4.2 K, carried by the standard four probe potentiometric method with sensitivity at the level of 0.01 nΩ×cm and an accuracy of 0.02%.

3. Experimental results

As mentioned above, studies were performed on samples in three states: quenched, annealed and aged to maximize the separation of the solid solution (SSS). To understand the states after heat treatment and structural-phase changes in the material it is performed isochronous annealing of non-irradiated steels.

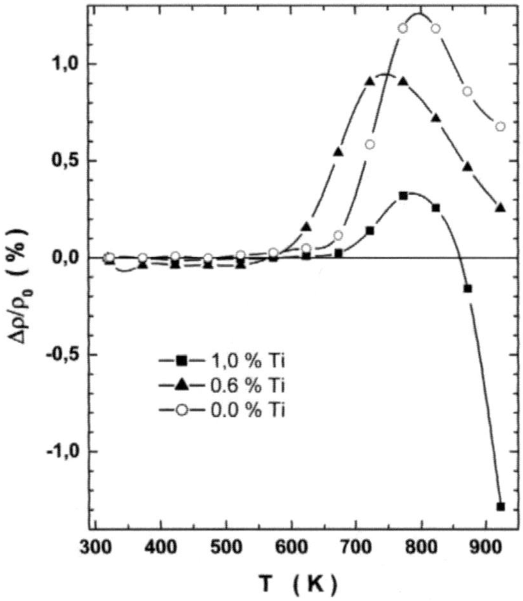

Figure 1. The dependence of the residual resistivity of steels in the quenched state on isochronous annealing temperature.

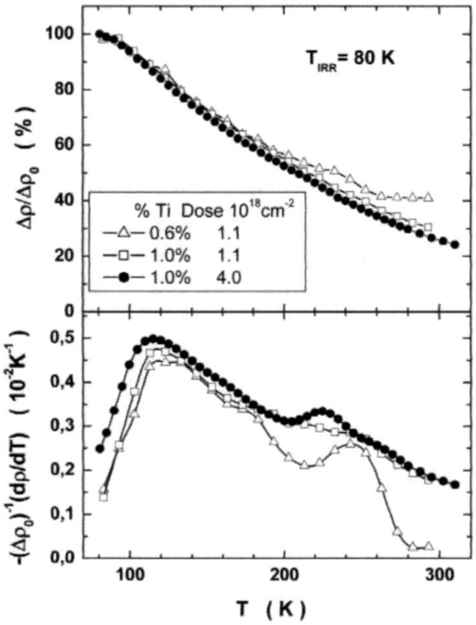

Figure 2. The dependence of the the resistivity steels in quenched state electron-irradiated to different doses on isochronous annealing temperature. In the lower part of the figure - differentiated curves.

Figure 1 shows the changes in the electrical resistivity of steels in the quenched state at the isochronous annealing. It can be seen that the increase of annealing temperature leads to an increase in resistivity at 700 - 800 K, and at higher temperatures up to 950 K begins to decline in the resistivity and the alloy with 1% Ti below the reference value. This indicates the reduction of the titanium (and nickel) concentration in the solid solution. It should be noted an important feature, the maximum growth of the electrical resistivity is higher, the lower the concentration of titanium. At the isothermal annealing at 923 K in steel containing titanium, the formation of intermetallic precipitates such as Ni_3Ti, is shown by electron microscopy studies. To determine the processes occurring during isochronous annealing in this temperature range in the aged at 923 K samples were also carried isochronous annealing in the range 600-920 K.

These isochronous annealing showed that the growth of the resistivity in the 700-800 K with the subsequent downturn discovered earlier in the isochronous annealing of quenched samples, reproduced with the same dependence on the concentration of Ti. Thus, there are two observed at annealing processes - at 700-800 K and in the vicinity of 900 K and above. The increase in the resistivity at 700-800 K is associated with the processes of separation and approach to equilibrium state, as has been shown for the fcc alloys Fe-36% Ni [7]. With increasing temperature above 800 K in the investigated alloys it is occurs the homogenization of austenitic solid solution and lowering resistivity in accordance with the equilibrium states. Thus, the presence of titanium in solid solution suppresses the thermal separation process, and decreases the equilibrium curve of the solid solution separation of stainless steel.

During the low-temperature electron irradiation at 80 K dose dependence of growth of residual electrical steels studied are linear at least up to a dose $10^{18}cm^{-2}$. This reflects a continuation of the resulting radiation-induced defects.

In Fig. 2 it is presented the results of isochronous annealing of quenched samples irradiated by electrons. Dependencies obtained for steel with 1% Ti at different doses are practically identical. In the region of room-temperature electrical resistivity of the irradiated samples is not reaches the initial value. In the case of steel alloyed by 0,6% Ti exceeding of the value over the initial resistivity in the region of room temperature is maintained for more than steel with 1% Ti.

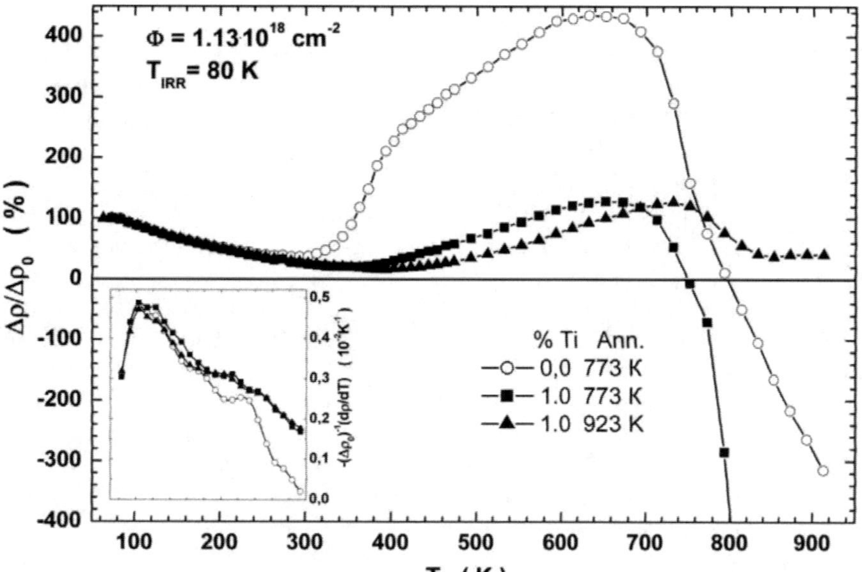

Figure 3. Resistivity dependences on the isochronous annealing temperature of steels annealed before electron irradiation at different temperatures. At the bottom part of the figure - differentiated curves.

Maximum value of annealing is at the first peak at 110-120 K. The second peak poorly defined, is about 170-180 K. The third peak of annealing is situated at 220-240 K. The amplitude of the first peak tends to increase with increasing dose and concentration of titanium. In the area of 220-240 with increasing temperature the resistivity decreasing of the steel with 0,6% Ti much less than of the steel with 1% Ti. The peak annealing, located at 240 K is well defined.

Figure 3 shows the data of isochronous annealing of steels annealed before irradiation at 773 K for maximum separation (SSS) or aged at 923 K for the formation of intermetallic precipitates. It can be seen that the peaks of the annealing substantially the same ones as in the case of quenched samples. The first peak of annealing almost the same amplitude and position on the temperature scale for all samples. The second peak is clearly visible only in the alloy without titanium. The third annealing peak in the area of 220-240 clearly visible on an alloy without titanium and is located near 240 K. In the region of room-temperature it is observed partial annealing, as well as in quenched samples.

With further increase in the annealing temperature it is observed an increase in resistivity at 350-400 K on all steels. However, the resistivity increase in the alloy without titanium substantially greater. The maximum value of the resistivity reaches at the samples annealed for SSS state, at 660 K, which is significantly lower than for the non-irradiated samples. At higher temperatures, at steels annealed for SSS the resistivity falls below the initial value. This drop is not to become at the steel with 1% Ti, annealed at 923 K, the electrical resistivity close to the initial level. These data indicate that during the annealing of the irradiated samples it is occure the further separation of solid solution and thus homogenization of a solid solution above 800 K, as well as for the unirradiated samples. Accordingly, this results in a decrease in the resistivity below the original values for samples annealed for SSS state.

4. Discussion

In this work were shown, that the increase in resistivity at low temperature irradiation (80 K) is linearly dependent on the radiation dose. This means that with increasing radiation dose is a linear growth of preserved concentrations of radiation-induced defects. It is possible that under irradiation it occurs the recombination of close pairs. Thus, dumbbell configuration of self-interstitials (SIA), unlike pure metals may be different: Fe-Fe, Fe-Ni, Fe-Cr, Ni-Cr, and others. From this follows that the migration energy of self-interstitials may be more high.

The activation energy of the processes corresponding to the detected annealing peaks may be estimated by known formula $E_A=kT_0\ln(\nu\Delta t)$ [8], where ν - frequency of the Debye, Δt - annealing time, T_0 - the peak annealing temperature. Evaluation provides for a peak at about 110 K (0,36±0.02) eV. For peak around 170 K estimation gives (0,55±0,05) eV. For a peak at 220-240 K estimation gives the activation energy of about (0,65±0,07) eV

Considering the annealing peak at 110-120 K which is broad and has a large amplitude can be assumed that it is associated with the annealing of self-interstitials on the sinks and vacancies. The presence of titanium leads to the formation of complexes SIA - Ti atom. Perhaps these complexes dissociate or migrate at 170 K. The third peak annealing at temperatures of 220-240 K is due to the migration of vacancies. And these vacancies in steels containing Ti react with the Ti atoms to form complexes vacancy-atom of Ti. Besides the vacancies annealed at sinks and form vacancy clusters (VC). On the formation of complexes indicates that this peak is clearly visible in the steel without titanium and 0.6% titanium. In the area of room-temperature dissociation of complexes and VC of small multiplicity and growth of VC greater degree of multiplicity occures. This is indicated by the beginning of the separation of the solid solution, leading to an increase in resistivity. At higher temperatures of isochronous annealing dissociation of the remaining VC, which leads to further separation of the solid solution to temperatures near 700 K.

Thus we can say that radiation defects due to the interaction with the Ti atoms form a series of more complex defects compared to defects in non-alloy steel. Some of these types of defects have higher thermal stability.

5. Conclusion

The data obtained on the effect of Ti doping on the accumulation and annealing of radiation defects and radiation-induced structural phase changes in the austenitic stainless steel 16Cr15Ni3Mo (0-1) Ti at low temperature (80 K) electron irradiation.

1 At the non-irradiated alloys the impurity of Ti inhibits the thermal separation of the solid solution at 700-800 K. At temperatures above 850 K, the formation of the second phase (Ni_3Ti) occurs.

2. The accumulation of radiation defects at 80 K leads to a linear increase in resistivity with increasing radiation dose, which means saving generated point defects.

3. In the irradiated at 80 K alloys migration of self-interstitials, with an activation energy of 0.36 eV, to sinks and formation of complexes self-interstitials - Ti atom occurs in the temperature region 110-120 K. Dissociation of these complexes occurs near 170 K. The binding energy of complexes - 0.19 eV.

4 The migration of of vacancies occurs at 220-240 K with an activation energy of 0.65-0.70 eV. Thus there are leads to formation of vacancy clusters and complexes vacancy - Ti atom. Dissociation of these complexes and vacancy clusters starts at around room temperature and occurs to 700 K. This results in a radiation-induced separation of the solid solution.

Obtained data indicate that radiation defects interact with titanium atoms and form a number of more complex defects compared to defects in the non-alloyed steel. Doping of titanium has an effect on radiation-induced processes.

Acknowledgements. The research was carried out within the state assignment of FASO of Russia (theme “Spin” No. 01201463330 and project of Ural Branch of RAS № 15-17-2-3), supported in part by RFBR (project No. 13-02-00359-a).

References

[1] Votinov S N, Prokhorov V I, Ostrovsky Z E 1987 *Irradiated Stainless Steels* Moskow, Science 128 p.
[2] Mariasz P J J. 1984 *Nucl. Mater.,* **122-123** 472
[3] Garner F A 1984 *J. Nucl. Mater.* **122-123** 459
[4] Sagaradze V V, Pavlov V A, Alyab’ev V M, Goshchitskii B N, and others 1988, *Fiz. Met. Metalloved,* **65** 970-977.
[5] Perminov D A, Druzhkov A P, Arbuzov V L 2011 J. *Nucl. Mater.*, 414, pp. 186-193.
[6] Arbuzov V L, Danilov S E, Kazantsev V A, and Sagaradze V V 2014 *The Physics of Metals and Metallography,* , **115**, 10, 1017–10.
[7] Danilov S E, Arbuzov V L, Kazantsev V A 2011 *Journal of Nuclear Materials* 414 200–204
[8] Konobeevsky S T 1967 *Influence of irradiation on materials*, Moskow Atomizdat 402c.

Separation of radiation defects in Ni and Ni-C alloys under electron and neutron irradiation

S E Danilov V L Arbuzov, B N Goshchitskii, A E Kar'kin, V D Parkhomenko

Institute of Metal Physics, Ural Branch of Russian Academy of Sciences, S.Kovalevskaya str. 18, Ekaterinburg, 620990, Russia

E-mail: danilov@imp.uran.ru

Abstract. Complex investigations of radiation damage of Ni and Ni- 880 at. ppm C alloy under electron and neutron irradiation in the region of room temperature hardened and deformed state. In pure nickel, with the deformation microstructure, both in electron and in the neutron irradiation is observed separation of radiation-induced defects. When electron irradiation in the alloy Ni-C separation effect is observed, and when neutron irradiation there is no. This is due to the interaction of carbon atoms with radiation defects. The main sinks for radiation-induced defects are the areas with a high concentration of defects in cascades of atomic displacements.

1. Introduction

Operation of nuclear power plants of various types with great sharpness raises the problem of the radiation resistance of structural materials. At irradiation temperatures below 1/3 of the melting point the radiation damage is manifested in the accumulation of small clusters of radiation defects and the individual single vacancies. Nickel is a model material for the austenitic stainless steels. In nickel at room temperature vacancies are immobile and interstitial free to migrate.

Vacancies and vacancy clusters (VC) play an important role in radiation damage to metals and alloys, including radiation-induced structural phase transitions [1]. When cascadeless electron irradiation, VC formed by the free migration of vacancies. When cascade neutron irradiation vacancy clusters, besides formed directly in the displacement cascades. VC are efficient sinks of point defects and affect the radiation-induced diffusion processes [2,3]. In the austenitic stainless steels to reduce radiation damage used plastic deformation. In [4,5] theoretically studied the role of dislocation sinks on vacancy pore formation. Comparison of cascade neutron and cascadeless electron irradiation deformed and quenched metals and alloys reveals the role of the free migrated radiation defects and cascades of atomic displacements.

We was experimentally discovered in nickel the effect of the separation of radiation defects [6]. A substantial part of self-interstitials annihilated when migrating to the dislocation sinks and there lead a greater accumulation of vacancies in the deformed nickel. Carbon in solid solution (tens - hundreds of atomic ppm) can have a major effect on the properties of irradiated alloy as a result of its strong interaction with point defects [7,8]. The aim of this work is to identify the role of carbon atoms in the separation effect of radiation defects.

2. Materials and methods

For nickel doping carbon at 880 at. ppm by vacuum remelting, it is taken the nickel with additions of iron carbide. Solubility of carbon in the solid nickel decreases with decreasing temperature to 2400 at.ppm at room temperature [9]. Nickel alloys were annealed at 1173 K in high vacuum for one hour followed by rapid cooling. Carbon in Ni-C alloys are in solid solution and thermally stable up to 850 K.

Electron microscopic studies were carried out on the microscope JEM 200CX. To delete a of vacancies created by deformation the sample was annealed at 450 K. The dislocation structure is completely preserved. Measurement of resistivity was conducted with standard four-probe measurement with error - 0.02%.

Irradiation of nickel and nickel alloys samples was performed of neutron dose (En> 0.1 MeV) F = $(1, 10, 50)\cdot 10^{18}cm^{-2}$ and electron doses (Ee = 5 MeV) in the range $(0\text{-}5)\cdot 10^{18}cm^{-2}$. The temperature of the neutron irradiation is 300-320 K and 270-300 K of electron irradiation. For dose dependence on irradiation with fast neutrons were used a set of samples exposed to different doses. When electron irradiation was performed it is sequential irradiation with different doses of the same sample.

3. Experimental results

To investigate the microstructure were carried out electron microscopic studies. Recrystallized by slow cooling pure nickel has a dislocation density of at 10^7-10^8 cm^{-2}. At the rapid cooling of pure nickel, a dislocation density is obtained of about 10^8-10^9 cm^{-2} and at C-Ni alloys at $5\cdot 10^9 cm^2$. At a plastic deformation it is formed cellular structure. The cell size decreases with increasing degree of deformation up to 90% from 700 nm to 200 nm. After low-temperature annealing for 0.5 hours at 450 K of deformed nickel the cellular structure is is stored. The volume of the cell have a density of dislocations $2{,}3\cdot 10^{10}cm^{-2}$ in the cell walls consisting of plexuses dislocation, the dislocation density is $4{,}8\cdot 10^{10}cm^{-2}$ and higher. At neutron irradiation it is observed vacancy clusters the size of 2-3 nm.

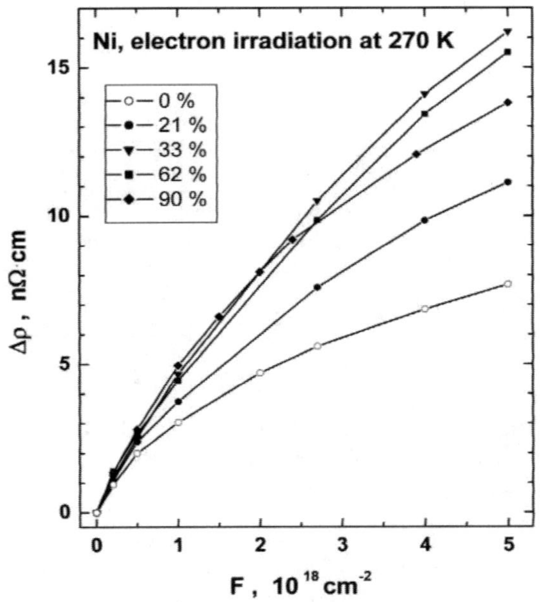

Figure 1. The dose dependence of resistivity growth for the pure nickel with varying degrees of deformation under electron irradiation at 270 K. Before irradiation, the samples were annealed at 450 K to remove deformation vacancies.

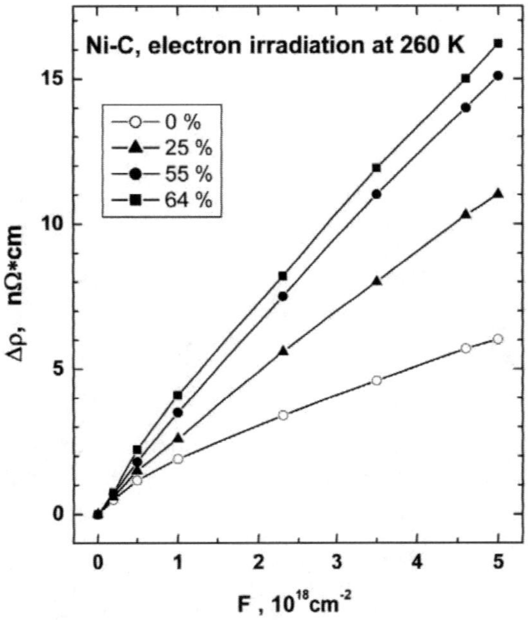

Figure2. Dose dependence of resistivity growth for the alloy Ni-C with varying degrees of deformation under electron irradiation at 260 K

The increase in resistivity when the nickel is deformed is the sum of the contributions from the dislocations and vacancies. These contributions may be divided. Changes in resistivity after annealing at 450 K, is determined by deformation vacancies. Using the value of the specific contribution of vacancies in pure nickel 0,23-0,28 nΩ·cm/at.ppm [7], we can estimate the amount of vacancies. At the degree of deformation of 40%, it is 67 at. ppm. The contribution given by the dislocations in deformed nickel corresponds to a change in the resistivity at recrystallization stage. At the specific contribution of dislocations in Ni, is equal to $9{,}4\cdot10^{-10}$nΩ cm^3 [10] the density of dislocations in nickel with a degree of deformation of 40% is approximately $5.6\cdot10^{10}$ cm^{-2}; as a result of electron microscopy it is $9\cdot10^{10}$ cm^2.

Fig. 1 shows the dose dependence for pure nickel with different degrees of deformation. The greatest effect is observed when the separation degree of deformation of about 40%. This is in agreement with electron microscopic data on the dislocation density in the cells. Figure 2 shows the dose dependence of the alloy Ni-C with varying degrees of deformation. It can be seen that under electron irradiation of Ni-C alloy also observed the effect of separation. As the level of deformation up to 64% the effect is increased. Maximum effect, as in the case of pure nickel at 40% deformation was not observed.

To determine the effect of carbon on accumulation vacancies on Figure 3 shows the dose dependence for deformed and non-deformed alloy of pure Ni and Ni-C. The figure shows that the increase in the resistivity of the Ni-C alloy during irradiation less than pure nickel.

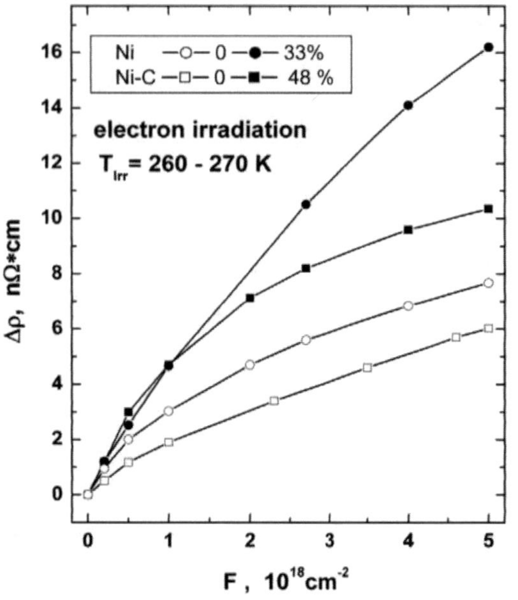

Figure 3. Dose dependence of the resistivity growth of deformed and non-deformed Ni and Ni-C alloy irradiated by electrons.

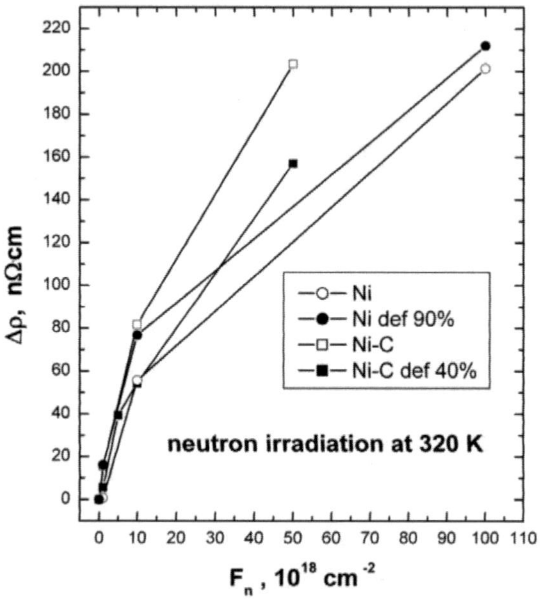

Figure 4. Dose dependence of growth of residual resistivity of deformed and non-deformed Ni and Ni-C alloy irradiated with neutrons.

Figure 4 shows the results of neutron irradiation of Ni and Ni-C alloy in the deformed and non-deformed state.It is seen that the magnitude of increase resistivity under neutron irradiation is considerably higher than under the electron irradiation. On pure nickel effect of radiation defects separation is observed. At the Ni-C alloy separation effect is not observed. Increase resistivity in the deformed alloy is less than in the undeformed alloy.

From the change in resistivity under electron irradiation to a dose of $5\cdot10^{18}$ cm^{-2} concentration of radiation vacancies is 30 at.ppm in undeformed nickel and 63 at.ppm in the deformed by 40% nickel. In the alloy Ni-C concentration of vacancies is 23 at.ppm in undeformed and 41 at.ppm in the deformed by 48% nickel. At neutron irradiation it is generated vacancies and VC. In [11] it is shown that the contribution to the resistivity of vacancies does not depend on whether they are located individually or in clusters. Under neutron irradiation at a dose of $2\cdot10^{19}cm^{-2}$ the amount of vacancies will be in pure nickel in the recrystallized state 222 at.ppm, in a deformed state 307 at.ppm and in the alloy Ni-C - 327 at.ppm in the recrystallized state and 217 at.ppm in the deformed state.

4. Discussion

The efficiency of radiation defects sinks is determined by their capacity. Sinks capacity of dislocations in the volume of subgrain 1.5-2 times more than the sinks capacity of the walls subgrain [6]. Hence the initial sinks capacity in deformed nickel and Ni-C alloy is associated with the value of the initial residual resistivity ρ_0. During irradiation due to the accumulation of vacancies total capacity of sinks will only increase, and at higher doses go into saturation. The amount of accumulated during irradiation vacancies is determined by the resistivity growth $\Delta\rho$. Thus it must be performed a linear relationship between the magnitude $\Delta\rho$ and ρ_0. Due to the fact that the initial resistivity of pure nickel alloys and Ni-C differ, we used the value of the difference between the deformed and undeformed state for nickel and Ni-C alloy ρ_0(def) - ρ_0(nodef). = $\Delta\rho_0$. Figure 5 shows the results of such a comparison for nickel alloys and Ni-C irradiated by electrons to a dose of $5\cdot10^{18}$ cm^{-2}.

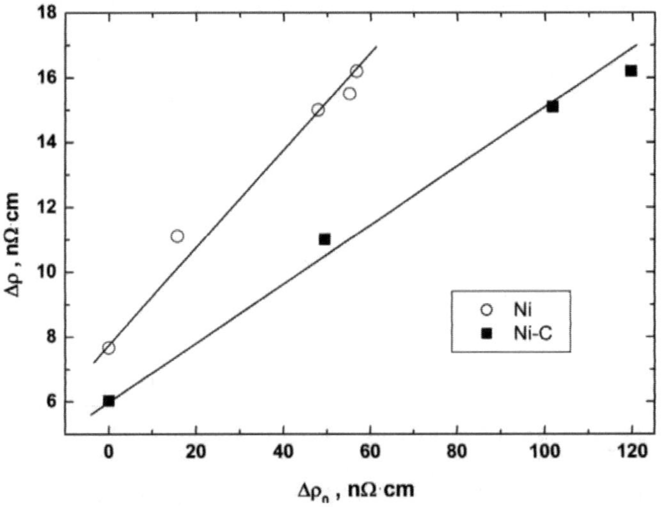

Figure 5. The dependence of resistivity increase under electron irradiation ($5\cdot10^{18}$ cm^{-2}) on the increase in the resistivity of plastically deformed the nickel and Ni-C alloy.

As can be seen from the figure, there is a linear relationship. However, the separation effect of radiation defects in the alloy, Ni-C less than pure nickel.

At our irradiation temperatures the vacancies and the carbon atoms are immobile and mobile are just self-interstitials (SIA). Currently, is not enough studies of the interaction of atoms and point defects in nickel. As the data modeling [12], the carbon atoms in nickel can be effectively trapped at the SIA and vacancies. In [7] it is assumed that the complex SIA - atom C annealed at 70 K. Thus we can assume that in this case, under irradiation it is move are not only SIA but also the complexes SIA - atom C. In its migration such complexes can interact with other atoms C to form carbon clusters. Thus, the formation of carbon clusters are carbon output from the solid solution and decrease resistivity. The contribution to the resistivity of carbon - 3.5 ·μΩ cm/at.% [13].

The number of generated displacements per atom in the alloy and pure nickel at any dose of irradiation is the same. At neutron irradiation in pure nickel separation effect is observed, as at the electron irradiation. In the alloy Ni-C situation is reversed - the increase in the resistivity of the deformed alloy less than undeformed about half. The values of the resistivity increments under neutron irradiation in Ni-C alloys greater than in the pure nickel. When migrating complexes CMA-C atom it can be formed carbon clusters, as under the electron irradiation. When migrating complexes

SMA-C atom in the areas of cascades of atomic displacements it is a high probability clustering, containing in the structure a few atoms and CMA and stable at irradiation temperature. This leads to increased accumulation of radiation defects and therefore to the large increase in resistivity in Ni-C alloys compared to pure nickel. In this case, a large concentration of dislocation sinks in deformed Ni-C samples compared to undeformed increases annihilation migrating complexes atom C-SMA at dislocations and reduce the growth of resistivity.

5. Conclusion

Complex investigations of radiation damage of Ni and Ni-alloys 880 atomic ppm C under electron and neutron irradiation in the region of room temperature quenched and deformed state. According to the research it may be concluded:

1. In pure nickel, with the deformation microstructure, both under electron and in neutron irradiation in the vicinity of room temperature is observed the separation of radiation defects. Under electron irradiation separation of radiation defects have the maximum at degree of deformation of about 40% and are determined by the dislocation microstructure and its spatial distribution.
2. When the alloy Ni-C irradiated by electrons it is also observed the effect of the separation of radiation defects. The growth of resistivity is constrained by the presence of carbon atoms through the formation of migrating complexes C atom - own interstitials, leading to the removal of carbon from the solid solution.
3. Upon irradiation Ni-C alloys with neutrons, the effect of radiation defects separation are not observed. This is due to the effect of carbon on the formation of stable clusters consisting of carbon atoms and self-interstitials in the areas of displacement cascades.

Acnowlegements

The research was carried out within the state assignment of FASO of Russia (theme "Spin" No. 01201463330 and "Flux" № 01201463334), supported in part by RFBR (projects No. № 13-02-00321-a and № 14-02-00150).

References

[1] Trushin Yu V 2002 *Radiation processes in multicomponent materials* Saint Petersburg Ed. Phys. Techn. Inst.
[2] Danilov S E, Arbuzov V L, and Kazantsev V A 2014 *The Physics of Metals and Metallography* **115**, 2, 126–133.
[3] Arbuzov V. L, Goshchitskii B N, Danilov S E, Kar'kin A V and Perminov D A 2008 *The Physics of Metals and Metallography*. **106**, 3, 266–275.
[4] Samsonidze, GG, Trushin Yu V 1988 *Sov. Phys. Tech. Phys*, **33** 1, 24-29
[5] Trushin Yu V, Orlov A N 1986 *Sov. Phys. Tech. Phys*., **31** 7,.763-767.
[6] Arbusov V L, Goshchitskii B N, Danilov S E, Kar'kin A E, Pecherkina N L and Sagaradze V V 2014 *Technical Physics Letters*, **40**, 2, 95–97.
[7] P. Ehrhart 1991 *Atomic Defects in Metals*, **25** Landolt-Börnstein Group III Condensed Matter, SpringerLink 437.
[8] Natesan K and Kassner T F 1973 *Metall. Trans*. **4** 2557-2566.
[9] Elliott R P 1965 *Constitution of Binary Alloys, First Supplement* McGray Book Company, 456 p.
[10] Yang Yi, Huang H, Zinkle S 2010 *J Nucl. Mater*. **405** 261-265.
[11] Arbuzov V L, Goshchitskii B N, Danilov S E, Karkin A E, Sagaradze V V, Zuev Yu N 2010 *Physics of Metals and Metallography* **109** 300-308
[12] Sorensen S M, Chen C W 1977 *Radiat. Eff.* **33** 109-118.
[13] Hugel J. 1973 *J. Phys. F*, **3**, 1723.

Cascade-probabilistic methods and solution of Boltzman type of equations for particle flow

A I Kupchishin[1,2], A A Kupchishin[2] and N A Voronova[2]
[1] Al-Farabi Kazakh National University, Almaty, Kazakhstan
[2] Abai Kazakh National Pedagogical University, Almaty, Kazakhstan

E-mail: ankupchishin@mail.ru

Abstract. A new method for calculating the spatial and temporal distributions of secondary particles and passing through a substance called cascade-probabilistic (HF) was proposed, developed and described. Original problems in the application of the original CP-technique in radiation physics, cosmic ray and positron physics was resolved. The general expression for the Boltzmann equation for the steady and unsteady cases was recorded based on an analysis of all the studies. In case of one-dimensional and three-dimensional (coordinate and angles θ and φ) Boltzmann equations solution goes into the previously obtained expressions for N_m. In addition, all previously derived analytical expressions for flow of different particles: protons, neutrons, electrons, concentration of defects, etc. are obtained after simplifying. At the same calculations give satisfactory agreement with experimental data.

1. Introduction

The intensive development of nuclear physics, especially nuclear energy, space research, the creation of various types of accelerators and nuclear reactors, obtaining materials and composites with desired properties using radiation caused interest in the study and quantitative description of the processes of interaction, education and distribution of primary and secondary particles in condensed and gaseous media [1 – 3]. As it is known, that during the passage of particles through matter happen quite complex and diverse events that are important both theoretical and practical significance for nuclear physics, solid state physics and other fields of science and technology. First of all, it refers to the dosimetry and protection from radiation, cosmic ray physics, radiation physics, solid state physics, neutron physics and reactors, radiation chemistry, biology and medicine, radioisotope analysis, monitoring and defectoscopy, nuclear geophysics and geology and especially lately radiation technology to manufacture materials with desired physical and chemical properties [4 – 7].

Naturally, in order to understand and describe these phenomena, on the one hand, it is necessary to know what happens to the particles (both primary and secondary, to generate different kinds of collisions). Thus one of the main problems in this case is to establish spatio-temporal and energy distributions of the incident and the secondary particles in the environment. In addition, at the time, and also after the passage of particles through the substance almost all physical and chemical properties of the substance and its structure are changed. In the description of these processes, there is a problem of choice of theoretical research method. The most famous and widely used theoretical methods of calculation – is the Monte Carlo method, the Boltzmann equation, Focker-Planck equation, and a variety of specialized methods and models. Without belittling the well-known numerical methods and models, apparently, it can be said that the indisputable advantages over them possess analytical methods, even with their help manage only approximately describe a phenomenon [1 – 3].

2. The history of CP-method

The history of CP-method associated with name of the great mathematician of the last century, S.D. Poisson, particularly with the distribution of random variables, named in his honor. Poisson distribution for certain values of parameters contained in it becomes a simple cascade-probability function. Initially, it was used in calculation of angular distributions and flow of secondary particles and meteorological factors without sufficient rigorous study. So, Goudsmit and Saunderson were able to create an approximate theory of multiple scattering in 1940, when describing angular distribution of electrons. Later, [1] using the simplest CP-functions calculations of integral multiplicities, coupling coefficients and meteorological factors and energy distribution of nucleon component of cosmic radiation in the Earth's atmosphere, as well as the neutron flow from the sun are made. In addition, energy spectrum of π-mesons in the earth's atmosphere was calculated. Despite the fact that in these studies achieved good agreement between theoretical calculations and experimental data, yet most CP-functions for different particles was not found. It is unclear role and place of these studies by comparing obtained results with CP-functions with similar distributions calculated by Monte Carlo methods and Boltzmann equation. Some clarity on these issues to a certain extent have been done in [1], in which CP-functions were derived for different stable, unstable particles and antiparticles as recurrence relations, and from Boltzmann equation, detailed mathematical analysis were done, basic physical properties, as well as the calculated spatial distribution of radiation defects in metals irradiated by electrons, protons and alpha particles were established. CP-method has been successfully used in calculation of average positron thermalization, angular distributions of annihilation photons and simulation of defects in depth with experimental data on positron annihilation [8 – 10].

3. Results

On the basis of general principles of cascade-probabilistic (CP) method and analysis of our fifty years research we recorded general solution of equations of unsteady cascade process (Boltzmann type) for the m-th component of any particle (including generation and elimination of phase space and their passage through substance to predetermined depth).

As it is known that Boltzmann equation – the equation of kinetic theory of gases, proposed L. Boltzmann for determination particle distribution function of an ideal monatomic gas [1, 2]. Equation in dimensionless variables for the m-th component of particles:

$$\frac{\partial f_m}{\partial t}+(\upsilon,\nabla_V f_m)+(F,\nabla_\upsilon f_m)=\frac{1}{\varepsilon}L(f_m,f_m). \tag{1}$$

Here $f(x, \upsilon, t)$ – distribution function density of number of particles in phase space $x\otimes\upsilon$, x – three-dimensional space coordinate, υ – speed, t – time, F – density of external mass forces, ε – dimensionless parameter (proportional to ratio of average distance that particles pass without collision, characteristic to scale of phenomena under consideration). Collision operator L in the simplest case is as follows:

$$L(f_m,f_m)=\int\left[f_m(\upsilon')f_m(\upsilon_1')-f_m(\upsilon)f_m(\upsilon_1)\right]\left|\upsilon-\upsilon_1\right|d\omega d\upsilon_1,$$

where υ_1 and υ – velocity of molecules before collision, υ_1' and υ' – velocity of molecules after collision, $d\omega$ – element of area in a plane perpendicular to $\upsilon_1-\upsilon$ vector.

In deriving Boltzmann equation it is assumed that evolution of $f_m(x, \upsilon, t)$ function is determined by its value at the time t and binary collisions between gas molecules and reaction time of two molecules of collision gas is much less than time during which they move as free particles. From mathematical point of view, conclusion of Boltzmann equation is a specific algorithm for developing L operator on the basis of the known laws of motion of two colliding with other gas molecules.

In equation (1) scope of variable t – half- $t \ge 0$, range of variation υ–all space R^3, range of x – subdomain Ω in R^3 (Ωmay coincide withR^3). Physically, function $f(x, \upsilon, t)$must be non-negative and such that

$$\int f_m(x,\upsilon,t)\upsilon^2 d\upsilon<\infty,$$

The simplest boundary condition on $d\Omega$ is as follows:

$$f_m(\upsilon - 2n(n,\upsilon); x; t) = f_m(\upsilon, x, t),\ x \in \partial\Omega, \upsilon \in R^3,$$

where n – normal to dΩ. There are several different precise performances of Cauchy problem for equation (1), but existence of a whole solution of (1) natural from physical point of view of assumptions about L operator had been proved for none of them. Passing in (1) distribution function to the flow of particles in a spherical coordinate system based on an analysis [1], it is offered general solution of non-stationary equation of system of Boltzmann equations for m-th component of particles (including generation and elimination of particles from phase volume) within cascade-probabilistic method in the form of:

$$\begin{gathered} N_m(x,y,z,\theta,\varphi,E,t) = \sum_{ink}\ldots\sum\int\ldots\int_{V_n t_n \Omega_n E_n} N_{0e}(t=0,x',y',z',\theta',\varphi',E')\frac{1}{\lambda_0\cos\theta_0}\bullet \\ \bullet\,\chi_0(x',y',z',\theta',\varphi',E',t')\frac{d\sigma}{\sigma_0 d\Omega dE_0}dV_0 J_n\ldots\bullet \\ \bullet\frac{1}{\lambda_n\cos\theta_n}\chi_{0in}(x^n,y^n,z^n,\theta^n,\varphi^n,E^n,t^n)\frac{d\sigma}{\sigma_n d\Omega dE_n}dV_n J_n, \end{gathered} \tag{2}$$

where, $N_m(x, y, z, \theta, \varphi, E, t)$ – intensity of m-th particles at x, y, z, under zenith θ and azimuth φ; χ_i – probability pass a way r_i at angles θ_i, φ_I with energy E_i after n interaction; $\frac{d\sigma}{d\Omega_i dE_i}$ – differential cross section for interaction; $dV_n(x_n, y_n, z_n)$ – volume element; J_n – Jacobian transformation; i, n, k – number of generations of interactions and reaction channels.

The integral $\chi_0 \div \chi_n$ switches to CPF and further to Poisson distribution. At present, we calculate optimal coefficients and simple sixty fold integrals, which allow solving huge number of problems with ~ 15% error. It is planned to calculate parameters for hundredfold and more integral. All previously derived formulas for probabilities of flows and various primary and secondary particles (stable and unstable), concentrations of defects in irradiated materials charged, neutral particles, and others were found from obtained solutions. Further, mathematical analysis of expression for flow of m-th component was performed and it was shown that asymptotic value of intensity of particles in phase space v (x, y, z, t), having a real physical meaning can be found from it [1]. Moreover, considered processes can take place both to passage of micro (nano) particles (portions) and with birth of new stable and volatile particles. It is shown that in general case, solution can be represented as a sum of multiple integrals.

The mathematical analysis of the found expression for the particle flow was performed.

1.From (2) it follows that $x = x'$, $y = y'$, $z = z'$ and $t = 0$

$$N_m(x,y,z,\theta,\varphi,E,t) = N_{0e}(t=0,x',y',z',\theta',\varphi',E'),$$

where N – flow of particles, x, x', y, y', z, z' and t – coordinates and time, θ and φ – zenith, azimuthal angles. In this case, the flow is equal to flow of primary particles.

2. From obtained equation (2) with $x \to \infty$, or $y \to \infty$, or $z \to \infty$, or $t \to \infty$, follows that

$$N_m(x,y,z,\theta,\varphi,E,t) \to 0.$$

This is easy to show, because each cascade-probability function has an exponent (with negative exponent), which for infinite values of parameters in each term of sum tends to zero.

3. If (2) to put i = 0, n = 0 and k = 0, z = h, the case of vertical incidence of one type of particles on substance (such as plates), we obtain:

$$N_m = N_0 e^{-\frac{h}{\lambda_0}}.$$

4. In case of one-dimensional and three-dimensional (coordinate and angles θ and φ) Boltzmann equations solution (2) goes into the previously obtained expressions for N_m (formulas 1.36 и 1.40 from [1]).

5. In addition, all previously derived analytical expressions for flow of different particles: protons, neutrons, electrons, concentration of defects, etc. are obtained after simplifying (2). At the same calculations give satisfactory agreement with experimental data [1].

4. Conclusion

1. General solution of the nonstationary Boltzmann type equations for the m-th component of particles (including the generation and elimination of particles from the phase volume) within the cascade-probabilistic method was proposed.

2. It is shown that for x = x' , y = y' , z = z' and t = 0

$$N_m(x,y,z,\theta,\varphi,E,t) = N_{0e}(t=0,x',y',z',\theta',\varphi',E'),$$

and for x →∞, or y→∞, or z→∞, or t→∞ on the basis of mathematical analysis of expression for particles flow.

3. After simplifying (2) it is possible to get all the previously derived analytical expressions for particles flows that give satisfactory agreement with experiment.

Acknowledgments
This work was supported by the grants from the Ministry of Education and Science of the Republic of Kazakhstan (a. N 32, 349).

References

[1] Boos E G, Kupchishin A I, Kupchishin A A, Shmygalev E V and Shmygaleva TA 2015 Relations to Markov chains (Almaty: Abay KazNPU. RINCT, al-Farabi Kazakh National University, LLP "Kama") 388
[2] Vinogradov Ed IM 1977 Mathematical Encyclopedia "Soviet Encyclopedia" V.1 (M.) 1152
[3] Boltzmann L 1956 Lectures on theory of gases/translated from german (M.)
[4] Kupshishin A I, Potatiy K V and Chukurova R M 1987 Cryst. Latt. Defects and Amorphous Mater. **I3** (34) 157 – 162
[5] Klimenko I V, Zhuravleva T S, Lenenko N D and ZhuravlevaYu V 2009 Zhurnal Physicheskoihimii 83 346 –350
[6] Harnutova E P, Petrov E I 2004 Journal Materials Research **19**(7) 1924
[7] Lappan U, Fuchs B, Geifiler U, Scheler U, Lunkwitz K 2003 Rad. Phys. Chem. **67** (4) 447
[8] Boos E G, Kupchishin A I 1988 The solution of physical problems by cascade-probabilistic method (Alma-Ata: Nauka) Part 1 112 Part 2 144
[9] Kupchishin A A, Kupchishin A I and Shmygaleva T A 2007 Computer modeling of radiation-physical problems (Almaty: Kazakh University) 432
[10] Kupchishin A I, Lisitsin A A and Kupchishin A A 2014 The interaction of high-energy radiation with matter (Tomsk: Tomsk Polytechnic University) 153

IOP Conf. Series: Materials Science and Engineering **110** (2016) 012036 doi:10.1088/1757-899X/110/1/012036

Optical research of biomaterials of Sorbulak

O V Esyrev[1], A A Kupchishin[2], A I Kupchishin[1,2] and N A Voronova[2]

[1] al-Farabi Kazakh National University, Almaty, Kazakhstan
[2] Abai Kazakh National Pedagogical University, Almaty, Kazakhstan

E-mail: ankupchishin@mail.ru

Abstract. Within the framework of optical research it was established that on the unpolluted samples of sedge stems occurs structuring of material, whereas on contaminated and irradiated blurring of its structure takes place. Sampling of sedges and rushes for research was carried out in areas near the first dam Sorbulak. For comparison, samples of same materials were taken far away from populated areas. Irradiation was carried out with high-energy electrons with energy of 2 MeV and integral dose of $3 \cdot 10^5$ Gr. Irradiation leads to a more pronounced structuredness of material. There is a significant difference in the structural elements (epidermis, vascular bundles, parenchymal cells, etc.). There are traced dark spots and bands associated with the presence of huge amounts of heavy metals against the background of a green matrix.

1. Introduction

It is known fact that one of the most important problems of environmental biotechnology for today is how to clean polluted water. Detailed analysis of the scientific and technical information indicates that the chemical and technological monitoring to ensure the normal operation of enterprises is not enough for environmental protection. It is necessary to conduct on-going biotechnological monitoring of flora, which includes biomaterials. In the environment, plants such as sedges, reeds and many others, are the most vulnerable to anthropogenic changes. Toxic contamination is a striking example that represents a dangerous threat of our time [1 – 4].There are many environmental problems that our world is facing today. For instance, pollution mechanisms of reservoirs, the impact of toxic substances on their flora and fauna, purification of water bodies from pollutants. Thus, the attention is focused on the contaminated areas. The wastewater, which is subjected to special purification systems in water treatment facilities, is collected in so-called “lake reservoirs”. The study of environmental issues of reservoirs occurs extremely rare.Being a part of the environment of the regions aquatic environment determines to a large extent the well being of the entire population. For this reason ecological problems of region aquatic environment and the location of wastewater treatment systems require immediate solutions. It is necessary to believe that investigation of ecology of reservoirs will be successful in sequential comparison with already known positions of ecology of freshwater environments [5 – 9].

Heavy metals (HM) are among the most main polluting substances of reservoirs. Therefore, in the environmental and hydro biological monitoring of pollutants HM is givenprimary importance. Reduction of heavy metals may occur as a result of physical processes of mass transfer, physicochemical processes of complexation, sorption, and accumulation. Many HM are biologically active, do not undergo transformation in hydrobiont organisms, and slowly leave the biological cycle. They are absorbed by the cell walls and accumulate within the cells of microorganisms such as bacteria, mycelial fungi, yeasts, and algae. The sorption capacity of most of the HM is relatively high. For this reason, these organisms can be considered as effective and cost efficient sorbents for water purification from impurities. Sedges, reeds, various algae are the most promising indicators of changes in the environment due to the increased concentration of biogenic elements. Higher aquatic plants (macrophytes) absorb, concentrate and convert those chemical elements into a chelate form during the growth and development. Thus, they reduce the concentration of toxic metals in the water and improve its quality. The search for natural physical factors, which have the effect on improving the water quality, should remain.

2. The experimental procedure

Sampling of sedges and rushes for research was carried out in areas near the first dam Sorbulak. For comparison, samples of same materials were taken far away from populated areas. Irradiation was carried out with high-energy electrons with energy of 2 MeV and integral dose of 3•105 Gr. This large dose was taken for determining the responses of plant facilities to irradiation. The reflection and transparency of contaminated and non-contaminated samples of biomaterials were examined to study their structure by using optical-microscopic method. The measurements were performed on an automated digital microscope (ACM) Leica DM 6000 M. This microscope is a precision system for research with high-resolution digital cameras and software for analysing and storing images. Selection of different colours in the image is used to detect small or subtle elements of the sample by contrast. Studies of the anatomical structure of elaeagnus leaves were carried out according to the methodological guidelines of Prozinoy M.N. [10]. Surface and pressed samples along with cross-sections were manufactured. First, glycerol was used to lighten preparations. Then, microphotographs were taken on MS-300 microscope (magnification X63). A large number of temporary preparations were produced. Measurement of morphometric parameters was performed for quantitative analysis using ocular micrometre MOV-1-15 (with the lens X9, magnification x10.7). The concentration of heavy metals in the sample is calculated as follows,

$$C = A\ V\ K / P,\ (\text{mg/kg}),$$

where A – concentration of the element in the sample solution, mcg/ml; V – volume of the sample solution, mL; K – dilution factor; P – mass of weighed portion sample, g; where: $A = \gamma/\ell$ st.s. ℓ p.l.s. where γ – concentration of standard solution; ℓ st.s. – peak length of standard solution; ℓ p.l.s. – peak length of sample.

3. Experimental research

Optical microphotographs of unpolluted, polluted, and irradiated samples of sedge plants are shown in figures 1 – 3. As follows from the figures, changes in the morphology of objects clearly appear on contaminated samples. As a result, a huge number of dark spots were found. Structuring of the material is observed on non-contaminated samples of sedge stems; whereas, blurring of its structure is occurred on contaminated and irradiated samples. Thus, it was revealed that irradiation leads to a more pronounced structuring of material.

Analysis of leaf biometrics of check and contaminated samples of sedge (Figures 5 and 6) shows a substantial difference in the structural elements between check and contaminated materials (epidermis, vascular bundles, parenchymal cells, etc.). Dark spots and bands associated with the presence of a huge number of heavy metals are clearly seen on the background of green matrix. Figures 5 and 6 reveal that cells of the sedge epidermis have tabular shape without any intercellular spaces. They also have a thickened outer wall, which is covered on the surface by cuticle along with a waxy coating, living (check sample), and dyingoff (contaminated sample) hairs, which act as screen for reflecting part of the solar rays. Moreover, there are some important parts among the main cells of the same type of epidermis. Those are the guard cells of stomata, the ferruginous and highly sensitive hairs (trichomes), the hydathode, the motor cells etc. The water and nutrients, such as foliar application for plants, penetrate through existing pores and pectin strands in the outer walls of cells of epidermis. This epidermis protects all the inner tissues of plants from desiccation, mechanical damage, infiltration of various infections. It also regulates gas exchange and transpiration of the whole plant through stomata. There is an accumulation of various substances, such as glycosides, tanning agents, alkaloids, as well as heavy metals with phytoncide properties, in the epidermal cells of contaminated leaf of elaeagnus. Glandular hairs form essential oils, resin, and mucus. Since covering hairs synthesise plant hormones, enzymes and substances necessary for normal functioning of plants, their existence and functioning is essential for the whole organism.

The analyses of the presence of heavy metals in non-contaminated and contaminated samples of sedges were conducted as described in methodology part of this work. The results indicate that sedge differently adsorb environmental elements, particularly heavy metal (Table 1).

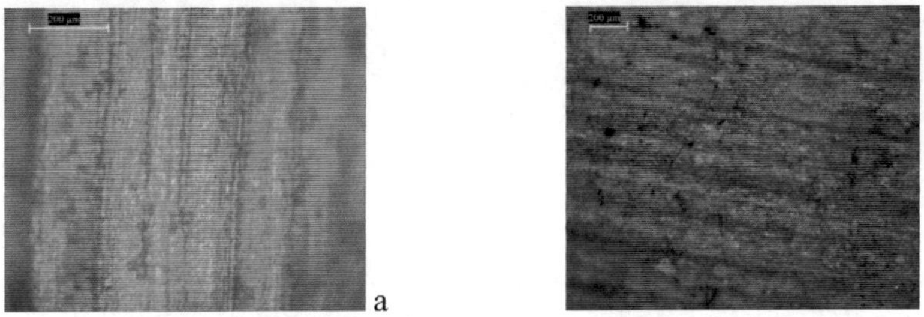

a b

a – non-contaminated; b – contaminated

Figure 1. Optical microphotographs of sedge.

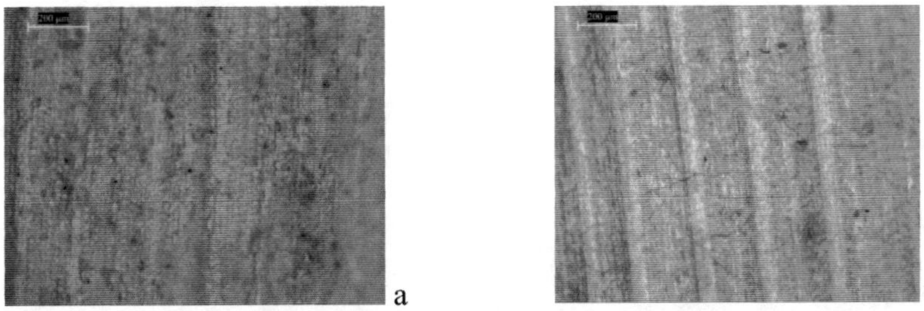

a b

a – non-contaminated; b – contaminated

Figure 2. Optical microphotographs of the irradiated sedge.

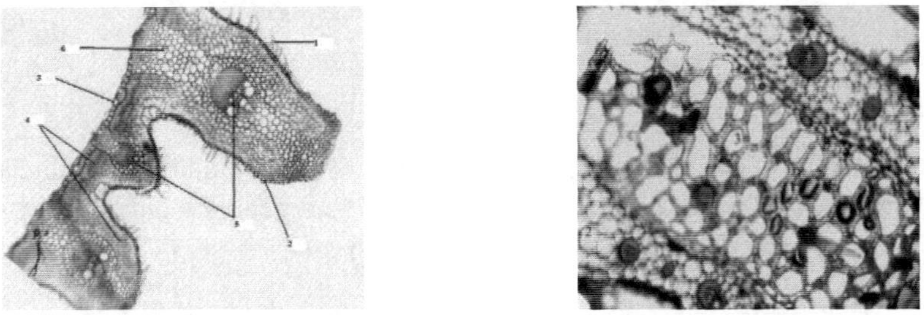

a – non-contaminated:
1 – trichomes, 2 – upper epidermis,
3 – the lower epidermis, 4 – mesophiles, 5 – vascular bundles, 6 – parenchymal cells

b – contaminated:
1 – vascular bundles,
2 – parenchymal cells

Figure 3. The anatomical structure of sedge.

Table 1. Heavy metals content in the material of sedge (Lake Sorbulak)

(mg/kg)	Pb	Cd	Zn	Cu	Fe	Ni	Co	Mn	Cr	Sr
Sedge (check sample)	6.40	0.80	32.00	9.60	648.0	20.40	0.40	70.0	25.20	49.60
Sedge (contaminated sample)	7.00	1.00	32.00	6.40	720.0	12.60	0.48	3800.0	8.40	89.60

The lowest concentration is observed for cadmium, from contaminated to non-contaminated materials of sedge, 0.8 and 1.0 mg/kg, respectively, and the highest is for manganese, 70 and 3800.0 mg/kg, respectively).The increase of concentration is also observed for ironfrom contaminated to non-contaminated materials of sedge 648.0 and 720.0 mg/kg, respectively. Selective adsorbing ability of certain heavy metals can be successfully used for cleaning various reservoirs from many harmful elements and compounds. We have found that industrial and domestic wastewater initially contains significant concentrations of metals. Thus, it is the main sources of heavy metals in the aquatic environment of Sorbulak.

As is known, the absolute content of heavy metals in plants are divided into four groups: the elements of increased (Sr, Mn, Fe, Zn), medium (Cu, Ni, Pb, Cr), low (Mo, Cd, Se, Co), and very low (Hg) concentration. In our case, to a first approximation, this pattern is confirmed for sedges as well (the first group).

4. Conclusion

1. It was established that changes in the morphology of objects clearly occur on the contaminated samples. A great number of dark spots were found. Structuring of material is observed on non-contaminated samples of sedge stems; whereas, blurring of its structure was revealed on contaminated and irradiated samples. It was discovered that irradiation leads to a more pronounced structuring of material.

2. Analysis of leaf biometrics of check and contaminated samples of sedge shows a substantial difference in the structural elements between check and contaminated materials (epidermis, vascular bundles, parenchymal cells, etc.). Dark spots and bands associated with the presence of a huge number of heavy metals are clearly seen on the background of green matrix.

Acknowledgments
This work was supported by the grants from the Ministry of Education and Science of the Republic of Kazakhstan (a. N 32, 349).

References

[1] Izrael Yu A 1984 *Ecology and control of the natural environment* (L.: Gidrometeoizdat) 375
[2] Tyumenev S D, Dostai Zh D 2009 *Management of water supply and sanitation of Almaty* (Almaty: Kazakh National Technical University named after Satpayev K.I. Institute of Geography) p 224
[3] Esyrev O V, Kupchishin A I, Kupchyshin A A, Hodarina N N, Aryutkin K N and Tyumenev S 2013 *News of higher educational institutions. Physics*, **56(1/2)** 122 – 126
[4] Gorelik S S, Rastorguev L N and Skakov Yu A 1970 *X-ray and electron-optical analysis* (M.: Metallurgy) 370
[5] Kent A D, Yannarell A C, Rusak J A, Triplett E W ana McMahon K D 2007 Synchrony in aquatic microbial community dynamics *ISME J1* 38 – 47
[6] Wanjugi P and Harwood V J 2013 *Environ Microbial* 517 – 526

[7] Brahmi M, Belhadi N H, Hamdi H and Hassen A 2010 *J. Environ. Sci. (China)* 1218 – 1224
[8] Buonanno M, Randers-Pehrson G, Bigelow A W, Trivedi S, Lowy F D et al. 2013 *In Vitro Studies. PLoS ONE* **8**(10): e76968. doi:10.1371/journal.pone.0076968
[9] Lee E, Lee H, Jung W, Park S, Yang D, Lee K 2010 *Korean J Chem Eng.* **26** 1301 – 1307
[10] Prozina M L 1960 *Botanical micro machinery* (M.) 280

Catastrophic models of materials destruction

A I Kupchishin[1,2], B G Taipova[2], A A Kupchishin[2], N A Voronova[2], V I Kirdyashkin[2] and T V Fursa[3]

[1] al-Farabi Kazakh National University, Almaty, Kazakhstan
[2] Abai Kazakh National Pedagogical University, Almaty, Kazakhstan
[3]National Research Tomsk Polytechnic University, 634050, Tomsk, 30 Lenin Avenue, Russia

E-mail: ankupchishin@mail.ru

Abstract. The effect of concentration and type of fillers on mechanical properties of composite material based on polyimide were studied. Polyethylene terephthalate (PET, polyester), polycarbonate (PCAR) and montmorillonite (MM) were used as the fillers. The samples were prepared by mechanically blending the polyimide-based lacquer solutions with different concentrations of the second component. The concentration of filler and its class, especially their internal structure and technology of synthesis determine features of physical and mechanical properties of obtained materials. Models of catastrophic failure of material satisfactorily describe the main features depending on tension σ from deformation ε.

1. Introduction

Currently, modern industrial production is characterized by different image types of used raw materials, methods of processing and an extremely wide assortment it as-received production [1, 2]. Modern technology puts new challenges to material science. When creating new materials one of the main objectives is to improve the complex of technological and operational properties. Obtaining composite materials (CM) is solution of problem. One of the main ways to create CM is injection of fillers [3]. The optimal choice of polymers for mixing provides obtain material with properties that do not possess any of the used components. Moreover, the main goal is directed on material strength, change in a wide range of physical-mechanical, chemical, and optical properties, and etc. Reaction of polymer (matrix) with the filler particles at the interface characterizes compositional modification of material properties. In this regard, significant attention is paid to thermoplastic matrix. Type of filler and its concentration significantly affects to properties of polymer composites [4, 5]. As it is known, physical and mechanical properties of material are significantly determined by contained amount of defects in it and probability of occurrence of various rearrangements with their participation. Deformation of solid body while subjected various loads (mechanical, thermal, electrical, radiation, etc.), leads to an electron-stimulated processes, causing internal (local and extended) tension. It is divided to an elastic, wherein shape and dimensions of the body substantially reduced and plastic in which the material does not return to its original state. In general, fairly complicated processes are occurred during long-term operation and high mechanical and thermal loads in solids. Elongations, commensurate with length of sample are observed in the polymers, and even exceed it.

There has not yet created general theory describing laws and mechanisms of processes in polymers under influence of ionizing radiation so far. Study of changes in physical and mechanical properties of polymer composites under influence of high-energy particles, causing significant changes in the structure, has scientific and practical interest [6]. The effect of different fillers on physical and mechanical properties of a series of composite films was studied and models of destruction were suggested in this paper.

2. The experimental procedure

Polyethylene terephthalate (PET, polyester), polycarbonate (PCAR) and montmorillonite (MM) were used as the fillers. The samples were prepared by mechanically blending the polyimide-based lacquer solutions with different concentrations of the second component. Composites were previously irradiated, and then subjected to mechanical testing. The samples were irradiated on electronic linear accelerator ELU-6 in the air at 25 °C with energy of 4 MeV. The average current density of beam on the sample was 0.5 mA/cm2. Pulse width corresponds to 5 ms with a frequency of 200 Hz. The irradiation dose was calculated according to standard procedure [7]. The absorbed dose was 250 kGy.

Studies of strength characteristics of these materials is carried out by mechanical tensile testing on tensile testing machine type RMU-0.05-1 with the speed of clips 36,09 ± 0,05 mm/min. Moving capture associated with the measurer does not exceed 0.1 mm. Uniaxial tension was measured in a constant load and temperature (20 ± 2) °C, relative humidity (45 ± 5)%. Computerized installation had appropriate software in standard Windows - applications [8]. The dependence of the elongation ε on the tension σ (up to the limit of material strength) are measured.

Workspace samples in the form of films were 50 mm (length) 5 mm (width). The thicknesses of the materials were equal: from polyimide – 35 microns, from composite materials – (70 ÷ 140) mm.

3. Results

Table 1 show the relative elongation and tension unirradiated polymer compositions based on PI. It can be seen that the introduction of the filler PET 5 wt.% leads to improved mechanical properties of the material.

Table 1. Dependence of the elongation from tension for the non-irradiated composites

PI		PI_{AB} + 1 wt.% MM		PI_{AB} + 2 wt.% PCAR		PI_{AB} + 5 wt.% PET	
ε (%)	σ (MPa)	ε (%)	σ (MPa)	ε (%)	σ (MPa)	ε (%)	σ (MPa)
36	51.7	45.8	36.7	59.5	34.2	229.2	72

The table 1 shows that the CM with MM filler has characteristic σ drop by 29% as compared with polyimide and increase plasticity by 27%.

Introduction concentration of 2 wt.% polycarbonate reduces the strength by 34% and to increase the plasticity by 65% compared with the pure polyimide (table 1).

The strength increases by 37% in samples with concentration of 0.5 wt. % PET; further increasing concentration (up to 5 wt.%) does not cause any significant changes in the strength, but causes a monotonic increase in elongation 6.7 times in comparison with the pure polyimide.

Changing strength and plastic properties of composite material based on montmorillonite due to the fact that the filler comprises interlayer water. In preparation technology of CM mixing polyimide with montmorillonite is carried out at 170 °C, promoting release of -HOH- radical from filler to the matrix, which does not completely react with the polyimide macromolecules. As a result, water accumulates in the matrix and causes destruction [9]. Changes associated with the introduction of polycarbonate determined by the characteristics of mesh matrix of polyimide on the one hand and with high rigidity chain polycarbonates on the other.

Improved mechanical properties of composite materials with a filler of PET due to the fact that there is structuring circuits polyimide with plastic carcass of terephthalate. Further, dependency of elongation from the tension for irradiated materials was studied (Table 2).

Table 2. Dependence of the elongation from tension for the non-irradiated composites

PI		PI_{AB} (D = 250 kGy)		PI_{AB} + 1 wt.% MM (D = 250 kGy)		PI_{AB} + 5 wt.% ПЭТФ (D = 250 kGy)	
ε (%)	σ (MPa)	ε (%)	σ (MPa)	ε (%)	σ (MPa)	ε (%)	σ (MPa)
36	51.7	28.4	39.6	37.3	37.8	137.2	70.3

The electron irradiation dose of 250 kGy causes increase in plasticity for 1 wt. % MM by 3,6% and strength decreases by 27% from initial non-irradiated polyimide (Table 2). For compositions with 5 wt.% PET electron irradiation leads to increase in the plasticity of 3.8 times and increase in strength by 36%. The data show that samples of PCM following radiation exposure nature of interaction of structure of the matrix with the filler and between reaction products varies. The electron irradiation of the composite material with the filler of montmorillonite causes ionization of interlayer water contained in the structure of MM and its interaction with radicals of polyimide, which leads to hydrolytic degradation. Also, the irradiation results in the release of water from the matrix of polyimide. Ionized water reacts with dedicated -HOH- radicals of polyimide, forming accumulations of water in the matrix. This process also leads to degradation of polyimide and, as a consequence - in degradation in mechanical properties of CM. Increasing content of PET final structure of film becomes more porous with a pore size of 40 nm. Effects of electron irradiation lead to the predominance of cross linking process, which leads to structuring of polyimide chains to carcass of polyethylene terephthalate. This improves mechanical properties of composite material.

To describe physical and mechanical properties we proposed catastrophic models of materials destruction [10].

In terms of the physical meaning of the most correct it is to establish not depending on σ from ε, and ε from σ, as a function of deformation and the arguments - tension. The dependence of σ from ε has no physical meaning.

In this model, change in elongation is represented as:

$$dl = l d\sigma * (\sigma_0)^{-1},$$

where l – length of sample, σ – tension, σ_0 – generalized module of strength.

Integrating the function l from l_0 to l and argument σ from 0 to σ, and considering deformation $\varepsilon = \Delta l / l$, we get:

$$\varepsilon = e^{\frac{\sigma}{\sigma_0}} - 1, \quad (1)$$

Or

$$\sigma = \sigma_0 \ln(\varepsilon+1), \quad (2)$$

and σ_0 – is the tension at which the ε + 1 increases e-times.

Expanding the expression (1) in a row, we get

$$\varepsilon = \frac{\sigma}{\sigma_0} + \frac{\sigma^2}{2!\sigma_0^2} + \frac{\sigma^3}{3!\sigma_0^3} + \ldots \quad (3)$$

When $\sigma << \sigma_0$ this ratio becomes classic Hooke's law.

Hooke's law describes a linear (elastic) of the curve of σ from ε, but often does not cover whole course according to σ = σ (ε).

Similarly, we obtained the dependence of ε from σ for exponentially square, parabolic and other models:

$$\varepsilon = e^{\frac{\sigma^2}{E_1^2}} - 1, \qquad (4)$$

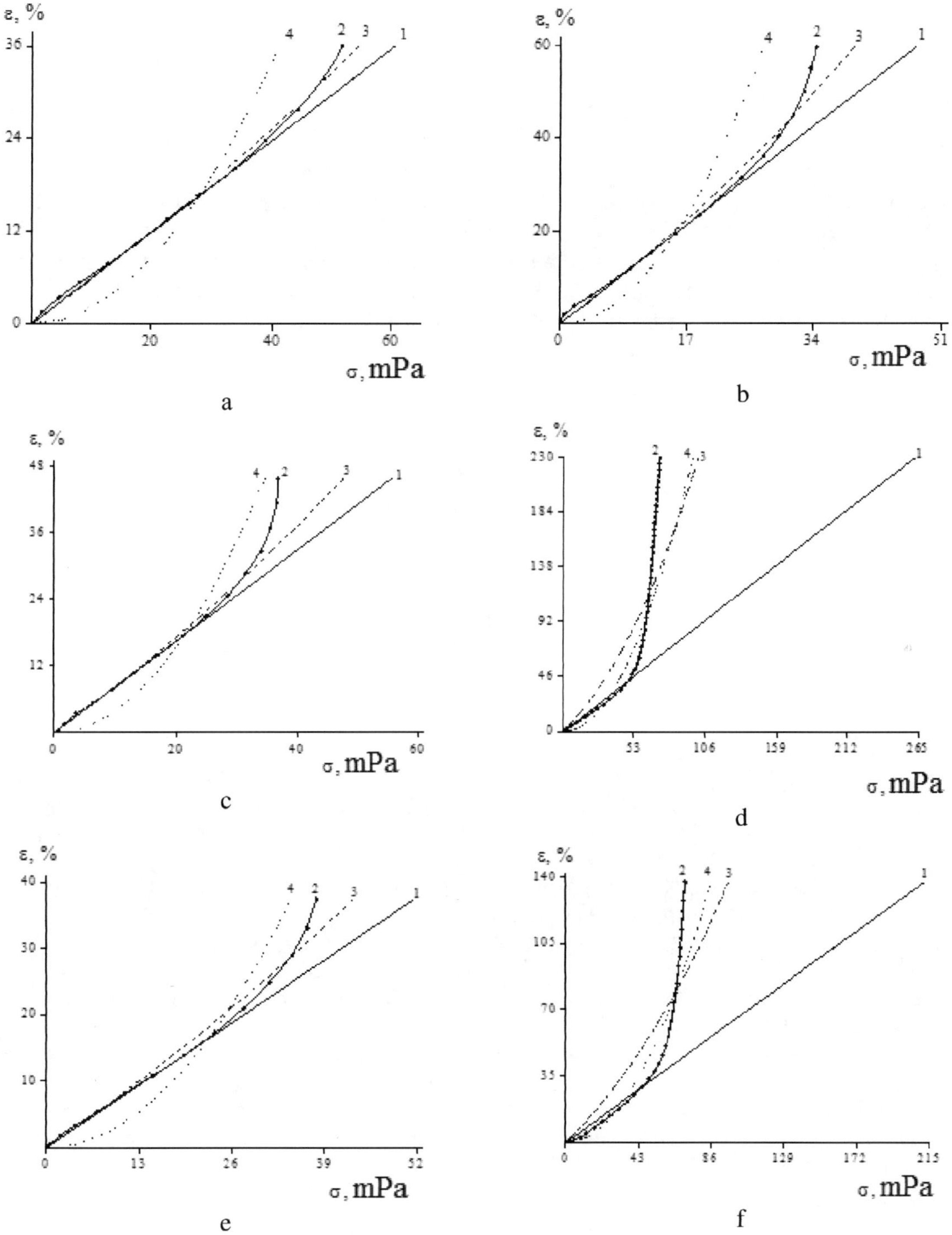

a – PI_{AB}; b – PI_{AB} + 2 wt.% PCAR; c – PI_{AB} + 1 wt.% MM; d – PI_{AB} + 5 wt.% PET; e – PI_{AB} + 1 wt.% MM irradiated (D = 250 kGy); f – PI_{AB} + 5 wt.% PET irradiated (D = 250 kGy)

1 – Hooke's law, 2 – experiment, 3 – exponential model; 4 – parabolic model

Figure 1. Dependence of elongation from tension for composite materials.

$$\varepsilon = \sigma^2 \cdot \left(E_2^2\right)^{-1}. \quad (6)$$

Figure 1 shows the experimental and calculated dependences of ε from σ for different models and composites.

Calculations based on ε from σ unirradiated composite materials (Figure 1) produced by the formulas (1), (3), taking into account only the first term (Hooke's law), and (6) for different concentrations of the second component.

Calculations show that, depending on ε from σ for non-irradiated and irradiated composite materials produced by different formulas for different concentrations of the second component, it is better to describe the experimental data exponential model.

4. Conclusions

Based on obtained results the following conclusions:

Study of effect of fillers allows creating a material with desired properties. The concentration of filler and its class, especially their internal structure and technology of synthesis determine features of physical and mechanical properties of obtained materials.

Various models are proposed under uniaxial loading materials. The best agreement with experiment gives the exponential model.

Acknowledgments

This work was supported by the grants from the Ministry of Education and Science of the Republic of Kazakhstan (a. N 32, 349) and by The Ministry of Education and Science of the Russian Federation in part of the science activity program.

References

[1] Shevchenko V G 2010 *Fundamentals of physics of polymer composite materials* (M.) p 98
[2] Lappan U, Fuchs B, Geifiler U, Scheler U and Lunkwitz K 2002 *Polymer* **43** 4325
[3] Panova L G 2010 *Fillers for polymeric composite materials* (Saratov: sarat. state. tech. uni.) p 68
[4] Kudaikulov S K, Iskakov R M, Kravcova V D, Umerzakova M B, Abadie M and Zhubanov B A 2006 *Polymers for special purposes* (Almaty) p 310
[5] Kudashov S V, Urmancev U R, Selezneva G V, Rahimova N A and Zheltobruhova V F 2012 *Journal of applied chemistry* **85(11)** 1860 – 66
[6] Komarov F F, Taipova B G, Kupchishin A I and Muradov A D 2013 *Advanced Materials* **4** 53 – 58
[7] Serikov L V, Urmazov T A and Shian L N A method of dosimetry of ionizing radiation. A.p. USSR N 1544030 from 14.12.87
[8] Discant G A, Zamanova S K, Kupchishin A I, Kolesov G Y and Muradov A D 2005 *Bulletin of Kazakh National University. Physical series* **2(20)** 75 – 79
[9] Edited Koritski U V 1987 *Handbook of Electrotechnical materials* (M.) **2** 93 – 99
[10] Kupchishin A I, Taipova B G, Kupchishin A A and Kozhamkulov B A 2015 *Mechanics of Composite Materials* **51(1)** 159 – 164

IOP Conf. Series: Materials Science and Engineering **110** (2016) 012038 doi:10.1088/1757-899X/110/1/012038

Computer simulation and study of the IR spectra of irradiated polymer materials

V Z Gabdrakipov[1], A A Kupchishin[1], A I Kupchishin[1, 2], K B Tlebaev[1] and V A Gyngazov[3]

[1]Abay Kazakh National Pedagogical University
[2]Al-Farabi Kazakh National University
[3]National Research Tomsk Polytechnic University, Tomsk, Russia

E-mail: ankupchishin@mail.ru

Abstract. Within the framework of MDDP the calculations of IR spectra have been performed for the initial polyethylene terephthalate (PET) and its state with the cation which satisfactorily describing experimental data. The calculations were made using the MDDP (modified neglect of diatomic differential overlap), implemented in the program MOPAC 2012. A feature of the method is that the overlap integrals are replaced by the δ-function, single-site Coulomb and exchange integrals are calculated by atomic parameters Slater-Condon. In recent years the studies of the structure and physico-chemical properties of polyethylene terephthalate and composites made on it basis have got the intensive development. Comparison of calculations with experimental data gives their satisfactory agreement.

1. Introduction

It is known that many physical and chemical properties of polymers, including optical, depend on the type and internal structure of the material, its energy characteristics, manufacturing conditions, the parameters of external loads, etc. In recent years the studies of the structure and physico-chemical properties of polyethylene terephthalate and composites made on it basis have got the intensive development.

At present PET is one of the basic polymeric materials, from which are made nuclear membranes having wide application in industry [1, 2].

The basis for preparing these membranes is the irradiation polyethylene terephthalate films (thickness of 10 to 300 microns) by heavy ions (argon, xenon, Gold et al.) followed by etching and obtaining pores with diameter of 10 to 300 nm and with a density of 10^9 cm^{-2}.

In the theoretical studies, one of the main methods for studying the structure of molecules is the method of IR-spectroscopy, because, as a rule, the absorption spectra of substances in the 500–500 cm^{-1} are investigated where you can explore the bands are the most characteristic for molecular structures [3 – 5].

This method is rather informative method of analysis and identification of organic compounds [6 – 9]. In this paper, the studies of the IR spectra in the unirradiated and irradiated polyethylene terephthalate films have been carried out.

2. Calculation procedure

The calculations were made using the MDDP (modified neglect of diatomic differential overlap), implemented in the program MOPAC 2012. As models were used PET chains for the ground state and the state with the cation. A feature of the method is that the overlap integrals are replaced by the □-function, single-site Coulomb and exchange integrals are calculated by atomic parameters Slater-Condon. Two-center electron-electron integrals are asymptotically Coulomb, and at zero internuclear distance, they tend to the corresponding single-center integrals. The skeleton repulsion is calculated from the selected formula (asymptotically Coulomb forces) with optimized parameters. The problem with this is that in one integral is summed nuclear-nuclear repulsion and attraction of the nuclei; in the electronic cores of another nucleus. In addition, it is also taken into account repulsion of the cores themselves.

3. Results

In the result of calculations it was obtained that for the initial state and variant with the cation, the location of the peaks in the frequency spectrum is the same. Meanwhile, their amplitudes vary considerably. Comparison of the experiments with the initial and with the cation gives good agreement.

Figures 1 and 2 show the calculated and experimental infrared (IR) spectra of polyethylene terephthalate (non-irradiated and irradiated).

Tables 1 and 2 show the results of calculations and interpretation of the absorption bands on the basis of the calculation is given. We took into account only those bands that were in the spectral range 2000 – 400 cm-1 and the moment of transition which was greater than 1.0 (approximately 15% of the total).

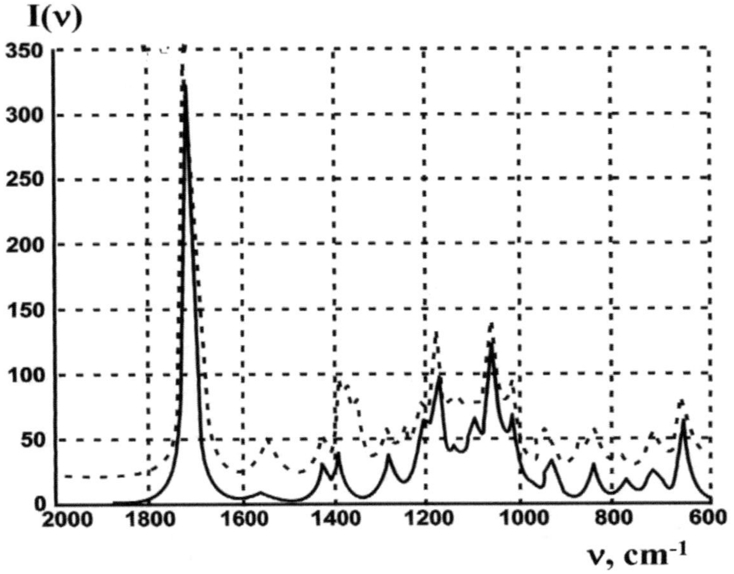

Figure 1. The IR spectra of PET (initial – solid curve and with cation – dotted line).

The following notation is used: *d* – bending vibration, *ν* – stretching vibration, (C = O) – a carbon atom of a carbonyl group, (C = C) – the carbon atoms of the vinyl group, *ar* – vibration of benzene rings, *i* – the group is inside molecule.

Since the calculation only gives a line spectrum, then for comparison with the experiment, he is subjected by additional processing – calculate the convolution integral, with so called instrument function that simulates broadening of the observed lines due to the finite width of the entrance slit and the thermal motion:

$$I(\nu) = \sum_i \int I_0(\nu_i) F(\nu - \nu_i) d\nu,$$

where I_o (ν_i) is the intensity of the i-th band line spectrum, F($\nu - \nu_i$) is the instrument function, which is usually taken either as Gauss function or Lorentz; $I(\nu)$ is the resulting intensity. As a result the weak bands are hiding under the "tails" of the more powerful. The convolution integral is calculated using specially written programs and package MATLAB. We used the function of the Lorentz with a half width 10 cm^{-1}. For comparison of theoretical calculations with the experiment were used the data of the work [10] for forming membranes in PET films irradiated xenon ions with energy of 450 MeV and a dose of $3\cdot10^9$ cm^{-2}.

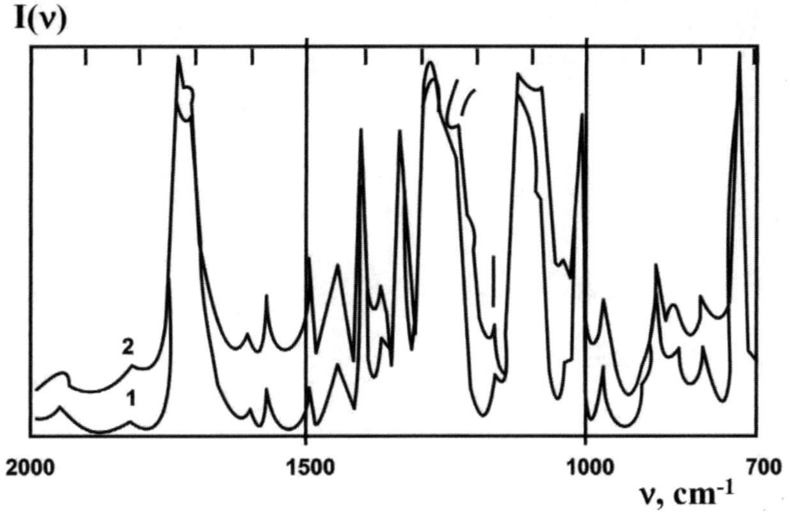

Figure 2. The experimental IR spectra of PET (initial and irradiated).

Table 1. Parameters IR spectra for initial polyethylene terephthalate.

Frequency (sm^{-1})	Moment of transition	Type of transition	Frequency (sm^{-1})	Moment of transition	Type of transition
715	1.35	d COC	1179	2.79	v CC + d CCH
776	1.20	d HCH	1189	1.44	v CC +v CO
937	1.78	d CCC ar	1210	2.65	d HCH
1040	1.28	d CCC ar	1292	2.24	Vv CC + v CCar
1054	1.06	d COCH	1393	1.34	d CCC ar
1058	1.72	d CCC ar	1429	1.35	d CCC ar
1061	5.04	d COC	1710	3.03	v C=O
1165	1.31	d CCH + vd CCC	1.712	2.87	v C=O
1066	0.93	vd COC + d CCH	1714	11.15	v C=O
1092	1.16	v CC + v CO	1717	2.70	v C=O
1101	1.76	d CCH	1719	6.42	v C=O
1131	1.10	d COC	1720	2.17	v C=O
1143	1.36	d CCH	1721	5.83	v C=O
1173	3.23	d CCC + d CCO			

The film thickness was 38 micrometers, density - 1.39 g/cm^3 (Figure 2). In the first approximation, it was believed that xenon ions in the first stage of passage knock out an electron from a macromolecule of polyethylene terephthalate that corresponds to PET with cation. The figure shows that the majority of experimental and theoretically calculated bands in the frequency range 1800 – 700 cm^{-1} is the same (in particular frequency bands with 1719; 1429; 1393; 1292; 1189; 1101; 937; 776; 715). However, it should be noted that by the intensity peaks are not always quantitatively described in terms of this model due to changes in the ratio of the middle and end of fragments.

Table 2. Parameters of IR spectra of PET with the cation.

Frequency (sm^{-1})	Moment of transition	Type of transition	Frequency (sm^{-1})	Moment of transition	Type of transition
714	1.11	d CCO	1146	1.15	v CC ar + d CCH ar
721	1.15	d CCO + d CCH	1177	2.23	v CC + d CCH
729	1.13	d CCO	1179	1.24	v CC ar + d CCH
790	1.03	d OCC + d OCCH	1181	1.19	d HCH
878	1.48	v CO + v CC	1185	1.44	v CC
944	1.22	d CCCC ar	1189	1.12	d HCH
951	0.94	d CCCC ar	1190	0.95	v CC +d CCO
953	1.15	d CCCCCC ar	1209	1.58	d HCH
1006	0.97	d CCO + d CCH	1226	1.14	d HCH
1013	0.92	d OCH	1252	1.93	v CC + d CCO
1021	1.87	v CO	1307	1.17	v CO + v CC
1022	1.56	v CO + d HCH	1355	4.75	v CO + v CC
1043	1.05	v CC ar	1376	4.30	v CO + v CC ar
1045	1.02	v CO + d HCH	1396	3.04	v CO +v CC ar
1054	1.28	v CO	1430	1.07	v CC ar + d CCC ar
1060	2.25	v CC + v CO + d COC	1528	1.17	v CC ar
1062	2.10	d OCH + d HCH	1545	2.10	v CC ar
1062	2.01	v CO	1693	7.00	v C=O + v CO
1067	1.71	v CC + d CCO	1700	6.78	v C=O
1074	1.27	v CC ar	1711	6.49	v C=O
1090	0.95	v CC + v CO	1718	4.01	v C=O
1099	1.09	v CC + v CO	1721	5.31	v C=O
1102	1.33	v CO + d OCH	1729	5.52	v C=O
1114	1.07	v CO +d OCH	1733	5.64	v C=O
1138	1.73	v CC + d CCC ar			

However, in our view, the model and the results are satisfactory for predicting features of the vibrational spectra of molecules of polyethylene terephthalate and other polymeric materials.

4. Conclusion

1. IR spectra calculations for polyethylene terephthalate for the initial state and a state with a cation were made.
2. Comparison of calculations with experimental data gives their satisfactory agreement.

Acknowledgments
This work was supported by the grants from the Ministry of Education and Science of the Republic of Kazakhstan (a. N 32, 349) and The Ministry of Education and Science of the Russian Federation in part of the science activity program.

References
[1] Charlesby A 1954 Atomic radiation and polymers (Pergamon, New York)
[2] Lunkwitz K, Lappan U, Fuchs B, Scheler U 2004 J. Fluor. Chem. 125 (5) 863
[3] Smith A A 1982 Applied Infrared Spectroscopy: Per. from English (M: Mir) 328
[4] Bellamy L 1963 Infrared spectra of complex molecules (M: Izdat.in.lit) 489
[5] Tarutino L I, Pozdnyakov FO 1986 Spectral analysis of polymers (L: Chemistry) 534
[6] Hummel D O 1997 Polymers, additives, surfactants and their mixtures: Computer-based search with special FTIR libraries: Pap.12th Eur. Symp. Polym. Spectrosc. (ESOPS-12), Lyon, Macromol. Symp., **119** 65 – 77
[7] Tlebayev K B, Kupchishin A I, Kusainov A T, Pivovarov S P and Aryutkin K N 2011 Radiation-thermal processes in polytetrafluoroethylene (Almaty: Ed. KazNPU. Abaya Ltd. "Kama") 271
[8] Kamanina N, Iskandarov M O and Nikitich A A 2003 Technical Physics Letters **29** 16 29 – 37
[9] Tarasevich BN 2012 Basics of IR spectroscopy with Fourier transform Training in Training (Manual, M.) 200
[10] Takahashi S, Yoshida M, Asano M et al. 2004 NIM. **B 217** 435

Computer simulation radiation damages in condensed matters

A I Kupchishin[1,2], A A Kupchishin[2], N A Voronova[2], V I Kirdyashkin[2] and V A Gyngazov[3]

[1]Al-Farabi Kazakh National University, Almaty, Kazakhstan
[2]Abay Kazakh National Pedagogical University, Almaty, Kazakhstan

E-mail: ankupchishin@mail.ru

Abstract. As part of the cascade-probability method were calculated the energy spectra of primary knocked-out atoms and the concentration of radiation-induced defects in a number of metals irradiated by electrons. As follows from the formulas, the number of Frenkel pairs at a given depth depends on three variables having certain physical meaning: firstly, C_d (E_0, h) is proportional to the average energy of the considered depth of the PKA (if it is higher, than the greater number of atoms it will displace); secondly is inversely proportional to the path length λ_l for the formation of the PKA (if λ_l is higher than is the smaller the probability of interaction) and thirdly is inversely proportional to E_d. In this case calculations are in satisfactory agreement with the experimental data (for example, copper and aluminum).

1. Introduction

In recent years in connection with the intensive development of nuclear power, studies of various physical processes in the near-Earth space, particularly in the Earth's radiation belts, the study of the influence of the electron, gamma and nucleon irradiations on the properties of materials has become one of the most important directions in solid state physics [1 – 5]. In the bombardment of solid bodies by charged particles, such as electrons, along the path of their movement are generated not only light particles such as secondary electrons (when energy is lost mostly by ionization of atoms), and atoms and ions of the medium which are the progenitor of the atom-atom cascades. The energy spectrum of primary knocked-out atoms (PKA) at various depths in the material depends on the energy, charge, mass of the incident particles and the type of target (atomic mass, density), integral and differential cross sections of particle-atom and atom-atom collisions and energy losses on ionization and excitation.

When propagating through medium the primary knocked-out atoms generate secondary knocked-out atoms, and then tertiary, etc. Ultimately, in the solid after the first process stage are formed the radiation defects such as vacancy-interstitial atom, divacancy and two knocked out atoms and so on [1, 2, 5 – 9].

2. The main results

Let us consider the multiple formation of PKA within the cascade-probabilistic method [10] and the possibility of using cascade-probability function in these processes. To define of the free path for the formation of the primary knocked out atoms (PKA) for electrons λ_l we use the data of calculations of the cross sections σ_d (E_l). We remark that when calculating of the total free path R was taken into account the following. When decelerating of the electrons with initial energy E_0 there comes such moment when their energy becomes less E_{th} (minimum energy for the formation of Frenkel pairs), and therefore further formation of PKA P_1 (E_2) is not possible. Therefore, this "residue" was subtracted from the total free path. Let us compute the spectrum of PKA within the cascade-probabilistic method. We use a simple cascade-probability function [10].

Let the electron after going depth i-times have been elastically interacted with the formation of PKA. Their power spectrum P_i (E_2) in the elementary act (normalized) can be written as:

$$P_i(E_2)=\frac{d\sigma_i(E_1,E_2)}{dE_2\sigma_i(E_1)},\quad \int_{E_d}^{E^i_{2max}} P_i(E_2)dE_2=1,$$

where $d\sigma/dE_2$ is the differential and $d\sigma_i(E)$ is the total cross section; E_1 is the energy of an electron at a depth h. As the second-knocked out atoms are taken into account by the cascade function ν (E_2) then the energy spectrum at a depth h for PKA can be too described by the function.

Here it is assumed that the electron and PKA retain the original direction of motion, the spectrum of PKA $W(E_0, E_2, h)$ at a depth h is determined by the formula:

$$W(E_0,E_2,h)=\sum_{i=0}^{\infty}\int_0^h\left(\frac{h'}{\lambda_1}\right)^i*\frac{1}{i!}e^{-\frac{h'}{\lambda_1}}P_i(E_2)e^{-\frac{h-h'}{\lambda_2}}(\lambda_2)^{-1}dh'\cdot(\lambda_1)^{-1}, \quad (1)$$

where λ_2 is the free path of the atom-atom-atom displacements h′ is the depth of generating PKA. When electron irradiation with energy up to 10 MeV arise low energy PKA (their average energy is 200 eV) and therefore are not able to reach greater depths. Consequently, in (1) it is necessary to choose a real physical area of integration. It must obviously be of a size comparable to the mean free path of PKA. The energy of an electron at a depth h was defined as $E_1 = E_0 - \Delta E$, where ΔE is the ionization loss of electrons. These losses were calculated using a modified Bethe-Bloch formula. As a result of it was obtained that the value of $W(E0, E2, h)$, depending on the E_2, is a decreasing function for all values of depths (Figure 1) and at low electron energies spectrum of PKA is much milder, at large E_0 is the tougher.

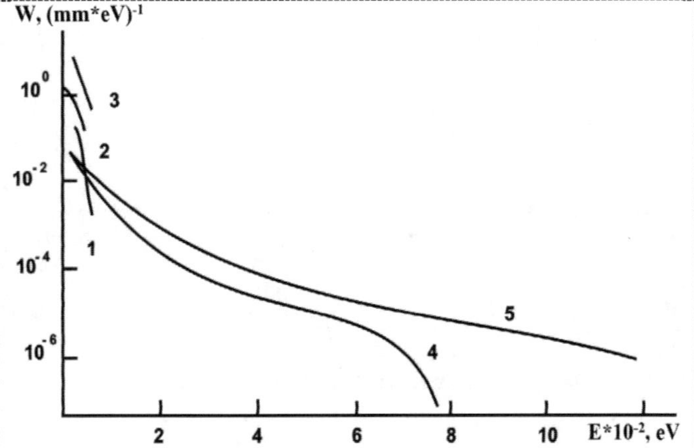

1 – 0,1 mm; 2 – 0,4; 3– 7; 4 – 4; 5 – 2

Figure 1. The spectra of PKA at different depths in the copper when it is irradiated with the electrons with the energy E_I = 1 (1, 2) and 10 (3-5) MeV.

Let define the average energy $\overline{E_2}$ of PKA:

$$\overline{E_2}=\int_{E_d}^{E_{2max}} E_2\frac{d\sigma}{dE_2}\cdot\left(\int_{E_d}^{E_{2max}}\frac{d\sigma}{dE_2}\right)^{-1}. \quad (2)$$

From the calculations of the average energy of PKA it follows that it is slowly changing with the change of E_1 and does not exceed 200 eV (Figure 2).

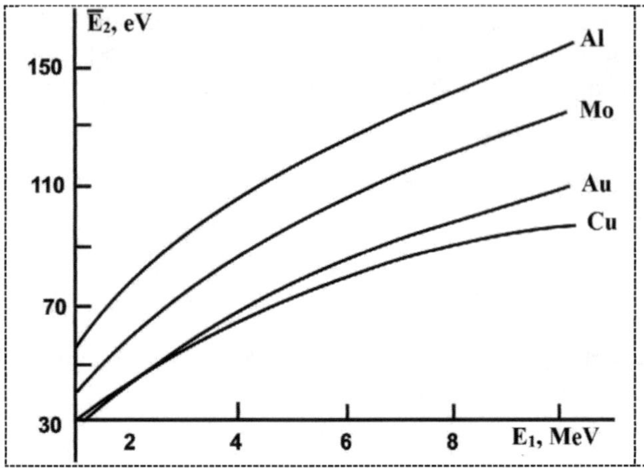

Figure 2. The dependence of the average transferred energy from energy of the electrons E_1 for different elements.

Further by means of the spectrum of PKA it can calculate the concentration distribution of Frenkel pairs in depth of the irradiated sample [10]:

$$C_d(E_0,h) = \int_{E_d}^{E_{2max}} \nu(E_2) W(E_0,E_2,h) dE_2, \tag{3}$$

where $\nu(E_2)$ is the cascade function; E_d is the threshold energy of displacement; E_{2max} is the maximum energy of the PKA.

As follows from the formulas (1), (3), the number of Frenkel pairs at a given depth depends on three variables having certain physical meaning: firstly, C_d (E_0, h) is proportional to the average energy of the considered depth of the PKA (if it is higher, than the greater number of atoms it will displace); secondly is inversely proportional to the path length λ_1 for the formation of the PKA (if λ_1 is higher than is the smaller the probability of interaction) and thirdly is inversely proportional to E_d.

Let us calculate the relative concentration of Frenkel pairs N_d. The calculations will produce for thin foils, i.e. when the loss $\Delta E << E_0$, which corresponds to the initial part of C_d (E_0, h). For the total number of displacements writes:

$$N_d = C_d(E_0,h)\Delta h \Phi t S,$$

where Δh is the thickness of the foil; S is the area the cross section of the sample; F is the flow; t is the time. Taking into account the total number of atoms in the sample for $N_d T = n_0 \Delta h S$ the relative concentration of point defects obtains [10]:

$$C_d(E_0,h) = 0{,}4\Phi t \overline{E_2}(h) \cdot (n_0 \lambda_1(E_0) E_d)^{-1}. \tag{4}$$

To define energy dependence of the total number of displacements per one incident electron, let us integrate (3) in depth from 0 to h_d (the maximum depth on which the electron forms another PKA), i.e.

$$C_d(E_0) = \int_0^{h_d(E_0)} C_d(E_0,h) dh. \tag{5}$$

Let us compare our calculations with the experimental values on a gain of residual electrical resistivity $\Delta\rho$ at low temperatures [2] (T ≃ 20, 4 K). In this case $\Delta\rho$ was determined by the formula:

$$\frac{\Delta\rho}{\Phi t} = \Delta\rho_f C_d(E_0) \cdot (n_0 h d(E_0))^{-1} \cdot 100\%, \tag{6}$$

where $\Delta\rho_f$ is the increase in the electrical resistivity of Frenkel pairs.

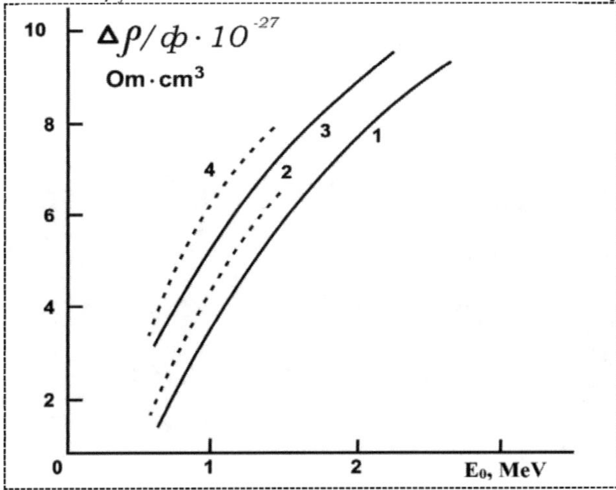

Figure 3. Dependence of the growth of the residual electrical resistivity in the Cu (1) and Al (3) on the energy of electrons: 1, 3 – calculation; 2, 4 – Experiment [2].

When calculatings of $\Delta\rho/\Phi t$ were used the following values for Al and Cu: $\Delta\rho_{Al}$ = 3, 4·10^{-6} ohm·cm/at. %, $\Delta\rho_{Cu}$ = 1, 3·10^{-6} ohm·cm/at. %. The dependences $\Delta\rho(E_0)/\Phi t$ are shown in Figure 2. In the interval of electron energies E_0 = 0.6 – 1.3 MeV (corresponding to the experiment [2]) calculated curves differ from the experimental no more than 20%.

3. Conclusion

1. With the use of the simple cascade-probability function an analytical expression for the energy spectrum of primary knocked-out atoms (PKA) and the concentration of radiation-induced defects have been received. The calculations of these functions for copper and aluminum were produced.
2. The calculation of the dependence of the electrical resistance on the energy of the electrons gives satisfactory agreement with experiment.

Acknowledgments

This work was supported by the grants from the Ministry of Education and Science of the Republic of Kazakhstan (a. N 32, 349) and The Ministry of Education and Science of the Russian Federation in part of the science activity program.

References

[1] Kupchishin A I and Potatii K B 1985 The processes of generating of radiation defects by relativistic electrons in metals: Pr. III All-Union. Meet., Radiation effects in metals and alloys (Alma-Ata: Science) 68 – 73
[2] Luccasson P G, Wolker R M 1961 Disc. of the Faraday Soc. **31** 57 – 66
[3] Pogrebnjak A D, Il'jashenko M V, Kul'ment'eva OP et al. 2001 Vacuum **62** 21–26
[4] Noli F, Misaelides P, Pavlidou E et al 2005 Applied Surface Science **252** 23 8043 – 8049
[5] Misaelides P, Noli F, Tyurin YN et al. 2005 Nuclear instruments and Methods in Physics Research. **B 240** 23 371–375
[6] Misaelides P, Hatzidimitrou A, Noli F and al 2004 Surf. Coat. Technol. **180-181** 290 – 296
[7] Zhang R F, Argon A S and Veprek S 2009 Phis.Rev. **79** 1 – 13
[8] Tyuirin Yu N and Pogrebnjak A D 2001 Surf. And Coat. Tech. **142–144** 293–299
[9] Pogrebnyak A D, Gritsenko B P, Kylyshkanov M K et. al. 2006 Tech. Physics Letters (Moscow) **32(12)** 1060–1063
[10] Boos E G, Kupchishin A I, Kupchishin A A, Shmygalev E V and Shmygaleva T A 2015 Relationship with Markov chains (Almaty, Abay KazNPU, RI and NCT& M Al-Farabi Kazakh National University, Ltd. "Kama") 388

IOP Conf. Series: Materials Science and Engineering **110** (2016) 012040 doi:10.1088/1757-899X/110/1/012040

Positron sensing of distribution of defects in depth materials

A I Kupchishin[1,2], A A Kupchishin[2], N A Voronova[2] and V I Kirdyashkin[2]

[1] Al-Farabi Kazakh National University, Almaty, Kazakhstan
[2] Abay Kazakh National Pedagogical University, Almaty, Kazakhstan

E-mail: ankupchishin@mail.ru

Abstract. It was developed a non-destructive method of positron sensing, which allows to determine the distribution of defects in the depth of the material. From the analysis we can conclude that the angular distribution curves of annihilation photons (well as and on the characteristics in experiments on the lifetime, 3γ - angular correlation, Doppler effect) is influenced by three main factors: a) The distribution of defects in the depth of the material, their dimensions as well as parameters of the interaction of positrons with defects. With increasing the concentration of defects the intensity $J_\gamma(\alpha, \xi)$ varies more; b) Modification of the energy spectrum of slow positrons due to the influence of defects, wherein the spectrum of positrons becomes softer, and the average energy of the positron annihilation is reduced; c) Deformation of the momentum distribution of the electrons in the region of defect. The energy spectrum of electrons is also becomes softer, and the average energy of the electrons (on which positrons annihilate) is less. The experimentally were measured spectra of photons in the zone of annihilation and were calculated the distribution of defects in depth for a number of metals.

1. Introduction

As is known, a change in the physical properties of the irradiated and deformed metals, alloys, and other substances is caused not only changes in the crystal structure of the solid body, but also the electronic structure of defects [1 – 4]. When irradiation of materials by the different particles the defects is substantially irregularly distributed over the depth of the solid body and these distributions are significantly influence on the physical and chemical properties of the substance [5 – 7]. A non-destructive method of positron sensing of defect distribution in the depth of a matter is offered by us. The positrons have a unique ability to localize in areas of the crystal with excess electronic charge inherent in conventional defects in the crystal structure that allows us to solve the linkages between electronic and atomic structure. The impact of defects on the parameters of positron annihilation in metals was found by I.Y. Dekhtyar et al. [8, 9] for nickel and its alloys with iron. It is shown that plastic deformation narrows the angular correlation curves; at the maximum the count rate increases. It is caused by the appearance of excess negative charge in the region of the defect during the formation of the stress field around the dislocations. The resulting local dipoles cause polarization of S- and d-bands and positrons are effectively captured by dislocation cores, which are confirmed by numerous experimental data on angular correlation and positron lifetime. At the same time the energy spectrum of electrons in the region of defects is undergoing significant changes, the probability of annihilation of positrons with electrons on the defects is increased by reducing their annihilation in the atomic core and due to the long lifetime of positrons in these areas. To obtain quantitative results on the change in the electronic structure of metals and alloys are currently used different models capture positrons as well as analytical ratios linking the intensity of annihilation photons with a pulsed electron spectrum.

However, these models and methods are obtained with a number of assumptions and approximations, and therefore cannot be implemented, for example, in studying the characteristics such as the distribution of defects in depth, sizes. In addition, for conducting experiments on the methods of depth-sensing and variable thickness of the absorber it is required a strict mathematical apparatus of two-photon annihilation of positrons in defective materials.

2. The main results

A detailed analysis of the previous relationships [10] for the flows of positrons in defective materials, as well as annihilation photons in experiments 2-γ angular correlation allows in a first approximation to obtain a formula for the intensity $J_\gamma(\alpha, \xi)$ depending on the concentration $n_i(h)$ defects in the form:

$$J_\gamma(\alpha,\xi)=\sum_{i=0}^{n}\int_{0}^{\xi} n_i(h)\Phi_i(\alpha,h)dh, \tag{1}$$

where α is the emission angle ξ is the maximum depth, $n_i(h)$ is the concentration of defects in the *i*-type, *h* – depth.

$$\begin{aligned}\Phi_i(\alpha,h)= & \int_{\Omega_1}\int\int_{\Omega_2}\int\int_{\Omega_3}\int\int_{\Omega_4}\int\int_{E_1}\int\int_{E_2}\int_{E_3}\int_{E_4}\int_{y}\int_{z}\varphi_i'(t,x,y,z,E_i,\Omega_i)\times \\ & \times\Phi_i(E_2,\Omega_2)d\psi_{вз}\psi_{ан}(E_1,E_2,\theta_{12})n_\gamma(E_1\div E_4,\Omega_1\div\Omega_4)\times \\ & \times\psi_{\gamma_1\gamma_2}(E_3,E_4,x,y,z)\sigma_{12}(E_3,E_4)d\Omega_1 d\Omega_2 d\Omega_3 d\Omega_4\times \\ & \times dE_1 dE_2 dE_3 dE_4 dydz.\end{aligned} \tag{2}$$

Here φ_i' is a modified stream of positrons; $\Phi_i(E_2, \Omega_2)$ is the concentration of electrons; $d\psi_{вз}$, $\psi_{ан}$ is the probability of interaction and positron annihilation; $\psi_{\gamma1\gamma2}$ is the probability of departure and the passage of annihilation photons; n_γ is the normalized energy spectrum of the annihilation photons in the elementary act; $\psi_{\gamma1\gamma2}$ is the probability of escape 2 gamma rays from the sample and passing to the detectors D_1 and D_2; σ_{12} is the detection efficiency, and $E_1 \div E_4$ и $\Omega_1 \div \Omega_4$ is the energy and the solid angle of positrons, electrons and photons; y, z is the coordinate. From the analysis of equations (1) and (2) we can conclude that the angular distribution curves of annihilation photons (well as and on the characteristics in experiments on the lifetime, 3γ - angular correlation, Doppler effect) is influenced by three main factors: 1. The distribution of defects in the depth of the material, their dimensions as well as parameters of the interaction of positrons with defects. With increasing the concentration of defects the intensity $J_\gamma(\alpha, \xi)$ varies more. 2. Modification of the energy spectrum of slow positrons due to the influence of defects, wherein the spectrum of positrons becomes softer, and the average energy of the positron annihilation is reduced. 3. Deformation of the momentum distribution of the electrons in the region of defect. The energy spectrum of electrons is also becomes softer, and the average energy of the electrons (on which positrons annihilate) is less.

The deformation of the energy and angular spectra of positrons and electrons, as well as an increase in the concentration of radiation-induced defects lead to the fact that the momentum distribution of the annihilating pairs varies considerably (becomes softer), the average energy of the pair in the C-system is reduced. Consequently, in the laboratory frame annihilation gamma-rays will be emitted at an angle close to 180 °, i.e. all this will lead to a narrowing of the angular correlation curves. The equation of the type (1) is a Fredholm integral equation of the second kind and it can be attributed to the number of incorrect equations which are difficult to solve directly. In this connection must be properly set up an experiment that could be used this ratio. At the same the detectors D_1 and D_2 arranged motionless on line 0 - 0 and the sample moves through the zone of annihilation in steps of $\Delta x = 0.01$ mm in the chamber using a special screw so that each measurement was conducted in the same position of the sample (it means that the projection on a plane perpendicular to the direction of movement of the screw) to reduce measurement errors.

The experimental intensity distribution curves J (x) of the annihilation photons in the annihilation zone are shown in Figure 1. It is seen that in the irradiated samples are observed a strong increase in the intensity of annihilation photons than annealed, and the effect is at the maximum of the curve can be up to about 20 %. The effect of increasing J (x) is associated with a more effective deceleration and capture of positrons in defects. All curves are asymmetric, with the right half of them falls substantially lower than the left. This is due to the fact that the moving sample sinking gradually overlaps the source of positrons and ultimately reduces the flow of the annihilation gamma rays to almost zero.

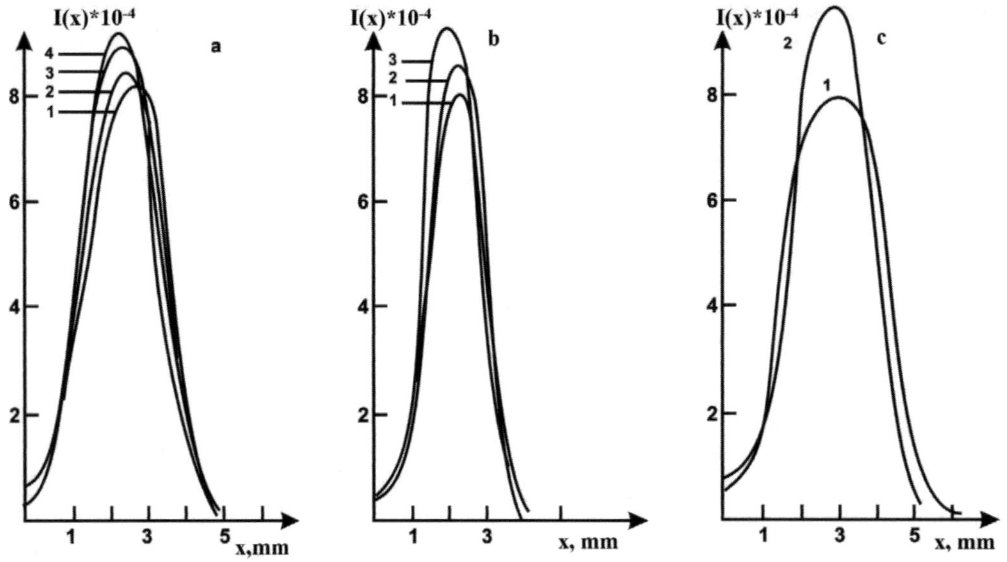

Figure 1. The experimental curves of the intensity distribution of photons in the zone of annihilation. a is the titanium annealed (I), 2 is the same, irradiated by α-particles 29 MeV to integral dose 10^{15} cm^{-2}, 3 – $5{\cdot}10^{15}$ cm^{-2}, 4 – 10^{16} cm^{-2};
b is alloy (Ti + 1,2 atm.% Zr) annealed (I), 2 is the same, irradiated by α- particles 29 MeV to integral dose 10^{15} cm^{-2}, 3 – $1.5{\cdot}10^{15}$ cm^{-2};
c is molybdenum annealed (I), 2 – irradiated by proton 30 MeV to a dose of $6{\cdot}10^{17}$ cm^{-2}

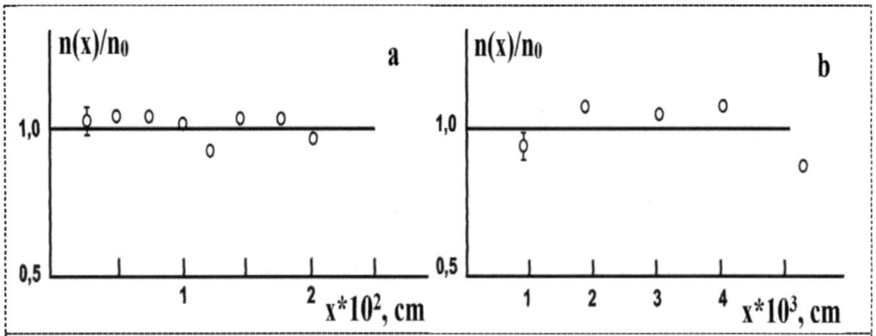

Figure 2. The dependence of the concentration of radiation-induced defects on the depth in titanium irradiated with alpha particles with an energy of 29 MeV (a) and molybdenum irradiated by protons with an energy of 30 MeV (b).

We have proposed a method of positron sensing vacancy defects in the depth of the material. In the case where the defect concentration (on which occurs the process of annihilation) is insignificant, the dependence *n (x)* on the depth can be found in the first approximation from the expression:

$$n(x) = \frac{d\left(J_2 - J_1\right)}{dx} \frac{n_0}{\Phi'(\alpha, x)}. \tag{3}$$

The experimental measurements of the curves intensity distribution of annihilation photons in the annealed and irradiated samples were performed on an automated installation of the angular correlation with program management.

In this way, inputted defects change the interaction of positrons with materials. Using this method one can determine the concentration of defects at a given depth. According to the experimental results dependencies J_1 (h) and J_2 (h) with taking into account the actual zone of positron annihilation as they pass through the sample, we obtained depth profiles of defects in titanium alloys irradiated with alpha particles with energy of 10 - 30 MeV, with application of the relations (13).

The calculation results are shown in Figure 2. It can be seen that the concentration of defects uniformly distributed over the depth of the samples of titanium and molybdenum. Thus, the method of positron depth sensing allows to determine the distribution of defects in the depth of the material. Comparing these results with those curves calculated by the method of variable thickness of the absorber and the cascade-probabilistic method [10], have them satisfactory agreement.

3. Conclusion

1. It is proposed and developed a method of positron-sensing defect distribution in the depth of the material.
2. For the first time were received the experimental intensity distributions of photons directly in the zone of annihilation for annealed and irradiated samples of Ti and Ti + 1,2 % Zr (α-particles) to different doses and on Mo, irradiated by proton on the basis of which are determined the distributions of defects in depth, consistent with theoretical calculations.

4. Acknowledgments

This work was supported by the grants from the Ministry of Education and Science of the Republic of Kazakhstan (a. N 32, 349).

References

[1] Kupshishin A I, Potatiy K V and Chukurova R M 1987 Cryst. Latt. Defects and Amorphous Mater. **I3** (**34**) 157–162
[2] Luccasson P G, Wolker R M 1961 Disc, of the Faraday Soc. **31** 57–66
[3] Howell R H 1978 Phys. Rev. **B18** (7) 3015–3025
[4] Mitchell J B, Logan G M and Echer C J 1973 J. Hucl. Mater. **48** 139–142
[5] Lisitsyn V M, Korepanov V I and Yakovlev A N 2000 *Proceeding of 11th conf. On Radiation Phys. And Chem. Of Condensed Matter.* (Tomsk, TPU) 125–128
[6] Lisitsyn V M, Korepanov V I and Yakovlev A N 2006 News of higher educational institutions. Physics **10** 98–101
[7] Korepanov V I, Lisitsyn V M and Lisitsyna L A 1997 Russian Physics Jour. **39** (11) 1082–92
[8] Dekhtyar J 1975 Electronic structure of defects in metals Metal, electronic, D lattice, Naukova Dumka 228–252
[9] Dekhtyar I Ya 1974 Physics Reports **9** C 5 243–353
[10] Boos E G, Kupchyshyn A I, Kupchyshyn A A, Shmygalev E V, Shmygaleva T A 2015 Contact Markov chains (Almaty KazNPU. Abaya, research institutes and NHT M Kazakh National University. Al-Farabi, LLP "Kama") 388

Modification of thin oxide films of MOS structure by high-field injection and irradiation

D V Andreev[1], **G G Bondarenko**[2], **V V Andreev**[1] **and A A Stolyarov**[1]

[1] Bauman Moscow State Technical University, Kaluga Branch, 2, Bazhenov Str., Kaluga 248000, Russia

[2] National Research University Higher School of Economics (HSE), 20, Myasnitskaya Str., Moscow 101000, Russia

E-mail: andreev@bmstu-kaluga.ru

Abstract. We have investigated processes of modification and changing of the charge state of MOS structures having a multilayer gate dielectric based on a thermal SiO_2 film doped with phosphorus under conditions of different modes of high-field electron injection and an electron irradiation. We have determined that negative charge, accumulating in the phosphosilicate glass (PSG) ultra thin film of the MOS structures having the two-layer gate dielectric SiO_2-PSG under conditions of both high-field tunneling injection of electrons and electron beam, could be used for a modification of devices having the same structure (e.g. correction of threshold voltage, increase of charge stability and breakdown voltage of MOS structure). We have shown that when thickness of PSG film increased, a raising of the electron traps density occurred, but the value of the cross-section of electron traps was the same. It was established that in order to obtain MOS structures with high thermal stability, one has to anneal them at 200 °C after performing irradiation treatment.

1. Introduction

Doping of the gate dielectric, based on thermal silicone dioxide, by phosphorus can be applied especially in order to stabilize the characteristics of devices [1–6]. Phosphosilicate glass (PSG) film can take place also in case of using polycrystalline silicon gates doped by phosphorus [7]. The fact of the PSG film presence considerably changes the character of charge state changing of MOS structures under both the high-field tunneling injection of electrons in dielectric and irradiation in comparison with the structures based on thermal SiO_2 films which are not doped with phosphorus. Currently the gate dielectric based on SiO_2 passivated by ultra thin PSG film is widely used for field effect devices based on SiC [3, 4]. The main problem in creation of the dielectric films for the semiconductor devices based on the MOS structures allowing the parameters of the devices to be controlled by the high-field injection modification and the irradiation modification after their manufacturing is fabrication of the required optimal structure of the dielectric film providing the effective trapping of the charge carriers on the traps and having the high injection and irradiation stability as well as low charge defectiveness [2, 3, 6–8]. Everything mentioned above is a cause of demand for an integrated and comprehensive study of the technological process of SiO_2 films doping with phosphorus and also the SiO_2-PSG structure study in order to optimize the parameters of the dielectric film needed to increase the stability and reliability of MOS devices.

In this work we have considered a method of MOS structures modification based on SiO_2 film doped by phosphorus under the high-field tunneling injection of electrons into the gate dielectric and modification by the irradiation of crystals with low-energy electrons. Also, we have studied the influence of SiO_2 phosphorous doping modes on the characteristics of MOS structures.

2. Materials and experimental techniques

The tested MOS capacitors were fabricated on Si $\langle 1\,0\,0 \rangle$ n-type wafers with a resistivity of 4.5 Ω·cm. As the gate dielectric, we used both a thermal silicon dioxide film and SiO_2 film passivated by phosphosilicate glass. Silicon dioxide having the thickness of 7÷50 nm was thermally grown by using a dry oxidation at 850÷1000 ^{0}C. PSG film having the thickness of 3÷22 nm was produced by doping with phosphorus from $POCl_3$-O_2 at 900 °C. In order to obtain experimental samples having different PSG thickness, the doping time by phosphorus was ranged from 1.5 to 7 min. After that the wafers were exposed to nitrogen annealing at 1000 °C. As the gates we used both a polysilicon (Si*) film doped with phosphorous up to 20 Ω/□ and an aluminium film. The gates had the square dimensions of 10^{-4}÷10^{-2} cm^2.

The thickness of the PSG film was calculated as the difference between thicknesses of the insulating film before and after selective etching in a mixture of HNO_3 (33 ml), HF (44 ml), and distilled water (923 ml) [8].

In order to realize the injection modifications of MOS structures, we used the FowlerNordheim high-field injection of electrons from the silicon substrate [9–11] by a constant current pulse of 0.1 μA/cm^2 to 10 mA/cm^2. During the injection process we monitored a value of voltage dropping on MOS structure that allowed us to obtain information about a change of the charge state of the dielectric film straightly at the time of the modification process. So as to determine the amount of thermostable components of an accumulated negative charge in the dielectric film after the injection and radiation treatments, the MOS structures were annealed in a range of temperatures of 150÷250 °C within the time from 3 to 30 minutes. We monitored the changes of the charge state of MOS structures by using the C-V technique and multilevel current stress technique [11]. The latter technique was to apply a special form of a current pulse to a sample, providing a charge of MOS capacity and subsequent high-field FN electron injection into the dielectric which generally carried out in the constant-current mode. During FN injection we measured the shift of the voltage drop on the MOS structure ΔV_I, characterizing a change of charge state of MOS structure [2, 6, 11]. From the C-V characteristics, we determined voltage shift (ΔV_{ss}) characterizing the surface state charge.

In order to study the influence of electron irradiation on characteristics of the MOS structures, we used a scanning electron microscope Carl Zeiss EVO 40. Using the microscope, we carried out the irradiation of the MOS structures by electrons with energy from 15 to 20 keV having a beam current of 8 nA and a fluence up to $5 \cdot 10^{14}\,\mathrm{cm}^{-2}$.

3. Experimental results and discussion

In this work we implemented a comparative study of the charge state modification of MOS structures by both the high-field tunnel injection of electrons into the gate dielectric and the electron irradiation of the MOS structures. A distinctiveness of the MOS structures having the two-layer dielectric SiO_2-PSG, being under high-field electron injection, is the accumulation of negative charge in the PSG film [1, 6, 7, 12]. In accordance with the assumptions made in [6, 12], the electrons injected into the insulator are captured by the positively charged groups located in the PSG film. Apparently these electron traps make a major contribution to the value of an accumulated negative charge in the PSG film and these traps have the capture cross section $\sigma_1 = 1.4 \cdot 10^{-15}$ cm^2 [8, 12, 13]. As shown in [12], along with the traps of the first type in the two-layer dielectric SiO_2-PSG there exist electron traps having capture cross section of $\sigma_2 = 3.2 \cdot 10^{-16}$ cm^2 which are observed under the injection of electrons from Si. One of possible explanations for the presence of electron traps of the second type can be the impact of molecules of $POCl_3$ on the SiO_2 structure [6] and, as consequence of that, a reconstruction of the silicon oxide structure near the SiO_2-PSG interface occurs what leads to the appearance of the dangling bonds of oxygen. These dangling bonds might be electron traps.

Fig. 1 shows the dependencies of the shift of a voltage drop on the MOS structure in time of the injection process modification by constant current (1), the voltage shift provided by the integrated charge in the surface states (2) and a change of these dependencies after annealing at 200 °C ($1', 2'$) on the density of injected electrons. In order to implement the mode of high-field electrons injection from the silicon substrate to the gate of MOS structure, the pulse of constant current having density of 1 μA/cm^2 was applied. In figure 1 the density of injected electrons was defined as $N = Q_{inj}/q$, where q is the electron charge; Q_{inj} is the density of injected charge.

We used the irradiation with 15 to 20 keV electrons what guaranteed us a track length of more than the thickness of the gate. We ascertained the fact that the irradiation of MOS structures having SiO_2-PSG film by electrons possessing energy enough to pass through the gate dielectric film significantly increased the density of surface states at the Si-SiO_2 interface as well as the accumulation of a negative charge in the volume of gate dielectric in PSG film. Fig. 2 shows the dependencies of the shift of voltage drop on the MOS structure under conditions of the irradiation modification (1), the voltage shift provided by the integrated charge located on the surface states (2) and also a change of these dependencies after annealing at 200 °C ($1', 2'$) on the fluence of electrons (in that case the energy of electrons is 18 keV) (2,2′). In order to measure ΔV_I and ΔV_{ss} in case of irradiation modification, the process of irradiation was interrupted. During the interruption short impulse of current of 1 μA/cm^2 was applied to MOS structure (to determine ΔV_I) [14]. Then we measured ΔV_{ss} by using of C-V characteristics. Measurements had negligible effect to the process of irradiation modification. As distinct from the high energy irradiation by electrons [15], in conditions of the low energy irradiation a change of the charge state of MOS structure is basically caused by the ionizing processes. The observed physical processes in case of the low energy irradiation in many respects are identical effects occurring in case of the high field electron injection [1].

Fig. 1 and fig. 2 show that in time of the irradiation by electrons, lower density of negative charge is accumulated in the gate dielectric in comparison with the high-field electron injection. Curves $1'$ in fig. 1 and fig. 2 characterize the thermostable component of negative charge. Fig. 1 and fig. 2 demonstrate that the thermostable component of negative charge accumulated under the injection modification (curves $1'$) highly exceeds the charge accumulated during irradiation by electrons. Therefore the modification by injection is more effective and under certain conditions of the injection causes less degradation of MOS structures (figures 1 and 2, curves 2). However the irradiation by electrons allows carrying out the group crystal treatment right on a semiconductor wafer. That significantly increases productivity of the treatment.

During the process of electrons injection from silicon if the thickness of PSG layer grows then the range of possible values of the threshold voltage of MOS transistors increases. However in order to ensure acceptable values of the surface state density, the value of injected charge at the time of adjusting of the threshold voltage should not exceed 0.3 μC/cm^2. The range of current exposure during changing of the charge state of MOS devices should be limited between $10^{-7} \div 10^{-5}$ A/cm^2. The decrease of the current stress density is accompanied by technical difficulties of implementation and is impractical as a result of the significant increase at the time of injection of the desired amount of charge. An increase of the injection current density leads to significant increment probability of the sample breakdown probability as well as to the build-up of the surface states density at the Si-SiO_2 interface [12] during the injection of electrons from Si.

Fig. 3 shows the dependences of the shift of voltage drop on the MOS structure at the time of the tunnel injection of electrons from silicon by a current pulse 1 μA/cm^2 on the value of

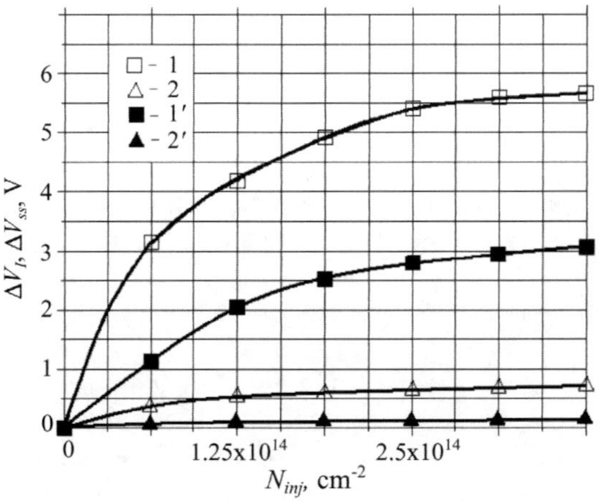

Figure 1. The dependencies of the shift of voltage drop on the MOS structure during the modification by the constant injection current (1), shift of the voltage caused by the integrated charge located on the surface states (2) and a change of these dependencies after annealing at 200 °C (1′,2′) on the number of injected electrons.

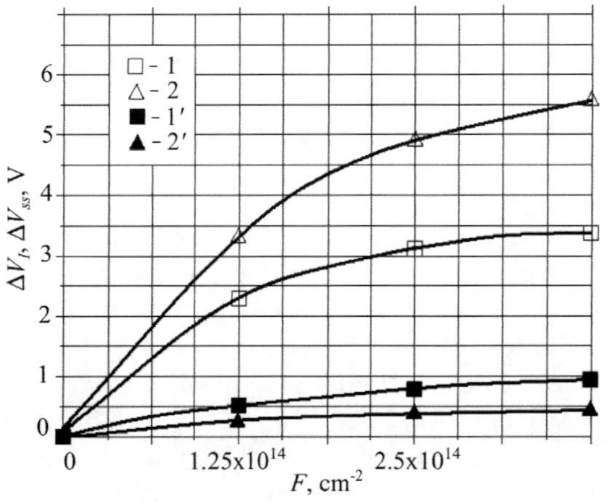

Figure 2. The dependencies of the shift of the voltage drop on the MOS structure in time of the radiation modification (1), the shift of the voltage caused by the integrated charge located on the surface states (2) and a change of these dependencies after annealing at 200 °C (1′,2′) on the fluence of electrons curves (2,2′).

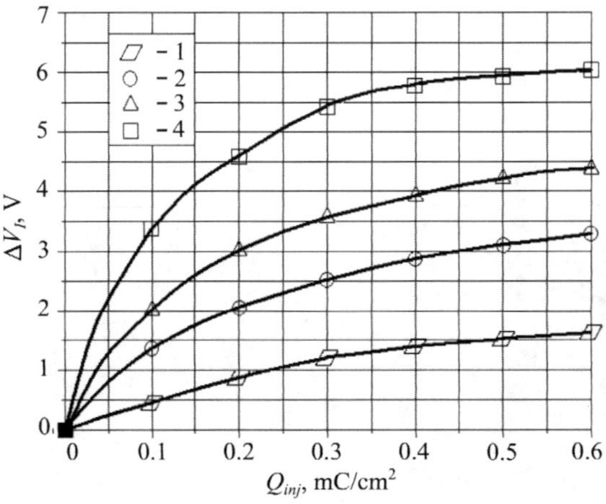

Figure 3. The dependencies of the shift of the voltage drop on the MOS structure having the thickness of the gate dielectric SiO_2-PSG of 40 nm at the time of the modification by constant injection current on density of injected charge for a different thickness of PSG film: 1 — 5 nm; 2 — 9 nm; 3 — 13 nm; 4 — 22 nm.

the injected charge for MOS structures in case of different PSG film thickness in case the total thickness of the gate dielectric SiO_2-PSG equals 40 nm. Fig. 3 shows the fact that the increase of PSG film thickness leads to extension of a range of ΔV_I possible changes on the same amount of injected charge.

Interrupting the process of injection at a certain distance of time (process of interruption does not affect the experimental curves) we were measuring the shift of flat-band voltage ΔV_{FB} by using C-V characteristics and also the change of voltage applied to the sample under the injection by the same current density at the negative polarity at the gate $\Delta V_I\,(-)$. The analysis of the obtained dependencies revealed that the centroid of the negative charge located in the PSG film closer to the SiO_2-PSG interface. We ascertained the fact that the accumulation of negative charge for all curves in fig. 3 was mostly caused by presence of electron traps having the cross section $1.4 \cdot 10^{-15}$ cm^2.

We found that both the density of accumulating negative charge and the density of the thermostable part of accumulating negative charge (after annealing) increased in case of raising of doping time of SiO_2 by phosphorus what leads to the increasing of the PSG film thickness. But the increasing of the PSG film thickness is not acceptable for thin gate dielectric films because this might be a reason of deterioration of the gate dielectric charge stability and deterioration of the gate dielectric-semiconductor interface. If the concentration of phosphorus in the PSG film is more than 1.5% then it might leads to polarization effect and decrease of charge stability of the gate dielectric [5, 8, 12].

We estimated the relaxation time constant of negative charge having high-temperature stability in the range of operating temperatures for MOS devices with due regard for the curves of the thermally stimulated depolarization currents [8]. The estimated value of the relaxation time is larger than $4 \cdot 10^8$ sec. Therefore, in order to obtain MOS devices having high-temperature stability after correction of the structures charge state, it is needed to anneal the MOS devices at temperatures not lower than 200 °C. Also, modification of the characteristics needs to be performed taking into account the relaxation of the charge part during annealing.

Accumulation in the volume of gate dielectric a desired density of the thermostable component of negative charge allows us to adjust the threshold voltage of MOS transistors and increase the breakdown voltage of the gate dielectric as well as reduce the probability of breakdown by healing "weak spots" due to the accumulation of negative charge [10, 12, 16]. Fig. 4 shows the histograms of the MOS structures distribution by the charge-to-breakdown (Q_{BD}) for the samples having SiO_2 gate dielectric (1) and for the samples having two-layer SiO_2-PSG gate dielectric (2).

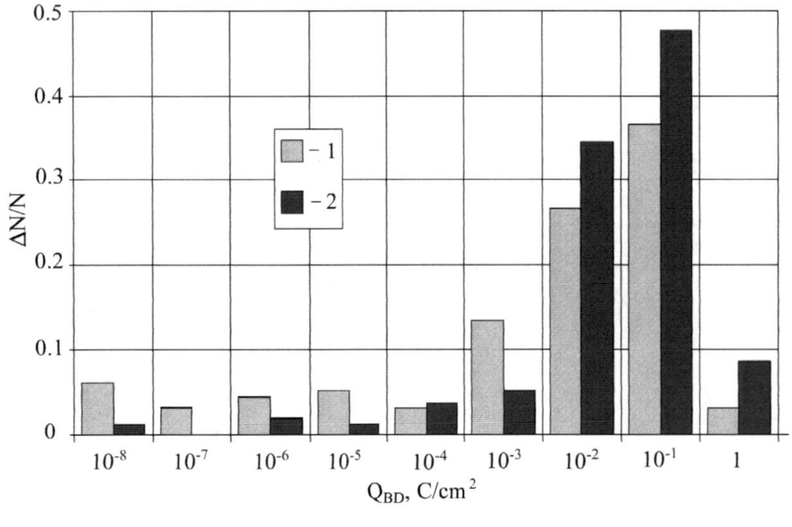

Figure 4. The histograms of the MOS structures distribution by the charge-to-breakdown for the samples having SiO_2 gate dielectric (1) and for the samples having two-layer SiO_2-PSG gate dielectric (2).

Fig. 4 shows the fact that the two-layer gate dielectric SiO_2-PSG allows to increase the average amount of charge injected into the dielectric before its breakdown and also allows to reduce the number of defective structures having a small value of the charge injected before breakdown. This effect is explained by the healing of "weak spots" in the gate dielectric due to the accumulation of negative charge and consequently by reducing the value of the local injection currents flowing in places of defects and leading to breakdown of the gate dielectric. As a result of the trapping of injected charge the potential barrier interfering the localization of currents in fields of long damages at the substrate interface raises.

4. Conclusions

We have established that the negative charge, accumulated in the PSG film in MOS structures having two-layer gate dielectric SiO_2-PSG at the time of both the high-field tunnel injection of electrons and the electron irradiation, could be used for modification of MOS devices. We have shown that the application of the high-field injection of electrons in order to modify the MOS structures charge state was more preferable than the usage of electron irradiation because in this case we could make an individual adjustment of the characteristics of each device. Also, by using selected modes of the high-field injection one may significantly reduce related degradation processes. We have ascertained the fact that both the density of the accumulated negative charge and the density of its thermostable component increased with the growth of a doping level of SiO_2 by phosphorus. The growth of a doping level of SiO_2 by phosphorus led to the increase of the PSG film thickness. We have shown that the application of two-layer SiO_2-PSG gate dielectric allowed to increase the average value of charge injected into the dielectric before charge-to-breakdown and to reduce the number of defective structures having a small value of the charge-to-breakdown. This effect is explained by the healing of "weak spots"in the gate dielectric due to the accumulation of negative charge and consequently by reduction of the value of the local injection currents.

Acknowledgments

This study was supported by the Ministry of Education and Science of the Russian Federation, state task of the Moscow State Technical University, project no.1117. Support from the Basic Research Program of the National Research University Higher School of Economics is gratefully acknowledged.

References

[1] Strong A W, Wu E Y, Vollertsen R, Suñé J, Rosa G L, Rauch S E and Sullivan T D 2009 *Reliability wearout mechanisms in advanced CMOS technologies* (Wiley-IEEE press) p 624
[2] Bondarenko G G, Andreev V V, Maslovsky V M, Stolyarov A A and Drach V E 2003 *Thin solid films* **427** 377–80
[3] Sharma Y K, Ahyi A C, Issacs-Smith T, Shen X, Pantelides S T, Zhu X and Feldman L C 2012 *Solid-State Electronics* **68** 103–7
[4] Sharma Y K, Xu Yi, Jennings M R, Fisher C, Mawby P, Feldman L C and Williams J R 2014 *International Journal of Fundamental Physical Sciences* **4** 37–42
[5] Balk P and Eldridge J M 1969 *Proc. of the IEEE* **57** 1558–63
[6] Bondarenko G G, Andreev V V, Stolyarov A A and Tkachenko A L 2002 *Vacuum* **67** 507–11
[7] Levin M N, Tatarintsev A V, Makarenko V A and Gitlin V R 2006 *Russian Microelectronics* **35** 329–36
[8] Andreev V V, Baryshev V G, Bondarenko G G, Stolyarov A A and Shakhnov V A 1997 *Russian Microelectronics* **26** 378–83
[9] Afanas'ev V V and Stesmans A 2007 *J. Appl. Phys.* **102** 081301–28
[10] Lombardo S, Stathis J H, Linder P, Pey K L, Palumbo F and Tung C H 2005 *J. Appl. Phys.* **98** 121301
[11] Andreev V V, Bondarenko G G, Maslovsky V M and Stolyarov A A 2012 IOP Conf. Series: Materials Science and Engineering **41** 012017
[12] Andreev V V, Bondarenko G G, Maslovsky V M and Stolyarov A A 2014 *Acta Phys. Pol.* A **125** 1371–73
[13] Bondarenko G G, Andreev V V, Drach V E, Loskutov S A and Stolyarov M A 2006 *Thin Solid Films* **515** 670–3
[14] Andreev V V, Bondarenko G G, Maslovsky V M, Stolyarov A A and Andreev D V 2015 *Phys. Status Solidi* C **12** 299–303
[15] Kaschieva S, Todorova Zh and Dmitriev S N 2004 *Vacuum* **76** 307–14
[16] Andreev V V, Bondarenko G G, Maslovsky V M, Stolyarov A A and Andreev D V 2015 *Phys. Status Solidi* C **12** 126–130

IOP Conf. Series: Materials Science and Engineering **110** (2016) 012042 doi:10.1088/1757-899X/110/1/012042

Formation of alpha and beta tantalum at the variation of magnetron sputtering conditions

E O Nasakina[1], M A Sevostyanov, A B Mikhaylova, A S Baikin, K V Sergienko, A V Leonov and A G Kolmakov

Institution of Russian Academy of Sciences A.A. Baikov Institute of Metallurgy and Material Science, Leninsky Avenue, Moscow, Russia

E-mail: nacakina@mail.ru

Abstract. Nano- and microdimensional surface layers of α and β tantalum on flat NiTi, Ti, glass, etc. substrates were created. Structure and composition of samples were defined by SEM, AES and x-ray diffractometry. With increase in deposition time surface layer thickness not linearly increases. The transitional layer provide high adhesion of a surface layer to a substrate. Irrespective of summary sputtering time the β phase is formed in the beginning and at sputtering time more than 20 min on it α tantalum is deposited, while temperature remains below 150 °C. Keywords: composite materials, surface layer, tantalum, alpha and beta phase, nitinol, corrosion resistance.

1. Introduction

Development of the composite materials allowing to effectively combine desirable operational characteristics of the modified surface layers and the main material at the present time is perspective in many areas of human activity: in optics (filtering, reflecting, absorbing, etc. environments), electronics (conductors, semiconductors, dielectrics), mechanical engineering, building and household (tribological, durable, functional, protective, resistant to action of aggressive environment, decorative and other coatings), medicine (biocompatible), etc. An effective method for layered composites formation is physical vapor deposition, including magnetron sputtering, since at relatively small expenditure of time and resources it allow to receive qualitative products of various nature and geometry and control properties of the created materials. In particular as a new surface on a substrate from other material tantalum is interesting due to high corrosion resistance, radiopacity, conductivity, wear resistance, etc. Thus in case of thin films of tantalum, according to literary data, formation of both alpha and beta phases, which differ in properties, is possible [1-19].

2. Materials and methods

Creation of layered composites was carried out by formation of tantalum surface layers on nitinol, glass, titanium and copper substrates (basis) with use of a magnetron sputtering method. Working and residual pressure in the vacuum chamber was 0,4 and $4x10^{-4}$ Pa, respectively. For cleaning, activation and polishing of a substrate surface bombarding by argon ions was carried out. The magnetron worked at a direct current at ~ 850 мA, ~ 700 V, deposition distance 200 mm and various time. Phase structure

[1] To whom any correspondence should be addressed.

of the deposited films was characterized by the x-ray diffractometer "Ultima IV" of Rigaku firm with Cu Kα radiation. The surface morphology and layer-by-layer element structure research was investigated by the TESCAN VEGA II SBU scanning electronic microscope with an INCA Energy energy dispersive spectrometer system and JEOL JAMP-9500F Auger spectrometer in combination with ion etching through argon ion bombardment at 30°.

3. Results and discussion

Figure 1 shows dependence of tantalum surface layer thickness from time of magnetron sputtering on a nitinol substrate. Increasing the sputtering time to 20 min increased the thickness of both the surface layer (consisting only of the deposited substance) and the transition layer (containing elements of both the substrate and deposited substance). Further increasing the sputtering time caused an increase only in the thickness of the surface layer and up to 30 minutes more intensively, than at bigger time. This situation remains at all used materials. It occurs, most likely, because at the beginning of layers formation atoms and ions of deposited substance, overcoming a sputtering distance, collides with particles of working gas, with each other and with substrate surface and don't appear at each site of its surface in equal volume and at the beginning interact with it chaotically and unevenly (including with the sites which are already was formed by the previous deposited particles). And further (at increase of sputtering time, and so of the time of influence on a surface) particles continue to collide, mixing up, try to reach thermodynamically more advantageous position and state, and more uniform distribution of the deposited substance at the surfaces takes place. By consideration of cross section of samples it is visible that at smaller time of a sputtering the surface shows big heterogeneity.

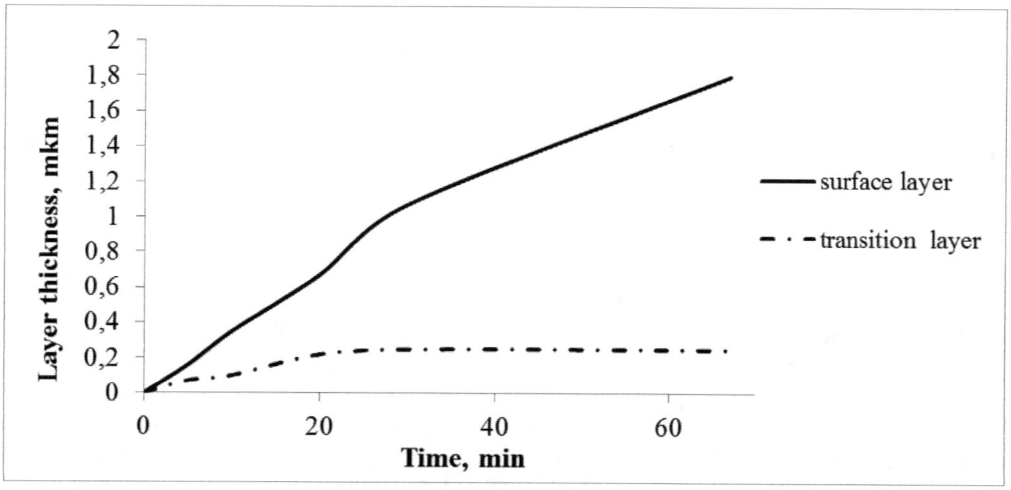

Figure 1. Dependence of surface layer thickness on time of magnetron sputtering on a nitinol substrate

The X-ray diffraction patterns of our samples with nitinol basis (figure 2) demonstrate that depending on sputtering time tantalum is formed in two various crystal states – an alpha (a cubic crystal lattice) and a beta (a tetragonal lattice with the small content of oxygen) [3 – 14].

In the case of a sample with a Ta surface layer obtained by sputtering for 5 min, the major phase was nitinol, but many peaks β-Ta in the 2-theta range from 33 to 81 degrees were observed that also corresponds to various crystal orientation. In the composites formed after sputtering for 10 and 20 min, β-Ta with O were a major phase, and only 2 main peaks were observed, but nitinol was also present. After 29 min, the strongest peak was that from α-Ta, and there were β-Ta with O and nitinol, due to the averaging of results over the entire probing depth; at further increase in time α-Ta dominated, and very weak peaks of β-Ta and nitinol were observed. Thus it turns out that irrespective of summary sputtering time the beta phase is formed in the beginning and at sputtering time more than

20 min on it alpha tantalum is deposited. The same regularities are observed in case of other substrates which are united by availability of oxygen in a surface.

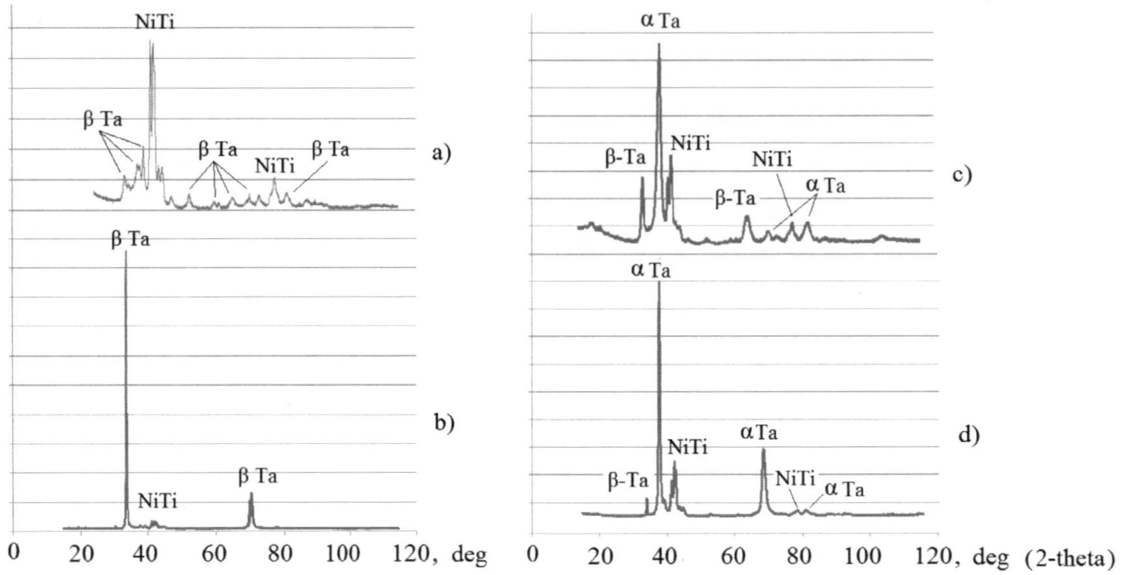

Figure 2. X-ray diffraction patterns of a composite on the basis of a nitinol received at magnetron sputtering time: a) 5 min, b) 10 and 20 min, c) 29 min, d) 30 min

Several theories of tantalum formation in α or β phase is developed which are generally connected with working temperature and pressure (defining mobility and energy of atoms) and the substrate nature. However different authors achieve often contradictory results.

It is noted that the alpha phase is formed at temperatures more than 400 °C promoting deposited atoms mobility increase: initially at heating of a substrate or as a result of the annealing following sedimentation (then deposited β-Ta transforms in α-Ta) [1, 3, 6, 11]. However at a temperature about 400-500 °C also β phase received (for example, in the form of the particles distributed in α) [3, 6], and α is also formed without heating [9, 12]. It is specified that with growth of temperature the size of grains, impurity amount in a surface layer (for example, the dissociation of oxides enhanced, i.e. the O contents lowered) and its amorphousness decreases.

Presence at the working atmosphere of the high oxygen content according to [13] leads to fast formation of oxides and, therefore, promotes formation of a tantalic layer in a beta state whereas in [12] oxygen environment did not prevent the formation of alpha tantalum. At a deposition on silicon and glass substrates in [10, 13, 14] 0,5 - 0,7 Pa sputtering pressure led to α-Ta formation and smaller or bigger pressure - β-Ta, but in [12] already at 0,28 Pa α phase was formed.

In [14] also alpha tantalum formed at 0,3 and 1,4 Pa pressure, but at sputtering on earlier deposited α-Ta (110) layer. Also it was specified that (110) is the most low-energy lattice for BCC materials and provokes formation on itself of the same structure. Being a zone of a new surface nucleation the substrate surface specifies the character of its structure formation. It was shown that on amorphous carboniferous or oxidic surfaces the beta tantalum is formed, for example, on titanium without natural oxide or TaN substrates the α-Ta is formed [7, 9, 11, 14].

And though availability of oxygen on a substrate surface not always prevented the formation of α-Ta, nevertheless it is considered that it promotes formation of β-Ta. Therefore its creation in this work in an initial time period on all substrates is quite expected, despite ionic etching.

In [11], as well as in this research, it was shown that longer time promotes layering of α-Ta on earlier formed β-Ta, but authors connected it with a considerable warming up of a surface (more than

350 °C) whereas in this work temperature of a substrate didn't rise higher than 150 °C and so couldn't influence formation of alpha phase. Thus it is noted that α-Ta is thermodynamically more stable phase. Therefore it is possible to assume that the alpha phase is formed in this work with increase in time as a result of more uniform filling of a surface (since with growth of sputtering time possibility of particle choice of more suitable state and position increases) and also because of a possible local surface heating and absence on the surface of an oxygen which is already expended for formation of an underlayer β-Ta.

4. Conclusion

Nano- and microdimensional surface layers of α and β tantalum on flat NiTi, Ti, glass, etc. substrates were created. With increase in deposition time surface layer thickness not linearly increases. The transitional layer provide high adhesion of a surface layer to a substrate. Irrespective of summary sputtering time the β phase is formed in the beginning and at sputtering time more than 20 min on it α tantalum is deposited, while temperature remains below 150 °C.

References:

[1] Cheng Y, Cai W, Li H T and Zheng Y F 2006 Surface modification of NiTi alloy with tantalum to improve its biocompatibility and radiopacity *J. Mater. Sci.* **41** 4961–64

[2] Zein El Abedin S, Welz-Biermann U and Endres F 2005 A study on the electrodeposition of tantalum on NiTi alloy in an ionic liquid and corrosion behaviour of the coated alloy *Electrochem. Commun.* **7 (9)** 941–46

[3] Zhang M, Yang B, Chu J and Nieh T G 2006 Hardness enhancement in nanocrystalline tantalum thin films *Scripta Mater.* **54** 1227–30

[4] Chakraborty B R, Halder S K, Maurya K K, Srivastava A K, Toutam V K, Dalai M K, Sehgal G and Singh S 2012 Evaluation of depth distribution and characterization of nanoscale Ta/Si multilayer thin film structures *Thin Solid Films* **520** 6409–14

[5] Zhou Y M, Xie Z, Xiao H N, Hu P F and He J 2009 Effects of deposition parameters on tantalum films deposited by direct current magnetron sputtering *Vacuum* **83** 286–91

[6] Dorranian D, Solati E, Hantezadeh M, Ghoranneviss M and Sari A 2011 Effects of low temperature on the characteristics of tantalum thin films *Vacuum* **86** 51-55

[7] Bernoulli D, Müller U, Schwarzenberger M, Hauert R and Spolenak R 2013 Magnetron sputter deposited tantalum and tantalum nitride thin films: An analysis of phase, hardness and composition *Thin Solid Films* **548** 157–61

[8] Zhou Y M, Xiea Z, Xiao H N, Hu P F and He J 2011 Effects of deposition parameters on tantalum films deposited by direct current magnetron sputtering in Ar–O2 mixture *Appl. Surf. Sci.* **258** 1699–1703

[9] Zhou Y M, Xie Z, Ma Y Z, Xia F J and Feng S L 2012 Growth and characterization of Ta/Ti bi-layer films on glass and Si (111) substrates by direct current magnetron sputtering *Appl. Surf. Sci.* **258** 7314–21

[10] Navid A A, Chason E and Hodge A M 2010 Evaluation of stress during and after sputter deposition of Cu and Ta films *Surface & Coatings Technology* **205** 2355–61

[11] Myers S, Lin J, Martins Souza R, Sproul W D and Moore J J 2013 The β to α phase transition of tantalum coatings deposited by modulated pulsed power magnetron sputtering *Surface & Coatings Technology* **214** 38–45

[12] Cacucci A, Loffredo S, Potin V, Imhoff L and Martin N 2013 Interdependence of structural and electrical properties in tantalum/tantalum oxide multilayers *Surface & Coatings Technology* **227** 38–41

[13] Navid A A and Hodge A M 2012 Nanostructured alpha and beta tantalum formation—Relationship between plasma parameters and microstructure Mater. Sci. Eng. A **536** 49– 56

IOP Conf. Series: Materials Science and Engineering **110** (2016) 012042 doi:10.1088/1757-899X/110/1/012042

[14] Navid A A. and Hodge A M 2010 Controllable residual stresses in sputtered nanostructured alpha-tantalum *Scripta Materialia* **63** 867–70

[15] Elangovan T, Murugeshan S, Mangalaraj D, Kuppusami P, Khan S, Sudha C, Ganesan V, Divakar R and Mohandas E 2011 Synthesis and high temperature XRD studies of tantalum nitride thin films prepared by reactive pulsed DC magnetron sputtering *Journal of Alloys and Compounds* **509** 6400–07

[16] Kuz'michev A I 2008 *Magnetron sputtering systems. Book 1. Vvedenie v fiziku i tekhniku magnetronnogo raspyleniya (Introduction to the Physics and Technology of Magnetron Sputtering)* (Kyiv: Avers)

[17] Simon G and Toma M 1991 *Applied Technology surface treatment of metallic materials* (Chelyabinsk: Metallurgy)

[18] Bunshah R.F. at al. 1982 *Deposition technologies for films and coating* (Park Ridge, New Jersey (USA): Noyes Publikations)

[19] Zabolotnyi V T 1997 *Ionnoe peremeshivanie v tverdykh telakh (Ion Intermixing in Solids)* (Moscow: MGIEM(TU))

IOP Conf. Series: Materials Science and Engineering **110** (2016) 012043 doi:10.1088/1757-899X/110/1/012043

Numerical simulation of hydrodynamic flows in the jet electric

V D Sarychev, A Yu Granovskii and S A Nevskii[1]

Siberian State Industrial University, Novokuznetsk, 42 Kirov Street, 654007, Russia

E-mail: nevskiy.sergei@yandex.ru

Abstract. On the basis of concepts from magnetic hydrodynamics the mathematical model of hydrodynamic flows in the stream of electric arc plasma, obtained between the rod electrode and the target located perpendicular to the flat conductive, was developed. The same phenomenon occurs in the welding arc, arc plasma and other injection sources of charged particles. The model is based on the equations of magnetic hydrodynamics with special boundary conditions. The obtained system of equations was solved by the numerical method of finite elements with an automatic selection of the time step. Calculations were carried out with regard to the normal plasma inleakage on the solid conducting surface and the surface with the orifice. It was found that the solid surface facilitates three swirling zones. Interaction of these zones leads to the formation of two stable swirling zones, one of which is located at a distance of two radii from the axis and midway between the electrodes, another is located in the immediate vicinity of the continuous electrode. In this zone plasma backflow scattering fine particles is created. Swirling zones are not formed by using the plane electrode with an orifice. Thus, the fine particles can pass through it and consolidate.

1. Introduction

Investigation of the interaction of electromagnetic fields, flows of particles, plasma with a solid body and its application in various areas of technology and production evokes interest to the problems associated with the study of electrical gas discharges. For practical application of the gas discharge device it is necessary to determine the dependence of plasma characteristics on external controlled parameters (pressure, type of gas and its consumption, electric current, voltage, geometric dimensions, incoming power of the electromagnetic field, etc.); to control processes of plasma maintenance and generation; select optimal operating conditions according to specific plasma technologies. Solution of these problems is possible through a comprehensive experimental study and mathematical modeling of electrical discharges. Basics of theoretical description of the physical processes, occurring in the gas-discharge plasma, are described in [1], where it is noted that magnetogasdynamic (MGD) description of plasma is effective within the limits of applicability of the continuous medium approximation as well as the fundamental kinetic approach. The technology of plasma arc is most widely used in engineering, especially in metallurgy, powder spraying, welding, etc. In the process of welding by consumable electrode a conductive thermal plasma is formed. Its properties, mode parameters greatly affect the processes of formation of the metal bath of the substrate, where the properties of the

[1] Corresponding author: Nevskii S.A.

modified surface are formed. Therefore, the numerical modeling of hydrodynamic and thermal fields is an important tool in the creation of surface modification technologies [2]. In [3 – 8] numerical studies with regard to the above mentioned processes are carried out. In [3] the influence of the cathode shell on the temperature field is studied. It is found that different potential boundary conditions on the back surface of the anode plate do not substantially affect the heat flow in the cathode or within the arc but influence the current density and the temperature distribution near the anode plate. When a single computational domain comprises cathode and anode arcs, additional conditions at the boundary "plasma-electrode" to account the energy flows through it should be used. For the electric current of 200 A calculation of isotherms gives a satisfactory agreement with the experiment. In [4] the formation of self-organizing structures of electrodes is studied. As a result a three-dimensional two-temperature simulation model is developed, which shows gradual emergence of a point structure alongside the increasing level of anode cooling: from one blurred spot for low levels of cooling to subsequent coating of the anode zone by small spots for intensive cooling. Paper [5] is devoted to comparison of turbulent flow models for a free burning of a high-intensity argon arc. Navier-Stokes equations are modified in the paper by taking into account the radiation transport, electric power consumption and electromagnetic forces. For modeling of turbulence, which has a value for the arc boundary due to the sharp temperature gradient between the arc column and the surrounding gas, zero- and two- viscosity equation is used. It is found that this model best describes the temperature profiles in the arc.

Works [6 - 8] are devoted to the study of processes occurring in the molten bath during plasma exposures. In [6] the problem of investigation of heat transfer and the fluid flow during pulsed arc welding of stainless steel is solved by the method of finite elements. The proposed model allows to study evolution of the weld pool in time at a constant and pulsed current. In [7] the numerical simulation was supplemented by experiments using high-speed shootings of the weld pool surface in the infrared. Processing and analysis of the images showed good agreement with the numerical model.

In [8] on the basis of the equilibrium magnetogasdynamic model the calculation of the flow and heating of gas (argon) in the diaphragmatic plasmatrone's channel depending on the diaphragm position and size, consumption and gas swirling intensity is performed. Peculiarities of changes of the arc characteristics in the channels of variable section for swirling and non-swirling flows are revealed. It is shown that diaphragming of the channel in the conditions of an intensive gas swirling ensures in the area up to the diaphragm localization of plasma flow in the narrow near-axial zone having a clear thermal interface with an external gas flow (effect of the swirling thermal insulation of plasma); the length of the stabilization area depends on the sizes of the diaphragm aperture and its position in the channel. In [9] a numerical study of the microwave discharge in the argon flow, axially injected into the coaxial with a shortened inner electrode, was performed. It was established that an important role in the formation of a microwave discharge plays the inner electrode's section, and the cold gas "blown" through its channel cooling the wall, stabilizes burning of the microwave discharge and contributes to the formation of a sharp leading heat edge.

Thus, we must conclude that at present time the modern trends of modeling of plasma effects target creation of a unified mathematical model that describes the processes in the area adjacent to the cathode as well as in the molten bath. Therefore, the aim of our work is to simulate hydrodynamic flows in the diaphragmatic electric arc.

2. Problem formulation

The model based on the equations of Navier-Stokes, thermal conduction and Maxwell's equations is used for the simulation of the plasma flow and temperature distribution in it.

Let us consider the cylindrical electrode with one of its ends having a hemispherical shape, another electrode we shall assume to be flat (Figure 1). On the border of GD (plasma/anode) the heat flow for the anode is set by the sum of the heating from plasma, heating from electron condensation $\left|\vec{j}\cdot\vec{n}\right|\varphi_a$ (where φ_a is the work function of anode electrons) and radiation losses $\varepsilon\sigma_B T^4$ (where ε – anode emissivity, σ_B – Stefan-Boltzmann constant). At the border of EB (plasma/cathode) the heating from plasma, ion heating $j_i V_i$, radiation losses $\varepsilon\sigma_B T^4$ and thermoionic cooling $j_e\varphi_c$ (energy required for electron emission from the cathode, which is given by the product of electron current density and work function of cathode material) contribute into the heat flow. The boundary conditions for Maxwell's and magnetohydrodynamic equations are provided in Table 1. The material properties accepted for calculations are given in Table 2. Kinetic parameters of plasma (specific heat, thermal conductivity, viscosity, conductivity) depend on the temperature and are given in a tabular format.

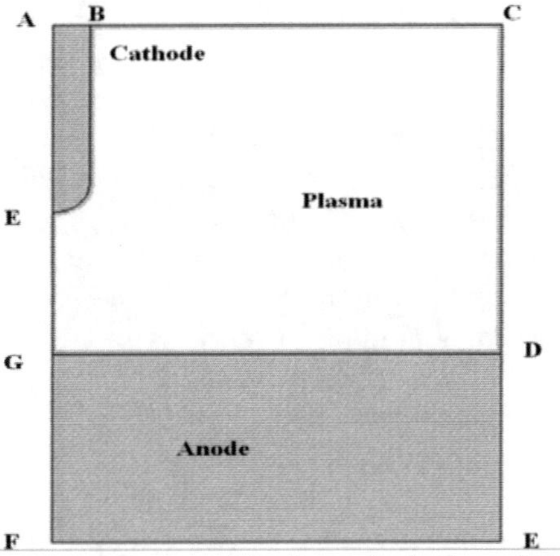

Figure 1. The geometry used in the calculation of the plasma arc

Table 1. Boundary condition

	AB	BC, CD	DE	EF	AE, FG	EG
T,q	$-\vec{n}\cdot\vec{q}=0$	T_0	T_0	T_0	$\vec{n}\cdot\vec{q}=0$	$\vec{n}\cdot\vec{q}=0$
(u,v,w)	-	$p=p_0$	-	-	-	$\vec{n}\cdot\vec{v}=0$
$(V,\vec{A})$	$\vec{j}\cdot\vec{n}=\dfrac{I_c}{\pi R_c^2}$	$\vec{j}\cdot\vec{n}=0$ $\vec{A}\times\vec{n}=0$	$\vec{j}\cdot\vec{n}=0$ $\vec{A}\times\vec{n}=0$	$V=0$ $\vec{A}\times\vec{n}=0$	$\vec{j}\cdot\vec{n}=0$ $\vec{B}=0$	$\vec{j}\cdot\vec{n}=0$ $\vec{B}=0$

Table 2. Material properties of the cathode, anode and gas

Symbol	Nomenclature	Value
ρ_m	Density anode and cathode metal	7500 kg m^{-3}
C_m	Specific heat of metal	602 J kg^{-1} K^{-1}
k_m	Thermal conductivity of metal	26 W m^{-1} K^{-1}
σ_m	Electrical conductivity of metal	7.7×10^5 Ω^{-1} m^{-1}
φ_c	Surface work function of cathode	4.52 V
V_i	Plasma's ionization potential	15.68 V
φ_a	Surface work function of anode	4.65 V
ε	Surface emissivity	0.9

3. Results and discussion

System (1), (2) with regard to the boundary conditions is solved by the finite element method for the case of aperture's absence and its presence. The cathode is a cylinder limited by hemisphere of the same radius, the radius is 0.8 mm, the anode in the first case is a plane and in the second case – a plane with an aperture of 2 mm radius. Distance from the cathode to the anode (EG) is 3.8 mm. On the cathode the current density is set, so that the total current is 100 A. On the anode it is assumed that the potential is zero. The initial conditions correspond to the quiescent gas and electrode material. The temperature at the initial time for the entire area is constant and equals to 300K. The initial density is chosen to be constant for each zone: for the cathode and anode – steel density, for the interelectrode gap – air density. Step on the spatial coordinate is 0.1 mm. The time step is automatically chosen basing on the Courant-Friedrichs-Lewy condition.

First, we shall analyze the case of diaphragm's absence. In moments of times up to 1 μs the flow in the plasma jet is non-stationary. From the cathode to the anode propagates the temperature wave and in the flow zone three swirling zones are generated (Figure 2 a). One of them, in the vicinity of the cathode and the symmetry axis, is due counter-propagating jets. These jets are generated because of the action of the radial component of the electromagnetic force directed toward the axis. Near the symmetry axis approximately at the midway between the cathode and the anode the second swirling zone is located. In the subsequent moments the interaction of these zones results in the formation of one stable swirling zone at a distance of two radii from the axis. The third swirling zone is formed near the anode at the moment of time 0.1 μs and is observed during the calculation time.

Thus, the plasma transits from essentially non-stationary state into a quasi-stationary. It is induced by the viscous and electrodynamic forces. The flow has a complex structure and includes the zones: close to the electrode, the near-axial flow with a shock transition, two swirling zones. The velocity distribution on the axis is shown in Figure 2 b, which implies that the velocity of plasma arc, starting from zero speed (E – cathode surface) reaches maximum ~500 m/s at a distance of 1 mm from the cathode surface, and then drops to zero and changes the sign. Analytical models and numerical calculations with large spatial steps do not reveal these features.

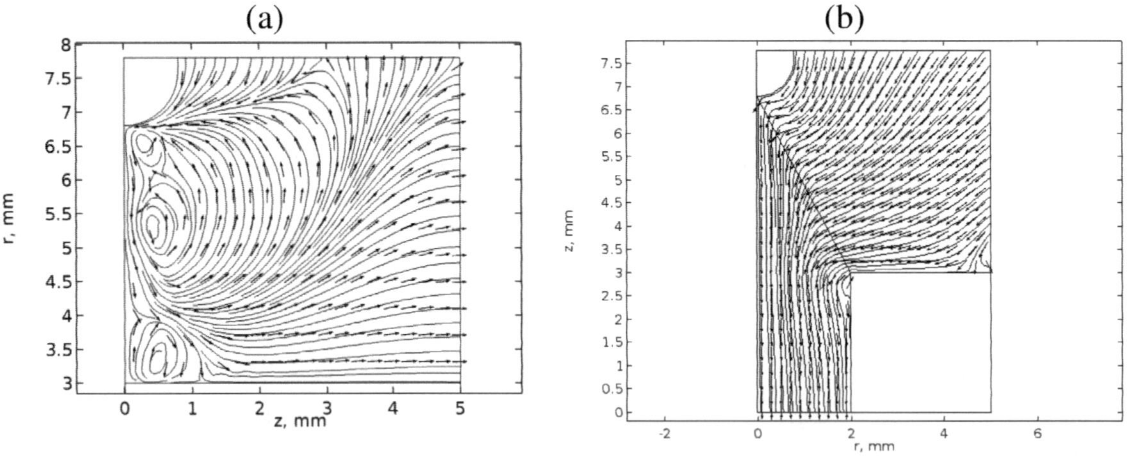

Figure 2. Velocity field of arc. (a) – streamlines at t=1.1 μs; b – streamlines at diaphragm

Analysis of the hydrodynamic situation reveals two important results. The formation of the first swirl near the cathode and its existence 1 μs creates a favorable situation for the generation of small particles 100 nm from the molten material. The second, the swirling zone near the anode creates a plasma counter-current, which does not allow the fine particles to reach the anode surface and participate in the coating formation. Therefore, we propose to use an orificed anode. The calculations with an orifice are given in Figure 2b, which shows that the swirling zones are not formed and it can be used to develop welding technology.

4. Conclusion

Mathematical model of hydrodynamic flow in the plasma during electric arc welding is developed. It is shown that at times $t < 1$ μs three swirling zones are formed. Formation of the first swirling zone, and its existence ~ 1 μs, create conditions for generation of fine particles. However, the presence of plasma backflow near the anode does not allow them to get on the surface of the modified material. The diaphragm in the electrode leads to the fact that the swirling zones are not formed and the fine particles can be consolidated.

5. Acknowledgements

The research grant was funded by the Russian Science Foundation (No. of project 15-19-00065)

References

[1] Finkelburg W, Maecker H 1956 *Elektrische bogen und thermisches plasma* (Berlin: Springer)
[2] Shih T -M, Zheng Y, Arie M, Zheng J -C. 2013 *Numerical Heat Transfer; Part A: Applications* **64** 435
[3] Xi Chen, He-Ping Li 2001 *International Journal of Heat Mass Transfer* **44** 2541
[4] Trelles J P 2014 *Plasma Sources Science and Technology* **23** 054002
[5] Kim Y-J, Lee J-C 2013 *Journal of the Korean Physical Society* **62** 1252
[6] Li H-P, Benilov M S 2007 *Journal of Physics D: Applied Physics* **40** 2010
[7] Traidia A, Roger F, Chidley A 2011 *World Academy of Science, Engineering and Technology* **73**
[8] Lelevkin V M, Semenov V F 2002 *Technical Physics Letters* **28** 722
[9] Gritsinin S I, Davydov A M, Kossyi I A, Kulumbaev E B, Lelevkin V M 2013 *Plasma Physics Reports* **39** 579

Forming patterns and mechanical properties of austenitic chromium-nickel steel due to strain aging

N V Kamyshanchenko, V V Krasilnikov[1], I S Nikulin, A V Gal'tsev, V A Belenko and I N Gal'tseva

Belgorod State University, Pobeda St. 85, Belgorod 308015, Russia

E-mail: kras@bsu.edu.ru

Abstract. The work presents the results of studies of forming patterns and mechanical properties of martensite transformation, found in the chromium-nickel steels of 08X18H10T grade, subjected to pre-heat treatment followed by deformation aging. Internal energy state is determined by using acoustic emission. The observed patterns improve the mechanical parameters of steels quenched and plastically deformed at low temperature and then subjected to temper under load in the optimum temperature being associated with obtaining a more stable condition of the structure through the processes of relaxation of internal stresses, high dispersion and uniform distribution of carbides and intermetallic particles, increasing the density of dislocations as well as through other processes occurring during deformation aging martensite. Start your abstract here…

1. Introduction

At present, some traditional ways of increasing the strength of steel and alloys almost have been failed or proved to be uneconomical, in particular, opportunities of doping become limited due to deficit of a number alloying elements. On the other hand, the increasing demands of industry for the quality of alloys and steels with special properties and high strength characteristics makes professionals to look for efficient ways to get simple constructional and corrosion - resisting materials. Promising ways to improve the strength characteristics of alloys are a better use of the structural - phase method for hardening, combined deformation, deformation-thermal, radiation and other technologies. This paper is devoted to studying the regularities of deformation aging of austenitic steel of grade 18 - 10.

2. Materials and equipment

Investigations are fulfilled on austenitic chromium-nickel corrosion - resisting steels of 12X18H10T and 08H18N10T grades. For the mechanical testing, the samples of complex shape with a thickness of a test portion of 1.2 mm, a length of 14.0 mm and a width of 2.1 mm have been prepared. The axis of the samples coincides with a direction the sheet rolling; samples have been pre-annealed in a vertical tube furnace up to 1320 - 1370 K for 15 minutes. Mechanical tests are carried out by stretching the samples at a tensile machine of «soft» type. The rate of deformation during tensile test is 10^{-3} s^{-1}. A sensitive strain gauge is used to test the modulus of elasticity, limit of proportionality and the strength. The test is carried out at elevated temperatures with the use of a tubular resistance furnace with air

[1] To whom any correspondence should be addressed.

atmosphere. At lower temperature, liquid nitrogen is used as the cooling medium. Fluctuations of temperature do not exceed $\pm 3^0$ C in the short-term tests and $\pm 5^0$C in long – term ones.

The metallographic studies are carried out by the optical microscope MIM - 10. X-ray studies of samples subjected to heating after deformation are performed on diffractometer Dron - 1 under chromium radiation with automatic recording the interference line (311) of austenite to chart tape. Calculated under the experimental conditions the relative error of measurement of the crystal lattice period is $\Delta\alpha / \alpha = \pm 0{,}016\%$. Stress relaxation tests are carried out by the tensile machine with recording time up to 30 minutes. The value of relaxation $\Delta\sigma_{рел}$ is represented by the quantity of voltage drop from the initial value σ_0 to a certain one σ_t reached over a period of time. Measurement of the microhardness H is done with the hardness testing instrument PMT - 3M under load 100 g. In the tests, elastic limit with a tolerance of ~ 0.001% $\sigma_{0.002}$, conventional yield strength, ultimate strength σ_B and specific elongation $\delta\%$ of the specimen up to rupture are defined. Acoustic emission (AE) is measured by a piezoelectric tranducer with gain coefficient of 40 dB. The noise level reduced to the input is 10 mkV. The sensitivity of the apparatus is 10^{11} V / m and a working range of 0.1 - 0.8 MHz. AE sensor output signal is input to the main amplifier with gain coefficient controlled up to 60 dB followed by computer registration with a built-analog-digital converter (ADC) allowing to take the readings with a frequency of 3 MHz.

3. Processing technology of steel investigated

After manufacturing, mechanical polishing and annealing the samples are divided into 3 batches. The first batch is subjected to heat treatment according to GOST 5582 – 75 (State Standard). The second batch after the heat treatment (GOST 5582 - 75) is subjected to plastic deformation up to $\leq$ 20% at a temperature (77 K) of the intensive transformation of austenite in the α phase of the martensite type followed by tempering at 720 K. Samples of the third batch after the heat treatment (GOST 5582 - 75), plastic deformation up to $\leq$ 20% at the temperature of 77 K followed by tempering at the optimal temperature regime of 675 - 875 K are investigated by both a no-load, and under-voltage (0.1 - 0.75) $\sigma_{0.2}$. Table 1 shows some of the parameters characterizing the state of the steel after the appropriate treatment.

A large number of reviews, monographs, and original papers (see, for instance, [1]) have been published on studying patterns of martensite transformation of austenitic steels. It is relevant for these steels to decrease viscosity and plasticity, fatigue spalling carbides, loss of gas density etc. Uniform distribution of carbide phase has a decisive influence on the formation of a high-complex of mechanical, corrosion and other properties. Resolving the problem of obtaining alloys with good mechanical properties is largely dependent on the technique of mastering the use of domestic reserves. One of the characteristics of this class of steels after heat treatment is a small stability of austenite which becomes α - phase of the martensite type under the influence of plastic deformation. It is known that at low temperatures with a increasing the deformation of steels of austenite class, the α - martensite content increases simultaneously. In the process of austenitic-martensitic transformation occurs a substantial change in the state of the residual austenite by volume and phase cold working [2]. In samples of 2 batch, it is manifested the dependence of strength on the degree of deformation with a clear indication of saturation (figure 1).

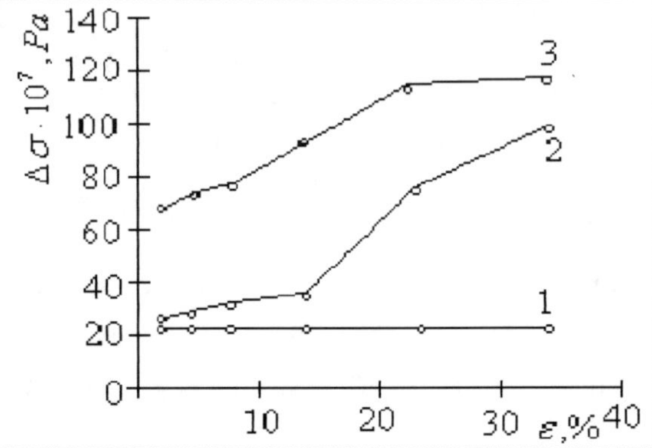

Figure 1. The increment $\Delta\sigma_{0.2}$ (curve 2) and $\Delta\sigma_t$ (curve 3) dependences of quenched steel on a degree of deformation (ε) at 77K and following heating-up to 770K with a heat rate of 0.175 K/s. A rupture is at 300K. Curve 1 is the value σ_t of heat treated steel (GOST 5582 – 75) deformed at 300 K.

Formation of martensite phase in steel at low temperature deformation is accompanied by elastic microstress initiation and dislocation structure [3]. Metallographic studies show that the austenitic steel structure in the quenched state is saturated by carbide inclusions. Intense precipitating carbides in austenite occur at stabilizing temper, which leads to a reduction in percentage of not only carbon but also the alloying elements in austenite and, consequently, it is accompanied with small changes in specific volume. Based on the above said, it is possible to suggest a slight decrease of quenching stresses. Further straining at 77 K up to $\varepsilon \leq 20\%$ after the heat treatment leads to an increase of strength parameters, the reason for which it is necessary to look at formation of both the martensite phase of austenite and greatly influence of the residual austenite and carbide inclusions (Table 1).

Table 1

Batch №	Processing technology	σ_t, x10^7N/m^2	$\sigma_{0.2}$, x10^7N/m^2	Rockwell hardness	Structure
1	Heat processing (GOST 5582 - 75)	63.8	22.5	54 – 56	Austenite, carbide
2	Heat processing, plastical deformation at 77K up to $\varepsilon \leq 20\%$ with next temper at 720K.	102	64	60 – 32	Martensit e (80%), austenite, carbide
3	Heat processing, plastical deformation at 77K up to $\varepsilon \leq 20\%$ with next temper at 720K under stress $\sigma_L = 0{,}5\sigma_{0,2}$	175	106	64 – 66	Martensit e (90%), austenite, carbide

Next temper at elevated temperatures appreciably effects on the strength characteristics, the magnitude and nature of which depends on temperature and heating rate (figure 2, figure 3).

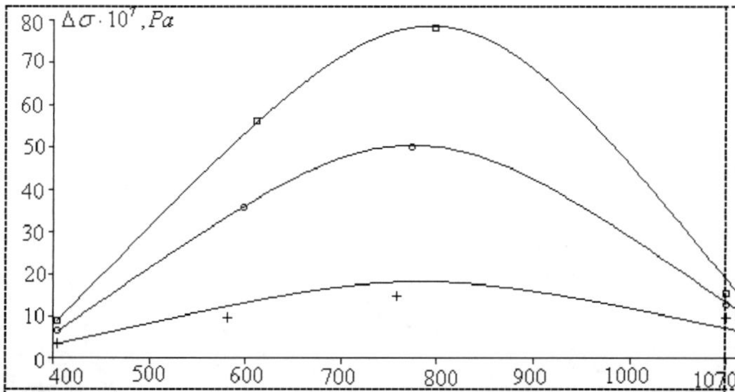

Figure 2. The dependence of the yield strength increment on aging temperature and the value of a residual deformation at the heat rate of 0.175 K/s.
$\varepsilon_1 = 10\%(+), \varepsilon_2 = 20\%(\circ), \varepsilon_3 = 30\%(\square)$
The rupture is at 300K.

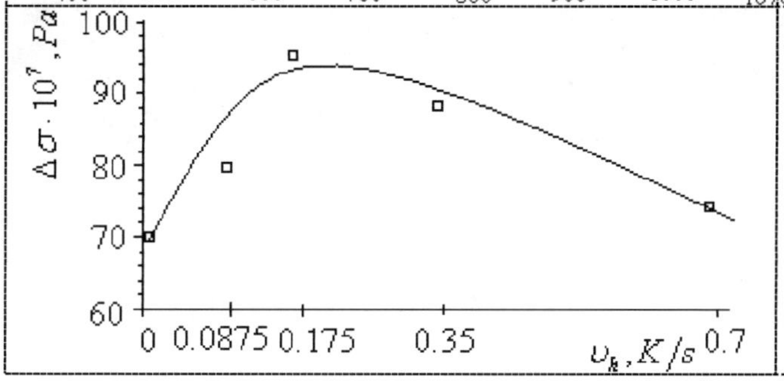

Figure 3. The dependence of the yield strength increment on the heat rate υ_H at the optimal annealing temperature 770K and the residual deformation $\varepsilon = 20\%$. The rupture is at 300K.

4. Determination of the internal state of the investigated steel using acoustic emission.

As the content of the martensite phase in the investigated steel is around 80 - 90% due to mechanical-thermal treatment, then the most likely source of acoustic emission (AE) in the field of elastic deformation is dislocation processes and processes related to the cracking of carbides formed as a result of strain aging martensite. Average energy of single pulse recorded with the deformation $\Delta\varepsilon$ of the sample is determined by the formula

$$W_u = \frac{W(\varepsilon)}{\dot{N}(\varepsilon)} \tag{1}$$

and is ~ 10^{-16} J. This value of a single pulse is at the range of 10^{-10} - 10^{-19} J, which corresponds to dislocation processes and processes of microcrack formation. In figure 4, the example of the dependence of the average single pulse energy W_u on a deformation quantity calculated by equation (1) is represented. Typically, bursts in the histogram correspond to the AE activity bursts that are associated with the unsteadiness of distribution of internal crystal lattice imperfections.

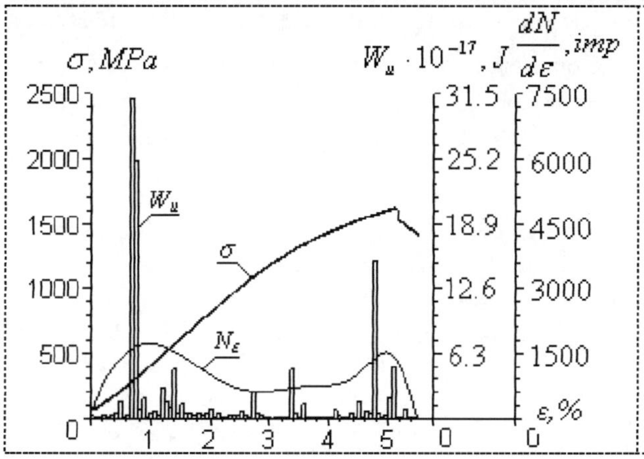

Figure 4. The dependence of AE activity N_ε and the average single pulse energy W_u on a relative deformation quantity ε% of third batch at following regime of temper under stress: T=770K, $\sigma_L = 0.5\sigma_{0.2}$. The time of averaging in constructing histogram of changing W_u is Δt=0.56 c. Mechanical tests were fulfilled at temperature of 300K.

On the other hand, nonmonotonic behavior of AE activity is driven by changes in the mechanisms of deformation in the mechanical tests (figure 4, smooth curve) [4, 5]. It is found that the value of AE activity in the area of elastic deformation is most pronounced in the case of temper under stress at $\sigma_L = 0.1\sigma_{0.2}$ and $\sigma_L = 0.15\sigma_{0.2}$. The quantity of AE activity monotonically decreases with increasing stress of temper in the elastic-plastic region. This pattern of behavior of AE activity in the elastic area, depending on the temper stress in the strength state, is associated with the ratio of processes of strain aging martensite: the formation of carbides, the increase in the density of mobile dislocations and pinning it by impurity atoms, as well as the relaxation of internal stresses. The behavior of the mechanical characteristics correlates with the behavior of some parameters of AE (figure 5).

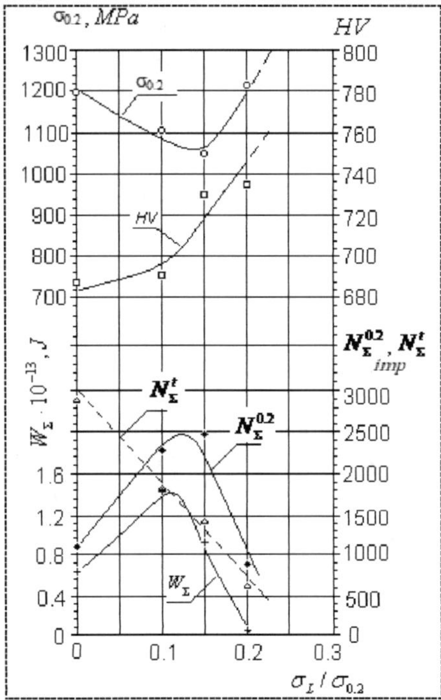

Figure 5. Changing mechanical characteristics and AE parameters on relative load quantity in deformation aging process at 770K of third batch samples..

Thus, improving the mechanical characteristics of austenitic steel plastically deformed at low temperature promotes lowering the relaxation and acoustic parameters as a result of temper under load in the optimum temperature regime. The observed patterns are due to the achievement of a more stable state of the processed steel structure through the processes of relaxation of internal stress, high dispersion and uniform distribution of carbides and intermetallic particles, increasing the density of dislocations and through other processes occurring during strain aging martensite. The dependence of the value of activity level $\dot{N}$ and the total AE N_Σ^t on a value of the applied stress and the tempering temperature is manifested, which gives an indication of the existence and magnitude of internal stresses in the investigated steel.

References

[1] Kurdyumov G V 1960 *Phenomenon of quenching and leaving steel* (Moscow: Metallurgisdat)
[2] Novikov N V and Gordynskii N I 1975 *Metal Science and Heat Treatment* **N2**, 67–68
[3] Benus V C 1975 *Dynamics of dislocations* (Kyiv: Naukova dumka)
[4] Kamyshanchenko N V, Neklyudov I M, Kuz'menko I N, Roganin M N and Gal'tsev A V *Physics of hardness and plasticity of materials. Abstracts of XV Int. Conf.* (*Samara, Russia, 26-29 June* 2006) pp 12-14
[5] Gusev O V 1982 *Acoustic emission in deformation of refractory metal single crystals* (Moscow: Nauka

Decay kinetics and energy transfer in ternary phosphate glass doped Eu and Eu/Dy

E F Polisadova[1] **and H A Othman**[2]

[1]National Research Tomsk Polytechnic University, Tomsk, Russia
[2]Faculty of Science, Menoufiya University, Shebin El-Kom, Egypt

E-mail: elp@tpu.ru

Abstract. Different series of glass samples with different glass compositions and different concentrations of rare earth ions (Eu and Eu/ Dy co-doped) were prepared. Excitation and photo-luminescence spectra were measured, for which the observed peaks are attributed to the *f-f* transitions of the rare earth ions. Decay kinetics of the characteristic emission peaks were registered and investigated using Inokuti–Hirayama model (IH model),when the emission decay deviated from the exponential pattern to study energy transfer in the prepared samples. The IH model is used to determine the energy transfer parameter which is correlated to the glass composition factor.

1. Introduction

Rare earth ions find wide applications in sensitizing solid state and glass lasers, infrared quantum counters as well as infrared to visible converters. Energy transfer between ions in solids can be accomplished either radiatively or nonradiatively and the mechanism and kinetics involved have been extensively addressed [1, 2]. Many workers have treated the radiative and nonradiative transitions in the rare earth ions and succeeded in realizing energy transfer process in doubly doped crystals and glasses [3–6]. In particular, the excellent work by Blasse and coworkers has resulted in the elucidation of energy transfer, energy migration and concentration quenching in various rare earth ions [7–10]. One important consequence of the energy transfer process is the phenomenon of sensitized luminescence in solids. It is an important factor in increasing the laser efficiency of the materials. Hence this phenomenon has a significant application in the research and development of new laser materials and it is possible to remarkably reduce the threshold energy of laser oscillation in these materials. A large number of studies on sensitized luminescence or fluorescence and concentration quenching have been made, since the wide spread interest results from the breadth of research areas encompassed by the phenomena of energy transfer [11,12].

2. Experimental

For the current work, two glass systems namely; series (B) of composition-$50P_2O_5$- xZnO - (50-x) Li_2O:5wt%Eu_2O_3 and series (C) of composition -$50P_2O_5$ - xZnO - (50-x) Li_2O:1wt%Eu_2O_3+3wt% Dy_2O_3 is prepared from high purity chemicals (Aldrich chemicals), using the melt-quench technique, where x changes in steps of 10, and ranging from 0 to 50 mol%. Photoluminescence and excitation spectra were recorded at room temperature using a Jasco spectrophotometer FP-777. Emission decay kinetics were recorded using a system of monochromator MDR-3 and GW-Instec GDS-2204 oscilloscope for signal detection, and a Xenon pulsed lamp of about 150μs pulse duration for excitation.

3. Results and discussion

Excitation spectra of samples under study (fig.1.) , were measured in the range 300-500 nm. Where spectrum of the Dy^{3+} doped glasses results from the f-f transitions as follows; from $^6H_{15/2}$ to $^4P7/2$ (350nm), $^6P_{3/2}$ (356), $^4F_{7/2}$ (388), $^4G_{11/2}$ (425) and $^4H_{15/2}$ (452). While for the Eu^{3+} ions the first excited state namely $^7F_1$ is thermally populated at room temperature so the excitation spectrum consists of transitions from the $^7F_0$ and $^7F_1$to the higher excited states. Peak centered at 394nm which results from the transition $^7F_0$ to $^5L_6$ of Eu^{3+} is the most intensive and was chosen as excitation wavelength in the measurements of the stationary photolumeniscence spectra of the two glass systems under study.
Photoluminescence spectra shown in Fig.2., consist of the $^5D_0$ to $^7F_1$ transition at 592 nm and to $^7F_2$ at

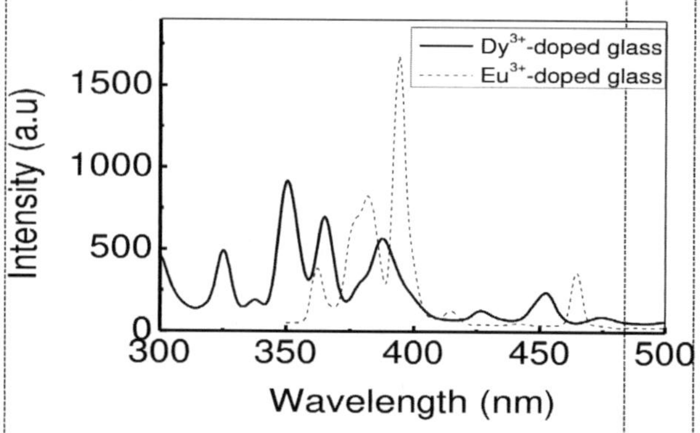

Figure 1. Typical excitation spectra for Dy^{3+} and Eu^{3+} ions.

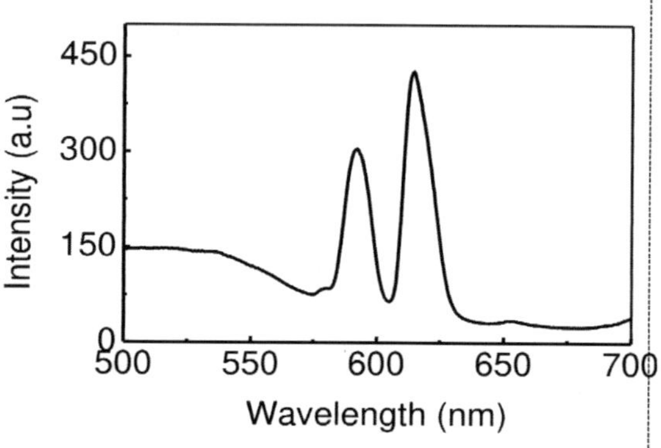

Figure 2. Emission spectrum of Eu_2O_3 doped glass (λ_{ex}=394 nm).

614 nm transitions of the Eu^{3+} ions, in case of singly doped glasses. While in case of Dy/Eu co-doped glasses (Fig.3) the 394 nm excitation line pumps the $^4F_{7/2}$ of Dy^{3+} Level results in a transition from the $^4F_{9/2}$ to $^6H_{13/2}$ component of the ^6H_j ground state of Dy^{3+} ion, at 575 nm in addition to the characteristic emission peaks of the Eu^{3+} ions.

Studying the decay patterns of the Eu^{3+} emission at its characteristic $^5D_0$ to $^7F_1$ transition, for different glass composition shows that: in case of Eu^{3+} singly doped glasses emission signals decay are ideal exponentials as shown in Fig.4.a. Using a single exponential fitting, the decay times are calculated for this series and are presented in Fig.5. In the case of double doping, decay patters (Fig.4.b.) deviated from the ideal exponential and the calculated decay times presented in Fig.5.for this series show a drastic decrease in its values compared to that in case of singly doped glasses. The above mentioned observations lead us to suggest an energy transfer from the Eu^{3+} excited ions to Dy^{3+} ions, this leads to deexcitation of the Eu3+ ions via non-radiational transition so increasing the decay time of the psecies and that this energy transfer depends on the composition of the glass matrix. When the interaction between rare earth ions is not important, decay curves can be described using single –exponential function to obtain the lifetime of the emitting level.

When energy transfer between the dopant ions appears and the decay curves show a non-exponential behavior, expression derived by Inokuti and Hirayama (Widely known as IH model) [13], for the decay of luminescence of a donor–acceptor system can be used.

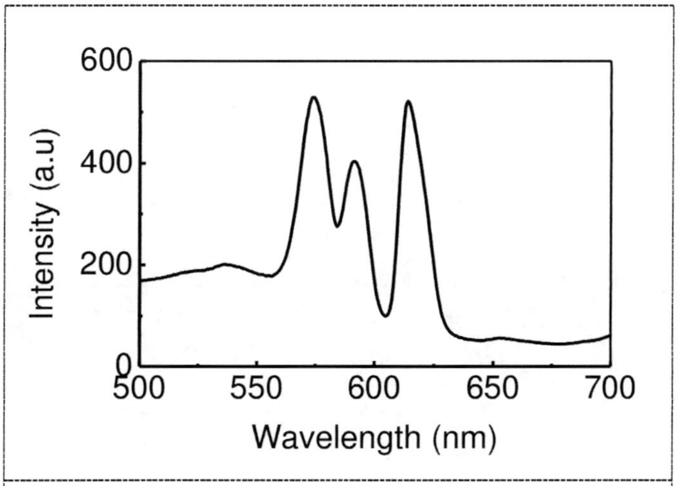

Figure 3. Emission spectrum of Eu/Dy co-doped glass (λ_{ex}=394 nm).

In the IH model, emission intensity as a function of time is given by

$$I(t) = I_o \exp\{-\frac{t}{\tau_o} - Q(-\frac{t}{\tau_o})^{\frac{3}{s}}\}$$

Where, t is the time after excitation, τ_0 is the intrinsic decay time of the donors in the absence of acceptors. Q is the energy transfer parameter. The value of s depends on the type of energy interaction, with s=6, 8, 10 for dipole-dipole, dipole- quadrapole, and quadrapole- qudrapole interaction respectively.

To study the nature of energy transfer processes between Eu^{3+}, and Dy^{3+} ions responsible for the non-exponential decay pattern. The behavior of the decay curves has been analyzed in the frame work of the Inokuti–Hirayama model. A good fitting using IH model with s= 6, reveals that the energy transfer is dipole-dipole in nature and the obtained values of the energy transfer parameter (Q) which are presented in fig.6., show dependence on the composition of the glass matrix. Lifetimes of the $^4F_{9/2}$ level are found to decrease with the composition factor x. suggesting that the consequent replacement of Li_2O by ZnO enhances the process of energy transfer between the rare earth ions. P_2O_5 based glasses consists of PO_4 tetrahedra connected via bridging oxygen ions with P-O bonds, and one dangling double bonded oxygen ion (non-bridging oxygen), from our previous work, using IR investigation of similar glass composition [15] we found that the addition of ZnO to glass composition creates additional non-bridging oxyges. So we could conclude that the replacement of Li_2O by ZnO results in an increase of non-bridging oxygens. The existence of which increases the covelancy of RE-O bonds, and enhance the RE site symmetry which may lead to enhancement of energy transfer.

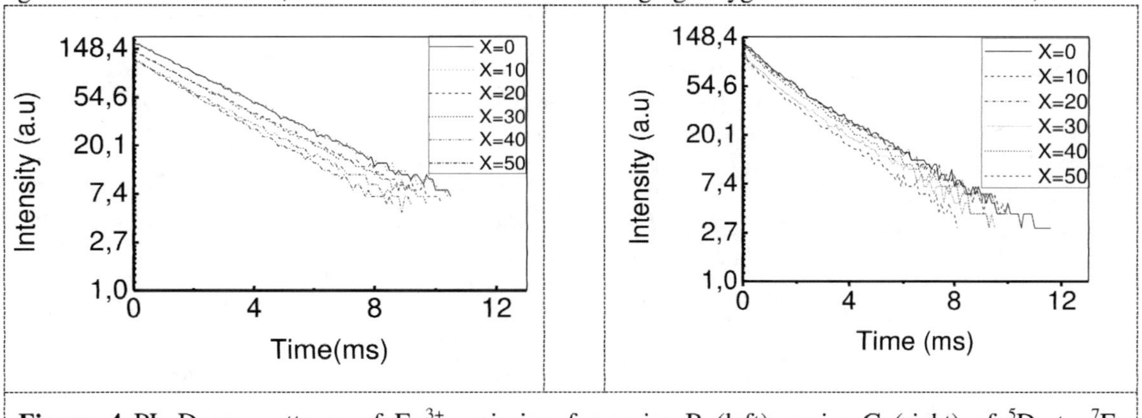

Figure 4 PL Decay patterns of Eu^{3+} emission for series B (left), series C (right) of $^5D_0$ to $^7F_1$ transition ions.

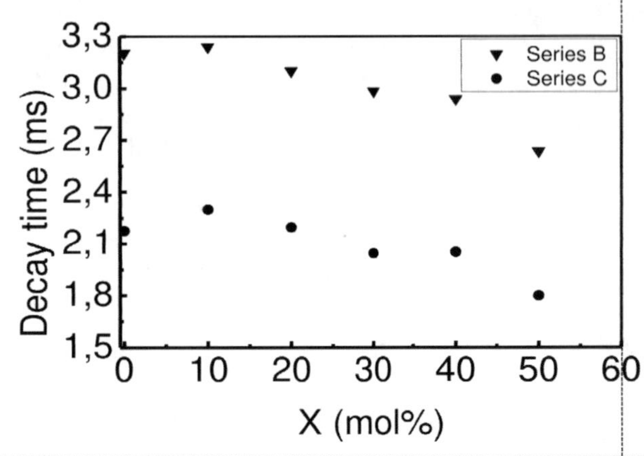

Figure 5. Decay time variation with the composition factor X (mol%) for series B, and C.

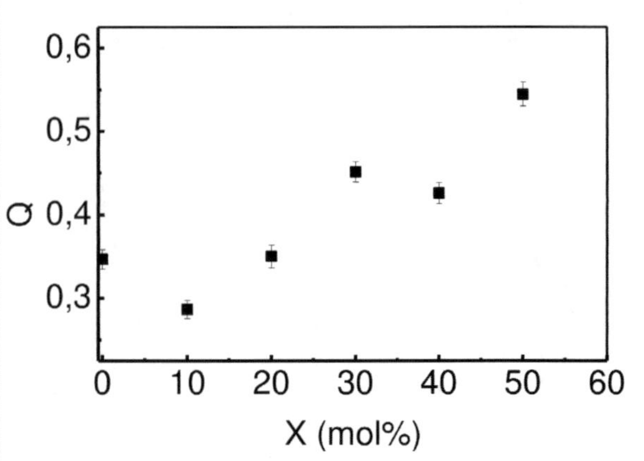

Figure 6. Variation of the energy transfer parameter, (Q) ,with composition.

4. Conclusion

It could be concluded that co-doping of RE ions affects the emission characteristics of the glass with introducing the emission lines of the co-doped ion and that the possibility of energy transfer between the RE ions effectively affect the emission decay time, where the decay time of theEu^{3+} $^5D_0$-$^7F_1$ transition decreased from about 3 ms in case of Eu^{3+} singly doped glasses to about 2 ms. In the case of co-doping Where the IH model could give insight about the mechanism of the energy transfer and the good agreement of the model proposed by Inokuti and Hirayama gives the opportunity to calculate the energy transfer parameter, which we found to depend on the glass compostion.

5. References

[1] Aderhold J, Davydov V Yu, Fedler F, Klausing H, Mistele D, Rotter T, Semchinova O, Stemmer J and Graul J 2001 *J. Cryst. Growth* **22** 701

[2] Reisfeld R 1975 *Struct. Bond.* **22** 123

[3] Joshi B 1995 *J. Non. Cryst. Solids* **180** 217

[4] Ostroumov V, Privis Yu, V.A. Smirnov, I.A. Scherbakov 1986 *J. Opt. Soc. Am.* B **3** 81

[5] Krupke W, Shinn M, Marion J, Caird J, Stokowski S 1986 *J. Opt. Soc. Am.* B **3** 102

[6] Peng B, Izumitani T 1995 *Opt. Mat.* **4** 797

[7] Eyal M, Reisfeld R, Jorgensen C, Jacoboni C 1986 *Chem. Phys. Lett.* **129** 550

[8] Blasse G 1979 *Handbook on the Physics and Chemistry of Rare earths* vol 4, ed Gschneidner K, Eyring L (North Holland Publishing Company) p 237

[9] Blasse G 1988 *Solid. Stat. Chem.* **18** 79

[10] Berdowski P, Buijs M, Blasse G 1985 *J. Physique* ***46*** 31

[11] Buijs M, Vree J, Blasse G 1987 *Chem. Phys. Lett.* **137** 381

[12] Franck J, Livingston R 1949 *Revs. Mod. Phys.* **21** 505

[13] Kelloy R 1970 *J. Lumin.* **2** 435.

[14] Inokuti M, Hirayama F 1965 *J. Chem. Phys.* **43** 1978

[15] Polisadova E, Othman H, Lisitsyn V 2013 *Functional materials* **20** 290

Hydrogen absorption by Zr–1Nb alloy with TiN_x film deposited by filtered cathodic vacuum arc

E B Kashkarov[1], N N Nikitenkov[1], M S Syrtanov[1] and M N Babihina[1]

[1]Tomsk Polytechnic University, 30 Lenin Avenue, Tomsk, Russian Federation

E-mail: egor_kashkarov@mail.ru

Abstract. This paper describes the opportunity of titanium nitride (TiN_x) films application as protective coating for Zr-2.5Nb alloy from hydrogenation. Dense TiN_x films were prepared by filtered cathodic vacuum arc (CVA). Hydrogen absorption rate was calculated from the kinetic curves of hydrogen sorption at elevated temperature of the sample (T = 673 K) and pressure (P = 2 atm). Results revealed that TiN_x films significantly reduced hydrogen absorption rate of Zr-2.5Nb.

1. Introduction

Zirconium-based alloys such as Zr–1Nb, Zr–2.5Nb are widely used in nuclear industry as cladding materials for Russian PWR of VVER type reactors. Zirconium has a good combination of properties such as low neutron absorption, stress-corrosion cracking resistance, creep behavior, reduced hydrogen uptake and corrosion [1–3]. Hydrogen released during the process of water radiolysis which appeared under the influence of radiation during reactor operation cycle. Hydrogen degradation can manifest itself in appearance of hydride phases resulting in substantial loss of plasticity, increase in ductile-brittle transition and sometimes in decrease in mechanical strength [4-9]. The presence of high temperatures, pressure, mechanical loads lead to the mechanical properties degradation [5-10].

Most attempts to prevent hydrogen embrittlement of zirconium alloys focused on the surface modification and deposition of thin surface layers by ion-plasma deposition, ion implantation, pulsed electron or ion beams surface modification [11-13]. TiN_x is a promising candidate as a diffusion barrier coating. It has high melting temperature, high hardness, excellent exhibited ion-irradiation tolerance and high thermal conductivity [14–17]. TiN thin films have found a numerous industrial application because high erosion and corrosion resistance which favorably affects the fretting wear of the coating [18]. It was previously found that TiN_x films deposited by direct current magnetron sputtering (dcMS) significantly reduced hydrogen uptake by Zr–1Nb alloy and it is promising films to protect zirconium alloys from hydrogen embrittlement [19, 20]. However, the sputtered TiN films usually exhibits a lower density and adhesion with substrate compared to arc evaporated films due to its low density and degree of ionized sputtered atoms. Additionally, magnetron sputtering always exhibits lower deposition rate due to target poisoning with an introduction of reactive gas. Cathodic vacuum arc (CVA) deposition has some advantages such as a high ion to neutral ratio of arc evaporation process which resulted in a densification of the growing film and a high level of adhesion to the substrate due to increased adatom mobility [21]. The formation of macro-particles during arc evaporation leads to a defect and rough surface morphology of coating [22, 23]. It was demonstrated that the application of plasma filters and negative pulsed bias to a substrate significantly reduced the microdroplets content in coatings [24, 25].

Hydrogen absorption kinetics by Zr–1Nb alloy with TiN_x films deposited by filtered CVA have not been discussed. In this paper we described the influence of TiN_x deposited by filtered CVA onto Zr–1Nb alloy on hydrogen absorption rate and element distribution before and after hydrogen saturation.

2. Materials and Methods

Zr–1Nb samples with fixed size of 20×20×0.5 mm were previously polished to the average roughness R_a of 0.045 μm using sandpaper and diamond paste. The chamber was evacuated to the base pressure of lower than 2.5×10^{-3} Pa. Before deposition samples were subjected to ion bombardment in argon glow discharge at 1500 V for 3 min. The cathodic vacuum arc evaporation system equipped with coaxial plasma filter (designed by authors [24]) was used. Ti (99.99 % purity) was used as the cathode material. The process parameters are shown in Table 1.

Table 1. Process parameters for TiN_x coatings deposition

Arc current, A	Pulsed bias, V	Pulse frequency, kHz	Pressure (Ar), Pa	Deposition time, min
75	150	100	0.15	30

Hydrogenation of the samples was performed by Sievert method (from hydrogen atmosphere) on the automated complex Gas Reaction Controller LPB (USA) during 120 min at 673 K temperature. The heating rate and hydrogen pressure in the chamber was 6 K/min and 2 atm respectively. The hydrogenation temperature was chosen due to the working temperature of zirconium fuel cladding in the operation process of nuclear reactors. Hydrogen absorption rate was measured from the absorption curves after hydrogen saturation.

The phase identification and structural investigations were performed by X-ray diffraction with CuKα radiation (1.5410 A wavelength) using Shimadzu XRD-7000 (Japan) diffractometer in asymmetric mode at 40 kV, 30 mA and θ=3° fixed angle. Study of the elemental composition of samples was performed by high-frequency glow discharge optical emission spectroscopy (GD-OES) with GD-Profiler 2 technique (Horiba, Japan). Hydrogen concentration was measured according to the principle of melting in inert gas at hydrogen analyzer RHEN602 (LECO, USA).

3. Results and discussion

Figure 1 shows that the intensity of hydrogen pick-up by Zr–1Nb significantly higher than Zr–1Nb with TiN_x. Hydrogenation of initial zirconium alloy characterized by a linear region at the beginning of the process and gradually reaches saturation to 2500 sec in comparison with Zr–1Nb/TiN_x which has not reached saturation during hydrogenation. Thus, TiN_x film deposited by filtered CVA reduces the intensity of hydrogen uptake by Zr–1Nb alloy. The average hydrogen absorption rate was calculated on the linear

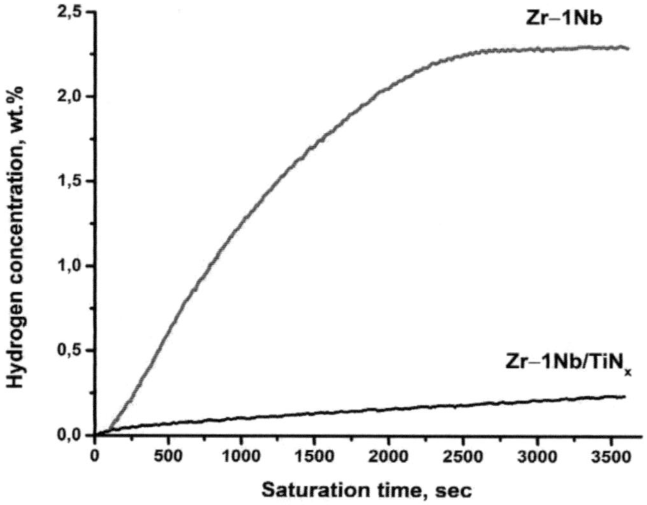

Figure 1. Hydrogen pick-up curve

regions and presented in Table 2. Hydrogen absorption rate by Zr–1Nb decreased in ~24 times as a result of TiN_x coating deposition.

Table 2. Average hydrogen absorption rate

Sample	Average hydrogen absorption rate, wt.%/sec
Zr–1Nb	1.4×10^{-3}
Zr–1Nb/TiN_x	5.7×10^{-5}

Table 3 shows that after hydrogen saturation the total hydrogen concentration in the volume decreased from 22100±300 ppm for Zr–1Nb to 1150±100 ppm for Zr–1Nb with TiN_x film. The principle of melting in inert gas at hydrogen analyzer allows to measure hydrogen concentration with the accuracy of less than 1 ppm. However, the concentration of hydrogen is generally distributed unevenly throughout the sample volume. Therefore we cut 3 pieces from different sites of the samples to more accurately determine the total hydrogen concentration. The small amount of hydrogen contained in initial Zr–1Nb and Zr–1Nb/TiN_x samples.

Table 3. Hydrogen concentration in the samples before and after hydrogen saturation

Hydrogen concentration	Zr–1Nb	Zr–1Nb/TiN_x
Before saturation	14±3 ppm	19±5 ppm
After saturation	22100±300 ppm	1150±100

Figure 3 shows diffraction patterns of Zr–1Nb with TiN_x coating before and after hydrogen saturation. In-plane XRD measurements were carried out in an asymmetrical layout. The incidence angle was 3 degree. The zirconium hydride reflections have been observed after hydrogenation. Analysis of this pattern shows that zirconium hydride reflections belong to the δ hydride. The most intensive reflection (200) of TiN coatings has been detected in both diffraction patterns. This fact indicates on the presence of the texture in the coatings. It could be associated with TiN_x coatings deposition parameters.

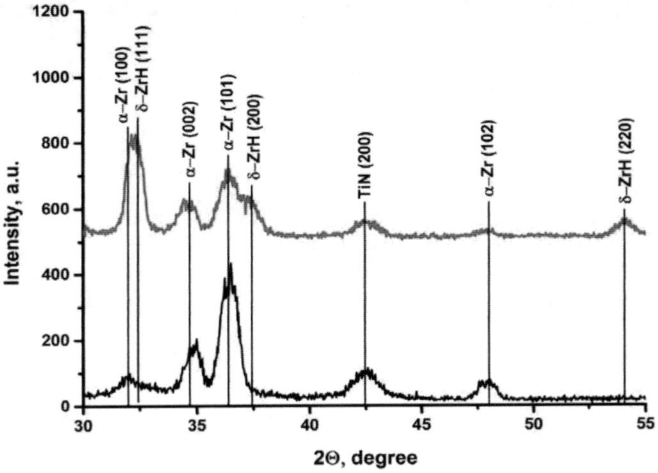

Figure 3. XRD analysis of Zr–1Nb with TiN_x before (below) and after (above) hydrogenation

Figure 4 shows GDOES profiles of Zr, Ti, N, O, H elements before and after hydrogen saturation. The results revealed that hydrogen is present in the TiN_x film and non-uniformly distributed in the sample volume after hydrogen saturation. Furthermore the thin oxide film was observed on the surface of the samples before and after hydrogenation. The intensity of hydrogen signal monotonically decreased with increasing the analyzing depth (sputtering time) and nearly reaches the minimum behind the TiN_x–Zr–1Nb interface (Fig. 4a). It also can be seen that Ti/N ratio decreased at the film depth. After hydrogen

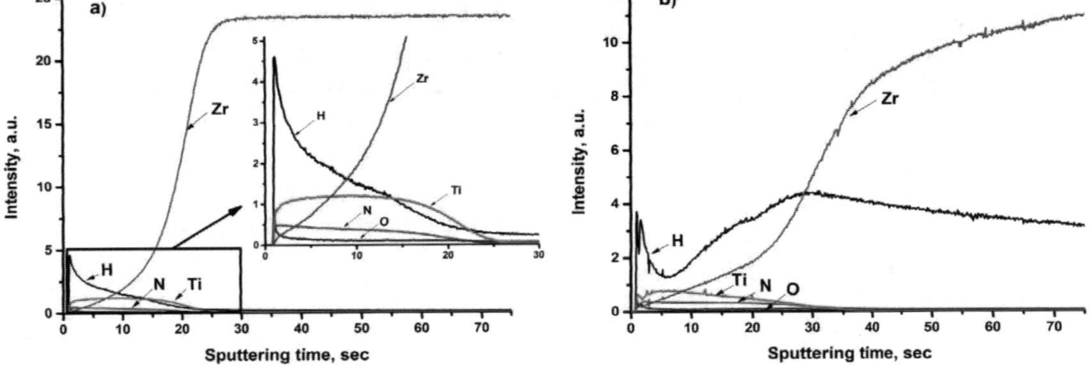

Figure 4. GDOES profiles of elements of Zr–1Nb/TiN_x before (a) and after (b) hydrogen saturation

saturation the intensity of hydrogen signal increased and reached the maximum in TiN_x–Zr–1Nb interface than monotonically decreased with increasing the depth (Fig. 4b). It indicates that hydrogen accumulates in the TiN_x film towards the interface which is probably associated with the decreasing of nitrogen content. Hydrogen penetrates outside the TiN_x film through Zr–1Nb alloy due to diffusion process induced by the difference of concentration in the interface and the sample volume.

4. Conclusion

Titanium nitride films deposited by filtered cathodic vacuum arc significantly reduced hydrogen uptake and total amount of hydrogen in the volume of Zr–1Nb alloy. Hydrogen absorption rate decreased in more than 20 times. δ-phase zirconium hydride formed during hydrogen saturation of Zr–1Nb with TiN_x film. Hydrogen predominantly accumulates in the TiN_x–Zr–1Nb interface where the nitrogen content is low.

References

[1] Sabol G P 2004 Proc. 14th Intern. Symp. 3
[2] Murty K L 2006 Prog. Nucl. Energy **48** 325
[3] Montgomery R, Waeckel N, Yang R 2001 Proc. 29th Annual NRC Water Reactor Safety Meeting 409
[4] Chernyaeva T P, Ostapov A V 2013 *Voprosy atomnoy nauki i tehniki* **87** 16
[5] Silva K R, Dos Santos D S, Robeiro A F, Almeida L H 2010 *Defect and Diffusion Forum* **722** 297
[6] Zielinski A, Sobieszczyk S 2011 *Intern. Journ. of Hydrogen Energy* **36** 8619
[7] Bai J B 1993 *Scripta Metallurgica et Materialia* **29** 1259
[8] Mani Krishna K V, Sain A, Samajdar I, Dey G K 2006 *Acta Materialia* **54** 4665
[9] Singh R N, Kishore R, Singh S S, Sinha T K, Kashyap B P 2004 *Journ. of Nuclear Materials* **25** 26
[10] Anders A 2007 *Jour. Phys. D: Appl. Phys.* **40** 2272
[11] Lider A M, Chernov I P, Cherdantsev Yu P, Pushilina N S, Stepanova O M, Garanin G V 2010 *Proc. 16-th Symp. on High Current Electronics* **2** 541
[12] Ivanova S V, Glagovskiy E M, Khazov I A, Orlov V K, Shlepov I A, Nikitin K N, Dubrovskiy Yu V, Denisov E A 2008 *Proc. IV Int. Conf. on Interaction of Hydrogen Isotopes with Structural Materials* 51
[13] Nikitenkov N N, Sutygina A N, Shulepov I A , Sivin D O , Kashkarov E B 2015 *IOP Conf. Series: Materials Science and Engineering* **81** 012018
[14] Wang H, Araujo R, Swadener J G, Wang Y Q, Zhang X, Fu E G, Cagin T 2007 *Nucl. Instrum. Methods B* **261** 1162.
[15] Oyama S T 1996 *Chemistry of Transition Metal Carbides and Nitrides* (Berlin: Springer)
[16] Toth L E 1971 *Transition metal carbides and nitrides* (Academic Press)
[17] Pierson H O 1996 *Handbook of Refractory Carbides and Nitrides: Properties, Characteristics, Processing and Applications* (Noyes Publications)
[18] Hultman L 2000 *Vacuum* **57** 1
[19] Nikitenkov N N, Kashkarov E B, Tyurin Yu I, Nikiforova N I 2014 Izvestiya vuzov. Physika **57** (11/2) 134
[20] Kashkarov E B, Nikitenkov N N, Tyurin Yu I, Syrtanov M S, Zhang Le 2015 *Iop Conf. Series: Materials Science and Engeneering* **81** 012017
[21] Boxman R L, Martin P J, Sanders D M 1995 *Handbook of Vacuum Arc Science and Technology: Fundamentals and Applications* (NJ: Noyes) 367
[22] Karpov D A 1997 *Surf. Coat. Technol.* **96** (1997) 22
[23] Coll B F, Sanders D M 1996 *Surf. Coat. Technol.* **81** 42
[24] A.I. Ryabchikov, D.O. Sivin, A.I. Bumagina 2014 *Applied Surface Science* **310** 130
[25] Ryabchikov A I, Ryabchikov I A, Stepanov I B 2005 *Vacuum* **78** 331

Hydrogenation of Zr–2.5Nb alloy after plasma-immersion titanium implantation

A N Sutygina[1], E B Kashkarov[1], N N Nikitenkov[1], Yu I Tyurin[1] and M S Syrtanov[1]

[1]Tomsk Polytechnic University, 30 Lenin Avenue, Tomsk, Russian Federation

E-mail: egor_kashkarov@mail.ru

Abstract. The study results of the influence of plasma-immersion ion implantation of titanium in Zr–2.5Nb on hydrogenation are presented. The titanium implantation was carried out in two modes: with active plasma filtering (APF) and passive plasma filtering (PPF). The results of total hydrogen concentration, absorption rate, XRD analyses and depth distribution of elements revealed that modified surface layer after titanium ion implantation is formed hydrogen diffusion barrier reduces hydrogen absorption by Zr–2.5Nb.

1. Introduction

Zirconium and its alloys are important structural materials of nuclear reactors. Zirconium is used in the nuclear reactors as the fuel assemblies cases, channels of control and protection system, the technological channels and the channel pipes.

In these reactors there is a problem of water radiolysis under the influence of radiation resulted the release of hydrogen. Zirconium alloys absorbs and accumulates hydrogen that leads to the hydrogen corrosion and embrittlement of the material. The presence of high temperatures, pressure, mechanical loads lead to the mechanical properties degradation [1-7]. All these factors reduce the operation time of zirconium elements.

One of the ways to solve these problems is the surface modification and deposition of thin surface layers (ion-plasma metal deposition, ion implantation and surface modification by pulsed ion beam) [8, 9]. One of the most perspective methods for modifying microstructure and elemental composition is plasma immersion ion implantation (PIII). PIII is the modification method of the surface by accelerated ions from the plasma arc discharge by applying a high voltage pulse current on the sample. PIII allows the ion implantation in the volume target and modify the deep layers of materials [10]. PIII leads to formation intermetallic phases and solid solutions of high concentration. As a result wear resistance, adhesive strength, corrosion resistance are increased after PIII [11]. This is due to the high concentration of implanted elements and the high target temperature reached during ion implantation [12].

Titanium implantation into steel, nickel, aluminum increases the hardness, corrosion resistance and improved tribological properties. Furthermore titanium has a low capture cross section of thermal neutrons. Titanium implantation into Zr–2.5Nb improves oxidation and corrosion behaviors [13, 14]. Therefore in this work titanium is implanted into Zr–2.5Nb. At the moment there are few studies about titanium implantation into Zr–2.5Nb and its influence on hydrogenation.

In according with this, the aim of the work is the study of the influence of titanium PIII with active plasma filter (APF) and with passive plasma filter (PPF) on hydrogenation of Zr–2.5Nb alloy.

2. Materials and Methods

Zr–2.5Nb samples with fixed size of 1 mm in thickness and 30 mm in diameter were previously polished to the average roughness R_a of 0.045 μm using sandpaper and diamond paste. The chamber was evacuated to the base pressure of lower than 2.5×10^{-3} Pa. Before deposition samples were subjected to ion bombardment in argon glow discharge at 1500 V for 3 min. For PIII – the cathode vacuum arc evaporation system equipped with coaxial plasma filter (designed by authors [15, 16]) was used. Ti (99.99 % purity) was used as the cathode material. Process parameters for PIII of Ti are: 0.15 Pa total pressure, 70 A arc current, 1500 V pulsed substrate bias with 100 kHz pulse repetition frequency and 3 A peak pulse current. The implantation time was 15 min. The part of the samples was implanted in active filtration mode (APF) another part in passive filtration mode (PPF).

Hydrogenation of the samples was performed by Sievert method (from hydrogen atmosphere) on the automated complex Gas Reaction Controller LPB (USA) during 120 min at 673 K temperature. The heating rate and hydrogen pressure in the chamber was 6 K/min and 2 atm respectively. The hydrogenation temperature was chosen due to the working temperature of zirconium fuel cladding in the operation process of nuclear reactors. Hydrogen absorption rate was measured from the absorption curves after hydrogen saturation.

The phase identification and structural investigations were performed by X-ray diffraction with CuK_α radiation (1.5410 A wavelength) using Shimadzu XRD-7000 (Japan) diffractometer in asymmetric mode at 40 kV, 30 mA and θ=3° fixed angle. Study of the elemental composition of samples was performed by high-frequency glow discharge optical emission spectroscopy (GD-OES) with GD-Profiler 2 technique (Horiba, Japan). Hydrogen concentration was measured according to the principle of melting in inert gas at hydrogen analyzer RHEN602 (LECO, USA).

3. Results and discussion

Figure 1 shows the total hydrogen concentration in the samples volume after hydrogen saturation. The total hydrogen concentration decreases in ~19 times and ~2.2 times in comparison with the initial Zr–2.5Nb after hydrogen saturation of Ti-implanted sample with APF and PPF respectively.

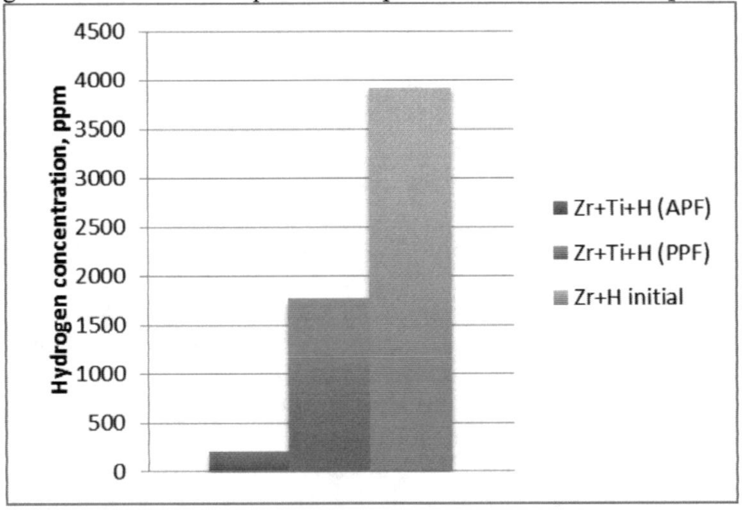

Figure 1. Hydrogen concentration of the initial and Ti implanted Zr–2.5Nb alloy after hydrogen saturation

Figure 2 shows that the average hydrogen absorption rate of Zr–2.5Nb after Ti implantation decreased from $13.65\cdot10^{-4}$ to $0.74\ 10^{-4}$ cm^3 $H_2/(sec\cdot cm^2)$ after APF and to $3.2\ 10^{-4}\cdot cm^3$ $H_2/(sec\cdot cm^2)$ after PPF.

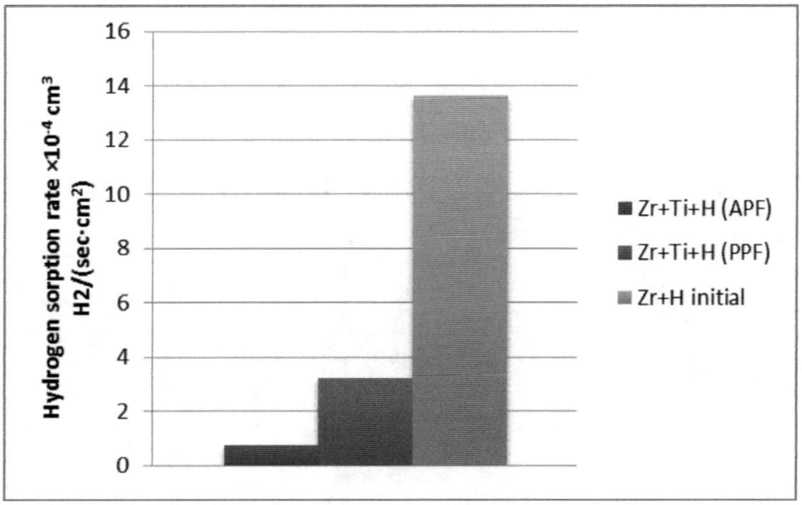

Figure 2. Average hydrogen absorption rate by Zr–2.5Nb after Ti implantation with APF and PPF modes

The hydrogen absorption rate was determined by differentiating the kinetics curve which characterized the dependence between hydrogen pressure in the chamber and time of saturation on the linear region and is calculated by the following formula:

$$q = \frac{V}{t \cdot S} \frac{P_0 - P}{P}, \tag{1}$$

where V – chamber volume (175 cm^3), t – saturation time, S – effective surface area of the samples, cm^3, P and P_0 – final and initial pressure, atm [17].

Figure 3 shows the results of X-ray diffraction analysis. Ti reflexes and different compounds after titanium implantation are not detected. Only reflections of α-Zr with hexagonal lattice were observed after PIII with APF and PPF modes.

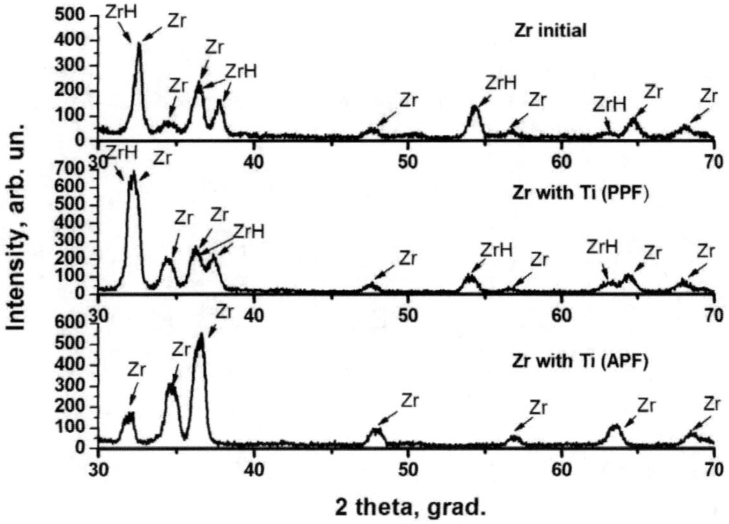

Figure 3. XRD analysis of Zr–2.5Nb after titanium implantation and hydrogen saturation

Zirconium hydride phases with tetragonal lattice are formed after hydrogen saturation in initial samples and Ti-implanted sample at PPF. Hydride phases in the sample modified with APF not observed due to a low hydrogen concentration (206 ppm). Moreover, some concentration of hydrogen is in dissolved state.

Figure 4 shows the GDOES profiles of elements distributions in the modified layers after hydrogen saturation. Ti penetration depth into Zr–2.5Nb with APF is ~250 nm and with PPF is ~350 nm. The implantation with APF reduces the ion energy and dose of implanted material.

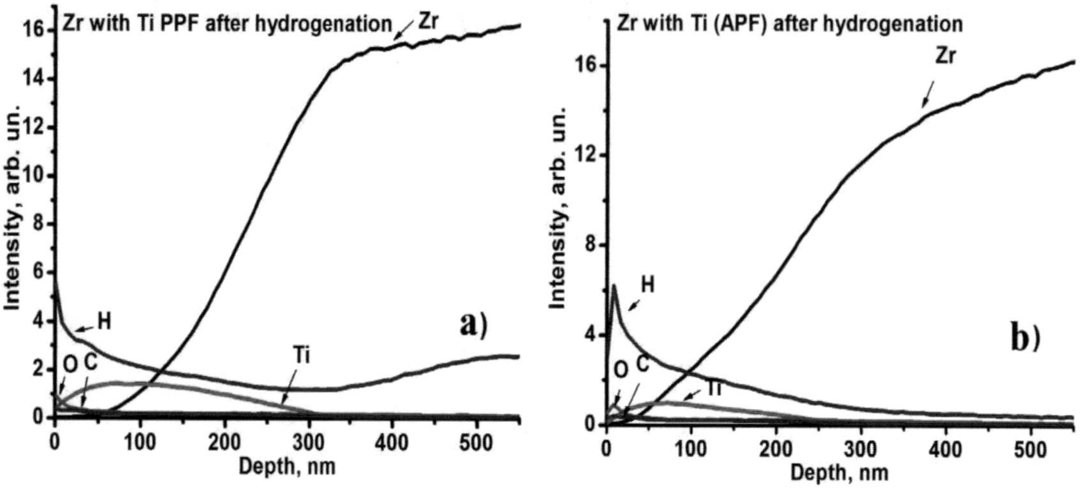

Figure 4. GDOES depth distribution of elements in Zr–2.5Nb after Ti implantation with APF (a) and PPF (b) after hydrogen saturation

Titanium implantation with APF significantly reduced hydrogen penetration into Zr–2.5Nb. The intensity of hydrogen signal monotonically decreased with increasing the depth and nearly reaches the intensity of hydrogen for the initial Zr–2.5Nb behind the modified layer with APF (Fig. 4a). However, the intensity of hydrogen increased behind the modified layer with PPF (Fig. 4b). This indicates the larger hydrogen concentration in the sample volume which confirmed by measured hydrogen concentration. Active plasma filtration more effectively reduced the amount of microdroplets formed in cathodic vacuum arc spots. Microdroplets can be pathways for hydrogen penetration into zirconium because it predominantly consists of metallic titanium [18] which has a high hydrogen absorption rate. Thus, hydrogen accumulates in the modified layer which is formed a diffusion barrier preventing hydrogen penetration into zirconium alloy.

Conclusion

PIII of titanium into Zr–2.5Nb reduces hydrogen absorption rate and total hydrogen concentration in the volume. The different radiation defects formed during titanium implantation can accumulate hydrogen primarily. When hydrogen fills the modified layer it begins to penetrate into Zr–2.5Nb volume by diffusion process. Titanium implantation with APF more effectively reduces hydrogen absorption by zirconium alloy in comparison with PPF. Intermetallic compounds were not observed in the experimental conditions.

References

[1] Chernyaeva T P, Ostapov A V 2013 *Voprosy atomnoy nauki i tehniki* **87** 16
[2] Silva K R, Dos Santos D S, Robeiro A F, Almeida L H 2010 *Defect and Diffusion Forum* **722** 297
[3] Zielinski A, Sobieszczyk S 2011 *Intern. Journ. of Hydrogen Energy* **36** 8619
[4] Bai J B 1993 *Scripta Metallurgica et Materialia* **29** 1259
[5] Mani Krishna K V, Sain A, Samajdar I, Dey G K 2006 *Acta Materialia* **54** 4665
[6] Singh R N, Kishore R, Singh S S, Sinha T K, Kashyap B P 2004 *Journ. of Nuclear Materials* **25** 26
[7] Anders A 2007 *Jour. Phys. D: Appl. Phys.* **40** 2272

[8] Lider A M, Chernov I P, Cherdantsev Yu P, Pushilina N S, Stepanova O M, Garanin G V 2010 *12-st International Congress on Radiation Physics, High Current Electronics, and modification of Materials. 16-th Simposium on High Current Electronics* **2** 541
[9] Ivanova S V, Glagovskiy E M, Khazov I A, Orlov V K, Shlepov I A, Nikitin K N, Dubrovskiy Yu V, Denisov E A 2008 *Proc. IV Int. Conf. on Interaction of Hydrogen Isotopes with Structural Materials* 51
[10] Dushyant G 2011 *International Journal of Advancements in Technology* **2** 2333
[11] Joseph V, Mantese I, Brown G, George A. 1996 *Collins Cambridge Journals* **21** 52
[12] Nikitenkov N N, Sutygina A N, Shulepov I A , Sivin D O , Kashkarov E B 2015 *IOP Conf. Series: Materials Science and Engineering* **81** 012018
[13] Peng D Q, Bai X D, Pana F, Sun H, Chen B S 2006 *Applied Surface Science* **252** 2196
[14] Zhanga X Y, Wan Q, Baib X D, Jina Z X 2006 *Vacuum* **80** 1003
[15] Ryabchikov A I, Ryabchikov I A, Stepanov I B 2005 *Vacuum* **78** 331
[16] Ryabchikov A I, Ryabchikov I A, Stepanov I B, Usov Yu P 2007 *Surf.Coat.Technol.* **201** 6523
[17] Jeshin P, Wonbaek K, Misook W 2007 *Materials Transactions* **48** 1012
[18] Matthews A, Lefkow A R 1984 *Thin Solid Films* **126** 283

Wear resistance of a metal surface modified with minerals

S V Kislov[1], V G Kislov[1], P V Balasch[1], A V Skazochkin[2,4], G G Bondarenko[3] and A N Tikhonov[3]

[1] Research and Production Association "Geoenergetika", Kaluga, Russia

E-mail: geoen@mail.ru

[2] Kaluga branch of the Russian Presidential Academy of National Economy and Public Administration, Kaluga, Russia

E-mail: avskaz@rambler.ru

[3] National Research University "Higher School of Economics", Moscow, Russia

E-mail: gbondarenko@hse.ru

Abstract. The article describes the advantages of the new technology of mineral coating of metal products for the friction pair of mechanical systems. It presents the research results of the wear rate of the samples made of 12X13 steel (X12Cr13) with mineral layers, in the experiments with a piston ring sliding inside a cylinder liner with grease. The wear rate of the samples with mineral layers is lower almost by two factors than that of the samples made of grey foundry iron and untreated samples. As the result of slip/rolling abrasion tests of parts with mineral layers under conditions of high contact pressure, a suggestion was made concerning probable mechanics of surface wear.

Introduction

The core indicators of the quality of machine's parts, assemblies and mechanisms are reliability and lifespan, which are greatly defined by the properties of the surface layers of parts and junctions: wear resistance, corrosion resistance, friction rate, contact rigidity, seating stability, connection tightness. The reason for a short lifespan of parts and other structural components may be significantly associated with tear-and-wear and corrosive deterioration of their surface layers. To eliminate or slow down the processes taking place on the border between metal and environment, which have a negative effect on the operability of materials, various types of surface treatment are used.

Besides multiple methods involving coatings, films and protective layers, there is a rapidly developing modern technology of changing physical and chemical properties of surface layers of metal parts by means of modification. In particular, industrial application of natural minerals for hardening surface layers of structural materials is finding its niche, due to the effects, combination of which leads to enhanced wear resistance, anti-friction, anti-corrosion and other specific properties of a friction pair.

[4] To whom any correspondence should be addressed.

The basis for a new original technology of creating mineral layers enhancing wear resistance of metal parts was created in Russia in 2009-2013 [1]. The mineral layers protect parts of all-purpose mechanical systems from wearing off, extend their lifespan and reduce energy consumption due to significant elimination of mechanical losses in machines and mechanisms [2]. The properties of mineral layers created by RPA “Geoenergetika”, have been confirmed by experiments and field tests conducted by partners from the Federal Institute for Materials Research and Testing, BAM, Germany [3].

The technology of creating mineral layers was introduced to the Russian and European markets in 2012-2015 and has already found application in various industries: metallurgy, machine building, power engineering, ship construction.

The mineral layers created enhance wear resistance of friction pairs from 3 to 10 times, feature unique tribotechnical parameters and can be applied to water lubricated metal parts, in environments with abrasive particles, and under conditions of unlubricated friction [2, 4, 5].

The created layers have unique properties, because parts coated by mineral layers of RPA “Geoenergetika”, demonstrate great performance:

- in seawater, salt mist, hydrogen sulphide, increased humidity, abrasive dust and other aggressive environments;
- under thermal cyclic loads in a wide temperature range.

Technology of mineral layers’ creation:

- does not require special basins, furnaces, vacuum chambers or specific conditions;
- has no fundamental limitations of mass and dimensional properties;
- does not change the geometry of the treated parts, does not require alteration of designs;
- is environment-friendly, has no harmful effect on human health and environment.

The present work investigates the wear rate of the samples before and after application of mineral coatings, as well as in comparison with some other samples.

Experiment details, results and their discussion

The technological process of mineral layers’ creation consists in embedding nanocomposite mineral powders into the crystal lattice of the treated metal by means of ultrasound and electrospark treatment, as well as ball and roller knurling. The details of the technological process of mineral layers’ creation, as well as structural changes on the surface and subsurface layers of metal samples (chromium steels, titanium alloys) are described in the work [4]. As the result of the combination of technological procedures changing the surface, microhardness of the surface layers is increased. Microstructural researches have demonstrated that a modified layer is observed, with granular structure, clearly different from the remaining mass [4]. The enhancement of hardness and strength properties is conditioned by the reduction of grain size.

This work investigates deterioration of samples with mineral layers and without special treatment, by means of an experiment including sliding of a piston ring in a cylinder liner lubricated with machine oil (the methodology is described in works [6, 7]) and an experiment of slip/rolling under conditions of high contact pressure (a thorough description of the method can be found in work [8]).

For all the experiments modified mineral layers according to NPO “Geoenergetika” technology were created on the surface of metal samples produced by lathe turning from 12X13 steel (Russian equivalent of EU X12Cr13 steel) rods in the shape of piston rings [1]. The thickness of the mineral layers created varied from 5 to 20 micron.

Wear rate was compared individually for a piston ring with untreated surface and a piston ring with modified surface. Synthetic oil based on polybutyleneglycol (GLYMOT PBG B20) and machine oil were used in the experiments. The mode involving machine oil lubrication was chosen because lube is typically present in most practical situations.

“PU”, “PV”, “PW” and “PT” samples in figure 1 designate various combinations of surface treatment conditions and metal from which the samples are produced.

Wear rates of the rotating disks were compared with wear rates of GGL20HCN European brand grey foundry iron disks with high carbon content (3.7 weight percent of carbon and 2.0 weight percent of silicon) which is normally used for production of cylinder liners and brake disks. The curly bracket

IOP Conf. Series: Materials Science and Engineering **110** (2016) 012048 doi:10.1088/1757-899X/110/1/012048

and "X12Cr13" title in figure 1 denote the results acquired for a 12X13 (X12Cr13) steel piston ring without any treatment.

Abrasion slip-rolling experiments were conducted under conditions of high contact pressure P = 2.25 GPa. Experiments involving sliding with machine oil were conducted under much lower contact pressure, of about 100 MPa.

Figure1 presents wear rates of the surface of PT series rings (PT-502, PT-503, PT-504 and PT-201, PT-202, PT-203) made of 12X13 steel with mineral layers which were compared with wear rates of the surface of untreated rings made of 12X13 steel, 20X13 steel and grey foundry iron.

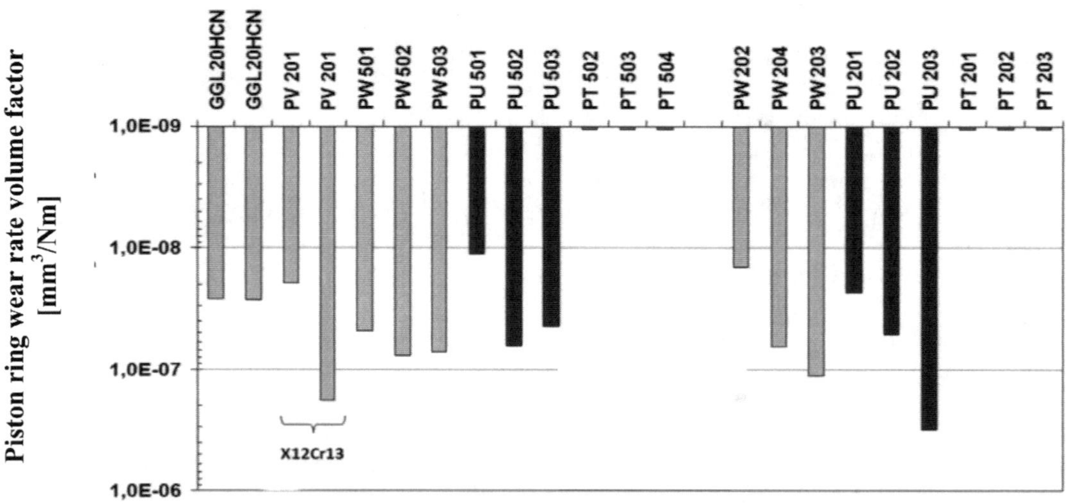

Figure 1. Wear rate volume factors for piston rings sliding under mixed lubrication conditions over cast iron and 12X13 steel treated by RPA "Geoenergetika" in machine oil (F_N=50 H; v =0,3 m/s; T_{oil} = 170°C; s = 24 km)

Evidently, creating mineral layers on the surface of PT series rings provided wear rate which is by two factors lower than that of standard samples made of grey foundry iron and untreated 12X13 and 20X13 steel rings.

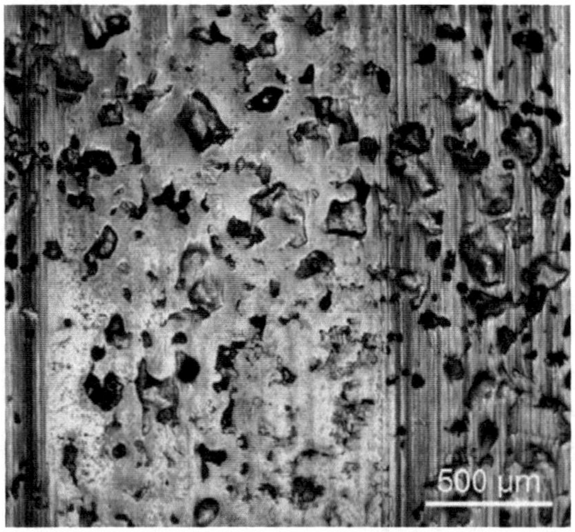

Figure 2. The picture of a wear track on a treated 12X13 steel surface performed by means of an optical microscope (T = 120 ^{0}C, P = 1.5 GPa, 10 million cycles, slip-rolling).

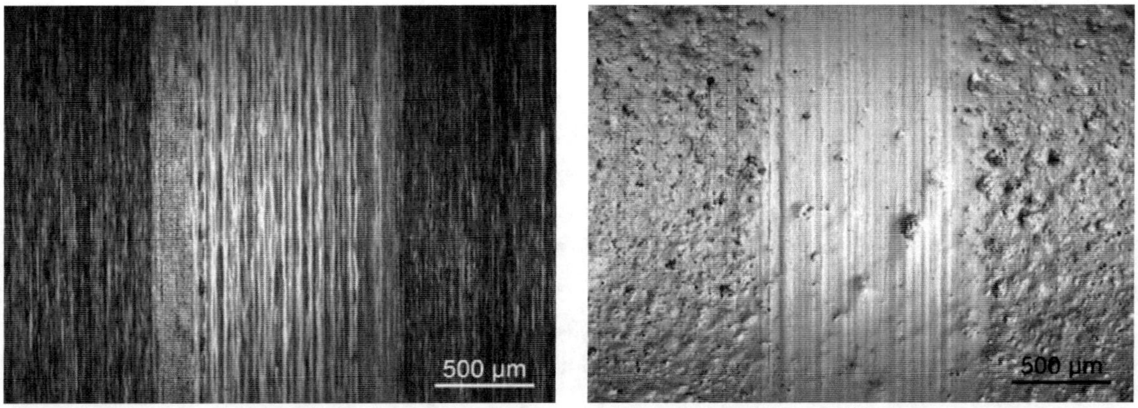

a b

Figure 3. The picture of a wear track after slip-rolling performed by means of an optical microscope (a - surface of the sample made of ШХ15 steel without mineral coating; b - cylindrical sample made of ШХ15 steel with mineral coating; T = 120 ^{0}C, P = 1.5 GPa, 10 million cycles, slip-rolling).

Figures 2 and 3 clearly demonstrate that after 10 million cycles of slip-rolling abrasion test the surfaces with mineral coating, exposed to friction, became smoother. It is possible that the technology of creating a thin modified layer on a metal surface, leading to its local strengthening and enhanced hardness [4], preconditioned good slip-rolling friction resistance of the samples made of 12X13 and ШХ15 (Russian equivalent of 102Cr6 German steel) steels, which was probably defined by general residual compression stresses.

The thin layer of the hardened surface of steel samples modified by minerals has demonstrated good resistance to slip-rolling procedure [7, 8] after ten million cycles at the temperature of 120 ^{0}C and pressure P = 1.5 GPa. Under such conditions the effective thickness of the oil film h_{min} at 120 ^{0}C was 0.027 micron [8] which is the evidence of boundary lubrication mode, so the wear of the surfaces is mainly conditioned by physical and chemical interaction taking place at the spots of actual surface contact. Besides, the pictures in figure 2 show microcavities randomly scattered over the working surface of the sample. It is due to these created microcavities that the oil is detained when lubrication is suspended, which may be listed among the advantages of the technology [4] because performance capabilities and equipment safety are enhanced during start-and-stop operations, e.g. spindle bearings in case of insufficient lubrication.

Conclusions

1. The mineral layers created on the surface of the samples presented by 12X13 steel rings provided wear rate which is almost by two factors lower than that of the samples made of grey foundry iron and untreated 12X13 and 20X13 steel samples during experiments involving piston ring sliding in a cylinder liner with machine oil lubrication.
2. When testing slip-rolling abrasion of parts with mineral layers under high contact pressure conditions, boundary lubrication mode is possible. That is why the surface wear is only conditioned by physical and chemical interaction at the spots of actual surface contact.

Acknowledgements

The authors are grateful to Dr. Mathias Woydt, head of Tribology and Wear protection division of Federal Institute for Materials Research and Testing, BAM, Germany, for performing sample measuring operations and consulting.

References

[1] Kislov S V, Kislov V G and Lazarev S Yu 2011 Creation method of combined mineral surface layer for metal products protecting it from action of aggressive medium and with designated tribotechnical parameters *Patent on invention №2421548 from 20.06.2011, patent application 2009142259/02 from 18.11.2009*

[2] Elagina O Yu, Gantimirov B M, Tomskij K O, Mazurkevich AN and Kislov S V 2013 Research of tribotechnical parameters of coat by metal to metal friction without oil *Perspective materials* **7** 56-60

[3] Kislov S, Kislov V, Ostrovsky P and Woydt M 2012 Localized hardening and structuring of steel by a simple process. *53 Tribologie Fachtagung «Reibung, Schmierung und Verschleiß» der GfT Göttingen, Herstellung: Druckservice Zillekens* **2** 49/1-7

[4] Kislov S V, Kislov V G, Skazochkin A V, Bondarenko G G and Tihonov A N 2015 Effective mineral coats for surface hardening of metal products *Metals* **4** 56-63

[5] Polovinkin V N 2009 *Nanotechnology in ship construction* Academician A.N. Krylov Reserch institute St-Peterburg 132

[6] Woydt M and Kelling N 2003 Testing the tribological properties of lubricants and materials for the system "piston ring/ cylinder liner" outside of engines *Industrial Lubrication&Tribology* **55 5** 213-222

[7] Löffelbein B, Woydt M and Habig K-H 1993 Sliding friction and wear of ceramics in neutral, acid and basic aqueous solutions *WEAR* **162-164** 220-228

[8] Scholz C, Spaltmann D and Woydt M 2011 Slip-rolling resistance of thin films and high toughness steel substrates under high Hertzian contact pressures *WEAR* **270 7–8** 506–514

Extraction of Alumina from high-silica bauxite by hydrochloric acid leaching using preliminary roasting method

D V Valeev[1], E R Mansurova[1], V A Bychinskii[2] and K V Chudnenko[2]

[1] Baikov Institute of Metallurgy and Materials Science, Russian Academy of Science, Leninskii pr. 49, Moscow, 119991, Russia
[2] A.P. Vinogradov Institute of Geochemistry of the Siberian Branch of the RAS, 1A Favorsky St., Irkutsk, 664033, Russia

E-mail: dmvaleev@yandex.ru

Abstract. A process of dissolution Severoonezhsk deposit boehmite-kaolinite bauxite by hydrochloric acid, as well as the processes that occur during open-air calcination, were investigated. A dehydration process has been studied, and the basic phase transformation temperatures were identified. Temperature and time of calcination influence on bauxite dehydration speed were determined. It is shown that the preliminary calcination increases the extraction ratio of alumina into solution up to 89%. Thermodynamic modelling of physical and chemical processes of bauxite decomposition by hydrochloric acid and the basic forms of aluminium speciation in solution were obtained.

1. Introduction

Russia is the world leader in aluminum production, but the bulk of the raw material (~70% of bauxite and alumina) are imported from abroad (Australia, Jamaica, Guyana, Guinea). Shipments of aluminum ore from distant deposits are questioning production profitability due to constant increase in expenses for freight ships and transportation by rail. The total length of some logistic chains (mine - alumina refinery - aluminum plant) may exceed 15 thous. km. High political risks and the issue of raw materials security in the country make this problem even more acute [1].

Huge reserves of low-grade and non-traditional for alumina production aluminum-containing materials, such as high-silicon bauxites [2–3], kaolin [4], anorthosite [5] and coal fly ash [6–9], can serve as an efficient source of raw materials. Researches in this area are essential and extensively developing in countries with developed aluminum industry, e.g. Canada [10] or China [11-14].

Due to development of new corrosion-resistant materials (alloys based on Ta, Zr, Nb), durable equipment for alumina refinery can be produced [15–16].

The use of acid decomposition offers great opportunities for efficient complex processing of alternative raw materials to produce alumina [17], rare or rare-earth metals [18–19], silicon-based materials [20], and a reagent for the water purification [21] that will lead to a significant increase in competitiveness of domestic enterprises. Therefore, the main task of this work was to develop an efficient method of acid leaching which could be applied to raw materials available in Russia.

2. Experimental

Acid leaching was carried out in a round-bottom flask placed in a heating flask. The bauxite samples used in the experiments were of two types, both milled to a particle size of 0.1 mm: the original sample and the sample calcined at a temperature of 700°C. Calcination was performed in muffle furnace (HTC 03/18/3N/PE, Nabertherm, Germany) in open corundum crucibles, heating time to temperature of 560-750°C was 30 min., sample weight was 10 gr., and the time - 3.5 h.

Bauxite was leached by 20% hydrochloric acid with mechanical stirring at 110°C for 1-5 hours. Then the solutions were separated from the solid precipitation (Si-stoff) using a vacuum filter. The Si-stoff was washed with distilled water and then dried in an oven at 110°C for 2 h. The resulting solutions, washing water and Si-stoff were analyzed for Al, Fe and Si by ICP-OES (Optima-4300 DV of PerkinElmer Instruments, USA). The Si-stoff was characterized by X-ray diffraction (Ultima IV, Rigaku, Japan) for identification of crystalline phases. To determine the change in mass of bauxite and temperature of phase transitions of minerals included in bauxite, we used the simultaneous thermal analyzer STA 409 Luxx (Netzsch, Germany).

The study of Al-Si-Fe-Cl-C-N-H-O-*e* system was studied by the method of minimizing thermodynamic potentials. Researches were performed with the help of «Selector» software [22]. The basis of the simulation was based on the concept of ion-complex structure of acidic aluminum-containing solution.

The list of substances potentially possible in equilibrium included 71 dependent component, including: 8 condensed phases, 51 components of aqueous solution and 12 components of gas phase. All thermodynamic properties were taken from the «Selector» database: a_Sprons98, s_Sprons98, g_Reid, s_Yokokawa [23–25].

3. Results and discussion

In the present work, domestic high-silicon Severoonezhsk deposit bauxites (Arkhangelsk region) have been used as a raw material. To date, the main direction of processing these bauxites is the production of refractory materials. Processing of the raw material for alumina will let to meet the needs of basic aluminum smelter of the European part of Russia (Kandalaksha, Nadvoitsy, Volkhov and Volgograd aluminum Smelter) in metallurgical alumina.

According to XRD (figure 1a) the main mineral components in Severoonezhsk bauxite are boehmite AlO(OH) ~ 41.5%, kaolinite $Al_2Si_2O_5(OH)_4$~ 25%, gibbsite $Al(OH)_3$~ 8%, hematite Fe_2O_3 ~ 4%, goethite FeOOH ~ 5%, anatase TiO_2 ~ 3%. The chemical composition of bauxite is 50 % Al_2O_3; 20.0 % SiO_2; 7.94 % Fe_2O_3; 2.94% TiO_2; 0.88 % CaO; 0.86% Cr_2O_3; 0.52% MgO; 16.86% loss after calcination. Due to a high silicon content, this bauxite is considered to be a low-grade and therefor has a low silicon module ($\mu_{Si} = 2.5$).

The results of leaching of bauxite by 20% hydrochloric acid at 110°C show that the extraction ratio of aluminum is not more than 23% in case of a process duration of 5 h. (the obtained solutions are composed of 10.6 g/l Al, 7.69 g/l Fe , and 0.01 g/l Si). The process is uneven, the first two hours are characterized by the most intensive leaching due to a complete dissolution of the easily dissolved by hydrochloric acid minerals such as gibbsite, goethite and hematite. Kaolinite and boehmite thus dissolve slightly, which is confirmed by X-ray diffraction (figure 1b).

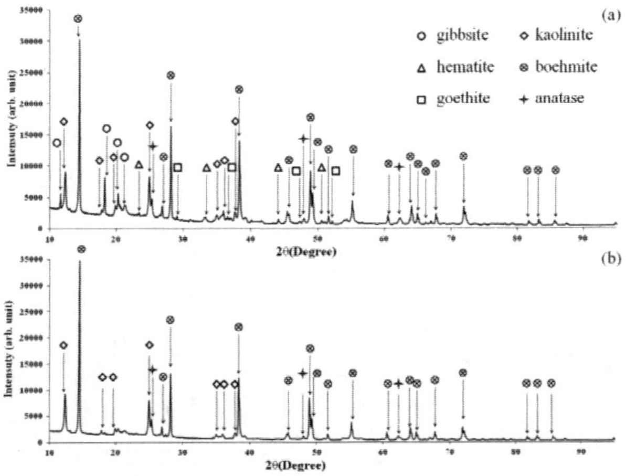

Figure 1. XRD patterns of the original bauxite (a) and the solid residue after leaching for 2 hours (b).

In order to increase the extraction ratio of aluminum bauxite precalcination was used, that allowed us to transfer low-soluble boehmite and kaolinite into the acid-soluble forms – γ-Al_2O_3 and $Al_2Si_2O_5$, respectively.

This is confirmed by thermodynamic modeling of the solubility of certain minerals by hydrochloric acid. Subsequently, the calculation results were compared with experimental data. It is found that the solubility of aluminum-containing minerals in chloride solutions at 25°C decreases (figure 2). Thermodynamic calculations have shown that increasing solubility of aluminum-containing minerals caused by change in the ionic composition of the solution. In particular, there is a relative decrease in the concentration of cationic and hydroxide complexes of Al, while the concentrations of chloride and chlorohydrate complex compounds of aluminum increase [26].

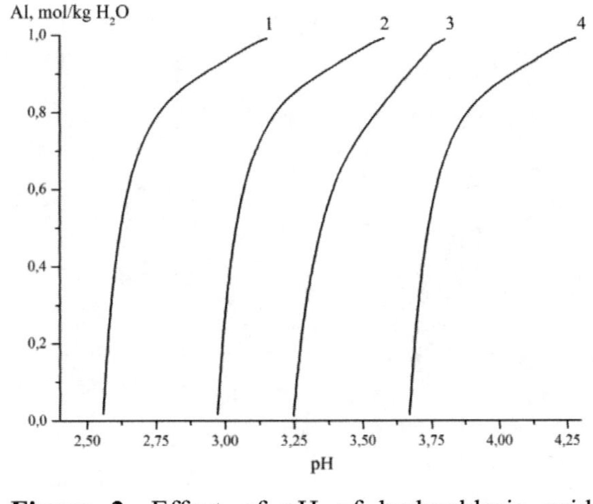

Figure 2. Effect of pH of hydrochloric acid solution on the solubility of aluminum-containing minerals: 1 – kaolinite; 2 – metakaolinite; 3 – boehmite; 4 – γ-Al_2O_3.

To determine the temperature of dehydration and related phase transitions, thermogram of the original bauxite during a linear heating to 1300°C in air was taken (figure 3).

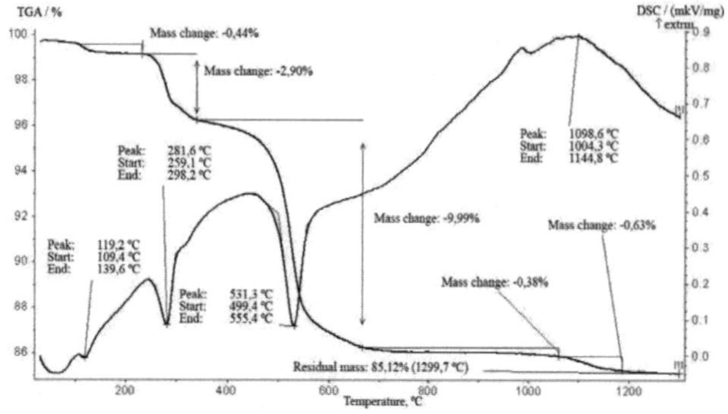

Figure 3. Effect of temperature on the weight change (TGA) and differential scanning calorimetry curve (DSC) of the original bauxite.

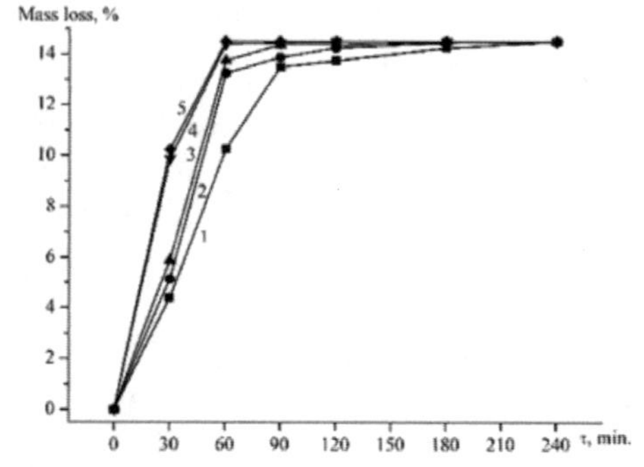

Figure 4. . Effect of temperature ageing time on mass loss of bauxite: 560 °C (1), 600 °C (2), 650 °C (3), 700°C (4), 750 °C (5).

It shows two distinct endothermic peaks at 260-300°C and 500-560°C. According to XRD at the first stage (at 300°C), constitutional water is removed in the form of OH-ions from gibbsite and goethite, and at the second (at 560°C) - from boehmite and kaolinite to form γ-Al_2O_3 and irregular metakaolinite $Al_2Si_2O_7$. Thus, after calcination of bauxite at optimal conditions the phase composition is the following: γ-Al_2O_3 – the main phase, metakaolinite $Al_2Si_2O_7$, hematite Fe_2O_3 and a small amount of anatase TiO_2.

Figure 4 shows the dependence of the dehydration rate on the temperature and duration of calcination. To remove the constitutional water completely, it is required to hold it from 30 min. at 700°C to 3 h. at 560°C. Temperature increase up to 750°C does not lead to a substantial increase in the dehydration rate. The maximum concentration of removed water is 14.5% of the total mass of the bauxite. Based on the data the optimal calcination parameters are temperature of 700°C and time of 30 min. Calcined bauxite was leached by 20 % hydrochloric acid under the same conditions as the original. Figure 5 shows that the extraction ratio of aluminum in the solution has reached 89% after 4 h., which leads to an increase in the content of the main components in solution: 42.2 g/l Al, 9.25 g/l Fe, and 0.02 g/l Si.

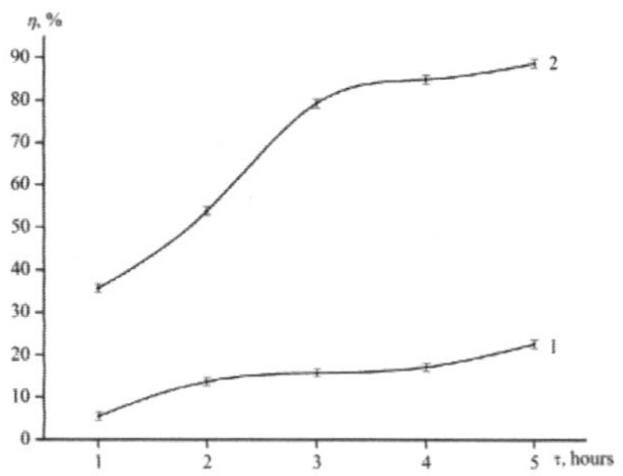

Figure 5. Effect of leaching time on the degree of extraction of aluminum (η, %): 1 - original bauxite; 2 - calcined bauxite.

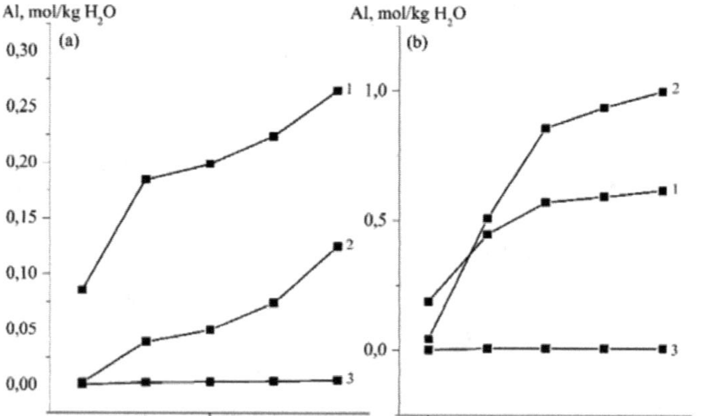

Figure 6. Distribution of chemical forms of aluminum in hydrochloric acid solutions after leaching the original (a) and calcined (b) bauxite: 1-Al^{3+}; 2-$AlCl_4^-$; 3 - $AlCl_3$.

Physicochemical modeling of the process of decomposition of bauxite by hydrochloric acid to determine the main forms of aluminum in solution is based on the experimental data (figure 5). As a result, the equilibrium component composition of hydrochloric acid solutions was determined. The basic compounds of leaching of original bauxite at 110°C are Al^{3+} and $AlCl_4^-$, $AlCl_3$ concentration is estimated to be $10^{-3} \div 10^{-4}$ mol/kg H_2O (figure 6a). Other forms do not have a significant effect on the solubility of aluminum since their concentration less than 10^{-6} mol/kg H_2O.

Using calcined bauxite increases the concentration of aluminum in the solution due to the formation of chloride aluminum forms. As shown in figure 6b, the basic form of the aluminum in solution becomes $AlCl_4^-$.

4. Conclusion

It is found that precalcination of Severoonezhsk bauxite at 700°C for 0.5 h. leads to a significant increase of the extraction ratio of aluminum in solution (89%), using leaching by hydrochloric acid, in comparison with leaching of uncalcined bauxite (23%). It has been shown that increasing of the extraction ratio of aluminum caused by the change of the phase composition of bauxite after calcination. This is a result of dehydration of low-soluble minerals, which are a part of bauxite, and occurs in two steps (at 260-300°C and at 500-560°C). As a result, soluble in hydrochloric acid phases – γ-Al_2O_3 and metakaolinite – are formed. The equilibrium composition of hydrochloric acid solutions was determined by the method of thermodynamic modeling. It is found that the increase in the aluminum concentration in the solution is due to the formation of chloride form $AlCl_4^-$, which along with Al^{3+}, is a main form of aluminum in solution.

5. References

[1] Senyuta A S, Panov A V, Suss A G and Layner Y A 2013 *Proc. Light Metals 2013 - At the TMS 2013 Annual Meeting and Exhibition* (San Antonio: Wiley) p 203
[2] Yusheng Wu, Laishi Li and Mingchun Li 2014 *Light Metals* 121
[3] Tariq M, Iqbal M M, Aziz A, Shafiq M, Sajid M and Mohammad B 2014 *J. Chem. Soc. Pak.* **35(4)** 624
[4] Suss A G, Damaskin A A, Senyuta A S, Panov AV and Smirnov A A 2014 *Light Metals* 105
[5] Knudsen C, Wanvik J, Svahnberg H 2012 *Geol. Surv. Denm. & Greenl. Bul.* 2012 **26** 53
[6] Shemi A, Ndlovu S, Sibanda V and Van Dyk L D 2014 *Int. J. Min. Proc.* **127** 10
[7] Ma Z, Wang C-Q, Li G-X, Duan Y-W and Peng F 2015 *Xian. Huag. / Mod. Chem. Ind.* **35** 34
[8] Tripathy A K, Sarangi C K, Tripathy B C, Sanjay K, Bhattacharya I N, Mahapatra B K, Behera P K and Satpathy B K 2015 *Int. J. Min. Proc.* **138** 44
[9] Yao Z T, Xia M S, Sarker P K and Chen T 2014 *Fuel* **120** 74
[10] Bazin C, El-Ouassiti K and Ouellet V 2009 *Hydrometallurgy* **88** 196
[11] Zhu P-W, Dai H, Han L, Xu X-L, Cheng L-M, Wang Q-H and Shi Z-L 2015 *J. Zhej. Univ.: Scien. A* **16** 161
[12] Dong L, Li Y, Yan J and Shu X Q 2014 *Adv. Mat. Res.* **878** 149
[13] Liu K, Xue J and Luo W 2014 *Proc. EPD Congress 2014 - TMS 2014 143rd Annual Meeting and Exhibition* (San Diego: Wiley) p 251
[14] Wu C-Y, Yu H-F and Zhang H-F 2012 *Tran. Nonfer. Met. Soc. Chin. (Engl. Edit.)* **22** 2282
[15] Gillesberg B 2011 *Valve world Magazine* **16** 79
[16] Robin A and Rosa J L 2000 *Int. J. Refr. Met. & Hard Mat.* **18** 13
[17] Zhao A-C, Zhang T-A, Lyu G-Z and Liu Y 2014 *Dong. Dax. Xueb. / J. North. Univ.* **35** 75
[18] Wang W, Pranolo Y and Cheng C-Y 2013 *Separ. Purific. Tech.* **108** 96
[19] Korovin V and Shestak Y 2009 *Hydrometallurgy* **95** 346
[20] Xiao J, Li F, Zhong Q, Bao H, Wang B, Huang J and Zhang Y 2015 *Hydrometallurgy* **155** 118
[21] Belyak A A, Zhavoronkova V I, Smirnov A D, Lainer Yu A and Milkov G A 2013 *Wat. Sup. & San. Techn.* **5** 32

[22] Chudnenko K V 2012 *Thermodynamic Modeling in Geochemistry: Theory, Algorithms, Software, Application* (Novosibirsk: Geo)
[23] Johnson J W, Oelkers E H and Helgeson H C 1992 *Comp. & Geoscien.* **18** 899
[24] Reid R, Prausnitz J and Sherwood T 1977 *The properties of gases and liquids, 3d edition* (New York: McGrawHill)
[25] Yokokawa H 1988 *J. Nat. Chem. Lab. Ind.* **83** 27
[26] Valeev D V, Bychinskii V A and Chudnenko K V 2015 *Russ. J. Inorg. Chem.* **60** 1428

6. Acknowledgments

The reported study was partially supported by RFBR, research project No. 15-38-50545 mol_nr.

IOP Conf. Series: Materials Science and Engineering **110** (2016) 012050 doi:10.1088/1757-899X/110/1/012050

Optical absorption of BaF_2 crystals with different prehistory when irradiated by high-energy electrons

E P Chinkov, S A Stepanov, V F Shtan'ko and T S Ivanova

Department of Lasers and Lighting Engineering, Institute of High Technology Physics, National Research Tomsk Polytechnic University, 30 Lenin Avenue, Tomsk 634050, Russia

E-mail: stepanovsa@tpu.ru

Abstract. The spectra of stable optical absorption of BaF_2 crystals containing uncontrollable impurities after irradiation with 3 MeV electrons are studied at room temperature. The dependence of the efficiency of stable color accumulation in the region of emerging cross-luminescence on the absorption coefficients measured near the fundamental absorption edge in unirradiated crystals of various prehistory is traced.

1. Introduction

Alkaline-earth metal fluorides are transparent over a wide spectral range, and they are used as an optical constructional material to transmit powerful UV and VUV radiation fluxes. BaF_2 crystals are of special relevance among them since cross-luminescence transitions at 195 and 220 nm and a decay time of less than 1 ns has been revealed and interpreted for the first time for these crystals [1]. This type of luminescence is also called as core-valence luminescence or Auger-free luminescence. BaF_2 crystals are currently the most widely used material for solid state scintillation detectors of x-ray, gamma radiation and other high energy particles.

The presence of intense inertial (600 ns) luminescence component at 300 nm due to the formed self-trapped excitons (STE) [2] restricts application of BaF_2 crystals as fast scintillators. Different methods are used to suppress the STE luminescence.

On the one hand, the studies were conducted to change the intensity relationship for the fast and slow luminescence components in pure BaF_2 crystals in favor of the fast component. For example, in [3], a special filter unit was used to reduce the intensity of the slow luminescence component. In [4], a gamma detector was developed based on the BaF_2 scintillator heated to 220 °C since the intensity and the decay time of the slow component are strongly dependent on temperature, whereas those of the fast component are not [5].

On the other hand, it was found that doping of BaF_2 crystals with some types of impurities results in suppression of the luminescence intensity of the slow component at 300 nm [6, 7]. Despite a large number of papers devoted to this problem, the effect of various impurities on the luminescent properties of BaF_2 crystals is still not clear. The problems are encountered when choosing the type and concentration of the dopant, and the technique for crystal growing.

The short-lived and unstable STE absorption occurs immediately after exposure of BaF_2 crystals to ionizing radiation [2, 8]. Pulsed spectroscopy with space-time resolution [9] was used to show that the efficiency of STE formation, and hence, the intensity of the slow luminescence component is determined to a large extent by the coordinates of the irradiated sample in the growth region with small and large-scale distortion of the crystal lattice periodic potential.

As the total dose increases, the long-lived or residual absorption is accumulated. There are a number of papers describing the effects of radiation damage in scintillators [10, 11]. Earlier studies of the BaF_2 crystal radiation resistance show conflicting results. The residual absorption is highly sensitive to the crystal quality. Doping with impurities (especially heterovalent dopants) increases the radiation sensitivity of crystals of a fluorite structure [12]. Most of the differences from sample to sample are considered to be associated with different amounts of uncontrollable impurities.

Compared to pure BaF_2 crystal, the presence of uncontrollable impurities is to be followed by change in the transmission cutoff in the VUV spectral region. However, so far no attempts have been made to relate the changes in BaF_2 crystal transmission to their radiation sensitivity.

The aim of the research is to study the impact of changes in optical transmittance near the fundamental absorption edge in BaF_2 crystals of various prehistory on their radiation resistance.

2. Experimental methods

The objects of study were BaF_2 crystals grown by the Bridgeman-Stockbarger technique in fluorinating atmosphere in Crystallography Institute of the Russian Academy of Science (Moscow) and in State Optical Institute (St. Petersburg).

The BaF_2 crystals of different lots used in this research, designated by numbers in table 1, were not intentionally doped. The grown samples were of a cylindrical shape. To eliminate the impact of microsegregation of 6 pieces from the central part, 2 plates were cleaved out from each of the pieces. One parallel cleavage was analyzed to estimate the impurity content, and the second one was used to measure the optical transmittance in the VUV spectral region and radiation-induced optical absorption.

Table 1. Impurity contents in samples.

Element (10^{-4} mol%)	No.1	No.2	No.3	No.4	No.5	No.6
O	1	0,5	0,7	0,7	1	0,9
Pb	3	3	3	4	7	10

Crystals No. 1 have been used previously [8, 9] in the study of the short-lived absorption. It is known that hygroscopicity of BaF_2 crystals is low. To prevent entry of oxygen into the crystals the so called scavenger is added to remove oxygen from the melt. For BaF_2 crystals, the scavenger is typically PbF_2 [10]. When changing the technology of growing, lead and oxygen impurities remain in BaF_2 crystals, which reduces the transmission in the VUV spectral region. The results of the preliminary oxygen impurity proton-activation analysis, as well as the lead impurity spectroscopic analysis of BaF_2 crystals are shown in table 1. The data of the spectroscopic analysis indicate the presence of other impurities (Ca, Sr, Fe, Al, Mg, Mo, Zn, W), the concentration of which varies from sample to sample within the range (10^{-5}–10^{-4}) mol%.

The transmission spectra of the unirradiated BaF_2 crystal in the VUV spectral region were measured at room temperature with the spectrophotometer of BMP-2 type in State Optical Institute (St. Petersburg) by V.M. Reyterov.

All samples were irradiated with the dose ~10^5 Gy at room temperature. The radiation source was a continuous electron beam accelerator. The accelerated electron energy was equal to 3 MeV. The sample thickness was within 2 mm to ensure uniform in-depth coloring. The absorption spectra of the irradiated BaF_2 crystals were measured at room temperature with the spectrophotometer of SF-256 type in the spectral range from 6.2 to 1.55 eV.

3. Results

Figure 1 shows the spectra of the induced optical absorption for BaF_2 crystals subjected to irradiation, which contained uncontrollable impurities. The spectra were measured at room temperature within 1 hour after the end of irradiation. The analysis of the data presented in figure 1 reveals the following patterns.

In irradiated BaF_2 crystals, the induced absorption is of complex spectral distribution in the range from 6.2 to 1.55 eV.

In all the irradiated crystals, the induced optical absorption is found to grow in the spectral region of the detected cross-luminescence at 5.64 eV. The greatest increase in absorption at 5.64 eV occurs in sample No. 4.

In the spectral position, the observed bands do not fit the absorption bands F (2.03 eV) and F_2 centers (1.71 eV) in the BaF_2 crystal reported in [13].

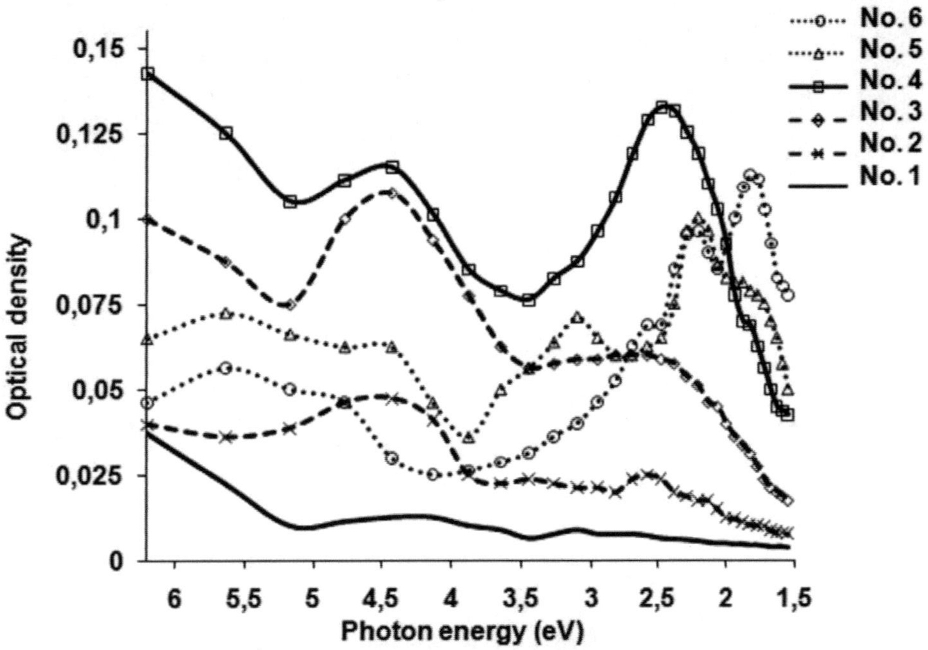

Figure 1. Optical absorption spectra of BaF_2 crystals of various prehistory at irradiation to the same dose.

The spectra of the induced absorption in samples No. 2, No. 3 and No. 4 contain bands in the region of ~ 4.3 eV, as well as a broad structureless band at 2.48 eV. Apparently, these bands are due to formation of metal colloidal nanoparticles [14] which are to be more stable defects than the aggregates of F centers. The spectra of the induced absorption in samples No. 5 and No. 6 contain absorption bands in the region of 2.25 eV and about 1.77 eV. An extremely large width of the absorption bands, and the shift of the maxima to longer wavelengths can be due to the distribution of the colloidal particle in size and/or due to nonspherical particle shape [15].

The concentration of impurities in BaF_2 crystals of various prehistory changes within a rather narrow range (table 1). Nevertheless, the data in figure 1 shows that the crystal colorability varies dramatically from sample to sample.

Oxygen, lead, yttrium, and sodium ions are commonly referred to uncontrollable or "biographical" impurities in crystals of a fluorite structure. The presence of these impurities leads, firstly, to the shift of the transmission cut-off of fluorite crystals to longer wavelengths, and secondly, to occurrence of selective absorption bands in the VUV spectral region [10, 11, 16-18].

At higher concentration of the doped, for example, oxygen impurity, the colorability of fluorite crystals is determined by the intensity of the selective absorption in the VUV spectral region of unirradiated samples [10, 11]. At low concentration of the doped impurity this dependence cannot be typically traced.

The area under the absorption curve of an individual sample was assumed to be the crystal colorability efficiency.

Figure 2 shows the dependence of the colorability for BaF_2 crystals of various prehistory on the absorption coefficients, measured in unirradiated samples near the fundamental absorption edge at 150 nm, as well as in the region of the oxygen impurity selective absorption at 190 nm. These results indicate that the greater the absorption coefficients in the VUV spectral region in unirradiated crystals, the higher the colorability of BaF_2 crystals by high-energy electrons.

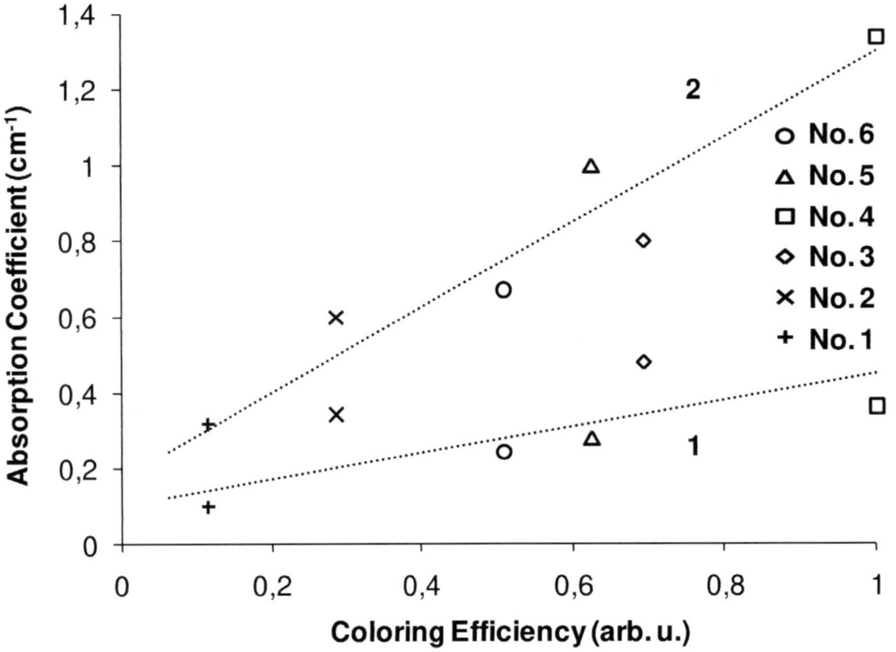

Figure 2. Influence of absorption coefficients of BaF2 crystals in the fundamental absorption edge of unirradiated sample on coloring efficiency (curve 1 - 190 nm, curve 2 - 150 nm).

It is well known that imperfection of the anion sublattice, mainly determines the radiation sensitivity of crystals of a fluorite structure. One of the techniques to control the number of excess anionic vacancies or interstitial fluoride ions is co-doping of crystals [17].

For example, doping of the CaF_2 crystal with yttrium or sodium leads to a long-wavelength shift of the transmission cutoff in the VUV spectral region and increase in the radiation sensitivity, as compared to those in an undoped crystal [16–18]. If the concentration ratio of these impurities is optimum, the transmission cutoff in the VUV spectral region of the crystal is comparable to that of a pure crystal, the radiation resistance of the co-doped crystal increases [17].

When oxygen enters the crystal through substituting fluorine ions, the compensation of an excess charge is followed by occurrence of anion vacancies [13]. We can assume that in BaF_2 crystals the compensation of the impurity oxygen (O^{2-}) excess charge occurs in substituting the barium ions in the lattice by positively charged ions of other impurities (R^{3+}). Ions of rare earth elements or transition metals can be regarded as these impurities. If the concentration ratio of these impurities is optimum, dipoles of ($O^{2-} – R^{3+}$) type are formed. The anion vacancies are not formed and the radiation resistance of the co-doped BaF_2 crystal increases. In BaF_2 crystals of various prehistory not intentionally doped, the concentration ratio of the uncontrollable impurities compensating the electroneutrality breakdown can be random. Therefore, the effect of the impurities on radiation resistance of the crystals is unpredictable.

4. Conclusion

The spectra of the stable absorption of BaF_2 crystals of various prehistory are measured after irradiation with high-energy electrons at room temperature.

It was found that in the spectral region of the detected cross-luminescence, the colorability efficiency of BaF_2 crystals of various prehistory can be determined by the absorption coefficient near the fundamental absorption edge of the unirradiated crystal.

The results can be the basis for a method to perform the in-process control of the radiation resistance of materials of a fluorite structure.

References

[1] Aleksandrov Yu M, Makhov V N, Rodnyi P A, Syreishchikova T I and Yakimenko M N 1984 *Sov. Phys. Solid State* **26** 2865
[2] Williams R T, Kabler M N, Hayes W and Stott J P H 1976 *Phys. Rev.* B **14** 725
[3] Han H, Zhang Z, Weng X, Liu J, Guan X, Zhang K and Li G 2013 *Rev. Sci. Instrum* **84** 073503
[4] Biasini M, Cassidy D B, Deng S H M, Tanaka H K M and Mills A P 2005 *Jr. Nucl. Inst. and Meth.* A **553** 550
[5] Schotanus P, Van Eijk C W E, Hollander R W and Pijpelink J 1985 *Nucl. Inst. and Meth.* A **238** 564
[6] Hamada M M 1994 *Nucl. Inst. and Meth.* A **340** 524
[7] Dorenbos P, Visser R, Doll R, Anderssen J and Van Eijk C W E J 1992 *Phys.: Condens. Matter* **4** 5281
[8] Shtan'ko V F and Chinkov E P 1998 *Phys. Sol. State.* **40** 1119
[9] Shtan'ko V F and Chinkov E P 1998 *Phys. Sol. State.* **40** 1326
[10] Woody C L 1991 *GEM TN*-91-38
[11] Lecoq P, Annenkov A, Gektin A, Korzhik M, Pedrini C 2006 *Inorganic Scintillators for Detector Systems: Physical Principles and Crystal Engineering* (Berlin: Springer) p 251
[12] Nikl M, Sato H, Mihokova E, Mabuchi T, Nawata T, Yoshikawa A, Pejchal J, Kawaguchi N, Ishizu S, Fukuda K, Suyama T 2009 *Proc. SPIE* **7361** 73610S
[13] Hayes W 1974 *Crystals with the Fluorite Structure: Electronic, Vibrational, and Defect Properties* (Oxford: Clarendon Press) **13** 9780198512776
[14] Alcala R, Orera V M 1976 *Journal de Physique* **37** 520
[15] Hughes A E, Jain S C 1979 *Advances in Physics* **28** 717
[16] Arkhangel'skaya V A, Reiterov V M, Trofimova L M 1980 *J. of Applied Spectroscopy* **32** 67
[17] Lisitsyna L A, Chinkov E P,. Reiterov V M and Trofimova L M 1983 *J. of Applied Spectroscopy* **38** 673
[18] Radzhabov E, Nepomnyashchikh A and Kirm M 2007 *Phys. stat. sol.* A **204** 670

Kinetic characteristics of the luminescence decay for industrial yttrium-gadolinium-aluminium garnet based phosphors

V M Lisitsyn[1], S A Stepanov[1], D T Valiev[1], E A Vishnyakova[2], H A Abdullin[3], A A Marhabaeva[3] and A T Tulegenova[3]

[1]Department of Lasers and Lighting Engineering, Institute of High Technology Physics, National Research Tomsk Polytechnic University, 30 Lenin Avenue, Tomsk 634050, Russia
[2]International Training and Laser Center, Lomonosov Moscow State University, GSP-1, Leninskie Gory, Moscow, 119991, Russia
[3]Department Solid State and Nonlinear Physics, al-Farabi Kazakh National University, 71 al-Farabi Ave., Almaty, 050040, Republic of Kazakhstan

E-mail: stepanovsa@tpu.ru

Abstract. The spectral and decay kinetic characteristics of pulse cathodoluminescence and photoluminescence of phosphors based on yttrium-gadolinium-aluminum garnet were investigated using pulsed optical time resolved spectroscopy.

1. Introduction

LEDs based on InGaN heterostructures emitting in the near-UV (blue) spectral range are considered to be most promising to convert a portion of this radiation into light with the desired color characteristics using the phosphor. These heterostructures emit in the range of 430 ... 490 nm which corresponds to photon energies from 2.5 to 2.9 eV. To produce white light, radiation in the yellow-red spectral region is to be added to the blue radiation of the heterostructure [1–3]. The average photon energy in the visible spectral region is about 2.2 eV. Therefore, conversion of the blue light into yellow-red light is accompanied by low energy loss that provides high light efficiency of the LED light source and insignificant energy losses in device heating. Luminous efficiency currently achieved for "white" LEDs is 160 lm/W, while the theoretically maximum possible values are 240 ... 260 lm/W.

Further increase in light efficiency is mainly determined by the ability to improve the efficiency of phosphors. Radiative characteristics of phosphors to a large extent depend on the history: synthesis technology, quality of raw materials and composition of the resulting material defects. Therefore, optimization of phosphor properties to obtain the maximum light efficiency for the LED is challenging [1, 4, 5].

Lack of information on the processes of energy transfer to the luminescence centers, the characteristics of the luminescence decay kinetics hinders phosphor improvement. This paper addresses the study of the spectral and decay kinetic characteristics of the luminescence for a group of industrial phosphors based on yttrium-aluminum gadolinium garnet as promising phosphors for emitting in the visible range of spectrum.

2. The tasted samples and the experimental method

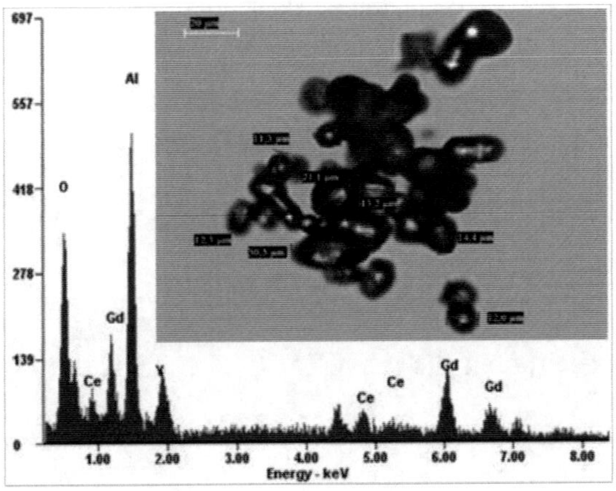

Figure 1. Elemental and particle size distribution of the FL-5049 phosphor

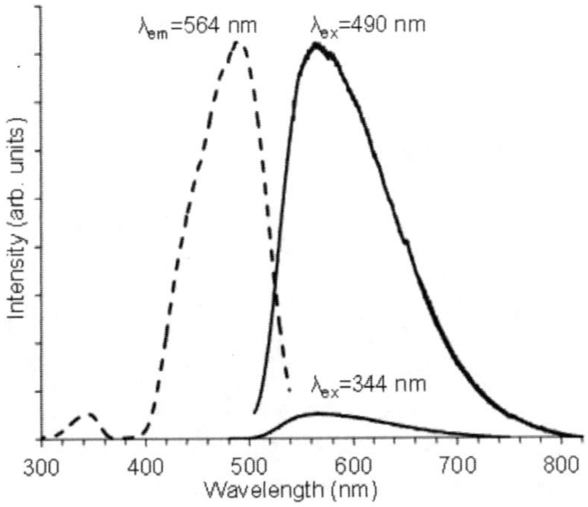

Figure 2. Excitation and luminescence spectra for FL-5049 phosphor

FL5049 $(Y\text{-}Gd\text{-}Ce)_3Al_5O_{12}$ phosphor doped with cerium has been provided by the INCOTEX Holding synthesized by Fultor Enterprises Co. Ltd, Dalian, China. Industrial phosphors manufactured by AWB3 (Billion Light Co., Taipei County, Taiwan) and LEUD 560 for were used to conduct the investigation as well. The elemental composition of the samples was determined using the scanning electron microscope Quanta 3D 200i. The images of the investigated powders were made using the Leica DM 6000 M optical microscope. The results obtained for the elemental composition of FL-5049 phosphor are presented in figure 1 as an example. The composition of the studied phosphors is provided in table 1.

FL-5049, AWB3 and LEUD 560 phosphors are similar in the composition of basic elements, but differ in their proportions.

Table 1. Phosphor composition

	FL5049		AWB3		LEUD 560	
Element	*Wt* %	*At* %	*Wt* %	*At* %	*Wt* %	*At* %
O	20.70	54.73	25.27	56.72	29.77	60.99
Al	16.64	26.09	21.17	28.18	19.53	23.73
Y	9.69	4.61	15.90	6.42	28.63	10.55
Ce	9.61	2.90	2.74	0.70	5.13	1.20
Gd	43.36	11.66	34.92	7.98	16.94	3.53

The sizes of the FL-5049 phosphor particles in measurement with the Leica DM 6000 M microscope vary from 12 to 30 μm. The luminescence and excitation spectra of the phosphor were measured with the AGILENT CARY ECLIPSE spectrofluorimeter and an optical complex based on two MDR-204 crossed monochromators. The xenon lamp was used as an excitation source. All the measurements were performed with allowance for the optical path spectral sensitivity.

Figure 2 shows the excitation and luminescence spectra for FL5049 phosphor. The luminescence spectrum of FL-5049 phosphor is a broad band with a maximum at 564 nm. The shape of the luminescence spectrum is independent of the excitation radiation. The luminescence is excited by radiation in the region of 400–540 nm, and it is weaker in the region of 310–360 nm. The excitation and luminescence spectra for the tested samples are characteristic of phosphor-based YAG with dopants [5]. The excitation and luminescence spectra for AWB3, LEUD 560 and FL-5049 phosphors are similar.

To investigate the time-resolved luminescence, four excitation sources were used: pulsed nitrogen laser with λ = 337 nm and pulse duration τ = 10 ns (LR 337), pulse energy of 50 mJ; LED heterostructure with λ = 395 (Chip 395) and 450 nm (Chip 450) (pulse energy of 15 mJ, τ = 50 ns), and high-current pulsed electron accelerator (PEF) with an average electron energy of 250 keV (τ = 10 ns). The pulse luminescence of phosphors was investigated in the spectral range of 300–800 nm at room temperature within 10^{-8} –10^{-3} s after the end of the excitation pulse. The luminescence and change in the optical absorption were recorded using FEU-106, FEU-84-6, digital oscilloscopes LeCROY (350 MHz bandwidth) and Tektronix TDS-2014 (100 MHz bandwidth), and monochromator MDR-204, MDR-3.

3. Experimental results

The results obtained for the luminescence decay kinetics in the range of 10^{-8} – 10^{-3} for the FL-5049 phosphor are shown in Fig. 3.

Excitation of FL-5049 phosphor by UV radiation flux (λ = 390 nm) does not cause sufficiently intense luminescence, as it can be expected from the excitation spectrum presented in Fig. 2. The luminescence is excited by heterostructure radiation (LED chip λ = 450 nm), and excitation by radiation with λ = 337 nm is not so efficient. However, high-power laser radiation is sufficient to excite intense luminescence. The short-time luminescence spectra are similar to those measured under steady state conditions: the maximum bandwidth accounts for 564 nm, and the band-width is equal to 0.44 eV.

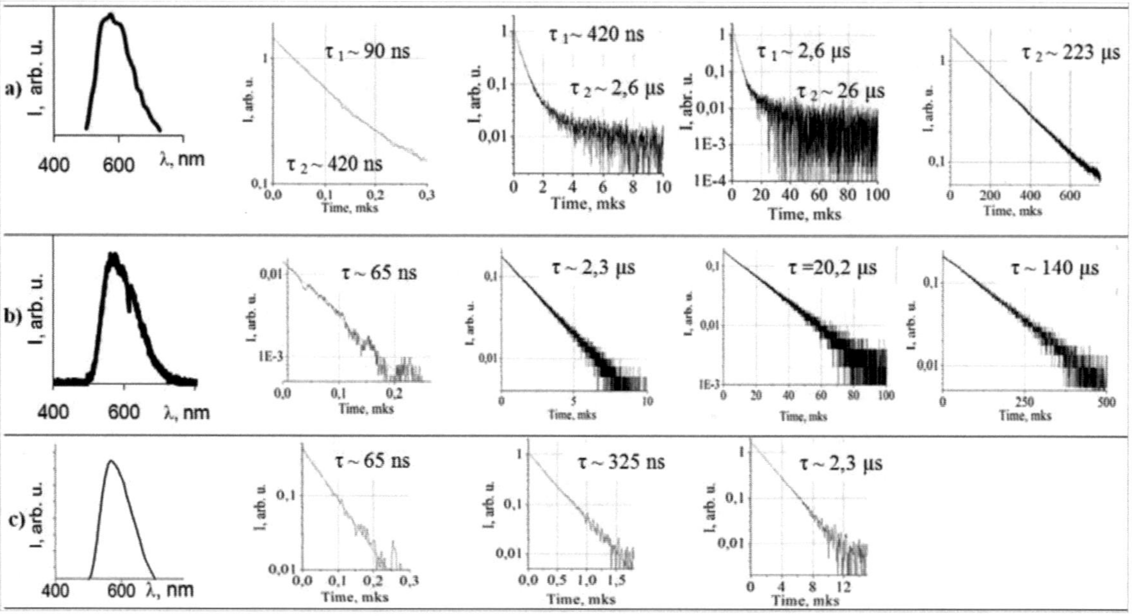

Figure 3. Spectral and decay kinetic characteristics of the luminescence for FL-5049 phosphor under different types of excitation: by electron beam (a), heterostructure radiation with λ = 450 nm (b), laser radiation flux with λ = 337 nm (c)

The luminescence decay kinetic curves in the maxima of the luminescence spectral bands for FL-5049 phosphor after excitation by PEF, radiation of LED Chip (450 nm) and laser radiation (337nm) are shown in figure 3. The measurements were performed in the time range from 10 ns to 1 ms. Within this time range, the luminescence intensity decreases by almost 10^6 times. To measure the luminescence decay within this time range, the measurement was performed with different temporal resolution (time window measurement). This allows identification of the main components of the

IOP Conf. Series: Materials Science and Engineering **110** (2016) 012051 doi:10.1088/1757-899X/110/1/012051

luminescence decay kinetics, but complicates kinetic curve matching. In the obtained decay kinetics, several components can be described by an exponential function. There are components of the nanosecond range: 65, 90 and 325 ns, and 420 ns. In the microsecond time range, there are components with characteristic times from 2.3 to 2.6 μs, from 20.2 to 26 μs and from 140 to 223 μs depending on the type of excitation. Note that the short-time decay component was observed in [6]. The contribution of each component to the total luminescence light sum is difficult to determine due to the uncertainty of the measurement of the amplitude values at the beginning of the decay kinetic curves. But it can be argued that the characteristic relaxation times are similar under different types of excitation.

From the results presented in table 2 it can be seen that the values of the characteristic luminescence decay times have are substantially similar for different excitation modes in all the investigated phosphors. The only difference is that no component with τ_2 can be observed under excitation by Chip 450. Lack of the components with τ_4 and τ_5 after excitation by laser radiation (LR 337) is due to the fact that the luminescence intensity of the phosphor is small and insufficient for recording in the large time domain. Note the difference in the quantitative characteristics of the luminescence decay in the studied phosphors under excitation by different sources. These differences are particularly distinct in the long lifetime components.

Table 1. Summarizes the values of the characteristic luminescence decay time τ_i for phosphors studied under different types of excitation

Phosphor	Excitation source	Luminescence decay time				
		τ1, ns	τ2, ns	τ3, μs	τ4, μs	τ5, μs
FL 5049	e-beam	90	420	2.6	26	223
	Laser radiation (λ=337 nm)	65	325	2.3		
	LED Chip (λ=450 nm)	65	-	2.3	20.2	140
LEUD560	e-beam	90	420	3.9	21	219
	LED Chip (λ=450 nm)	65	-	2.3	20.2	142
AWB3	e-beam	65	300	3.9	32	220
	LED Chip (λ=450 nm)	65	-	2.3	20.2	141

4. Discussion

Excitation of FL5049, AWB3 and LEUD 560 phosphors by energy pulses from different light sources: laser pulse with λ_{ex} = 337 nm, electron excitation with electron energy of 250 keV and LED chip with λ_{ex} = 450 nm cause burst of luminescence. The LED chip with λ_{ex} 395 nm does not excite the tested phosphors. Two characteristic properties of the pulsed luminescence in these phosphors have been found.

1. The pulse luminescence spectra of FL-5049, AWB3 and LEUD 560 phosphors under the excitation modes used in the experiment are of the same shape similar to that measured under steady-state conditions. Consequently, the same luminescence centers are excited by photon energies of 2.5 eV (LED Chip λ_{ex}=450 nm), 3.68 eV (LR 337 nm) and electrons (PEF).

2. The same centers are excited in FL-5049, AWB3 and LEUD 560 phosphors which are of insignificantly different quantitative elemental composition.

It is reasonable to explain the observed results as follows. During synthesis of phosphors based on yttrium-gadolinium-aluminum garnet doped with cerium, a large concentration of intrinsic lattice defects enters the crystals. High degree of stoichiometry in the crystal of complex composition cannot be provided during growing from salt mixtures at high temperatures. The crystal contains defects in the form of vacancies, interstitial ions, a variable dopant ion ratio, and a large diversity of defect complex. Apparently, in highly imperfect crystals (crystals containing large concentrations of various

IOP Conf. Series: Materials Science and Engineering **110** (2016) 012051 doi:10.1088/1757-899X/110/1/012051

defects accidentally or intentionally introduced during synthesis to provide activation), a set of processes initiated by external excitation should be different from those occurring in perfect crystals [8, 9]. In these crystals, the spatial and energy structure of the formed complexes and nanodefects differ from those in perfect crystals. In highly imperfect crystals doped with gadolinium yttrium-aluminum garnet, a nanodefect substructure is formed. Therefore, in studying a set of the processes of emission excitation and generation, the possibility of processes in nanodefects is to be considered.

If we assume the existence of the nanodefect substructure in phosphor crystals, the cumulative research results can be interpreted as follows. Exact coincidence of spectral characteristics (position and FWHM of the luminescence bands), satisfactory agreement of the luminescence decay kinetics characteristics indicate a similar structure of the luminescence centers. The emission center in these crystals can be Ce^{3+} ion [7]. In turn, the emission center is a component of the nanodefect. Therefore, radiative characteristics do not dependent (or weakly dependent) on the change in elemental composition. The emission center may be a complex with a cerium ion in its composition.

The investigated phosphors provide high light yield of the conversion of the energy excited by laser, electron beam and LED Chip 450 radiation into luminescence. Consequently, the excitation energy is transferred to the luminescence centers with high efficiency, though energy from different sources is absorbed as a result of totality of different processes. The bandgap absorption of yttrium aluminum garnet crystals starts from 5.2 eV [11]. The photon flux of photons with energies of 2.5 eV, 3.68 eV excite either the luminescence center or the trap site which effectively transfers the energy to the luminescence center. The efficiency of energy transfer to the luminescence center is determined by spatial and energy correlation of the "trap-luminescence center" pair.

When excited by high-energy electron fluxes, the absorbed energy is spent mostly on creation of electronic excitations: electron-hole pairs and excitons in the host. It is highly probable that the resultant electronic excitations are trapped by nanodefects since their capture cross section is at least two orders of magnitude higher than that of the point defect. The electronic excitation captured by the nanodefect cannot escape. It is highly likely that the energy is transferred to the luminescence center located in the same nanodefect.

References

[1] Nakamura S 1997 *Blue laser Diode* (Berlin: Springer)
[2] Schubert F E 2006 *Light-emitting Diodes* (UK: Cambridge University Press)
[3] Zorenko Yu, Gorbenko V, Konstankevych I, Voloshinovskii A, Stryganyuk G, Mikhailin V, Kolobanov V, Spassky D 2005 *J. Lumin.* **114** 85
[4] Soshchin N P 2007 *SID-2007 Proceeding ECNUP* 277
[5] Briskina Ch M, Rumyantsev S I, Ryzhkov M V, Soshchin N P, Spasskiy D A 2012 *Svetotekhnika*[in Russian] **5** 37
[6] Zorenko Yu V, Savchin V P, Gorbenko V I, Wozniak T I, Zorenko T E, Puzikov V M, Dan`ko A Ya, Nizhankovskii S V 2011 *Fiz. tverd. tela* [in Russian] **53** 1542
[7] Nishimira S., Tanabe S., Fujoka K., Fujimoto Y. // Optical Materials. - 2011.- N. 33.-P.688-691.
[8] Lisitsyna L A, Lisitsyn V M 2013 *Fiz. tverd. tela* [in Russian] **55** 2183
[9] Lisitsyn V M, Valiev D T, Tupitsyna I A, Polisadova Ye F, Lisitsyna L A, Stepanov S A, Oleshko V I 2013 *Izv. Vuz. Fizika* [in Russian] **56** 75
[10] Tanner P A., Fu L, Ning L 2007 *J. Phys.: Condens.Matter* **19** 216213

Mathematical modelling of energy conversion of pulsed electron beam in BaF_2 crystal

S A Stepanov, V F Shtan`ko, E P Chinkov and T S Ivanova

Department of Lasers and Lighting Engineering, Institute of High Technology Physics, National Research Tomsk Polytechnic University, 30 Lenin Avenue, Tomsk 634050, Russia

E-mail: stepanovsa@tpu.ru

Abstract. This paper presents the results of mathematical modeling of spatiotemporal distribution of energy in a BaF_2 crystal in an area of dissipation of energy of an pulsed electron beam. Fluence is varied from 0.188 to 0.626 J/cm^2. Electron beam pulse duration was 24 ns, and the maximum electron energy of 280 KeV. Based on the obtained of energy distributions temperature change of the material during irradiation were calculated.

1. Introduction

Through a combination of unique electrical and optical properties of BaF_2 is a promising material for a wide range of constructional applications. Using pulsed electron beams allows generating short-lived defect concentration sufficient for detection and research of the mechanisms of defect formation in solids. However, the impact on ionic crystals pulsed electron beams (PEB) has some specific features compared with low-intensive sources of radiation.

In an area of energy dissipation of pulsed electron beam is formed strong electric field and the temperature rises along with the generation of non-equilibrium electron-hole pairs, which affects the efficiency of formation and decay of an electron-hole excitation and also on the postradiation processes. Effect of temperature increase in an area of energy dissipation is not enough studied, making it difficult to correct account of this factor in the analysis of research results.

Effect of temperature increase in the area of energy dissipation is not enough researched, making it difficult to correct account of this factor in the analysis of research results.

2. Calculation of distribution of the injected electrons in BaF_2 crystal

Feature of interaction of accelerated electrons with matter manifests itself in a change driving directions, which is conditioned mainly by the elastic scattering on the nuclei of atoms, while the change in the energy of the electrons occurs mainly is due to inelastic scattering by electrons. Therefore, changing the magnitude and direction of the velocity vector of the electron can be considered independently [1].

Ionization energy losses of electron per unit path length determined by Bethe-Bloch formula [2, 3], accurate enough in energy range of electrons $10^4 \le E \le 10^7$eV.

$$\left(-\frac{dE}{dz}\right)_I = \frac{2\pi e^4 N_A}{m_0 c^2 \beta^2}\frac{Z}{A}\rho\cdot\left\{\ln\frac{\tau^2(\tau+2)}{2(\bar{I}/m_0c^2)} + 1 - \beta^2 + \frac{1}{(\tau+1)^2}\left[\frac{\tau^2}{8} - (2\tau+1)\ln 2\right] - \delta\right\} \quad (1)$$

Where Z, A – atomic number and mass of a element; N_A – Avogadro's number; e, m_0 – charge and electron rest mass; c – speed of light in vacuum; $\beta = \upsilon/c$; $\beta^2 = 1-(\tau+1)^{-2}$, where υ – electron velocity; $\tau = E/(m_o c^2)$ – kinetic energy of electrons in units $m_o c^2$ ($m_o c^2 = 0{,}511$ MeV); $\bar{I}$ – average ionization potential of the atom; δ – correction for the effect of density.

The average ionization potential $\bar{I}$ (in eV) is defined by the Sternheimer formula [4]:

$$\bar{I} = Z \cdot (9{,}76 + 58{,}8 \cdot Z^{-1.19}) \tag{2}$$

For substances with a complex chemical composition, the braking ability can be determined with good accuracy by summing a braking ability of the constituent elements, thus calculation is performed with the replacement $Z \rightarrow Z_{ave}$, $A \rightarrow A_{ave}$, $\bar{I} \rightarrow I_{ave}$. For BaF_2 crystal obtained Z_{ave} =24.67, A_{ave} = 58.44, I_{ave} = 397.34 eV.

To calculate the distribution of injected electrons in the sample and the absorbed energy density applied a variation "enlarged" collisions of Monte Carlo method using the angular distribution of Moliere-Bethe [3]. In the calculations the thickness of sample was chosen equal to the maximum extrapolated path of electrons R_o. In normalized coordinates $\tau = z/R_o$, for BaF_2 crystal was calculated a universal curve of distribution injected electrons, which depends only on the properties of this material (figure 1).

3. Calculation of energy release in BaF_2 under the influence of PEB

In the calculations used spectrum of PEB [3], which was determined experimentally by attenuation in thin foils, as described in [5]. Modelling of distribution of energy release produced considering the action of the retarding field of injected space charge [3].

Formula (3), together with the calculated distribution universal injected electrons (Fig. 1), allows the calculation of the density distribution of the space charge $\rho(z_i, t_m)$ in absolute coordinates for PEB of arbitrary shape with a different beam current density at any moment of excitation.

$$\rho(z,t) = \sum_{i=1}^{k} \rho_i(z) = \frac{e(1-\gamma_0)}{\int_0^1 \varphi(\tau)d\tau} \cdot \sum_{i=1}^{k} \frac{N_{0i}}{R_{0i}} \varphi(\tau) \tag{3}$$

Where γ_0 - the proportion reflected electrons; N_0 - number of electrons in monoenergetic pulse; e - electron charge; $\rho(z)$ - function of the distribution of the space charge in absolute units; $\varphi(\tau)$ - function of the distribution of the injected electrons, R_0 - distance extrapolated path of the electrons.

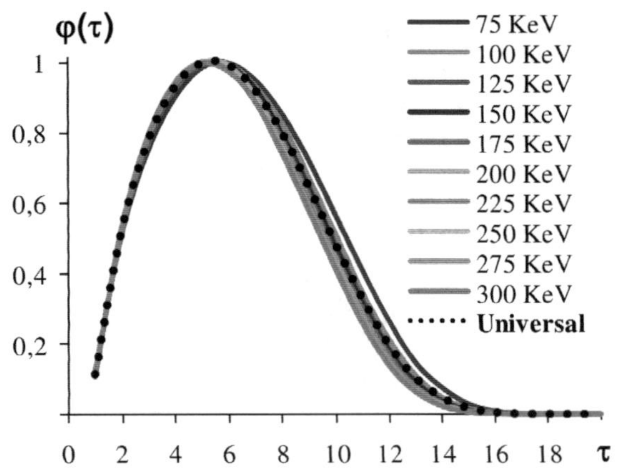

Figure 1. Normalized distribution of injected electrons in BaF_2 at different energies of incident electrons E_o

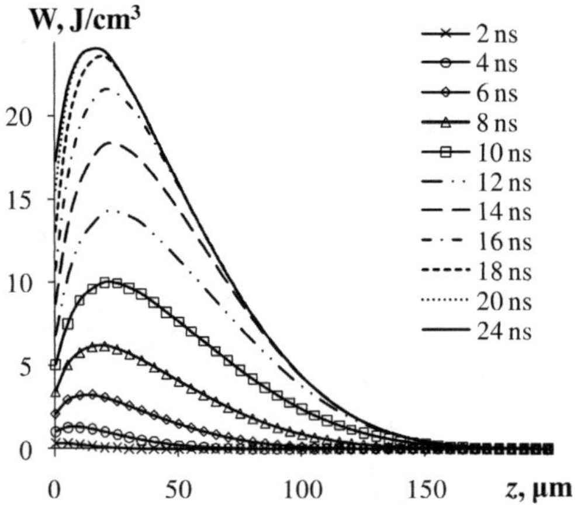

Figure 2. Dynamics the forming of profile energy release in BaF_2. Fluence of PEB is 0.188 J/cm^2.

IOP Conf. Series: Materials Science and Engineering **110** (2016) 012052 doi:10.1088/1757-899X/110/1/012052

The dynamics of the formation of the energy profile for BaF_2 shown in figure 2. In contrast to the monoenergetic beam (figure 1), profile excitation of the sample at the beginning of irradiation pulse is shifted from the irradiated surface and after reaching maximum of a extrapolated path, determined by the electrons with maximum energy ($t \approx 10$ ns), returns to the irradiated surface.

Accounting a spectral-time parameters of pulsed electron beam leads to complex spatiotemporal energy release function W(z,t) in the sample. The time dependence of energy release to the surface layers of the sample has a a complicated form with multiple peaks. With increasing z coordinate layer - distribution W(z, t) acquires a dome shape. It should be noted that with increasing distance from the irradiated surface of the material decreases time of irradiation of a layer.

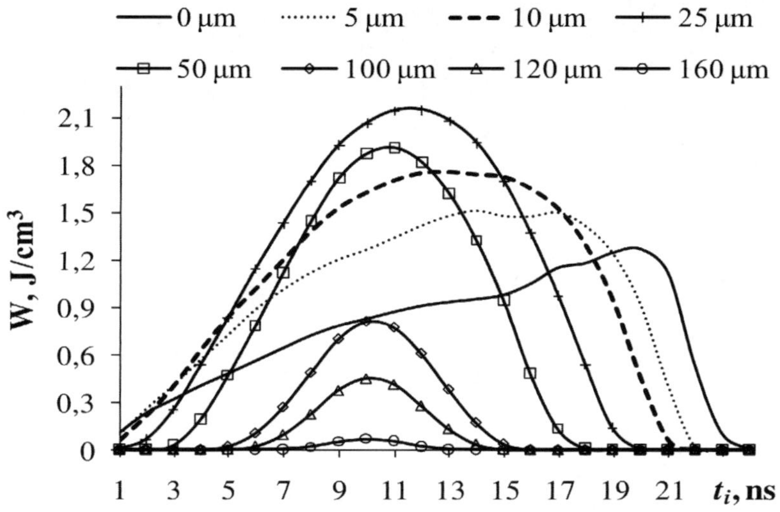

Figure 3. Calculated time dependences of energy release at different distances from the irradiated surface BaF_2. Fluence of PEB is 0.188 J/cm^2

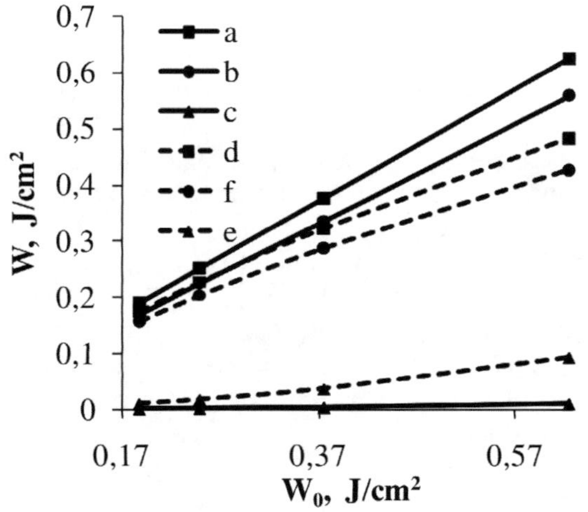

Figure 4. Dependence of influence of a injected space charge field and the vacuum gap on the initial fluence W_0. a - fluence PEB on the irradiated surface (d- with vacuum gap); b - absorbed energy (f- with vacuum gap); c- energy stored in the electric field of the injected charge (e- with vacuum gap)

Results in figure 4 shows that at short-circuited geometry of irradiation, the ionization energy losses of electrons is much higher loss due to the inhibitory effect of the electric field of the injected charge. Reducing the absorbed energy relative to the incident energy on the surface caused mainly by reflected electrons. In the case of geometries with a vacuum gap before the irradiated surface due to deceleration of the electrons in the electric field of the space charge decreases the energy density incident on the surface (dashed curves in figure 4), which leads to a further decrease in the proportion of the absorbed energy.

4. Calculation of a temperature change in field dissipation of energy of PEB in BaF_2 crystal

Calculation of heating the material in the field of energy dissipation PEB, was based on the resulting spatiotemporal energy release in the material W_j (z, t), discussed in more detail in [6].

Figure 5 shows the results of calculating distribution of heat (ΔT) in BaF_2 crystal at the initial temperature of the material (T_M) 293 K. Irradiation is accompanied by an increase in temperature caused by thermalization of nonequilibrium electrons and holes. Profile of the temperature distribution is determined by the spatial distribution of the absorbed energy.

Figure 6 shows the calculated dependence ΔT at the maximum of the energy release from the T_M for BaF_2 crystal at the end time irradiation of PEB. In the area of dependencies $\Delta T(T_M)$ where $T_M < 80$ K, the determining factor of high heat is the low heat capacity of the material. With increasing T_M above 80 K, values of $\Delta T(T_M)$ are becoming almost constant. In this area the heating caused mainly by the high energy release.

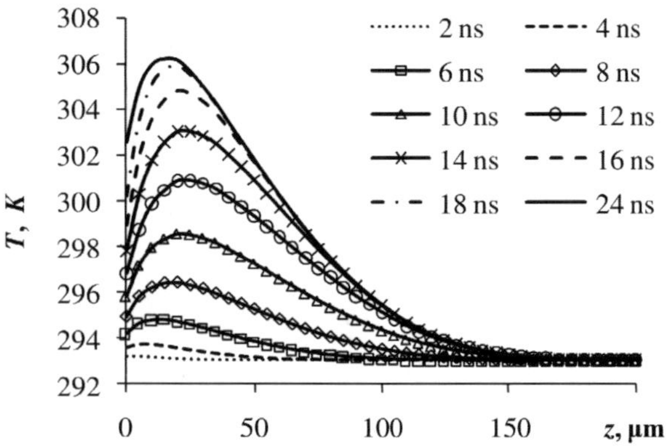

Figure 5. Calculated a formation dynamics of a the profile of temperature distribution (T) in BaF_2. Fluence of PEB is 0.188 J/cm^2. The initial temperature of the material is 293 K

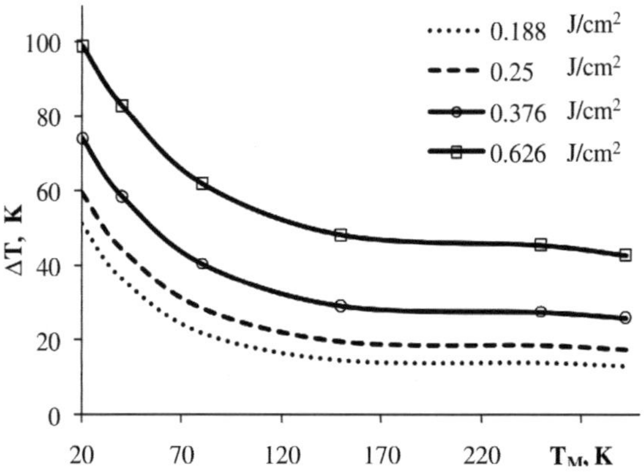

Figure 6. Calculated dependence ΔT at the maximum of the energy release from the T_M for BaF_2 crystal at the end time irradiation of PEB Fluence of PEB is indicated for each curve

5. Conclusion

At research of temperature dependences of the spectra of pulsed cathodoluminescence, values of decay component, as well as the temperature dependence of the yield of products of irradiation, an important factor is the temperature of the material in which the measurements are carried out. Consequently, as shown in this paper, the presence of spatiotemporal inhomogeneity of temperature of the crystal, as well as a significant heating of the material during irradiation PEB, even at low fluence, will make a significant distortion in the measurement results. This effect of inhomogeneous change of a temperature must be taken into account in the research of the temperature dependence of the yield of products of irradiation and kinetic processes postradiational.

References

[1] Tabata T, Ito R 1979 *Nucl. Instrum. and Methods* **158** 521
[2] Bethe H A 1953 *Phys. Rev.* **89**
[3] Shtanko V F, Tolmachev V M, Chinkov E P and Stepanov S A 2012 *Tech. Phys.* **57** 225
[4] Sternheimer R M 1966 *Phys. Rev.* **145** 247
[5] Krastelev E G, Meskhi G O and Yablokov B N 1976 *Prib. Tekh. Eksp.* **3** 39
[6] Stepanov S A, Shtan`ko V F and Chinkov E P 2014 *J. of Phys.: Conf. Series* **552** 012045

Luminescent properties of lithium-phosphate-borate glasses doped with Tb^{3+}/ Eu^{3+} ions

D T Valiev[1], **S A Stepanov and Liu Cong**

Department of Lasers and Lighting Engineering, Institute of High Technology Physics, National Research Tomsk Polytechnic University, 30 Lenin Avenue, Tomsk 634050, Russia

E-mail: dtdamirka@gmail.com

Abstract. The luminescence of Li_2O-B_2O_3-P_2O_5-CaF_2 scintillation glass doped Tb^{3+}, Eu^{3+} under different types of excitation sources are investigated. Changing the europium concentration of 0.5 to 1 wt% leads changes in luminescence intensity of Tb^{3+} ions. The luminescence spectrum of the Tb^{3+} ions are depend on the concentration of Eu^{3+}. It was found, that the luminescence decay kinetics of terbium ion in the band 543 nm depending on the concentration of europium and from type of excitation. The difference in the nature of the luminescence decay kinetics of glasses under pulsed photo- and electronic excitation discussed.

1. Introduction

To date, the most promising pair of rare-earth ions (REI) that act as activators, ions is terbium (Tb^{3+}) and europium (Eu^{3+}), which are very intensively studied in a variety of amorphous systems of complex composition [1]. Interest in such systems is due to the widespread use of phosphors and scintillator materials, activated a couple of Tb-Eu, in scientific and technological fields [2]. For example, glass doped with terbium, are used in radiography as thermal neutron detectors due to their unique optical properties [3].

The trivalent ions of terbium (Tb^{3+}) cause luminescence in the visible range due to optical transitions $^5D_4 \rightarrow {}^7F_J$ (J = 2, 3, 4, 5, 6), the most intense of which is the transition $^5D_4 \rightarrow {}^7F_5$, corresponding to a wavelength of 544 nm radiation. Intense luminescence of terbium ions is observed in many hosts, for example, aluminum oxide films [4]. In order to improve the luminescent characteristics of terbium ions are added to the host material other ions, which, through various mechanisms can transfer excitation energy to luminescence centers. Among the REI doped, Eu^{3+}/Tb^{3+} co-doped compounds have received considerable interesting due to their following potential luminescent properties in phosphors: the Eu^{3+} ions emit red emission at 614 nm ($^5D_0 \rightarrow {}^7F_2$ transition) and Tb^{3+} ions emit green emission at 544 nm ($^5D_4 \rightarrow {}^7F_5$ transition), which make up the disadvantage of low color rendering index due to the lack of a green emission in most white LEDs fabrication process.

Most published studies are limited to the study of optical properties of materials containing different promoters. Remain insufficiently studied spectral and decay kinetic characteristics of scintillation glasses. A more detailed discussion of this issue will improve the understanding of the mechanisms of interaction between the ions of the activators and help with the decision of applied

[1] The reported study was supported by RFBR, research project No. 14-02-31297 mol_a

problems related to the study and development of new materials with predetermined properties, as well as to seek opportunities synthesis of radiation-sensitive material by introducing these adjuvants.

2. Experimental details

2.1. Glass preparation

In the present work glass system of the composition Li_2O-B_2O_3-P_2O_5-CaF_2 (LBPC) doped with Tb 5 wt%:X wt% Eu where X= 0.5, 0.7 and 1 (Table 1). The samples were synthesized at the Institute for Single Crystals of the National Academy of Sciences of Ukraine (Kharkov). For glass synthesis the following compounds were chosen: $LiPO_3$, H_3BO_3, CaF_2, Eu_2O_3 and Tb_2O_3. All the reagents were high purity used without further purification. The resulting glasses are found to be homogeneous, transparent and moisture resistant. The composition of the samples is given in Table 1.

Table 1. Composition of the tested scintillating glasses

Sample	Composition, weight %				
	Eu	Tb	CaF_2	H_3BO_3	$LiPO_3$
LBPC:Tb5Eu0.5	0.5	5	10	10	to 100
LBPC: Tb5Eu0.7	0.7	5	10	10	to 100
LBPC: Tb5Eu1	1	5	10	10	to 100

The major component of all the prepared glasses was lithium metaphosphate. It was chosen due to its low melting point and high solubility to different oxides and salts. This property facilitates the modification of glass composition and incorporation of a large amount of dopants. Lithium metaphosphate was made by melting lithium dihydrogen phosphate (LiH_2PO_4) powder at 900 °C in a platinum crucible for 40 minutes. The molten mass was poured into a glassy carbon crucible and cooled to room temperature. The obtained glass is then crushed and used a precursor for the batch preparation.

The melting temperature of REE oxides is in the range 1690-2400 °C. lithium metaphosphate well mixes with the REE oxides in the batch, wherein the melting point of the mixture reduces. Lithium has a high cross section of thermal neutrons capturing, lack of heavy element atoms in the host is also reason to choose lithium methaphosphate and the possibility of forming homogeneous optically transparent glass with the volume of a perfect working surface which need no mechanical treatment directly in crucibles.

The initial batch was melted and heat-treated in platinum and glassy carbon crucibles in a muffle furnace SNOL 7.2/1300. The temperature and melting time were optimized. Complete dissolution of the components occurs at temperatures above 900 °C. The melt is frequently stired at regular intervals to make it homogeneous. The initial temperature of the glassy carbon crucible in which glass casting was performed was varied to control the cooling rate of the melt. The technique for making glass castings was developed. The batch was first preheated at 800 °C in a muffle furnace. The temperature was raised up to 900 ... 1100 °C (depending on REE concentration). The melting time was 90 minutes. Melting at 900 °C allows addition of up to 2 wt. % of REE, and at 1100 °C up to 8–5 wt. % of REE can be added. The homogeneous melt was poured into the heated up to 200 °C glassy carbon crucible and cooled at room temperature. The prepared samples were colorless glass plates with a thickness of 3 mm.

2.2. Measurements

The integral luminescence spectra of pulsed cathodoluminescence (PCL) ("spectrum per pulse") were recorded after electron beam excitation in the range of 300–900 nm with fiber optic spectrometer AvaSpec-2048. The integration time was from 1 ms to 10 ms. As a source of photoluminescence excitation ultraviolet LED light source (λ = 365 nm) and pulse nitrogen laser (λ = 337.1 nm) with 7 ns of pulse duration were used.

The PCL was studied by time-resolved spectrometry. For excitation was used electron source was a compact high-current accelerator of GIN-600 type with a vacuum diode. Duration of electron pulse at FWHM 10-15 ns, the average energy of the accelerated electrons was ~250 keV. The sample was placed in vacuum, the excitation energy density varied from 6 to 300 mJ/cm^2. The luminescence decay kinetics was recorded with a photomultiplier PMT-84-6 using monochromator MDR-3 and a digital oscilloscope LeCROY (350 MHz). Time-resolved measurement of the spectra was performed by scanning along the wavelengths and by recording the luminescence decay kinetics; the emission spectrum was reconstructed by the emission intensity at a given time I_t relative to the moment of the excitation pulse action (the time required to achieve the maximum amplitude of the electron beam current is assumed as t = 0).

3. Results and discussion

The optical transmission spectra of glasses are shown in figure 1. For all the investigated samples the transmission edge lies in the region of 350 nm. In the visible region the samples are transparent, i.e. the transmission coefficient is about 90%.

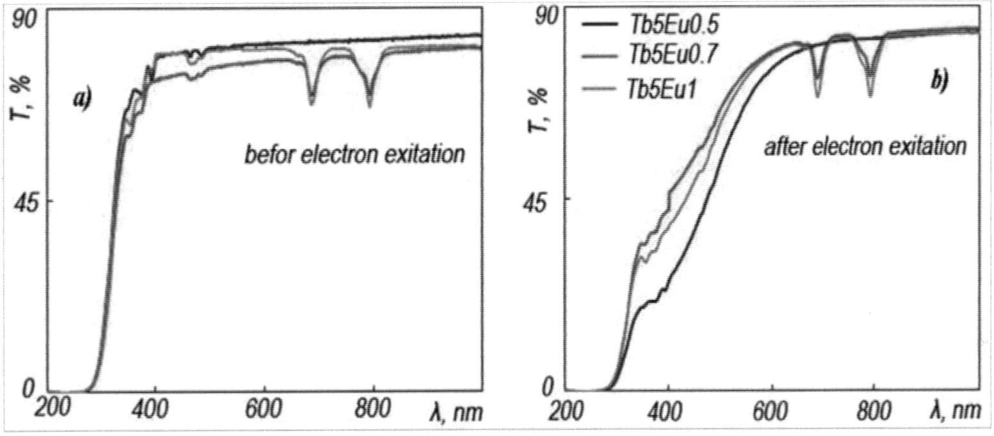

Figure 1. Optical transmission spectra LBPC: Tb, Eu glasses (a) with different concentration of Eu^{3+} ions; (b) glass irradiated with a pulse train (100 pulses) with the density of excitation energy of 69 mJ/cm^2

In the sample with the europium concentration of 0.7, 1 wt%, absorption peaks with maxima at 350, 367, 378 and 485 nm are recorded in the transmission spectrum which is apparently associated with f–f transitions in the terbium ion [5].

The transmission coefficient at 350 nm is 70%. The absorption peaks at 690 and 795 nm is recorded in samples which contain 0.7 and 1 weight% of europium ions (Figure 1). Increasing the concentration of europium ions results in the absorption of the bands (Figure 1a).

After irradiation with electron pulse train (100 pulses) it was observed a decrease of the samples transmittance. The absorption peaks in the range 350–378 nm could not be distinguished in the spectra of the irradiated samples (Figure 1, b). It was also found that the samples with the lowest concentration of europium are most sensitive to changes in the optical properties under electron irradiation.

When excited glass samples accelerated electron beam intense luminescence is observed whose spectrum consists of a series bands (Figure 2 a) with a peak wavelength of 489, 544, 588, and 620 nm, corresponding to radiative transitions in the ion Tb^{3+}:$^5D_4 \rightarrow {}^7F_6$, $^5D_4 \rightarrow {}^7F_5$, $^5D_4 \rightarrow {}^7F_4$, $^5D_4 \rightarrow {}^7F_3$.

For all samples the luminescence spectra is dominant in intensity in the "green" emission with λ_{max} = 544 nm. With increasing concentration of europium ions an increase in the intensity of luminescence of all major emission bands. Characteristic is the manifestation of luminescence band at

700 nm in a sample with a minimum concentration of europium Tb5Eu0.5. This band is due to the radiative transitions in europium ion $^5D_0 \rightarrow ^7F_4$. The increase of Eu^{3+} leads to quenching of luminescence of europium in the band 700 nm.

When excited by a nitrogen laser glasses doped with europium and terbium ions was observed that an intense luminescence whose spectrum consists of a series bands corresponding to radiative transitions in the ion Tb^{3+} (Figure 2 b). The photoluminescence spectrum qualitatively similar to that of pulsed cathodoluminescence spectrum (PCL). However, in the spectrum of the radiation transition in europium ion $^5D_0 \rightarrow ^7F_4$ at 700 nm is observed. Apparently this is due to the selectivity of the laser radiation. When excited glasses doped with europium heterostructure with λ = 365 nm (figure 2 c), the spectrum can be noted band 453 nm ($^5D_3 \rightarrow$ $^7F_3$), which does not appear in the spectra of the PCL and under the excitation of nitrogen laser.

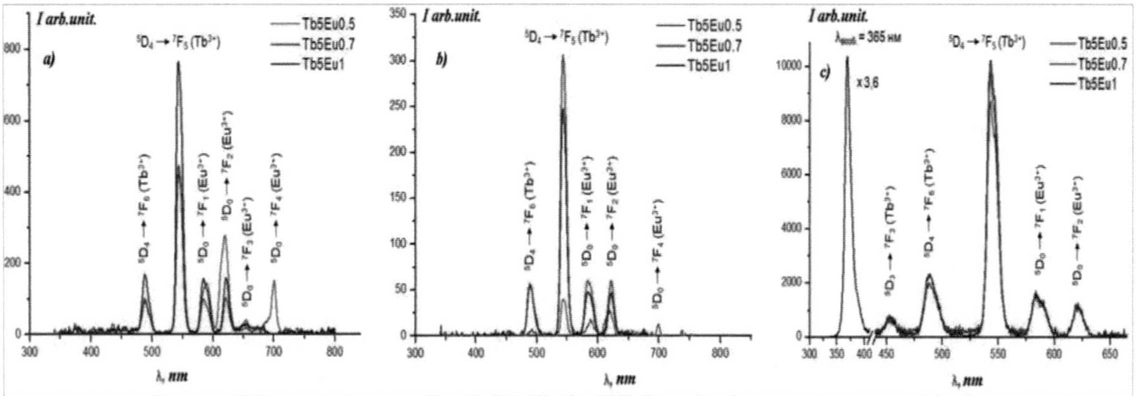

Figure 2. Luminescence spectra of LBPC:Tb5EuX (X=0.5; 0.7; 1) glasses measured by "spectrum per pulse" technique use different source of excitation: electron beam (a), nitrogen laser (λ = 337.1 nm) (b), LED chip (λ = 365 nm) (c)

Figure 3 shows the luminescence decay kinetics in the bands of 544 nm, the nature of which is related to the radiative transitions in terbium ions after exposed high-energy flow with nanosecond duration [6]. From the data obtained it can be noted that there are two components in decay kinetics: fast with the decay time $\tau_1 \sim 0.4$ ms and the slow component $\tau_2 \sim 1.8$ ms. Time characteristics are similar for all test glasses doped with REI in this type of excitation.

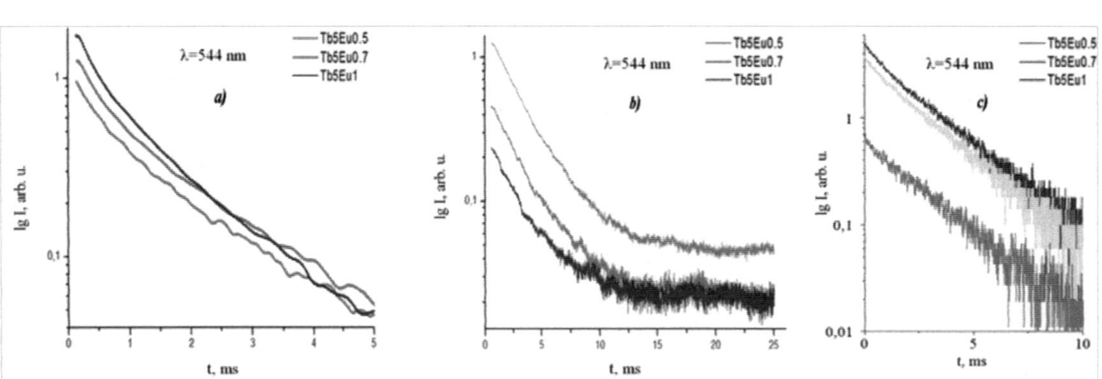

Figure 3. The luminescence decay kinetics curve of the of LBPC:Tb^{3+}, Eu^{3+} glasses (λ=544 nm) under electron (a), nitrogen laser (λ = 337.1 nm) (b) and LED chip (λ = 365 nm) excitation

Under photoexcitation (nitrogen laser pulses) the characteristic decay time of the long time component is comparable to the decay time of luminescence excited heterostructure (LED chip λ_{ex}=450 nm) τ is ~ 2.6 ms (table 2).

Table 2. Decay time of luminescence tested samples under different type of excitation

Type of excitation	Tb5Eu0.5		Tb5Eu0.7		Tb5Eu1	
	τ_1, ms	τ_2, ms	τ_1, ms	τ_2, ms	τ_1, ms	τ_2, ms
e-beam excitation	0,4	1,8	0,4	1,8	0,3	1,5
nitrogen laser (λ_{ex}=337 nm)	-	2,6		2.5		2.5
LED chip (λ_{ex}=450 nm)	0,8	2,7	1,2	2,9	0,6	2,6

The reason for the difference luminescence decay kinetics under pulsed photo- and electron excitation may be due to non-radiative transfer of excitation. When selective UV laser irradiation host glass excited weaker than in the case of high-current electron beam irradiation. As a result of the excitation of the host, are short-lived electronic excitations thus possible to increase the efficiency of nonradiative energy transfer to the host.

4. Conclusion

The performed studies demonstrate that lithium-phosphate-borate glass doped with Tb^{3+} and Eu^{3+} ions with different concentration has sufficiently effective luminescence upon excitation with a strong current electron beam, pulses of nitrogen laser and LED chip. The increase of europium content from 0.5 to 1wt% is cause the quenching of the emission bands at 700 nm of ions Eu^{3+}. On the other hand it leads to an increase in the intensity of luminescence bands in all major terbium ions. The bands of emission Tb^{3+} are dominating in luminescence spectrum. The luminescence decay time of terbium ions longer under photoexcitation, it is possible due to non-radiative energy transfer between ions co-activator. It was found, that the luminescence decay kinetics of terbium ion in the band 544 nm depending on the concentration of europium.

References

[1] Wang R, Zhou D, Qiu J at al. 2015 *Journal of Alloys and Compounds.* **629** 310
[2] Linganna K, Sreedhar V B, Jayasankar C K 2015 *Materials Research Bulletin.* **67** 196
[3] Chen D, Yu Y, Wang Y, Huang P, Weng F 2009 *J. Phys. Chem.* **113** 6406.
[4] Fu J, Kobayashi M, Sugimoto S, Parker J M , 2009 *J. Am. Ceram. Soc.* **92** 2119.
[5] Guo H, Wang X, Chen J, Li F 2010 *Opt. Express* **18** 18900
[6] Valiev D, Polisadova E, Stepanov S at al. 2015 *Journal of Luminescence.* **162** 128

Carbon saturation of silicon target under the action of pulsed high-intensity ion beam

N E Aktaev and G E Remnev

National Research Tomsk Polytechnic University, Tomsk, Russia

E-mail: nurkenaktaev@gmail.com

Abstract. The action of the pulsed high-intensity ion (carbon) beam on the silicon target is investigated by means of the theoretical model. The forming of the carbon concentration profile in depth of the silicon sample is modelled. It is argued, that there are two ways of the profile forming: short-pulsed ion (carbon) implantation and diffusion of the carbon atoms adsorbed on the silicon surface. It is shown, that the carbon atoms adsorbed on the silicon surface and diffused into the silicon target play the main role in the concentration profile forming.

1. Introduction

The investigation of influence of the pulsed high-intensity ion (carbon) beams (PHIIB) on the various materials is of interest to many scientific groups [1-6]. By decreasing of the PHIIB energy current (lower than 10^7 W/cm^2 [3]) it may be realized the short-pulsed implantation regime (SPIR) without surface melting. The main feature of the SPIR is characterized by rapid heating and cooling-down of the target material. The rapid process leads to high temperature gradient which promotes the diffusion process from the surface into the depth the silicon simple. The diffused particles are the carbon atoms adsorbed on the silicon surface. The sources of the adsorbed atoms can be [2]: i) the explosive plasma formed on the surface of graphite anode; ii) the impurity atoms from the residual atmosphere of the working chamber. In works [6,7] it was proposed that the beam ions penetrating into the simple by means of the ion implantation mechanism do not play a key role in the formation of the concentration profile. The assumption is based on the discrepancy between the integral implantation dose (IID) and the number of impurity ion (NIA) of a carbon: the NIA is much more than number corresponding of the IID in the experiments [6-8].

In present paper we attempt to quantitative theoretical analyze the diffusion effect at the short-pulsed ion implantation under the actions of the pulsed high-intensity ion beams. The analyze will be performed by mean of the model taking into account the rapid heating, cooling-down of the target material and the high temperature gradient producing by the PHIIB.

2. The Model

To analyze the diffusion process at the short-pulsed implantation the mathematical model is developed. The model allows the calculation of heating of the silicon target and the transport diffusion processes in the surface layer of the target. When there is the pulse action of PHIIB on the silicon target, the time evolution of the temperature field $T(x,t)$ in the depth x is determined within the framework of the Stefan task solution

$$c_p \rho_m \frac{\partial T(x,t)}{\partial t} = \lambda \frac{\partial^2 T(x,t)}{\partial x^2} + \frac{j_{beam}(t)}{e}\left[Q_n(x) + Q_e(x)\right], \tag{1}$$

where c_p is the specific heat capacity, ρ_m is the density, e is the elementary electronic charge, $j_{beam}(t)$ is the current density function. The energy specific losses at the nuclear $Q_n(x)$ and the electron $Q_e(x)$ decelerations can be written as

$$Q_{n(e)}(x) = \frac{1}{\sqrt{2\pi}\Delta R_p} \int_0^{E_0} \frac{S_{n(e)}(E)}{S_{n(e)}(E) + S_{e(n)}(E)} \exp\left\{ -\frac{\left[x - R_p(E_0) + R_p(E)\right]^2}{2\Delta R_p^2} \right\} dE, \tag{2}$$

where R_p and ΔR_p are the average projected range and standard deviation respectively, E_0 is the energy of beam ions, $S_n(E)$ and $S_e(E)$ are the sections of the nuclear and electron decelerations respectively.

The diffusion of carbon under the action PHIIB is described by the relation

$$\frac{\partial N(x,t)}{\partial t} = \frac{\partial}{\partial x}\left\{ D\left[T_{cd}(x,t)\right] \frac{\partial N(x,t)}{\partial x} \right\}, \tag{3}$$

where $N(x,t)$ is the concentration of carbon ions on the depth. The diffusion functional is written as

$$D\left[T_{cd}(x,t)\right] = \frac{\omega(r) a_{lat}^2}{2} \exp\left[-\frac{E_{diff}}{k_B T_{cd}(x,t)} \right], \tag{4}$$

where a_{lat} is the carbon lattice constant, k_B is the Boltzmann constant. The diffusion activation energy $E_{diff} = 2.18$ eV was defined as the difference between zero level and minimum of the potential energy $W(\tilde{r})$ determined by the Tersoff potential [10]

$$W(\tilde{r}) = \frac{1}{2} \sum_{i,j(\neq i)} f_C(r_{ij}) \left[f_R(r_{ij}) + f_A(r_{ij}) \right], \tag{5}$$

where f_C, f_R, f_A are the cutoff, repulsion and attraction functions respectively, r_{ij} is the distance between particles. The absolute value of the angular frequency of the collective motion is estimated as $\omega(\tilde{r}) = \left[\tilde{m}^{-1} d^2W / d\tilde{r}^2\right]^{1/2}$, where $\tilde{m}$ is the generalized mass of the "atom-lattice" system. The deformation dependence of the potential energy for carbon and silicon is presented in figure 1.

The temperature dependence of the cooling-down process $T_{cd}(x,t)$ is characterized by the relation

$$\frac{\partial T_{cd}(x,t)}{\partial t} = \frac{\lambda}{c_p \rho_m} \frac{\partial^2 T_{cd}(x,t)}{\partial x^2}, \tag{6}$$

The boundary condition can be written as

$$T_{cd}(0,t) = T_0 + (T_{ini} - T_0)\exp\left(-\frac{\lambda}{c_p \rho_m d_{Si}^2} t\right), \tag{7}$$

where T_{ini} is the temperature of the sample surface after the PHIIB action, $d_{Si} = 400$ μm is the thickness of the silicon sample, $T_0 = 300\,\text{K}$ is the ambient temperature.

The concentration implantation profile is calculated as follows

$$N_I\left[\tilde{X}(x,E),t\right] = \frac{N_{dose}}{\sqrt{2\pi}\Delta R_p(E)} \frac{k_{PRSN}}{\left[1+\left(\frac{\tilde{X}(x,E)}{a^*}\right)^2\right]^q} \exp\left[-v^* \operatorname{arctg}\left(\frac{\tilde{X}(x,E)}{a^*}\right)\right], \tag{8}$$

where $k_{PRSN} = 0.2$ is the normalization factor,

$$\tilde{X}(x,E) = \left[x - R_p(E)\right]\Delta R_p^{-1}(E), \tag{9}$$

$$a^* = \Delta R_p \sqrt{\left[16(\eta-1) - \beta_1(\eta-2)^2\right]/16}\,, \tag{10}$$

$$v^* = -\frac{\eta(\eta-2)\sqrt{\beta_1}}{\sqrt{16(\eta-1)-\beta_1(\eta-2)^2}}, \tag{11}$$

$$\eta = \frac{6(\beta_2 - \beta_1 - 1)}{2\beta_2 - 3\beta_1 - 6}, \tag{12}$$

and parameters are $\beta_1 = 2.6896$, $\beta_2 = 11.006$.

3. Results and discussion

The temperature profile calculated after the action (600 pulses) of one pulse is shown in figure 2. The curve at the current density 20 A/cm^2 (open circles) corresponds to the experimental setup (the range of the beam current density of the TEMP-4 is 15-20 A/cm^2). It can be seen from figure 2 that the melting processes can be neglected when mass transfer procedure is modelled.

IOP Conf. Series: Materials Science and Engineering **110** (2016) 012054 doi:10.1088/1757-899X/110/1/012054

The figure 3 shows the profile formed by the diffusion of the carbon adsorbed on the surface layer into the silicon target and implantation profile. The concentration value N_{norm} is normalized to the amorphous carbon concentration $N_{Amorf} = 1.1 \cdot 10^{23}$ cm^{-3} (of course, $N_{norm} = N / N_{Amorf}$).

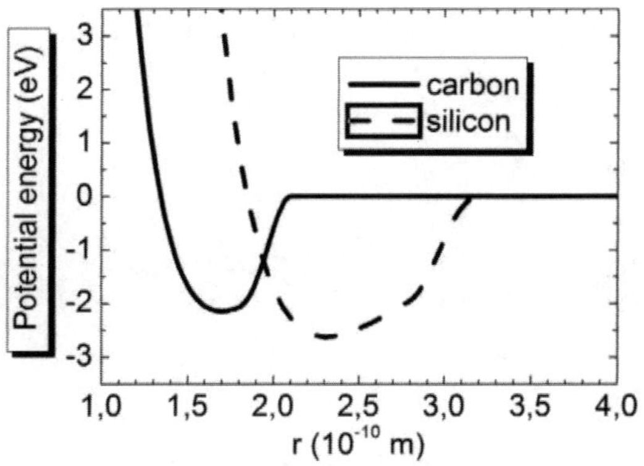

Figure 1. Deformation dependence of potential energy for carbon (solid) and silicon (dash)

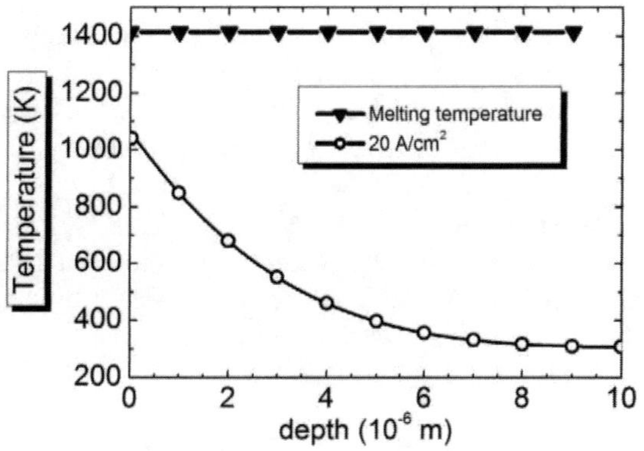

Figure 2. Temperature distribution in depth of silicon target

Inside picture is the same as a dependence (square) on the main picture but it have a wide range of the depth. The discrepancy between the concentration value of the diffusion profile and implantation profile at least tree orders. Consequently, it can be concluded that the carbon saturation in the sub-surface silicon layer occurs by means of the diffusion of carbon adsorbed on the silicon surface. The contribution of the implantation ions to concentration profile is forming insignificant.

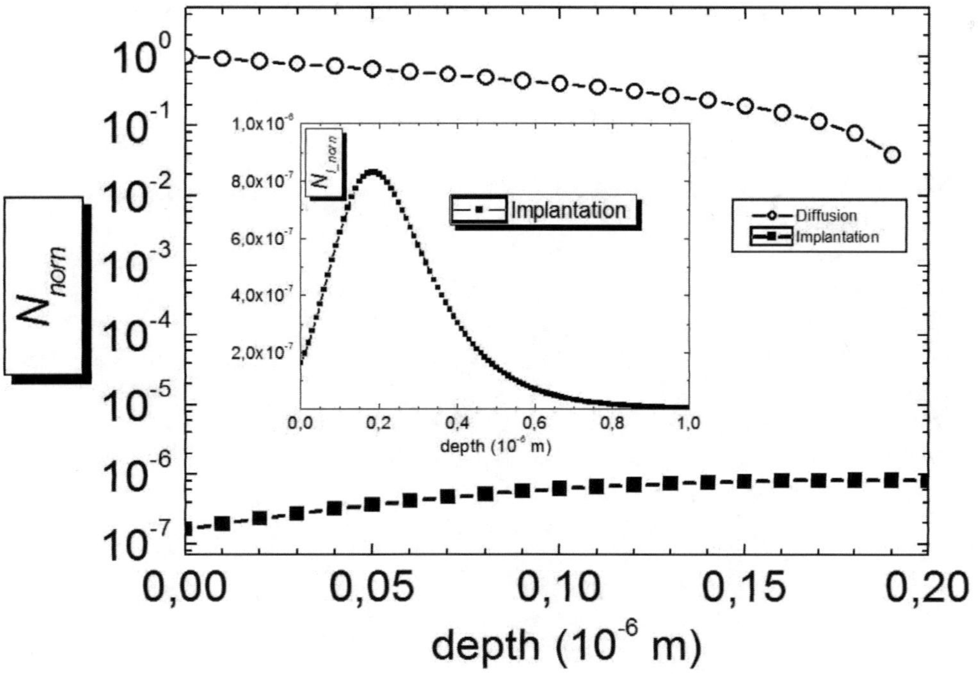

Figure 3. Concentration profiles: diffusion (circles), implantation (squares).

4. Conclusion

The influence of the pulsed high-intensity ion beam on the silicon is studied by use the developed theoretical model. The input parameters of the model were the settings of the experimental setup of the TEMP-4. It is shown, that at the short-pulsed implantation regime of the TEMP-4 the silicon surface does not melt. However, the regime leads to the high temperature gradient which promotes the diffusion process from the surface into the depth the silicon simple. Thus, it is shown that the carbon atom diffused from the surface make the main contribution to the forming of the concentration profile. The concentration of the implanted carbon ions less more than tree orders compared with the concentration of the diffused carbon atoms.

Acknowledgment

This work was funded within the framework of realization of Strategic Programme on National Research Tomsk Polytechnic University Competitiveness Enhancement in the Group of Top Level World Research and Academic Institutions.

References

[1] Neri J M, Hammer D A, Ginet G, Sudan R N 1980 *Appl. Phys. Lett* **37** 101.
[2] Remnev G E, Shulov V A 1993 *Laser Particle Beams* **11** 707.
[3] Davis H A, Remnev G E, Stinnett R W, Yatsui K 1996 *MRS Bull* **21** 58.
[4] Bayazitov R M, Khaibullin I B, Batalov R I, Nurutdinov R M, Antonova L Kh, Aksenov V P, Mikhailova G N 2003 Nucl. *Instr. Meth. Phys. Res. B* **206** 984.
[5] Renk T J, Provencio P P, Prasad S V, Shlapakovski A S, Petrov A V, Yatsui K, Weihua J, Suematsu H 2004 *Proc. IEEE* **92** 1057.
[6] Remnev G E, Uglov V V, Shymanski V I, Pavlov S K, Kuleshov A K 2014 *Appl. Surf. Science* **310** 204.
[7] Zhu X P, Zhang F G, Tang Y, Lei M K 2011 *Laser and Particle Beams* **29** 283.
[8] Remnev G E, Uglov V V, Shymanski V I, Konarski P, Samtsov M P, Pavlov S K, Lapchuk N M 2013 *Vacuum* **89** 118.
[9] Remnev G E, Isakov I F, Opekounov M S, Kotlyarevsky G I, Kutuzov V L, Lopatin V S, Matvienko V M, Ovsyannikov M Yu, Potyomkin A V, Tarbokov V A 1997 *Surf. and Coat. Tech.* **96** 103.

IOP Conf. Series: Materials Science and Engineering **110** (2016) 012055 doi:10.1088/1757-899X/110/1/012055

Propagating self-sustained annealing of radiation-induced interstitial complexes

P M Bokov[1] and P A Selyshchev[2]

[1]Radiation Science Department, South African Nuclear Energy Corporation (Necsa), Building 1900, P.O. Box 582, Pretoria 0001, South Africa

[2]Department of Physics, University of Pretoria, Private bag X20, Hatfield 0028, South Africa

E-mail: pavel.bokov@necsa.co.za, selyshchev@gmail.com

Abstract. A propagating self-sustained annealing of radiation induced defects as a result of thermal-concentration instability is studied. The defects that are considered in the model are complexes. Each of them consists of one atom of impunity and of one interstitial atom. Crystal with defects has extra energy which is transformed into heat during defect annealing. Simulation of the auto-wave of annealing has been performed. The front and the speed of the auto-wave have been obtained. It is shown that annealing occurs in a narrow region of time and space. There are two kinds of such annealing behaviour. In the first case the speed of the auto-wave oscillates near its constant mean value and the front of temperature oscillates in a complex way. In the second case the speed of propagation is constant and fronts of temperature and concentration look like sigmoid functions.

1. Introduction

The irradiation of materials with high-energy particles typically causes deterioration of their operational characteristics [1]. This leads to the necessity for frequent replacement, decontamination and disposal, which in itself is a difficult and expensive task. But the problem is particularly acute for bulky parts and assemblies, and especially for the reactor vessel. Therefore, exploring opportunities to extend the service life through the restoration of the operational characteristics of structural materials is extremely important, especially given the current state of nuclear power when the projected service life for many nuclear power plants is approaching the end.

It is known that one of the ways of restoring material characteristics after irradiation is annealing. However, restoring the material characteristics of bulky structures is difficult with conventional annealing techniques. The idea of extending the life of nuclear reactor vessels by annealing is though being actively discussed. Traditional annealing by heating to the necessary temperature of the entire reactor vessel requires a large amount of energy, a long period of time and can consequently cause irreversible phase-structural changes of the material.

For this reason, it is of interest to study the particularities of annealing by means of a temperature auto-wave, as it was proposed in [2]: annealing as a result of passing a short pulse of high temperature occurs locally and is devoid of the drawbacks discussed above. It is more precise, although associated with the setting up of a more complex auto-wave mechanism of temperature propagation. This mechanism is based on the self-sustaining annealing of radiation defects. Consider a radiation damaged crystalline sample, which contains the radiation-induced lattice defects. Due to these defects, the crystal is not in a state of thermal equilibrium. In the formation of defects, the regular bonds are broken and new, irregular ones are established. As a result, stresses appear in the crystal and its elastic energy increases. Although the state of the crystal becomes metastable, its relaxation to equilibrium can take a long time, due to the high values of activation barriers.

The energy of a radiation-damaged sample after irradiation is significantly higher than its energy prior to irradiation. The magnitude of the difference between these energies is of the order of energy of the defect formation. For instance, the formation of a Frenkel pair requires from 5 to 7 eV. The energy stored in defects does not contribute to the thermal energy of the crystal, but it can be converted into thermal energy in the process of the annealing of defects, for instance, in the case of the recombination of vacancies and interstitial atoms. As a result, the temperature of the crystal increases, and since the intensity of defect annealing depends exponentially on temperature, the intensity of annealing is further boosted. Thus, a positive feedback loop is established. Under certain conditions, this positive feedback, which we will call thermal-concentration feedback, leads to the development of self-oscillations of temperature and of defect concentration [3, 4]. However, if the annealing is initially localized, then by thermal conduction and diffusion of the defects it may spread in space and again, under certain conditions, form a self-sustained wave.

Despite the topicality, the practical importance and a huge number of publications, the kinetics of annealing of radiation defects is studied insufficiently. This applies primarily to inhomogeneous annealing of, especially, multicomponent materials, and to nonlinear relations that can be realized in the process of annealing and lead to a wide variety of relaxation processes. It should also be noted that the study of the auto-wave annealing is of considerable academic interest because it allows the establishing of mechanisms of development of auto-waves in active media of a different nature.

2. Problem statement

Let us investigate the propagation of the annealing auto-waves caused by the thermal-concentration feedback using as an example the simplest model that describes the dynamics of the annealing inactive complexes. The complex binds an impurity atom, which is a trap for the defect, and one interstitial atom. Let the concentration of traps be n_t = const, and $n_0 = n_0(t)$ be the number of available traps that are not bound to interstitial atoms, and $n_1 = n_1(t)$ be the number of traps bound with one interstitial atom. Both n_0 and n_1 change during the annealing process and are functions of time, but their sum remains constant: $n_t = n_0(t) + n_1(t)$.

Since the concentration of the complexes is comparable with the concentration of traps [1, 4], the formation energy of the interstitial atoms is several times higher than one of the vacancies, then for high densities of dislocations, which are typical for irradiated materials, and for the characteristic temperatures of annealing, we can neglect the contribution of the vacancies. Note that this approximation reduces the energy accumulated in the defects and slows the development of the auto-wave annealing process in its initial stage.

Then the evolution of defect concentrations n_i, n_1 and of temperature T can be described by the system of equations:

$$\frac{\mathrm{d}n_i}{\mathrm{d}t} = -\frac{n_i}{\tau_i} - \gamma_i n_0 n_i + \frac{n_1}{\tau_1}, \tag{1}$$

$$\frac{\mathrm{d}n_1}{\mathrm{d}t} = +\gamma_i n_0 n_i - \frac{n_1}{\tau_1}, \tag{2}$$

$$c\rho\frac{\partial T}{\partial t} = \kappa\frac{\partial^2 T}{\partial x^2} + \theta_i\frac{n_i}{\tau_i} + \theta_b\gamma_i n_0 n_i - \theta_b\frac{n_1}{\tau_1} - h(T - T_e). \tag{3}$$

The first term in (1) describes the absorption of interstitial atoms by dislocations, the second and third terms in (1) and the first and the second terms in (2) describe the creation and the thermal destruction of complexes. Equation (3) describes the change in the crystal temperature with respect to thermal conductivity, heat sink and heat dissipation during annealing of defects and complexes. The last term in equation (3) describes the dissipation of heat. Here $\gamma_i = \mu D_i$ is the complex creation rate constant, $\tau_i^{-1} = z_i \rho_d D_i$ and $\tau_1^{-1} = \mu D_i^0 n_{\mathrm{at}} \exp\left(-\left(E_{mi} + E_b\right)/kT\right)$ are the inverse lifetimes of interstitial atoms with respect to absorption by dislocations and of complexes with respect to thermal destruction. $D_i = D_i^0 \exp\left(-E_{mi}/kT\right)$ is the diffusion coefficient, ρ_d is the density of dislocations and z_i is the preference of dislocation absorption of interstitial atoms. c and ρ are the specific heat capacity and the density of the crystal and κ is the heat transfer coefficient. E_{mi} is the migration energy, θ_i is the energy converted to heat in the process of one annealing of one interstitial atom, which is approximately equal to the energy of its creation and θ_b is the release (absorption) of energy in destruction (creation) of the complex, E_b. T_e is the environment (heat sink) temperature, k is the Boltzmann constant. n_{at} is the number of atoms in the unit volume. The lifetime of a complex with respect to thermal destruction, τ_1 can be obtained from the principle of detailed balance: $\gamma_i n_0 n_i^e = n_1^e/\tau_1$.

The system (1) to (3) is essentially nonlinear, as the typical lifetimes of defects depend exponentially on temperature. Taking into account that the lifetime of the interstitial atom is much smaller than that of the complex, i.e. $\tau_i \ll \tau_1$, we set $\mathrm{d}n_i/\mathrm{d}t = 0$. Also take into consideration that the loss of interstitial atoms as a result of complex formation dominates. Then instead of (1) to (3), we obtain a system of two differential and one algebraic equation:

$$\frac{\mathrm{d}n_1}{\mathrm{d}t} = -\frac{n_1}{\tau_1}, \tag{4}$$

$$c\rho\frac{\partial T}{\partial t} = \kappa\frac{\partial^2 T}{\partial x^2} + (\theta_i - \theta_b)\frac{n_1}{\tau_1} - h(T - T_e), \tag{5}$$

$$\frac{n_i}{\tau_i} = \frac{n_1}{\tau_1}. \tag{6}$$

Let the coordinate axis Ox be perpendicular to the surface and be directed into the medium. Consider the boundary conditions of the second kind: we assume zero heat flux across the boundary Γ: $\mathrm{d}T/\mathrm{d}x|_\Gamma = 0$. The initial concentration of complexes is given by: $n_1(0, x) = n_{1,0}$. We will assume that the material temperature is initially equal to the environment temperature, T_e, except for a narrow boundary layer of width Δ in which, as a result of external heating, an elevated temperature, T_0, is attained and maintained (possibly for some finite period): $T(t, x) = T_0$ if $0 \le x \le \Delta$ and $T(t, x) = T_e$ if $x > \Delta$.

The problem of radiation defect annealing is similar to the well-known combustion problem [5], however unlike the latter the structure of radiation damages is non-stationary. Defects are continually annealed at a non-zero rate and the auto-wave can propagate to final distances only.

Non-linear system (4) – (6) was solved numerically for a finite interval of length L with the Crank-Nicolson finite difference scheme. This discretisation scheme yields a system of nonlinear equations, which was solved with Newton's iterations at each time step. The Jacobian matrix used in the Newton method was obtained analytically and the system of linear equations was also analytically reduced to a tridiagonal form, which can be efficiently solved with the Thomas algorithm.

Calculations were performed for parameters of steel: c = 460 J/(kg·K), ρ = 7800 kg/m^3, κ = 74.4 W/(m·K), z_i = 1.1, $D_i^0 = 10^{-5}$ m^2/s, E_{mi} = 0.3 eV, $\rho_d = 10^{12}$ m^{-2}, θ_i = 5.9 eV, $\theta_b = E_b$ = 1 eV and $\mu = 8 \cdot 10^{-8}$ m. Geometrical parameters and initial conditions were chosen as follows: L = 10 cm, Δ = 1 mm, T_e = 300 K, T_0 = 425 K, $n_{\mathrm{at}} = 6 \cdot 10^{28}$ m^{-3} and $n_{1,0} = 8 \cdot 10^{26}$ m^{-3}. However, in order to investigate mechanisms of propagation of a self-sustained wave of radiation defect annealing, the parameters were varied in wide ranges.

3. Results and discussions

The obtained results are presented in figures 1 and 2. Note that in this paper, we will report defect concentration normalized by a constant factor, $C = (\theta_i - \theta_b)n_1/(c\rho)$. The dimension of parameter C

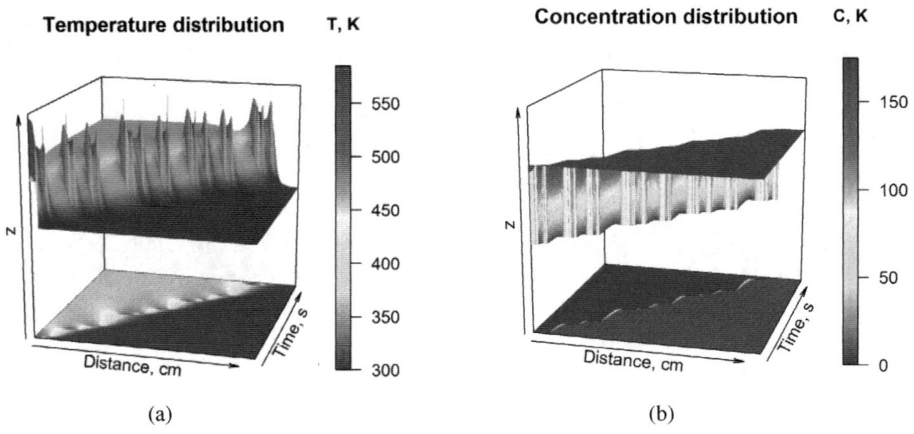

Figure 1. Variation of temperature (a) and of normalized concentration (b) during the self-sustaining annealing without dissipation.

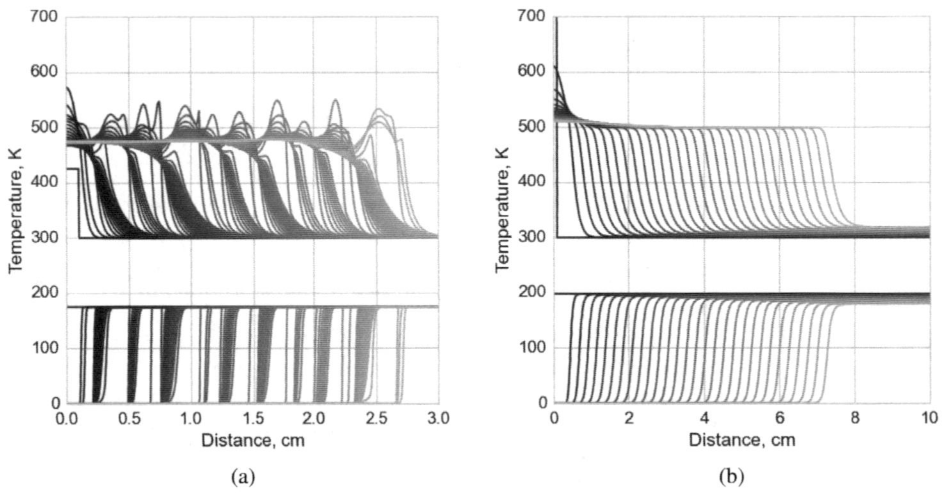

Figure 2. Propagation of temperature (top curves) and of normalized concentration (bottom curves) fronts of self-sustaining annealing with a variable (a) and a constant speed (b), without dissipation.

is kelvin and it has a physical meaning of temperature increase of an isolated piece of matter when all defects (complexes) are annealed.

From the results it follows that the development of thermal-concentration instability causes a rapid annealing of defects (propagating self-sustaining annealing of defects), when the defect density and the temperature, remaining almost constant throughout the depth of the sample, change dramatically in a narrow region that is moving with some constant average speed. In this area, the defect density falls to zero, and the temperature increases to some maximum value, which is then, taking into account the dissipation, reduced to ambient temperature.

The rapid change of the derivative of temperature on a short interval means a high contribution of thermal conductivity. Hence, in the annealing zone one can neglect the contribution of convective heat flow changes in the heat transfer equation (5) as compared to the contribution of the thermal conductivity (in the reaction zone the temperature differences are small, because of the strong dependence of lifetime on temperature the defects are essentially annealed at a maximum temperature).

It was found that the propagation of annealing has specific features. Unlike the classical model auto-waves (presented in figure 2b), the wave does not propagate at a constant speed for realistic values of the annealing parameters. The speed oscillates in a complicated manner near its average value, which is of the order of a few centimetres per second. The front of the defect concentration looks like a step or sigmoid function, while the temperature front oscillates in a complicated way in the area of the annealing, as is illustrated in figures 1a, 1b and 2a.

Such behaviour may be associated with pre-heating of the material in the vicinity of the annealing zone by thermal conduction. Because of the exponential dependence of the rate of annealing on the temperature, a certain combination of parameters leads to a bifurcation and a stationary regime of heat propagation is replaced by an oscillating one. This is supported by the fact that, for some parameters, the temperature front also has the shape of a step or sigmoid function, and the speed of the annealing front in this case is almost constant as illustrated in figure 2b.

In the case when heat dissipation is introduced, the temperature decreases slowly down to the environment temperature. The temperature front of the self-sustaining annealing has the shape of a moving hump with a steep climb and a more gentle (almost linear) descent, and not the shape of a sigmoid function as in the system without dissipation of energy.

Due to the instability of defects, along with intensive annealing in the front of the auto-wave, a more gradual (because of high activation energy) spontaneous annealing of defects takes place all over the damaged volume; their density decreases and the condition for the self-sustaining defect annealing is violated, especially if this is accompanied by heat removal.

Comparison of a traditional and of a self-sustaining propagating annealing of defects demonstrates that the heating of a sample of finite size, to a certain temperature, by the usual heat conduction requires a larger (asymptotically infinite) time, while heating time by propagating self-sustaining annealing is limited. Thus the material will be exposed to an elevated temperature in a local spatial region for a significantly shorter time, which to some extent can prevent the heat-induced phase-structural transformations of the material and thereby prevent the change of its properties in the process of annealing. The self-sustaining annealing of defects requires significantly less of externally supplied energy. Nevertheless, the propagation of self-sustaining annealing of defects occurs when the stored energy of defects is sufficiently large, and requires special conditions of initiation, which are not always achievable.

4. Conclusions

It was found that thermal-concentration instability constitutes the mechanism of spreading self-sustaining annealing of defects, which occurs in a narrow region in a short time. It is shown that there are two types of such annealing behaviour: in the first case the annealing propagation speed oscillates around a constant mean value and the thermal front in the field annealing oscillates in a complicated way; in the second case the velocity is constant, and the temperature and concentration fronts are of the form of a sigmoid function.

References

[1] Ivanov L I and Platov Yu M 2004 *Radiation Physics of Metals and its Applications* (Cambridge: CISP)
[2] Selyshchev P A 2014 Propagation of self-reinforcing annealing of radiation defects, *Proc. of the XXIV Int. Conf. Radiation Physics of Solids.* (Sevastopol, 07–12 July 2014) (Moscow: GNU NII PMT) pp 589–94
[3] Andronov A A, Witt A A and Haykin S E 1981 *Oscillation Theory* (Moscow: Nauka)
[4] Selyshchev P A 2008 *Self-organization in Radiation Physics* (Moscow, Izhevsk: R&C Dynamics)
[5] Vasil'ev V A, Romanovskii Yu M and Yakhno V G 1980 *Autowave Processes* (Moscow: Nauka)

Functional properties of multilayer vacuum-arc TiN/ZrN coatings

S V Plotnikov[1], A D Pogrebnyak[2], Y O Tleukenov[1] and N K Erdybaeva[1]

[1] *Ust-Kamenogorsk, 070003, Kazakhstan,* D. Serikbayev East Kazakhstan State Technical University, 23/2-54 Serikbayev Street
[2] *Sumy, 40000, Ukraine,* Sumy State University, Rimsky-Korsakov Street, 2

E-mail: erick_090@mail.ru

Abstract.

Nanostructured multilayer Ti/ZrN coatings were synthesized by vacuum-arc deposition with a number of layers 134-533 and an average thickness 20-125nm of layers. A good planarity was revealed resulting in a range of nanometer layer from plasma streams in a reactive environment. Phase-structural changes mechanisms were established as a model of critical operating coatings' conditions of in the surface layers under the action of an aggressive oxygen atmosphere at high temperature (700 0C). The thickness parameter effect on its hardness of the multilayer system was shown. It was found that the maximum hardness of 42 GPa and the lowest abrasion of coating 1,3 ×10–5 mm3×H–1×mm–and counterbody 1,9 ×10–6 мм3×H–1×мм–1 inherent in TiN/ZrN system with the smallest layer thickness of 20 nm in the period. The results are explained by the influence of the size factor interphase boundaries magnified in a multilayer system with a nanometer thick layers.

1. Introduction

One of the ways to improve the functional properties of coatings based on refractory compounds nitrides is creating a multi-layer systems, in which the properties superior to the properties of each element individually can be attained by combining the various layers of nitrides. One of the most popular uses of such coatings is a cutting tool, and the friction units of machine parts [1].

Analysis of the literature indicates about perspectives of using for improving the mechanical properties of multilayer coatings in which the layers consisting of nitrides of refractory materials have a thickness of about 20 - 30 nm [2, 3]. Coatings consisting of two layers of hard refractory metals nitrides have high hardness because, tension created in layers prevents the movement of dislocations.

In this regard, unification of TiN and ZrN layers with high hardness in a multilayer coating allows greatly improve the strength of the material for the preservation of high hardness such a comprehensive cover by the formation of a large number of strongly phase boundaries.

2. Experimental details

Coatings were prepared by vacuum arc of the two evaporators. One of which contains titanium brand BT-1-00, the second - zirconium, obtained by electron-beam melting (EBM). Samples with size

(15×15×2,5 mm) from steel 12X18H9T (R_a = 0,09 m) were used as substrates for the deposition of coatings. Samples were obtained at PN = P_N= 3×10^{-3} Torr, and a constant bias potential U_b = -150 V.

During the experiment, three sets of coating samples were formed which differ in layers thickness in the period.

In the first series of samples, pair of TiN/ZrN layers had a thickness of about 40 nm λ at a total coating thickness of about 13 microns. In the 2nd series of samples, TiN/ZrN layers had a thickness of about $\lambda \approx 70$ nm at a total coating thickness h ≈ 14 microns. The third series of samples' TiN/ZrN layers were formed with a thickness of $\lambda \approx 250$ nm with a total coating thickness h ≈14 m.

A number of layers for the different series are: - for the 1st a number of layers 533, for the 2nd a number of layers 233, and for the third a number of layers 134.

Annealing of samples was carried out in a vacuum chamber previously evacuated to a pressure of 10^{-5} Torr in oxygen, self-inflicted to the pressure to 5×10^{-3} Torr.

The surface morphology, fractograph of fracture, friction track were studied by scanning electron microscope FEI Nova NanoSEM 450. Phase-structural condition research was carried out by diffractometer DRON-3M in the emission of Cu-K_α.

The tribological tests were performed in air on a "ball - drive". «Tribometer» was used as the friction machine, CSM Instruments. Coatings were deposited onto the polished surface of cylindrical samples (R_a = 0,088 microns), made of steel 45 (diameter 42 mm, height of 5 mm). A ball with 6.0 mm diameter was used as a counterface, made of sintered and certified material - Al_2O_3. The load was 6.0 N, sliding speed was 10 cm/s. The tests conform to international standards ASTM G99-959, DIN50324 and ISO 20808.

The hardness of the coatings was measured with a hardness tester Model DM 8 by micro-Vickers method at a load of 0.2 N on indenter.

3. Results and discussion

Fig. 1 shows a typical radiograph fragment of the multilayer coating sample. It is seen that layers containing TiN and ZrN phase with cubic (structural type NaCl) crystal lattice without expressly preferred orientation (texture) crystallites are formed after the precipitation on surface layers (informative layer depth of about 4 microns) are formed. TiN layers increases slightly with an increase in the specific contribution period λ, which is manifested in the intensity ratio change of TiN and ZrN phase peaks (see. Fig 1 a and b). The grating period of the nitride phase in the fibers changes with an increase in the duration of the deposition and, respectively, the thickness of layers and the total period of the multilayer system λ: the grating period decreases with increasing thickness of TiN from 0.4241502 nm at deposition time of 20 seconds ($\lambda \approx 70$ nm) to 0.4238870 nm at the deposition time of 40 s ($\lambda \approx 250$ nm). The decrease is less significant in ZrN from 0.4581055 nm at 20 seconds to 0.4581046 nm at 40 seconds.

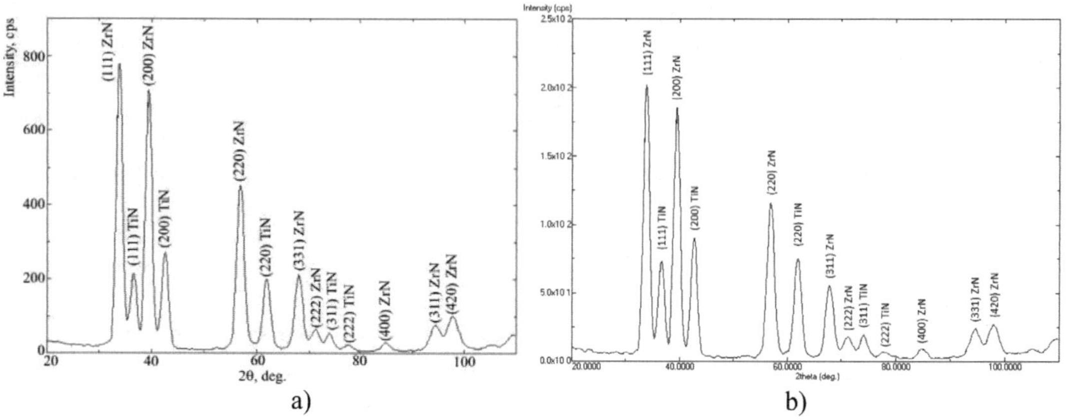

Figure 1. Lands of coatings X-ray diffraction spectra with $\lambda \approx 70$ nm (a) 2nd series and $\lambda \approx 250$ nm (b) the 3[d] series

Annealing in atmosphere of oxygen leads to surface layers acidulation and the formation as the main phases in the dioxides fibers - TiO_2 with tetragonal type rutile (base, containing up to 95% in fibers based on Ti) and anatase (containing 5% and less.) Anatase (DB card number 5000223) is detected on the diffraction spectrum (Fig. 2a) by the most powerful first-line angle $2\theta \approx 25.36$ deg. The whole diffraction spectrum (Fig. 2a) is shown for rutile (DB card number 9007531). Only one type of dioxide ZrO_2 (Arkel having a cubic system, DB card number 5000038) is formed in the layers of zirconium nitride by oxidation.

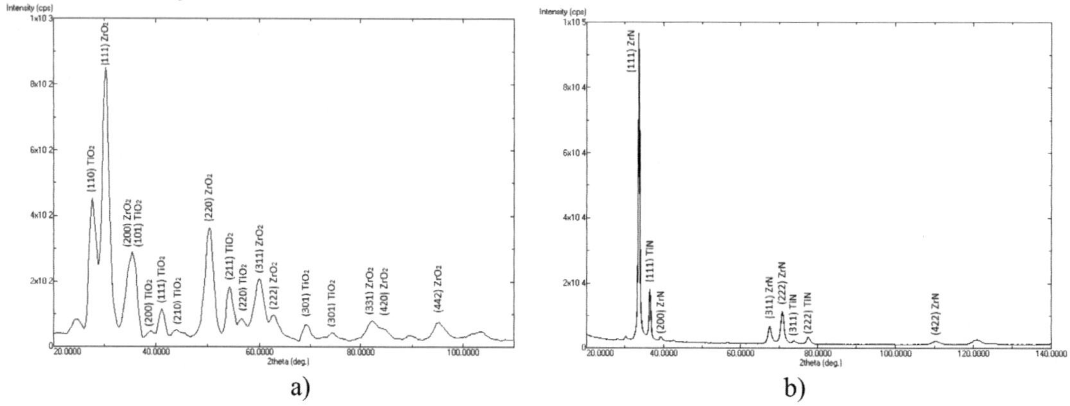

Figure 2. The plot of X-ray diffraction spectra of TiN/ZrN coatings with $\lambda \approx 70$ nm obtained after an hour annealing at 700 °C, a) – from the surface without grinding, b – after grinding with oxidized surface of 5 microns.

However, as fiberwise X-Ray analysis performed after grinding of a surface layer thickness of about 5 micrometers, the presence of oxides observed in the surface layer of the coating. Oxide formation is not observed in thicker layers after grinding (Fig. 2b), forming nitrides during the deposition have a preferred orientation of crystallites with [111] axis perpendicular to the plane of growth. From this it follows that the predominant orientation of TiN and ZrN crystallites with [111] axis is formed in the initial growth stage, which loses its priority and thus becomes non-oriented crystalline state in the process of increasing the total coating thickness and the relaxation of compressive stress growth.

In this connection, from the results of X-ray analysis it is shown in Figure 2, that oxygen from the atmosphere penetrates the subsurface layer of the coating during annealing, promoting the oxidation up to the formation of stable dioxide phase of metal due to disorientation of the crystallites in the surface layers and low compressive stress. Texture growth is developed in deeper layers under the influence of compressive stresses [111] [4.5], which reflects the initial growth stage, because of the high packing density (111) plane complicates the diffusion depth of such layers. Oxygen does not penetrate in an amount sufficient to form oxide phases.

A detailed analysis of the changes in the surface layers when oxidation allows for data transmission microscopy. Figure 3 shows the results of electron microscopic chips of samples for 3 sets at different magnifications.

It can be seen, that there is a good planarity of precipitation even for the greatest number of the thinnest layers of Series 1 (Figure 3, 1). Also the high denseness of the coating and the lack of it in the form of inhomogeneities in the droplet phase are characteristic and visible when deposited on the surface.

Energy dispersive spectrum is characteristic for all 3-series and shows the stoichiometry of the coating composition of the coating TiN/ZrN which is shown in Figure 4.

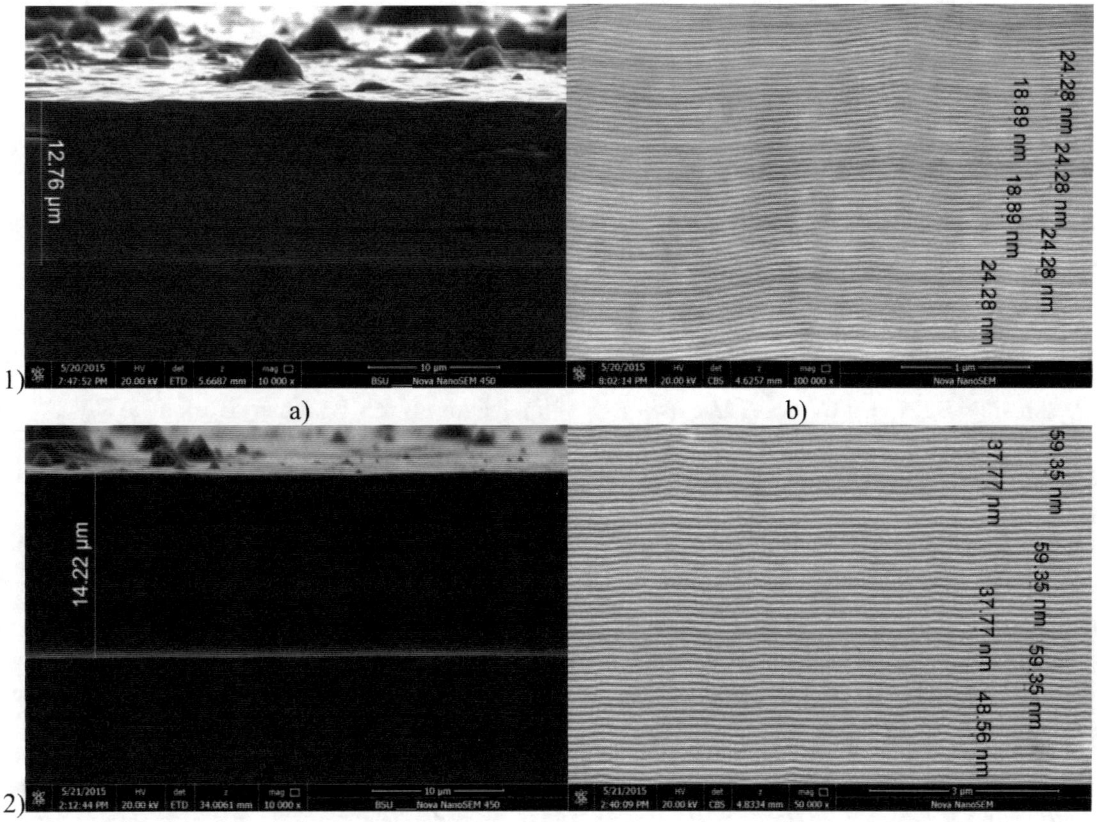

3) 

a) b)

Figure 3. Electron micrographs of multilayer TiN/ZrN coatings cross-section: 1) for the first series with 533 the number of layers; 2) for the second series with 233 the number of layers; 3) for the third series with 134 the number of layers.
a) general view of the sample cleavage; b) an enlarged detail of the multilayer coating side surface.

4. Conclusion

The use of vacuum-arc method allows obtaining a two-phase multi-layer system TiN/ZrN with nanometer thickness range of good planarity. With increasing thickness of the composition TiN/ZrN, deposited under the action of U_s = -150 V, the transition occurs from the preferred orientation with [111] axis to the non-oriented state. The presence in the initial layers growth of coating with [111] texture prevents oxidation during high-temperature annealing at 700 ^{0}C in oxygen atmosphere.

References

[1] Cavaleiro, Albano, De Hosson, Jeff Th M, Springer 2006 *Nanostructured coatings*, pp. 648.
[2] Anishik B M, Uglov V V, Zlotskii S V, Emelyanov V A, Ponomar B N, Uhov V A 2003 Mnogosloinie nanostrukturirovannie pokritiya Ti/ZrN: struktura I mehanicheskie svoistva. Perspertivnie materiali. Vol. **4** pp. 75-78.
[3] Siao-Fan Chen, Yu-Chu Kuo, Chaur-Jeng Wang, Sung-Hsiu Huang, Jyh-Wei Lee, Yu-Chen Chan,Hsien-Wei Chen , Jenq-Gong Duh, Tsung-Eong Hsieh 2013 The effect of Cr/Zr chemical composition ratios on the mechanical properties of CrN/ZrN multilayered coatings deposited by cathodic arc deposition system. *Surface & Coatings Technology,* pp. 247–252.
[4] Sobol O V, Andreev A A, Gorban V F, Stolbovoi V A, Pinchuk N V, Meilehov A A 2015 Vliyanie potenciala smesheniya na phazavi sostav, strukturu, substrukturu I mehanicheskie harakteristiki mnogosloinoi systemi TiN/ZrN, poluchennoi vakuumno-dugovim ispareniem. *J. Nano I electronnoi phyziki.* Vol. **2** pp. 6.
[5] Sobol O V, Andreev A A, Gorban V F, Stolbovoi V A, Pinchuk N V, Meilehov A A 2015 *Vliyanie tolshini bisloev TiN/ZrN na strukturu I svoistva mnogosloinih pokritii, poluchennih vakuumno-dugovim ispareniem*. Vol. **2** pp. 124-129.

Anomalous high-velocity outbursts ejected from the surface of tungsten microdroplets in a flow of argon-air plasma

I P Gulyaev[1,2,5], A V Dolmatov[2], P Yu Gulyaev[2], V I Iordan[1,3], M Yu Kharlamov[4] and I V Krivtsun[4]

[1] Khristianovich Institute of Theoretical and Applied Mechanics SB RAS, 4/1 Institutskaya str., Novosibirsk, Russia
[2] Ugra State University, 16 Chekhova str., Khanty-Mansiysk, Russia
[3] Altai State University, 61 Lenina pr., Barnaul, Russia
[4] Paton Electric-Welding Institute of the NAS of Ukraine, 11 Bozhenko str., Kyiv, Ukraine

E-mail: gulyaev@itam.nsc.ru

Abstract. For the first time, a phenomenon of high-velocity outbursts ejected from the surface of liquid tungsten microparticles in a flow of argon-air plasma under atmospheric pressure was observed. As tungsten particles sized 50 to 200 μm moved in a plasma flow, stratified radiating spheres up to 9 mm in diameter formed around such particles. The spheres were sources of high-velocity outbursts whose ejection direction coincided with the direction of the plasma flow. The velocity of the anomalous outbursts amounted to 3-20 km/s. In the outburst images, the distribution of glow intensity along outburst tracks exhibited a wavy decaying behavior with a wavelength of 5-15 mm. Possible physical factors that could be the cause of the phenomenon are discussed.

1. Introduction

The phenomenon of high-velocity discharges was first observed in our experiments aimed at the study of the formation and fractionation process of melt droplets, and at the measurement of the velocity and temperature of such droplets during a plasma-arc wire spraying process implemented for depositing coatings on a Plazer 30-PL-W spraying facility [1-3]. A specific feature of the method was using, as the spray material, a conducting wire fed into the plasma flow behind the exit plane of the plasma-gun double nozzle; this wire simultaneously served the function of the electric-arc anode. Thus, in the experiments, completely melted spray particles came off from the wire end into the plasma flow. The optical study was performed using an original spectral-brightness pyrometry diagnostics facility developed by researchers from Ugra State University (Khanty-Mansiysk) and Institute of Theoretical and Applied Mechanics SB RAS (Novosibirsk) [3-5]. The latter system comprised a digital video camera, which registered images of individual spray particles in the form of particle tracks, and a visible range spectrometer, which registered the integral optical emission spectrum of the two-phase jet.

[5] To whom any correspondence should be addressed.

2. Experimental

In our experiments, spraying of steel and tungsten wires 1.6 to 2 mm in diameter was performed. A detailed description of the operating conditions of the spray facility was given elsewhere [2, 3]. According to our estimates, the mass-mean temperature of the argon plasma was about 13 thousand degrees, and the velocity of the plasma-torch flow was greater than 2.5 km/s. The diagnostic system included an HD1-1312 (Photon Focus AG) video camera, which was equipped with a 1312x1082 CMOS photosensor, and an LR-T1 (Aseq Instruments) photospectrometer, with spectral range 300-1000 nm. The optical path of the video camera contained a bandpass optical filter with central wavelength 575 nm (FWHM 40 nm). The frames were registered at a rate of 55 fps. The objective lens of the video camera provided for a filming spatial scale of 51.9 µm/pixel. The focal plane of the video-camera optical system was coincident with that of the photospectrometer optical system. The region in which the measurements on the spectrophotometer were conducted was a 10-mm diameter circle located at the center of the camera frame.

A typical frame taken from the two-phase flow during the spraying of steel wire is shown in Fig. 1a. In the right-hand side of the frame, a bright optical emission due to the electric arc and due to the most heated region of the plasma is observed. Individual spray droplets had sizes ranging from 50 to 200 µm, and they formed tracks several millimeters long. Typical velocities of such droplets were 40-80 m/s. A similar pattern of a two-phase jet was also observed during the plasma spraying of other materials (alloys of the NiAl, NiCrBSi and NiCoCrAlY systems, ZrO_2 and Al_2O_3 oxides, and others) [3-6].

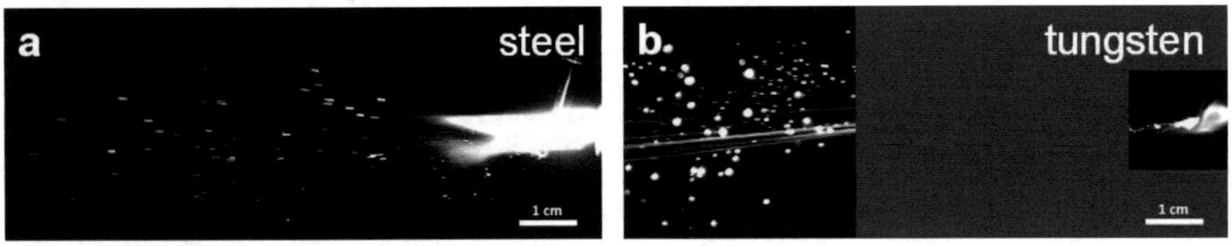

Fig. 1. Images of a two-phase plasma jet seeded with melted spray droplets: a – steel, b – tungsten. The exposure time of the frames is 30 µs (left frame) and 10 µs (right frame).

In the registered images of the two-phase plasma jet with moving tungsten droplets (see Fig. 1, b), two new phenomena were observed: (i) appearance of bright tracks crossing the entire frame in the direction of the plasma jet; 2) formation of bright luminous regions, shaped as spheres, around some of the tungsten droplets. In the registration of the jet with tungsten droplets the frame format was 800x800 pixel (the physical size of the observation region was 42x42 mm), and the exposure time was 10 µs. Thus, the velocity of an object that traversed the observation region in the exposure time exceeded 4.2 km/s (42 mm / 10 µs). Below, we give a description of observed phenomena and an analysis of their assumed nature.

3. Results

Numerous registered frames provide confirmation to the fact that the sources of the high-velocity outbursts were melted spray particles that moved in the plasma jet. For instance, a bright luminous region 1-2 mm in size is observed around such particles in Fig. 2. Simultaneously, the frame involves ordinary particle tracks about 100 µm wide and 0.5-1 mm long. Interestingly, all the high-velocity outbursts are directed along the direction of the local motion of plasma in the two-phase jet. A study of the integral emission spectrum of the jets revealed no additional bands or lines in experiments with tungsten in the working spectral region of the video system, 555 to 595 nm. Besides, from the continuous portion of the emission spectrum, 520 to 650 nm, dispersed-phase temperatures in the jet section located 9 cm downstream from the nozzle exit were determined by the spectral pyrometry method [7, 8]; for the flows with steel and tungsten droplets, those temperatures proved to be respectively 2730-3090 K and 3650-3890 K. The recorded frames show that, in this spectral region, the optical emission due to plasma is insignificant, and the measured temperatures can therefore be assumed to refer to the hottest dispersed-phase particles.

Fig. 2. A group of tungsten particles being the sources of high-velocity tracks. The scale is 51.9 μm /pixel, and the exposure time of the frame is 10 μs.

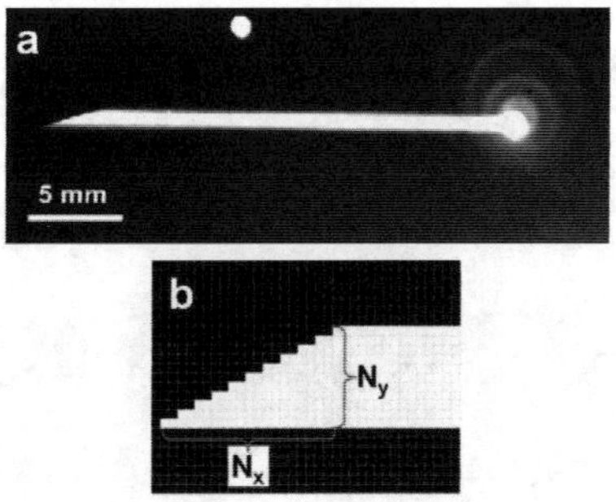

Fig. 3. Example of a radiating particle and a high-velocity outburst. The exposure time of the frame is 10 μs (a). Schematic representation of a track-tip raster image (b)

Consider in more detail an individual particle with a high-velocity outburst (Fig. 3). A region sized approximately 1.5 mm in immediate vicinity of the particle is occupied by the radiation which was most likely emitted by the brightest part of the gas discharge. Around the particle, a series of nearly spherical luminous contours about 2 mm, 4 mm, 6 mm, and 9 mm in diameter is observed; with increasing contour diameter, the contour brightness decreases in value while the contour center gets displaced toward the external region of the plasma flow (in Fig. 3, in upward direction).

As it is seen in all the images of the completely registered high-velocity outbursts, the track ends of these outbursts are all linearly skewed (see Fig. 3, a). The latter regularity can be explained by the operating algorithm of the CMOS photosensor, which accumulates photosignal even during period when a charge is being stored in the FD (floating-diffusion) zone waiting for read out [9]. In fact, the charge accumulation times in neighbor rows were different, and this circumstance enables the determination of the velocity of the moving objects in the frames from the inclination of their cross-sections. The clock frequency of the A1312 photodetector (Photon Focus AG) was 40 MHz; hence, the readout times for the signals in neighbor rows differed by $\tau = 25$ ns. A schematic representation of a raster image of the track end of an outburst is shown in Fig. 3, b with the aim of introducing designations for the track tip length, N_x, and for the track tip width, N_y (both quantities are measured in pixels). The velocity of an outburst can be calculated by the formula $V = (s \cdot N_x)/(\tau \cdot N_y)$, where s is the image spatial scale. For the track in Fig. 3a, the calculation performed using values $N_x = 52$, $N_x = 17$, s =51.9 μm, $\tau = 25$ ns yields for the outburst velocity a value of V =6350 m/s. Taking into account the fact that the length of the entire track is 22 mm, it can be argued that the frame in Fig. 3a was registered 3.5 μs after the outburst was formed. The calculated velocities of all other measured tracks predominantly fall into the range from 3 to 10 km/s, the maximum values reaching 20 km/s. It is worth noting that the obtained values well agree with the lower estimate for velocity 4.2 km/s.

Figure 4 (left) shows images of a several tracks of high-velocity outbursts; in particular, the upper track is the track shown in Fig. 3a. Evidently, the track widths of the various high-velocity outbursts range from 0.2 to 1 mm. Also, Fig. 7 (right) shows the distribution of brightness along the tracks. It is significant that the track brightness varies along the track length, and it exhibits a wave pattern. The wavelength of the waves varies from one track to another, and it falls into the range from 5 to 15 mm, the wavelength of the individual tracks increasing in magnitude with distance from the parent particle. In the upper graph of Fig. 4, a sawtooth structure with a period of 1-1.2 mm is seen. On close inspection, such a structure with a characteristic size of 4-5 mm can also be seen on the middle graph.

IOP Conf. Series: Materials Science and Engineering **110** (2016) 012057 doi:10.1088/1757-899X/110/1/012057

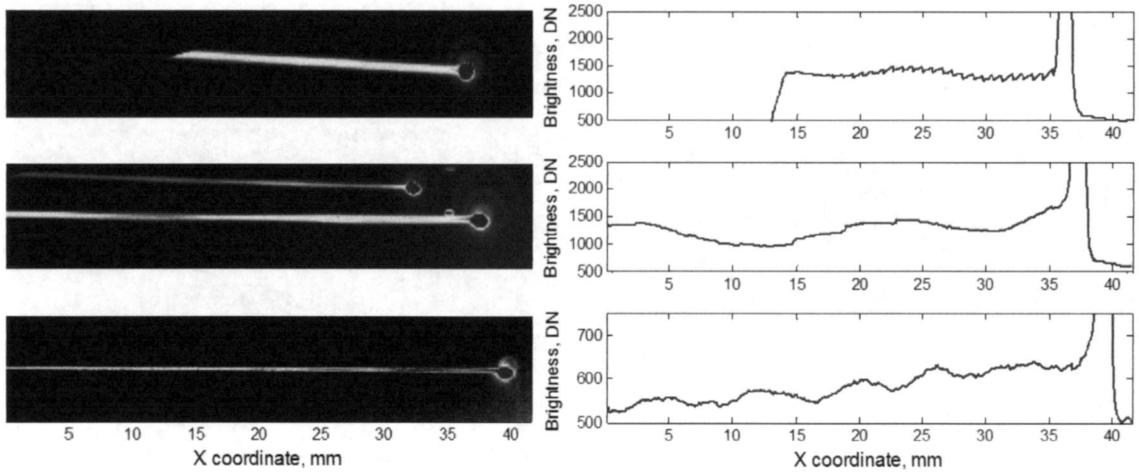

Fig. 4. Images of individual tracks of high-velocity outbursts (left) and the distribution of luminosity along the tracks (right).

Consider spherical discharges around tungsten particles which present sources of the high-velocity outbursts. Figure 5 shows, to identical scale, images of 17 different independent discharges. We assume that the various sizes (ranging from 1 to 10 mm) of the luminous regions in those images refer to various stages of their expansion process. It is seen that, irrespective of its size, a luminous region has a stratified structure that involves a series of bright rings. A high-velocity outburst arises after, in the expansion process, the spherical region reaches a size of 8-10 mm, the outburst always being ejected in the plasma flow direction (in the frames, from right to left).

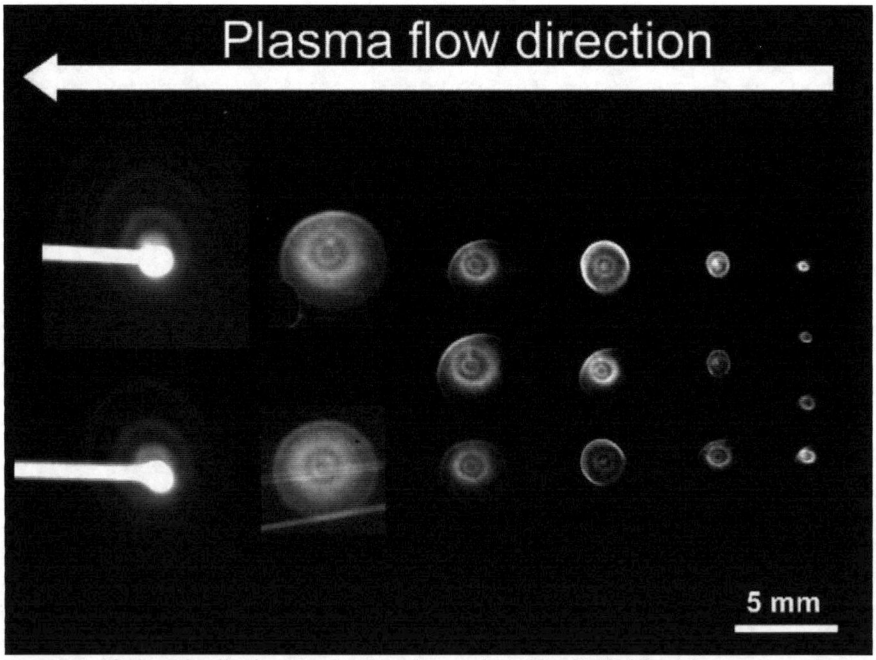

Fig. 5. Images of electric discharges in the plasma jet. The individual images are independent fragments taken from different frames; these images therefore bear no relation to each other.

4. Discussion

As it was noted above, the mechanism underlying the observed phenomenon still remains unclear. However, the obtained experimental data suggest that, in our experiments, we observed electric gas discharges followed by directed ejection of charged particles (electrons). Apparently, for realization of gas discharges, the formation of regions with strong electric fields is necessary. Under conditions of a dispersed plasma flow, such regions may arise due to the thermoionic emission of electrons from the surface of metal particles. As a result, a spray particle acquires a positive charge while the electron cloud around the particle turns out to be charged negatively. The occurrence of the observed phenomenon in the experiments with tungsten droplets can be explained by the high melting temperature of tungsten. Indeed, according to the Richardson-Dushman law, the thermoionic emission of electrons from the surface of melted W, Ta and Mo heated to a high temperature is 2-4 orders of magnitude more intense in comparison with the group of Cr, Fe, Co and Ni metals [10].

The nature of the stratified structure of the luminous spheres is probably similar to the nature of the alternating bright and dark regions in the near-cathode zones of glow discharges [11]. For instance, O.A. Nerushev et al. [12] examined, under laboratory conditions, a steady stratified spherical glow discharge 5-10 cm in diameter that burned in rarefied atmosphere of molecular gases with an admixture of high-molecular substances. The factors that cause the occurrence of directed high-velocity outbursts still remain unclear; however, it can be hypothesized that those factors are most probably related with the electron motion.

5. Conclusions

A new phenomenon was observed while studying the interaction of melted tungsten droplets with low-temperature argon-air plasma using an optical system based on a digital video camera. Around each of the metal particles, whose sizes ranged from 50 to 200 μm, there formed a spherical luminous region sized up to a few millimeters; this region exhibited a stratified structure. After the expansion of such a sphere to a size of 8-10 mm, there occurred an outburst of spray material from the surface of the metal particle in the direction of the plasma flow. The velocity of the outbursts amounted to 3-20 km/s, and the radiation intensity in the outburst wakes exhibited a wavy behavior.

We relate the observed phenomenon to the high rate of the thermoionic emission of electrons from the surface of liquid tungsten particles and to the formation of a space-charge region around such particles. The discovered phenomenon calls for further study; we believe that, as a result of such a study, the exact nature of the phenomenon will be established.

6. Acknowledgements

The authors express their gratitude to Dr. O.I. Demyanov and Dr. V.I. Tkachuk (Paton Electric-Welding Institute) for the fruitful joint experiments on the plasma-arc spray facility.

This work was supported by the Russian Foundation for Basic Research (project Nos. 14-08-90428 and 15-48-00100) and National Academy of Sciences of Ukraine (project No 06-08-14).

References

[1] Kharlamov M Yu, Krivtsun I V, Korzhik V N and Petrov S V 2011 Formation of liquid metal film at the tip of wire-anode in plasma-arc spraying *Paton Weld. Journal* **12** 2-6

[2] Gulyaev I P, Dolmatov A V, Kharlamov M Yu, Gulyaev P Yu, Iordan V I, Krivtsun I V, Korzhik V N and Dem'yanov O I. 2015 Plasma-arc wire spraying: Optical investigation of process phenomenology *J. Therm. Spray Technol* (in press).

[3] Gulyaev I P, Gulyaev P Yu, Korzhik V N, Dolmatov A V, Iordan V I, Krivtsun I V, Kharlamov M Yu and Demianov A I Experimental investigation of process of plasma-arc wire spraying *Paton Weld. Journal* **3/4** 36-41

[4] Dolmatov A V, Gulyaev I P and Jordan V I, The optical control system of dispersed-phase properties in thermal spray process 2015 *IOP Conf. Ser.: Mater. Sci. Eng.* **81(1)** 012041.

[5] Vaschenko S P, Gulyaev I P, Gulyaev P Yu, Dolmatov A V and Kuzmin V I Express-Control of

Particles Motion and Heating in Thermal Spray 2015 *Proc. 17th Int. conf. «Hardening, coating and repair technologies: Theory and Practice» (Saint Petersburg, Russia)* 61-68 (in Russian)

[6] Gulyaev I P, Ermakov K A and Gulyaev P Yu 2014 New high-speed combination of spectroscopic and brightness pyrometry for studying particles temperature distribution in plasma jets *European researcher* **3/2** 564-570.

[7] Magunov A N 2010 Spectral pyrometry of objects with a nonuniform temperature *Technical Physics. The Russian Journal of Applied Physics* **55** 991-995.

[8] Dolmatov A V, Gulyaev I P and Imamov R R 2014 A spectral pyrometer for the control of temperature in thermal synthesis processes *Ugra State University Bulletin* **33** 32-42 (in Russian).

[9] Meynants G 2014 Global shutter imagers for industrial applications *Proc. of the SPIE 9141 Optical Sensing and Detection III* 914108.

[10] *Handbook of Ion Sources* 1995 ed B. Wolf (CRC Press) p 560

[11] Raiser Y P 1991 *Gas Discharge Physics* (Springer) p 449

[12] Nerushev O A, Novopashin S A, Radchenko V V and Sukhinin G I 1998 Spherical stratification of a glow discharge *Phys. Rev. E* **58** 4897-4902.

Control of dispersed-phase temperature in plasma flows by the spectral-brightness pyrometry method

A V Dolmatov[*1] **, I P Gulyaev**[2]**, P Yu Gulyaev**[1] **and V I Iordan**[3]

[1] Ugra State University, Khanty-Mansiysk, Russia
[2] Institute of Theoretical and Applied Mechanics SB RAS, Novosibirsk, Russia
[3] Altai State University, Barnaul, Russia

E-mail: adolmatov@bk.ru

Abstract. In the present work, we propose a new method for measuring the distribution of temperature in the ensembles of condensed-phase particles in plasma spray flows. Inter-relation between the spectral temperature of the particles and the distribution of camera brightness signal is revealed. The established inter-relation enables an in-situ calibration of measuring instruments using the objects under study. The spectral-brightness pyrometry method was approbated on a Plazer plasma-arc wire spraying facility at the Paton Institute of Electrical Welding (Ukrainian Academy of Sciences, Kiev) and on the Thermoplasma 50-1 powder spraying facility at the Institute of Theoretical and Applied Mechanics (Russian Academy of Sciences, Siberian Branch, Novosibirsk). The work was supported by the Russian Foundation for Basic Research (Grants Nos. 14-08-90428 and 15-48-00100).

1. Introduction

Studies of multiphase plasma flows present a vital task due to the broad spectrum of relevant scientific and applied problems. Interaction of a dispersed phase with plasma is under study in such fields as combustion, jet propulsion, and high-temperature synthesis of materials [1, 2]. Control of heat- and mass-transfer processes of plasma with condensed particles is a key problem in the development of metal- and ceramic-coating spraying processes [3].

In diagnostics of multiphase plasma flows, the aggressive medium and the non-uniformity of the high-temperature flow leave no alternative to optical pyrometry methods [4]. However, using such methods is largely restricted by the short observation time of objects in the rapidly moving dispersed phase. The brightness pyrometry approach calls for its further development toward the control of moving objects. Spectral pyrometry methods have low temperature sensitivity, which circumstance strongly restricts the capabilities of those methods in constructing hyper-spectral images with a time constant of order 100 μs [5, 6]. Thus, new pyrometry methods need to be developed for solving the problem of measurement of the temperature of condensed-phase particles in plasma spray flows.

[*] To whom any correspondence should be addressed

2. Principles of spectral-brightness pyrometry

The basis for the spectral-brightness pyrometry (SBP) is the idea of *in-situ* calibration of imaging means to be performed through comparison of statistical characteristics of the distributions of digital monochrome camera signals and digital spectrometer signals registered while measuring one and the same ensemble of dispersed-phase particles. To this end, the camera measurement volume and the spectrometer measurement volume in the dusted plasma flow are to be made coincident (Fig. 1).

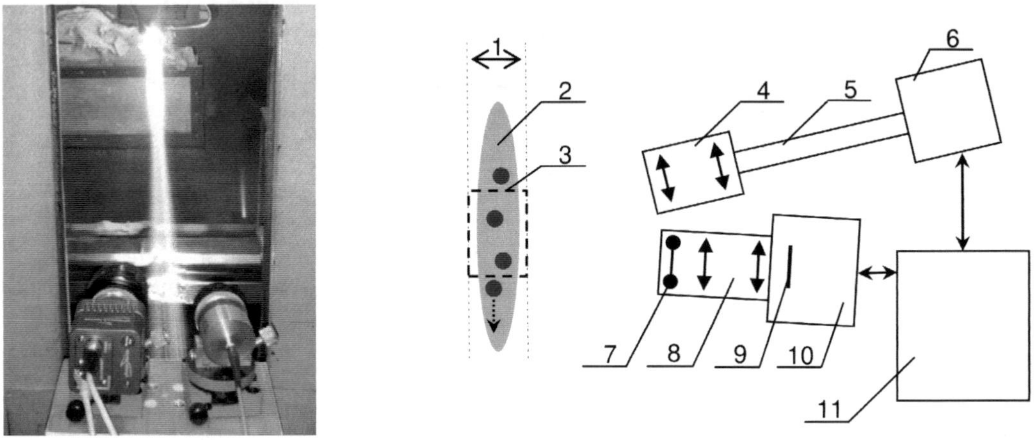

Fig. 1. Instrumentation (a) and diagram (b) of a facility for measuring the characteristics of plasma-flow dispersed phase by the SBP method (1 – depth of field of the measuring devices; 2 – dusted plasma flow; 3 - measurement volume common to the camera and to the spectrometer; 4 – spectrometer lens; 5 – optical fiber; 6 – digital spectrometer; 7 – narrow-band optical filter with central wavelength λ_b; 8 – camera lens; 9 – 2D image photo sensor; 10 – digital video camera; 11 – PC with installed signal-processing software)

Powder materials intended for spraying coatings normally exhibit good uniformity of their optical properties. Providing that the spectral emissivity of the material in the working temperature range exhibits no sharp variations, then, for a narrow spectral region, one can use the grey-body approximation and assume the emissivity factor of dispersed-phase particles, or their coefficient of radiation, to be unknown constant $\varepsilon = const$. Then, under the condition that some reference thermodynamic temperature T_s can be put in correspondence to each reference signal level b_s of the brightness pyrometer, for an arbitrary signal level b the temperature in the Wien region can be found by the formula

$$T = \left(\frac{1}{T_s} + \frac{\lambda_b}{C_2} \ln \frac{b_s}{b} \right)^{-1}, \quad (1)$$

where λ_b is the wavelength of the brightness-pyrometer light filter, μm; and $C_2 = 14388$ μm·K [7].

We assume that, as the detector in the digital spectrometer, a charge-accumulating multi-element photo sensor is used. We denote the signal measured by the spectrometer and corrected for distortions in the electronic and optical paths of the instrument as $q(\lambda, T)$. During observation of an object with uniform temperature T, linearization of its optical emission spectrum plotted in the variables $x = C_2/\lambda$ and $y = \ln(q \cdot \lambda^4)$:

$$y = -\frac{1}{T} x + const \quad (2)$$

allows identification of the shape of the object and determination of its thermodynamic temperature from the slope of the plotted curve:

$$T = -(dy/dx)^{-1}, \tag{3}$$

Consider a digital image of the measurement volume of the camera and spectrometer. Let each image element be characterized by an arbitrary signal level b_k, where k is the number of the element in the 2D photo sensor of the camera (here, linear indexation is used). Then, according to the radiation law of real bodies in the Wien approximation, some temperature T_k corresponds to this signal:

$$b_k = A \cdot \varepsilon \cdot \frac{C_1}{\lambda_b^4} \exp\left(-\frac{C_2}{\lambda_b \cdot T_k}\right) = A_1 \cdot \exp\left(-\frac{x_b}{T_k}\right). \tag{4}$$

In formula (4), A is the sensitivity of the digital camera; $C_1 = 37418$ W·μm^4/cm^2; $x_b = C_2/\lambda_b$; and $A_1 = const$ at fixed parameters of the camera measuring tract. The spectrometer integrates radiation coming from the entire projection of the measurement volume; hence, the integral signal of the spectrometer q_{sum} can be expressed in terms of temperatures that correspond to the respective image elements:

$$q_{sum}(\lambda) = \sum_k q(\lambda, T_k) = \frac{B \cdot \varepsilon \cdot C_1}{\lambda^4} \cdot \sum_k \exp\left(-\frac{C_2}{\lambda \cdot T_k}\right) = \frac{B_1}{\lambda^4} \sum_k \exp\left(-\frac{x}{T_k}\right). \tag{5}$$

In formula (5), B is the sensitivity of the spectrometer, and $B_1 = const$ at all parameters of the spectrometer measuring tract being fixed. Following the passage to auxiliary coordinates (2), relation (5) assumes the form

$$y(x) = \ln(B_1) + \ln\left(\sum_k \exp\left(-\frac{x}{T_k}\right)\right). \tag{6}$$

On the basis of the spectral pyrometry method, we define the reference temperature T_s as the temperature of a grey emitter whose thermal emission spectrum has a value of its derivative at the point x_b identical to that in spectrum (6). The temperature T_s may happen to not belong to the set of image-element temperatures; yet, this temperature can be expressed in terms of this set with the help of formulas (3) and (6):

$$\frac{1}{T_s} = \frac{\sum_k \frac{1}{T_k} \cdot \exp(-x_b/T_k)}{\sum_k \exp(-x_b/T_k)}. \tag{7}$$

Using formula (4), we replace the set of temperatures $\{T_k\}$ in formula (7) with the levels $\{b_k\}$ of the image-element signals registered by the digital camera. Following this, we obtain the following formula for determining the reference signal level b_s that corresponds to temperature T_s:

$$b_s = \exp\left(\frac{\sum_k b_k \cdot \ln b_k}{\sum_k b_k}\right). \tag{8}$$

Then, the set of temperatures $\{T_k\}$ for the image elements in the measurement volume can be determined by using formula (1). Evidently, here the unknown value of the emissivity ε of the material exerts no influence on measurement results.

For adaptation of the spectral-brightness pyrometry method to the measurement of the distribution of dispersed-phase temperature in a plasma flow, we adopt the following model. At fixed operating conditions of the spray facility, the spraying process is assumed to be a steady one, with the expected local values of controlled dispersed-phase characteristics being time-independent quantities. Here, there is no need in tracking the motion and heating of individual particles along their trajectories;

instead, it just suffices to collect sufficient statistical data from a great number of particles at each point of the measurement-volume. The latter allows us to evaluate the spatial distribution of dispersed-phase temperature from the local mean value of this temperature over a coordinate. The stationary model of the spray flow as applied to the entire measurement volume makes it possible to represent the whole ensemble of dispersed-phase particles with a sufficiently large volume of sampled data for estimating the reference signal level $\tilde{b}_s$ and the reference temperature $\tilde{T}_s$. The exposure time of the digital camera τ_b and the exposure time of the spectrometer τ_q can be chosen different and, for the same particle ensemble under observation, it suffices that the registration of a frame series be time-matched with the signal integration process in the spectrometer. The latter matching allows reaching a high signal-to-noise ratio in measured spectra of dispersed-phase particle ensembles by setting the time τ_q to about 1 second and by adopting a value of τ_b on the order of 10 μs for registration of rapid-particle tracks in the images of a frame series with total observation time $t = \tau_q$. In estimating the reference signal level $\tilde{b}_s$ and reference temperature $\tilde{T}_s$, the fact should be taken into account that the object under observation is not an image element of the measurement volume but, instead, a particle with relative area

$$F = \frac{\pi \cdot d^2}{4} \tag{9}$$

and equivalent signal level of stationary particle S_p [8]. That is why in formulas (7) and (8) we have to pass from summation over image elements to summation over ensemble particles:

$$\frac{1}{\tilde{T}_s} = \frac{\sum_n d_n^2 \cdot \frac{1}{T_n} \cdot \exp(-x_b / T_n)}{\sum_n d_n^2 \cdot \exp(-x_b / T_n)}, \tag{10}$$

$$\tilde{b}_s = \exp\left(\frac{\sum_n d_n^2 \cdot S_{p_n} \cdot \ln S_{p_n}}{\sum_n d_n^2 \cdot S_{p_n}} \right). \tag{11}$$

In formulas (10) and (11), n is the number of an ensemble particle that was identified in the frame series, and d_n and S_{p_n} are respectively the relative diameter and the equivalent signal level of the stationary particle with number n.

3. Approbation of the method

In the original implementation of the SPB method, the following instruments and means were used: an HD1-1312-1082 digital camera (Photon Focus, Switzerland), an LR-1T digital spectrometer (ASEQ Instruments, Canada), a synchronization circuit of the camera and spectrometer designed around an Arduino microcontroller assembly, a SL-series set of narrow-band optical filters (PhotoOptic-Filters, Russia), and MATLAB software [9]. The measuring complex was approbated while doing research on the plasma-arc wire spraying setup at the Paton Institute of Electrical Welding (Ukrainian National Academy of Sciences, Kiev) and on the determination of performance characteristics of technological processes for the Thermoplasma 50-1 plasma powder spray facility designed around the PNK-50 plasma torch with inter-electrode insert (Institute of Theoretical and Applied Mechanics, Novosibirsk) [10, 11].

After registration of a frame series and optical emission spectrum of the dispersed-phase, MATLAB means were used to reveal the possible presence of plasma-induced lines in the working spectral region of the digital camera. If such lines were revealed, then the optical filter was changed for another one, and the registration was performed once again. Automated change of the spectral region of the digital camera is possible in using, in the complex, KURIOS-WB1(/M) acousto-optic or liquid-crystal tunable filters with a re-adjustment time of 40 ms.

The thermal emission spectrum over the working region in variables (2) was used to evaluate the dispersed-phase reference temperature in the measurement volume. For suppressing the effect due to the transmittivity of plasma-flow medium on the registered signal level, digital filtration and morphologic-analysis methods were employed to remove from measured data the variable background with a heterogeneity scale one order of magnitude greater than the particle size. Then, individual tracks were recognized in the images; from the tracks, such characteristics of the particles as their position in the flow, velocity, direction of motion, diameter, and equivalent signal level of stationary particle were determined [12]. Afterwards, for the ensemble of the particles, the reference signal level was calculated by formula (11), and the temperature of all the particles, by formula (1). Selection of the particles over the coordinate of the spray flow projection has allowed us to obtain local sampled datasets for dispersed-phase properties and construct the maps of mean values and dispersions (Fig. 2).

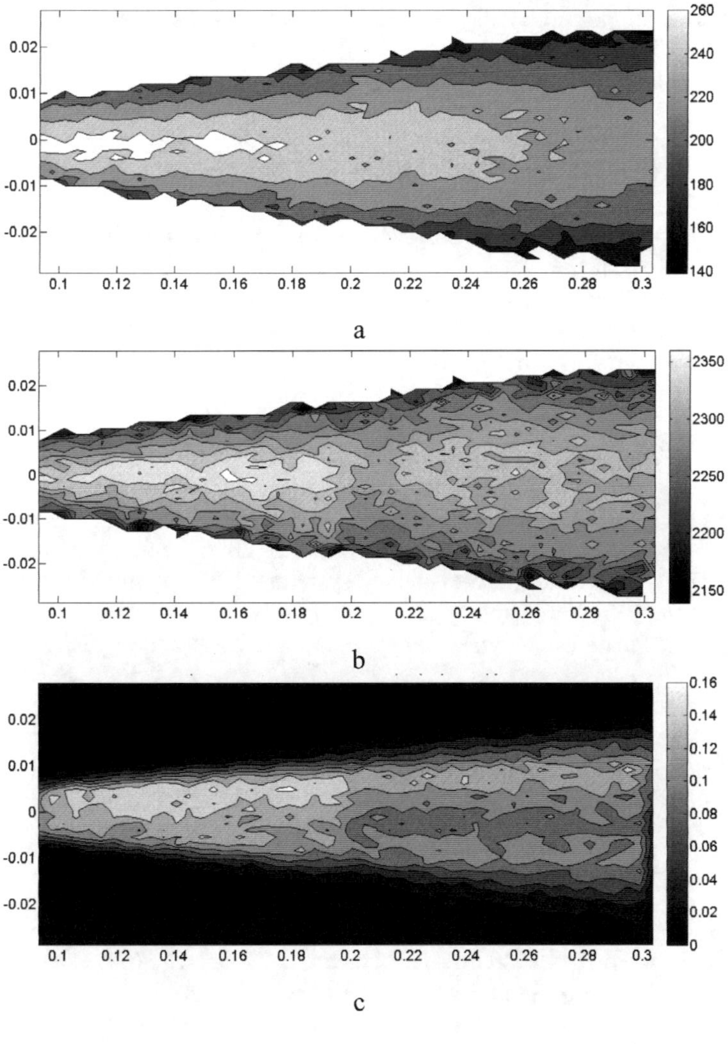

Fig. 2. Maps of dispersed-phase properties in the spray flow of the PNK-50 plasma torch (a – mean particle velocity, m/s; b – mean particle temperature, K; c – relative concentration of particles, %).

The volume of sampled data in the evaluation procedure for $\tilde{b}_s$ and in the plotting of the maps of dispersed-phase properties amounted to $5 \cdot 10^5 \div 10^6$ particles that were identified in 10000 image frames. A parallel data-processing algorithm implemented in MATLAB software has ensured an analysis rate of 3500-5000 particles per second. In the course of the analysis, the inaccuracy in measuring the dispersed-phase temperature in plasma flow by the spectral-brightness pyrometry method was estimated to be at a level of 0.75-1.5%.

4. Conclusions

The spectral-brightness pyrometry method can be used to control the spatial dynamics of the thermodynamic temperature in fast processes. The sensitivity of the method is comparable with the sensitivity of the brightness pyrometry approach. Moreover, the inter-relation between the signals in the brightness and spectral channels permits an in-situ calibration of measuring instruments while observing the objects under study. The latter makes the developed pyrometric method an invariant one with respect to the emissivity values of substances in condensed state. An additional analysis of the spectrometer signal is used for verification of the thermal nature of the optical emission in the working band of the camera optical filter, and it helps in excluding inadequate measured data. The inaccuracy in determining the temperature of individual particles by the spectral-brightness pyrometry method proved to be less than 2%. This work was supported by the Russian Foundation for Basic Research under Grants Nos. 14-08-90428 and 15-48-00100.

5. References

[1] *Self-propagating High-temperature Synthesis* 2001 ed. Evstigneev V V (Novosibirsk: Nauka) 284 (in Russian)

[2] Puzryakov A F 2003 *Theoretical Foundations of plasma spraying* (Moscow : Publishing House of the MSTU. NE Bauman) 360 (in Russian)

[3] *Thermal spray: Textbook* 2007 ed. Baldaev L H (Moscow: Market DS) 344

[4] Dubnishscev Yu N, Arbuzov V A, Belousov P P and Belousov P Ya 2003 *Optical methods for flow research* (Novosibirsk: SUI) 420 (in Russian)

[5] Magunov A N 2009 *Spectral Pyrometry (Review)* Instruments and Experimental Techniques **4 (52)** 451-472

[6] Mauer G, Vassen R and Stover D 2008 Study on detection of melting temperatures and sources of errors using two-color pyrometry for in-flight measurements of plasma sprayed particles *Proc. of Int. Thermal Spray Conf.*

[7] Karimova F F, Orlovskaya S G, Kalinchak V V, Shkoropado M S 2006 *Investigation of temperature fields of radiating objects* Physics of aero-dispersed systems **43** 39-45 (in Russian)

[8] Gulyaev I P, Ermakov K A, Gulyaev P Yu *New High-Speed Combination of Spectroscopic And Brightness Pyrometry For Studying Particles Temperature Distribution In Plasma Jets* 2014 European researcher **3-2 (71)** 564-570

[9] Boronenko M P, Gulyaev I P, Gulyaev P Yu, Demyanov A I, Dolmatov A V, Jordan V I, Korzhyk V N, Krivtsun I V and Kharlamov M Yu 2014 *Estimate of Dispersed Phase Velocity and Temperature in Plasma-Arc Spraying Jets* Fundamental Research **11-10** 2135-2140 (in Russian)

[10] Gulyaev I P, Gulyaev P Yu, Korzhik V N, Dolmatov A V, Jordan V I, Krivtsun I V, Kharlamov M Yu and Demjanov A I *Experimental investigation of process of plasma-arc wire spraying* The Paton Welding Journal 36-41

[11] Vaschenko S P, Gulyaev I P, Gulyaev P Yu, Dolmatov A V, Kuzmin V I *Express-Control of Particles Motion and Heating in Thermal Spray* 2015 Proc. 17th Int. conf. «Hardening, coating and repair technologies: Theory and Practice» (Saint Petersburg, Russia) 61-68 (in Russian)

[12] Dolmatov A V, Gulyaev I P, Jordan V I 2015 *The optical control system of dispersed phase properties in thermal spray process* IOP Conference Series: Materials Science and Engineering 012041

Effect of potassium iodide on luminescent and photovoltaic properties of organic solar cells P3HT-PCBM

N Kh Ibrayev[1], D A Afanasyev and K A Zhapabaev

Institute of Molecular Nanophotonics, Academician E.A. Buketov Karaganda State University, Karaganda, Kazakhstan

E-mail: niazibraev@mail.ru

Abstract. It has been investigated spectral-luminescence properties of polymer films, doped with potassium iodide (KI). Using of KI didn't lead to the gradual changes of optical density of polymer films and the range of band gap semiconductor polymer P3HT. The fluorescence intensity of P3HT decreased and changed by use of KI. Using of 1% KI in polymer leaded to decrease of fluorescence lifetime. Influence of heavy atom on photovoltaic effect of organic solar cells has been investigated. 1% of KI in polymer film leaded to decrease of Isc and slightly decrease of Uoc. Investigation shows that magnetic field does not affect on photovoltaic properties of cells P3HT-PCBM. Magnetic field increased of open circuit voltage and short circuit current of solar cells with 1% of KI. Study of electrical impedance of cells revealed the magnetic sensivity of solar cells with KI additives. The lifetime of free charge carriers increased in the magnetic field for solar cells with KI additives.

1. Introduction

Polymer solar cells based on poly [3-hexylthiophene] (P3HT) and fullerene derivative [6,6]-phenyl-C61-butyric acid methyl ester (PCBM) could be an alternative to the traditional silicon solar cells. Efficiency of polymer cells based on P3HT-PCBM has reached 5% [1]. One of the possible ways to increase efficiency of polymer cells is increase of singlet-triplet conversion in polymers. It is reachable by changing spin-orbit interaction, by using effect of heavy atom. There is a lot of information about influence of heavy atom on photophysics organic molecules [2]. But there are not many studies of influences of external heavy atom effect on the process of transformation of solar energy into electrical energy in polymer cells.

Addition of external heavy atom in semiconductor polymers could change not only the process of photophysical reactions but also probability of creation of free charge carriers in polymer composites. However, the level of influence of external heavy atom on the electronic processes in semiconductor polymers is not thoroughly investigated.

In the present work, we report about the investigation of influence of non-organic impurity (potassium iodide) in polymer on process of transformation of solar energy into electrical in solar cells based on polymer-fulleren mix P3HT+PCBM which caused their use as a model system for investigation of electronical processes in organic solar cells.

[1] Author to whom any correspondence should be addressed.

2. Experemental

Polymer films based on P3HT with the addition of PCBM or inorganic impurities KI were prepared for studying of the spectral-luminescent properties of polymer composites. The weight ratio was 1: 1 and 100: 1 respectively for the components in composites for mixtures P3HT-PCBM and P3HT -KI.
The glasses with conductive ITO (indium tin oxides) layer with area of 2x1.5 cm were used for the assembly of solar cells. The resistivity of ITO is 10 Ω*cm. The transmission of glass with the ITO is not less than 72 % at the maximum absorption of polymer P3HT. Solar cell was fabricated with structure ITO/PEDOT:PSS/P3HT:PCBM/Al following with method of described in detail in [4, 5].
Registration the absorption spectra of the films was carried out using a spectrophotometer Cary 300. The excitation and luminescence spectra of the films were measured on a fluorescence spectrophotometer Cary Eclipce.
The current-voltage characteristics (CVC) of the cells was measured with light cell by a conductive layer (ITO) radiation of the xenon lamp power of 100 mW/cm^2 on simulator sunlight Solar Cell IV Measurement (Photo Emission Tech Inc.). The measurements were performed at 25 ° C. The real (Z') and imaginary (Z") electrical resistance of solar cells was measured at impedancometry Z500PRO in potentiostatic mode for a given zero constant potential.
Neodymium magnet is used to measure the effect of an external magnetic field on the current-voltage characteristics and the impedance of cells. The magnetic field of magnet is 0.55 T.
The kinetics of fluorescence of polymer films was measured using a pulsed spectrofluorometer with picosecond resolution and recording mode with time-correlated photon counting (Becker & Hickl). Fluorescence excitation of samples was carried out using a pulsed semiconductor laser with a wavelength of λ_{gen} = 488 nm and full width at half maximum of τ = 80 ps.

3. Result and discussion

Spectral-luminescent properties were investigated for the P3HT polymer films and P3HT-PCBM, PCBM-KI polymer composites. Absorption spectrum of the pure P3HT is shown in Figure 1. The broad absorption band with a maximum at 520 nm was observed for the P3HT.
Additional absorption band appears after addition PCBM in the film. This band is in the ultraviolet region of the spectrum with a maximum around 230 nm (Figure 1, a, curve 2). The optical density of the P3HT absorption decreases 2-fold compared neat a P3HT film. This is due to a decrease in amount of polymeric material in the film. Adding KI does not lead to significant changes in the optical density of the composite film P3HT-KI. There is some change in shape of the absorption spectrum of P3HT – KI film (Figure 1, a, curve 3) in comparison with the film of the P3HT (curve 1).

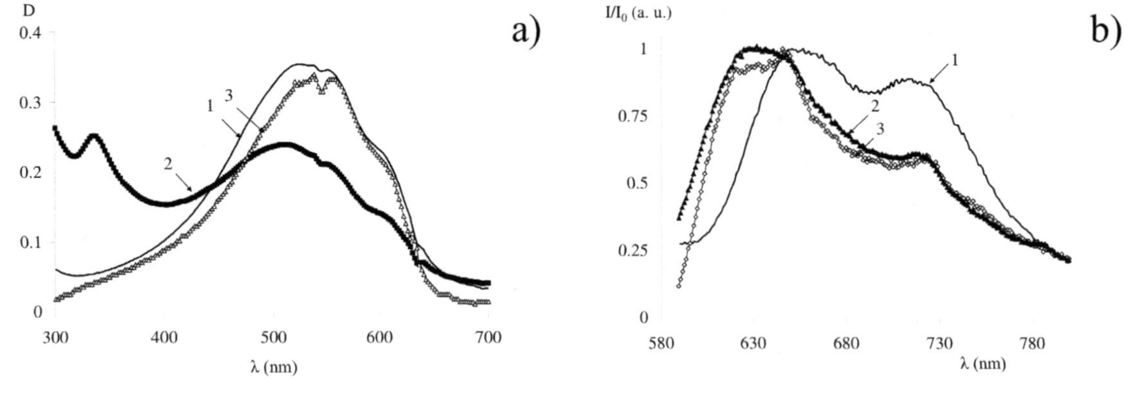

Figure 1. The absorption spectra (a) and normalized fluorescence spectra (b) of the polymer films P3HT (1), P3HT–PCBM (2) and P3HT–KI (3)

Fluorescence spectra of the P3HT, P3HT-PCBM and P3HT-KI films are shown in Figure 1, b. Broad band with a maximum at 650 nm and the shoulder at 720 nm (Figure 1, b, curve 1) is observed in the fluorescence spectrum of the P3HT film. Addition of PCBM molecules leads to a significant decrease in fluorescence intensity (10 times) and to change of the spectral curves form (Figure 1, b, curve 2). The fluorescence spectrum of the P3HT-PCBM film is shifted to shorter wavelengths compared with the fluorescence spectrum of the P3HT film. Decreasing fluorescence intensity of more than 10 times occurs for the P3HT-KI film compared with fluorescence of P3HT films. Short-wavelength shift of the fluorescence spectrum was observed for the P3HT-KI film similar to shift of the fluorescence spectrum of the P3HT-PCBM film (Figure 1, b, curve 3).
The band gap energy (E_g) of P3HT with organic and inorganic impurities was determined from the absorption spectra of the films (Figure 2). Results of calculation of E_g polymer composites shows in Table 1. These results show that the organic and inorganic impurity is not significantly in changing of the band gap of the composite films.

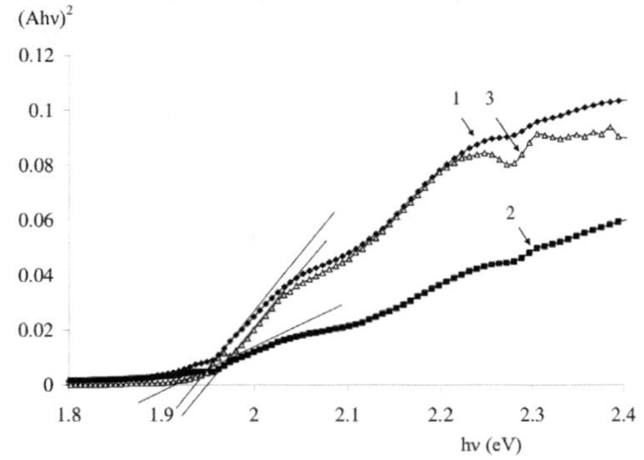

Figure 2. Plot of $(Ahv)^2$ versus hv for all the samples along the sequence: 1) P3HT; 2) P3HT - PCBM; 3) P3HT - KI

Figure 3 shows the fluorescence kinetics of P3HT, P3HT-PCBM and P3HT-KI films. Registration signal is at the maximum of the fluorescence spectrum at 650 nm wavelength. Figure 3 shows, that addition of 1% KI in the polymer leads to the same fluorescence quenching as addition of 50% fullerene in P3HT film. The lifetime of fluorescence was calculated from the linear part of the logarithmic curve (Figure 3, b). The results of the fluorescence lifetime measurement are shown in Table 1. The largest decrease in the lifetime of the film was observed for P3HT-KI. Decrease of the lifetime of the compared with decrease P3HT-KI film is observed for the film P3HT-PCBM.

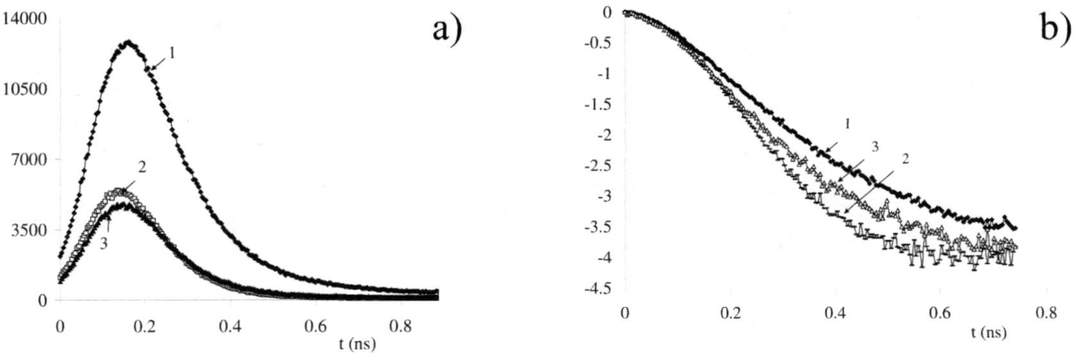

Figure 3. Fluorescence kinetics *I(t)* (a) and logarithmic curves of fluorescence kinetics (b) for the P3HT film (1), P3HT- PCBM films (2) and P3HT- KI films(3)

Table 1. The size of the band gap, fluorescence lifetime and the lifetime of the free charge carriers in polymer films doped with organic and inorganic impurities

Material of films	Bandgap (eV)	τ_{Fl} (ns)	τ_y (s)
P3HT	1,93	0,228	
P3HT–PCBM	1,92	0,184	0,012
P3HT–KI	1,94	0,145	0,012
P3HT–KI (B = 0,55 T)	-	-	0,014

For the P3HT-PCBM film change in fluorescence lifetime associated with the formation of electron-hole pairs at the interface between P3HT and PCBM. For a sample of P3HT-KI quenching of fluorescence due to the dynamic fluorescence quenching through KI. The iodine atom (I) is active not only in quenching of luminescence, but also it increases the probability of intersystem crossing from the singlet (S_1) to the triplet (T_1) state [3].

Solar cells have been made for determination of the free charge carriers lifetime in the polymer films. It was made planar heterojunction P3HT-PCBM instead of bulk heterojunction. KI added to the polymer P3HT. P3HT- KI weight ratio was 100: 1.

Heavy atom can lead to increased intersystem crossing in organic molecules [2]. Dynamic fluorescence quenching film P3HT + KI indicates growth intersystem crossing $S_1 \rightarrow T_1$ in the film of P3HT + KI. An additional factor pointing to the increase of the concentration of molecules in the T_1 state can be a magneto-sensitive processes of transformation of light energy into electrical energy. Figure 4 shows CVC of solar sells with planar heterojunction based on P3HT – PCBM and P3HT+KI – PCBM.

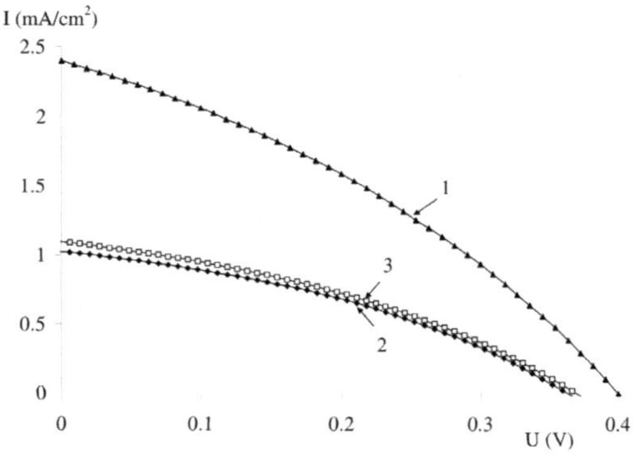

Figure 4. Influence of the external magnetic field on the current-voltage characteristics of solar cells: 1) P3HT- PCBM; 2) P3HT+KI – PCBM (B=0 T); 3) P3HT+KI – PCBM (B=0,55 T)

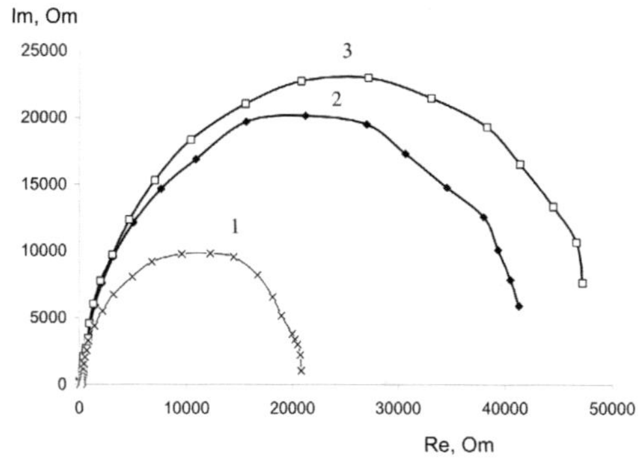

Figure 5. Impedance spectra of solar cells: 1- P3HT- PCBM; 2 - P3HT+KI – PCBM (B=0 T); 3- P3HT+KI – PCBM (B=0,55 T)

Studies of the influence of magnetic field on the CVC of solar cells showed absence of influense of MF on the planar heterojunction of P3HT-PCBM. Analogous results are observed in [6, 7]. It indicate about absence of the formation stage of the polaron pair in the P3HT polymer and effective transformation of excitons into free charge carriers at the interface of the heterojunction. Addition of 1% KI in the film leads to a significant reduction of I_{sc} and a small decrease of the U_{oc}. Influence of an external magnetic field on the current-voltage characteristics of the solar cells is fixed on the cell containing the impurity of KI. Growth of the U_{oc} and I_{sc} occurs in the magnetic field (Figure 4, curve 3).

Measurement of the electrical impedance of cells shows the effect of an external magnetic field on the electrical resistance of the cells with the addition of KI. The effect of the external magnetic field is not fixed on the cells without KI. The value of both the real part and the imaginary part of the electrical resistance of the P3HT+KI – PCBM cell increases in an external magnetic field (Figure 5). Thus, surprisingly, that the lifetime of the free charge carriers in the magnetic field increases (Table 1).

4. Conclusion

Spectral-luminescent properties of polymer films doped the heavy atom of potassium iodide (KI) have been investigated. Addition of KI in polymer films doesn't lead to significant changes in optical density polymer films and doesn't change the bandgap of the polymer. Fluorescence spectrum of P3HT film have broad band with the maximum at 650 nm and the shoulder at 720 nm. Addition of KI in polymer film leads to decreases of intensity of the fluorescence and change the form of the spectral curve. Investigation of the fluorescence kinetics showed a decrease in fluorescence lifetime in the films of P3HT-PCBM (τ_{fl} = 0,184 ns). This is associated with the formation of electron-hole pairs at the heterojunction interfaces. Quenching of fluorescence intensity and decrease of fluorescence lifetime (τ_{fl} = 0,145 ns) occurs in the film P3HT + KI. It is associated with the dynamic fluorescence quenching polymer P3HT in using KI. The increase in the probability of intersystem crossing from the singlet (S_1) to the triplet (T_1) state is observed in the film P3HT + KI.

Influence of the external magnetic field on the current-voltage characteristics of the solar cells with the addition of a heavy atom indicates an increase intersystem crossing between S_1 and T_1 electronic state of the polymer P3HT. Addition of KI in the amount of 1% leads to a decrease in Isc and slightly decreases Uoc. The magnetic field is not affect on the characteristics of photovoltaic P3HT-PCBM solar cell. U_{oc} and I_{sc} is increase in the magnetic field for solar cells with the addition of 1% KI.

Measurement of the electrical impedance of cells shows the same magnetic sensitivity of cells with the addition of KI. The lifetime of the free charge carriers increases in a magnetic field. These results indicate a significant contribution of excited triplet states of the P3HT polymer in the process of transformation of light energy into electrical energy in an organic solar cell

References

[1] Ma W, Yang C, Gong X, Lee K, Heeger A J 2005 *Adv. Funct. Mater.* **15** 1617

[2] Solov'ev K N, Borisevich E A Intramolecular heavy-atom effect in the photophysics of organic molecules 2005 *Phys. Usp.* **48** 231

[3] Vladimirov Y A, Potapenko A J 1989 *Physico-chemical basis of photo-biological processes* (Moscow: Executive. wk.)

[4] Bisquert J, Garcia-Belmonte G, Munar A, Sessolo M, Soriano A, Bolink H J 2008 *Chemical Physics Letters* **465** 57

[5] Garcia-Belmonte G, Munar A, Barea E M, Bisquert J, Ugarte I, Pacios R 2008 *Organic Electronics* **9(5)** 847

[6] Xu Z, Hu B 2008 *Adv. Funct. Mater.* **18** 2611

[7] Hu B, Yan L, Shao M 2009 *Adv. Mater.* **21** 1500

Synthesis of nanostructured films of zinc oxide and a study of its structural and luminescent properties

B R Ilyassov[1, 2] and N Kh Ibrayev[1*]

[1]Institute of Molecular Nanophotonics, E.A. Buketov Karaganda State University, 100028 Karagandy, Kazakhstan.
[2]Laboratory of Solar Energy, PI NLA, Nazarbayev University, 010000 Astana, Kazakhstan

E-mail: niazibrayev@mail.ru

Abstract. ZnO nanorod array were synthesized on sol-gel prepared ZnO seed layer over quarts substrates by hydrothermal method. X-ray diffraction results show that nanorods are the typical hexagonal wurtzite structure without any other new phase and preferentially oriented along the c-axis perpendicular to the substrate surface. The scanning electron microscopy (SEM) of the surface of the ZnO film showed vertical ZnO nanorods with hexagonal crossection. The luminescence spectrum of ZnO nanorod array consisted of a narrow UV band and a broad band in visible region of spectrum and its temporal characteristics were measured.

1. Introduction

ZnO is a multifunctional semiconductor material having direct wide band gap of 3.37 eV and high excitonic binding energy of 60 meV at room temperature. These properties make it a promising semiconducting material for transparent electronics, solar cells and other optoelectronic devices [1-3]. Synthesis of ZnO nanostructures such as one-dimensional rods, wires, tubes etc., has got increasing attention for its specific properties and for the fabrication of nano-scale devices. Many different techniques were used for the synthesis of nanostructures; they include vapor-liquid-solid (VLS) epitaxy [4], chemical vapor deposition (CVD) [5], pulse laser deposition (PLD) [6], thermal decomposition [7], hydrothermal synthesis [8], etc. The hydrothermal method is a more simple and cost effective process for the synthesis of ZnO nanorods. It has very low synthesis temperature below 100°C, catalyst free process and is suited for the uniform synthesis of ZnO nanorods over large area of the substrates. The structural and optical properties of the nanorods obtained by hydrothermal process exhibits strong dependence on conditions such as seed layer crystallinity, choice and concentration of surface-active agent used and reaction time for synthesis. Some authors [9] had reported the effect of annealing substrates on the density of ZnO nanorod array. Other authors [10] had studied the influence of addition of NaOH on the diameter of ZnO nanorods prepared by hydrothermal method. Liu et al. [11] had synthesized ZnO nanorods on sputtered ZnO seed layer.

In this work we have synthesized ZnO nanorod array on sol-gel prepared ZnO seed layer [1] over quarts substrate. The structural properties of the nanorods were studied by X-ray diffraction (XRD). The nanorods in array showed c-axis orientation perpendicular to the substrate and had structure of wurtzite. The surface morphology of the nanorods was studied by Scanning electron microscopy

(SEM). The surface of the ZnO film exhibits vertical ZnO nanorods with hexagonal crossection. The optical properties of the samples were studied by UV-Vis spectrophotometer. The reflectance of the nanorod showed an absorption edge at the band gap of ZnO. The photoluminescence of nanorods showed UV emission at room temperature.

2. Experiment procedure

ZnO nanorods arrays were synthesized on quartz substrates by hydrothermal method [11], initially on the surface of the cleaned quartz substrates were coated thin seed layer of zinc oxide ZnO by sol-gel method [12]. The sol is prepared by dissolving zinc acetate dehydrate in isopropyl alcohol (with concentration of 20 grams per liter of solvent) at 20 °C with intensive stirring on a magnetic stirrer for 1-2 hours. Preparation of the clear solution at high concentrations of zinc acetate was achieved by adding a few drops of lactic acid. The resulting solution remains clear for a long time, it allows to obtain high-quality seed layers.

Uniform distribution of the sol on the substrate surface is achieved by dripping several drops of the solution onto the substrate mounted on a chuck of a spin-coater (VTC-200 Spin Coater, MTI), and then the substrate spun at ~ 3000 rpm for 30 seconds. Later, the substrate is placed on a hot oven and kept at a temperature of 300 °C for 10 minutes. Final annealing at a temperature of 450 °C for one hour leads to the formation of thin transparent layers on the substrate surface. This method allow us to synthesize ZnO multi-layered films on the substrate with area of 1-1.5 cm^2, having uniform properties and a high optical transmission factor in the visible range.

Hydrothermal synthesis of ZnO nanorods was carried out in the aqueous solution of zinc nitrate dehydrate and hexamine ($C_6H_{12}N_4$) in a glass beaker; the substrates with the seed layer of ZnO were placed at the fluoroplastic holder into the upright position. The synthesis was carried out in the temperature range 90-95 °C for 3 hours with intensive stirring. The resulting samples were washed with deionized water and dried.

3. Results and discussion

Studies of as-grown ZnO films on the scanning electron microscope (FEI Quanta-250) exhibited that the ZnO film is an array of hexagonal nanorods predominantly oriented perpendicular to the substrate (Fig. 1). X-ray patterns of as-grown ZnO films exhibited that ZnO nanorods have a hexagonal wurtzite structure with unit cell parameters a = 3.260 Å, c = 5.214 Å and predominantly oriented along the direction of the crystallographic c-axis (Fig. 2).

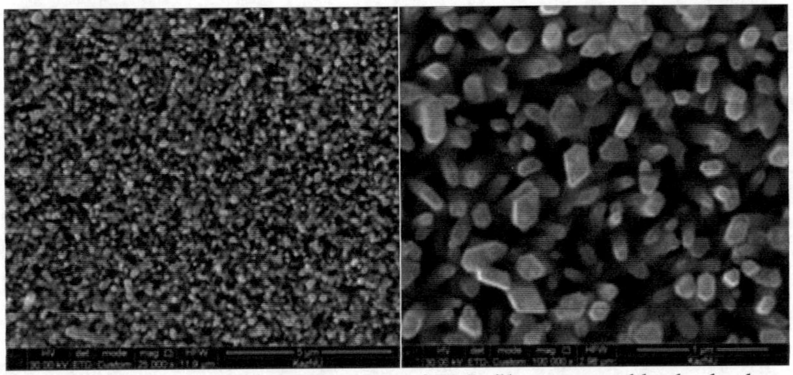

Figure 1 - SEM image of surface microstructure of ZnO film prepared by hydrothermal method at different magnification.

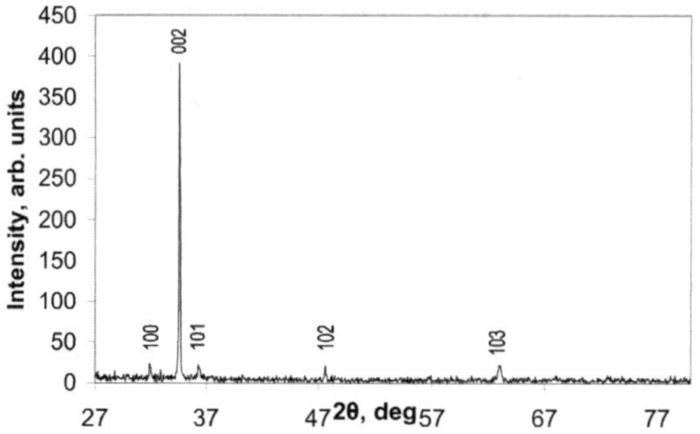

Figure 2 - X-ray pattern of ZnO films prepared by hydrothermal method.

The luminescent properties of zinc oxide nanorods array were investigated. The luminescence spectra of ZnO nanorods array was measured on a highly sensitive fiber-optic spectrophotometer AvaSpec-2048 at room temperature, photoexcitation of samples was performed the third harmonic of a neodymium laser LCS-DTL-374QT (λ = 355 nm, τ = 7 ns, E = 5 mJ). The luminescence spectra of the films ZnO (Fig. 3) exhibited two bands: a narrow band in the UV with a maximum at 370-372 nm and a broad band in the visible region. The broad band in the visible region had several peaks with a maximum of 524 nm in the green region, 572 nm and weak peaks in the orange and red region. UV band has exciton nature, the main luminescence centers are free and bound excitons [13]. Visible luminescence of zinc oxide nanostructures related to point defects mainly with oxygen vacancies on the surface and inside the crystal, for the less defective ZnO nanostructures it covers mainly the green region of the spectrum [14].

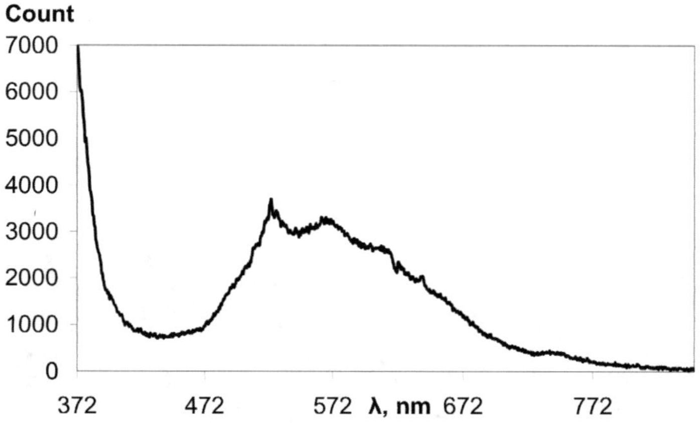

Figure 3 - The photoluminescence spectra of ZnO films prepared by hydrothermal method.

The temporal characteristics of luminescence of ZnO in subnanosecond and nanosecond time ranges were measured (Fig. 4 and 5). The samples were excited by a picosecond pulse semiconductor laser (BDL-SMC) with repetition rate of 50 MHz and pulse duration of 50 ps. The detection of luminescence carried out by a picosecond spectrometer using mode time-correlated single photon counting. All curve of kinetics of UV luminescence decay is not an exponential function (Fig. 4). However, the initial part of the kinetic decay curve (up to 10 ns) is described by an exponential. Lifetime calculated from this exponential part of curve is ~ 0.2 ns. All curve of kinetics of luminescence decay in the visible region is also not an exponential function (Fig. 5). Lifetimes of the emission bands with maxima at 524 nm and 572 nm calculated from the initial part of the exponential curve decay are the same and approximately 60 ns.

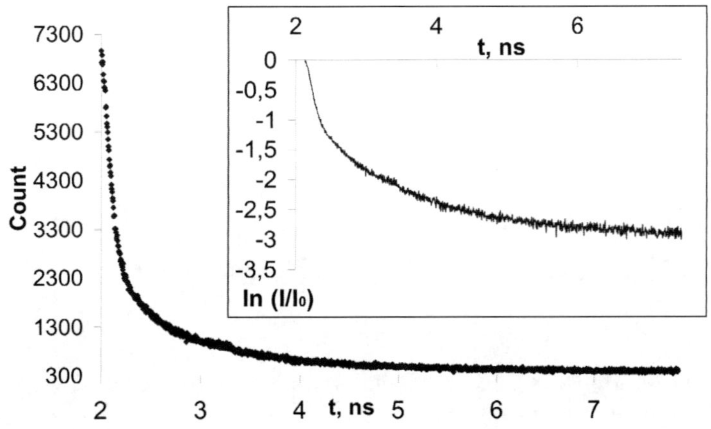

Figure 4 - The ultraviolet luminescence decay of zinc oxide films prepared by hydrothermal method.

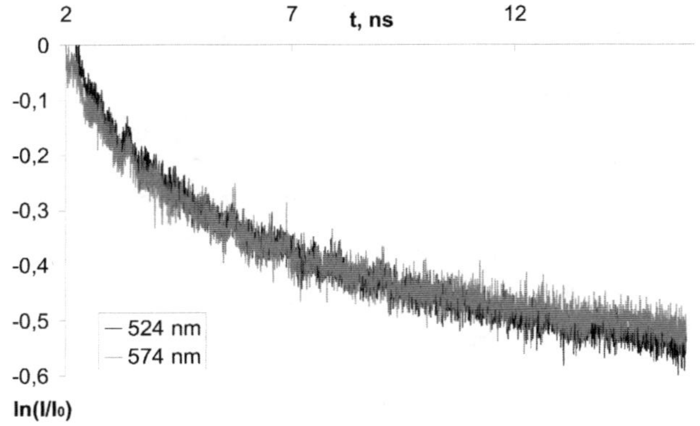

Figure 5 - Kinetics of luminescence decay of zinc oxide films in the visible region prepared by hydrothermal method.

References

[1] Yi G, Wang C and Park W 2005 *J. Semicond. Sci. Technol.* **20** S22
[2] Law M, Greene L E, Johnson J C, Saykally R, Yang P D 2005 *J. Nature Materials* **4** 455
[3] Park S H, Kim S H, Han S W 2007 *J. Nanotechnology* **18** 055608
[4] Huang M H, Wu Y, Feick H, Tran N, Weber E, Yang P 2001 *J. Advanced Materials* **13** 113
[5] Wang G Z, Wang Y, Yau M Y, To C Y, Deng C J 2005 *J. Materials Letters* **59** 3870
[6] Bae C H, Park S M, Ahn S E, Oh D L, Kim G T, Ha J S 2006 *J. Applied Surface Science* **253** 1758
[7] Shakti N and Gupta P S 2010 *J. Optoelectronics and Advanced Materials-RC* **4** 662
[8] Vayssieres L 2003 *J Advanced Materials* **15** 464
[9] Ma T, Guo M, Zhang M and Wang X D 2007 *J. Nanotechnology* **18** 035605
[10] Wu W Y, Ting J M, Kung W Y 2010 *J. Electrochem. Soc.* **157** K71
[11] Baruah S and Dutta J 2009 *J. Sci. Technol. Adv. Mater.* **10** 013001
[12] Znaidi L 2010 *J. Materials Science and Engineering B* **174** 18
[13] Meyer B K, Alves H, Hofmann D M and et al. 2004 *J. Phys. Stat. Sol. B* **241** 231
[14] Rodnyi P A and Khodyuk I V 2011 *J. Opt. Spectrosc.* **111** 776

Effect of nonequilibrium hydrogen release in the ultrafine-grained Zr-1Nb alloy under the electron beam exposure

E N Stepanova[1], G P Grabovetskaya[2], V S Sypchenko[1], I P Mishin[2], V N Kudiiarov[1] and T N Melnikova[1]

[1] Tomsk Polytechnic University, 30 Lenin Avenue, Tomsk, Russia
[2] Institute of Strength Physics and Materials Science of Siberian Branch of Russian Academy of Sciences, 2/4, Akademicheskii Avenue, Tomsk, Russia

E-mail: enstepanova@tpu.ru

Abstract. The evolution of structural and phase state and hydrogen release from the ultrafine-grained hydrogenated zirconium Zr-1Nb alloy during vacuum annealing and electron beams exposure were studied. The use of electron beam irradiation for hydrogen degassing is shown to decrease the temperature of active hydrogen release by 100–200 K and/or reduce the time required for hydrogen degassing from the alloy to concentrations corresponding to technical standards.

1. Introduction

Recently works on finding materials that can accumulate and store hydrogen are actively developed. It is known that hydride-forming metals and alloys in the ultrafine-grained (UFG) state can be effective solid-state hydrogen storages. However, the extraction of hydrogen from such materials generally occurs at elevated temperatures [1]. Therefore, the task is to reduce the temperature of the active hydrogen release from solid-state hydrogen storages. One of the possible problem solutions can be using the effect of nonequilibrium hydrogen release for hydrogen degassing. Electron beam irradiation of metals and alloys is known to activate hydrogen diffusion and desorption from the target material [2]. This leads to a significant lower-temperature shift of the peak in the curve for the temperature dependence of the dehydrogenation rate in metallic materials as compared to conventional thermal heating of the specimens in dynamic vacuum [3]. To activate the hydrogen release from the materials both continuous [3] and pulsed electron beams [4] can be used. In our opinion, pulsed electron beams irradiation is of particular interest. During pulsed electron beam irradiation the substantial heating occurs only near the surface of the material and the temperature in the bulk of the material varies slightly [5]. The latter is particularly important for the preservation of the UFG structure wherein recrystallization and grain growth are observed at temperatures of (0.3–0.4)T_{melt} [6]. Taking into account the foregoing, the purpose of this work was to carry out comparative studies of the effect of vacuum annealing and pulsed and continuous electron beams exposure on the evolution of structural and phase state and hydrogen release from the UFG Zr-1Nb-H alloy.

2. Experimental Procedure

The experiment was performed using commercial Zr-1Nb alloy (E110 grade). The UFG state in the alloy was obtained by the method combining preliminary hydrogenation and hot plastic deformation by pressing to 75–80 % at the temperature of 873 K [7]. Preliminary hydrogenation of the alloy to concentrations of 0.23 % (hereinafter, the hydrogen concentration is indicated in weight percent) was carried out by annealing in gas media using a Siverts' device at the temperature of 853 K. The hydrogen concentration in the samples was measured by RHEN602 gas analyzer with ±0.0001 % accuracy.

Electron-microscopic investigation of thin foils was performed with the use of EM-125K transmission electron microscope. The size of structure elements was determined using the relevant micrographs by means of the secant method. The volume fractions and lattice parameters of the phases were obtained with the accuracy of ± 1% and 0.0001 nm, respectively, using Shimadzu XRD7000 diffractometer with Cu-Kα radiation source.

Test samples for annealing and irradiation were cut from billets by the electrospark method. Before testing the surface of samples were subjected to mechanical grinding and electrolytic polishing.

Hydrogen degassing of the samples was carried out in three ways: by annealing in a dynamic vacuum and by pulsed and continuous electron beams irradiation. Annealing in a dynamic vacuum was performed using a Siverts' device at 873 K. The heating rate to the annealing temperature was 10 K/s. Irradiation of the samples was conducted using «Solo» set-up with pulsed electron source [5]. Samples were irradiated by pulsed electron beam with the energy of (7–10) keV, pulse duration of 50 μs and frequency of 2 Hz for (15–30) minutes. Beam diameter was equal to 10 mm. Continuous electron beam irradiation of the investigated samples was performed at the special set-up [4] at the following conditions: electron beam energy E = 30 keV, current on the sample I = 35 μA. The sample was fixed on the heater and was placed in a vacuum chamber and then was preheated to temperatures of 628, 653, and 685 K for obtaining the hydrogen release dependences. The electron beam was turned on after stabilization of required temperature. Temperature and current were measured from the sample surface directly in the process of irradiation. Hydrogen release intensities depending on irradiation time was recorded with the use of bundled software, permitting to control the sweep of MH-7304 mass spectrometer and execute the linear heating of the samples with the rate of 1–10 K/s (in this experiment heating rate was 1 K/s).

3. Results and Discussion

The typical electron micrographs and corresponding electron diffraction patterns of the UFG structure of the Zr-1Nb-0.23H alloy are shown in Fig. 1. It is evident that an entangled deformational contrast is observed and some structure elements are poorly distinguished in the bright-field image of the structure (Fig. 1, a).

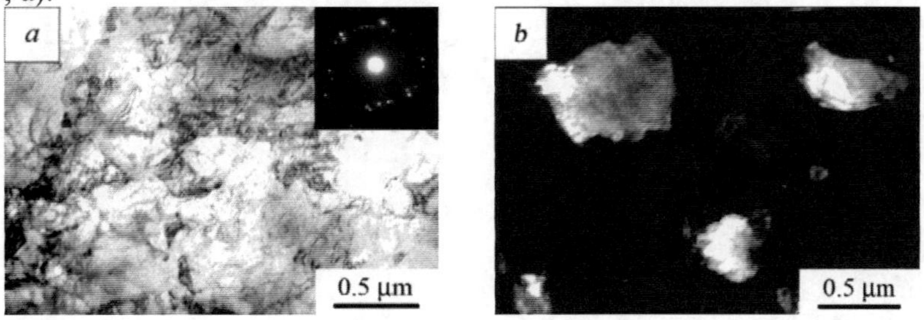

Figure 1. UFG Zr-1Nb-0.23H alloy microstructure, (a) bright field image, (b) dark field image

The electron diffraction patterns of the structure obtained for the area of 1.6 μm^2 show an appreciable number of reflections uniformly distributed over a circle (Fig. 1, a). This is indicative of the presence of a large number of grain-subgrain structure elements in a unit volume and of their substantial misorientation. The average size of the grain-subgrain structure elements of the Zr-1Nb-0.23H alloy determined from the dark-field image (Fig. 1, b) was 0.4 μm.

X-ray structure analysis demonstrated the Zr-1Nb-0.23H alloy in the UFG state has complicated phase composition. Alloy contains both main Zr_{α} phase and secondary phase Nb(Zr) (bcc) (reflections 1 – (100) and 2 – (200), Fig. 2, a). Particles of the secondary phase are located in the bulk and at grain boundaries of Zr_{α} phase. Furthermore, ZrH_86 and ZrH2_225 hydrides were detected in the UFG Zr-1Nb-0.23H alloy (reflections 3 – $(111)_{225}$, 4 – $(111)_{86}$, and 5 – $(200)_{86}$). It can be seen that $(220)_{86}$, $(311)_{225}$, and $(222)_{225}$ hydride reflections are superimposed on the (110), (103), and (112) reflections of the Zr_{α} phase. Apparently, some portion of hydrogen is present in solid solution of the Zr_{α} phase. This is indicated by the increase of its lattice parameters in Zr-1Nb-0.23H alloy after compaction in comparison with the initial fine-grained state.

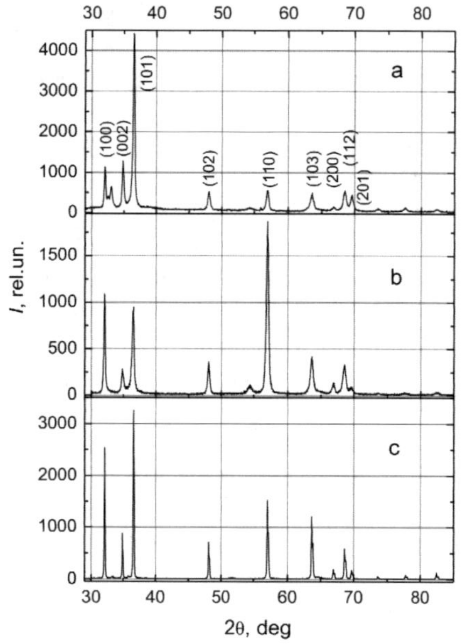

Figure 2. Portions of diffraction patterns for zirconium alloy samples: Zr-1Nb-0.23H (1); Zr-1Nb-0.115H (2); Zr-1Nb-0.00532H (3)

Temperature and time dependences of the hydrogen release from the UFG Zr-1Nb-0.23H alloy samples during dynamic vacuum annealing are shown in Fig. 3. It can be seen that hydrogen release from the alloy begins after heating to a temperature of ~623 K (Fig. 3, a).

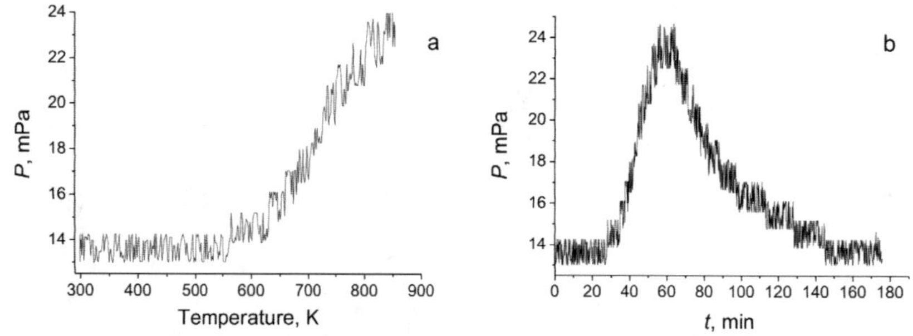

Figure 3. Temperature (a) and time (b) dependences of hydrogen release from the samples of the UFG Zr-1Nb-0.23H alloy during dynamic vacuum annealing

Maximum rate of hydrogen release is observed in the first 10 minutes after reaching the temperature of 873 K and during exposure at this temperature the rate of hydrogen release gradually decreases (Fig. 3, b). Measurement of residual hydrogen concentration showed its decrease from 0.23 to 0.115 % as a result of annealing by an indicated mode.

Table 1 shows the values of residual hydrogen concentrations in the Zr-Nb-H alloy after annealing at dynamic vacuum and pulsed electron beam exposure. It can be seen that after using modes of dynamic vacuum annealing hydrogen concentration in the Zr-Nb-0.23N alloy remains substantially greater than the concentration corresponding to the technical standards for this alloy (0.002–0.001 %).

At the same time use of pulsed electron beam allows not only to decrease the temperature and/or time of hydrogen degassing from the UFG alloy in comparison with annealing at dynamic vacuum, but in some modes to reduce the hydrogen concentration in alloy to the values closed to technical standards. In addition hydrogen concentration in the alloy decreases both in irradiated and non-irradiated parts of the sample. It should also be noted that after annealing at dynamic vacuum and pulsed electron beam irradiation the remaining hydrogen, apparently, is present mainly in the precipitations of the hydrides. This is indicated by the intensity growth of the (101), (110), and (103) reflections of the Zr_{α} phase, on which hydrides reflections are superimposed (Fig. 2, b).

Table 1. Hydrogen concentration in the UFG Zr-1Nb-0.23H alloy after annealing and electron beam irradiation

State of the UFG Zr-1Nb-H alloy		After annealing and in the irradiated part of the sample, %	In the non-irradiated part of the sample, %
Vacuum annealing	803 K, 60 min	0.178	
	823 K, 60 min	0.147	
	853 K, 60 min	0.087	
Pulsed electron beam irradiation	803 K, 15 min	0.0165	0.105
	823 K, 15 min	0.0266	0.119
	853 K, 15 min	0.0074	0.1
	803 K, 30 min	0.0036	0.0536
	823 K, 30 min	0.00187	0.00572
Continuous electron beam irradiation	657 K, 53 min	0.0227	
	698 K, 45 min	0.00532	

Four modes were used in the study of the continuous electron beam irradiation effect on hydrogen release from the UFG Zr-1Nb-0.23N alloy. Typical dependences of the intensity of hydrogen release from the alloy during continuous electron beam exposure on the temperature of preliminary (before the start of the electron beam irradiation) sample heating and the irradiation time are shown in Fig. 4.

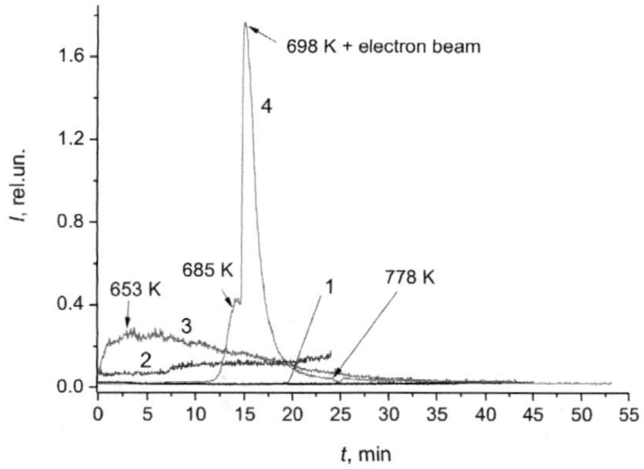

Figure 4. Typical dependences of the hydrogen release intensities on the preheating temperature and irradiation time: 1 – continuous electron beam exposure without preheating; 2 – heating to 628 K; 3 – preheating to 628 K and exposure of continuous electron beam; 4 – preheating to 685 K and exposure of continuous electron beam

It can be seen that hydrogen degassing performed by means of electron beam irradiation without preheating does not lead to any significant changes in the hydrogen content in the alloy (mode 1, Fig. 4, curve 1). In the case of the preliminary sample heating to 628 K, its temperature under electron beam exposure increases to 665 K and hydrogen starts to release from alloy (mode 2, Fig. 4, curve 2). However, the intensity of hydrogen release is insignificant. The preheating temperature rise by 25 K leads to a sharp increase in the intensity of hydrogen release from the alloy with a slight changing the sample temperature (from 653 to 657 K) after electron beam switching (mode 3, Fig. 4, curve 3). During further irradiation for 45 min sample temperature does not change and hydrogen release intensity gradually decreases (practically to zero). Hydrogen concentration in alloy after irradiation by mode 3 decreases 10 times (to 0.0227 %). Gradual sample heating to the initial temperature of hydrogen release (685 K) was used for the irradiation mode 4. As curing time at the temperature increases the intensity of hydrogen release begins to decrease (Fig. 4, curve 4). Beginning of electron beam exposure leads to a rise in sample temperature to ~ 698 K and significant growth of hydrogen release intensity (Fig. 4, curve 4). During the further 5 min of irradiation the decrease of hydrogen release intensity practically to zero is observed, and for subsequent 20 min of irradiation, it is not changed. For the total time of heating and electron beam exposure in the indicated mode (~ 45 min) hydrogen concentration in the UFG alloy decreases from 0.23 to 0.00532 %. In this case no hydride precipitations are observed in the material (Fig. 2, c).

Structure examinations of Zr-1Nb-0.23H alloy revealed that during vacuum annealing and exposure of pulsed and continuous electron beams growth of grain-subgrain UFG structure elements occurs. Thus, after annealing at 853 K, 60 min the average size of structure elements increased to ~1.3 μm. During irradiation by pulsed and continuous electron beams the rate of growth of the grain-subgrain UFG structure elements increased in comparison with annealing. In the case of pulsed electron beam exposure at 853 K the average size of grain-subgrain structure elements increased to ~1.2 μm for 30 min. Continuous electron beam irradiation at 698 K for 60 min results in increase of average size of structure elements to 0.67 μm. At the same time alloy annealing in dynamic vacuum for 60 min even at indicated temperature does not lead to change of the average size of structure elements. Observed fluctuating of growth rate of UFG structure elements of the Zr-1Nb-H alloy under electron beams irradiation is apparently connected with the activation by electron beam not only hydrogen diffusion, but also diffusion of other elements.

4. Conclusion

Thus, as exemplified by the hydrogenated to a concentration of 0.23% Zr-1Nb alloy, both pulsed and continuous electron beam irradiation can be used for hydrogen degassing from the metallic hydride-forming materials. It was found that exposure of pulsed and continuous electron beams allows to decrease the temperature of the active hydrogen release by 100–200 K and/or reduce the time required for hydrogen degassing from the alloy to concentrations corresponding to technical standards.

Acknowledgements
This work was supported by the Russian Foundation for Basic Research (project No. 15-08-03823).

References

[1] Il'in A A, Kolachev B A, Nosov V K, and Mamonov A M 2002 *Vodorodnaya tekhnologiya titanovykh splavov* (Moscow: MISIS) p 392
[2] Tyurin Yu I and Chernov I P 1999 *Dokl. Ros. Akad. Nauk* **3** 328
[3] Nikitenkov N N, Kudryavtzeva E N, Chernov I P, Tyurin Yu I, Grabovetzkaya G P, Melnikova E N 2008 *Proc. 9 Int. Conf. on Modification of Materials with Particle Beams and Plasma* Flows (Tomsk) p 316
[4] Stepanova E N, Grabovetskaya G P, Mishin I P, Chernov I P, Teresov A D 2013 *Izvestia visshih uchebnih zavedenii. Phizika* **56** 95
[5] Koval N N, Grigoryev S V, Ivanov Yu F, Ovcharenko V E, Kolubaeva Yu A, Teresov A D 2009 *IEEE Trans. Plasma Sci.* **37** 1998
[6] Kolobov Yu R, Valiev R Z, Grabovetskaya G P, Zhilyaev A P, Ivanov M B, Dudarev E F, Kashin O A, Ivanov K V, Naidenkin E V 2001 *Grain boundary diffusion and properties of nanostructured materials* (Novosibirsk: Nauka) p 232
[7] Grabovetskaya G P, Stepanova E N, Ratochka I V, Naidenkin E V, Lykova O N *2013 Inorganic Materials: Applied Research* **4** (2) 92

The influence of the preparation conditions on structure and optical properties of solid films of graphene oxide

E Seliverstova[1], N Ibrayev, R Dzhanabekova and V Gladkova
Institute of Molecular Nanophotonics, Karaganda 100028, Kazakhstan

E-mail: genia_sv@mail.ru

Abstract. In this study, we investigated the physico-chemical properties of graphene oxide monolayers at the interface water-air. Monolayers were formed by the spreading of dispersion of graphene oxide in acetone and THF. It was found than graphene monolayers are in the "liquid" state on the surface of subphase. Monolayers were transferred onto solid substrates according to Langmuir-Blodgett (LB) method. SEM images show that the films have an island structure. The films obtained from acetone solutions are more uniform, which makes them more promising in terms of their use as conductive coatings. Absorption spectrum of graphene LB films exhibits a broad band in the ultraviolet and visible region of the spectrum. The optical density of the film obtained from acetone solution is greater than the optical density of the film prepared from THF. In the visible region of the spectrum both films have high transparency.

1. Introduction

Currently, search and active use of new alternative energy sources in many developed countries accepted as vital, strategic resources needed to ensure the future development of these economies. However, despite of the promising results, alternative energy sources have not yet reached the level of optimal responsiveness of expectations of mass consumer. The most widely used silicon photovoltaic elements are very expensive. Therefore search of cheaper dye sensitized solar cells is carried on.

Intensive research worldwide is in pursuit of cheaper production methods of photovoltaic devices without compromising performance. Given that graphene production costs may be cheaper than other carbon nanomaterials, it's single or few layered structure is nearly fully transparent, their conductivity, plus graphenes can be easily chemically altered, their physicochemical properties tuned and their 3-d structure manipulated, means that graphene possess many useful properties for the production of photovoltaic devices. The technology could be realized by casting graphene sheets into composite devices using Langmuir-Blodgett casting techniques.

Widespread and low-cost graphene can be used in the counter electrodes instead of platinum. In the works of [1, 2] was showed that graphene had high specific surface area and high mobility of electrons, having a higher oxidation potential than Pt. The wide variety of existing forms of graphene can improve the necessary parameters of counter electrodes. For example, reduced graphene oxide [3], has a surface lattice defects, which implies a higher catalytic activity than the fully reduced defect-free graphene [4-6]. Additionally, graphene production costs are cheaper than other carbon nanomaterials. This combines many useful properties necessary for effective modification of photovoltaics device

[1] To whom any correspondence should be addressed.

design. Duo to this aspect study of conditions for producing of graphene oxide solid films with predictable properties that is presented in this paper, it is particularly relevant.

2. Experiment

Single layer graphene oxide was used for dispersion preparation (SLGO) (Cheaptubes, USA). The suspension of graphene oxide was prepared in acetone and tetrahydrofuran (THF). All solvents were purchased from Sigma Aldrich and used without purification. Concentration of graphene in solution was $6*10^{-4}$ mol L^{-1}. Suspensions were sonicated in ultrasonic bath during 50 min. The morphology of particles and films of graphene were obtained by scanning electron microscopy (TESCAN Mira-3) on the surface of glasses with FTO coating.

Solid films were obtained by Langmuir-Blodgett (LB) technology. Monolayers were formed on the surface of the subphase by the spreading of the solution in the Langmuir-Blodgett through. The deionized water was cleaned by AquaMax and was used as subphase. The resistivity of the deionized water was 18.2 MΩ/cm. The surface tension of water was equal to 72.8 mN/m at pH=5.6 and temperature 22 °C.

3. Results and discussion

Compression isotherms (π-A isotherms) of graphene oxide monolayers are shown on the Figure 1.

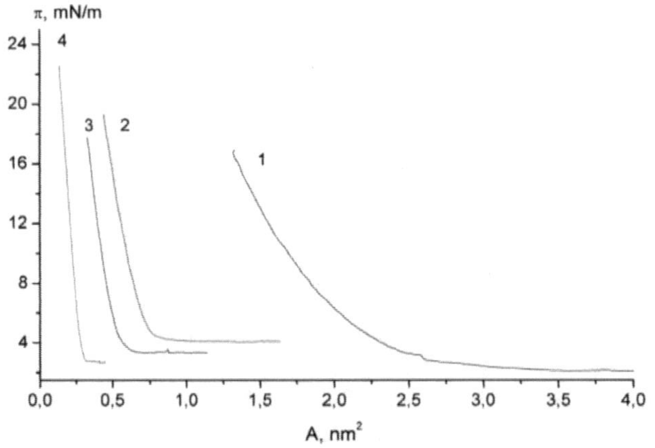

Figure 1. Compression isotherms of graphene oxide monolayers at subphase surface. Monolayer is formed from acetone solution at various volumes of solution, ml: 1 – 0,5; 2 – 1,5; 3 – 3,5; 4 – 5

Form minimal volume that was deposited onto subphase monolayer is in the "gaseous" state [7] at the surface pressures of 2-2.5 mN/m (curve 1). It means that the distance between the graphene particles on a surface of the water is large. With further compression of the monolayer the particles approaches and the film becomes "liquid". Moreover, this "liquid-expanded" state preserved at further compression of the monolayer up to its minimum area.

Spreading of greater volume of substances onto the surface of the subphase (1.5 ml) results in a dense packing of particles within monolayer (curves 2-4). In this case film on the water surface is also initially in the "gaseous" state (curve 2). Transitions of monolayer in "liquid-expanded" and "liquid-condensed" states were observed upon compression of film in the pressure range of 5-9 mN/m and 9-19 mN/m, respectively. From isotherms it also can be seen that the surface pressure equal to 18 mN/m the specific area of graphene has a smaller values than at the pressure equal to 9 mN/m. For monolayers that were formed from greater volumes of dispersion the increasing of specific surface area was registered with the growth of SLGO concentration. It should be noted that all monolayers had no collapse, despite the rather large amount of the spreaded solution.

When the monolayer is formed from the dispersion of graphene oxide from THF similar phase states were observed upon compression of the films on the surface. Increasing of particle concentration in monolayer leads to shift of curve to the side of the little area, and the extent of "gaseous" and "liquid" phases is decreased.

In could be concluded, from the comparison of two π-A isotherms registered for different solvents, that behavior of monolayers for both solvents is similar. But in monolayers of graphene oxide, that were spreaded from acetone, packing of SLGO is denser, though average size of particle in solutions is nearly equal (Figure 2).

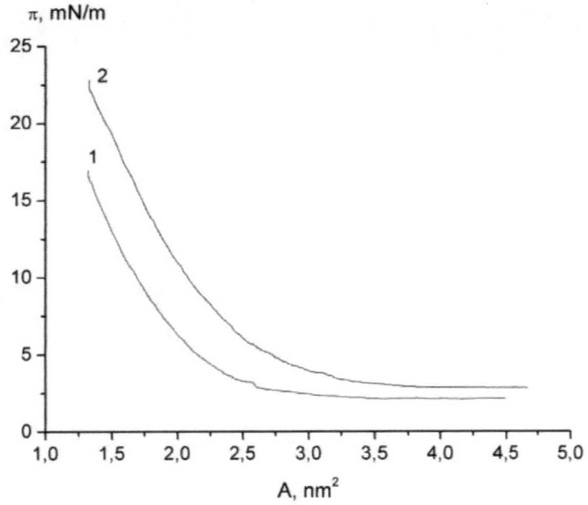

Figure 2. Compression isotherms of graphene oxide monolayers at subphase surface. Monolayer is formed from: 1 – acetone; 2 – THF. Volume of solution is 0.5 ml.

To maintain the required density of monolayers in the course of their transfer onto solid substrate the stability of films was studied [8]. The changing of surface pressure was fixed at constant area of monolayer during 120 minutes. For all monolayers a significant change of surface pressure was observed during 20 minutes at vaporation of solvents. The value of the surface pressure decreased on 2 mN/m times during all time of observation. This testifies that films formed from graphene oxide particles have high stability and could be transferred onto solid surface.

SEM-images were obtained for graphene oxide LB films (figure 3). Monolayers were deposited onto solid substrates according and Y-type (transfer during downward and upward stroke) at surface pressure of π=16 mN/m. The thickness of the films consisted of 4 monolayers.

IOP Conf. Series: Materials Science and Engineering **110** (2016) 012062 doi:10.1088/1757-899X/110/1/012062

Figure 3. SEM-images (registered with detectors of secondary electrons and reflected electrons) of graphene oxide LB films prepared from solution in: 1 – acetone; 2 – THF

As could be seen from the figure, the graphene films on a solid substrate have an island structure. In the pictures you can clearly distinguish the individual particles of graphene. Particles are distributed more uniformly across the substrate surface for the film obtained from acetone dispersion. At closer look at the surface of the film the dark areas of varying intensity can be seen. This fact is evidence about the presence of multilayer graphene particles in the film.

Absorption of the LB films was measured (figure 4). Absorption spectra have typical form and correspond to the spectra of other authors [9].

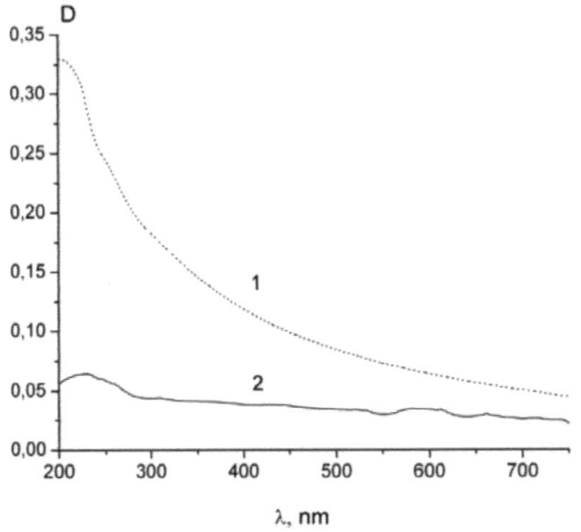

Figure 4. Absorption spectra of graphene oxide LB films prepared from solution in: 1 – acetone; 2 – THF

For films prepared from acetone solutions the optical density at the maximum absorption band is higher about in several times than absorbance of the films prepared from THF. In the visible region of the spectrum both films have high transparency.

4. Conclusion

In this study, we investigated the physical-chemical properties of graphene monolayers formed from dispersion of SLGO in acetone and THF at the interface of water-air. From the compression isotherms of monolayers it was found that for the pressures in the range of 0 to 2 mN mm^{-1} monolayer is predominantly in a gaseous state. With further compression of the monolayer the particles

IOP Conf. Series: Materials Science and Engineering **110** (2016) 012062 doi:10.1088/1757-899X/110/1/012062

approaches and the film becomes "liquid". But in monolayers of graphene oxide, that were spreaded from acetone, packing of SLGO is denser, though average size of particle in solutions is nearly equal.

Absorption spectrum of graphene LB films exhibits as broad band in the ultraviolet and visible region of the spectrum with a maximum at 230 nm. The optical density of the film obtained from acetone dispersion is greater than the optical density of the film prepared from THF solution. The transparency of the films is more than 90% in the wavelength range from 400 to 800 nm. SEM images show that the films have an island structure. In the pictures the individual particles of graphene are clearly distinguished. The pictures are clearly visible single sheets of graphene. The films obtained from acetone dispersion are more uniform, which makes them more promising in terms of their use as conductive coatings.

References

[1] Guo C X, Guai G H, and Li C M 2011 *Adv. En. Mat.* **1** 448
[2] He Z, Guai G H, Liu J, Guo C, Chye Loo J S, Li C M, Tan T T 2011 *Nanoscale* **3** 4613
[3] Hong W, Xu Y, Lu G, Li C, Shi G 2008 *Electrochem. Commun.* **10**. 1555
[4] Wang G., Xing W, Zhuo S 2009 *J. Power Sources* **194** 568
[5] Trancik J E, Barton SC, Hone J 2008 *Nano Letters* **8** 982
[6] Wang X, Zhi L J, Mullen K 2008 *Nano Letters* **8** 323
[7] Adamson A 1997 *Physical Chemistry of Surfaces* (New York: Wiley–Interscience)
[8] ArslanovV V 2000 *Uspekhi Khimii (Rus. Chem. Rev.)* **10** 1155
[9] Sutar D S 2012 *Thin Solid Films* **520** 5991

IOP Conf. Series: Materials Science and Engineering **110** (2016) 012063 doi:10.1088/1757-899X/110/1/012063

Spectra of explosive glowing of heavy metal azides at initiation by high-current electron beam

V I Oleshko and V V Lysyk

Department of Lasers and Lighting Engineering, Institute of High Technology Physics, National Research Tomsk Polytechnic University, 30 Lenin Avenue, Tomsk 634050, Russian Federation

E-mail: oleshko@tpu.ru

Abstract. Glowing spectra of products resulted by heavy metal azides explosive decomposition initiated by high-current electron beam were measured and identified. Intensive emission lines related to atoms of alkali metals were observed in spectra of samples under study. These atoms enter explosives during their preparation. Emission lines of elements being part of a sample holder were also presented in spectra of explosion.

1. Introduction

Pulsed optical spectroscopy with the high time resolution is one of the promising and modern methods for studying of explosive decomposition of energetic materials (EM) [1–4]. Pulsed laser and electron beams are used as an energy source for explosion initiation in this method as they have several characteristics necessary for studying kinetics of explosive decomposition of condensed EM. The main of them are short duration of initiating pulse (~10 ns), high power density of pulse and possibility of its varying in a wide range of 10^5–10^9 W/cm^2 by focusing of energy flux on target surface. These characteristics accompanied by pulsed measurements of optical, mechanical and electric parameters of explosion with high time and dimensional resolution allow obtaining of new information related to initial stages and developing of energetic materials explosive decomposition process.

At present, a high time (~ 1 ns) [1, 2] and dimensional (~ 5 μm) [5] resolutions are achieved. As for the spectral resolution, it is quite low and doesn't exceed ~ 5–20 nm in the majority of publications [1,2]. The cause is the low sensitivity of optical channel of photo-chronograph being used for registration of time-resolved spectra of explosive decomposition of energetic materials at the one pulse. Insufficient spectral resolution of the equipment doesn't allow to identify unambiguously products of explosive decomposition of EM. Therefore, the reliable information about formation dynamics and expansion velocity of atoms and molecules being produced at explosive decomposition of energetic materials cannot be obtained.

The presented work is aimed at measuring of emission spectra observed during explosive decomposition of heavy metal azides (HMA) in the spectral range of 200–980 nm with high spectral resolution sufficient for identification of explosion products.

IOP Conf. Series: Materials Science and Engineering **110** (2016) 012063 doi:10.1088/1757-899X/110/1/012063

2. Experiment

Experimental setup that was used for registration of spectra of explosive decomposition is depicted on figure 1. A pulsed nanosecond electron accelerator based on a pulsed high voltage generator GIN-600 connected to a vacuum diode was taken as a source of initiation. Average electron energy in a beam was ~ 250 keV, maximal current was 3 kA and pulse duration at FWHM amounted ~ 12 ns. The energy density of high-current electron beam (HCEB) was varied in the range of 0.1–1 J/cm^2 by changing cathode-anode distance and using diaphragms.

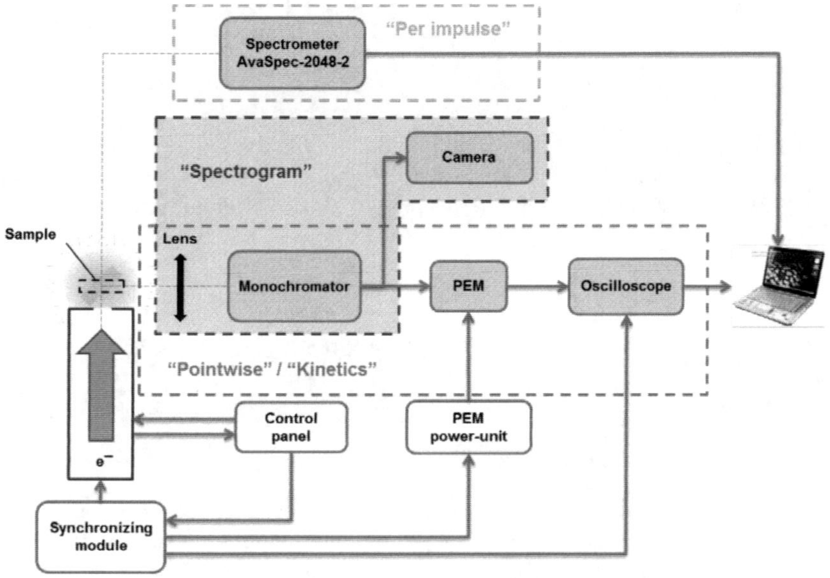

Figure 1. Scheme of experimental setup.

Two methods were used for registration of explosion spectrum. In the first one the image of explosion plasma from near-obstacle area was projected with a lens onto entrance slit of monochromator MDR-23 and photographed with digital camera SONY α500 mounted on monochromator instead of exit slit. Spectral resolution of the method amounted ~ 0.2 nm. Spectral range of measurements was 400–700 nm.

Time-integrated fragments of explosion spectrum ($\Delta\lambda$ ~ 25 nm) were registered in spectral response range of camera SONY α500 per one excitation pulse. Atomic spectral lines registered in explosion plasma were identified using well-known atomic spectra of spectral lamps LSP-1. Those reference spectra were photographed along with explosion spectrum onto the same spectrogram.

In the second method integrated spectra of explosion were registered by means of fiber-optical spectrometer AvaSpec DUAL. Spectral range of measurements was 220–1100 nm, spectral resolution amounted ~ 1.5 nm.

Silver and thallium azides were chosen as subjects of research. Samples were pressed in the form of 250–300 μm thick tablets with diameter of 3 mm. Silver azide whiskers (threadlike monocrystals) with characteristic dimensions of 0.1 mm × 0.05 mm × 5 mm were also taken. Samples were placed and initiated in a vacuum chamber. Schema of samples' placement in experimental chamber are shown on figure 2.

Sample under study 1 was fixed on sample holder 2 and excited by electron beam through anode aperture 3 and 10 μm thick aluminum foil 4. Sample holders were prepared from different metals (aluminum, duraluminum, copper) and from M18 alloy. An obstacle 5 (20 mm × 10 mm × 2 mm aluminum plate) was placed at distance L from samples's surface. It was stopping the plasma jet formed during sample explosion. The purpose of using an obstacle was increasing of intensity of explosion products glowing and dimensional localization of plasma.

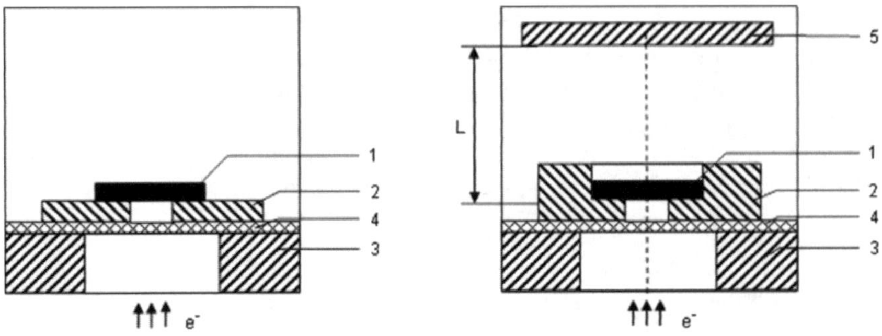

Figure 2. Schema of explosive initiation by electron beam that were used for measuring spectra of explosion decomposition products: 1 – sample, 2 – sample holder, 3 – anode, 4 – 10 μm thick aluminum foil, 5 – obstacle.

3. Results and discussion

Figures 3 and 4 depict spectrogram fragments of explosion plasma of AgN_3 and TlN_3 pressed samples obtained by means of digital camera SONY α500 in different spectral ranges.

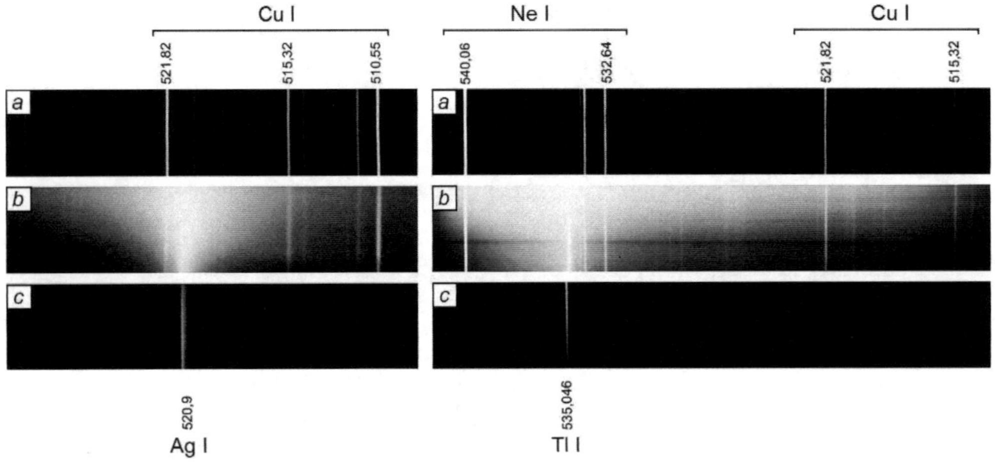

Figure 3. Explosion spectrograms of pressed AgN_3 (on the left) and TlN_3 (on the right) measured by means of optical spectrometer MDR-23 and digital camera in "green" spectral range: a – fragments of lamp LSP-1 (Cu) spectra, b – fragments of explosives samples explosion spectra, measured near obstacle on the background of LSP-1 (Cu) spectra, c – fragments of explosive samples explosion spectra at free expansion of explosive decomposition products.

It can be seen that two types of glowing can be observed in spectrum of explosion. These are line spectra of atoms being a part of explosive material and a continuous spectrum representing glowing of dense low-temperature plasma. Plasma pressure and temperature grow at the deceleration of explosion products on a solid obstacle and this leads to broadening of spectral lines of atoms being a part of explosive (figures 3 b, 4 b) It has been established that intensity ratio of continuous and line emission in spectra of explosion plasma depends on a mass of a sample and expansion conditions of explosive decomposition products (free expansion into vacuum or deceleration on solid obstacle). In case of low mass of the sample ($m \approx 5$ mg), the temperature and pressure in plasma decrease rapidly while the explosion products expand freely. This leads to violent intensity decrease of continuous spectrum. As a rule line spectra of explosion is observed in such conditions. With increasing of explosive mass ($m \approx 20$ mg) the continuous spectrum intensity and the volume occupied with dense plasma increase. The above is confirmed by spectrograms depicted in figures. For example, the spectra of TlN_3 and AgN_3 ($m \approx 5$ mg) explosion are line (figures 3 c and 4 c). At the explosion of HMA samples with mass more than 20 mg line emission of atoms is observed on the background of continuous spectrum (figures 3 b, 4 b).

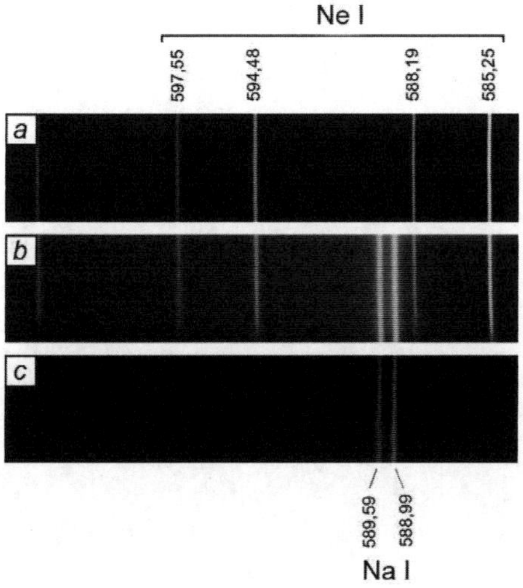

Figure 4. Spectrograms of pressed TlN_3 explosion measured by means of optical spectrometer MDR-23 and digital camera in "yellow-red" spectral range:
a – fragment of lamp LSP-1 (Cu) spectra, b – fragment of explosives samples explosion spectra, measured in near obstacle on the background of LSP-1 spectra, c – fragment of explosive sample explosion spectra at free expansion of explosive decomposition products.

It can be seen that two types of glowing can be observed in spectrum of explosion. These are line spectra of atoms being a part of explosive material and a continuous spectrum representing glowing of dense low-temperature plasma. Plasma pressure and temperature grow at the deceleration of explosion products on a solid obstacle and this leads to broadening of spectral lines of atoms being a part of explosive (figures 3 b, 4 b) It has been established that intensity ratio of continuous and line emission in spectra of explosion plasma depends on a mass of a sample and expansion conditions of explosive decomposition products (free expansion into vacuum or deceleration on solid obstacle). In case of low mass of the sample ($m \approx 5$ mg), the temperature and pressure in plasma decrease rapidly while the explosion products expand freely. This leads to violent intensity decrease of continuous spectrum. As a rule line spectra of explosion is observed in such conditions. With increasing of explosive mass ($m \approx 20$ mg) the continuous spectrum intensity and the volume occupied with dense plasma increase. The above is confirmed by spectrograms depicted in figures. For example, the spectra of TlN_3 and AgN_3 ($m \approx 5$ mg) explosion are line (figures 3 c and 4 c). At the explosion of HMA samples with mass more than 20 mg line emission of atoms is observed on the background of continuous spectrum (figures 3 b, 4 b).

The boost of continuous spectrum intensity can be observed in case of plasma jet deceleration on solid obstacle. The cause of this intensity growth is additional plasma heating by shock wave, formed in near-obstacle region. In this case the continuous spectrum is observed only near the obstacle. On a distance from obstacle intensity of continuous spectrum decreases. And starting from some point the line spectrum of explosive decomposition products can be registered. FWHM of spectral lines decreases the farther explosion products are from the obstacle. Regularities noticed above are typical for explosion plasma as well as for spectra of plasmadynamic jets forming during the pulsed evaporation of metals by HCEB [6].

It has been established that the strongest lines appearing in the emission spectra of AgN3 and TlN3 samples are the lines of silver Ag I (520.91, 546.55, 547.16 nm) and thallium Tl I (535.05 nm) respectively. These lines are responsible for green color of plasma observed at explosion of these explosives in vacuum.

Two densely spaced ($\Delta\lambda = 0.6$ nm) intensive emission lines of sodium atoms have been found in the "orange" spectral range in explosion spectra of monocrystals and pressed powder compacts of AgN_3 and TlN_3. These are Na I 588.99 and 589.59 nm (figure 4).

Emission spectra of explosive decomposition products of pressed samples of AgN_3 and TlN_3 measured by means of optical spectrometer AvaSpec DUAL are presented on figure 5.

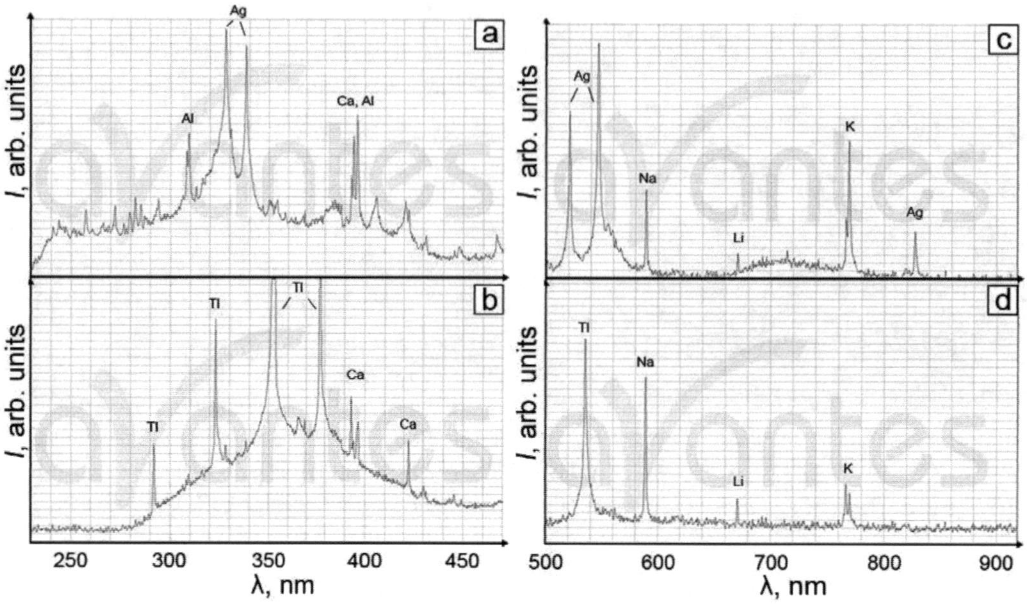

Figure 5. Explosion spectra of pressed samples of AgN_3 (a, c) and TlN_3 (b, d) measured in UV and visible spectral ranges by means of optical spectrometer AvaSpec DUAL.

The most intensive lines in spectra of explosive glowing of silver azide (figure 5 a, c) and thallium azide (figure 5 b, d) are the atom lines of metals being a part of HMA: Ag I (328.07, 338.29, 520.91, 546.55 and 547.16 nm) and Tl I (291.83, 322.98, 351.92, 352.94 and 535.05 nm). Intensive emission lines of alkali metal atoms have been found in "orange" and "red" spectral ranges of explosion spectra of AgN_3 and TlN_3 pressed samples and AgN_3 whiskers: Na I (588.99; 589.59 nm), K I (766.49, 769.89 nm) and Li I (670.78 nm).

Analysis of obtained experimental results demonstrates that alkali metals are pollution agents that enter explosives during the process of their producing. Appropriate spectral lines are always presented in explosion glowing spectra of HMA irrespective of the sample holder type. The presence of these elements in samples under study has been confirmed by measuring of impurity concentration by means of inductively coupled plasma method.

The role of sample holder played in forming of explosion spectrum has been established. In case of exploding of HMA samples placed on duralumin holder aluminum Al I (308.22, 309.27, 394.41, 396.15 nm) and calcium Ca I (393.37, 396.85, 422.67 nm) lines appear in explosive glowing spectra (figure 5). Atomic lines Al I 396.15 and Ca I 396.85 nm (as well as sodium lines Na I 588.99 and 589.59 nm) has been registered as one line due to insufficient spectral resolution of spectrometer AvaSpec DUAL. It is known that calcium is the main pollution agent of metals surface [7]. Therefore calcium lines do appear along with intensive aluminum lines in spectra of such surfaces. Atomic lines registered in explosion plasma of AgN_3 and TlN_3 samples are presented in table 1.

Table 1. The list of identified spectral lines registered in HMA explosion plasma.

Material	Identified line	
AgN_3	Ag I	(328.07, 338.29, 520.91, 546.55 + 547.16, 827.35)
	Na I	(588.99 + 589.59)
	Li I	(670.78)
	K I	(766.49, 769.89)
	Al I	(308.22, 309.27, 394.41, 396.15)
	Ca I	(393.37, 396.85)

IOP Conf. Series: Materials Science and Engineering **110** (2016) 012063 doi:10.1088/1757-899X/110/1/012063

TlN_3	Tl I	(291.83, 322.98, 351.92 + 352.94, 377.57, 535.05)
	Na I	(588.99 + 589.59)
	Li I	(670.78)
	K I	(766.49, 769.89)
	Ca I	(393.37, 396.85)

4. Conclusion

Emission spectra of products of HMA explosive decomposition are measured with high spectral resolution sufficient for their identification. It is found that there are several types of glowing can be found in explosion spectra of samples under examination: these are line spectrum of atoms being a part of explosive material and continuous spectrum representing glowing of dense low-temperature plasma. Line and continuous spectra intensity ratio depends on a mass of explosive and conditions of plasma jet expansion.

The most intensive lines in spectra of explosive glowing of AgN_3 and TlN_3 relate to atom lines of silver and thallium. And these lines are responsible for green color of explosion plasma. Intensive emission lines are observed in "orange" and "red" region of explosion spectra of HMA. These lines relate to alkali metals Na, Li and K which enter explosives during their preparation.

Information obtained in this work is of great interest to modeling process of explosive decomposition of initiation and brisant explosives under external energetic pulse action and to developing explosive-based light sources for atom and molecular spectroscopy of energetic and inert materials. The results show possibility of developing rapid and economical atomic emission method for determination of elemental composition of explosives as well as solid bodies of different types.

This work was performed with the financial support of Russian Ministry of Education and Science, government order «Science», theme # 3.1329.2014.

References

[1] Aduev B P, Aluker E D, Belokurov G M and Drobchik A N 2000 *Combustion, Explosion, and Shock Waves* **36** **5** 78–89

[2] Svaiko V N, Krechetov A G, Aduev B P and Gudilin A V 2005 *Technical Physics* **75** **6** 59–62

[3] Oleshko V I, Korepanov V I, Lisitsyn V M and Tsipilev V P 2004 *Technical Physics Letters* **30** **22** 17–22

[4] Oleshko V I, Korepanov V I, Lisitsyn V M and Tsipilev V P 2007 *Combustion, Explosion, and Shock Waves* **43** **5** 87–9

[5] Oleshko V I, Lisitsyn V M, Skripin A S and Tsipilev V P 2012 *Technical Physics Letters* **38** **9** 37–43

[6] Oleshko V I 2012 Spectral element analysis with using of high-current electron beams. *Tomsk. TPU*, p 96

[7] Kurakin S I, Diamant G M and Pugachev V M 1990 *Proceedings of AS USSR Inorganic Materials* **26** **11** 2301-4

Formation of cluster systems in condensed matters and IR spectra of liquids

G Melnikov[1], N Ignatenko[1], P Krasnych[1], V Melnikov[1] and E Cherkasov[1]
[1]*Kursk, 305040, Russia,* South-West State University (SWSU)

E-mail: inmkstu@bk.ru, melnikovga@mail.ru

Abstract.

Modern approaches to the interpretation of IR spectra of polyatomic liquids are based on cluster models of the structure of matter. First of all it concerns the far infrared region of the spectrum (20-300 cm^{-1}) where rotationally libration motions in the structure of clusters are found. This work is a continuation of research conducted by the authors earlier [G. Melnikov at al. 2015 *IOP Conf. Ser .: Mater. Sci. Eng.* **81** p 012032]. The authors have adopted a model in which the appearance of spectral bands is explained by to libration oscillations vibrations of dimers with different configurations in the structure of clusters.

Introduction

The quantization process in cluster systems is complex and there are numerous experimental and theoretical studies [1-6] devoted to the search of the laws of quantum transitions in the formation and evolution of clusters in liquids. The principles of quantization of oscillation vibration frequencies of dimers were discussed in detail by Kirchhoff and Lida [1] and later were used by Fraser et al. [2]. To describe the structure and rotational spectra of benzene-water dimer Gutowsky H. et al. for transition frequencies used the following ratio [3].

$$\nu = 2(J+1)\{B - D_{JK}K^2 - D_{Jm}m^2 - D_{JKm}Km - H_{JKm}m^2K^2\} - 4(J+1)^3 D_J \tag{1}$$

where $m = 0, \pm 1, \ldots$ – quantum number that specifies internal rotation of the dimer; D, H – constant values.

The problem of quantization of symmetric top rotation and rotational spectrum of the dimer SiF4-NH3 is considered in [4, 5]. In [6] the following ratio is proposed to quantize oscillation vibration frequencies in dimers.

$$\nu = 2\bar{B}(J+1) - 4D_J(J+1)^3 - 2D_{JK}(J+1)K^2 \tag{2}$$

where $\bar{B}$, D_J, and D_{JK} - constants.

The Equations (1) and (2) demonstrate that if rotational- librational motion of dimers in a cluster is simulated as a diatomic molecule, the energy of this motion can be written as

$$E_{rot} = B_v J(J+1) - D_v\left[J(J+1)\right]^2 + H_v\left[J(J+1)\right]^3 + \ldots \tag{3}$$

where $B_v, D_v, H_v, ...$ - are series of powers: $(v+0.5)$.

The frequency spectrum of rotational oscillations vibrations in cluster systems is represented as an expansion in powers of x similar to Dunham series [7], and the constant Dv is called centrifugal tension constant

$$T(v,J)=\sum_{l,j} Y_{l,j}\left(v+\frac{1}{2}\right)^{l}\left[j(j+1)\right]^{j} \tag{4}$$

Practical use of different expansions in powers $(v+0.5)$ like Dunham series for predicting rotational and vibrational-rotational-librational frequencies of cluster systems is rather difficult, because the expansions contain several empirical constants, the calculation of which requires a great deal of experimental data on the spectra of objects. This paper discusses some alternative ways of predicting frequencies in the IR range.

The rule of sequential formation of clusters in the structure of liquids

As a of Density fluctuations in a disordered condensed medium result in the formation of cluster systems that have composition, geometrical dimensions, formation energy and lifetime. In the series of papers [8-11], the authors have shown that under the conditions of dynamic equilibrium a bound state of two particles (a dimer) forms up in the structure of the most probable cluster. These particles make libration oscillations with a frequency ω determined by the condition

$$\omega_i=\sqrt{\frac{2\Delta H_{\dim}}{J_{\dim}}}\cdot\sqrt{\hat{Z}_i} \tag{5}$$

Equation (5) shows that the frequency is determined by the number of particles in the cluster, the number and characteristics of possible dimer configurations that are composed of free molecules.

When the spontaneous collapse of the cluster takes place, the number of particles in its composition changes and becomes equal to $\hat{Z}_j$, then the frequency of the dimer librational oscillations in the newly formed cluster will be

$$\omega_j=\sqrt{\frac{2\Delta H_{\dim}}{J_{\dim}}}\cdot\sqrt{\hat{Z}_j} \tag{6}$$

The change in frequency is accompanied by the emission or absorption of the quantum of energy with a frequency

$$\omega_{ij}=\sqrt{\frac{2\Delta H_{\dim}}{J_{\dim}}}\cdot\sqrt{\hat{Z}_j}\left(1-\sqrt{\frac{\hat{Z}_j}{\hat{Z}_{j+1}}}\right) \tag{7}$$

Assuming that the root in the parenthesis of (7) remains constant, you can set this value as equal to the square root from the "golden section" constant, i.e. set $\frac{\hat{Z}_j}{\hat{Z}_{j+1}}=\frac{5}{8}=0.625$, then the formula (7) can be written as a ratio

$$\omega_{ij}=0.2094\sqrt{\frac{2\Delta H_{\dim}}{J_{\dim}}}\cdot\sqrt{F_j} \tag{8}$$

where $F_i=1, 2, 3, 5, 8, 13, 21, 34, 55, ...$-are Fibonacci numbers.

IOP Conf. Series: Materials Science and Engineering **110** (2016) 012064 doi:10.1088/1757-899X/110/1/012064

Figure 1 shows a diagram of the spontaneous decay of a cluster with the most probable number of particles $\hat{Z}_i = 8$ in accordance with the Fibonacci law.

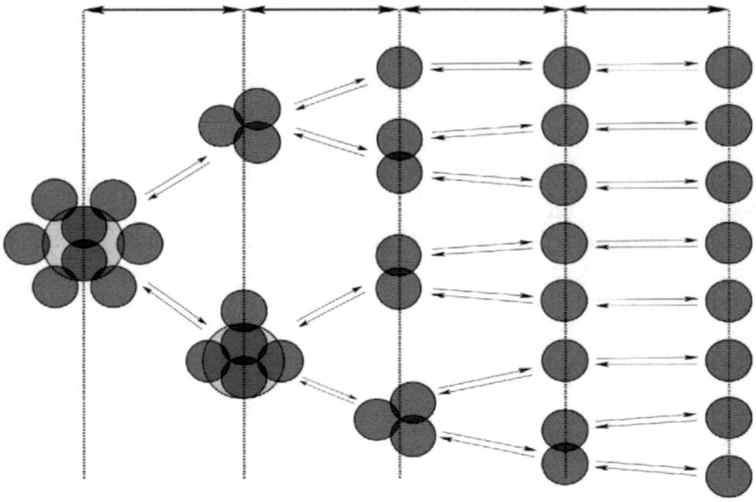

Figure 1. A diagram of the spontaneous decay of a cluster ($\hat{Z}_i = 8$) according to the Fibonacci law

Polyatomic molecules with complex geometric shapes can form a number of possible configurations of a dimer, that will have different moments of inertia relative to the main axes and energies of configuration formation. In the formula (8) to designate the value before the multiplier $\sqrt{F_i}$ we'll introduce the notation Figure 1 shows a diagram of the spontaneous decay of a cluster with the most probable number of particles $\hat{Z}_i = 8$ in accordance with the Fibonacci law

$$\omega_{\min,i} = 0.2094\sqrt{\frac{2\Delta H_{\mathrm{dim,j}}}{J_{\mathrm{dim,j}}}} \tag{9}$$

where $\omega_{\min,i}$ - is minimal frequency for the selected *i-th* configuration of the dimer.

The set of frequencies in IR spectra of a liquid is determined according to formulas (8) and (9) by simple relationship

$$\omega_i = \omega_{\min,i}\sqrt{F_i} \tag{10}$$

The existence of spectral bands in IR spectra suggests certain rules of quantization of librational oscillations of dimers in the structure of clusters.

In order not to use the cumbersome and time-consuming calculations by Dunham scheme, we'll use an alternative representation of Fibonacci series. It is possible to apply de Moivre-Binet formula, which is written as [12-13]:

$$F_n = \frac{1}{\sqrt{5}}\left[\left(\frac{1+\sqrt{5}}{2}\right)^n - \left(\frac{1-\sqrt{5}}{2}\right)^n\right] \tag{11}$$

with $n \geq 0, n = 0,1,2,...$ it is possible to demonstrate that there is asymptotic behavior, which allow us to represent Fibonacci series as a power function of the golden section base $\phi = 1.618039...$ [12-13].

$$F_n = \frac{1}{\sqrt{5}}\phi^n \qquad (12)$$

With various integer values of n we'll obtain integral values of the Fibonacci sequence, despite the fact that $\sqrt{5}$ *and* ϕ - are irrationals number, then according to formulas (10) and (12) the rules of frequencies quantization in IR spectra of liquids can be written

$$\omega_{mn} = \omega_{\min} \cdot 5^{-\frac{1}{4}} (\phi)^{\frac{1}{2}J_L} \qquad (13)$$

where $J_L = 0, 1, 2, ...$ - librational quantum number.

Fibonacci numbers can also be expressed in terms of Chebyshev polynomials according to recurrence relations or in terms of trigonometric functions [14-15]. Considering the formula (9) the ratio (13) can be written in final form as

$$\omega_J = 0.1423\sqrt{\frac{2\Delta H_{\mathrm{dim,J}}}{J_{\mathrm{dim,i}}}} (\phi)^{\frac{1}{2}J_L}, \qquad (14)$$

Calculation results and their comparison with experimental data

Let's check the validity of the obtained relations (10) and (14) by applying them to liquid benzene as the most studied object. It has been established that benzene molecules can form three types of dimer configurations [16-19]: Sandwich (S), T-shaped and parallel-displaced (PD) configurations that differ in the value of the equilibrium distances between the centers of molecule masses R_{dim}, the energy of configuration formation (dimer's binding energy) ΔH_{dim} and the moment of inertia J_{dim} in relation to the axis going through the center of masses in the configuration.

It is hardly possible to calculate the exact moment of inertia of the dimeric configuration and its formation energy for liquids with complex multi-nuclear molecules. So the value $\omega_{\mathrm{min},i}$ in the formula (14) shall be considered as an empirical constant.

Table 1 shows the frequencies of libration oscillations of benzene dimers in the structure of the clusters calculated by formula (10). They have been calculated dimers containing in their structure different numbers of particles with different values of $\omega_{\mathrm{min},i}$. By applying the Fibonacci series and the formula (10) we could predict the position of spectral bands in benzene IR spectrum within a wide frequency range ($20 cm^{-1} - 6000 cm^{-1}$), moreover it was found that benzene can have clusters with the number of particles varying from 2 to 6700, however, the most probable number of particles is determined by Erlang distribution and ranges from 3 to 21 depending on the parameters of the state of a liquid [8].

The accepted model of IR spectra formation in organic liquids allows to predict the position of spectral bands with an accuracy of about 5%. The use of formula (14) to calculate frequencies in IR spectra of liquids with different molecular structure and crystals has demonstrated that complex spectra of such objects always contain spectral lines that correspond to cluster formations with the number of the particles corresponding to a Fibonacci series.

Table 1. Libration oscillation frequencies of dimers (cm^{-1}) in the cluster structure of liquid benzene and its Raman.

Calc. (10)	[20,21]	Calc. (10)	[20,21]	Calc. (10)	[20,21]
11.57 $\omega_{1\min}$	**11.57**	277 $\omega_{4\min}\sqrt{5}$	**275.77**	1605 $\omega_{3\min}\sqrt{610}$	**1606,8**
				1830 $\omega_{2\min}\sqrt{1597}$	**1815**
				1893 $\omega_{4\min}\sqrt{233}$	**1885**
20.0 $\omega_{1\min}\sqrt{3}$		286 $\omega_{1\min}\sqrt{610}$		2042 $\omega_{3\min}\sqrt{987}$	**2045**
25.9 $\omega_{1\min}\sqrt{5}$	**20.25**	298 $\omega_{3\min}\sqrt{21}$	**301.6**		
				2338 $\omega_{2\min}\sqrt{2584}$	**2325**
32.7 $\omega_{1\min}\sqrt{8}$	**31.4**	341 $\omega_{2\min}\sqrt{55}$	–	2408 $\omega_{4\min}\sqrt{377}$	**2407.7**
				2598 $\omega_{3\min}\sqrt{1597}$	**2595**
41.7 $\omega_{1\min}\sqrt{13}$	–	363 $\omega_{1\min}\sqrt{987}$	**374,12**	2974 $\omega_{2\min}\sqrt{4181}$	
46.0 $\omega_{2\min}$	–	380 $\omega_{3\min}\sqrt{34}$	**388.8**		**2950**
		434 $\omega_{2\min}\sqrt{89}$	**462**	3063 $\omega_{4\min}\sqrt{610}$	**3044**
53.0 $\omega_{1\min}\sqrt{21}$	**53.5**	447 $\omega_{4\min}\sqrt{13}$	**447.6**		**3065**
65.0 $\omega_{3\min}$	**61.0**	482 $\omega_{3\min}\sqrt{55}$	–	3304 $\omega_{3\min}\sqrt{2584}$	
67.5 $\omega_{1\min}\sqrt{34}$	**70**				**3309**
		552 $\omega_{2\min}\sqrt{144}$		3783 $\omega_{2\min}\sqrt{6765}$	
79.7 $\omega_{2\min}\sqrt{3}$	**79.06**	568 $\omega_{4\min}\sqrt{21}$			**3788**
		588 $\omega_{1\min}\sqrt{2584}$	**600**		
85.8 $\omega_{1\min}\sqrt{55}$	**86.6**	613 $\omega_{3\min}\sqrt{89}$	**626.3**	3895 $\omega_{4\min}\sqrt{987}$	
92.0 $\omega_{3\min}\sqrt{2}$					**3893**
102.8 $\omega_{2\min}\sqrt{5}$	**100**	702 $\omega_{2\min}\sqrt{233}$	**690**	4203 $\omega_{3\min}\sqrt{4181}$	
109 $\omega_{1\min}\sqrt{89}$	–	723 $\omega_{4\min}\sqrt{34}$	**713.5**		**4208**
124 $\omega_{4\min}$	**128**	780 $\omega_{3\min}\sqrt{144}$	**790**	4955 $\omega_{4\min}\sqrt{1597}$	**4959**
130.1 $\omega_{2\min}\sqrt{8}$	**128**	893 $\omega_{2\min}\sqrt{377}$	**952**	5346 $\omega_{3\min}\sqrt{6765}$	
139 $\omega_{1\min}\sqrt{144}$	–		**972.5**		**5346**
145 $\omega_{3\min}\sqrt{5}$	**144.63**	992 $\omega_{3\min}\sqrt{233}$	**992.5**		
166 $\omega_{2\min}\sqrt{13}$	**168**	1136 $\omega_{2\min}\sqrt{610}$	**1148.5**		
175 $\omega_{4\min}\sqrt{2}$	**171**	1170 $\omega_{4\min}\sqrt{89}$	**1176.8**		
211 $\omega_{2\min}\sqrt{21}$	**210**	1262 $\omega_{3\min}\sqrt{377}$	**1249**	6303 $\omega_{4\min}\sqrt{2584}$	–
215 $\omega_{4\min}\sqrt{3}$	**213.9**	1445 $\omega_{2\min}\sqrt{987}$		8018 $\omega_{4\min}\sqrt{4181}$	–
268 $\omega_{2\min}\sqrt{34}$	**264**	1488 $\omega_{4\min}\sqrt{144}$	**1479**		

IOP Conf. Series: Materials Science and Engineering **110** (2016) 012064 doi:10.1088/1757-899X/110/1/012064

Conclusion

The distribution of clusters by the number of particles in their composition suggests the existence of clusters with a most stable configuration and with the number of particles determined by a Fibonacci series. Successive collapse of a selected cluster having the corresponding to a Fibonacci series takes place according to the rule saying that a newly emerged cluster also has the number of particles corresponding to a Fibonacci series with the preserved ratio between the number of particles in two successively transformed clusters, which is equal to the "golden section " $\phi = 0.618...$. A change in the dimer 's libration oscillation frequency with the change of the number of particles in the cluster structure is accompanied by emission of a photon having the frequency that is proportional to the square root of the number of particles in the cluster.

Mathematical construction of the adopted model is based on the assumption that there is an effective potential of inter- particle interaction and it is possible to calculate the most probable number of particles in the structure of the cluster by the Erlang distribution.

References:

[1] Kirchhoff W. H. and Lide D. R., Jr. 1965 *J. Chem. Phys.* **43** 2203

[2] Fraser G. T., Lovas F. J., Suenram R. D., at al. 1986 *J. Chem. Phys.* **84** 5983

[3] Gutowsky H. S., Emilsson T. and Arunan E. 1993 *J. Chem. Phys.* **99** 4882

[4] Ruoff R.F., Emilsson T., Jaman A. I., at al. 1992 *J. Chem. Phys.* **96** 1

[5] Guardiola R., Navarro J. 2008 *J. Chem. Phys.* **128** 144303

[6] Arunan E., Gutowsky H. S. 1993 *J. Chem. Phys.* **98 (5)** 4294

[7] Dunham J.L. 1932 *Phys. Rev.* **41** 721

[8] Melnikov G.A., Verveyko V.N., Melikhov Yu.F., at al. 2012 *High Temperature* **50** 214

[9] Melnikov G.A., Ignatenko N.M., Melnikov V.G. **.** 2013 *Journal of Nano- and Electronic Physics* **5** 04044(4)

[10] Melnikov G.A., Ignatenko N.M., Melnikov V.G., at al. 2014 *Proceedings of the South-West state university. Technics and Technologies* **4** 96

[11] Melnikov G.A., Ignatenko N. M., Melnikov V.G., Cherkasov E.N. 2015 *IOP Conf. Ser.: / Materials Science and Engineering* **81** 12032

[12] Naylor M. 2002. *Mathematics Magazine* **75** 163

[13] Melker A. I. 2012 *Materials Physics and Mechanics* **13** 85

[14] Harne S. 2014 *International Journal of Contemporary Mathematical Sciences* **9** 553

[15] Morgado J. 1995 *Portugaliae Mathematica* **52** 363

[16] Park Y. Ch., Lee J. Sh. J. 2006 *Phys. Chem.* **A110 p** 5091

[17] Smith T., Slipchenko L. V., Gordon M. S. 2008 *J. Phys. Chem.* **A112** 5286

[18] Schnell M., Erlekam U., Bunker P. R., at al. 2013 *Phys. Chem. Chem. Phys.* **15** 10207

[19] Yaya A., Tiburu E. K., Onwona-Agyeman B., Dodoo-Arhin D., Efavi J. 2014 *Journal of Computations & Modelling* **4** 27

[20] Broude V.L. 1961 *Uspekhi Fizicheskikh Nauk* **LXXIV** 577

[21] Bertie J. E., Keefe C. D. 2004 *Journal of Molecular Structure* ***695–696*** 39

IOP Conf. Series: Materials Science and Engineering **110** (2016) 012065 doi:10.1088/1757-899X/110/1/012065

The quantization of the radii of coordination spheres cubic crystals and cluster systems

G Melnikov[1], S Emelyanov[1], N Ignatenko[1] and G Ignatenko[1]

[1]*Kursk, 305040, Russia,* South-West State University (SWSU)

E-mail: inmkstu@bk.ru, melnikovga@mail.ru

Abstract.

The article deals with the creation of an algorithm for calculating the radii of coordination spheres and coordination numbers cubic crystal structure and cluster systems in liquids. Solution has important theoretical value since it allows us to calculate the amount of co-ordination in the interparticle interaction potentials, to predict the processes of growth of the crystal structures and processes of self-organization of particles in the cluster system. One option accounting geometrical and quantum factors is the use of the Fibonacci series to construct a consistent number of focal areas for cubic crystals and cluster formation in the liquid.

Introduction

The calculation of the radii of coordination spheres in the crystals is needed to construct the theory of many-atomic interactions, to decrypt X-ray patterns of crystal having complex structures, and of solid solutions with embedded interstitial atoms [1, 2]. In theoretical terms in the lattice theory and cluster model of liquid to describe the intermolecular interaction it is necessary to calculate the sum of the characteristic [3,5]. For disordered condensed media the application of thermodynamic relations for the calculation of the pressure in a system of interacting particles is also related with the calculation of the radii of successive coordination spheres and coordination numbers [6].

In quasicrystalline theories the structure of cluster systems is modeled by lattices of cubic type (fcc, bcc, SC), and it is believed that structural properties of arrays of clusters and bulk samples are similar. In coordination spheres the structure of such a cluster, or of a cubic crystal lattice is represented as a system of successive coordination spheres with single center in f randomly selected atom [8].

Studies of the structure of cubic system crystals allowed us to obtain the well-known rule that says that the radius of the N-th coordination sphere is proportional to the square root of the number of this sphere

$$R_n = R_1\sqrt{N} \tag{1}$$

The radius of the first coordination sphere R_1 defines the type of the crystal lattice, it is related with the geometric size (effective diameter) of the particles and the nature of the interaction between the particles in the crystal.

From the series of natural numbers $N = 1, 2, 3, \ldots$ it is necessary to exclude the numbers that correspond to the terms of extinction in the analysis of X-ray diffraction measurements. One of the main problems of theoretical crystallography is to find an algorithm for calculating the radii of coordination spheres in crystal structures and cluster systems.

Algorithms for calculating the radii of coordination spheres in crystals

The paper [7] proposed analytical expressions that allow us to calculate the radii of coordination spheres and corresponding coordination numbers for cubic lattices by the number of coordination spheres. For the fcc lattice the variation range is written as formula

$$R_k = \frac{1}{\sqrt{2}} a\sqrt{k} = R_1\sqrt{k} \tag{2}$$

where $k = 1, 2, 3, \ldots$ - takes the values of natural numbers, a - is the lattice constant, R_1 -is the radius of the first coordination sphere

From the series of natural numbers it is necessary to eliminate the numbers in accordance with the formula [7]

$$k_{excl} = 2\left[4^q(8i-1)\right],\ q = 0, 1, 2, \ldots\ , i = 1, 2, 3, \ldots \tag{3}$$

In an ideal fcc lattice there is no coordination spheres with numbers $k_{excl} = 14, 30, 46, 62, 78, 94, \ldots$ (with $q = 0$).

The radii of the coordination spheres of an ideal SC-lattice are written in the

$$R_k^{SC} = a\sqrt{k} = R_1\sqrt{k} \tag{4}$$

From the series of natural numbers the coordination spheres with the following numbers should be excluded:

$$k_{excl}^{SC} = 4^q(8i-1),\ q = 0, 1, 2, \ldots\ , i = 1, 2, 3, \ldots \tag{5}$$

In an ideal SC lattice there are no coordination spheres with numbers: $k_{excl} = 7, 15, 23, 28, 31, 39, 47, \ldots$

In [8] an algorithm was suggested to calculate the radii of coordination spheres and coordination numbers on their surface depending on the sphere number for diamonds. The suggestion was made based on experimentally obtained facts that in the direction $[\bar{2}11]$ the diamond lattice has cubic three-layer package with a period $0.728A$, and in the direction $[111]$ there is a periodicity with a constant of $2.059A$.

The radius m of the coordination sphere r_m are determined by the diamond lattice parameter $a = 3.566A$ and the direction vector $\vec{R} = \langle \mu v w \rangle, (\mu, v, w = 0, \pm1, \pm2, \ldots)$ and are defined as

$$r_m = \frac{a}{4}\sqrt{\mu^2 + v^2 + w^2} = \frac{a}{4}R_m, \tag{6}$$

where $R_m = \sqrt{4m},\ m = 2(i-1),\ \ i = 1, 2, 3, \ldots,\ R_m = \sqrt{4m-1},\ m = 2i-1,\ \ i = 1, 2, 3, \ldots$

According to the proposed algorithm, the authors of [8] calculated the radii of 192 coordination spheres and the number of particles on each of them. All coordination spheres are considered filled [9]. Location of the nearest neighbors in the amorphous carbon film is similar to the arrangement of them in the diamond lattice, but the distribution of layers over coordination numbers is seriously disrupted [10].

IOP Conf. Series: Materials Science and Engineering **110** (2016) 012065 doi:10.1088/1757-899X/110/1/012065

Principles of quantization of the radii of coordination spheres in cluster systems

Currently, there are several approaches to modeling the structure of cluster formations in condensed matters. Quasicrystal models represent the structure of a cluster by means of cubic lattices or lattices of the diamond type. Quasicrystal models give good results in modeling the structure of metal clusters and can identify the most stable cluster formations with the "magic" number of particles in their composition [11].

The paper [12] proposes a model of "double barrier of nucleation", according to which the mechanism of formation of the cluster begins with the formation of the cluster crystalline core o in the form Bravais unit surrounded by liquid crystalline shell. The appearance of the crystal nucleus means overcoming the first energy barrier of nucleation. Then there is a consistent core cell proliferation that ends in an irreversible process of crystal growth.

In [13] the structure of nanoscale objects and metal clusters is modeled by face-centered cubic (fcc), body-centered cubic (bcc) or hexagonal close packings, and inter-configuration of atoms in a cluster corresponds to the structure of the bulk sample. For fcc lattice the radii of successive coordination spheres are defined by the formula

$$R_n = \sigma_0 \sqrt{n} \tag{7}$$

where σ_0 - the diameter of a solid sphere (atom), n - the number of the coordination sphere.

For the bcc lattice:

$$R_n = \frac{\sqrt{3}}{3} \sigma_0 \sqrt{n} \tag{8}$$

The number of balls that are located on the n-th sphere is determined by the condition

$$u^2 + v^2 + w^2 + uv + uw + vw = n \tag{9}$$

where u, v, w are integers.

The model of the close packing of hard balls in the ideal crystal lattices with cubic structure cannot describe the sequential arrangement of coordination spheres in real crystals. The replacement of solid balls by soft spheres in the model does not basically change the calculation results, so geometrical factors should be supplemented physical principles of the quantization of coordination spheres.

In their studies [14-16] the authors of this article proposed a model of cluster formation in a disordered condensed media. According to this model the core of the cluster is a dimer around which a system of successive coordination spheres is built up. The result is a "shell structure" of the cluster. The appearance of defects in the cluster shell leads to unification and redistribution of the mutual arrangement of coordination spheres in comparison with crystalline phase of matter and determines temperature dependence of coordination numbers in the structure of the cluster. The studies of the suggested model showed that quantization process must be taken into account when the radius of the coordination sphere is changed in the structure of the cluster, similar to the quantization process of electron orbits in complex atoms.

For a system of successive coordination spheres of spherical particle the radius of such spheres can be represented by the ratio

$$R_n = R_1 \sqrt{pF_n} \tag{10}$$

where R_1 - the radius of the first coordination sphere, F_n - Fibonacci numbers, $p = 1, 2, 3, \ldots$ - natural numbers.

Now based on de Moivre-Binet formula (Binet's formula), that allows us to present a series of Fibonacci numbers as the power function of the base of the golden section $\phi = 1.618039\ldots$ [17].

$$F_n = \frac{1}{\sqrt{5}}\phi^n \tag{11}$$

Using formulas (10) and (11) we'll obtain the law of quantization of the radii of coordination spheres as a function of the principal quantum number.

$$R_n = \frac{2}{3}R_1\sqrt{p}(\phi)^{\frac{n}{2}} \tag{12}$$

where $n \geq 1, n = 1, 2, 3, ...$ - principal quantum number, $\phi = 1.618039...$ - golden section, R_1 - the radius of the first coordination sphere, the value of which determines the radius of the number of successive coordination spheres.

In [18, 19] the ratio is suggested that gives trigonometric interpretation of the Fibonacci series of numbers.

$$F_n = \prod_{l=1}^{n-1}\left(1 - 2i\cos\frac{\pi l}{n}\right) \tag{13}$$

According to formula (13) Fibonacci series is defined by two integers n and l that can be regarded as principal quantum number n and orbital quantum number l of the electron as it moves on the surface of coordination sphere. The relationship between the integers n, and l in formula (13) coincides exactly with the rules of the quantization of orbital angular momentum of electrons in complex atoms: $l = 0, 1, 2, ..., (n-1)$. However, formula (13) eliminates quantum state with quantum numbers $n = 1$ and $l = 0$ in the second coordination sphere (and subsequent ones); the electron is just in excited state. Fibonacci numbers can be expressed in terms of Chebyshev polynomials [19, 20].

The results of the calculations. Discussion

Direct verification of relation (10) for disordered condensed media can be based on the analysis of radial distribution function (RDF) of particles. Table 1 shows the results of calculating the radii of nine consecutive coordination spheres for argon near its melting point and compares these values with experimental data [21]. The calculation error is substantially does not exceed 5%, which proves the applicability of formula (10) for predicting radii of coordination spheres in liquid noble gases.

Calculating the radius of the coordination spheres in benzene, the molecules of which are structured on benzene ring, agree with experimental data for coordination spheres radii that were derived from the construction of radial distribution function [22, 23].

Papers [10, 24, 25] provide some data on the radii of three successive coordination spheres in n-paraffins series. The data was obtained from the RDF analysis and specify the width of the peaks corresponding to these radii. Coordination spheres radius calculation based on formula (10) for liquid n-hexane near its melting temperature demonstrated the applicability of this formula for predicting sequential arrangement of coordination spheres in the structure of this class of liquids.

Conclusion

In The authors have shown that it is possible to consider two factors - geometric and physical in a model that takes into account "golden" section in the location of successive coordination spheres and allows us to formulate the rule of formation of space coordination spheres: the radius of the n-th coordination sphere reduced to the radius of the first coordination sphere is equal to the square root of the Fibonacci number that stands at the (n + 1) - th place in the Fibonacci series: 1, 1, 2, 3, 5, 8, 13, 21, 34, 55....

The Formulated rule of the formation of coordination spheres in cluster systems means that the ratio of the radius of a subsequent coordination sphere to the radius of the previous sphere remains constant and is equal to the "gold" section square root . $R_{n+1} / R_n = \sqrt{1.618...} = 1.272...$.

Quantization process of coordination spheres follows from the trigonometric representation of Fibonacci numbers and the square root of these numbers. According to formula (13) Fibonacci series is defined by two integers n and l that can be regarded as the principal quantum number n and orbital quantum number l of the electron as it moves on the surface of the coordination sphere. The relationship between the integers n and l in the formula (13) conforms exactly with the rules of quantization of the orbital angular momentum of an electron in complex atoms and molecules.

Table 1. The radii of coordination spheres in liquid argon (85K) derived from the RDF and the calculated through Fibonacci numbers (p=1).

Sphere number	F_i	$R_n = R_1\sqrt{pF_n}$ (10)	Experiment	
		Argon (T=85 K)	[21]	[10]
1	1	3.677	3.677	3.71 – 3.80
2	2	5.200	–	
3	3	6.37	6.9462	
4	5	8.22	–	
5	8	10.40	10.079	
6	13	13.26	13.2795	
7	21	16.85	16.5483	
8	34	21.44	19.75–22.95	
9	55	27.27	26.2185	

Analysis of the results of applying Fibonacci numbers and the proposed rule (10) for calculating the radii of successive coordination spheres for crystals with fcc structure and liquids with different molecular structure leads to the conclusion that the universalism of rule (10) is connected with general properties of supramolecular structure of condensed matters.

References:

[1] Medvedev N, Naberukhin Y I 1978 *Phys. Chem. Liq.* **8** 167
[2] Voloshin V P, Naberukhin Yu I 1993 *J. Phys.: Cond. Mat.* **5**. 5685
[3] Medvedev N, Voloshin V, Luchnikov V, Gavrilova M 2006 *J. Comput. Chem.* 27 1676
[4] Shor Y, Yahel E, Makov G 2012 *J. Non- Cryst. Solids* **358** 2687
[5] Brouwers H J H 2013 *Phys. Rev.* E **87** 032202
[6] Burstein A 1974 *Journal of Physical Chemistry* **78** 2684
[7] Bokov O G, Shevlakov V V *Ultrasound and thermodynamic properties of the substance* 1982 Kursk: KSU 17
[8] Andreev V, Malik R, Podzyarey G 2001 *Journal of Structural Chemistry* **42** 233
[9] Kulikovsky V, Metlov K, Kurdyumov A et al. 2002 *Diamond and Related Materials* **11** 467
[10] Skryshevsky A F 1980 *Structural analysis of liquids and amorphous solids* M .: Higher School - 328
[11] Harris I A, Kidwell R S, Northby J A 1984 *Phys. Rev. Lett.* **53** 2390
[12] Tsvetkov E G, Kidyarov B I 2007 *Notes of the Russian Mineralogical Society* **136** 66
[13] Sabutz A V, Sekerzhitsky S S, Liopo V A 2013 *Vesnic of Brest University. Physics. Mathematics* **1** 37
[14] Melnikov G, Verveyko V, Polyansky A 2011 *Int. J. Thermophys.* **32** 901 DOI 10.1007/s10765-

011-0973-3
[15] Melnikov G, Ignatenko N, Melnikov V, Cherkasov E, Apalkova O 2014 *J. of Nano-and electronic Physics* **6** 03016
[16] Melnikov G, Ignatenko N, Melnikov V, Cherkasov E 2015 *IOP Conf. Series: Materials Science and Engineering* **81** p 012032, doi:10.1088/1757-899X/81/1/012032
[17] Naylor M 2002 *Mathematics Magazine* **75** 163
[18] Melker A 2012 *Materials Physics and Mechanics* **13** 85
[19] Benjamin A T, Walton D 2009 *Mathematics Magazine* **82(2)** 117
[20] Harne S 2014 *International Journal of Contemporary Mathematical Sciences* **9** 553
[21] Yarnell J L, Katz M J, Wenzel R G 1973 *Physical Review A* **7** 2130
[22] Bogdan T V 2006 *Russian Journal of Physical Chemistry* **80** 14
[23] Fileti E E, Canuto S 2004 *Journal of Computational Methods in Sciences and Engineering* **4** 559
[24] Bochynski Z, Drozdowski H 1995 *Acta Physica Polonica* **A 88** 1089
[25] Belyakova O A, Slovokhotov Yu L 2003 *Russian Chemical Bulletin, International Edition* **52** 1

IOP Conf. Series: Materials Science and Engineering **110** (2016) 012066 doi:10.1088/1757-899X/110/1/012066

Surface and sorption properties of TIO_2 nanotubes, synthesized by electrochemical anodization

T M Serikov[1], N Kh Ibrayev[1] and Zh Smagulov[1]

[1] Karaganda, 100028, Republic of Kazakhstan, Institute of Molecular Nanophotonics, E.A. Buketov Karaganda State University

E-mail: serikov-timur@mail.ru

Abstract. Electrochemical anodizing of titanium metal in the fluorine-containing electrolyte was used for production of nanotubes of titanium dioxide. The surface morphology of the films was investigated by scanning electron microscopy. It was found that with increasing anodizing voltage the inner diameter, growth rate and nanotubes interporous distance is increasing. The distribution of pore volume studied by the method of adsorption and desorption of nitrogen. The specific surface area of titanium dioxide films was measured for determine the conditions of anodizing on their changes. The sorption capacity of the films was studied.

1. Introduction

Currently titanium dioxide is a major semiconducting material in the dye sensitized solar cells (DSSC). It has been achieved about 12-13% efficiency at the best lab samples of solar cells manufactured on the basis of TiO_2 nanoparticles [1]. The use of nanotubes instead of nanoparticles can be the one of the ways to increase the efficiency of DSSC-cells. This is due to the 3-dimensional electron transport carried out in the films made of nanoparticles TiO_2. The connection between the nanoparticles, which is formed by thermal annealing, affects the electron transfer ability between the nanoparticles. Unformed chemical bonds lead to a decrease in the efficiency of electron transport. When using nanotubes it is a one-dimensional electron transport along the walls. It reduces the transfer of electrons from the charge generation points to the electrodes, as well as, if electrodes are properly designed the defects hindering electron transport are minimal [2].

The amount of light absorbed by a photovoltaic cell depends on the number of dye molecules adsorbed on the semiconductor surface. Increasing the specific surface area of the semiconductor oxide layer allows adsorb a greater number of the dye molecules and leads to increased absorption of solar light and increases the charge carriers concentration in the semiconductor layer.

There are many different methods, which results structures of different morphology, are used last years for obtaining nanostructures on the basis of titanium oxide [3-9]. They are mainly nanoparticles, nanotubes, nanowires and nanorods. There have been synthesized nanotubes of titanium dioxide [10-14] by various methods, the specific surface area of which exceeds hundreds of meters per unit area.

TiO_2 films with vertically oriented and highly ordered arrays of nanotubes were firstly obtained by electrochemical anodization of titanium foil in the fluorine-containing electrolyte by D. Gong et al. [15]. Since then, many works have been devoted to the study of the growth mechanism and the formation of TiO_2 nanotubes [16-19].

IOP Conf. Series: Materials Science and Engineering **110** (2016) 012066 doi:10.1088/1757-899X/110/1/012066

This paper presents the investigation results of electrochemical anodization conditions on a surface area of titanium oxide films, pore volume, sorption properties and geometric characteristics of TiO_2 nanotubes.

2. Experimental part

The titanium foil with a thickness of 60 microns was subjected to chemical polishing in a solution consisting of H_2SO_4 - 60%, HF - 25%, $C_3H_5(OH)_3$ - 15%. The polishing process occurs at a voltage of 20 V with vigorous stirring for 10 minutes. After polishing the samples were washed with distilled water, dried and then they were excised from the plate the required dimensions. Before anodization the samples were sonicated into 2 M solution of HCl, and then in acetone for 10 minutes to remove impurities and degreasing

As the basis of electrolyte the $C_2H_6O_2$ was used (content of NH_4F is 0,5 wt%, H_2O is 3 wt%). Anodic oxidation of titanium was carried out in an electrochemical cell in potentiostatic mode at a temperature of 5-7 ^{0}C. The platinum foil was as a cathode. The distance between the anode and cathode was 3 cm.

The electrochemical oxidation process consists from two steps. First is anodizing for 2 hours. Separation the formed film at the first step is carried out in an ultrasonic bath of 1 M hydrochloric acid solution. Then the sample was washed with copious amounts of deionized water and dried at a temperature of 80 ^{0}C. The duration of the second step of anodizing was 24 hours. Anodizing voltage was varied from 20 V to 70 V.

After the second stage of the anodizing the obtained TiO_2 nanotubes were sonicated to remove the byproducts and the surface oxide layer. Then they were washed with copious amounts of distilled water and dried in a stream of nitrogen. Crystallization of the amorphous phase of TiO_2 was performed by the thermal treatment of samples in a muffle furnace at 450 ° C for 3 hours with a temperature increase of 5 ^{0}C per minute.

For measurement of the specific surface area the nanotubes film were separated from the titanium foil by H_2O_2 (37%) and placed in the adsorber. Specific surface area was measured by the BET method. Pore size distribution and pore volume dependency on their diameter was obtained from the adsorption isotherm and desorption of nitrogen in measuring complex Sorbi-MS (Russia). Thermal training of samples was carried out at 100 ^{0}C for 180 minutes in the preparation block «SorbiPrep».

The images of samples surface were obtained on the scanning electron microscope MIRA 3LMU (Tescan, Czech Republic).

3. Results and discussion

Surface morphology of porous titanium dioxide films, obtained by scanning electron microscope, is showed on figure 1. All samples were anodized in the electrolyte: ethyleneglycol + NH_4F 0.5 wt.% + H_2O 3 wt. % at a variation of the anodization voltage from 20 V to 70 V.

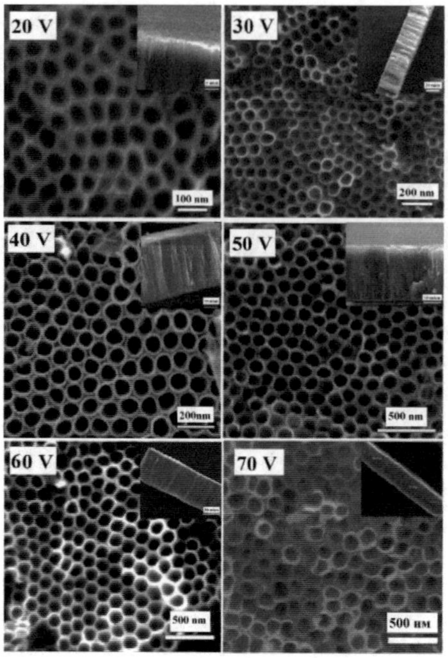

Figure 1. SEM images of the surface and cross-cleavage of titanium dioxide films

It can be seen that there are highly ordered titanium dioxide nanotubes, perpendicular to the plane of the titanium foil, on its surface. Table 1 shows the characteristics of nanotubes obtained by 24 hours anodizing. There are results of measurements of the specific surface area of the films and the total pore volume of titanium dioxide nanotubes. The total pore volume and pore volume distribution were obtained from the adsorption isotherm and desorption of nitrogen

Table 1. Geometric characteristics and specific surface area, total pore volume of TiO2 films

Voltage, V	The length of the nanotubes, μm	The average inner diameter of the nanotubes, nm	The average interporous distance, nm	Specific surface area S_{BET}, m^2/g	Specific surface per 1 μm S_{BET}, m^2/g	The total pore volume V_p, cm^3/g
20	17,1	55	60	71,3±0,8	4,20	0, 200
30	22,0	70	75	67,84±1,0	3,08	0.155
40	27,0	90	86	61,3±0,9	2,30	0.128
50	36,0	110	139	65,4±0,8	1,80	0.112
60	37,0	140	150	62,45±0,8	1,70	0.106
70	40,0	150	170	59,3±0,9	1,50	0.102

The data shows that while increasing the anodizing voltage the average interporous distance, inner diameter and growth rate of the nanotubes length are increasing too. When the anodization voltage is increased the specific surface area and pore volume of the films decreased. This is due to the fact that the number of nanotubes per 1 cm^2 is reducing due to increased internal nanotubes diameter.

Picture 2 shows the isotherms of nitrogen adsorption.

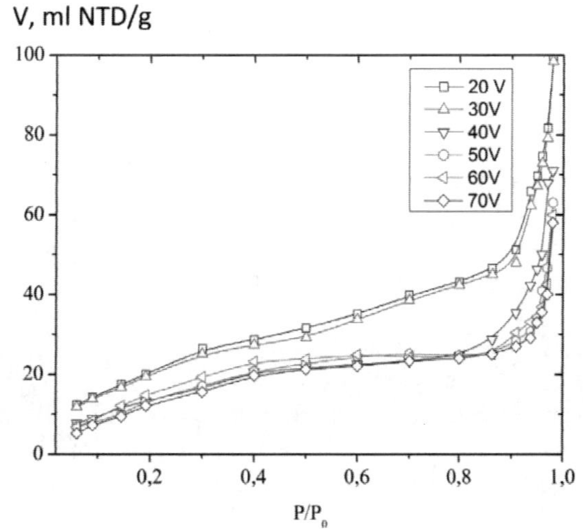

Picture 2. The isotherms of films adsorption, obtained at different voltages

According to the classification of the IUPAC, the obtained isotherms have II adsorption type, which corresponds with multilayer adsorption [20]. With pressure increasing the isotherm asymptotically approaches the straight line $P/P_0 = 1$. In the initial part of the isotherm it has a convex shape, which indicates the strong interaction of adsorbate with the adsorbent. In addition, the adsorption isotherm for all the samples have the same form, which indicates the identity of the porous structure of the films, which differ from each other in the volume of adsorbed gas.

Also the molecules sorption into the porous film of titanium dioxide was investigated. For this purpose sorption of the Z-907 (cis-Bis (isothiocyanato) (2,2'-bipyridyl-4,4'-dicarboxylato) (4,4'-di-nonyl-2'-bipyridyl) ruthenium (II)) dye molecules was carried from the ethanol solution into the pores of TiO_2 films, prepared in the electrolyte: ethyleneglycol + NH_4F 0.5 wt.% + H_2O 3 wt. % at a anodizing voltage of 20 V.

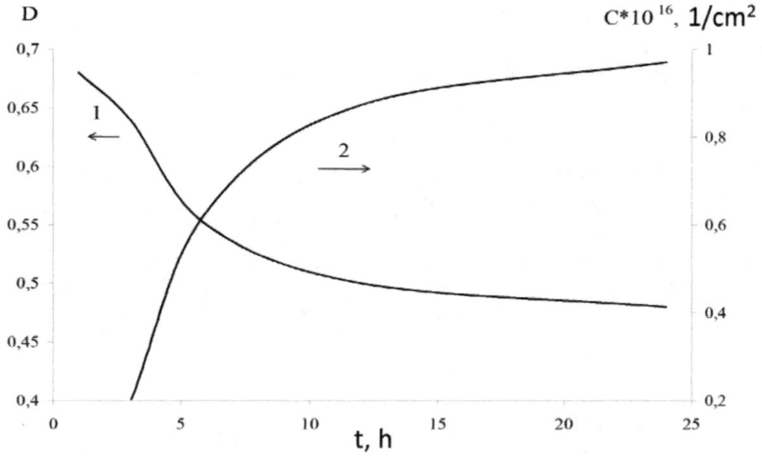

The presented data shows that with increasing time of sorption the optical density of the solution decreases, which indicates a decrease in the concentration of molecules in solution and flow physisorption of molecules on the surface of the TiO_2 film.

The number of molecules adsorbed per unit area was determined by the change in optical density D before and after the sorption.

$$C_S = \frac{N_A \cdot c \cdot V'}{S}(1 - \frac{D_1}{D_2}), \quad (1)$$

where: N_A - the Avogadro constant; c - the phosphor molecules concentration in solution; V' - the volume of solution; S - area of the adsorbent; D_1 and D_2 - optical density of the solution before and after adsorption.

It was established that with increasing adsorption time the concentration of molecules on the surface of the film increases, and then reaches saturation.

4. Conclusion

Thus, by the two stage electrochemical anodization there are highly ordered titanium dioxide nanotubes, perpendicular to the plane of the titanium foil, formed on the surface of the titanium foil. It was found that with increasing anodizing voltage the internal diameter, growth rate and nanotubes interporous distance is increasing. The specific surface area and total pore volume begin to decrease due to the falling number of nanotubes per unit area. With increasing time sorption the optical density of the solution decreases, the density of molecules on the surface of the film is increasing, and then reaches saturation.

5. References

[1] O'Regan B, Gratzel M 1991 Nature **353** 737
[2] Mukul D, Hongshan H 2012 *Scanning Electron Microscopy* **27** 537
[3] Yang J, Mei S, Ferreira J M F 2001 *Materials Science and Engineering: C* **15** 183
[4] Yan XM, Kang J, Gao L, Xiong L, Mei P 2013 *Applied Surface Science.* **265** 778
[5] Sahil Sahni, S Bhaskar Reddy, BS 2007 *Materials Science and Engineering: A.* **452** 758
[6] Pierson HO 1992 NJ 235
[7] Nam S H, Hyun J S, Boo J H 2012 *Materials Research Bulletin.* **47** 2717
[8] Zhu Y, Li H, Koltypin Y, Hacohen Y R, Gedanken A 2001 *Chemical Communications* **24** 2616
[9] Wu X, Jiang Q Z, Ma ZF, Fu M, Shangguan WF 2005 *Solid State Communications.* **136** 513
[10] Nogueira A F, Freitas J, Freitas J N, Winnischofer H 2006 Activity report **1** 1
[11] Abidaa B, Chirchia L, Barantonb S, Ghorbela A 2011 *Applied Catalysis B: Environmental* **106** 609
[12] Wen L, Gao J, Zhang F, Zhang G 2007 *Materials Transactions* **48** 2464
[13] Wang H, Li H, Wang J, Wu J, Li D, Liu M, Su P 2014 *Electrochimica Acta.* **137** 744
[14] Pugazhenthirana N, Murugesana S, Anandan S 2013 *Journal of Hazardous Materials* **263** 541
[15] Gong D, Grimes C A, Varghese O K, Hu W, Singh R S, Chen Z, Dickey E C 2001 *J. Mater. Res.* **16** 3331
[16] Macak J M, Zlamal M, Krysa J, Schmuki P 2007 *Small.* **3** 300
[17] Varghese O K, Gong D, Paulose M, Grimes C A, Dickey E C 2003 *J. Mater. Res.* **18** 156
[18] Zwilling V, Aucouturier M, Darque-Ceretti E 1999 *Electrochim. Acta* **456** 921
[19] Brunauer S, Deming L S, Deming W S, Teller E 1940 *Journal of the American Chemical Society* **62** 1723
[20] IUPAC Manual of Symbols and Terminology, Appendix 2, Pt.1, Colloid and Surface Chemistry 1972 *Pure Appl. Chem.* **31** 578

The impact of Ag nanoparticles on the parameters of DSS–cells sensitized by Z907

N Kh Ibrayev, A K Aimukhanov and A K Zeinidenov

Institute of Molecular Nanophotonics, Academician E.A. Buketov Karaganda State University, Karaganda, Kazakhstan

E-mail: niazibrayev@mail.ru

Abstract. Research of influence of Ag nanoparticles are in-process undertaken on absorption and on parameters CVC DSS-cells sensitized Z907. It is set that with the height of concentration Ag nanoparticles in tape to the concentration of 0.3% wt%. the absorbance of Z907 in a short-wave stripe grew to the value 1,6. It is set that under reaching the concentration of Ag nanoparticles in the cell of value the 0.3% wt%. efficiency of cell increased to 2.2%.

1. Introduction

To increase the work efficiency of solar elements the plasmons excitation can be used on the surface of metallic nanoparticles [1]. Three types of light traps plasmon are known, in their number dispersion in a far field, non communicative plasmon resonance in a near field, and superficial plasmon of polaritons on the border of border metal/semiconductor. A superficial plasmon arising up in metals can be used for the absorption increase of optical radiation or generation of photoelectric in different modifications of solar cells.

In basis of work principle of photodetectors using superficial plasmons, the generation of photoelectric lies in active material of photodetector under the action of variable electric-field of the plasmons excited by a radiation on the metallic nanoparticles surface. Thus an important circumstance is the local strengthening of electric-field of light wave as a result of excitation of superficial plasmon. Strengthening of the field is conditioned by that after the capture of photon the effective volume occupied by the electromagnetic field diminishes the surface of metal (its conversions in a superficial plasmon). Reduction of the field volume results in growth of spatial closeness of the field energy and its increasing.

The prosecutions of improvement of DSS-cells descriptions are constantly conducted. One of the perspective methods of this problem decision is the use in the construction of solar battery of the specially formed clusters from nanoparticles, which are possessing plasmon properties [2,3].

In this work the influence of Ag nanoparticles on volt-ampere characteristics of DSS-cells sensitized by Z907 was studied.

2. Experiment

Nanoparticles of silver (Ag NP) were prepared by citrate recovery of AgNO3 in aqueous solution [4]. In this method, the reducing agent and the stabilizer is a citrate ion derived by dissolving in water sodium trisodium salt of citric acid. When heating of the solution and the citrate ion oxidation the acetonedicarboxylic and itaconic acid are formed. These acids are adsorbed on the particle surface and their growth is monitored. The size measuring of the resulting structures was conducted using the submicron particles size analyzer Zetasizer Nano ZS. The average particle size of silver is equal to 4 nm. The scatter of the values obtained by dispersion of silver particles is minimal and ranges from 1 to 5 nm (Figure 1).

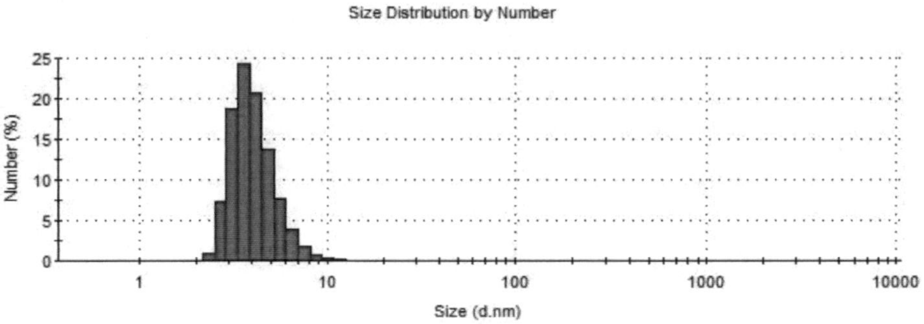

Figure 1. Distribution of Ag nanoparticles sizes in water solution.

Registration of absorption spectrums of the investigated standards was made on the spectrophotometer Cary 300 UV - Vis.

Paste on the basis of Degussa P25 powder was prepared by following methodology: 100 mg of colloid TiO_2 (size of particles of 25 nm) ground in a porcelain mortar with the small amount of deionized water (2 ml) and acetylacetone (0.2 ml). After formation of viscid paste gradually 3 ml of water was added additionally, continuing grinding

For preparation of tapes from titanic pastes glass samples were used, covered by the conducting layer of FTO (6 Ω/cm^2, Sigma Aldrich) cleared with the use of ultrasonic bath. After a 15 minute ultrasound treatment the samples were washed by the distilled water and ethanol. Then the cleaning of samples surface was conducted by an ionic etch in plasma of argon during 20 minutes. Then paste was inflicted on the surface of conducting glass plate by the stencil printing. Procedure of the stencil printing (causing, drying) recurred to the receipt of suitable thickness of semiconductor tape. After it semiconductor tapes gradually heated in a muffle stove at temperatures: 325°C for 5 mines; 375°C for 5 mines; 450°C for 15 mines; 500°C for 15 min. Introduction of silver nanoparticles in tape of TiO_2 took place on the stage of preparation.

The study of structure and element composition of titanic pastes was carried out using electronic microscope Mira 3 LMU, Tescan. Microstructure and element analyses of paste with Ag nanoparticles are shown on figure 2a.

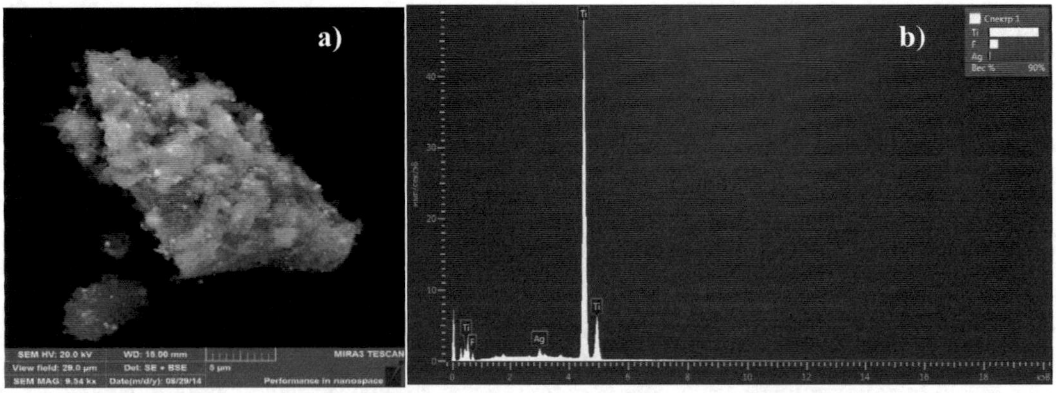

Figure 2. SEM images (a) and element analysis (b) of paste with Ag nanoparticles.

Figure (a) shows that paste is a porous structure with disseminations of Ag nanoparticles. The study of element composition of tapes showed the presence of titan atoms, oxygen and silver (figure 2b).

3. Results and discussion

Figure 3 (a) shows absorption spectrums of titan tapes with presence of Ag nanoparticles in tape at different wt% concentrations. Picture (a) shows that the increase of Ag nanoparticles concentration in tape results in the increasing of absorbance of titan tapes.

Adsorption of photoactive Z907 molecules by the surface of titan tapes from ethanol solution with the concentration of dye 10^{-3} mol/l was further conducted. Tape was maintained in solution of phosphor during 20 hours. Then the tape was extracted from solution and dried out in a drying closet (T=80°C) for 3 hours. Figure 3 (b) shows absorption spectrums of Z907 of titan tapes at Ag nanoparticles presence in different wt% concentrations. The figure shows that with the increasing the concentration of Ag nanoparticles in tape results in the increasing of dye absorbance.

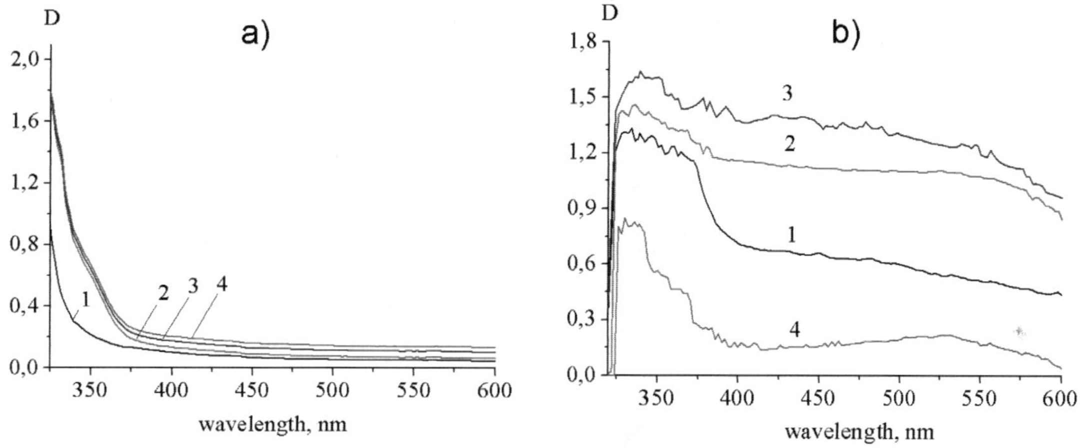

1-0%, 2-0.01%, 3-0.3%, 4-2%

Figure 3. Spectra of absorption of titan tapes (a) and Z907 in titan tape (b) at different Ag nanoparticles concentrations.

It was investigated that increasing the concentration of Ag nanoparticles in tape to the concentration of 0,3wt% the absorbance of Z907 in a short-wave stripe grew to the value 1.6. After further increase of Ag nanoparticles concentration the reductions of Z907 absorbance to the value 0.7 is observed.

CVC of DSS-cells with different Ag nanoparticles concentrations in tape is shown on a figure 4. Figure shows that CVC is nonlinear. Values of tension of idling (Uoc), current of short circuit (Icc), fill factor (FF) and efficiency determined from CVC.

Analysis of CVC on a figure 4 shows that adding of silver nanoparticles into DSS-cells substantially changes the parameters of CVC. Adding the silver nanoparticles 0.3 wt% into the porous layer of TiO_2 results in the increasing of short circuit current in 3.3 times. After this concentration the reductions of short circuit current was observed below of primary value.

IOP Conf. Series: Materials Science and Engineering **110** (2016) 012067 doi:10.1088/1757-899X/110/1/012067

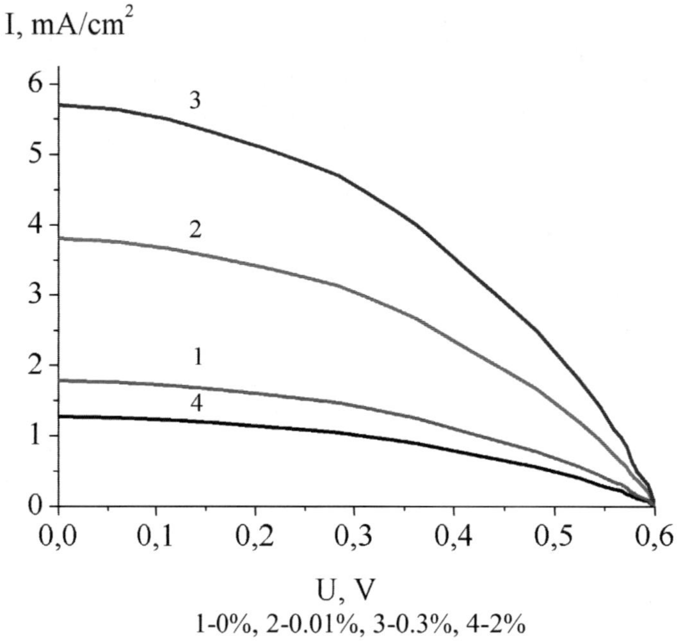

Figure 4. CVC of DSS–cells with Ag nanoparticles.

Dependence of value of cell short circuit current on silver nanoparticles concentration is presented on a figure 5. Figure shows that dependence passes through a maximum that is observed during the concentration of Ag 0.3 wt%. Efficiency of cells during a concentration the 0.3 wt% changed from 0.7% to 2.2%, Fill factor of CVC changed from 50% to 54%.

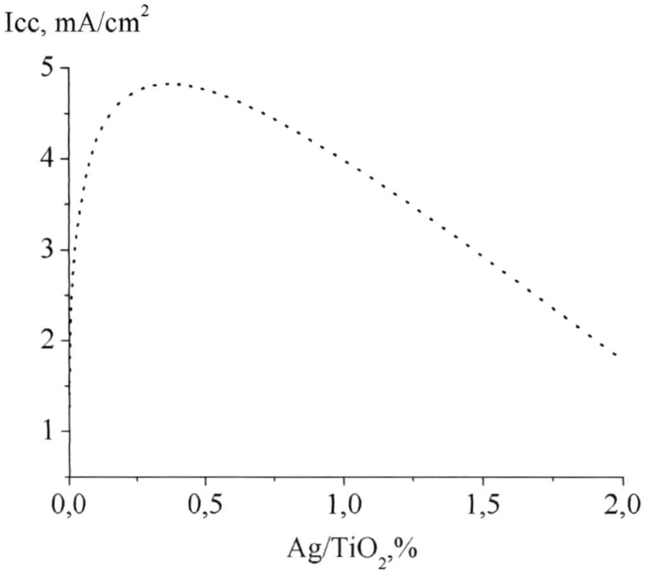

Figure 5. Dependence of value of cell short circuit current on a concentration of Ag nanoparticles.

4. Conclusion

Thus, the work was investigated the effect of plasmon silver nanoparticles on the photovoltaic properties of DSS-cells. It was found that with increasing the concentration of silver NPs in the film to the concentration of 0.3 wt% the Z907 absorbance in the short band increases in 1.3 times. And this leads to an increase the efficiency of DSS-cells in 3 times.

References

[1] Pillai S, Catchpole K R, Trupke T and Green M 2007 *Journal of Applied Physics* **101** 093105-1-093105-8
[2] Ishi T, Fujikata J and Makita K 2005 *Japanese Journal of Applied Physics* **44** L364-L366
[3] Liang Tang, Sukru E K and Salman L 2008 *Nature Photonics* **2** 226-229
[4] Sheng R S, Zhu L and Morris M D 1986 *Analytical Chemistry* **58** 1116-1119

Fluoride technology of obtaining REM magnetic alloys and master alloys

V L Sophronov[1], A N Zhiganov[1], Yu N Makaseev[1], I Yu Rusakov[1], V V Verkhoturova[2]

[1] Seversk Technological Institute, MEPhI, 636036, Russia, Seversk, Tomsk oblast, Kommunistichesky prospect street, 65
[2] Tomsk Polytechnic University, 634050, Russia, Tomsk, Tomsk oblast, Lenina prospect street, 30

E-mail: VLSofronov@mephi.ru

Abstract. Rare earth permanent magnets (REPM) based on neodymium-Fe-boron system are the most promising, since they have the highest magnetic and satisfactory mechanical characteristics. The paper covers physical-chemical principles and shows the results of experimental studies of the process of obtaining REM alloys and master alloys using fundamentally new fluoride technology based on ladle calciothermal REM fluorides and Fe reduction.

1. Introduction

Magnetic materials are alloys of, at least, one rare earth element with one 3d - transition metal [1].

Fluoride technology is one of the methods of obtaining magnetic materials to produce REPM. It includes the following stages: fluorination of raw materials with fluorine or another fluorinating agent, calciothermal reduction of fluorides with obtaining compact alloy ingots and (or) REM-Fe master alloys and magnets manufacturing based on metal powder industry method using the mechanical or hydride grinding and recycling [2, 3]. The main advantages of this technology are shown in [4].

2. Physical and chemical principles of the reduction process

2.1. Thermodynamics of the process

The thermodynamics of the process of metals fluorides reduction was discussed in the paper [5].

The obtaining Nd-Fe master alloys and Nd-Fe-B alloys can be represented by the equations:

$$NdF_3 + nFeF_3 + 1.5(1 + n)Ca = \mathbf{NdFe_n} + 1.5(1 + n)CaF_2, \quad (1)$$

$$NdF_3 + nFeF_3 + mFe + gB + 1.5(1 + n)Ca = \mathbf{NdFe_{(n+m)}B_g} + 1.5(1 + n)CaF_2, \quad (2)$$

where n, m, g – stoichiometric coefficients.

Table 1 shows the thermodynamic parameters (ΔH, q, T_a) for the process of REM fluorides combined reduction reaction 1.

Table 1 shows that part of the alloys and master alloys components could be added to the reaction charge in order to reduce thermicity of the process in the form of a powder or facings. The same effect is achieved by the addition of inert components, e.g. CaF_2, to the charge.

Table 1. The values of enthalpy, thermicity and adiabatic temperature of the reaction of obtaining $NdFe_n$ depending on the master alloy composition

Parameters	Composition of master alloy $NdFe_n$, % Wt. with neodymium content							
	100	90	80	70	60	50	40	30
$-\Delta H^{\circ}_{298}$, kJ/mol	170	410	650	1050	1530	2250	3130	4970
q, J/g	650	1310	1780	2320	2750	3160	3530	3820
T_a, K	1400	1860	2335	2700	3000	3300	3600	3900

Note. The thermodynamic calculations do not take into account the heat of Fe and neodymium mixing.

2.2. Kinetics of the process

The main kinetic characteristics of metallothermic reactions are: the activation energy E_a, the rate constant K_c, and the rate of the W process. Activation energy of the process of obtaining magnetic alloys and master alloys, described by equations 1 and 2, was determined by the equation [6]:

$$T_a - T_M = R(T_M - 273)^2/E_a, \tag{3}$$

where T_a – adiabatic temperature of the process;

T_M – maximum process temperature. It was determined empirically by direct measurement using a tungsten-rhenium thermocouples VR-1 and low-inertia locking device (loop oscillograph).

The data obtained are presented in Table 2.

Table 2. The maximum temperature and the activation energy of calciothermal process of obtaining alloys and master alloys of different composition

Parameter	Master alloy composition containing Nd, %Wt. (reaction 1)				Amount of Fe in the form of metal powder, % Wt. (reaction 2)		
	100	90	80	70	55	60	65
T_a, K	1400	1860	2335	2700	2520	2380	2240
T_M, K	–	1700	2100	2400	2260	2140	2020
E_a, kJ/mol	–	126,2	120,8	115,3	105,8	118,0	125,4

Table 2 shows that the activation energy of the process of coreduction of REM fluorides and transition metals with the formation of alloys and master alloys depends on their composition and is within 105.8-126.2 kJ/mol.

3. Experimental study

3.1 Installation description and the research methods

Study of the process of ladle calciothermal reduction of metal fluorides was carried out on the reduction reactor shown in Figure 1. It is made as a cylindrical unit (2) whith cover (1). Outside, the reactor is equipped with a cooling jacket (6). Its inner part is equipped with steel pad (3) with CaF_2 lining (4) and the prepared charge (5). The reaction in the reactor starts with electric-squib ignition.

The methodology of the study was as follows. Necessary amounts of trifluorides metals, Fe, FeB metal powders and Ca facing were weighed separately, laminated, stirred in a rotating drum mixer for 10-20 minutes and, then, were charged into the crucible. Ignition device, consisting of a metal spiral, was assembled on the surface of the charge. All the items were placed in a reactor and sealed. Initiation of the reduction melting reaction was carried out with a spiral, heated by the transformer.

For conducting calciothermal reduction the following materials were used: metal fluorides with a conversion coefficient of the native oxides into fluorides for: Nd_2O_3 – 94-96 % ($\alpha_{T\Phi H}$), for Fe_2O_3 – 92-94 % ($\alpha_{T\Phi Ж}$); Fe metal powder of the PZHV-1, 2 (GOST 9849) brands; FBO-20 with a particle size of not more than 0.25 mm; Metallic Calcium (TC 95824) in the form of facing.

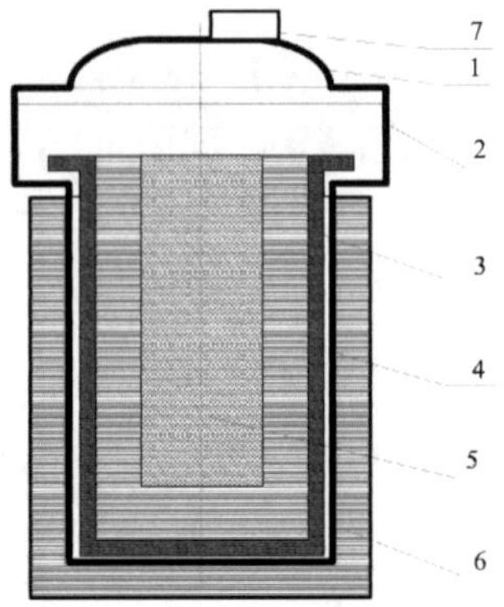

1 – reactor cover; 2 – reactor housing; 3 – steel pad; 4 – CaF_2 lining;
5 – charge; 6 – cooling jacket; 7 – emergency pressure relief valve

Figure 1. General view of the reduction reactor

3.2 Studies of the influence of various factors on the yield of the metal into the ingot

The influence of overreductant, reducing melting scale, alloy and master alloy composition, the degree of metals oxides fluorination, the wall thickness of the lining of the crucible to the reduction degree of fluorides, metal yield into an ingot and its quality were studied. The data are shown in Figures 2 and 3.

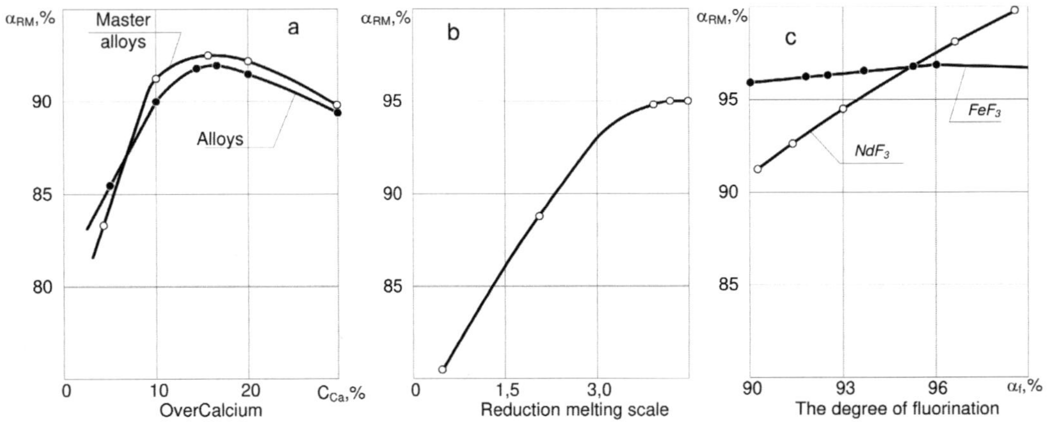

Figure 2. Dependence of the metal yield into the ingot on overcalcium (a), reduction melting scale (b), the oxides fluorination degree(c)

The data in Figures 2 and 3 show that the alloys and master alloys (metals) yield into the ingot and the reduction degree of REM fluorides and Fe are influenced by many factors.

Reduction melting scale. With increasing the scale up to 4 kg for the amount of metal, the yield of the metal into the ingot increases. This is due to decrease of heat loss from the reduction melting products, and, respectively, increase of the system duration of the stay in a molten state.

OverCalcium. With increasing of overCalcium up to 10%, metal yield into the ingot increases, and, then, with overreductant from 10 to 20%, practically does not depend on its quantity, and it starts decreasing with an excess of more than 20%.

The fluorination degree. Fluorination degrees of neodymium and Fe oxides have different effects on the metal yield into the ingot. The oxide content in Fe fluoride in up to 10% does not have serious effect on the metal yield into the ingot, but the same change of the content of neodymium oxide in fluoride reduces its yield up to 91-92%.

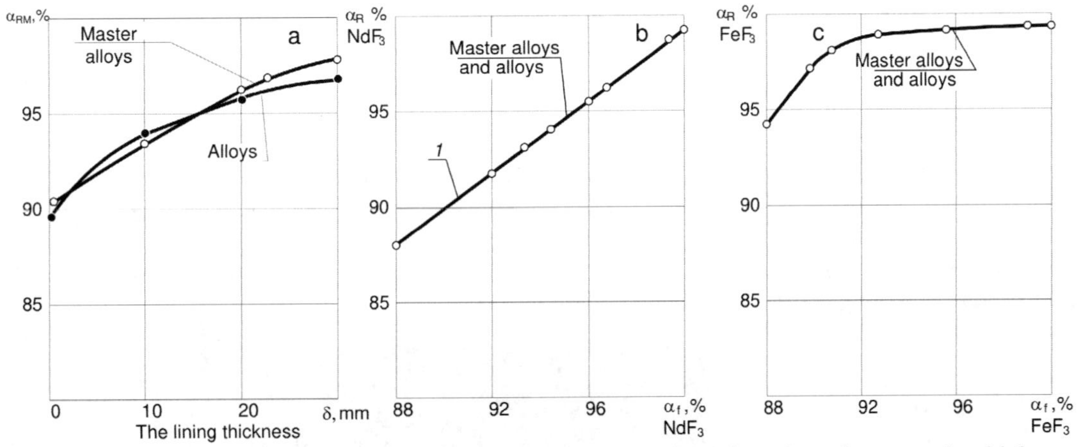

Figure 3. Dependence of Nd, Fe yield, as well as alloys and master alloys into ingot on the thickness of the lining layer (a), the fluorination degree of neodymium (b) and Fe (c)

3.3 The study of the alloys and master alloys quality

To assess the quality of materials 2 alloys and master alloys ingots were subjected to chemical (emission spectrometer with inductively-coupled plasma ICAP 6200 DUO), X-ray diffraction (X-ray diffractometer ARL'XTRA) and electron microscopy (scanning electron microscope Vega 3 SBH (TESCAN)) analysis, the results of which are shown in Table 3 and in Figures 4-7.

Table 3. Alloys and Master alloys chemical composition

N RM	$\alpha_{вп}$, %	M_{ingot}, kg	Calculated alloy composition, % (Wt.)		The results of chemical analysis, % (мWt.)				Yield into ingot, α, %	
			$C_{(Nd+Pr)}$	C_{Fe}	C_{Nd+Pr}	C_{Fe}	C_B	$C_{\Sigma(Nd,\ Pr,\ Dy,\ Tb,\ Fe,\ B)}$	FeF_3	$P3F_3$
1	97.8	4.89	32	67	31.1	67.5	1.1	99.7	98.5	95.0
2	98.4	4.92	36	63	34.8	63.6	1.1	99.5	99.4	95.2
3	98.0	4.90	40	60	39.7	59.9	-	99.6	99.5	95.0
4	97.8	4.89	40	60	39.8	60.0	-	99.8	99.5	97.0

Note: 1, 2 - Alloys chemical composition; 3, 4 - Master alloys chemical composition

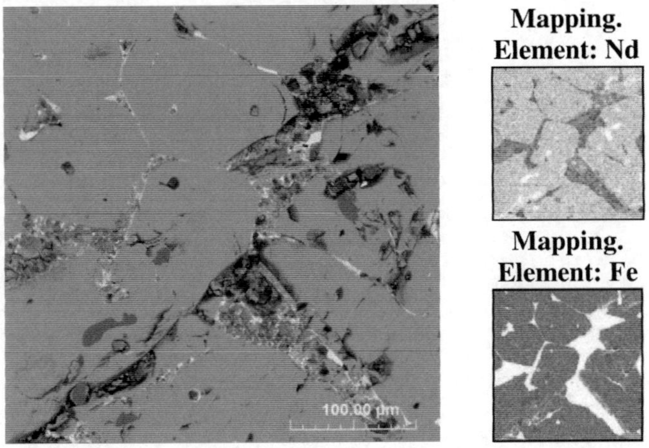

Figure 4. The results of electron microscopy analysis of alloy sample

The theoretical weight of alloys and master alloys at the initial charge is 5 kg.

Data analysis in Table 3 shows that the total content of REM, Fe and boron is greater than 99.5% for alloys and 99.6% for master alloys (REM and Fe), the degree of fluorides reduction is 98.5% (Fe) and 95 % (REE).

Figure 4 shows that the phase composition of the alloy is nonhomogeneous. Conducted mapping finds out that the grain slice consists essentially of Fe and neodymium. Dark spots on the image of grains slices are the phases, enriched with Fe. The grains boundaries and the grains pores are saturated with neodymium, containing a minor amount of oxygen.

Figure 5 shows that the surface of the sample is inhomogeneous The surface of the edges has the defects with the size up to 1 micron.

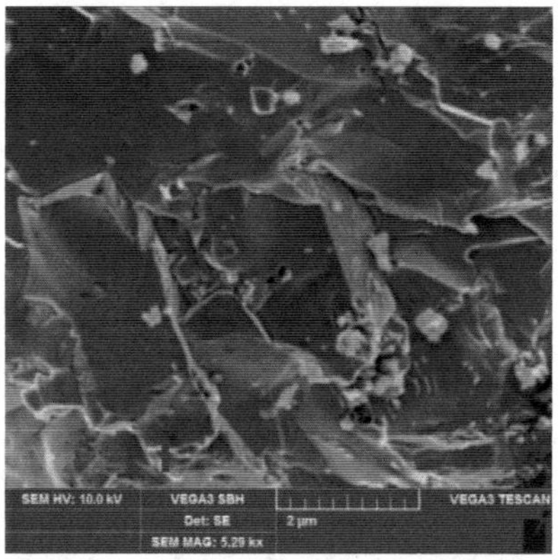

Figure 5. The results of electron microscopy analysis of master alloy

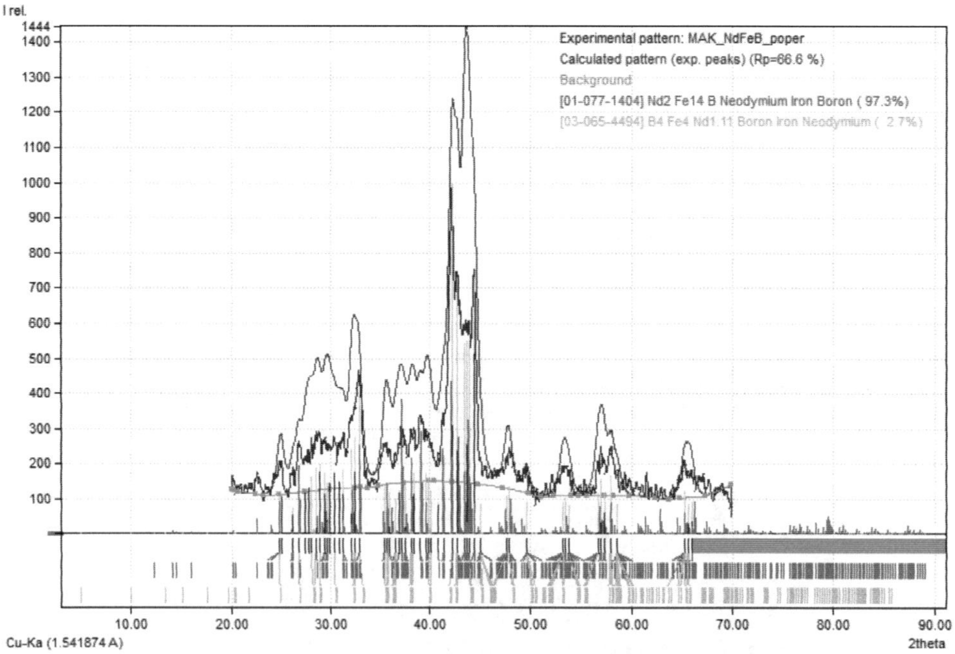

Figure 6. The results of X-ray diffraction analysis of alloy sample

Figure 6 shows that the content of the main magnetic phase $Fe_2Nd_{14}B$ in the sample is 97.3%, the rest is $Nd_{1,11}Fe_4B_4$ phase.

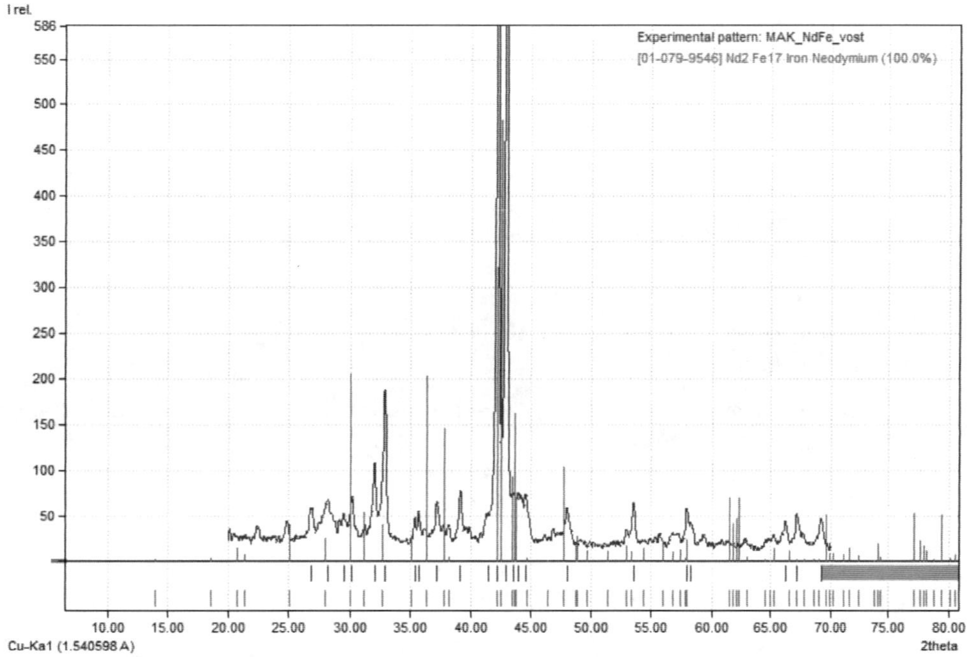

Figure 7. The results of X-ray diffraction analysis of master alloy sample

Figure 7 shows that the main Nd_2Fe_{17} phase content in the sample is 100%.

4. Conclusion

1 The thermodynamics of the REM and Fe fluorides reduction process is considered. It is shown that the alloys based on Nd-Fe-B system and Nd-Fe master alloys could be obtained with ladle calciothermal metal fluorides reduction.

2 The kinetics of the process of the complex reduction of REM fluorides and Fe is studied and the activation energy of the process is determined, which is in the range of (105,8-126,2) kJ / mol.

3 Experimental studies of the process of obtaining alloys and master alloys by ladle complex reduction of RE fluorides and Fe are carried out.

4 The quality of the alloys and master alloys obtained is proved with chemical, X-ray diffraction and electron microscopic analysis.

This work was supported by the Federal target program "Research and development on priority directions of scientific-technological complex of Russia for 2014-2020» (RFMEFI57814X0018).

References

[1] Tulupov I. F. 1977 Permanent magntes (M.: CRI for patent information, TEI and State Commettee on Inventions and Discoveries, p 1–48)

[2] Sofronov V, Buynovskiy A 2012 Fluoride technology of alloy production on the basis of rare-earth metals for the production of high-energy permanent magnets Non-ferrous metals **№ 1** P 14-18

[3] Sachkov V, Anufrieva A, Sofronov V 2014 Study of the process of obtaining the initial metal fluorides for the production of high-energy permament magnets by dry fluoride technology *Osaca Japan conference proceedings* p 519–520

[4] Sofronov V L, Buynovskiy A S, Makaseev Yu N 2014 Fluoride technology for obtaining magnetic materials based on Nd-Fe-B. Kinetics of fluorination processes of neodymium and iron oxides Procedia Chemistry **№ 11** p 158-164

[5] Bolshakov K. 1976 Chemistry and techology of Rare elements (M.: Higher School, p 360)

[6] Frank-Kamenetsky D. 1967 Diffusion and Heat Transfer in Chemical Kinetics (M.: Sc., p 492)

IOP Conf. Series: Materials Science and Engineering **110** (2016) 012069 doi:10.1088/1757-899X/110/1/012069

Study on the process of Fe (III) oxide fluorination

V L Sophronov[1], M E Kalaev[1], Yu N Makaseev[1], V I Sachkov[1], V V Verkhoturova[2]

[1] Seversk Technological Institute, MEPhI, 636036, Russia, Seversk, Tomsk region, Kommunistichesky avenue, 65
[2] Tomsk Polytechnic University, 634050, Russia, Tomsk, Tomsk region, Lenina avenue, 30

E-mail: VLSofronov@mephi.ru

Abstract. The article deals with a fundamentally new fluoride technology for obtaining fluoride materials, provides data on the kinetics of the process of fluorination of Fe oxide with fluorine, fluoride and ammonium bifluoride. The physical and chemical properties of obtained fluorides are shown: a study of the elemental composition, grain-size composition using the method of scanning electron microscopy and laser diffraction.

1. Introduction

The fluoride technology of obtaining alloys for permanent NdFeB magnets is based on the process of ladle calciothermal reduction of the mixtures of non-aqueous REM fluorides and Fe. The description of fluoride technology and its advantages are given in papers [1-3].

The methods of obtaining inorganic fluorides, including Fe (FeF_2 и FeF_3), could be divided into two fundamentally different groups - hydrometallurgical and thermal.

Hydrometallurgical methods are based on deposition of metal fluorides from saline solutions or hydroxide sludges using a fluorinating agent, with their subsequent drying and calcination.

Unfortunately, Fe fluorides are unstable in water vapor atmosphere. So, FeF_2 and FeF_3, in the presence of water vapor, at 520 K, go into Fe_3O_4 and Fe_2O_3, respectively. Thus, the formation of Fe fluoride crystallohydrates with the number of water molecules from 1 to 6 is possible, and, at deposition of metal fluorides from saline solutions with fluorhydric acid, the formation of Fe fluoride crystallohydrates with a number of hydrogen fluoride molecules, e.g. $FeF_3{\cdot}3HF{\cdot}3H_2O$ is possible. It is impossible to obtain non-aqueous fluorides from crystallohydrates by thermal treatment.

During heating of Fe (III) fluoride crystallohydrate the following processes could be observed:

$$FeF_3 \cdot 4{,}5H_2O \xrightarrow{70-100^\circ C} FeF_3 \cdot 3H_2O \xrightarrow{100-150^\circ C} Fe_2OF_4 \xrightarrow{200^\circ C} Fe_2O_3.$$

Therefore, non-aqueous fluorides, which are suitable for metallothermic reduction, could not be obtained by aqueous methods.

Thermal methods. For obtaining non-aqueous trifluorides the methods of Fe oxides fluorination are preferable. As fluorinating agents gaseous F_2 and HF or NH_4F and NH_4HF_2 melts could be used.

The results of the studies of the process of Fe oxides fluorination with gaseous elemental fluorine and in the melt of ammonium fluoride are shown in the paper.

2. Kinetics of fluorination process

2.1 The fluorination with elemental fluorine

Study of the kinetics of the process of Fe oxide fluorination was conducted using thermogravimetric installation with nickel retort, shown in Figure 1. The installation can operate in isothermal conditions, both in pure fluorine atmosphere and in the atmosphere of a gas mixture of fluorine-argon or nitrogen at the temperatures up to 600 ° C and permanent recording of the sample mass and temperature.

Fe (III) oxides of Ch.p brand (Ch.p. – chemical purity) were chosen as the objects of the studies. The research was carried out at the temperatures from 420 ° C to 550 ° C and in the following constant conditions: initial sample weight is 150 mg; gas flow (composition: F2- 96-97%, HF – in res.) is 2 l / h; argon flow is 4 l/h; particles size is less than 100 microns.

The results of kinetic studies and experimental data processing for the Fe oxide fluorination process are shown in Figures 2 and 3.

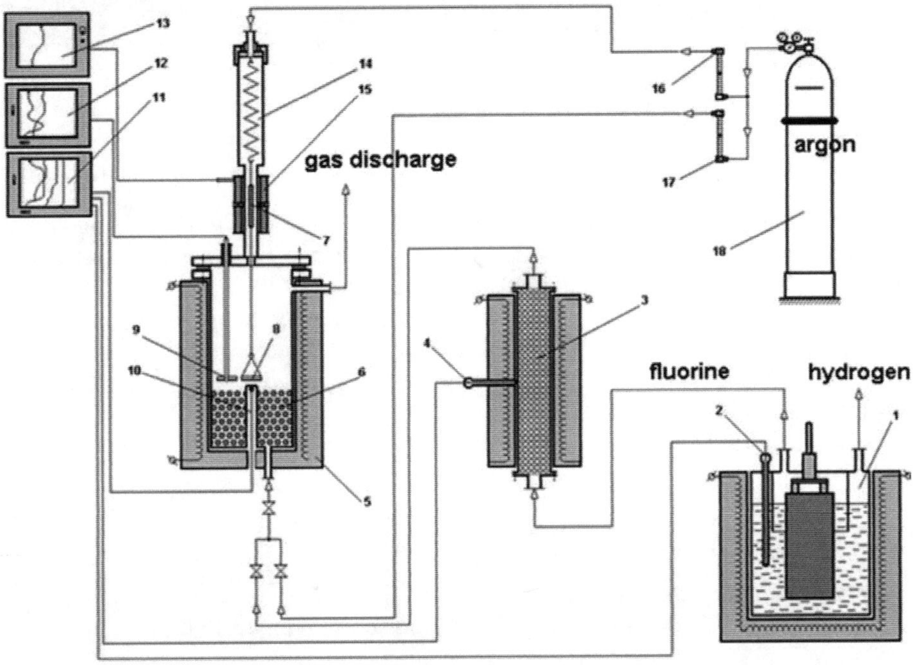

1 - electrolyser; 2, 4, 10 - thermocouples for recording the temperature of the electrolyte in elektrolyser, the sorbent agent in a column and the gas in the reactor, respectively; 3 – the column with the sorbent agent (NaF); 5 - fluorination reactor; 6 - nickel filler; 7 - nickel core; 8 - a cup with the product; 9 - thermocouple with a cup; 11 - multi-channel recording device of temperature mode of adsorption column and electrolyser; 12 - recording device of the product temperature; 13 - recording device of the product mass changing; 14 - weighing spring; 15 - measuring inductors; 16, 17 - flow rotameters; 18 – a tank with an inert gas

Figure 1. Experimental installation for thermogravimetric studies of fluorination processes

In Fe_2O_3 fluorination degree dependence on the process duration at various temperatures, shown in Figure 1, we could see that the degree of fluorination increases with increasing the temperature.

Thus, if the temperature increases from 420 ° C to 550 ° C, the degree of fluorination at the time of fluorination of 2400 sec. increases from 14% to 93%.

When the results of kinetic studies of Fe_2O_3 fluoridation were processed, several mathematical models were used to describe the chemical: for the kinetic region the equation of districting sphere and surface was used, for diffusion region Todes and Yander equations were used, for description of the processes in the initial period of the reaction Kazeev-Erofeev-Kolmogorov equation was used [4].

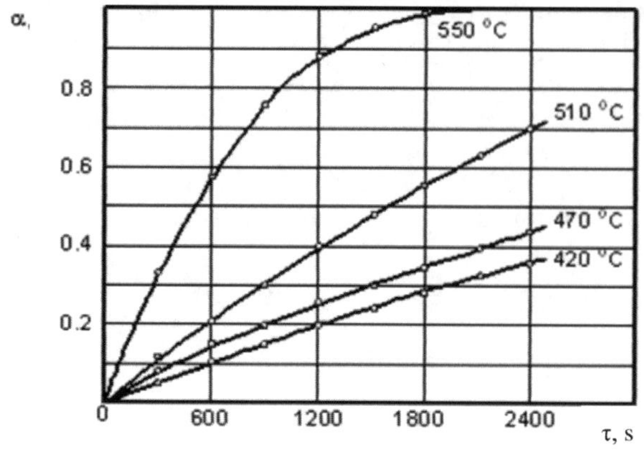

Figure 2. Fe_2O_3 fluorination degree dependence on the duration of the process at different temperatures

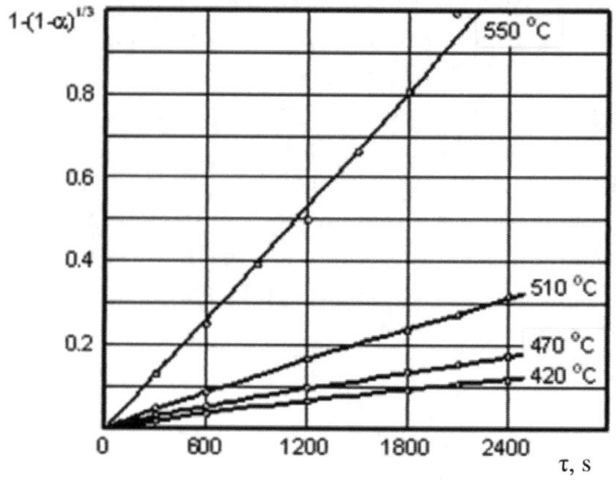

Figure 3. The results of experimental data processing using the districting surface equation

It should be noted that, at high temperatures (over 500 ° C), there is some melting and sintering of solid fluoride films based on the Fe_2O_3 powder particles, which decreases the fluorine access to the surface and reduces the degree of resulting product fluorination.

The results of experimental data mathematical processing concerning Fe oxide fluoridation using the different models have shown that the districting surface equation gives the smallest error in the linear approximation. The reaction order, determined using this model, is close to 1, and the apparent activation energy is $E_a = (43,0 \pm 3,0)$ kJ/mol.

To describe the kinetics of Fe oxide fluorination in the range of 420-550 ° C the following equation could be used:

$$1-(1-\alpha)^{\frac{1}{3}} = (0,087 \pm 0,007)\cdot \tau^{-\frac{(43000\pm 3000)}{R\cdot T}}.$$

2.2 Fluorination with ammonium fluoride

Fe fluoride was obtained using the reaction of Fe oxide and ammonium fluoride interaction according to the scheme:

$$Fe_2O_3 + 6NH_4F = 2FeF_3 + 6NH_3\uparrow + 3H_2O.$$

For this purpose, Fe oxide sample was mixed with NH_4F in a molar ratio of 1: (8-9) and placed into the reactor preheated to 300 ° C. The sample was being kept there for 1.5 hours, then the system was vacuum-processed to a residual pressure of $1\cdot 10^{-3}$ mm. of mercury and accommodated.

The products research was carried out using the installation of combined heat and differential calorimetric analysis SDT Q-600, produced by Intertech instruments company (hereinafter, thermal analyzer). Derivatographic analysis allowed us to determine the areas of chemical and phase transformations, while diffraction method (XRD) analysis - structural constitution.

Results of the study of the process of Fe oxide fluorination are shown in Figure 4.

X-ray diffraction analysis shows the absence of the fluorination reaction at the temperature of up to 120 ° C, but at the temperature of 175 ° C ammonium complex composition NH_4FeF_4 appears.

Heating the mixture of reactants up to 206 ° C leads to the formation of the complex with $(NH_4)_3FeF_6$ composition. This complex is thermally unstable and with further temperature increase decomposes into two compounds: NH_4FeF_4 and FeF_3. At the temperature of 365 ° C, the product of fluorination in 99% consists of NH_4FeF_4 complex. With further temperature increase up to 450-600 °C tetra-ammonium fluoride complex is destroyed in the air with forming Fe oxide.

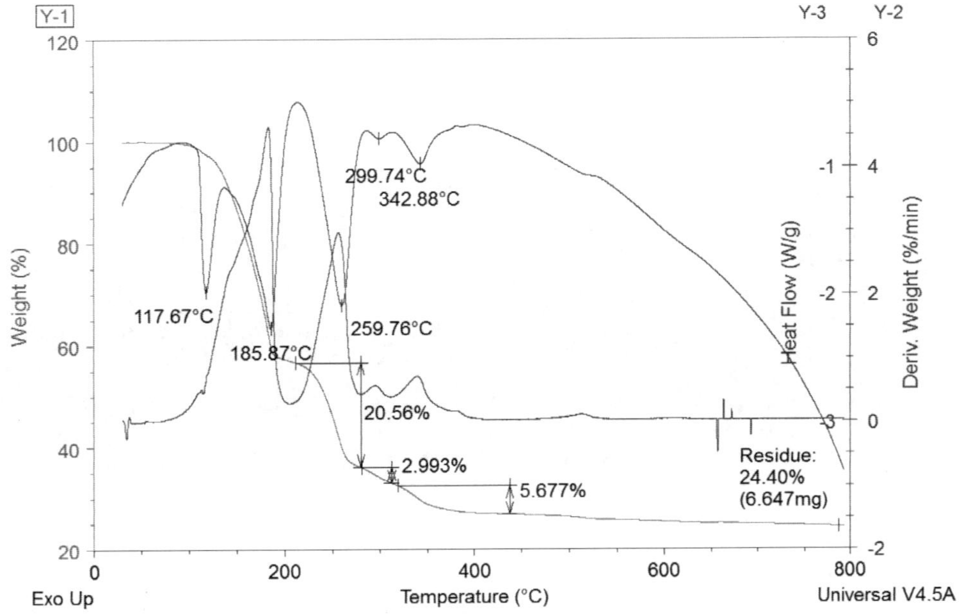

Figure 4. Thermogram of Fe oxide fluorination with 100% of ammonium fluoride excess (heat flux)

Fe oxide fluorination with ammonium difluoride is similar to the reaction with ammonium fluoride, as fluoride decomposes at the temperature over 120 °C, becoming a bifluoride. The study finds out that the reactions of non-aqueous Fe fluoride formation is a complex mechanism with forming intermediates.

At the first stage, the formation of ammonium bifluoride occurs:

$$nNH_4F \rightarrow {}^{n}\!/_{2}\, NH_4HF_2 + {}^{n}\!/_{2}\, NH_3\uparrow, t = 125°C.$$

At a subsequent stage, interaction of formed molten NH_4HF_2 with Fe oxide occurs according to the scheme:

$$Fe_2O_3 + nNH_4HF_2 \leftrightarrow 2NH_4FeF_4 + (n–2)NH_3 + (n–8)HF + 3H_2O.$$

At further heating, the formed compounds complex decomposes with forming non-aqueous Fe fluoride according to the scheme:

$$NH_4FeF_4 \rightarrow \mathrm{Fe}F_3 + NH_3\uparrow + HF\uparrow.$$

Identification of endothermic processes, recorded according to DSC data, is confirmed with the results of differential scanning calorimetry simultaneous analysis - mass spectrometry - x-ray diffraction analysis (DSC - MS - XRD). The interconnection of the temperature and the activation energy of the process of Fe oxide fluorination is determined by the process of ammonium fluoride melting and the formation of a heterophase liquid - solid system, that provides a significant change in the diffusion coefficients and, thus, - reaction acceleration, which correlates properly with the mechanism of ammonium fluoride interaction, proposed in K. Rajeshwar work [5]:

1) $n\mathrm{NH_4F}_{(s)}$ □ $(n/2)\mathrm{NH_4HF}_{2(l)} + (n/2)\mathrm{NH}_{3(g)}$;

2) $Me_2O_{3(s)} + nNH_4HF_{2(l)} \rightarrow 2NH_4MeF_{4(s)} + (n\text{-}2)NH_{3(g)} + (n\text{-}8)HF_{(g)} + 3H_2O\ (g)$

and decomposition of obtained complexes according to the scheme:

$$NH_4MeF_{4(s)} \rightarrow MeF_{3(s)} + NH_{3(g)} + HF_{(g)}.$$

Table 1. The results of differential scanning calorimetry of NH_4F-Fe_2O_3 system

Stage №	Temperatures region, °C	$E_a, \frac{kJ}{mol}$	T_{max}, °C	Process characteristics
I	30–75	38.5	44.8	HF desorbtion
II	30–140	163.2	81.7	H_2O desorbtion
III	115–145	510.0	135.0	NH_4F melting
IV	150–210	150.7	183.0	Complex formation
V	205–230	269.0	214.2	Complex decomposition
VII	305–338	273.0	338.8	Removal of the evolved NH_4F

XRD pattern of NH_4FeF_4 complex is shown in Figure 5 (a), where I is relative intensity, and 2θ is determined in angular degree. Subsequent thermal decomposition of this product leads to the formation of non-aqueous Fe fluoride. XRD pattern of obtained non-aqueous Fe fluoride is shown in Figure 5 (b).

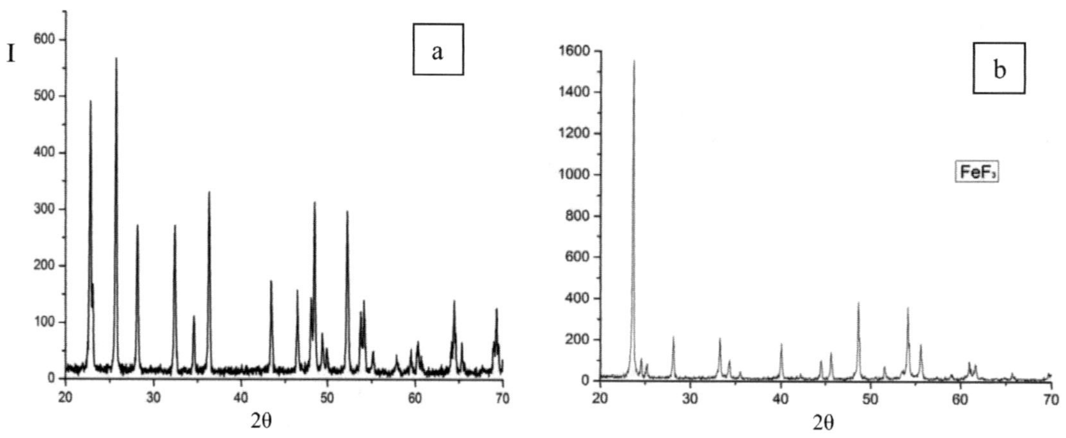

Figure 5. NH_4FeF_4 XRD pattern (a) and FeF_3 XRD pattern (b)

2.3 The study of fluorides physical and chemical properties

To determine fluorides physical and chemical properties modern analytical equipment and instrumental methods of analysis: X-ray fluorescence method of scanning electron microscopy and laser diffraction method were used.

To measure the moisture content (mass fraction of moisture) of Fe trifluoride weight hydrometer MH-50 was used. The moisture of the sample was 0.76 % (wt).

A study of the elemental composition of Fe fluoride. To determine the elemental composition of the Fe fluoride X-ray fluorescence method was used. The study was conducted using X-ray spectrometer (XRF-1800 model), produced by SHIMADZU company. Averaged results of three simultaneous determinations of the elemental composition of Fe fluoride samples: the content of Fe – 50.56%; F - 46.91% wt .; degree of Fe oxide fluorination with fluorine content - 93%. The magnitude of the mass fraction error for determined elements is (± 5)%.

The studies of Fe fluoride grain-size composition. The research was carried out using a scanning electron microscope (Quanta 200 3D SEM, produced by FEI company) and laser diffraction method (analyzer SALD-7101 (Shimadzu, Japan)).

Micrograph of Fe fluoride samples are shown in Figure 6 (a), and a bar graph of particle size distribution - in Figure 6 (b).

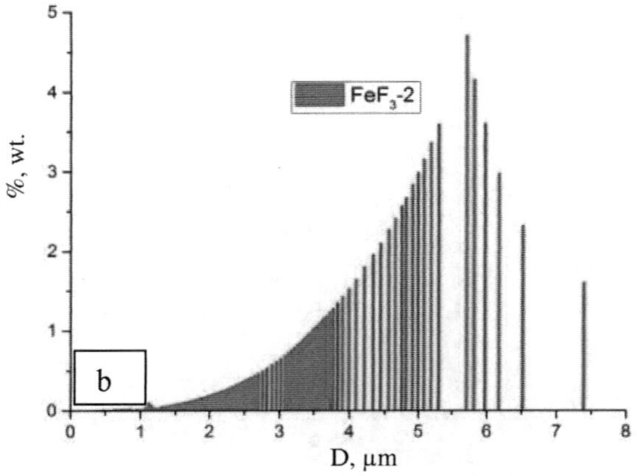

Figure 6. The micrograph (a) of Fe fluoride samples and the particles distribution based on Fe fluoride sample sizes (b)

Particles average diameter was 5.3 microns.

The XRD pattern of obtained non-aqueous Fe is identical to the XRD pattern presented in Figure 5b, which shows that the product is composed of FeF_3 in almost 100% .

Conclusion

1 The kinetics of Fe (III) oxide powder fluorination with elemental fluorine and ammonium fluoride is studied. The activation energies of the processes and reaction order of gaseous fluorine are determined.

2 The physical and chemical properties (moisture content, elemental composition, grain size and phase compositions) of Fe fluoride samples, that were obtained with the method of metal oxides interaction with fluoride and ammonium fluoride.

The work was supported by the Federal target program "Research and development on priority directions of scientific-technological complex of Russia for 2014-2020» (RFMEFI57814X0002).

References

[1] Sofronov V L, Buynovskiy A S, Makaseev Yu N 2014 Fluoride technology for obtaining magnetic materials based on Nd-Fe-B. Kinetics of fluorination processes of neodymium and iron oxides Procedia Chemistry № 11 P 158-164

[2] A.S. Buynovskiy, A.N. Zhiganov, V.L. Sofronov, V.I. Sachkov, N.V. Daneikina 2014 Current state of the rare earth industry in Russia and Siberia Procedia Chemistry № **11** P 126-132

[3] Sarychev G.A., Kosynkin V.D., Trubakov Yu. M. Current state of research in the field of technology of rare earth elements in Russia. *Proceedings of All-Russian conference on rare earth materials "REM 2013";* 2013

[4] Shcherbakov V.I., Zuev V.A., Parfenov A.V., 1985 Kinetics and mechanism of U, Pu and Np compounds fluorination with fluorine and halofluorides (M .: Energoatomizdat, P. 128)

[5] Krishnan Rajeshwar and Etalo A. Secco. // Can. J. Chem. 1977. Vol. 55. P. 2628

Detection and defect correction of operating process

Elena Vasendina[1], Inna Plotnikova[1], Anastasiya Levitskaya[1], Svetlana Kvesko[2].

[1] Institute of Non-Destructive Testing, National Research Tomsk Polytechnic University, Lenin Avenue 30, 634050 Tomsk, Russia

[2] Faculty of Innovative Technologies, National Research Tomsk State University, Lenin Avenue 36, 634050 Tomsk, Russia

E-mail: vasendina@tpu.ru

Annotation:
The article is devoted to the current problem of enterprise competitiveness rise in hard and competitive terms of business environment. The importance of modern equipment for detection of defects and their correction is explained. Production of chipboard is used as an object of research. Short description and main results of estimation efficiency of innovative solutions of enterprises are considered.
Key words: defect, production, parameter, chipboard, control, losses.

Introduction

Competition is a main feature of business environment. A government implements a policy of control and limitations of monopoly arbitrary action and support of fair competition for social production efficiency rise and social focus. In modern conditions of innovative development of economics problems of competitiveness rise of both production and enterprise itself are becoming vital. At the same time implementation of innovations is considered as one of the main ways of produced goods competitiveness rise, keeping up high rate development and earning power of an enterprise. That is why the problem of new equipment and technologies implementation at an enterprise is current and very important nowadays.

The quality of products produced at an enterprise and business costs are interrelated factors: the higher quality is the less nonconforming products are and the less level of business costs is. The less level of business costs is, the more possibilities for competitiveness in price are, which is often one of the main conditions of consumption. Thus, problems connected with the quality, can do harm to an enterprise [1].

Combining factors and conditions of production process, which correction can keep from nonconformity, cause problems connected with the quality of products.

Purpose of the work

The aim of the article is to show how modern methods allow identifying the sources of losses in the form of defect production, make analysis of its appearance at manufacturing of chipboard during three years (2013-2015).

Results

A technological process of chipboard manufacturing is a combination of a lot of stages [2, 3] where there is a possibility of nonconformity appearance which is described in detail in [4].

For production quality detection [5,6] a part of nonconformity production out of the whole number of chipboard is defined. Losses due to rejects in terms of value are defined as losses of an enterprise. Types of defects, detected during chipboard manufacturing and fixed monthly in a quality control department in reports for registration of defects, are set. They are: grinding defects, dust and tar pitch, end face, lengthwise edge, pulling off, shear of angle, short board, flocculent board, oil. Ways of chipboard use are identified in dependence on defect types. Prices of chipboard with defects are defined. Losses connected with reject rate amounted 13,8835 million rubles in 2014.

Analysis of reject rate frequency on defect names was done with the help of Pareto diagram shown in fig. 1.

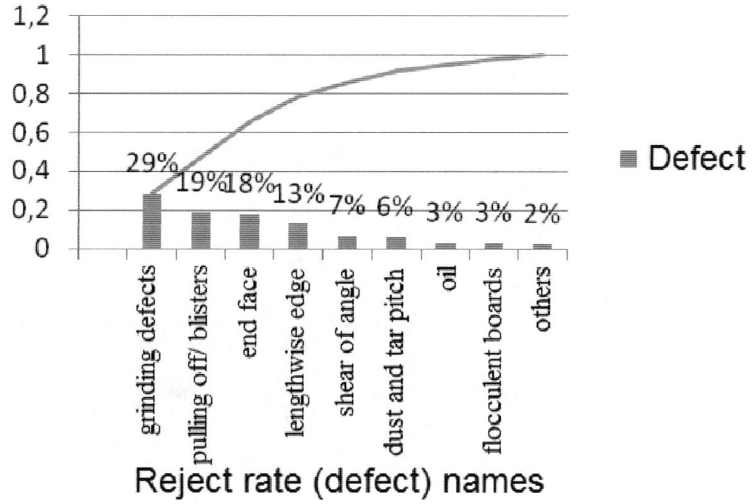

Figure 1. Reject rate (defect) names

It is obvious from the Pareto diagram that the most spread defects are: grinding defects, pulling off/ blisters, dust and tar pitch, short and flocculent boards.

For correction of defects such as pulling off and blisters (air blisters) a modern defectoscope UPU 3000 was used. The defectoscope detects some types of blisters such as: air blisters 1 / area, air blisters 2 / reject rate, air blisters 2 / maximum allowed length, air blisters 2 / category, cross-cut air blisters.

There is information on the defectoscope display, shown in fig. 2.

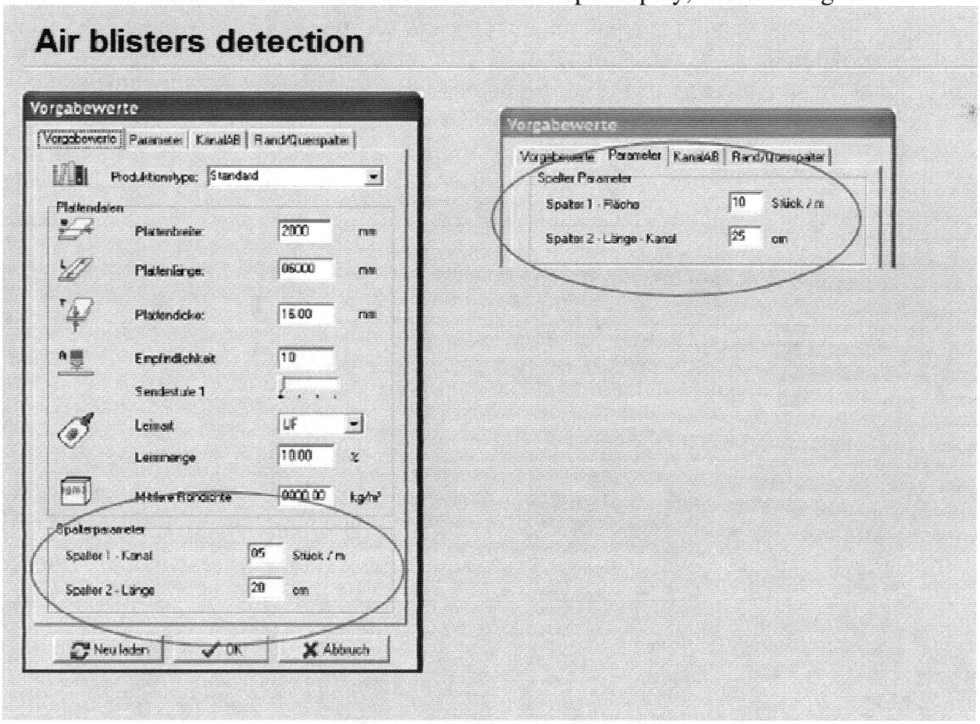

Figure 2. Air blisters detection

Understandings of air blisters are in fig. 3

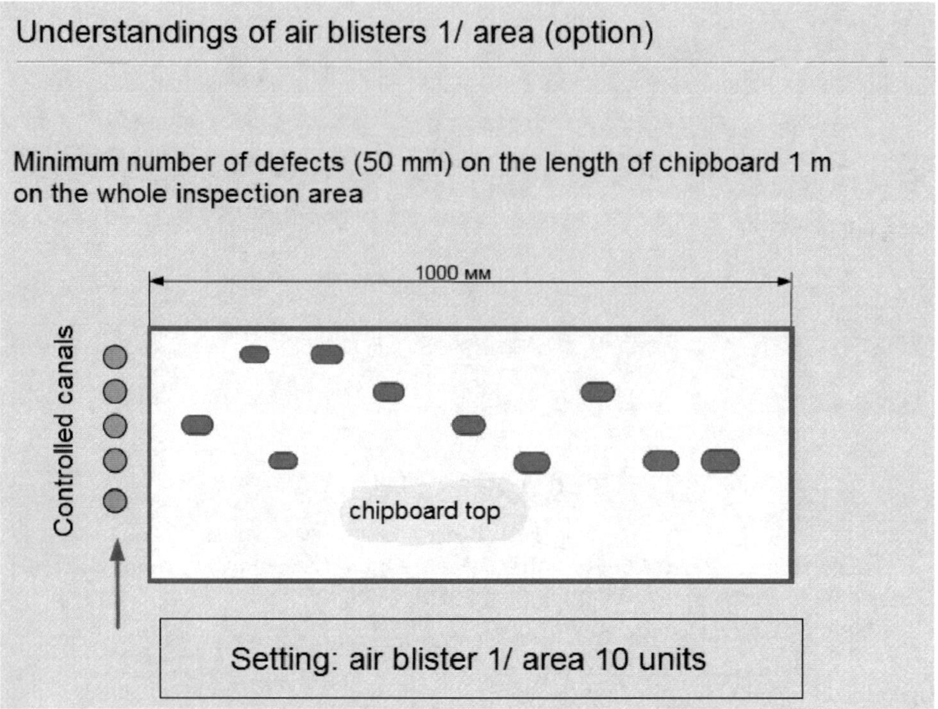

Figure 3. Understandings of air blisters 1/ area (option).

Furthermore there are quality indicators of process on reject rate before (2013) and after defectoscope installation in 2014 and 2015.

On the basis of the graph it can be stated that: a number of rejected rate is more in winter months rather than in warm ones. This can be explained by the fact that moisture is high in winter months and there are more blisters at manufacturing in winter months rather than in summer time. A number of reject rate decreased after installation of defectoscope UPU 3000. This is obvious when comparing January 2013 with January 2014 and 2015.

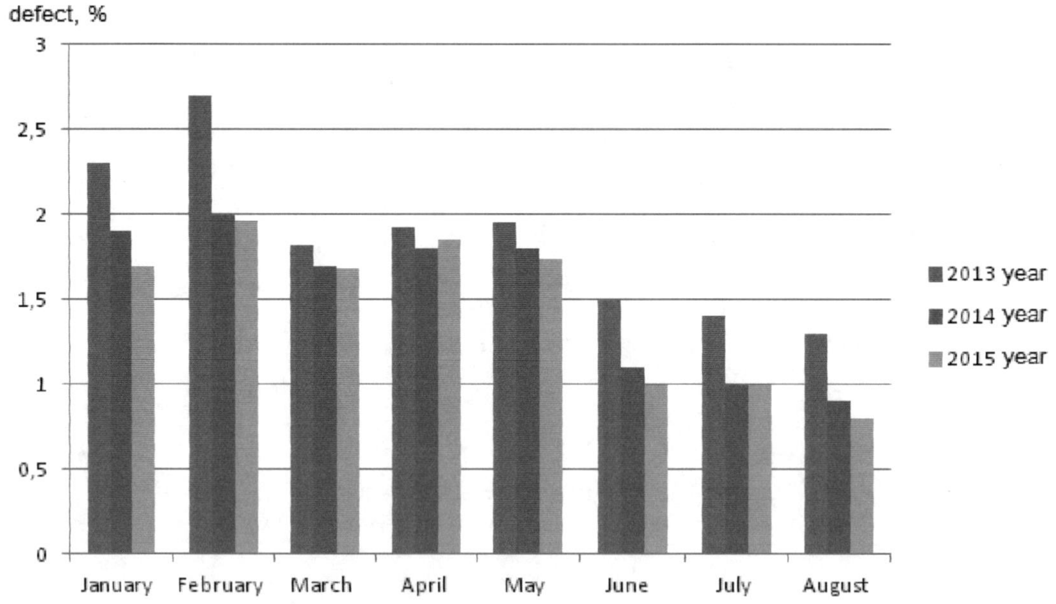

Figure 4. Quality indicators of process on reject rate.

IOP Conf. Series: Materials Science and Engineering **110** (2016) 012070 doi:10.1088/1757-899X/110/1/012070

Conclusion

Thus, having made analysis between the diagrams we can conclude that the appearance of the defect "pulling off, blisters" decreased after installation of the defectoscope. The defectoscope is intended for the defect "pulling off, blisters" prevention and not to allow chipboard with the defect to come in the market to consumers. It was proved that the defectoscope installation positively influenced the decrease of the reject rate indicator and prevention of production launch. The sum of losses caused by reject rate products was calculated; this value makes up about one per cent percentage wise or a share of one per cent from the volume of produced production.

On the basis of the research we can make a conclusion about the necessity of further modernization of technological process, study of the factors influencing the parameters of chipboard quality for reasonable fixation of their optimum values in normative and technical documentation.

References:

[1] Deming U J 2009 Vyhod iz krizisa: novaja paradigma upravlenija ljud'mi, sistemami i processami *v 2 kn.: per. s angl. (M.: Bizneskom vol 1*) 280

[2] GOST 10632-2014 Plity drevesno-struzhechnye. Tehnicheskie uslovija. EN 312:2010 NEQ.

[3] Volynskij V N Tehnologija drevesnyh plit i kompozicionnyh materialov 366

[4] Vasendina E A, Plotnikova I V , Red'ko L , Zyablova N N 2015 Study of types of defects in wood chipboard production [Electronic resource] (*IOP Conference Series: Materials Science and Engineering vol* 81) Radiation-Thermal Effects and Processes in Inorganic Materials 7

[5] Ishhenko T L 2009 Izmenenie prochnosti drevesnostruzhechnyh plit v period ih poslepressovoj vyderzhki (*Derevoobrabatyvajushhaja promyshlennost' vol* 2) 7-9

[6] Panov N G 2010 Povyshenie prochnostnyh svojstv drevesnyh materialov na osnove primenenija nanotehnologii (*Tehnologija i oborudovanie himicheskoj pererabotki biomassy dereva. Himija drevesiny) (SPb: SPbGLA*) 173-176

The influence of nickel layer thickness on microhardness and hydrogen sorption rate of commercially pure titanium alloy

V N Kudiiarov[1], E B Kashkarov[1], M S Syrtanov[1], I.S. Yugova[1]

[1]Tomsk Polytechnic University, 30 Lenin Avenue, Tomsk, Russian Federation

E-mail: viktor.kudiiarov@gmail.com

Abstract. The influence of nickel coating thickness on microhardness and hydrogen sorption rate by commercially pure titanium alloy was established in this work. Coating deposition was carried out by magnetron sputtering method with prior ion cleaning of surface. It was shown that increase of sputtering time from 10 to 50 minutes leads to increase coating thickness from 0.56 to 3.78 μm. It was established that increase of nickel coating thickness leads to increase of microhardness at loads less than 0.5 kg. Microhardness values for all samples are not significantly different at loads 1 kg. Hydrogen content in titanium alloy with nickel layer deposited at 10 and 20 minutes exceeds concentration in initial samples on one order of magnitude. Further increasing of deposition time of nickel coating leads to decreasing of hydrogen concentration in samples due to coating delamination in process of hydrogenation.

1. Introduction

Titanium alloys are indispensable as structural materials in aircraft, automotive, shipbuilding, nuclear, chemical and other industries [1, 2]. They have a corrosion resistance (due to the formation of the solid oxide film on the surface) and high mechanical strength. Moreover, titanium is widely spread in nature [3-7].

The mechanical properties of titanium depends on the content of impurities (H, N, O, C), which form the interstitial solid solutions with titanium and intermediate phases such as oxides, hydrides, carbides, nitrides. A content of oxygen, carbon and nitrogen increases the hardness and strength of the titanium, but reduces its plasticity. Therefore, the concentration of these impurities should not exceed 0.1 wt. %. The hydrogen content in titanium has the greatest impact on the mechanical properties. In the presence of hydrogen in titanium at the grain boundaries allocated thin brittle hydride phase plate, causing the hydrogen embrittlement. Hydrogen embrittlement leads to occurrence of internal stresses in the material. At lower temperatures, the titanium hydride becomes brittle and severe degradation of the mechanical and fracture behaviors of these alloys can occur [3]. Thus, the hydrogen content of titanium should not exceed 0.012 wt.% [3, 5].

In the process of production the hydrogen content in titanium alloys could exceed permissible value. Therefore, in the manufacture of constructions it is necessary to consider the risk of hydrogen embrittlement [7-11]. In order to study phenomenon of hydrogen embrittlement in titanium alloys require the preparation of experimental samples with different content of hydrogen for mechanical testing. However, hydrogen saturation of titanium alloys are often complicated by the presence of the surface oxide film, which significantly reduces hydrogen sorption rate, particularly at low temperatures. To increase the sorption rate could be used ionic cleaning method for removing a surface oxide film and subsequent deposition of nickel micron-thick layer. The nickel layer significantly reduces the formation rate of the oxide film, and also actively interacts with hydrogen even at low temperatures [12]. In this paper established the influence of the nickel coating thickness on the microhardness and hydrogen sorption rate by commercially pure titanium alloy.

2. Experimental Procedure

The rectangular flat samples of commercially pure titanium alloy VT1-0 with the size of 20×20×1 mm were prepared for this experiment. The samples were mechanically polished using sandpaper and annealed in vacuum at T = 750 °C for one hour. The deposition of nickel layer on the surface of the samples was carried out by the «Raduga Spectr» technique which was developed in General Physics Department of Tomsk Polytechnic University. Samples were fixed in a vacuum chamber by the samples holder. Ultimate pressure in the chamber was equal to $2.5\cdot10^{-3}$ Pa. Ionic surface cleaning of the samples is of prime importance before the thin protective films sputtering. It is necessary for the removal of surface contamination of various types, as well as oxide layer removal. As a result, the ion cleaning of the surface directly before the coating process improves adhesion of the coating material with the substrate. Surface cleaning was carried out directly in the vacuum chamber before the coating deposition process by argon glow discharge at the parameters shown in Table 1.

Table 1. Parameters of the ionic surface cleaning process

Gas	Voltage, V	Power, W	Current, A	Pressure, Pa	Time, min.
Argon	1800-2000	1000	0.5	0.06	3

The nickel coating was deposited by magnetron sputtering method. High purity nickel (99,99%) was used as a target for magnetron system. Deposition parameters are presented in Table 2. Sputtering time was varied in order to form nickel layers of different thicknesses.

Table 2. Parameters of nickel magnetron sputtering

Voltage, V	Power, W	Current, A	Pressure, Pa	Time, min.
550	2000	3	0.11	10 ÷ 50

Structural and phase composition was determined by X-ray diffraction using diffractometer Shimadzu XRD 7000 (Japan). Vickers microhardness of the samples was measured using microhardness tester KB30S (Pruftechnik, Germany). The thickness of the nickel layers was determined by spherical cross-section method using Calotest CAT-S (CSEM, Switzerland). Hydrogenation was carried out from hydrogen atmosphere at 500 °C, 2 atm for 60 minutes using special equipment Gas Reaction Controller (USA) [13].

The X-ray penetration depth was varied by changing the incidence angle in an asymmetrical layout. Decreasing of the incidence angle was conducted up to complete elimination of the signal from the substrate. Evaluation of the coating thickness was carried out using the geometry of the sliding beam by the relation:

$$h = \frac{\ln(1-R)\sin\alpha\cdot\sin(2\theta-\alpha)}{\mu(\sin\alpha+\sin(2\theta-\alpha))}$$

where α –incidence angle, θ – gliding angle (Bragg angle), μ – linear absorption coefficient, R – the proportion of radiation reflected from the layer with the thickness h. When determining the thickness of the nickel coating the value of R has a value of 0.95.

3. Results and Discussion

Table 3 shows the results of nickel layer thicknesses determination by different techniques as a function of sputtering time. It has been established that the increase in sputtering time from 10 to 50 minutes increases the thickness of the nickel layer from ~ 0,56 to 3.78 μm. The results of the thickness determination by X-Ray method correlate with the results obtained by spherical cross-section method.

Non-destructive XRD analysis allows estimating the nickel layer thickness before hydrogen saturation without coating destruction.

Table 3. Results of nickel layer thicknesses determination

Time of magnetron sputtering, min	Thickness (XRD), μm	Thickness (by the method of spherical cross-section), μm
10	0.63	0.56
20	2.52	2.48
30	3.35	3.26
40	3.69	3.62
50	3.84	3.78

Figure 1 shows the dependences of Vickers microhardness on load for different samples. For the sample of titanium alloy hardness under different loads is from 140 to 170 HV. Obtained hardness values could be increased with a decrease in the indenter size [14, 15].

It is seen that at a 1 kg hardness load of all commercially pure titanium samples is approximately the same in view of the error. The hardness of the samples increases under low loads after the deposition of the nickel layer. The obtained hardness values increased with increasing coating thickness and reduced to a titanium alloy hardness values with increasing applied load. For a sample of nickel, hardness is from 200 to 210 HV in the investigated range of loads. Thus, the titanium alloy substrate has no effect on coating thickness over 3.3 μm when measuring the hardness of the samples at a load of 0.1 kg. The maximum hardness 240-260 HV observed for samples with coating thickness of 3.62 and 3.78 μm.

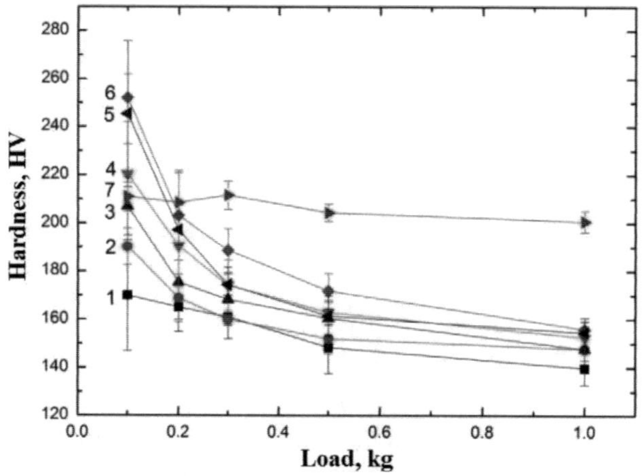

Figure 1. The dependences of Vickers microhardness on load for 1 - initial samples and for samples with different thicknesses of nickel layer: 2 - 0.56 μm; 3 - 2.48 μm; 4 - 3.26 μm; 5 - 3.62 μm; 6 - 3.78 μm and for 7 - nickel sample

Figure 2 shows the absorption curves for all samples (a) and the dependence of the hydrogen content on the deposition time of nickel (b). Comparing sorption curves is possible to say that the greatest hydrogen sorption rate is observed in samples with 10 and 20-minutes deposition of nickel. However, the graphs have a different nature: for a sample with a 10-minutes deposition the curve is close to linear form, for the sample with 20-minutes deposition the linear dependence is observed on two linear sections from 0 to 02 wt.% and from 1.5 wt.% up to 3.25 wt.%. Between these sections which are corresponding to different phases of titanium-hydrogen system observed the break related with increasing the hydrogen sorption rate. For the remaining samples, increasing deposition time leads to decreasing the sorption rate due to peeling of the coating. Figure 2b shows that during the same saturation time the samples accumulated different hydrogen concentration. The lowest concentration is in the initial sample. The greatest, as already mentioned, for the samples with 10- and

20-minutes deposition of nickel. For the remaining samples, the hydrogen concentration is inversely proportional to the time of deposition.

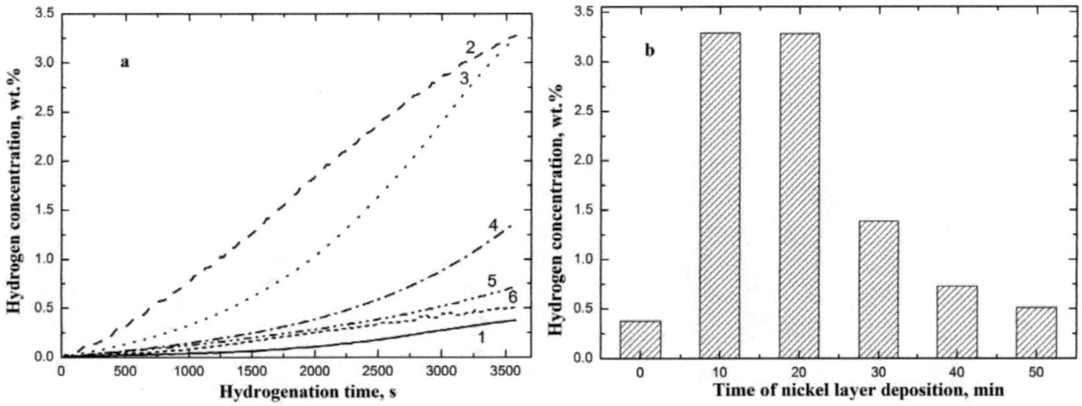

Figure 2. Hydrogen sorption curves (a) by titanium alloy with nickel layer after different time of magnetron sputtering: 1 - without nickel layer; 2 - 10 minutes; 3 - 20 minutes; 4 - 30 minutes; 5 - 40 minutes; 6 - 50 minutes, and hydrogen concentration (b) in titanium alloy with nickel layer after different time of magnetron sputtering hydrogenated at 500 °C, 2 atm, for 60 minutes

4. Conclusion

The influence of nickel layer thickness on microhardness and hydrogen sorption rate by commercially pure titanium alloy was established. Nickel coatings were deposited by the method of magnetron sputtering with ionic surface cleaning before deposition. It was showed that with deposition time increasing from 10 to 50 minutes the thickness of coatings increase from 0.56 μm to 3.62 μm. Microhardness values for all samples are not significantly different at loads 1 kg. Hydrogen content in titanium alloy with nickel layer deposited at 10 and 20 minutes exceeds concentration in initial samples on one order of magnitude. Further increasing of deposition time of nickel coating leads to decreasing of hydrogen concentration in samples due to coating delamination in process of hydrogenation.

References

[1] Chinkov E P, Baginskiy A G 2013 *Materialovedenie i tehnologiya konstruktsionnyh materialov* (Tomsk: Tomsk Polytechnic University) 230
[2] Malahov A I, Zhukov A P 1978 *Osnovy materialovedeniya i teorii korrozii* (Moscow: Vysshaya shkola) 192
[3] Arzamasov B N, Sidorin I I, Kosolapov G F 1986 *Materialovedenie* (Moscow: Mashinostroenie) 384
[4] Uhlig H H, Revie R W 1989 *Corrosion and corrosion control* (NY: John Wiley & Sons)
[5] Arzamasov B N, Brostrem V A, Bushe N A 1990 *Konstruktsionnye materialy* (Moscow: Mashinostroenie) 688
[6] Gormakov A N *Materialovedenie* 2003 (Tomsk: Tomsk Polytechnic University) 120
[7] Kolachev B A, Livanov V A, Buhanova A A 1974 *Mehanicheskie svoystva titana i ego splavov* (Moscow: Metallurgiya) 544
[8] Tal-Gutelmacher E, Eliezer D 2004 *Mater. Trans.* **45** 1594
[9] Tal-Gutelmacher E, Eliezer D 2005 *The Journal of The Minerals, Metals & Materials Society* **57** 46
[10] Boellinghaus T, Hoffmeister H 2000 *Corrosion* **56** 611
[11] Nelson H G 1996 *Hydrogen Effects in Metals* (PA: TMS) 699
[12] Kolachev B A 1966 *Vodorodnaya hrupkost tsvetnyh metallov* (Moscow: Metallurgiya) 256
[13] Kudiiarov V N, Gulidova L V, Pushilina N S, Lider A M 2013 *Advanced Materials Research*

IOP Conf. Series: Materials Science and Engineering **110** (2016) 012071 doi:10.1088/1757-899X/110/1/012071

740 690
[14] Iost A, Bigot R 1996 *J. Mater. Sci.* **31** (13) 3573
[15] Moshchenok V I, Doshchechkina I V, Lyapin A A 2008 *Razmernyy effekt v znacheniyah tverdosti materialov* (Kharkov: Bulletin of Kharkov National Automobile and Highway University)

Research of nanocomposite structure of boron nitride at proton radiation

Y V Borodin[1], D S Ermolaev[1], V Pak[2] and K Zhang[3]

[1]Tomsk Polytechnic University, 30 Lenina Av., Tomsk, 634050, Russia,
[2] Kemerovo State University, 6 Krasnaya St., Kemerovo, 650043, Russia,
[3]Harbin Institute of Technology, 11 Siling St., Nangang District, Harbin, 150001, China

E-mail: luryborodin@tpu.ru

Abstract. Using roentgen diffraction and electron microscopy, the influence of nanosecond irradiation by ion beams of high energy on forming of self-organized nanoblocks in near surface's layers of boron nitride (BN) has been studied. It was shown that low temperature transitions from hexagonal to wrutz boron nitrides is associated with changes of shape and sizes of self-organized particles consisting the nanoblocks. We have calculated the parameters of nanoblocks using the meanings of interplane distances and properties of subreflexes orders. The collective shifting deformations of layers in nanoblocks provides phase transition under the screen and forming the set of nanotubes with escaping of five order axes of symmetry. It has been realized that pentagons and stars arranged in points of entrance of five order axis of symmetry are associated with peculiarity of self-organization of the spiral-cyclic structures.

1. Introduction

Due to the development of nanophysics, the investigations of origin and behavior of the self-organized nanoparticles and nanoblocks of various invariant scales have been carried out. The increasing number of nanoblocks is associated with wave, beam and other impacts of nano- and pica second action providing a low temperature atom shifts and distribution of high electron density band.

The appearing of nanoblocks might be a result of irradiation of the crystal compounds by nanosecond ion beams. Self-organization as a rule is observed in high packed and layered materials. In BN, graphite, SiC this effect could be obtained due to the large number of high packed planes [1-3]. In this work the regimes of creating of nanotubes and filling them by the doped substances was shown. It has been realized the influence of nanosecond irradiation by electron and ion beams on organization of nanotubes. It seems essential to study the impact of nanosecond ion irradiation to shape and linear size of nanoblocks and nanotubes in near-surface region of BN.

2. Experimental methods

The BN plates obtained using pyrolylic methods from chemical clear B and N. It is well-known from the electronic-microscopic dates that in original BN hexagonal phase composes up to 98 percent of the volume, the rest is rhombohedral one. The irradiation was carried out by ion beams of 120–250 KeV and with density 80 A/cm^2 and time 20–100 ns using the screen made up from foil W of 0.5 mm. thickness restricting the irradiational zone. The roentgen investigations were carried out on diffractometr DRON-3 in Cu-K$_\beta$ flux. The method of non-focused lau grams .was used to reconstructing the thin subreflex structure. The sample is shifted from the center of goniometer to the side of roentgen source and diffraction pattern was observed using motionless sample. In the roentgen-diffraction investigations by Shtolz the non-focused picture method was used. Microscopic investigations were earned out on electronic microscope EM-125K; the image which allow us to have information about phase composition was obtained in area 0.5–1.0 μm.

3. Numerical results

In original BN and in irradiated by the nanosecond ion beams one the complicated distribution of integrated intensity of difractional stripes can be a result of complicated distribution of electron density in atomic planes (Figure l). Uneven electron density could be associated with distribution of old nanoblocks and forming the new one. According to the geometry of subreflexes the dependence of interangle distribution can be written as:

$$\Delta\varphi_0\ \tau^0, \Delta\varphi_0\ \tau^1, \Delta\varphi_0\ \tau^2; \quad (1)$$

$$\Delta\varphi_0\ \tau^0, \Delta\varphi_0\ \tau^1, \Delta\varphi_0(1/2\tau^2), \Delta\varphi_0(1/3\tau^2), \quad (2)$$

where Δφ0 – minimal interangle distance; τ = 1.618 -gold number.

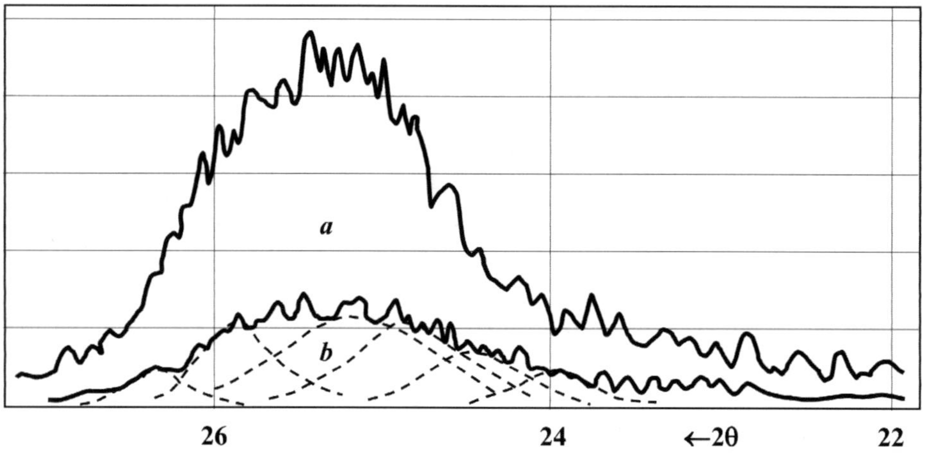

Figure 1. Thin diffraction reflex (002) structure of original (a) and irradiated (b) BN.

The experimental dates of parameters of nanoblocks along the planes [110] and [001] are presented in table 1 taking into account the expressions (1) and (2). Using the experimental dates of interplane distance between the high intensity stripes, the size of nanoparticles- is calculated from the following expression:

$$A = d_{max} N^{1/2} \tau^{1/6} K^{-1/3}, \quad (3)$$

where $N=H^2+K^2+L^2$ – is the sum of square indexes of high packed plane;

K - is a shape coefficient of nanoparticles ($K_{cub} = 1$; $K_{oct} = 0.4714$; $K_{tetr} = 0.1222179$).

The numerical meanings of nanoblocks were obtained according to the model of nanoparticles and their scaling in accordance with Fibonacci sequence. As shown in table 1, use of octahedral nanoparticles of 0.46 nm size can effects on that nanoblocks have the sizes closed to theoretical. The larger nanoblocks in terms of their sizes are formed along the plane (001) and part of these blocks are in self-organized sequences with self-similar transitions. However the experimentally observed nanoblocks of sizes 3.1, 3.9, 4.2 and 11.6 nm prevent the realization of such transitions. Using topographical observations by Shtolz, the distribution of nanoblocs of size 1–20 nm rightly arranged is observed.

The processing of origin BN by nanosecond ion beams of high energy is accompanied by forming of new wrutz BN phase. A dependence of BN volume has clear minimums of 1.6, 3.2, 4.8 and 7.9 μm depth. The changes in structure are shown also in a distance up to 15 mm under the protective screen from the irradiation zone. Low temperature transition $BN_G \rightarrow BN_W$ is concerned with reduction of nanoparticle's linear size on $\Delta a = 0.009$ nm, shifts deformations and formation of nanotubes.

The formation of nanotubes in BN in an enough distance from the irradiation area and modifications in arrangement of spherical packets of nanotubes are proved by the electron-microscopic investigations (Figure 2a). The correlation between the sizes of nanofube's packets and their density in sample and far from irradiation area was shown in table 2. It was noted that the size of the packets of nanotubes becomes smaller and their contribution is decreased and becomes 45 percents on depth of 0.2 μm and 15 mm from the area of irradiation. On a depth of 50 μm the packets of nanotubes in BN of five order symmetry as pentagons of size 32 nm become packets in form of stars of size 62.5 nm in a depth of 500 μm. The mechanism of nanotube's formation in BN is associated with particularities of self-organization of spiral-cyclic structures [4]. The experimental results of block formation in such self-organized structure with axis of five order symmetry can be observed on Figure 2b.

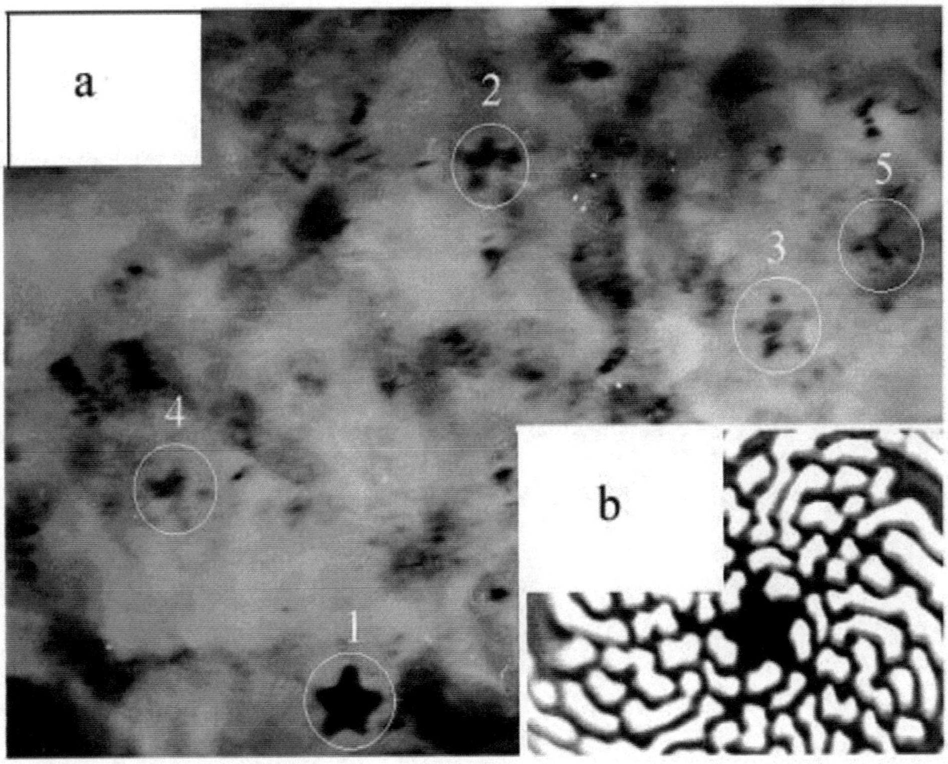

Figure 2. The formation of stars in the points of L5 axis escaping on electron-microscopic picture (a) and under the packing of bright blocks in self-organized structure (b); 1,2,3,4,5 – various levels in which the stars appears.

Table 1. The sizes of nanoblocks in BN_W

Plane	(110)			(001)		
$a_{exper.}$, **nm**	1.37	2.614	9.596	3.66	6.40	1.156
Step of Fibonacci sequence	3	4	7	5	6	7
$a_{calcul.}$	1.365	2.275	9.555	3.64	5.915	9.555

Table 2. The arrangement of spherical packets of nanotubes in BNw

Parameters of nanofube's packets	Original	The distance from the irradiation area, mm		
		0	10	15
Size, *d*, nm	8.4	82	64.5	6.6
d_{max} (d_{min})	600 (37.5)	250 (37.5)	87.5 (50)	87.5 (50)
Volume fraction, ΔV,%	~1.0	35–45	3–5	≤1.0

4. Summary

The irradiation of hexagonal BN by nanosecond ion beams accelerates in near-surface layer the phase transition and formation of nanoblocks characterized to wrutz-like BN. It was shown that a phase transition is reached on the distance far from the irradiation area and accompanied with formation of nanotubes with escaping of five order symmetry axis on the surface. The appearance of nanotubes is associated with collective shift deformations of nanoblocks and it occurs in the self-organization of spiral-cyclic structure.

References

[1] Kiang Ch H, Goddard W A, Beyers R et. al. 1996 *J. Phys. Chem.* **100** 3749–52
[2] Golberg D, Bando Y 1996 *Appl. Phys. Lett* **69** 2043– 47
[3] Borodin Y 2011 *Proceedings of the 6th International Forum on Strategic Technology, IFOST 2011* **1** 218–21
[4] Horiuchi S 1995 *Jap. J. Appl. Phys. Pt.2* **34** Nu 12 A 1612–15

Sorption properties of radiation-cross-linked polymer hydrogels containing ion-exchange fibers

M A Rezvova[1], V D Zhevnyk[1], V Pak[1], Y V Borodin[2] and E V Kachina[1]

[1]Kemerovo State University, 6 Krasnaya St., Kemerovo, 650043, Russia
[2]Tomsk Polytechnic University, 30 Lenina Av., Tomsk, 634050, Russia

E-mail: rezvovamaria@mail.ru

Abstract. Polymer hydrogel modification for soft contact lenses by ion-exchange fibers was studied in this work. The obtained results showed that the ion-exchange fiber modifiers have a number of advantages as compared with ion-exchange resin modifiers.

1. Introduction

Polymers are widely used in all fields of human activity. Medical devices polymers that exhibit high hydrophilicity, lack of toxicity, and many other useful properties are of particular interest. Currently, polymeric materials have been widely used in the development of new-generation wound healing. It is bandaging materials, wound dressings, adhesive bandages and other. Polymeric ion-exchange hydrogels are promising materials for the development of this field.

The ability to modify the hydrophilic polymeric material for soft contact lens (SCL) based on N-vinylpyrrolidone and methacrylic acid methyl ester with ion exchange resins (IER) to produce the polymeric sorptive materials for eye burn treatment [1-3] has been shown in previously published works. Soft contact lenses are one of the effective means for comprehensive treatment of recent chemical and thermo-chemical eye burns. The implemented functions in their use are neutralization of the damaging agent, acceleration of the cornea epithelialization, prevention of symblepharon, at a later date - the healing of post-burn lesions of the cornea. The appearance of new ion-exchange polymeric materials based on the material for soft contact lenses made possible to expand the indication range for their use.

Polymer hydrogel modification for soft contact lenses by ion-exchange fibers was studied in this work. Fibrous ion-exchangers have developed active surface and better kinetic characteristics compared with granular ion-exchangers (high baud rate, the greater availability of ionic groups for exchanging ions, including a large organic ions) [4]. Their ion-exchange capacity is high enough for practical use, and is not reduced by repeated regeneration cycles of acids and alkalis. High ion-exchange capacity, heat resistance and chemical resistance of synthetic ion-exchange fibers should also be noted.

The object of study is ion-exchange fiber brand MION AK-22 in OH^--form; MION K-5 in H^+-form and MION K-5, Na^+-form (fiber length: 0.5-1.0 mm Diameter: 20-50 microns) and material for soft contact lenses [5].

IOP Conf. Series: Materials Science and Engineering **110** (2016) 012073 doi:10.1088/1757-899X/110/1/012073

2. Results and Discussion

Values of water content and ion-exchange capacity (IEC) were determined for the study of ion-exchange fibers. Water mass in the fibers was related to the mass of air-dry samples in determining the water content of the hydrated samples. Water content measurement was carried out on reaching the maximum swelling of the fibers (24 hours). The water content average value was determined according to the results of five replicates for each type of ion-exchange fibers (it is shown in Table 1).

Table 1. The water content of the ion-exchange fibers, granular ion-exchange analogue and material for soft contact lenses.

Type of material	Water content (%)
Ion-exchange fiber MION K-5, Na^+	80 ± 3
Ion-exchange fiber MION K-5, H^+	65 ± 2
Ion-exchange fiber MION AK-22, OH^-	55 ± 1,5
Ion-exchange resin D-113, Na+	49,8 ± 1,2
Hydrogel for SCL	70 ± 2

Comparison between the ion-exchange resins water content and ion-exchange fibers water content showed the value of the characteristics for the ion-exchange fibers is higher than the same parameter for granular inoculants. Swollen hydrogel water content values and swollen ion-exchange fibers water content values are similar, therefore, the swelling of the fiber modified hydrogel is uniform and has no void formations. Thereby, fibrous materials are preferred for use as hydrogel modifiers according to this result. This kind of ion-exchange hydrogel will have greater strength and elasticity.

Determining of the alkali and acid sorption degree was carried out using the method of acid-base titration [6]. The fibers were placed in a volume of acid / alkali with a concentration of 0.1 g / mol at various periods of time. An aliquot of the acid / alkali was titrated and the mass difference in the original volume and the volume was calculated after sorption absorbed weight of acid or alkali.

Dependence of the degree of sorption on time is presented in graphical form (Figure 1, Figure 2, Figure 3).

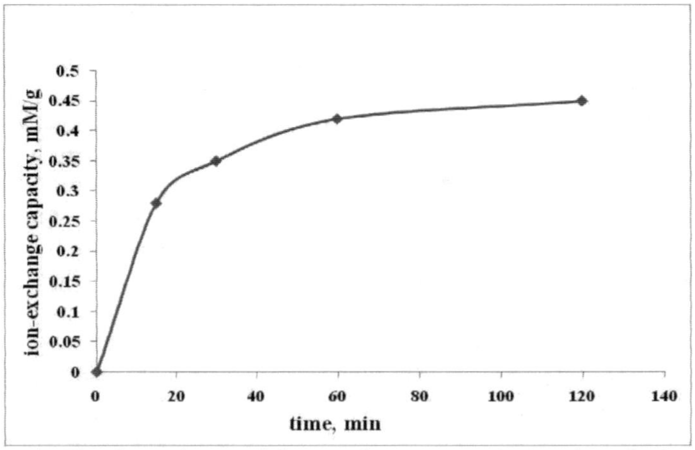

Figure 1. The dependence of ion-exchange capacity on time for ion-exchange fiber MION K-5, H^+ of relatively alkali.

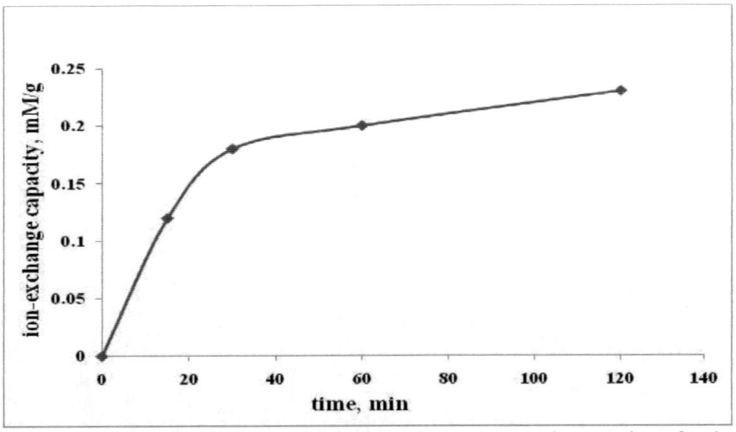

Figure 2. The dependence of ion-exchange capacity on time for ion-exchange fiber MION K-5, Na^+ of relatively acid.

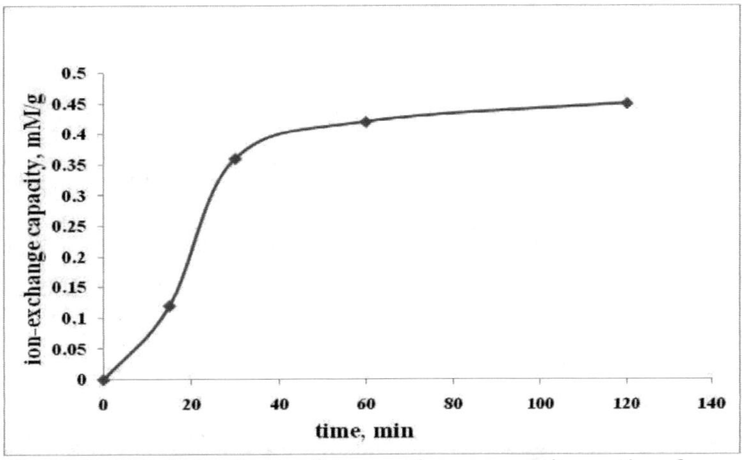

Figure 3. The dependence of ion-exchange capacity on time for ion-exchange fiber MION AK-22, OH^- of relatively acid.

The graphs show that ion-exchange fibers sorb significant proportion of acid/alkali by the maximum possible amount within a short period of time (about 40 minutes). This occurs by reason of a large active surface area of the fibers. The high speed of the process is achieved due to the short ion diffusion path inside the ion-exchange fiber.

Fiber MION K-5, Na^+ was investigated at the first stage of the ion-exchange polymer hydrogel synthesis. Composition of polymer hydrophilic material for SCL was used as a basic hydrogel [4]. Synthesis was carried out on the installation of ionizing radiation from Co^{60}RHM-γ-20 [7,8]. A prepolymer was obtained by irradiating a mixture of N-vinylpyrrolidone, methyl methacrylate and divinylcross-linker dose 10kGy. After the fiber (5% by weight) was added into the prepolymer under an inert atmosphere. The mixture was stirred placed in special molds and irradiated the absorbed dose 25kGy to fully cure. After hydration the samples were examined for the ion-exchange hydrogel water content and acid sorption. Water content of the samples obtained almost the same as the water content of the hydrogel base. Acid sorption is shown in Figure 4 compared with the ion-exchange hydrogel containing pellet filler, and ion-exchange resin D-113.

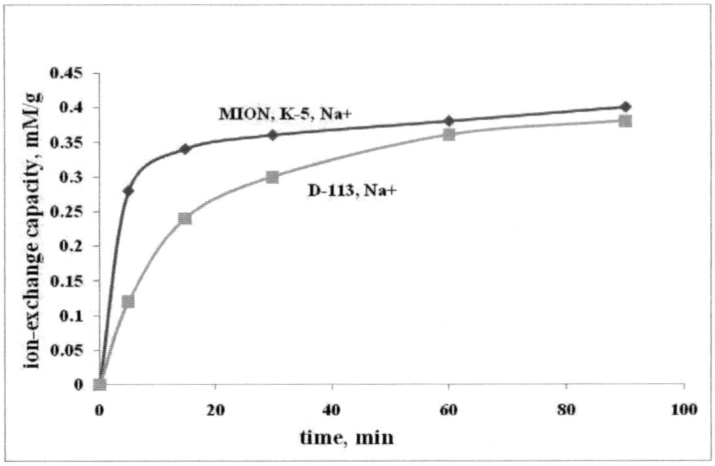

Figure 4. The dependence of ion-exchange capacity on time for ion-exchange fiber MION K-5and the ion-exchange resin D-113 of relatively acid.

3. Conclusion

Comparative evaluation of the ion-exchange fibers properties and granular analogs properties showed advantage of usefibrous materials as modifiers in the synthesis of ion-exchange polymeric hydrogels. Higher values of water content and active ion-exchange reaction were achieved due to the developed surface of the fiber. At the graph of hydrogels sorption degree it is seen that at the initial stage of the ion-exchange hydrogel, modified by fiber material,is more active, than modified by pellet filler.

References

[1] Zhevnyak V D, Pak V H and Le V M 2009 New technologies, materials and innovations in productionpp 154-156.
[2] Pak V H, Zhevnyak V D, Dikunova T V, Shraibman G N and Le V M 2010 *9th International Conference on "Modification of Materials with Particle Beams apd Plasma Flows"* 351-352
[3] Zhevnyak V D, PakV H, Le V M 2011 *J.Proceedings of Higher Education. Physics* **54** 252-255
[4] Chen L, Yang G and Zhang J 1996 A study on the exchange kinetics of ion-exchange fiber React. Funct. Polym. **29** 139-144
[5] Zhevnyk V D, Stalkovsky V V and Fomina M P 2011 RF Patent number 2119927 "Method of producing a polymer material for soft contact lenses"
[6] Vasiliev V P 2002 Analytical chemistry. Titrimetric and gravimetric methods of analysisp 368
[7] Ivanov V S 1988 Radiation chemistry of polymersp 320
[8] Shur A M 1969 Macromolecular compounds 656

IOP Conf. Series: Materials Science and Engineering **110** (2016) 012074 doi:10.1088/1757-899X/110/1/012074

Radiolysis of Potassium Picrate in 77 K

D G Yakubik[1], V Kh Pak[1], V A Anan'ev[1] and S A Ghyngazov[2]

[1] Kemerovo State University, Krasnaya st. 6, 650 043, Kemerovo, Russia
[2] National Research Tomsk Polytechnic University, Lenina st. 30, 634 050, Tomsk, Russia

E-mail: den@kemsu.ru

Abstract. The formation of paramagnetic centers in potassium picrate under irradiation at low temperature was investigated. The heating irradiated at 77 K potassium picrate crystal to room temperature results in paramagnetic centers - 2,6-dinitro-para-quinone radicals, ortho- and para-iminoxyl radicals and atomic oxygen. These products are formed under irradiation at room temperature.

1. Introduction

Potassium picrate (hereinafter PP) – nitroaromatic compound $C_6H_2(NO_2)_3OK$, being power explosive. Moreover it can undergo to all kinds of slow solid-phase decomposition under constant influence of the external power factor, such as heat, light, radiations, electromagnetic fields [1-4]. The goal of present investigation is to study the formation and transformations of the paramagnetic centers formed in PP at low-temperature irradiation.

2. Experimental

Polycrystalline PP was prepared by neutralization of a hot picric acid solution with a solution KOH. Reagents used were analytical reagent grade. PP single crystals were prepared by slow evaporation hot (60 °C) water solutions. They sizes was 15×3×4 mm and of 20-30 mg weight.

The crystals were irradiated by with ^{60}Co γ-rays at 77 K and 307 K. The dose rate was 0.16 Gy/s. The dose absorbed by the sample was calculated using the mass energy absorption coefficients.

Low-temperature annealing of the irradiated crystals of PP was made in a cryostat in a range 77-300 K with accuracy ±2K.

The ESR spectra of the irradiated crystals were recorded on a ''Rubin'' X-band ESR spectrometer at 77 and 300 K. At 77 K the crystal was immersed in liquid nitrogen contained in the finger of a Dewar. For absolute g-value determinations a calibration using DPPH (g = 2.0036) was used. The magnetic properties of studied centers were derived from series of experimental ESR spectra corresponding to rotations of the samples in three perpendicular planes in the static magnetic field.

The spin concentrations of the paramagnetic centers formed in irradiated samples were determined by comparing the intensities of the doubly integrated spectra with those obtained from reference sample – $CuCl_2 \cdot 2H_2O$ (the error in spin concentration is ±10%). To minimize experimental errors, the same condition of registration was used for all ESR measurements. The interpretation of ESR spectra was carried out according to standard procedures [5, 6].

IOP Conf. Series: Materials Science and Engineering **110** (2016) 012074 doi:10.1088/1757-899X/110/1/012074

3. Results and discussion

Figure 1 displays ESR spectrum of the γ-irradiation of potassium picrate single crystals at 77 K. Wide (1,45 mTl) isotropic asymmetric singlet with not resolved HFS was observed. When stored for 1 day its intensity do not change. Variation orientation of a single crystal concerning a magnetic field direction results in the change of the signal form, but resolution of a singlet line into separate components is not observed. A change of g-factor values was not revealed because of asymmetry of a singlet and its large width. Increasing magnetic field strength results in increasing signal intensity with slight change of the line shape. Probably this paramagnetic center (PC-0) is a combination of several defects stable at 77 K with different sensitivity to changes of the magnetic field. Thus, to identify the nature of the center based on the spectroscopic parameters it is extremely difficult. Therefore, to clarify its nature, it is necessary to carry out additional investigations.

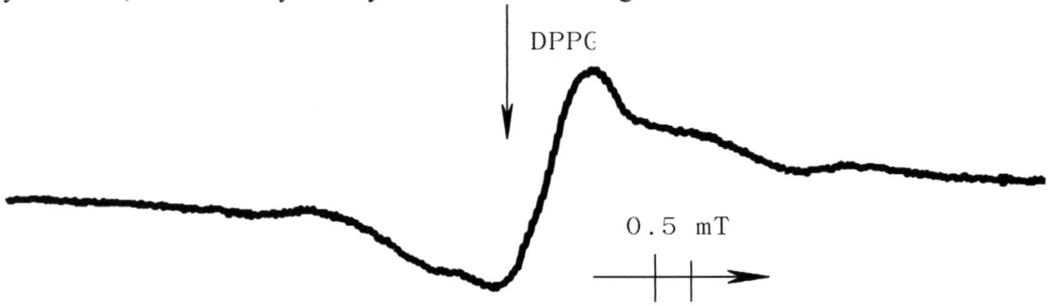

Figure 1. The ESR spectrum of the PP irradiated to 69 kGy at 77 K

The γ-irradiation of single crystals of PP at room temperature results in ESR-signal with HFS [1] is due to presence of four paramagnetic centers (figure 2):

PC-1 – The anisotropic triplet with splitting 0.15 – 0.53 mTl and the intensity ratio 1:2:1; has been identified as a 2,6-dinitro-p-quinone radical formed at eliminating on NO molecule from picrate-anion;

PC-2 – Two nonequivalent triplets, apparently, iminoxyl-radical formed at break off of oxygen from nitro-group in an ortho-position;

PC-3 – The isotropic singlet in half-width 0.25 mTl, it is presumably due to O^-;

PC-4 – The singlet with half-width 0.55 mTl, is preliminarily identified as a radical with nitro-group in para-position in a benzene ring.

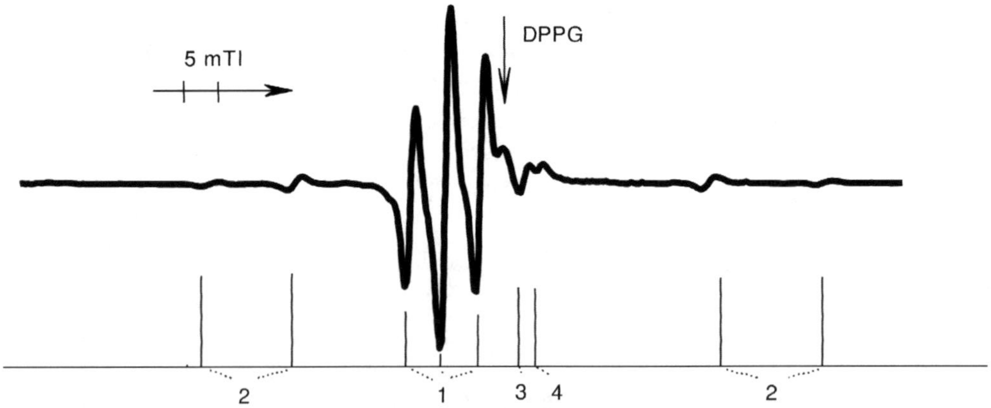

Figure 2. The ESR spectrum of the PP, irradiated to 238 kGy at room temperature: 1 – PC-1, 2 – PC-2, 3 – PC-3, 4 – PC-4

The spectroscopic parameters of paramagnetic centers in γ-irradiated at room temperature potassium picrate crystals [7] are given in table 1.

PC-1, PC-2 and PC-3 are stable at room temperature. When stored for 1 day the line intensity of PC-4 decreases slowly.

Table 1. ESR spectroscopic parameters of paramagnetic centers in γ-irradiated potassium picrate crystals.

Parameter	Paramagnetic center				
	PC-0	PC-1	PC-2	PC-3	PC-4
T_{xx} (mTl)	–	0.199	1.214	–	–
T_{yy} (mTl)	–	-0.291	-0.205	–	–
T_{zz} (mTl)	–	-0.205	-0.705	–	–
g_{xx}	2.0033	2.0086	2.0098	2.0028	20032
g_{yy}	2.0033	2.0046	2.0045	2.0028	2.0032
g_{zz}	2.0033	2.0019	2.0018	2.0028	2.0032
A_{iso} (mTl)	–	-0.342	3.125	–	–
ΔH_{max} (mTl)	1.450	0.150	0.245	0.550	0.250
ρ	–	0.127	–	–	–

The heating from 77 K to room temperature result in irreversibly and completely decay of the PC-0. From 77 to 100 K the signal intensity decreases sharply, further to 140 K - is not actually changing. At the temperature range 230-235 K the signal was vanished almost instantly and there is no new ESR signal is appeared.

When storing irradiated at 77 K PP crystal at room temperature during 1 hour, ESR signal is appeared again. It is identical to the PC-1 and PC-3 that in γ-irradiated at 307 K PP. During storage, the PC-1 concentration increased, tending to saturate for 24 hours. The maximal concentration is equal to (within experimental error) with that of PC irradiation at 307 K with the same dose of radiation (figure 3).

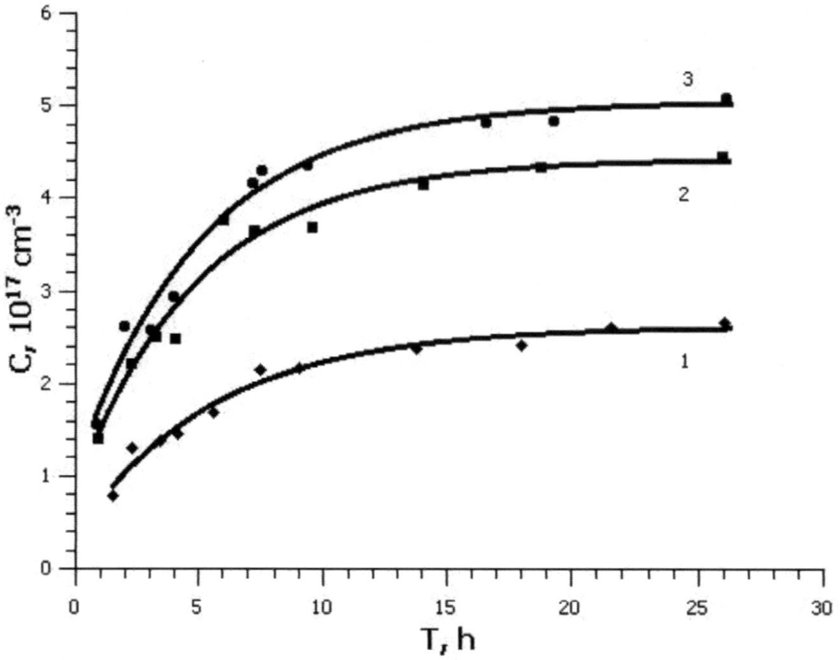

Figure 3. The accumulation of the PC-1 at different doses (1 – 30 kGy, 2 – 75 kGy, 3 – 98 kGy)

At the absorbed dose above 100 kGy, intensity ESR signal sharply increased and is displayed HFS signal PC-1. The analysis of spectroscopic characteristics displayed that a HFS is caused by a appearing iminoxyl radical in an ortho- and para-positions (figure 4).

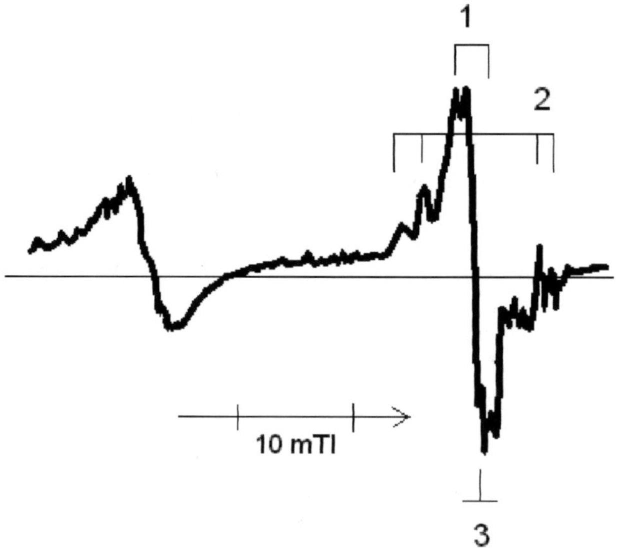

Figure 4. The ESR spectrum of the PP, irradiated to 453.2 kGy: 1- iminoxyl radical in a para-positions, 2 - iminoxyl radical in an ortho-positions, 3 - 2,6-dinitro-p-quinone radical.

IOP Conf. Series: Materials Science and Engineering **110** (2016) 012074 doi:10.1088/1757-899X/110/1/012074

Experiment of low temperature annealing showed that the nature of thermal decay of the paramagnetic centers in crystals PCs irradiated γ-rays to a dose of 300 kGy, is very different. ESR signal intensity of PC-1 in the range of 80-150 K decreases almost 10 times (figure 5), after remaining constant up to room temperature. The intensity of the EPR signal of PC-2 is reduced gradually. For PC-3 signal characteristic nonmonotonic change - at first it slowly increases, then gradually decreases (figure 6)

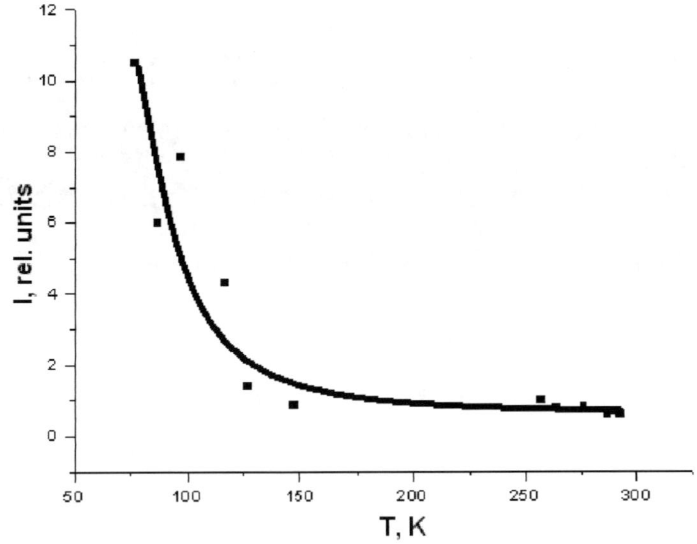

Figure 5. The temperature dependence of ESR signal intensity of PC-1

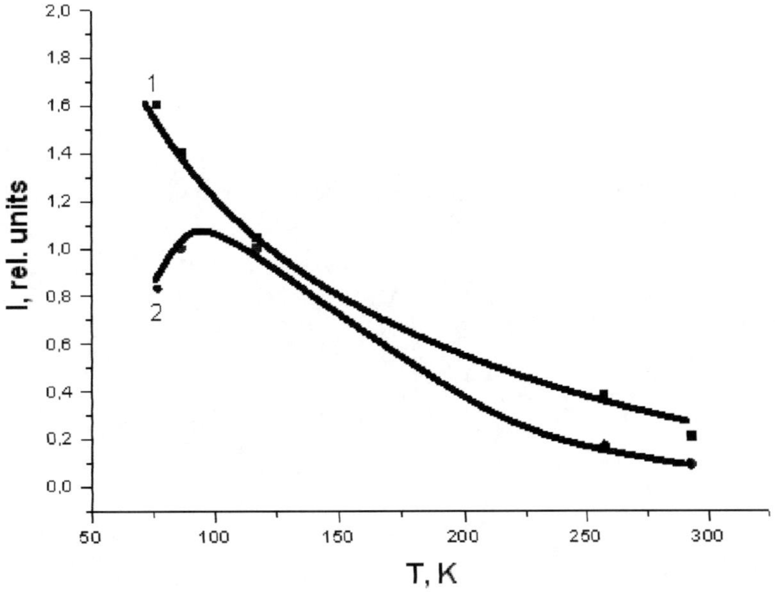

Figure 6. The temperature dependence of ESR signal intensity PC-2 (1) and PC-3 (2)

The obtained experimental data indicate the complex mechanism of the formation and transformation of paramagnetic centers formed in γ-irradiation of potassium picrate at 77 K.

IOP Conf. Series: Materials Science and Engineering **110** (2016) 012074 doi:10.1088/1757-899X/110/1/012074

Acknowledgements

This work was financially supported by The Ministry of Education and Science of the Russian Federation in part of the science activity program.

References

[1] Ryabykh S M, Martynova N V and Zhulanova V P 1989 *J Proceedings of the universities. Ser. Chemistry and Chemical Technology* **32** 44
[2] Andreev K K 1966 Thermal decomposition and burning explosives (Moskow: Science) p. 345
[3] Maksimov YY, Vorontsov E D and Pavlenko V G 1973 *J Kinetics and Catalysis* **14** 1139
[4] Makarevich G G, Pak V Kh, Pugachev V M and Ryabykh S M 1998 *High Energy Chemistry* **33** 245
[5] Vertz J E and Bolton J R 1986 Electron Spin Resonance. Elementary Theory and Practical Applications Springer New York, NY USA p. 500
[6] Zhidomirov G M et al. 1975 Interpretation of complex EPR spectra (Moskow: Science) p. 215
[7] Makarevich G G, Pak V Kh and Ryabykh S M 2004 *High Energy Chemistry* **38** 346

IOP Conf. Series: Materials Science and Engineering **110** (2016) 012075 doi:10.1088/1757-899X/110/1/012075

The Study of Radiolysis of $Ba(NO_3)_2$ by ESR

V Kh Pak[1], E P Dyagileva[2], S Yu Lyrshchikov[3], M B Miklin[1], V A Anan'ev[1]

[1] Kemerovo State University, Krasnaya str, 6, Kemerovo, 650043, Russia,

[2] Kemerovo State Medical Academy, Voroshilov str, 22a, Kemerovo, 650029, Russia

[3] National Research Tomsk Polytechnic University, Institute of Power Engineering, 30, Lenin Avenue, Tomsk, 634050, Russia

E-mail: miklin@kemsu.ru

Abstract. Irradiation of $Ba(NO_3)_2$ γ-rays at 308 K at increasing absorbed dose creates several new PC: 1, 2, 3, X. All of these are characterized by almost isotropic g-factor. Comparison of the observed parameters with literature data allows us to identify the PC as follows: 1 - O_3^-, 2 - O^-. PC 3 and X presumably attributed to the complex centers $[NO_2...O_2]$ or $[NO_2^-...O^-...O_2]$ and $[O^-...O_2]$, respectively.

1. Introduction

The paramagnetic centers (PC) NO_2, O_n^- (n = 1-3), and F-centers were found under the radiolysis of crystalline $Ba(NO_3)_2$ at 300 K [1, 2]. At 77 K PC NO_3, ONOO, and NO_3^{2-} were found [3-6]. The diamagnetic products peroxynitrite $ONOO^-$ and nitrite NO_2^- were formed by radiolysis of crystalline $Ba(NO_3)_2$ at 300 K [7-9].

X-ray irradiation at 300 K (absorbed dose ~ $5 \cdot 10^3$-$5 \cdot 10^4$ kGy was estimated by us from the data represented in [1]) results in ESR spectrum due to F-centers, O_2^-, O_3^-, and NO_2 [1]. The latter was registered at 85 K only. The annealing of F-centers was observed at room temperature, and at 358 K they completely decay within 3-5 minutes. This process simultaneously results in the increasing of O_2^- and O_3^- concentration.

A γ-ray irradiation at 300 K (absorbed dose ~ 1000 kGy) results in ESR spectrum due to O^-, O_3^-, and unidentified paramagnetic center [2]. The former is stable up to 570 K, the latter is annealed at 330-350 K. The ozonide ion decay under the heating of irradiated sample up to 440-480 K. Thus, there are the difference both PC observed in [5,6] and they thermal stability. It may be connected with the values of absorbed dose.

The goal of this investigation is to study the accumulation of PC under γ-irradiation of $Ba(NO_3)_2$ at 308 K.

2. Experimental

Single crystals of $Ba(NO_3)_2$ were grown by slow evaporation of aqueous solution of "reagent grade" salt. The samples were irradiated with ^{60}Co γ-rays at 308 K. The dose rate 0.69 Gy/s was measured with a Fricke dosimeter, assuming the radiation chemical yield of Fe^{3+} to be equal to 15.6 $(100\ eV)^{-1}$. The absorbed dose was calculated using the mass energy absorption coefficients. ESR spectra were registered by X-band spectrometer Bruker EMX Micro 6/1 at 300 K. The angular dependences of ESR spectra were constructed with using DPPH (g = 2.0036). The value of g-tensor were determinate with accuracy ± 0.0007.

IOP Conf. Series: Materials Science and Engineering **110** (2016) 012075 doi:10.1088/1757-899X/110/1/012075

3. Results

ESR spectrum due to all of observed PC was registered under irradiation by dose 400 kGy only (Fig. 1). Eigenvalues of PC' g-tensor represent in Table 1 were calculated from dependencies of the ESR spectra for different crystal orientation in the magnetic field. As seen, all of observed PC have almost isotropic g-tensors while in the earlier studies they have axial or rhombic symmetry (Table 1).

From our calculated g-factors values and published data it is follows that PC 1 is radical O_3^-, the PC 2 and PC X are radical O^-. A g-factor of PC 3 corresponds to radical NO_2, but its hyperfine structure differ from the one for NO_2 in other matrix. For the latter case, the assignment is clearly self-explanatory, because we observed only two from three lines (low intensity of spectrum is not possible to observe the remaining line) and the hyperfine splitting does not match the known values for this radical.

Irradiation by the doses from 10 to 150 kGy of barium nitrate single crystals results in the ESR spectrum due to ion-radical O_3^- only. It is stable at 300 K.

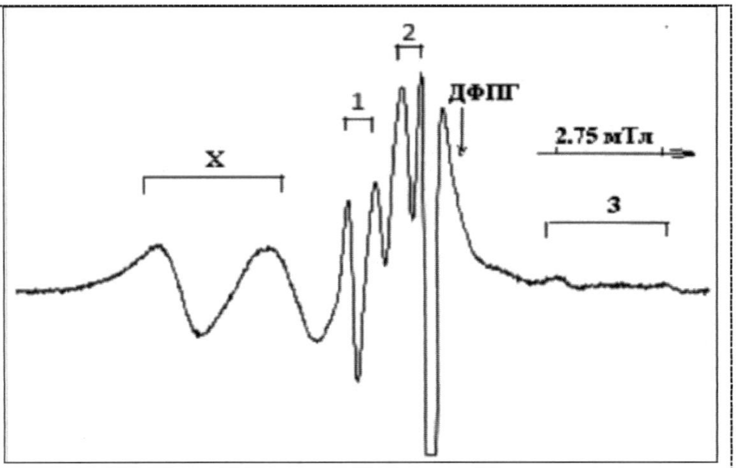

Figure 1. Spectrum ESR γ-irradiated (400 kGy) $Ba(NO_3)_2$.

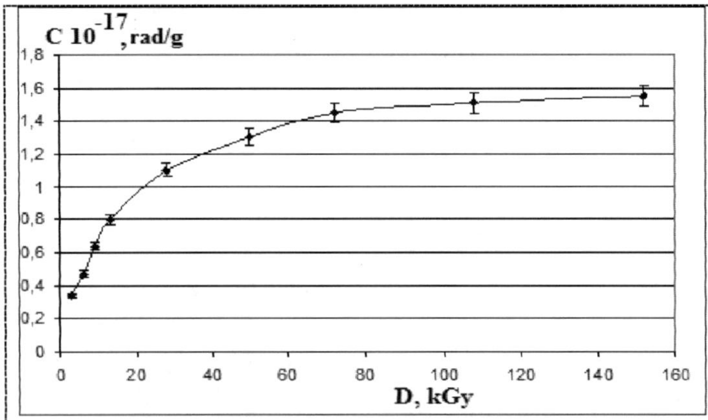

Figure 2. Accumulation radical O_3^- at γ-radiolysis of $Ba(NO_3)_2$.

Table 1

Eigenvalues of g- tensors for PC X, 1, 2, and 3.

Radical	g_{xx}	g_{yy}	g_{zz}	g_{ave}
X	2.0453	2.0448	2.0451	2.0451
1	2.0218	2.0170	2.0172	2.0190
2	2.0110	2.0107	2.0128	2.0115
3	1.9889	1.9798	1.9761	1.9816

The accumulation of O_3^- up to 15 kGy is almost linearly (Fig. 2). The value of initial radiation-chemical yield of O_3^- is equal to 0.016 rad/100 eV. Increase of the irradiation dose results in decrease of the rate of O_3^- formation and at the absorbed dose higher than 80-100 kGy the concentration become constant (1.6 10^{17} rad/g).

Irradiation of barium nitrate by dose > 120 kGy results in ESR spectrum due to O^-, and when dose > 1400 kGy one can be see center X (Fig. 3). At dose > 350 kGy in the ESR spectrum appears two small signals assignment as 3 in Fig. 1. The increase of the absorbed dose results in the decrease of its rate of formation X, but the accumulation of the paramagnetic center observes up to ~ 500 kGy.

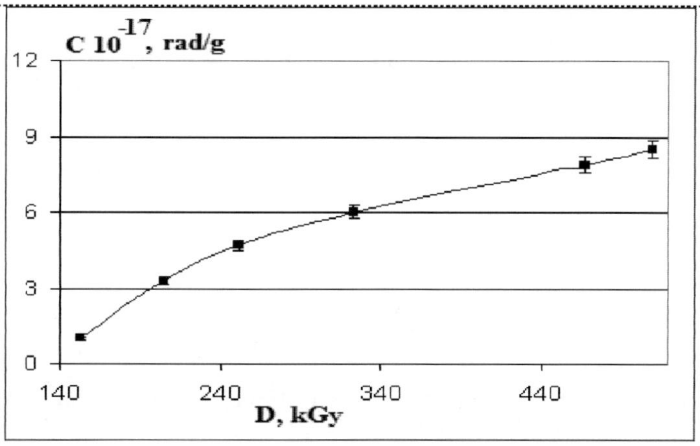

Figure 3. Accumulation radical X at γ-radiolysis of $Ba(NO_3)_2$.

The consistent irradiation of single crystal $Ba(NO_3)_2$ at 308 K to 40kGy and then at 77 to 1kGy result to appearance in ESR spectrum lines belonging to the radical NO2, located in four magnetically nonequivalent positions (Fig. 4).

4. Discussion

Comparison of the parameter of g-tensor observed radicals with literature data allows us to identify the PC as follows: 1 - O_3, 2 - O^-. PC 3 and X presumably attributed to the complex centers [O^- ... O_2] and [NO_2 ... O_2] or [NO_2^- ... O^- ... O_2], respectively.

At γ-irradiation is forming complex "cell" centers [NO_2^-...O] or [NO_2^-...O_2...NO_2^-]. It is supposed that initially analogical "cell" pairs are formed in the volume and on the surface of $Ba(NO_3)_2$ at photolysis [10] too.

The collapse of complex defects may result in the "free" radical O^-. Interaction O^- with a molecule of oxygen O_2 leads to O_3^-. Presumably, at low absorbed doses, is forming centers of the type [NO_2^- ...O_2...NO_2^-] and radicals are formed O_3^- only. Then, in the ESR spectra appear free radicals O^-.

Next capture free electrons and holes on the above-mentioned complex radiation defects can lead to radical products of transformation of these particles. Thus, paramagnetic particles can be formed from $[NO_2^-...O^-]$ or $[NO_2^-...O_2]$ to create free oxygen radicals or complex products, for example:

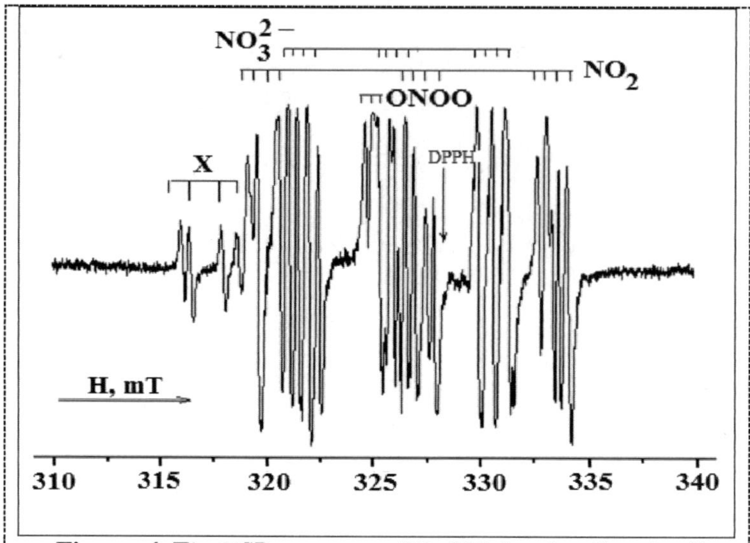

Figure. 4. The ESR spectrum of $Ba(NO_3)_2$ γ-irradiated at first at 300K (40 kGy) and then at 77 K (1 kGy), measured at 77K.

$$[NO_2^-...O] + e \rightarrow NO_2^- + O^-$$

$$[NO_2^-...O] + e \rightarrow [NO_2^-...O^-]$$

Formation of simple paramagnetic centers O_n^- output due to the possibility of diffusion such a radical in the interstitial space.

$$[NO_2^-...O_2] + O^- \rightarrow NO_2^- + O_3^-$$

Capture free holes can be result in formation radical NO_2

$$[NO_2^-...O] + h \rightarrow NO_2 + O$$

Acknowledgments

This works was supported state assignment of Ministry of Education and Science of Russian Federation No 2015/64.

References

[1] Tagaya K 1985 J. Appl. Phys. **2** 75
[2] El'vikh L P, Gorbovitskaya T I, Bugaenko L T and Tiliks Yu E 1977 High Energy Chem. **11** 348
[3] Harrinson L G, Ng Hok Nam H N1973 J. Chem. Soc. Faraday Trans I **69** 1432
[4] Ždansky K, Šroubek Z 1965 Phys. Stat. Sol. **10** 571
[5] Kikuchi R, Nogaito T, Tagaya K, Matsumoto K 1967 Mem. Inst. Sci. Ind. Res. **24** 53
[6] Bannov S I, Miklin M B1997 High Energy Chemistry **31** 233
[7] Anan'ev V, Kriger L, Miklin M 2015 IOP Conf. Series: Mater. Sci. Engineer. **80** 012023
[8] Chen T-H, Johnson E R 1962 J. Phys. Chem. **66** 2249
[9] Kulkarni S P, Garg A N 1986 Journal of Radioanalytical and Nucl. Chem. **98** 65
[10] Miklin M B, Ghyngazov S A, Pak V Kh, Anan'ev V A 2016 J. Photochem. Photobiol. A **315** 94

Influence of the technique for producing lithium-titanium ferrite ceramics on its electrical properties

A P Surzhikov, A V Malyshev and S A Lamonova

National Research Tomsk Polytechnic University, Tomsk, Russia

E-mail: malyshev@tpu.ru

Abstract. The patterns of electrical, dielectric, magnetic, and structural characteristics for the lithium-titanium ferrite ceramics, prepared in different technological conditions have been investigated for the first time. The determining factor of the obtained patterns is structural redistribution of iron ions over the sublattices and their diffusion from the grain boundary in the volume of the ferrite ceramics grain.

Keywords: Ferrite; Dielectric permittivity; Saturation magnetization.

1. Introduction

Modern development and updating of electronic equipment specifies increasingly strict requirements to the properties of ferrite materials. Lithium ferrites attract considerable attention of researchers due to feasibility of their application in microwave technology as circulators, isolators, and phase shifters due to high resistivity, low eddy current and fairly low cost.

It is well-known [1] that electrical properties of ferrite ceramics are sensitive to its composition and microstructure, which in turn are determined by technological conditions of sintering. Changes in temperature and sintering time cause redistribution of iron cations over the octahedral and tetrahedral sites in the spinel lattice. Therefore, the study of the interrelation between the conditions of ferrite ceramics sintering with its main physical characteristics is of current interest.

2. Experimental techniques

The studied samples were sintered under different conditions: sintering temperature T_s = 1283 K and 1373 K; sintering time t_s = 2, 4, and 7 hours. The technology of sample preparation was similar to that described in [2]. The sintering conditions were close to those currently used for production of 3SCH-18 ferrite.

The dielectric parameters of ferrite were obtained with a measurement setup based on automatic meter LCR-819 Goodwill, a special measuring cell and Origin to process the experimental results [3].

The saturation magnetization of the ferrite was measured at room temperature in pulsed magnetic fields with a maximum strength of 5 kOe using an original magnetometer [10].

The structural parameters of powder samples were studied with the X-ray diffractometer X'TRA.

3. Result and discussion

Figure 1 presents the experimental temperature dependence of the complex permittivity components. As can be seen in the Figure, increase in the values of the dielectric permittivity occurs at least within two stages. As the temperature grows, the slope of the curve increases, which implies the involvment of the second type of relaxators of higher concentration.

To approximate the experimental curves and to obtain the values of the dielectric parameters we used both the classical expression for the Debye relaxation polarization (1–4) and the expressions for the Wagner-Koops mechanism of interlayer polarization [8]. These models are most frequently used for mathematical processing of the measurement results obtained for dielectric ferrite ceramics [6, 7, 11].

The expressions for the complex permittivity components ε' and ε'' of the classical Debye model for two types of dipoles [7] are as follows:

$$\varepsilon' = \varepsilon_\infty + \frac{\varepsilon_2 - \varepsilon_\infty}{1+(\omega\cdot\tau_1)^2} + \frac{\varepsilon_{sf} - \varepsilon_2}{1+(\omega\cdot\tau_2)^2} \quad (1)$$

$$\varepsilon'' = \frac{(\varepsilon_2 - \varepsilon_\infty)\cdot\omega\cdot\tau_1}{1+(\omega\cdot\tau_1)^2} + \frac{(\varepsilon_{sf} - \varepsilon_2)\cdot\omega\cdot\tau_2}{1+(\omega\cdot\tau_2)^2} + \frac{\sigma_a}{\varepsilon_0\cdot\omega} \quad (2)$$

In these expressions, ε_∞ is the dielectric permittivity due to inertialess polarization; ε_{sf} is static effective dielectric permittivity; $\tau_1=\tau_{01}\cdot\exp[E_{a1}/(k\cdot T)]$ is the expression for the characteristic time of relaxation polarization for dipoles of the first type; E_{a1} and τ_{01} are the activation energy of relaxation for relaxation oscillators of the first type and the pre-exponential factor, respectively; τ_2 is the characteristic time of relaxation polarization for dipoles of the second type, ε_0 is vacuum permeability.

$\sigma_a = \sigma_{0a}\cdot\exp[-E_{a\sigma}/(k\cdot T)]$ and $E_{a\sigma}$ are the conductivity and the activation energy of the alternating current electrotransport; k is the Boltzmann constant; ε_2 is the dielectric constant in the absence of the relaxation oscillators of the second type; ω is angular frequency.

Classical expressions (1, 2) used for calculations did not allow satisfactory approximation of the experimental temperature dependence for ε' and ε'' components. Similar results were obtained in calculations when using the expressions for the Wagner-Koops polarization model [8, 9].

However, frequency-decreasing expressions (3, 4) used for τ_{0f} and ε_{sf} parameters in (1, 2)

$$\tau_{0f}=A\cdot\omega^{-q1} \quad (3)$$

$$\varepsilon_{sf}=B\cdot\omega^{-q2} \quad (4)$$

provide satisfactory approximation of the temperature dependence for ε' and ε'' components for two types of relaxation oscillators (Fig. 1). Symbols are the experimental data and lines are the calculated curves (the dependences for two of the six samples). Mathematical processing implied automatic adjustment of the parameters in expressions (1–4) for satisfactory coincidence of the calculated curves with the experimental temperature dependences of the components of the complex dielectric permittivity.

Introducing frequency decreasing expressions (3, 4), from the physical point of view, can be interpreted as reduction of the concentration of relaxation oscillators in the form of oppositely charged ion pairs involved in dielectric polarization [4]. We consider the process of relaxation polarization in the form of electron hopping between oppositely charged ions of iron located at different distances from each other. In this case, the concentration of oppositely charged ions involved in polarization will largely determine the ε' values of ferrite permittivity, and the dielectric polarization will depend on the frequency of the electric field and the temperature of the material.

The analysis of the dielectric parameters (table 1) shows that the values of activation energy for each type of the relaxation oscillators are virtually the same for all types of samples, which may indicate similar physical nature of each type of the relaxation oscillators.

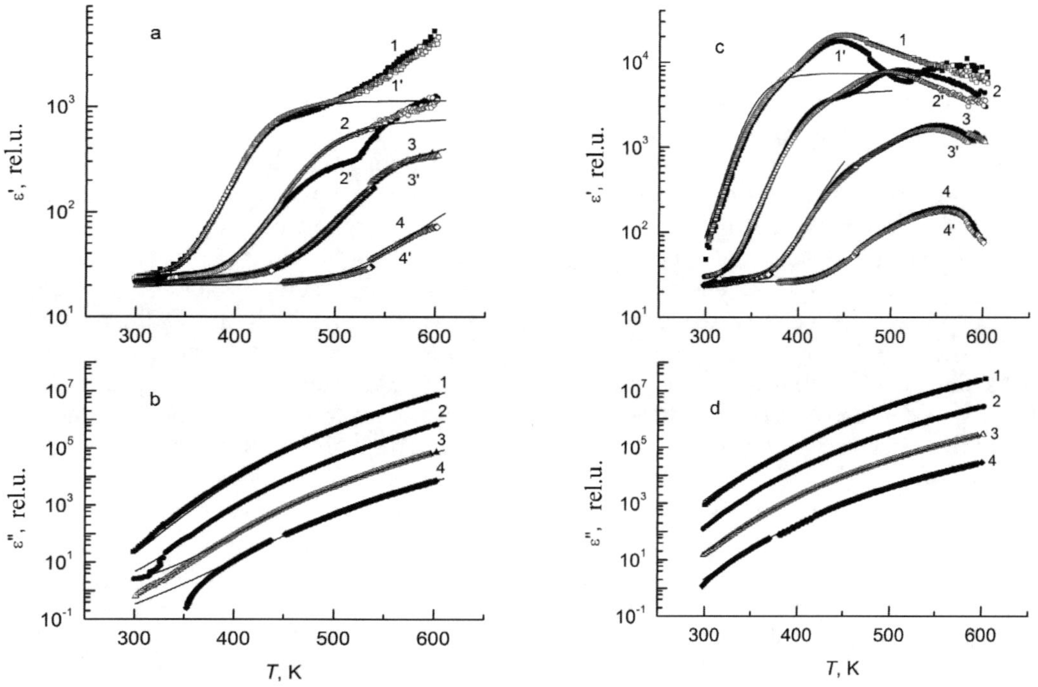

Figure 1. Temperature dependence of the real ε′ and imaginary ε″ components of the complex dielectric permittivity under heating (1–4) and cooling (1′–4′) of the ferrite samples, T_s= 1283 K, t_s = 4 h (a, b) and T_s = 1373 K, t_s = 4 h (c, d). The frequency of the test signal: 100 Hz (1, 1′), 1 kHz (2, 2′), 10 kHz (3, 3′), 100 kHz (4, 4′).

Table 1. Dielectric parameters of lithium-titanium ferrite samples.

	T_s =1283 (K)			T_s=1373 (K)		
	t_s =2 (h)	t_s =4 (h)	t_s =7 (h)	t_s =2 (h)	t_s =4 (h)	t_s =7 (h)
q1	–0.45	–0.5	–0.5	–0.4	–0.32	–0.32
q2	–0.15	–0.17	–0.17	–0.3	–0.2	–0.28
B (c^{q2})	3972	3433	3287	44010	26702	99560
A (c^{-q1})	$1.48\cdot10^{-6}$	$2.04\cdot10^{-6}$	$2\cdot10^{-6}$	$2.36\cdot10^{-7}$	$6.4\cdot10^{-8}$	$3.8\cdot10^{-8}$
E_{a2} (eV)	0.375	0.375	0.375	0.375	0.375	0.375
ε_∞	20	20	20	26	26	26
ε_2	25	25	25	23.8	23.8	23.8
τ_{01} (s)	$1.2\cdot10^{-7}$	$1.2\cdot10^{-7}$	$1.2\cdot10^{-7}$	$1.2\cdot10^{-7}$	$1.2\cdot10^{-7}$	$1.2\cdot10^{-7}$
E_{a1} (eV)	0.15	0.15	0.15	0.15	0.15	0.15
σ_{0a} $(\Omega\ m)^{-1}$	$3.18\cdot10^{4}$	$4.34\cdot10^{4}$	$3.66\cdot10^{4}$	$6.66\cdot10^{3}$	$7.10\cdot10^{3}$	$5.04\cdot10^{3}$
$E_{a\sigma}$ (eV)	0.714	0.720	0.715	0.574	0.557	0.520

E_{a2} is the activation energy of the relaxation process for the second type of relaxation oscillators dominant in its concentration; A and B are the frequency-independent parameters in expressions (1) and (2).

However, $E_{a\sigma}$ and σ_{0a} values in the ferrite samples sintered at T_s=1373 K are considerably higher than those in the samples sintered at T_s=1283 K. The electrical conductivity of ceramic materials in the radio frequency region depends on the height of the intergranular potential barrier, which results from the difference between the iron ion concentration in the grain volume and intergranular layer. Redistribution of concentrations of Fe^{2+} and Fe^{3+} ions in these regions relates to diffusion of oxygen ions from the intergranular layer into the grain volume and oxidation of Fe^{3+} ions to Fe^{2+} [2].

To identify unambiguously the electromigration mechanism in the grain volume, we need to measure the conductivity in a higher frequency range of the electric field. In this case, as the frequency increases, the intergranular layer capacitance shunting will cause decrease in the conductivity values. Most of the studies [11] consider the mechanism of electromigration during electron hopping between the localized states in ferrite crystal lattice ions. In this case, electromigration mechanisms and relaxation polarization equivalent, and therefore, the activation energy of electromigration processes at under high frequency electric field should be similar to the energy of relaxation polarization.

In our research, E_a and $E_{a\sigma}$ values are not identical due to the heterogeneous structure of ferrite ceramics and a relatively narrow range of the measuring frequency of the electric field.

The thermogravimetric analysis of the ferrite powder under sintering showed that charges lose their weight under sintering in isothermal curing due to ferrite reduction, i.e. to diffusion of oxygen ions from the material into the atmosphere. Cooling at temperature in the range of T=1273–1073 K results in replacement of the reducing medium with oxidizing atmosphere and mass gain due to the direct oxygen diffusion in the material volume, primarily in the grain boundaries on the powder particle surface. In this case, the effect of sintering time on concentration of Fe^{2+} ions found mostly in the volume of the ferrite grain is not significant since the values of the bulk diffusion coefficient of oxygen ions in the grain are considerably lower if compared to the intergranular layer. The duration of the cooling phase for the sample is limited as well.

The investigation of the sample magnetic characteristics at room temperature showed that the value of saturation magnetization M_s is crucial for samples sintered at t_s=2 h and T_s=1373 K (table 2). The values correspond to the nominal values of M_s for ferrite 3SCH-18 [5]. As T_s increases, M_s grows, whereas sintering time insignificantly affects the value of M_s.

Table 2. Values of M_s (G).

t_s (h)	T_s=1283 (K)	T_s=1373 (K)
2	143	172
4	156	158
7	149	165

It is known [1, 7] that the values of M_s in the spinel ferrite depend on the distribution of iron cations over sublattices. The more the number of iron cations Fe^{3+} in the octahedral (B) sublattice and the less the number of iron cations in the tetrahedral (A) sublattice, the higher the value of M_s.

In this case, the concentration of Fe^{3+} ions in the sublattice B is expected to reach the maximum under sintering at T_s=1373 K and t_s=2 h.

The values of basic structural parameters for ferrite samples of various types were determined via the diffractometer analysis of ferrite ceramics. The radiograms for each of the samples (powders) obtained with a diffractometer were processed in the program Powder Cell and enabled calculation of the basic structural parameters of ferrite (table 3).

Table 3. Structural parameters of ferrite ceramic samples.

t_s (h)	T_s=1283 (K)			T_s=1373 (K)		
	content of the main phase (%)	chemical formula	a (Å)	content of the main phase (%)	chemical formula	a (Å)
2	92.8	$Fe_{0.61}Zn_{0.2}Li_{0.19}[Fe_{0.494}Li_{0.229}Ti_{0.25}Mn_{0.025}]O_4$	8.365	98.9	$Fe_{0.58}Zn_{0.2}Li_{0.22}[Fe_{0.509}Li_{0.214}Ti_{0.25}Mn_{0.025}]O_4$	8.366
4	95.8	$Fe_{0.57}Zn_{0.2}Li_{0.23}[Fe_{0.514}Li_{0.209}Ti_{0.25}Mn_{0.025}]O_4$	8.364	89.1	$Fe_{0.63}Zn_{0.2}Li_{0.17}[Fe_{0.484}Li_{0.239}Ti_{0.25}Mn_{0.025}]O_4$	8.367
7	90.9	$Fe_{0.59}Zn_{0.2}Li_{0.21}[Fe_{0.504}Li_{0.219}Ti_{0.25}Mn_{0.025}]O_4$	8.366	92.3	$Fe_{0.63}Zn_{0.2}Li_{0.17}[Fe_{0.484}Li_{0.239}Ti_{0.25}Mn_{0.025}]O_4$	8.366

Where a is lattice constant.

In this case, the chemical formula for stoichiometric composition of 3SCH-18 ferrite is as follows: $Fe_{0.62}Zn_{0.2}Li_{0.18}[Fe_{0.489}Li_{0.234}Ti_{0.25}Mn_{0.025}]O_4$

As can be seen from the table, the content of the main phase of 3SCH-18 ranges from 89 wt. % to 98 wt. %. Apart from to the main phase, an additional non-magnetic phase of orthoferrite lithium $LiFeO_2$ was found, which "dilutes" the main phase, thereby affects the magnetization value, namely, it reduces magnetization.

The values of lattice constant (a) appeared to be close for all ferrite samples regardless of the changes expected during redistribution of iron cations in the sublattices as the radius of Fe^{2+} ions is greater than that of Fe^{3+} ions.

Table 3 also shows that redistribution of iron ions in the sublattices of the samples with different sintering parameters indicates a pattern when the sintering time with the greatest values of M_s corresponds to each value of T_s with the highest concentration of iron ions Fe^{3+} in the sublattice B.

If the mechanism of electrical conductivity and relaxation polarization of ferrite is considered in terms of the model of hopping over localized states between Fe^{2+} and Fe^{3+} ions located in the octahedral (B) sublattice [8, 9, 11], the concentration of relaxation oscillators and therefore ε' values will depend on the cationic distribution of iron ions. Thus, the transfer of iron ions into octahedral positions under varying modes of ferrite ceramics sintering will lead to an increase in ε' values. The comparison of the experimental data illustrates the tendency to this dependence (tables 1 and 3).

4. Conclusion

The comparative analysis of the data obtained through the diffraction analysis of powder samples (Table 3) and experimental data showed that M_s values correlate with the values of Fe^{3+} ion concentration in the octahedral (B) sublattice of the ferrite spinel and the concentration of the 3SCH-18 main phase for samples sintered both at T_s=1283 K and T_s=1373 K.

The obtained results can be used to choose a technological mode for lithium-titanium ferrite ceramics sintering in order to obtain an optimal set of operating parameters for ferrite products.

5. Acknowledgements

The research was supported by the Ministry of Education and Science of the Russian Federation in part of the "Science" program.

References

[1] Gorter E V 1955 *Uspehi Fiz. Nauk [in Russian]* **10** 279–346
[2] Surzhikov A P, Pritulov A M, Peshev V V and Malyshev A V et al 2001 *Russ. Phys. J.* **44** 11 1244–47
[3] Malyshev A V and Peshev V V 2007 *Russ. Phys. J.* **50** 2 161-164
[4] Surzhikov A P, Lysenko E N and Malyshev A V et al 2014 *Russ. Phys. J.* **57** 5 621–626
[5] Malyshev A V, Lysenko E N and Vlasov V A 2015 *Ceramics International* **41** 10 13671–75
[6] Smit J and Wijn H P J 1959 Ferrites. Physical properties of ferrimagnetic oxides in relation to their technical applications (Philips Technical Library: Eindhoven) p 299
[7] Krupichka S Physics of ferrites and related magnetic oxides 1976 [in Russian] (Moscow: Nauka) p 683
[8] Koops C G 1951 *Phys. Rev.* V **33** 1 121–124
[9] Watawe S C et al 2000 *J. of Magn. and Magn. Mat.* **214** 55–60
[10] Kreslin V Yu and Naiden E P 2002 *Pribory i tehnika experimenta [in Russian]* 1 63–66
[11] Lipare A Y et al 2003 *Materials Chemistry and Physics* **81** 108–115

Investigation of antioxidant capacity of the extracts of bilberry (VACCINUM MYRTILLIS L.) by voltammetry

A N Vtorushina, E D Nikonova

Lecturer, *Ecology and Human Safety Department, Tomsk Polytechnic University*
Student, *Ecology and Human Safety Department, Tomsk Polytechnic University*
Lenin av. 30, Tomsk Polytechnic University, Tomsk, Russia, 634050

E-mail: anl@tpu.ru

Abstract. This paper deals with the urgent issue of the search of new drugs based on plant raw materials that have an influence on various stages of oxidation processes occurring in the human body. The aim of this paper is to determine the antioxidant activity of the bilberry extracts that are used in the medicine practice by a cathodic voltammetry method.
We consider the influence of water and alcohol bilberry extracts on the process of oxygen electroreduction. From these extracts the most activity relation to the process of cathodic oxygen reduction showed alcohol (40%) bilberry extract. It was also stated that the alcohol extract of bilberry has a greater antioxidant activity than other known antioxidants such as ascorbic acid, glucose, dihydroquercetin.
Thus, after consideration of a number of plant objects, we showed the possibility of applying the method of cathodic voltammetry for the determination of total antioxidant activity of plant material and identifying and highlighting the most perspective sources of biologically active substances (BAS), as well as the ability of identifying extractants that fully extract BAS from plant raw materials. The activity data of extracts of plant raw materials gives an opportunity of establishing an effective yield phytopreparation based on bilberry that has an antioxidant effect.

1. Introduction

Nowadays therapy that includes antioxidants (AO) is becoming more common in the treatment of diseases associated with the development of radical chain oxidation processes in the body cells [1-3]. The use of natural AO for the treatment and prevention of free radical pathology has several advantages, such as: the absence of side effects for most of them; the low toxicity even during the extended use; formulations based on them are available for patients, etc. Many individual substances that are contained in plant raw material are currently synthesized and produced by chemical-pharmaceutical industry. However, the plant raw material is still very relevant in the application because it contains a large set of biologically active substances (BAS), which stimulates the immune system of the human body, providing a stabilizing effect during the flow of many vital activity processes. In recent years the interest in the methods of biologically active substances isolation and their activity evaluation in natural objects is increasing. Based on them the variety of food additives and drugs that stimulate the metabolic processes in the body are created [4-6]. Thereby, there is a great interest in a perennial plant from the family of Vacciniaceae - Vaccinium Myrtillus.

In national medicine Vaccinium Myrtillus is widely used because of its astringent qualities due to the presence of tannins. Tinctures and extracts of the fruits are used in cases of acute and chronic enterocolitis, putrid fermentation in the bowels, especially children, diarrhea, milder forms of diabetes. It externally applied as compresses for treating hemorrhoids. There are also indications of catarrhal and follicular tonsillitis, burns, canker sores, etc. treatment with the bilberry broth. Also the fruit of bilberries improves vision, especially night vision [7-9].

Thereby, the aim of the study is to investigate the influence of Vaccinium Myrtillus extracts on the electrochemical behavior of oxygen in water in the condition of the semi-infinite linear diffusion.

2. Methods and materials

We used a express and highly sensitive method of voltammetry (VA) having simple method implementation.

All experiments were carried out on a computerized voltammetric analyzer TA-2 produced by LLC Research and Production Enterprise RPE «Tomanalyt» (Tomsk) with three-electrode cell. Mercury-film electrode (MFE) was used as an indicator electrode, and argent chloride electrode served as an auxiliary reference electrode (ACE).

Direct cathode-current mode VA was used; speed of potential sweep (W) was 40 mV/s, working potential range from 0 to -1 V. The substances were mixed by vibration of the electrodes.

As a model reaction the oxygen reduction reaction was used. In the beginning we recorded the first wave of cathodic reduction of oxygen dissolved in a supporting electrolyte in the specified potential range. Then we successively added the solution of investigated substance and recorded the voltammograms of the first wave of oxygen cathodic reduction.

Oxygen reduction reaction was used as a model reaction. Records were in following order: record of first wave cathodic reduction of oxygen dissolved in a supporting electrolyte in the specified potential range. After substance addition, the solution was stirred about 20s. After the stirring is stopped, the potential was scanned negatively, causing oxygen reduction, giving a current first wave electroreduction of oxygen (ER O_2). Its value was proportional to the amount of oxygen in the bulk of the solution. Oxygen concentration was monitored by oxygen analyzer. Based on the ammetric measurements the concentration of oxygen in phosphate buffer at 25.0 ± 0.5°C was 2.56 ± $0.05\cdot10^{-4}$ $mol\cdot l^{-1}$.

Oxygen concentration in the electrolyte solutions was monitored by potentiometric oxygen sensor № 5972 produced by the center of computing automation and measuring «MERA-ELVRO», with automatic temperature compensation in the range from 0 to 40 C. All reagents were of analytical grade. Samples material were weighted on laboratory analytical balance VL 210 firm «Gosmetr» with a weighing error of ± 0.0002 g. Nanopure water was used for making solutions.

3. Experimental

Recent studies have shown that herbal medicines contained a complex of biologically active substances have influenced on the various body systems, including antioxidant defense system [10-12].

The list of antioxidant activity (AOA) determination methods is quite wide: chemiluminescent method, gas phase chromatography, spectrophotometric and fluorometric methods, etc [13-15]. For this study we used cathodic voltammetry method to determine the antioxidant activity of vegetable raw materials [16]. The method is based on a process of oxygen electroreduction (ER O_2) that has a mechanism similar to the restoration of oxygen in the cells and body tissues.

In this study we used vaccinum myrtillis' shoots. Hydroalcoholic extracts of vegetable raw materials were prepared by mixing water and ethanol in various proportions. Different ratios of water and alcohol (40, 70 and 95 percent) in order to obtain extracts of vaccinum myrtillis' shoots were examined.

The influence of the test substances on the electrochemical behavior of oxygen is considered in a supporting electrolyte – phosphate buffer with pH 6.86 (0.025M KH_2PO_4 and 0.025M Na_2HPO_4).

The total antioxidant activity of vaccinum myrtillis' extracts was determined by the relative decrease in the current ER O_2 (figure 1) in the presence of the test components in solutions.

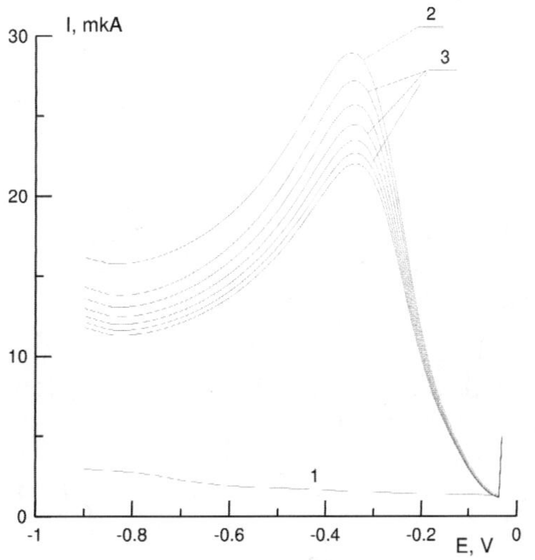

Figure 1. Voltammograms of the ER O_2 current in phosphate buffer (0.025M, pH 6.86) on the MFE without (2) and with 10^{-4} g/ml of vaccinum myrtillis aqueous extract in the solution (3), the residual current of the supporting electrolyte in the absence of O_2 and the substance in the solution (1). Potential scan rate is – 40 mv / s.

Lines (3) in figure 1 characterize the reduction of oxygen current with time of the reaction between the active oxygen forms and BAS vaccinum myrtillis extract in solution.

The dependence of relative changes of the limiting current ER O_2 on the time of the process in the presence of the sample was build according to obtained data. The kinetic criterion for characterizing the total AOA - K (μmol $l^{-1}min^{-1}$) was calculated with the help of the slope of linear part of the curve (Figure 2) [17].

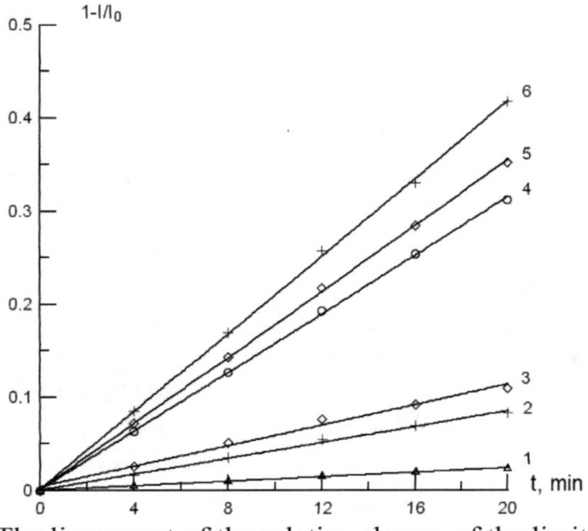

Figure 2. The linear part of the relative change of the limiting current of ER O_2 on the time of the process in the presence of 10^{-4} g/ml of vaccinum myrtillis extracts: water (1), 40% (6) 70% (4) and 95% (5) ethanol; dihydroquercetin (2), ascorbic acid (3).

4. Discussion of results

Experiments showed activity of all samples against of ER of O_2 process, but the activity of the water based samples are considerably smaller in comparing with alcoholic based samples which sufficiently close to each other in activity (table 1).

It is noteworthy that all samples moved the first wave potential of O_2 ER to a negative region, which means a loss of the thermodynamic process. The exception was the aqueous extract in the presence of its wave potential of ER O_2 shifted to a positive region, which gives the right make a conclusion about facilitating the process of ER O_2 by adding the sample solution in the background electrolyte solution. Such behavior of the cathode waves let us to suggest that in this case the probable EC mechanism of inhibition of the electrode process ER O_2 in the presence of AO includes subsequent chemical reaction of AO with active oxygen radicals.

Table 1. The total antioxidant activity of the vaccinum myrtillis extracts in relation to the process of ER O_2 (p=0.95, n=5).

Substance name	K (μmol $l^{-1}min^{-1}$)
The vaccinum myrtillis aqueous extract	0.153 ± 0.012
The vaccinum myrtillis extract with 40 % of ethanol	5.815 ± 0.026
The vaccinum myrtillis extract with 70 % of ethanol	4.351 ± 0.018
The vaccinum myrtillis extract with 95 % of ethanol	4.907 ± 0.037
Ascorbic acid	1.165 ± 0.053
Glucose	0.169 ± 0.037
Dihydroquercetin	0.650 ± 0.040

Thus the most active alcoholic extracts of vaccinum myrtillis is the extract with 40% of ethanol. It should be noted that the activity of considered ethanol extracts surpass the activity of such well known antioxidants like ascorbic acid and dihydroquercetin.

Study of the extracts effects on animals showed the highest activity of the extract with 70% of ethanol. In order to establish extractant that fully take out BAS there was a study of vaccinum myrtillis extract, obtained by treatment of 70% of ethanol. Fractionation was carried out by the extract with a number of solvents with increasing polarity. Due to this plant extract was dissolved in water (1:40) and extracted in a separating funnel sequentially with chloroform, ethyl acetate and butanol (table 2).

The first wave of ER O_2 was recorded in the presence of various fractions of the extract vaccinum myrtillis shoots with 70% of ethanol.

Table 2. AOA of vaccinum myrtillis extract shoots with 70% of ethanol and its fractions in the relation to the process of ER O_2 (p=0.95, n=5).

The sample	K, μmol $l^{-1}min^{-1}$
The chloroform fraction	3.432 ± 0.035
The ethyl acetate fraction	0.751 ± 0.014
The butanol fraction	5.604 ± 0.005
The aqueous residue	2.047 ± 0.005

The presented data show that the butanol fraction extract is the most active and the ethyl acetate is the most inactive. It can be said that butanol is best in removing of the bioactive substances from the extract of vaccinum myrtillis with 70% of ethanol.

5. Conclusion

Thus we showed the possibility of applying the method of cathodic voltammetry for the total antioxidant activity determination of vegetable raw materials, identification and selection of the most perspective sources of biologically active substances. We also demonstrated the possibility to identify extractants that fully extracting BAS from vegetable raw materials with a purpose of their concentration and the creation of pharmaceutical products based on them.

The obtained data on the activity of extracts of plant raw materials suggested in the paper give a possibility of creation of effective phytopreparation based on vaccinum myrtillis that has an antioxidant effect.

References

[1] Halliwell B and Gutteridge J M C 1984 *Biochemical Journal* **219** (1) 1
[2] Wolff S, Garner A and Dean R 1986 *Trends in Biochemical Sciences* **11** (1) 27
[3] Wang H, Provan G and Helliwell K 2000 *Trends in Food Science and Technology* **11** (4-5) 152
[4] Halliwell B 1996 *Free Radical Research* **25** (1) 57
[5] Atoui A, Mansouri A, Boskou G and Kefalas P 2005 *Food Chemistry* **89** (1) 27
[6] Chung H S, Chang L C and Lee S K 1999 *Journal Agricultural and Food Chemistry* **47** 36
[7] Sinelli N, Spinardi A, Di Egidio V, Mignani I and Casiraghi E 2008 *Postharvest Biology and Technology* **50** (1) 31
[8] Vendrame S, Guglielmetti S, Riso P, Arioli S, Klimis-Zacas D and Porrini M 2011 *Journal of Agricultural and Food Chemistry* **59** (24) 12815
[9] Molan A L, Lila M A, Mawson J and De S 2009 *World Journal of Microbiology and Biotechnology* **25** (7) 1243
[10] Riso P, Klimis-Zacas D, Del Bo' C, Martini D, Campolo J, Vendrame S, Moller P and Porrini M 2013 *European Journal of Nutrition* **52** (3) 949
[11] Dai J and Mumper R 2010 *Molecules* **15** (10) 7313
[12] Moyer R, Hummer K, Finn C, Frei B and Wrolstad R 2002 *Journal Agricultural and Food Chemistry* **50** 519
[13] Shilova I V, Krasnov E A, Korotkova E I, Nagaev M G and Lukina A N 2006 *Pharmaceutical Chemistry Journal* **40**(12) 660
[14] Campanella L, Bonanni A, Bellantoni D, Favero G and Tomassetti M 2004 *Journal of Pharmaceutical and Biomedical Analysis* **36** 91
[15] Blasco A, Rogerio M, Gonz′alez M and Escarpa A 2005 *Anal. Chim. Acta* **539** 237
[16] Avramchik O, Korotkova E, Plotnikov E, Lukina A and Karbainov Y 2005 *J. of Pharmaceutical and Biomedical Analysis* **37** 1149
[17] Vtorushina A N and Nikonova E D 2015 *IOP Conf. Ser.: Mater. Sci. Eng.* **81** 012091

IOP Conf. Series: Materials Science and Engineering **110** (2016) 012078 doi:10.1088/1757-899X/110/1/012078

Molecular dynamical modelling of endohedral fullerenes formation in plasma

A S Fedorov[1,2], E A Kovaleva[2], T A Lubkova[2], Z I Popov[1,3], A A Kuzubov[2], M A Visotin[2] and S Irle[4]

[1] Kirensky Institute of Physics, 50/38 Akademgorodok, Krasnoyarsk, 660036 Russia
[2] Siberian Federal University, 79 Svobodny pr., Krasnoyarsk, 660041 Russia
[3] National University of Science and Technology MISiS, Moscow, 119049 Russia
[4] Department of Chemistry, Graduate School of Science, Nagoya University, Furo-cho, Chikusa-ku, Nagoya, 464-8601 Japan

E-mail: visotin.maxim@gmail.com

Abstract. The initial stages of fullerene and endohedral metallofullerene (EMF) synthesis in carbon-helium plasma at 1500 K and 2500 K have been simulated with quantum chemical molecular dynamics (MD) based on density-functional tight-binding (DFTB). The cases of formation of large (>100 atoms) sp^2-carbon clusters with scandium atoms inside were observed. These clusters are considered as precursors of fullerenes or EMFs, and thus it is shown that formation of EMFs can be explained within the framework of "shrinking hot giant" mechanism. Also, the dependence of formation rates on plasma parameters, including temperature, buffer gas and metal atoms concentrations, has been studied.

1. Introduction

Since the discovery of buckminsterfullerene C_{60} [1], the family of fullerenes has attracted much attention due to their unusual electronic and optoelectronic properties, such as high electron affinities as well as novel chemical reactions [2]. It's well-known that encapsulations of metal atoms or small molecules into fullerene's cage lead to significant change of its chemical and physical properties [3,4]. The chemistry of these complexes is considerably different from that of bare fullerenes, in particular, their reduction-oxidation properties. The most promising and well investigated are endohedral metallofullerenes (EMFs, fullerenes with metal atoms inside) and endohedral metal carbide fullerenes (EMCFs, fullerenes containing metal atoms and a carbon molecule) [5-7].

These promising compounds have a lot of various applications due to their unusual electronic and magnetic properties, which strongly depend on their structure. Moreover, being very inert, carbon EMF cage can be used for medical purposes [8, 9]. Such structures may be used as carriers of radioactive isotopes [10], radiotracers or MRI contrast media [11]. The possibility of their using in anticancer therapy is also intensively investigated.

However, the yield of EMFs via plasma chemical synthesis is very low at the present moment. There are still many questions about endohedral metallofullerenes' formation mechanisms, as well as thermodynamic and kinetic stability. For example it can be mentioned that large fullerenes (C_n, n>70) are thermodynamically more favorable, whereas the fullerene mixture consists essentially of C_{60}. Thereby, two opposed mechanisms can be realized: growth of the fullerene and its shrinking via C_2 evaporation (so-called "fullerene road" and "shrinking hot giant" mechanisms, respectively) [12-14].

IOP Conf. Series: Materials Science and Engineering **110** (2016) 012078 doi:10.1088/1757-899X/110/1/012078

According to the recently developed "shrinking hot giant" mechanism, large fullerenes are formed first during carbon cluster aggregation process. Being the result of differently sized carbon clusters' coagulation and due to high thermodynamic stability, these giant fullerenes include large number of atoms as well as structural defects. Subsequently, these giant fullerene clusters shrink in size by the loss of the C_2 clusters [14] during the further process of plasma cooling, because the larger clusters have less kinetic stability [12].

This mechanism of fullerene formation was supported by a number of investigations, including quantum-chemical molecular dynamical simulations [12, 15]. One possible confirmation of this model is the fact that yields of C_{70} and higher fullerenes in fullerene mixture are much higher at high helium pressure (more than 0.10 MPa) [16] than at reduced pressure (0.013 MPa) [12].

In this work, plasma chemical synthesis of fullerenes and EMFs is simulated by quantum-chemical molecular dynamics. It is shown that formation of endohedral fullerenes can be explained within the framework of "shrinking hot giant" mechanism. Also, the dependence of formation rates on plasma parameters (including temperature, buffer gas and metal atoms concentrations) is studied.

2. Methods

All MD simulations were performed within the framework of non-consistent charge density-functional tight-binding (NCC DFTB) method [17, 18]. This method is often used for conducting MD for large systems (thousands of atoms) and for long run times (up to several nanoseconds), while having results comparable with conventional DFT [13].

In order to model large systems, we used boxes sized 100*100*100 Å. Each MD box contained 120 C_2 dimers and different numbers of He and Sc atoms (0, 120 or 240 helium atoms and 12 or 24 Sc, respectively). The spatial distribution of atoms and molecules as well as C_2 dimers orientations were uniformly random. Each box was simulated at two finite temperatures (1500 and 2500 K) by Berendsen thermostat [19] with coupling parameter of 100 fs. Taking into account that the system under consideration is large, only Γ point was used for the reciprocal space sampling. MD runs were conducted up to 0.6 ns with a time step of 1 fs. At least 5 different trajectories for each combination of parameters (i.e. temperature, number of He and Sc atoms) were simulated to get statistics for averaging the formation rates values.

In order to measure the dynamics of cluster aggregation and transformation, changes in orbital hybridization type of carbon atoms were tracked with time. Atoms having four, three and two carbon neighbors within a coordination sphere with radius equal to 1.9 Å were considered to be in sp^3, sp^2 and sp hybridization states respectively. Some atoms had only one neighboring C atom, for example atoms on the ends of chains or atoms in C_2 dimers. This state is further referred as 1b. The time evolution of number of atoms in certain hybridization states was averaged among 5 calculated trajectories for every parameters set.

3. Results and discussion

3.1. Fullerene formation mechanism

Qualitatively, almost the same picture is observed for all the trajectories with all the parameter sets:

At the beginning of MD run, constant collisions of C_2 dimers lead to formation of C_n chains. This initial chain building process is rapid and almost irreversible, as the driving force for their formation is the creation of carbon-carbon σ-bond between two carbon atoms with energy of nearly 3.4 eV. As a consequence, these long chains grow longer and longer, until almost no free C_2 molecules are present in the system. This stage is reached in a very short time in the order of tens of picoseconds.

After that, chains gradually begin to connect and form so-called "Y-junctions" containing a single sp^2-type carbon atom, and subsequently forming cyclic structures, containing pentagons, hexagons, and heptagons, as a result of thermal vibrations and bending. Formation of energetically favorable hexagons and heptagons from the interaction between two long carbon chains are also frequently observed. This can be considered as the initial stage of the growth of a carbon sheet fragments.

After formation of a cycle-containing nucleus, the cluster begins a transformation into a network of sp^2-type carbon atoms. The transformation is mainly due to bond formation between linear carbon chains attached to the nucleus in close proximity to each other. The driving force for this process is growing π-conjugation and stabilization, so the most kinetically stable pentagons and hexagons are formed. The pentagons, introducing a uniform curvature to the growing cluster, give it a bowl shape.

Concerning Sc atoms, we observe frequently that ends of carbon chains are terminated by Sc atoms, connected to the carbon chain by Sc-C σ-bond. The Sc atoms participate actively in the carbon sheet growth process too. We observed that Sc become di- and sometimes even trivalent, effectively replacing sp^2 carbon. Some of the carbon sheets formed also contain Sc at the ends of carbon side chains or on top of the sp^2-type carbon networks, with the position inside a carbon cage bowl being more stable.

Similar to nanotube to fullerene conversion, long carbon chains attached to the opening of the bowl reach over the opening and thereby form additional large macrocyclic rings, which in turn can collapse in a "zipper"-type cycloaddition reaction mechanism. As a result, clusters of more than 100 carbon atoms are likely to evolve into a form, which resembles a giant defect-rich fullerene. Figure 1 shows an example of such a structure, consisting of 125 C atoms and 3 Sc atoms inside the almost closed carbon cage.

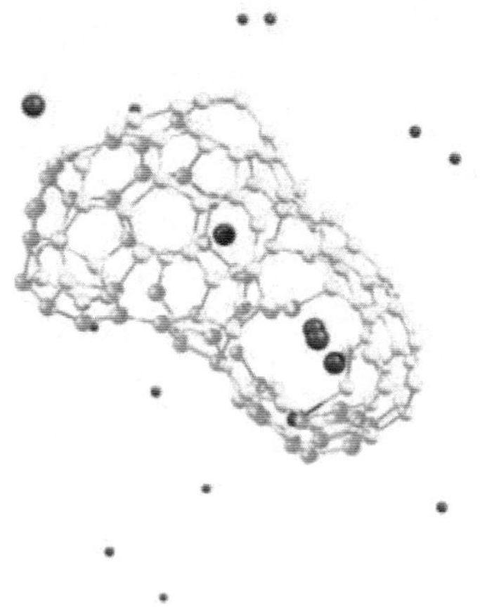

Figure 1. One of EMF cages after 0.4 ns simulation. Yellow balls represent carbon atoms, purple – scandium, cyan – helium.

3.2. Effect of plasma parameters

The rates of fullerene formation are proportional to the rates of sp^2-carbon network growth. Thus, we can compare the rates of sp^2-carbon network growth in order to estimate, how different plasma parameters affect fullerene and endohedral fullerene production in plasma chemical synthesis. The dynamics of carbon atom transformation from 1b state to sp^2 are shown in figure 2. It should be noted, that there are almost no atoms in the sp^3 state.

3.2.1. Effect of buffer gas concentration. Buffer gas atoms are known to play crucial role in fullerene growth. Comparing the rates of changes in the number of sp^2 carbon atoms (in figure 2 He concentration rises from top row to bottom), we can conclude that presence of helium atoms nearly doubles the speed of carbon cages growth. This fact can be explained by the heat transfer from the growing sp^2 nucleus to the buffer gas, which results in stabilization of the nucleus. However, the effect comes to saturation at high helium concentrations. The results for equal carbon and buffer gas concentrations (120 C_2 dimers and 120 He atoms) are almost the same as for 1:2 proportion.

3.2.2. Effect of plasma temperature. MD simulations at 1500K and 2500K reveal unusually weak temperature dependence of sp^2-carbon cluster formation rates. We speculate that the rates of cluster transformation to sp^2 sheet depend on internal vibrational energy of the cluster, which can be higher than the plasma temperature, because of the energy released during new bonds creation [20].

3.2.3. Effect of metal atoms concentration. Comparing the results for the He-free systems with different Sc atom numbers (see figure 2 (a) and (d)) we can see that increase in Sc concentration rises amounts of carbon atoms in 1b state. Analysis of the trajectories showed that this is due to Sc atoms hanging on the ends of carbon chains and thus not allowing them to connect to each other. As a result, the growth rates of unbranched chains and cycles are significantly reduced. In the cases with the buffer gas, however, the effect doesn't take place, because collisions with He atoms tear Sc from chain ends.

IOP Conf. Series: Materials Science and Engineering **110** (2016) 012078 doi:10.1088/1757-899X/110/1/012078

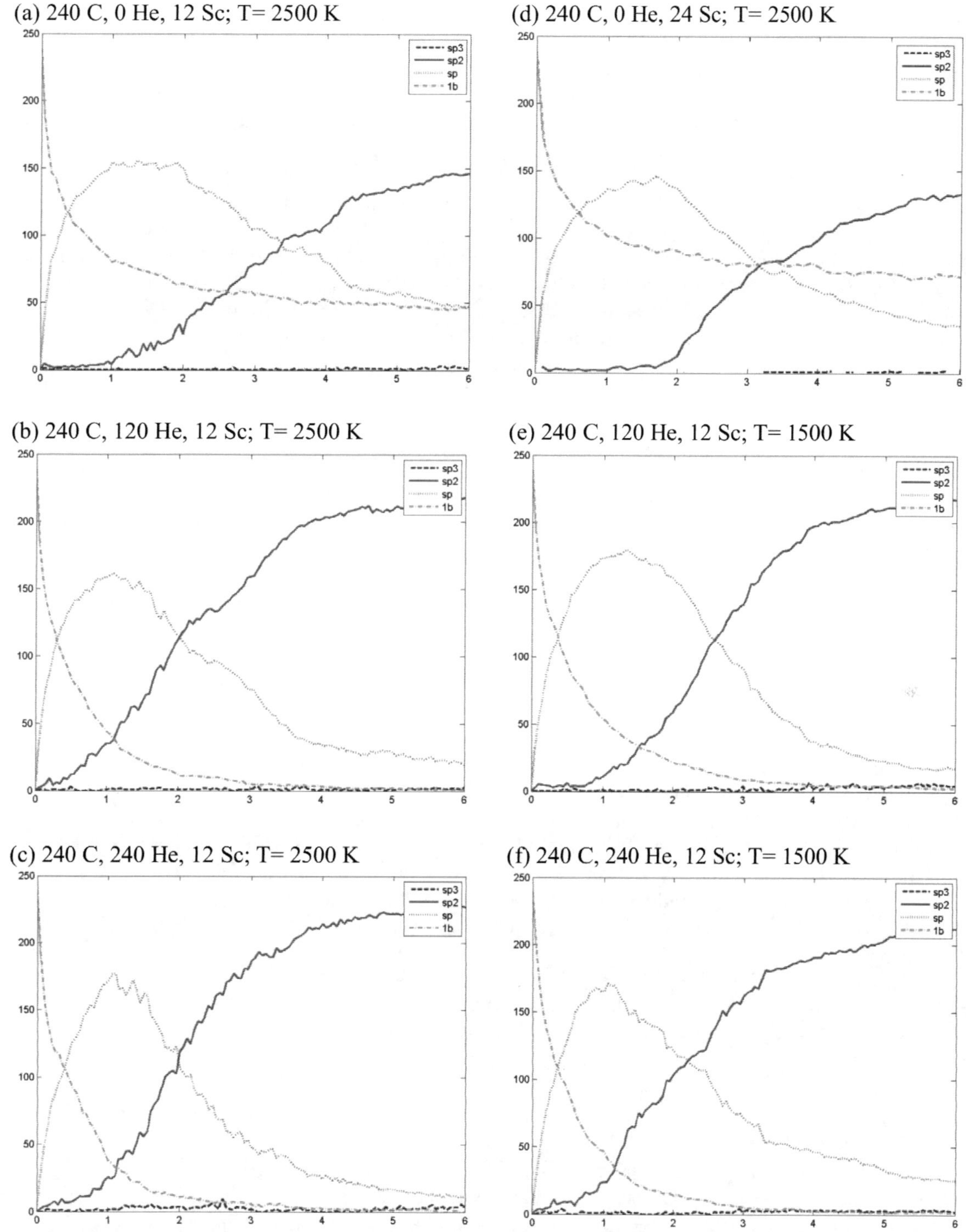

Figure 2. (a) – (f) Time evolution of the number of atoms in different hybridizations states. The numbers of carbon and helium atoms as well as the temperature used in the simulations are given above each plot. Horizontal time axes are in 10^5 fs.

4. Conclusions

In the present work we have performed DFTB MD simulations of the initial stages of fullerenes and scandium EMFs' synthesis in carbon-helium plasma at temperatures of 1500K and 2500K. During the simulations, which have started from uniformly distributed separate C_2 dimers, He and Sc atoms, most of the carbon have coagulated with formation of large (>100 atoms) sp^2-carbon shell-like clusters. It also should be noted, that there was almost no sp^3-associated clusters. The observed closed sp^2-carbon cage structures, some of which had several scandium atoms inside, can be considered as precursors of fullerenes or EMFs, which will obtain their stable structure after further annealing. These observations confirm the "shrinking hot giant" mechanism of fullerenes or EMFs formation in plasma.

Also the dependences of sp^2-carbon cage growth rates from the plasma parameters were investigated. It was shown that sp^2-carbon cage growth rates are almost 2 times bigger at 1:1 He and C_2 concentrations than without helium at all, although the further increase of He concentration has no visible effect. Also, an unusually weak dependence of sp^2-carbon cage growth rate on the temperature was found. Finally, it was shown that increase in Sc concentration lowers the growth rates of unbranched chains and cycles in the cases without helium, but has no effect in cases with buffer gas.

Acknowledgments

The work was supported by Ministry of Education and Science of Russia (Russian-Japanese joined project, Agreement № 14.613.21.0010, ID RFMEFI61314X0010).The authors are grateful to the Joint Supercomputer Center of Russian Academy of Sciences, Moscow and Siberian Supercomputer Center of SB RAS, Novosibirsk for the opportunity to use their computer clusters to perform the calculations.

References

[1] Kroto H W, Heath J R, O'Brien S C, Curl R F and Smalley R E. 1985 *Lett. to Nature* **318** 162
[2] Guldi D W and Nazario M 2002 *Fullerenes: from synthesis to optoelectronic properties* vol 4 (Dordrecht: Kluwer Academic Publishers)
[3] Akasaka T and Nagase S 2002 *Endofullerenes: A New Family of Carbon Clusters* (Dordrecht: Kluwer Academic Publishers)
[4] Shinohara H 2000 *Rep. Prog. Phys.* **63** 843
[5] Chaur M N, Melin F, Ortiz A L and Echegoyen L 2009 *Angew. Chem. Int. Ed. Engl.* **48** 7514
[6] Popov A A, Yang S and Dunsch L 2013 *Chem. Rev.* **113** 5989
[7] Inoue T, Tomiyama T, Sugai T and Shinohara H 2003 *Chem. Phys. Lett.* **382** 226
[8] Diener M D, Afford J M, Kennel S J and Mirzadeh S 2007 *J. Am. Chem. Soc.* **129** 5131
[9] Shultz M D, Duchamp J C, Wilson J D, Shu C Y, Ge J, Zhang J, Gibson H W, Fillmore H L, Hirsch J I, Dorn H C and Fatouros P P 2010 *J. Am. Chem. Soc.* **132** 4980
[10] Cagle D W, Kennel S J, Mirzadeh S, Alford J M and Wilson L J P 1999 *Natl. Acad. Sci. USA* **96** 5182
[11] Dunsch L and Yang S 2007 *Small* **3** 1298
[12] Irle S, Zheng G, Wang Z and Morokuma K 2006 *J. Phys. Chem.* B **110** 14531
[13] Zheng G, Irle S and Morokuma K 2005 *J. Chem. Phys.* **122** 14708.
[14] Irle S, Zheng G, Elstner M and Morokuma K 2003 *Nano Letters* **3** 1657
[15] Saha B, Shindo S, Irle S and Morokuma K 2009 *ACS Nano* **3** 2241
[16] Churilov G N, Kratschmer W, Osipova I V, Glushenko G A, Vnukova N G, Kolonenko A L and Dudnik A I 2013 *Carbon* **62** 389
[17] Aradi B, Hourahine B and Frauenheim T 2007 *J. Phys. Chem.* A **111** 5678
[18] Frauenheim T, Seifert G, Elstner M, Niehaus T, Kohler C, Amkreutz M, Sternberg M, Hajnal Z, Di Carlo A and Suhai S 2002 *J. Phys. Cond. Matter* **14** 3015
[19] Berendsen H J C, Postma J P M, Van Gunsteren W F, Dinola A and Haak J R 1984 *J. Chem. Phys.* **81** 3684
[20] Fedorov A S, Novikov P V, Martinez Y S and Churilov G N 2007 *J. Nanosci. Nanotechnol.* **7** 1315

IOP Conf. Series: Materials Science and Engineering **110** (2016) 012079 doi:10.1088/1757-899X/110/1/012079

Simultaneous spectrophotometric determination of chromium(VI) and iron (III) by H-point standard addition method

E V Larionova[1] and K A Bulygina[2]

[1]Lecturer, *Ecology and Human Safety Department, Tomsk Polytechnic University*
[2]Student, *Ecology and Human Safety Department, Tomsk Polytechnic University*
Lenin av. 34, Tomsk Polytechnic University, Tomsk, Russia, 634050

E-mail: evl@tpu.ru

Abstract. In this work the possibility of simultaneous spectrophotometric determination of chromium (VI) and iron (III) in alloys with help of the mixed organic reagent (diphenylcarbazide and 1,10-phenanthroline) is studied. We have applied H-point standard addition method to determine concentrations of chromium (VI) and iron (III) from the mixture. The pure signals of complexes of chromium (VI) with diphenylcarbazide and iron (III) with the 1,10-phenanthroline and their calibration plots are previously carried out. We established the possibility of simultaneous determination of chromium (VI) and iron (III) in the different concentration ranges by H-point standard addition method. Correctness of determination of concentration by means of the offered technique is proved by "added-found" method for a series of mixtures with different ratios of concentration of chromium (VI) and iron (III). It is founded that the error of determination of concentration doesn't exceed 33 %.

1. Introduction

Many common alloys consist of chromium and iron [1]. A quantity of iron in alloys is determined by ICP-AES [2], flame atomic absorption spectrometry [3], spectrophotometry [4], X-ray fluorescence spectrometry [5], voltammetry [6, 7]. Chromium can be analyzed by ICP-AES [8], sequential injection analysis [9], spectrophotometry [10], X-ray fluorescence spectrometry [11]. Simultaneous determination of chromium and iron can be realized by ICP-AES [12]. Analysis methods of chromium and iron often require additional stages of sample preparation, the chemical pretreatment stages, unconventional and high-cost equipments. Moreover for properties characterization of alloys it often requires to know the speciation of elements. It gives reasons for the development of simple and accessible techniques of simultaneous determination of Cr (VI) and Fe (III) in alloys.

In the industrial analysis the spectrophotometric determination of chromium (VI) with diphenylcarbazide (λ_{max} = 540 nm) and iron (III) with 1,10-phenanthroline (λ_{max} = 510 nm) are widely applied [13]. Spectrophotometric techniques of chromium (VI) and iron (III) determination are rather sensitive and selective. The problem of simultaneous determination of Cr (VI) and Fe (III) can be solved with use of the chromogenic and mixed organic reagents [14–20]. However in this case absorption spectra of chromium (VI) and iron (III) complexes overlaps therefore direct determination of concentration of the studied substances is impossible. To separate individual signals from epy overlapped absorption spectrum the different mathematical approaches, such as H-point method [16],

PLS regression [17], principal component analysis [16], artificial neural networks [18] and curve fitting [19, 20] were applied.

In this work we used the H-point standard addition method to determine concentrations of chromium (VI) and iron (III) from the overlapped absorption spectrum. The H-point method does not require application of the complicated procedure of resolution methods and have a sufficient accuracy provided additivity of overlapped signals and verification of the Beer's law.

Methodology of the H-point method was thoroughly described in the previous work [16]. The H-point standard addition method is a modification of the standard addition method. In this method a set of spectrophotometric signals of mixture of two components with the successive standard addition of second component is obtained. The method bases on the plotting of the calibration plots at two wavelengths. Two straight lines are obtained that have a common point with coordinates H($-C_H$, A_H), where $-C_H$ is the unknown concentration of the second component and A_H the analytical signal due to the first component.

Thus our goal is to study the possibility of simultaneous spectrophotometric determination of Cr (VI) and Fe (III) with mixed organic reagent (diphenylcarbazide and 1,10-phenanthroline) via the H-point method. For detection of absorption spectra we used UV Cary 60 Agilent Technology spectrophotometer in the range of wave length from 200 to 800 nm with cell length equal to 10 and 50 mm.

2. Results and discussion

We obtained the individual spectra of Cr (VI) with diphenylcarbazide and Fe (III) with 1,10-phenanthroline and their sum spectrum with mixed organic reagent (diphenylcarbazide and 1,10-phenanthroline). Figure 1 shows the received signals. As shown in figure 1, spectra of Cr (VI) and Fe (III) complexes are closely overlapped. Direct measurement of signals sizes and determination of concentration of Cr (VI) and Fe (III) from overlapped spectrum is impossible. As the signals are additive the application of the H-point method is valid.

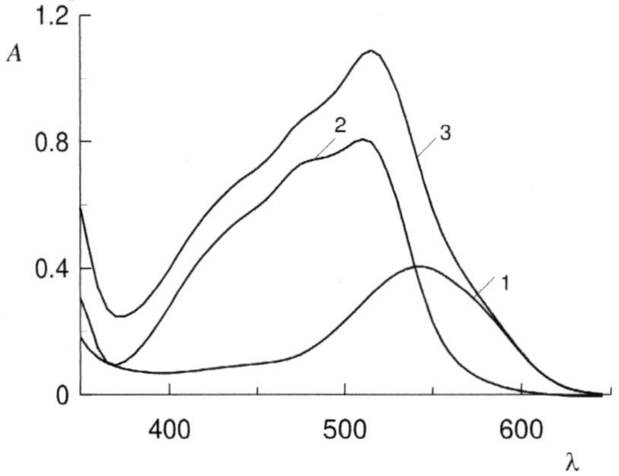

Figure 1. Absorption spectra of: a – (1) Cr(VI) with diphenylcarbazide; (2) Fe(III) with 1,10-phenanthroline and (3) Cr (VI) and Fe (III) with the mixed reagent (diphenylcarbazide and 1,10-phenanthroline).

We obtained the calibration plots of Cr (VI) with diphenylcarbazide in the two concentration ranges (0,8–4,0 and 0,02–0,1 mg/ml) and Fe (III) with 1,10-phenanthroline in the two concentration ranges (0,8–6,4 and 0,1–1,0 mg/ml). The appropriate correlation coefficients obtained (in the range 0.998–0.999) indicate the linearity in the all range of concentrations. The intercepts on the ordinates are negligible in the all graphs. Thus, the H-point method seems to be suitable for use in this system.

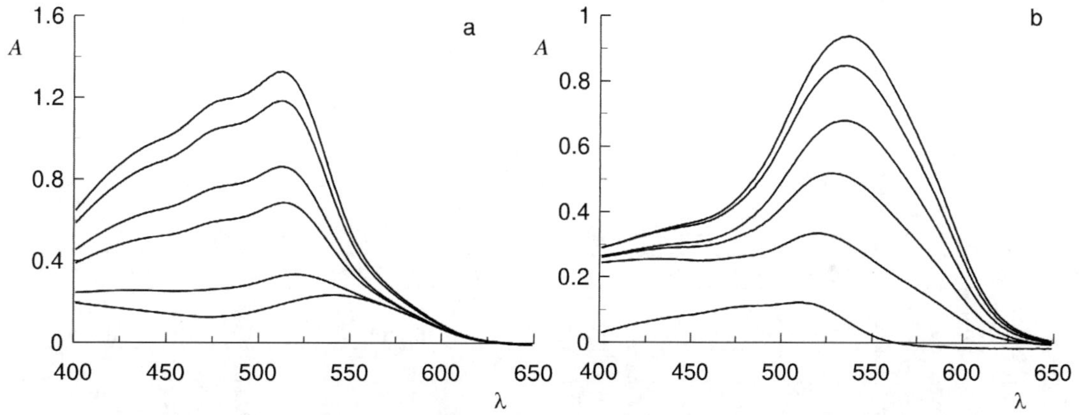

Figure 2. Absorption spectra of: (a) 0,8 mg/ml Cr(VI) with mixed reagent (diphenylcarbazide and 1,10-phenanthroline) with the standard addition of 0,0; 0,8; 2,4; 4,0; 5,6; 6,4 mg/ml Fe (III); (b) 0,8 mg/ml Fe(III) with the standard addition of 0,0; 0,8; 1,6; 2,4; 3,2; 4,0 mg/ml Cr(VI).

Firstly to apply the H-point method it is necessary to choose two wavelengths for plotting two calibration plots provided the systematic errors of concentrations determination are minimal [16]. In case of standard addition of Cr (VI) in the concentration range 0,8–4,0 mg/ml we determined the optimal two wavelength 470/530, in the concentration range 0,02–0,10 mg/ml application of the H-point method is not suitable. In case of standard addition of Fe (VI) in the concentration range 0,8–6,4 mg/ml we determined the optimal two wavelength 520/560, in the concentration range 0,1–1,0 mg/ml – 510/540. Then we plotted calibration graphs in case of standard addition of Cr (VI) in the concentration range 0,8–4,0 mg/ml with fixed Fe(III) concentration and standard addition of Fe (VI) in the concentration range 0,8–6,4 mg/ml and 0,1–1,0 mg/ml with fixed Cr (VI) concentration at chosen two wavelengths (Figure 3).

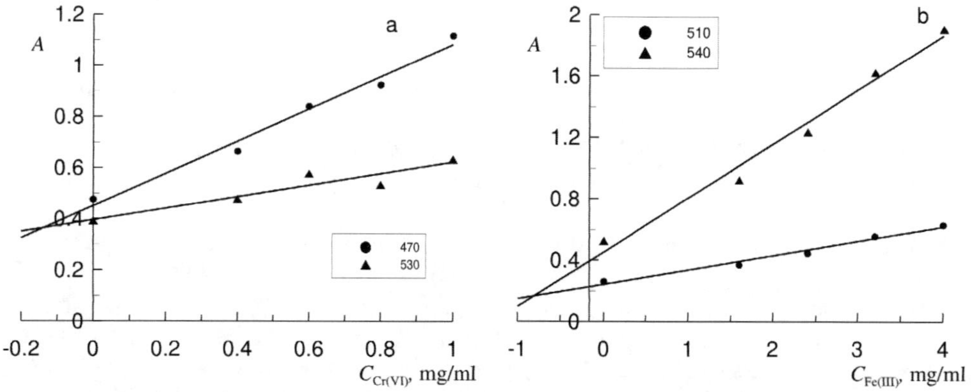

Figure 3. Plots of H-point standard addition method for: (a) fixed Fe(III) concentration and different concentrations of Cr(VI); (a) fixed Cr(VI) concentration and different concentrations of Fe(III).

For the evaluation of correctness of the proposed technique of simultaneous determination of Cr (VI) and Fe (III) we used several synthetic mixtures that had ratios of elements prepared according to some alloys [1]. Concentration of second component (Cr (VI) or Fe (III)) is calculated by the abscissa of intercepts of two calibration plots (Figure 3). Concentration of the first component (Cr (VI) or Fe (III)) is calculated by the ordinate of intercepts of calibration plots and individual calibration plot of first component. Tables 1 and 2 contains the results of concentrations of Cr (VI) and Fe (III) founded by "added-found" method with help of the proposed technique. As it can be seen the error of concentrations determination does not exceed 33 percent.

Table 1. Results of concentration determination of Cr (VI) and Fe (III) by "added-found" method.

Standard addition of C (Cr) (mg/ml)	Added		Found		Δ (%)	
	C_{Cr} (mg/ml)	C_{Fe} (mg/ml)	C_{Cr} (mg/ml)	C_{Fe} (mg/ml)	C_{Cr} (mg/ml)	C_{Fe} (mg/ml)
0,8–4,0	0,8	0,8	0,8	0,8	0	0

Table 2. Results of concentration determination of Cr (VI) and Fe (III) by "added-found" method.

Standard addition of C_{Fe} (mg/ml)	Added		Found		Δ (%)	
	C_{Fe} (mg/ml)	C_{Cr} (mg/ml)	C_{Fe} (mg/ml)	C_{Cr} (mg/ml)	C_{Fe} (mg/ml)	C_{Cr} (mg/ml)
0,8–6,4	0,8	0,8	0,8	0,7	0	13
0,1–1,0	0,1	0,06	0,1	0,08	0	33

3. Conclusion

The problem of simultaneous spectrophotometric determination of Cr (VI) and Fe (III) can be solved by application of mixed reagents and H-point method. For effective application of H-point method the optimal two wavelengths should be chosen, signals must be additive and the Beer's law must be verified. The offered technique can be used for the chromium and iron speciation in alloys. Application of mathematical treatment of signals expands the possibilities of spectrophotometric methods, allows avoiding the chemical pretreatment stages and using high-cost instrumentation and techniques.

References

[1] http://www.csudh.edu/oliver/chemdata/alloys.htm

[2] Xu J, Li M, Li S, Peng H, Zhao B and Yu D 2014 Determination of main components in nd-fe-B alloy with high precision by inductively coupled plasma atomic emission spectrometry *Zhongguo Xitu Xuebao/Journal of the Chinese Rare Earth Society* **32** 234

[3] Wang Y and Feng X 2007 Determination of micro chromium, copper, magnesium and iron in silicon aluminum alloy by flame atomic absorption spectrometry *Yejin Fenxi/Metallurgical Analysis* **27** 50

[4] Devanna N, Satheesh K P and Chandra Sekhar K B 2005 Derivative spectrophotometric determination of iron(II) using diacetylmonoxime benzoyl hydrazone *Asian Journal of Chemistry* **17** 1767

[5] Yao Q, Wang Y, Zhu Y, Yang D and Wu Q 2011 Determination of six elements in magnesium alloys by X-ray fluorescence spectrometry *Yejin Fenxi/Metallurgical Analysis* **31** 54

[6] Locatelli C and Torsi G 2003 Voltammetric determination of metals in copper alloys at trace and ultratrace concentrations *Collection of Czechoslovak Chemical Communications* **68** 1437

[7] Nazarov, B. F., Larionova, E. V., Stromberg, A. G., & Antipenko, I. S. 2007 Theory of the reversible electrode process in case of cyclic voltammetry at finite thickness film electrode *Electrochemistry Communications*, **9(8)** 1936

[8] Ma Q, Li Z and Zhai B 2013 Determination of chromium and zirconium in copper-chromium-zirconium alloy by inductively coupled plasma atomic emission spectrometry *Yejin Fenxi/Metallurgical Analysis* **33** 52

[9] Fan J, Sun Y, Wang J and Fan M 2009 An organic-reagent-free method for determination of chromium(VI) in steel alloys, sewage sludge and wastewater *Analytica Chimica Acta* **640** 58

[10] Xue C and Wang R 2009 Determination of chromium in aluminium alloy by potassium iodide-starch spectrophotometry *Yejin Fenxi/Metallurgical Analysis* **29** 75

[11] Zhang H, Ji S and Li C 2014 Determination of twelve elements coexisting with chromium, vanadium and titanium in titanium alloys by X-ray fluorescence spectrometry *Yejin*

Fenxi/Metallurgical Analysis **34** 30
[12] Cheng Y 2012 Determination of the silicon, vanadium, iron, aluminum, nickel, molybdenum and chromium in titanium alloy by ICP-AES *Xiyou Jinshu Cailiao Yu Gongcheng/Rare Metal Materials and Engineering* **41** 1871
[13] Marczenko Z and Balcerzak M 1998 *Spektrofotometryczne metody w analizie nieorganicznej* (Warszawa: Wydawnictwo Naukowe PWN)
[14] Martelli P B, Reis B F, Kronka E A M, F°H B, Korn M, Zagatto E A G and Araujo A N 1995 Multicommutation in flow analysis. part 2. binary sampling for spectrophotometric determination of nickel, iron and chromium in steel alloys *Analytica Chimica Acta* **308** 397
[15] Fernandes R N, Reis B F and Campos L F P 2003 Automatic flow system for simultaneous determination of iron and chromium in steel alloys employing photometers based on LEDs as radiation source *Journal of Automated Methods and Management in Chemistry* **25** 1
[16] Abdollahi H 2001 Simultaneous spectrophotometric determination of chromium(VI) and iron(III) with chromogenic mixed reagents by H-point standard addition method and partial least squares regression *Analytica Chimica Acta* **442** 327
[17] Safavi A, Abdollahi H and Mirzajani R 2006 Simultaneous spectrophotometric determination of fe(III), al(III) and cu(II) by partial least-squares calibration method *Spectrochimica Acta - Part A: Molecular and Biomolecular Spectroscopy* **63** 196
[18] Kompany-Zareh M, Massoumi A and Pezeshk-Zadeh S 1999 Simultaneous spectrophotometric determination of fe and ni with xylenol orange using principal component analysis and artificial neural networks in some industrial samples *Talanta* **48** 283
[19] Larionova, E. V., Vtorushina, A. N., Zabolotskaya, E. V., & Bulygina, K. A. 2015 Technique of simultaneous spectrophotometric determination of chromium (vi) and iron (III) via curve fitting *Proceedings - ISSAT International Conference on Modeling of Complex Systems and Environments 2015* 15
[20] Larionova, E. V., & Bulygina, K. A. (2015). Simultaneous spectrophotometric determination of chromium (VI) and iron (III) in alloys. *IOP Conference Series: Materials Science and Engineering* 81

IOP Conf. Series: Materials Science and Engineering **110** (2016) 012080 doi:10.1088/1757-899X/110/1/012080

Electron injection and polarization reversal processes in ferroelectrics analyzed with SEM: modelling of electron beam-stimulated effects

A V Pavelchuk, T K Barabash and A G Maslovskaya

Physics Engineering Department, Amur State University, 21 Ignatyevskoe Shosse, Blagoveshchensk 675000, Russia

E-mail: maslovskayaag@mail.ru

Abstract. The article is devoted to simulation of the injection and polarization switching processes arising in ferroelectrics analyzed with SEM techniques. Monte-Carlo simulation of electron trajectories was performed to estimate electron penetration depth depending on major scanning parameters. The charging kinetics caused by electron injection in ferroelectrics under nonequilibrium conditions of electron irradiation were also explored. The proposed model of polarization reversal process provides good agreement with experimental data.

1. Introduction

Recently the application of scanning electron microscope (SEM) techniques has become increasingly important in investigation of ferroelectrics owing to ability of getting specific responses and creating new modes of image formation. Indeed, the general modes usually used for SEM imaging are mostly considered to be the secondary electron image as well as the backscattered electron image modes. These modes permit one to explore a surface topography, composition inhomogeneity and field-induced contrast of ferroelectric and ferromagnetic materials. The earlier work [1] has reported the approach which allows a-c domains in $BaTiO_3$ crystals to be visualized due to a geometrical relief occurring on a sample surface covered by metal electrode. Afterwards numerous studies rapidly have shown the principal possibility of domain structure imaging on uncoated sample surface [2-3]. Moreover raised accelerating voltages (>10 keV) and injection of an electron beam into samples can be employed to observe evolution of domain structure with simultaneous registration of polarization switching current electron beam-induced [4]. Also the charging processes in ferroelectrics have been used to design controlled polarization switching method and create periodical submicron domain structures [5].

The theoretical studies of electron beam-material interaction and related phenomena usually consider numerous aspects. The basic approach consists in Monte-Carlo simulation of electron trajectories in irradiated target. This approach enables evaluating the geometry of electron trajectory distribution in the sample versus different values of energy of primary beam. In previous works methods, algorithms as well as numerical experiments have been discussed relative to mechanisms and principles describing electron beam interaction with irradiated materials [6-8]. In reality, multiple effects resulting from electron beam interaction with dielectric material can be analyzed by means of simulation, such as secondary electron emission, charging effect, heating process etc. [9-11]. In view of the importance of SEM advantages for analyzing as well as modifying ferroelectrics, there is a need for an improved understanding of the mechanisms and principles of electron beam interaction with ferroelectric materials in order to explain the registered processes. The following study was done to simulate electron injection and formation of polarization switching response focusing on conditions of observation of electron beam-induced polarization reversal processes in ferroelectrics being analyzed with SEM modes.

2. Monte-Carlo simulation of electron trajectories in irradiated target: overview of algorithm

The analysis of electron injection effect was based on conventional approach, which consists in Monte-Carlo simulation of large number N of single electron interaction with irradiated target. The procedure uses as basis the following principles. An electron with primary beam energy E_0 falls perpendicularly on surface of the sample. The landing position of the electron is determined to be P_0. The interactions between an electron and material are assumed to be separated into elastic and inelastic type. This implies that the random moving or energy loss can be realized. The type of interaction is specified stochastically. In order to evaluate the electron position at current point P_i scattering angle φ as well as azimuthal scattering angle ω need to be calculated. Between the random scattering events each electron travels the distance s with energy E_i. The distance can be calculated by the following equation [6]:

$$s = -\lambda \ln(\xi), \quad \lambda = \left(1/\left(N_0 \cdot \rho\right)\right)\sum_{k=1}^{M}\left(\omega_k \cdot A_k / \sigma_k\right), \tag{1}$$

where λ is mean free path in cm; $\xi \in (0,1)$ is uniform distributed random quantity; A_k is atomic weight in g/mole; ω_k is weight fraction of element k; σ_k is cross-section for each element k in cm^2; M is number of elements composed substance; N_0 is Avogadro's number in mole^{-1} and ρ is layer density in g/cm^3.

The total cross-section σ_k can be determined by means of Mott cross-section ratio [11]:

$$\sigma_k^{Mott} = \frac{4.7 \cdot 10^{-18}\left(Z_k^{1.33} + 0.032Z_k^2\right)}{\left(E + 0.0155Z_k^{1.33} \cdot E_i^{0.5}\right)\left(1 - 0.02Z_k^{0.3} \cdot \exp\left(-u_k^2\right)\right)}, \text{ cm}^2, \tag{2}$$

where $u_k = \lg\left(8E_i \cdot Z_k^{-1.33}\right)$, Z_k is atomic number of element k and energy E_i is expressed in keV.

In addition the algorithm groups all the electron energy loss events in a continuous energy loss function. In this way we can calculate the energy between collisions with use of the form of Bethe law:

$$\frac{dE_i}{ds} = -75800\frac{\rho}{E_i} \cdot \sum_{k=1}^{M}\omega_k \frac{Z_k}{A_k}\ln\left(\frac{1.166\left(E + 0.8J_k\right)}{J_k}\right), \text{ keV/cm}, \tag{3}$$

where $J_k = \left(9.76Z_k + 58.5Z_k^{-0.19}\right) \cdot 10^{-3}$ is Seltzer-Berger expression, which is mean ionization potential of element k in keV [6].

Note that ordinary differential equation (3) augmented by initial condition determined as previous value of energy is solved numerically for each inelastic scattering event. Besides the calculation of electron position and energy loss should be performed till the energy is less than threshold energy E_{th}, which is set to be 0.5 keV. If electron escapes the sample surface it is recorded as backscattered electron. This procedure is repeated for each electron interaction. With this assumption the Monte-Carlo simulation requires initializing the initial parameters such as geometrical size of the sample, material element composition, start electron beam energy E_0 and number of incident electrons N.

3. Simulation model of polarization reversal process electron beam-stimulated in ferroelectrics

The electron beam-induced polarization current (EBIC) mode is based on electron injection into ferroelectric sample at increased accelerating voltages (10-40 kV). This mode enables visualizing domain structure images and also exploring polarization switching response [4]. Injection of electron beam into sample leads to accumulation of charges in the irradiated layer, which stimulates appearance of charges on the top and bottom electrodes. This causes the occurrence of fields E_1 and E_2 in irradiated and unirradiated parts of sample respectively. The field intensity E_2 can activate polarization reversal processes in domains "tail-to-beam" orientated and, on the contrary, can stabilize the domain structure with "head-to-beam" orientation of spontaneous polarization vector $\mathbf{P_S}$. The diagram of charge and field distributions electron beam-induced in ferroelectric sample is shown in figure 1.

The electron beam irradiation of one of the sample surfaces results in drift of ferroelectric domain structure due to injection and charging-up effects. This gives rise to forming response of polarization switching. In order to express polarization switching current the Kolmogorov-Avrami model is generally applied [12]. Nevertheless this approach possesses a number of disadvantages regarding to electron beam-stimulated polarization current mode. Firstly this method can be referred to approximating one and it does not consider specifics of domain structure dynamics induced by an electron probe. Secondly the model does not include the operating parameters of the SEM modes and as a consequence it does not result in satisfactory fit with a form of current curves registered experimentally. The proposed model of polarization switching in ferroelectrics with 180° domain structure is based on main features of polarization reversal process experimentally observed. We assumed, that polarization switching is realized owing to growth of domains possessing wedgewise shape with the fixed base width $2y$ and the changing angle of domain boundary slope φ (figure 2 demonstrates the geometrical scheme).

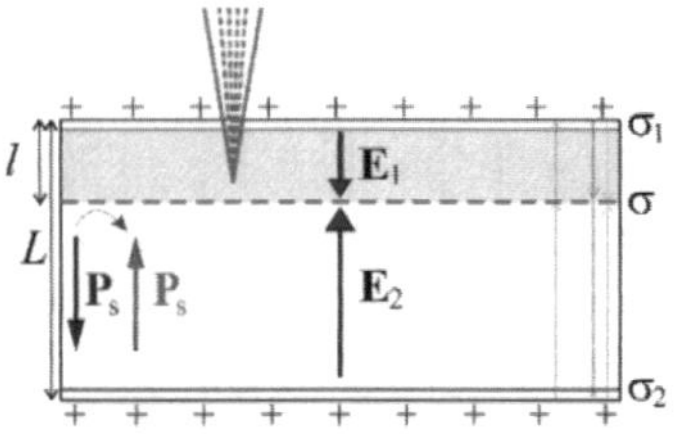

Figure 1. The scheme of charge and field distributions (σ, σ_1, σ_2 – surface charge densities).

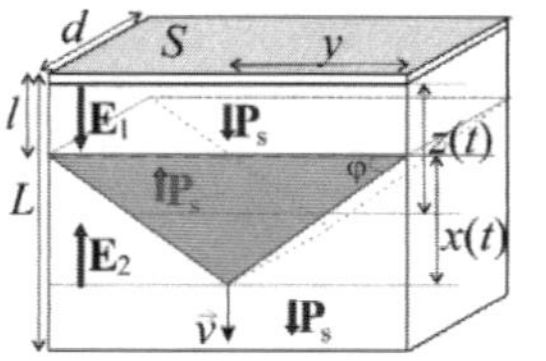

Figure 2. The diagram of domain growth (l – electron injection depth, L – sample thickness, $S=2y\cdot d$ – irradiated area).

The first stage of polarization switching process completes as long with the middle of a wedge $z(t)$ reaches the bottom electrode. The second stage suggests that the lateral domain growth is incorporated to mechanism of polarization switching. In general the repolarization process is due to growth of n wedge-shaped domains with an identical speed and width of the base $2y/n$. The polarization switching current is determined by superposition of deposits which are given by separate switching areas. To specify the self-similarity dynamics of domain structure we found fractional order differential equation:

$$\frac{d^\alpha s}{dw^\alpha} = \exp\left(-\frac{\tau_2}{\tau_1 w + \tau_3 Ls\left(1+\cos^2\varphi\right)/(4l)}\right), \quad s(w_0)=0, \ 0 \le s(w) \le 2(1-l/L), \qquad (4)$$

where $s=x/L$ and $w=t/\tau_1$ are dimensionless distance and time variables; $\tau_1=L/v_\infty$ is the characteristic time of boundary growth through a crystal in s; $\tau_2=\delta\varepsilon\varepsilon_0 L/(jl)$ is the characteristic time (in s) of accumulation of charge generating the field $E_2=\delta$ at given current density j (in A/m^2); δ is activation field in V/m; $\tau_3=2P_S/j$ is the characteristic time period (in s) during which the charge density σ is defined to be $2P_S$; α is fractal parameter; v_∞ is maximal velocity of domain boundary movement in m/s.

In this way we can obtain the polarization switching current in following form:

$$I(w)=\begin{cases}\dfrac{2P_S S}{\tau_1}\dfrac{d^\alpha s}{dw^\alpha}, & 0\le s\le 1-\dfrac{l}{L},\\[2ex] \dfrac{2P_S S}{s^2\tau_1}L^2\left(1-\dfrac{i}{N}\right)\dfrac{d^\alpha s}{dw^\alpha}, & 1-\dfrac{l}{L}<s\le 2\left(1-\dfrac{l}{L}\right),\quad i=\overline{1,N},\end{cases} \tag{5}$$

The differential equation in (4) can be solved numerically using fractional derivative approximation by Grunwald-Letnikov formula.

4. Results and discussion

In order to perform the simulation of electron trajectories in irradiated target one needs to specify parameters: primary beam energy E_0, chemical composition of material, probe diameter d_{pr}, threshold value of energy loss E_{th}, number of electrons N. The figure 2 demonstrates the results of modeling of electron transport in typical ferroelectrics. The analysis of similar computations suggests that the geometry of electron beam-ferroelectrics interaction can be characterized by hemispherical, semi-ellipsoidal (in preference for crystals of inorganic group – $LiNbO_3$, $LiTaO_3$, $BaTiO_3$ et al.) and "pear-shaped" forms (for group of organic crystals – TGS, DTGS, thiourea, Rochelle salt) depending on element composition and beam energy. Depth of electron penetration l is defined by a chemical composition as well as primary beam energy and does not depend on size of irradiated area on the sample surface. For instance, depth of electron penetration for lithium niobate crystal was estimated to be: l=1.2 μm at E_0=15 keV, l=2.4 μm at E_0=25 keV, l=5 μm at E_0=40 keV.

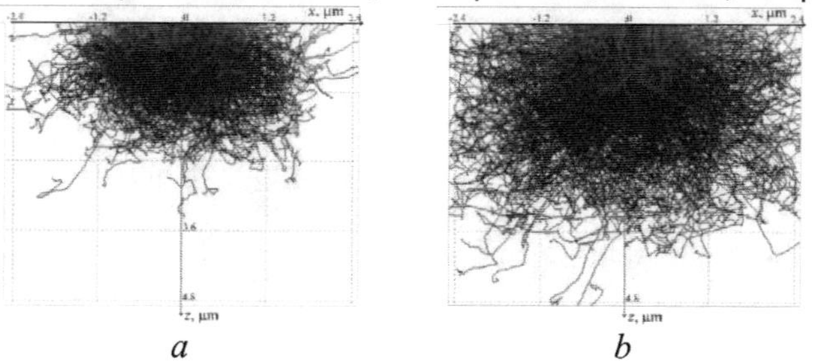

a *b*

Figure 2. The simulated electron trajectories at start beam energy E_0=25 keV, N=1000 electron histories, probe diameter d_{pr}=0.2 μm for samples: $LiTaO_3$ – *a*; $BaTiO_3$ – *b*.

Note, that EBIC mode assumes using samples covered by thin metal electrode. The figure 3 displays the simulated electron transport in TGS covered by metal electrodes at various values of electrode thickness (d_{el}). So we can conclude that metal electrode will be transparent for electron injection provided its thickness is approximately less than 0.1 μm (the results of electron transport coincide with modelling results for uncoated sample). In this case the electrons can freely penetrate into the sample and accumulate in surface layer which is necessary condition for beginning the polarization reversal process in EBIC mode.

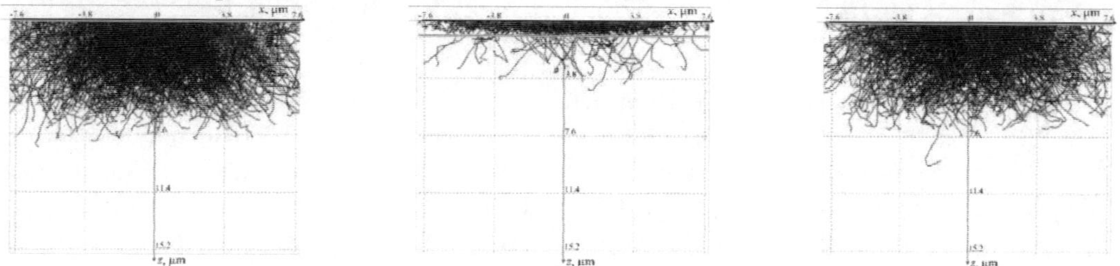

a *b* *c*

Figure 3. The simulated electron trajectories in TGS for the mode used: silver electrode at d_{el}=0.1 μm – *a*, d_{el}=1 μm – *b* and gold electrode at d_{el}=0.1 μm – *c* (E_0=25 keV, N=1000, d_{pr}=10 μm).

Then the following parameters should be also estimated to calculate the characteristics of polarization switching process: probe parameters (primary beam current I_{pr} and accelerating voltages

U), geometrical size of crystal (d, L, y), specific time parameters τ_1, τ_2, τ_3, spontaneous polarization P_S and dielectric permittivity ε. The simulation was performed for TGS crystal according to experimental data [4]. The curve of polarization switching current reproduces features experimentally observed. Also the computed response enables the time moment t_{st} of beginning of polarization reorientation to be evaluated. During the period t_{st} start charge q_{st} is accumulating in the irradiated layer and in the same time the field in unirradiated part of the sample reaches the coercive field value. The completing of polarization switching process corresponds to falling polarization current to zero level. The field in unirradiated part of the sample is more than coercive field for TGS at ε=50. The dependence of start time moment t_{st} of polarization switching versus injection current density j indicates that t_{st} decreases exponentially with increasing j. The computation of maximum value of polarization current I_{max} normalized by primary value of electron beam current I_{pr} allows us to conclude that when current density increases (j>0.8·10^{-3} A/m^2) the polarization current can exceed the value of probe current as presented in figure 4 *a*.

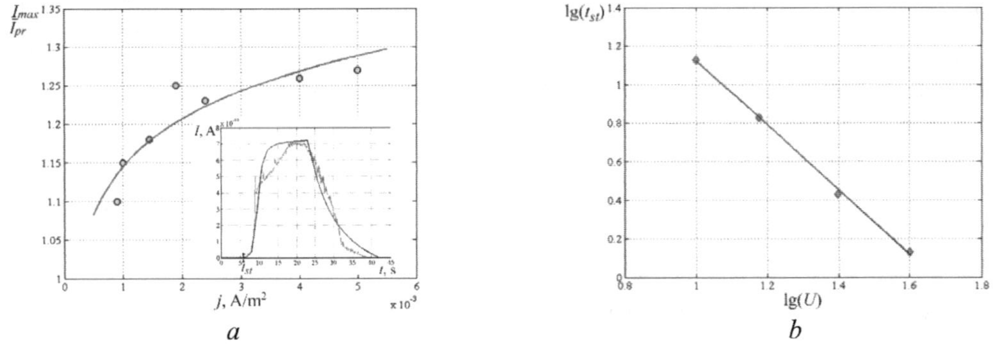

Figure 4. The dependence of relative of impulse current on current density – *a* (simulated polarization switching current compared with experimental data is shown in the insert figure), the start moment of polarization switching as a function of accelerating voltages (in log-log scale) – *b*.

Obviously, different values of accelerating voltages U correspond to different values of electron penetration depth l. The field in unirradiated part of the sample can reach the coercive field if $q_{st}l=const$. Taking into account the power relation between l and U: $q_{st}U^p=const$ the accumulated charge kinetics can be analyzed by means of calculation polarization switching start moment t_{st} depending on accelerating voltages U: $t_{st}U^p$=const. The corresponding function is shown in figure 4 *b*. As a consequence we can obtain the index of power p=1.7 at given values U=[10, 15, 25, 40] kV.

5. Concluding remarks

Finally, Monte-Carlo simulation of electron trajectories and computation of polarization switching current were performed to analyze the condition as well as characteristics of polarization reversal processes electron beam induced in ferroelectrics. The calculation of electron transport in ferroelectrics allows us to evaluate the electron penetration depth and features of geometry electron beam-material interaction region at given experimental parameters. Also the evidence indicates that the thickness of used metal electrode is approximately estimated to be less than 0.1 μm to observe polarization reversal process in ferroelectrics with EBIC mode of SEM. Also the characteristics of polarization switching processes in ferroelectrics were analyzed with use of proposed model of 180° domain structure dynamics and formation of polarization current. Our results theoretically support the experimental EBIC mode and permit us to study the effect of variation of probing parameters such as acceleration voltages and probe current on charging kinetics, start switching moment, the form and value of polarization current. The essential features of simulation results correspond to experimental data observed.

IOP Conf. Series: Materials Science and Engineering **110** (2016) 012080 doi:10.1088/1757-899X/110/1/012080

References

[1] Robinson G Y, White R M 1967 *Appl. Phys. Lett.* **10** 320
[2] Le Bihan R, Sella S 1970 *J. Phys. Soc. Japan.* **28** 377
[3] Ikeda S, Uchikawa Y J 1980 *Electron Microsc.* **29** 209
[4] Sogr A A, Maslovskaya A G, Kopylova I B 2006 *Ferroelectrics* **341** 29
[5] He J, Tang S H, Qin Y Q et al. 2003 *J. Appl. Phys.* **93** 9943
[6] Joy D C 1995 *Monte-Carlo modeling for electron microscopy and microanalysis* (New York: Oxford University Press)
[7] Napchan E 1992 *European Microscopy and Analysis* **2** 21
[8] Drouin D, Couture AR, Joly D et al. 2007 *Scanning* **29** 92
[9] Ohya K, Inai K, Kuwada H, Hayashi T, Saito M 2008 *Surface and Coating Technology* **202** 5310
[10] Maslovskaya A, Pavelchuk A 2015 *IOP Conf. Series: Materials Science and Engineering* **81** 012119
[11] Czyzewski Z, MacCallum D O, Roming A, Joy D C 1990 *J. Appl. Phys.* **68** 3066
[12] Ishibashi Y 1990 *Journal of the Physical Society of Japan* **59** 4158

Dynamics of combined electron beam and laser dispersion of polymers in vacuum

P A Luchnikov[1], A V Rogachev[2], M A Yarmolenko[2,4], A A Rogachev[3,4] and A A Bespal'ko[4]

[1] Moscow State University of Information Technologies, Radio Engineering and Electronics (MIREA), Moscow, Russia
[2] F. Skoryna Gomel State University, 104, Sovetskaya Street, Gomel 246019, Belarus
[3] Belarusian State University of Transport, Gomel, Belarus
[4] Institute of Non-Destructive Testing, National Research Tomsk Polytechnic University, 30, Lenina Avenue, 634050, Tomsk, Russia.

E-mail: xamdex@gmail.com

Abstract. The mechanisms of the impact of the laser assisting effect on the dispersion kinetics and on the structure of the deposited layers in electron beam dispersion of a polymer target were analyzed. The proposed model and analytical expressions adequately describe the kinetic dependence of the polymer materials dispersion rate in a vacuum on the intensity of laser processing of their dispersion zone.

1. Introduction

When a flow of charged particles (ions, electrons) impact on the surface of the polymer material in order to generate the active gas phase, complex physicochemical processes proceed, the character and the kinetics of which are determined, first of all, by the particles energy, the nature of the polymer, which ultimately affects the structure and composition of volatile dispersion products, their reaction activity and, accordingly, the structure and properties of the polymerized layers [1]. Further impact on the dispersion zone of the electromagnetic radiation allows expecting significant changes in the mechanism and the kinetics of those processes.

The impact of the assisting radiation on the polymer dispersion processes is multifactorial; it depends on the target nature and the radiation parameters [2–4]. It is manifested, first of all, in additional heat activation of macromolecules, bond breakage (if the photon energy is higher than the binding energy), photochemical activation of the formed volatile dispersion products and, accordingly, increase in their reactivity. Short-wave electromagnetic radiation may also cause photoelectron emission, which reduces the value of the electric charge adsorbed when processed by electrons and, accordingly, determines a decrease in intensity of the electric field decelerating the incident electron flow. This, in turn, increases the particles energy and, accordingly, the dispersion intensity, determines higher secondary electronic emission that affects the character and the kinetics of the secondary polymerization process.

[1] To whom any correspondence should be addressed.

IOP Conf. Series: Materials Science and Engineering **110** (2016) 012081 doi:10.1088/1757-899X/110/1/012081

2. The physical model

The quantitative analysis of patterns of the material laser dispersion is based on a comparison of the photon energy $\varepsilon = hc/\lambda$ (where h is the Planck constant, c is the speed of light, λ is the wave-length of laser radiation) with the binding energy of the atoms. When the materials are evaporated by laser pulse emission, the change in state of the target material is significantly determined by the absorption intensity of the energy by the material, which is in simple case described by Beer-Lambert law

$$I(x,y,z) = I_0(x,y)e^{-\alpha z}, \quad (1)$$

where I_0 is the intensity of the light that falls on the target surface; α is the absorption coefficient that depends on the nature of the material and the radiation wave-length; z is the radiation penetration depth.

The absorption intensity significantly determines the temperature condition of the process. And if the photon energy is high enough then the electron excitation caused by the photon absorption may lead to direct photochemical decomposition. If the radiation quantum energy is lower than the atoms binding energy and the heat flow density exceeds a certain threshold then the main dispersion mechanism is the thermochemical decomposition [5].

As noted, when the electron flow is exposed on the target surface, its charging occurs due to the introduction of the particles into the polymer by fixing them on traps in the surface layer. The analysis of the electrophysical processes that occur when dispersing the polymers shows that the electrostatic charge value on the surface is determined by the electrons density j, the secondary electron emission coefficient α, and, when volatile macromolecules fragments are formed, their ionization coefficient β. Note that the secondary emission coefficient α significantly depends on the energy W of the particles bombarding the surface. At sufficiently low electron energies, it has, as a rule, the value greater than one. It is known, for example, that when the surface of PTFE (polytetrafluorethylene) is irradiated by electron flow, its charging is observed at the energy $W > 1.5$ keV. At $W < 1.5$ keV, the secondary electron emission coefficient α is greater than one and the surface charging does not occur. When short-wave electromagnetic radiation with the photon energy $\varepsilon = h\nu > A_e$ (where A_e is electron work function) is exposed to the dispersing zone, the electron emission occurs with the flux density $J_f = J/h\nu$ (where J is the electromagnetic radiation energy flux density).

Given these factors and assuming that the electrical conductivity of the polymer is low enough, the charge surface density q change in time may be described by the following system of differential equations

$$\begin{cases} \dfrac{dq}{dt} = j - \alpha j + \beta \dfrac{dm}{dt} - eJ_f \\ \dfrac{dm}{dt} = -kUj \end{cases}, \quad (2)$$

where m is the mass of the polymer target; k is the value that depends on the polymer nature; $U=W/e$ is the effective accelerating voltage of the electron gun that disperse the polymer target; W is the electrons energy; e is the electron charge.

As already noted, the secondary electronic emission coefficient depends on the electron energy (or the accelerating voltage U) also in linear approximation, which can be calculated from the correlation

$$\alpha = \alpha_0 - dU, \quad (3)$$

where α_0 and d are constant values for this polymer.

In (2), we note that, according to the experimental data [6], the volatile products generation rate (the dispersion rate $V_d = \dfrac{dm}{dt}$) is proportional to the power of energy deposition on the target.

The appearance of charge on the surface determines the existence of an electric field decelerating the incident electrons. Then the effective acceleration voltage is

$$U = U_a - \frac{\alpha}{C}, \tag{4}$$

where U_a is the acceleration voltage of the electrons (ions) source; C is the capacitance of the charged polymer layer.

Differentiating (4), we obtain

$$\frac{dq}{dt} = -C\frac{dU}{dt} \tag{5}$$

Then, based on (2) and taking into account (3) and (5), we have

$$\frac{dU}{dt} = -\frac{1}{C}\{[1-\alpha_0 + (d + k\beta)U]j - eJ_f\}. \tag{6}$$

After integration of (6) with the initial condition $U(0) = U_a$, the effective voltage expression is as follows

$$U = \frac{j(\alpha-1)+eJ_f}{j(d+k\beta)} + \left[U_a - \frac{j(\alpha-1)+eJ_f}{j(d+k\beta)}\right]\exp\left(-\frac{d+k\beta}{e}jt\right). \tag{7}$$

One should note that the value of $U_0 = \frac{j(\alpha-1)+eJ_f}{j(d+k\beta)}$ is numerically equal to the electron source acceleration voltage whereby the dispersing process is stationary and the charge accumulation on the polymer surface does not occur.

The analysis of (6) shows that at the lumen output $J_f = \frac{j}{e}[U_a(d+\beta k)+1-\alpha_0]$ there is lack of charge and the standard disperse mode is realized. If the electromagnetic radiation intensity is higher than that value then, due to the photoeffect, a positive charge appears on the surface and the induced electric field is accelerating.

The linear velocity of the dispersion is obtained from (2) with (7)

$$V_d = \frac{k}{\rho}j\left[U_0 + (U_a - U_0)e^{-\frac{t}{\tau_0}}\right], \tag{8}$$

where $\tau_0 = \frac{C}{j(d+\beta k)}$ is the charging time constant; ρ is the target material density.

It is followed from (7) and (8) that, with sufficiently large dispersion time $t >> \tau_0$, the value of the electron gun accelerating voltage U_a virtually does not affect the energy of the particles bombarding the target and the dispersion rate. Therefore, to ensure high efficiency and stationarity of the dispersing process with such time values, the accelerating voltage of the electric gun should be set close to U_0, and to ensure high dispersion rate, intensive enough electron (ion) flows should be used. However, it is necessary to consider that as j increases, the charging time constant decreases, the efficiency of charge removal by electromagnetic assisting decreases and the charge is accumulated faster on the surface, which reduces the effective energy of electrons and, accordingly, the dispersion rate.

The stationary value of the dispersion rate while assisting the electron beam dispersion with the electromagnetic radiation may be expressed as follows $\Delta V_{d,0} = \frac{k}{\rho}\frac{j(\alpha-1)+eJ_f}{(d+\beta k)}$, where the value linearly increases with increase in the electron flux density. The difference in dispersion rates with and without laser assisting $\Delta V_{d,0} = \frac{k}{\rho} \frac{eJ_f}{(d + \beta k)}$ does not depend on the electron flux density but is determined only by intensity of the electromagnetic radiation.

The above kinetic features were considered in technological processes of forming composite film structures during deposition from the active gas phase with the selection of the conditions and parameters of coating condensation on the substrate by pulse laser-assisted electron beam deposition of initial components [2, 3].

3. Conclusion

The analysis of the possible mechanisms of laser assisting impact on the dispersing kinetics and the structure of the deposited layers in electron beam target dispersion revealed that the main impact factors on the target are the thermal effect produced by long-wave radiation, the photochemical activation and the photo emissive effect. The latter process in electron beam dispersion affects charging of the surface layer of the polymer sputtered and finally on the energy of the electrons that bombard the target. The proposed model and analytical expressions satisfactory describe the kinetics dependence of the dispersion rate of polymer materials in a vacuum on the intensity of laser processing of their dispersion zone.

Acknowledgments

This work was supported by the Ministry of Education and Science of Russian Federation (main part, project № 1079)

References

[1] Johnson S L, Papantonakis M R and Haglund R F 2010 Deposition of Polymer and Organic Thin Films Using Tunable, Ultrashort-Pulse Mid-Infrared Lasers *Laser-Surface Interactions for New Materials Production* ed A Miotello and P Ossi (Berlin: Springer Series in Materials Science) chapter 8 177–202

[2] Yarmolenko M A, Rogachev A A, Rogachev A V Luchnikov P A and Zhubo Liu 2012 *Nanomaterials and Nanostructures* **3** 21

[3] Zhubo Liu, Rogachev A V, Bing Zhou, Yarmolenko M A, Rogachev A A, Gorbachev D L and Xiaohong Jiang 2011 *Progress in Organic Coatings* **72** 321

[4] Kejie Zhang, Yarmolenko M A, Rogachev A A, Bing Zhou, Xiaohong Jiang, Ruiqi Shen and Xiaoheng Liu 2013 *Applied Surface Science* **273** 836

[5] Zhubo Liu, Rogachev A V, Zhou Bing, Yarmolenko M A, Rogachev A A Gorbachev D L and Jiang Xiaohong 2011 *Key Engineering Materials* **480** 30

[6] Bui M V and Rogachev A V 1997 *Relaxation-diffusion theory of interfacial processes* (Gomel: BelSUT) 196

IOP Conf. Series: Materials Science and Engineering **110** (2016) 012082 doi:10.1088/1757-899X/110/1/012082

Composition of Powders Produced by Electrospark Dispersion of Metal Granules in Water

S P Zhuravkov[1], G L Lobanova[1], A V Pustovalov[1], P E Slyadnikov[2] and L V Nadeina[3]

[1] Institute of High Technology Physics, National Research Tomsk Polytechnic University, 2A, Lenin Ave, Bld. 11, 634028, Tomsk, Russia
[2] The faculty of innovative technologies, National Research Tomsk State University, 36, Lenin Ave, 634050, Tomsk, Russia
[3] Institute of Natural Resources, National Research Tomsk Polytechnic University, 2A, Lenin Ave, Bld. 5, 634028, Tomsk, Russia

E-mail: zhursp@yandex.ru

Abstract. The results of experimental studies of metal powders composition produced by electrospark dispersion according to the scheme "metal electrodes – metal granule loading – distilled water" are given in the paper. With a help of X-ray diffraction analysis and other methods, it was found that chemical and phase composition of dispersion products is determined by affinity of metal to oxygen.

1. Introduction

Metal powders, used in different technological processes, must possess not only intended sizes of particles, but also, according to chemical composition and properties, they must comply with specific metal or alloy. The investigations of physico-chemical processes, developing under conditions of metal powder production by various methods, are relevant today.

The method of metals dispersion by electric spark was offered by B.R. Lazarenko and N.I. Lazarenko [1]. The method is that the granules of metal are poured between two electrodes in the power fluid bath (water or organic liquid), and then voltage impulses with a given amplitude value apply to the electrodes, and spark discharges are carried out between granules of metal. The metal dispersion under these conditions is caused by thermal processes developing in the discharge channel. Microdischarges observed as sparking occur between pairs of metal granules due to the influence of electrical energy impulses. These pairs of metal granules are located in equipotential zones of the interelectrode gap. Due to close gaps between particles involving in the process, base energy given off in discharge channels is used to heat metal in zones of anode and cathode spots, located on these particles surfaces [2]. It causes local heating of metal to melting and boiling temperatures. As a result of it, melting, evaporation and dispersing of molten metal occur. A dimple is formed on the surface of the eroded metal particle. Spherical or tear-drop shapes are characteristic of the electroerosion powder particles of metals, and these shapes verify their liquation nature, i.e. the metal liquid separation from locally heated channel zone of metal granules (Figure 1).

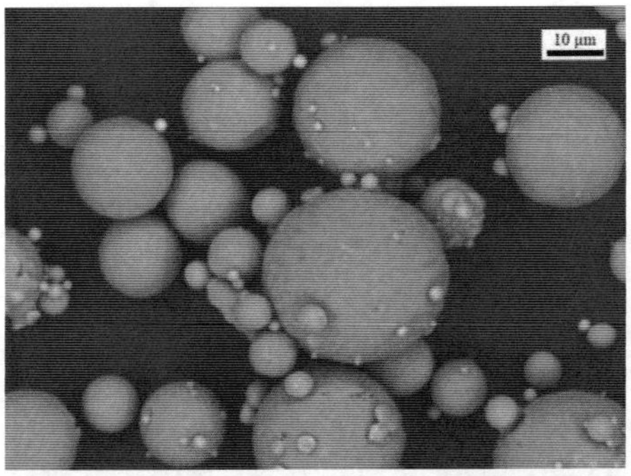

Figure. 1. Products of iron electrospark dispersion in water

The power fluid is also subjected to processes attending development of the discharge channel: dissociation and ionization of fluid occur. There are primary ionized particles of hydrogen and oxygen in the discharge channel in water [3]. Plasma property of the discharge channel is to accumulate energy, and then to use it slowly. These plasma properties lead to the discharge channel degeneration. It degenerates into a steam and gas bubble, in which recombination processes such as $H + H \rightarrow H_2$; $H + wall \rightarrow (1/2)H_2$; $O + O \rightarrow O_2$ and so on occur after completing processes in the channel [4]. Moreover, water is exposed to hot particles of molten metal spitting from erosion dimples, and gaseous products are formed in accordance with overall reaction $2H_2O+116$ kcal $\rightarrow 2H_2 + O_2$.

Therefore, hot metal melted by electrical erosion at first interacts with gaseous products of the steam and gas bubble. Then, flowing out into the power fluid, metal interacts with it. If water is used as power fluid, metal cools when it is in the hydrothermal oxidation zone. In consequence of these processes, oxide phases appear in composition of iron, niobium powders produced by electroerosion method [4]. The analysis of existence conditions of metallic and oxide phases of metals at different temperatures and fugitiveness of oxygen is given in the paper [5]. According to degree of affinity between metals and oxygen, the authors divided metals into three groups. Use of these approaches makes the processes be understood easier, so long as development of these processes is possible due to electroerosive production of different metals powders.

The aim of this paper is to study the effect of physicochemical processes on the phase composition of powders, taking into account that these process proceed due to electrospark dispersion of metals such as silver (Ag), zinc (Zn), iron (Fe), titanium (Ti) in water, as well as to obtain basic data to develop an electroerosion technology of the fine powders production with reproducible properties.

2. Experimental

Investigated metals, presenting different groups in accordance with degree of affinity between metals and oxygen [5], are shown in table 1.

Table 1. Classification of metals in conformity with level of oxygen affinity

Group name	Metal
Metals, which can be in equilibrium with non-dissociated water	Ag
Metals, which are unstable in medium containing water, and they corrode. They can exist only in reducing conditions in atmosphere of H_2 and CO	Zn, Fe
Metals, which are stable in special reducing conditions	Ti

The laboratory setup described in the paper [7] was used to produce electroerosion powders of metals. A porcelain jar (1 dm^3) was used as a reactor. Electrodes were made of metal which were in keeping with metal of granulated loading. They were fastened with a help of brackets, and dipped in the jar in the line of the jar's wall to the bottom. Electrode spacing was 60 mm. The interelectrode gap was filled with metal granules at a height of 10-15 mm. Distilled water was used as power fluid;

capacity size 500 cm^3. After finishing the dispersion stage, the lay of generated suspension was decanted in a receiving container. Electroerosion powders were fractionated with sedimentation or they were settled for thickening. Then they were dried into the vacuum oven, first at temperature no more than 40^0C, then they were brought up to constant-weight at 100^0C. Produced powders were investigated by the metal component fraction measurement and the phase composition study.

The metal component fraction was determined by method of metal chemical equivalent determination due to oxygen volume displacement and the weight of metal chemical analysis in the sample of powder. The investigation of phase composition and structural parameters of the sample was carried out by X-ray diffraction analysis, with a help of diffractometer Shimadzu XRD-7000 with CuKα-radiation (Nanocenter, Tomsk polytechnic university). The size analysis of coherent scattering regions (CSR) was carried out with a help of program of full-profile analysis – program POWDERCELL 2.4. The investigation of fine powders morphology was carried out by transmission electron microscopy method (TEM) using the electron microscope JEM-2100 (JEOL) (Nanocenter, Tomsk polytechnic university). The value of produced powders specific surface was determined with the analyzer of specific surface and porosity "Sorptometer M" (CJSC "Katakon") according to thermal nitrogen desorption. Due to findings, the software of "Sorptometer M" has calculated value of the investigated samples specific surface by Brunauer-Emmett-Teller method (BET method).

3. Discussion of Results

X-ray patterns of powders produced by electrospark dispersion of investigated metals (Ag, Fe, Zn, Ti) in water are shown in Figures 2,3,6,8.

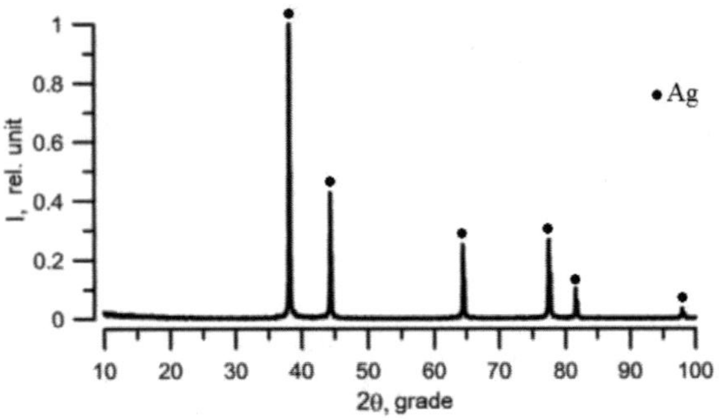

Figure 2. X-Ray diffraction of silver powder

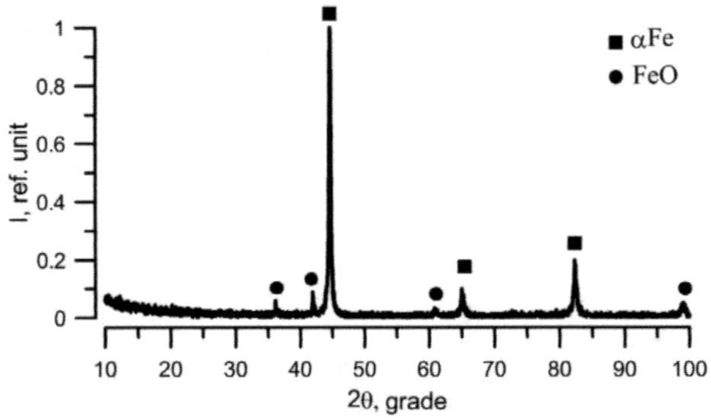

Figure 3. X-ray pattern of iron powder

X-ray pattern of silver powder produced by electrospark dispersion of metallic silver granules in water is shown in Fig. 2. According to data obtained by X-ray diffraction analysis, the product includes well crystallized metallic silver particles; according to phase composition, they are in line with metal of starter granules.

The calculation of crystal lattice parameters of Ag powders was carried out by Rietveld method. It was found that the structure of electroerosion Ag powder corresponds to cubic crystal lattice with near - standard parameters (a = 4,0871 Å). Some compounds of silver and water decomposition products (oxygen, hydrogen) were not identified in the produced powder composition.

According to data obtained by X-ray diffraction analysis (Fig. 3,6), in contrast to inert silver, the zinc and iron electroerosion powders are more inclined to oxidation. They contain substances mixture in their composition: not only Fe and Zn metallic phases but compounds of these metals and oxygen.

Having been washed with ethanol and acetone, iron powder dried after electro-impulse dispersion in water consists of two phases, which are in keeping with clear diffraction lines of metallic α-iron (αFe) and iron oxide (II) (FeO) in X-ray pattern (Fig.3). The content of metallic phase of iron in the powder is equal to 85-90%, and the specific surface value of this fraction is no more than 10 m^2/g.

IOP Conf. Series: Materials Science and Engineering **110** (2016) 012082 doi:10.1088/1757-899X/110/1/012082

Physical configuration of fine dispersed fraction particles (according to data obtained by transmission electron microscopy) is presented in Figure. 4.

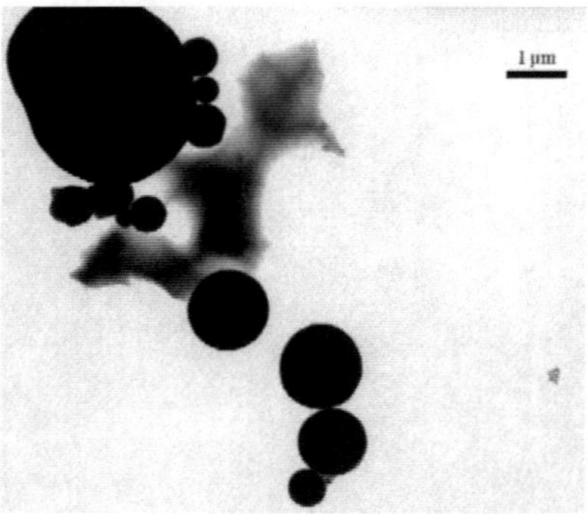

Figure. 4 Transmission electron microscopy of fine dispersed fraction of iron powder

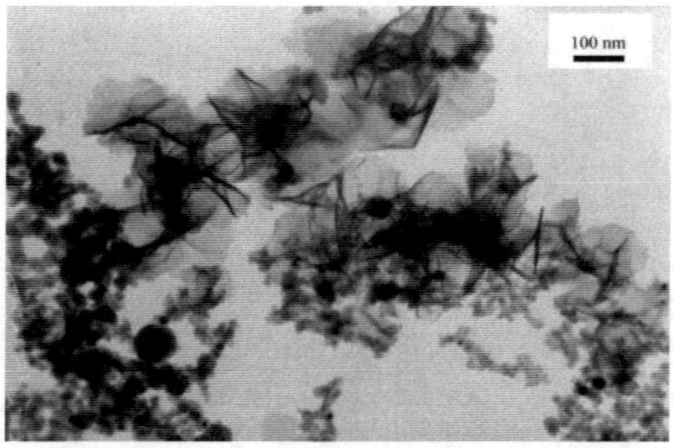

Figure. 5 Transmission electron microscopy of colloidal fraction of iron powder

Electroerosion powders, dried in vacuum oven from aqueous media, contain less metallic iron (60-65%) in their composition. The value of this powder specific surface increases to 40 m^2/g due to formation of a new phase (third) – hydroxides. They are formed due to the iron and oxygen interaction in humid atmosphere. Physical configuration of fine dispersed fraction particles is presented in Figure. 5.

X-ray pattern characterizing phase composition of electroerosion powder is given in Figure. 6. Zinc granules were used to produce this powder. According to data, obtained by X-ray diffraction analysis of the selected powder, products of dispersion is a mixture of at least two substances: metallic zinc (Zn) and zinc oxide (ZnO).

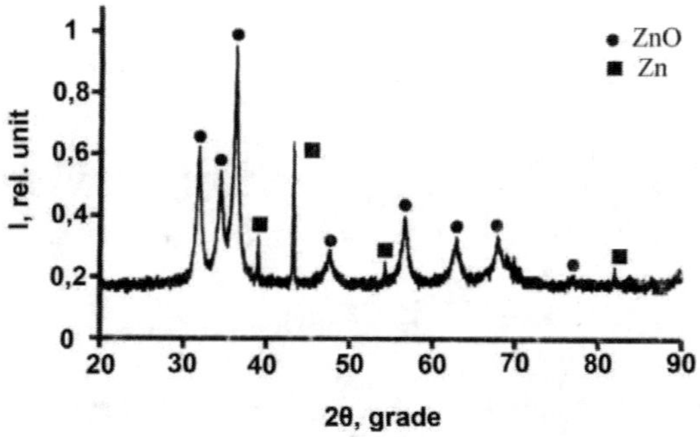

Fig. 6. X-ray pattern of zinc powder

Since high-temperature oxidation processes are such profound ones, so not only metallic zinc finely dispersed particles, produced due to thermal processes in discharge channels (Figure. 7a), but new formations held on their surface (Figure. 7b) are clearly observed in an electronic photography.

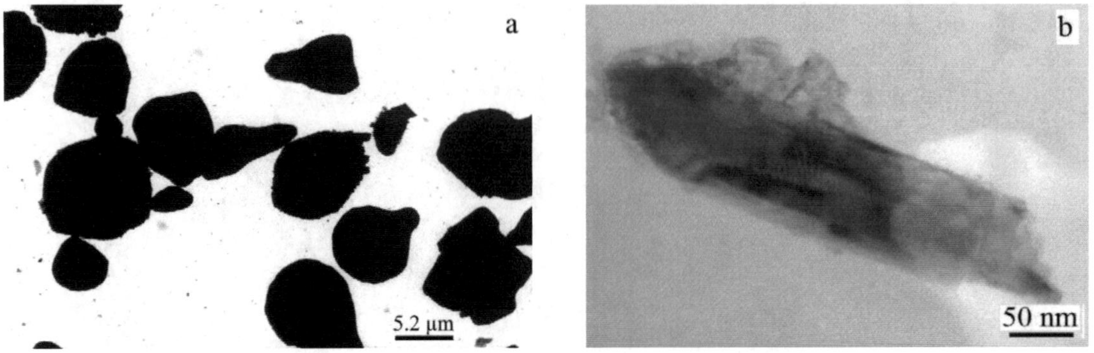

Figure. 7. Electronic photography of powders produced by electrospark dispersion of zinc granules: a) finely dispersed particles, b) single nanorod ZnO

These new formations are presented by eumorphic particles. The particles are long like nanorods, the length of them can be 150 nm or less, and 50 nm or less in cross direction corresponding to red zinc ZnO. The colour of this powder is grey. According to data of chemical analysis, the metallic zinc concentration in powder is no more than 70%.

Even more profound changes in phase and chemical compositions of electroerosion powders can be observed when the titanium electrospark dispersion is carried out. X-ray diffraction analysis data of powder produced from metallic titanium is presented by X-ray pattern in Figure. 8.

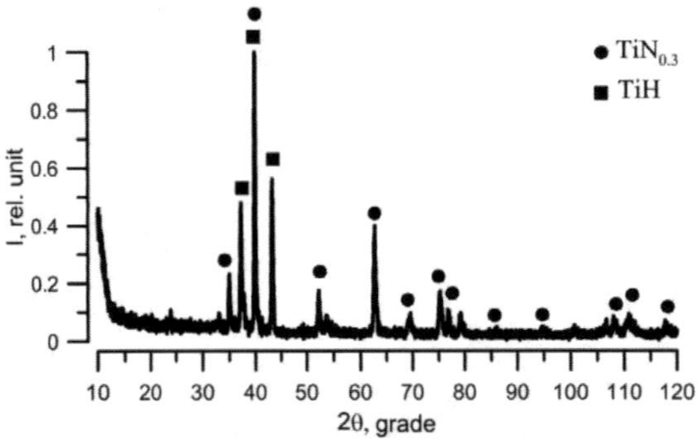

Fig. 8 X-ray pattern of titanium powder

According to data of this X-ray pattern, the lines, which could be classified as pure titanium, are shifted or closed by overlapping of lines which are typical of other substances. It is possible to come to conclusion that there is no pure metallic titanium presence in the powder. Twelve lines, including the most intensive one, are likely can be related to the phase $TiN_{0,3}$. Three lines can be related to the phase TiH: two independent lines possess the intensity of 68 and 96 %, and the third line is overlapped with the most intensive line of the phase $TiN_{0,3}$. Some lines can be related to the phase Ti_2O, but nine lines are overlapped with the lines of phase $TiN_{0,3}$, and only one line is independent.

Therefore, due to data of X-ray diffraction analysis, products of metallic titanium electrospark dispersion in water are likely to be presented as a mixture of at least two phases: $TiN_{0,3}$ and TiH. Observed behavior of metallic titanium, when in use electrospark dispersion in water, is in agreement with its reported chemical properties and high reactivity, when interacting with nitrogen and oxygen to form titanium nitride and titanium oxide at high temperatures [8].

Moreover, when in use metals electroerosion dispersion, high local temperatures increase sharply the velocity of interaction them with environment. Metals, which are prone to interaction with oxygen

IOP Conf. Series: Materials Science and Engineering **110** (2016) 012082 doi:10.1088/1757-899X/110/1/012082

and hydrogen, forming due to water decomposition in a spark, and other gases dissolved in water are able to absorb a great deal of oxygen (and other gases) in short time of the erosion cycle. Adsorbed oxygen can be in metals either in dissolved condition (oxides of the lowest oxidation level) or suboxides, and also endogenous non-metallics can occur. Both the first condition and the second one have an effect on phase composition of the produced metallic powders. The more affinity of original metal to oxygen and other gases in the composition of forming gas atmosphere is, the more effect is. When metallic powders are enriched with oxides, the enrichment has an effect on physical properties, for example, it decreases plasticity. Oxidation-reduction reactions between metals and atmosphere, forming in the process of water (power fluid) decomposition by a spark, are a source of oxygen in metals when in use electroerosion dispersion. When in use electroerosion, interaction of oxygen and metals is complicated by dissolution of oxides in metals, and it decreases their thermodynamic stability a lot.

4. Conclusions

Due to these investigations, it was found that only phase composition of products of the silver electrospark dispersion agrees with phase composition of original metal. The products, including original metal and oxides, are formed from oxygen hungry metals such as zinc and iron under experimental conditions. And the phases, which are in keeping with original metal, are not found in the composition of new products formed from titanium under these experimental conditions.

References

[1] Lazarenko B R and Lazarenko N I 1964 Author's certificate 70000 USSR. Powder production process and apparatus to produce it. Discoveries Inventions **22** 120

[2] Goloveyko A G 1966 Izvestiya of higher education establishments. Energy **6** 83-88

[3] Naugolnyih K A and Roy N A 1971 Electric discharges in water (Moskow: Nauka) 155

[4] Semkin B V, Usov A F and Kurets V I 1995 Principles of electro-impulse material destruction (St. Petersburg: Nauka) 276

[5] Asanov U A and Sakavov I E 1967 Izvestiya AS Kirgizskoy SSR **4** 71-75

[6] Oleynikov B V, Okrugin A V, Tomishin M D et al. 1985 Native metal formation in platform basic rocks (Yakutsk: Nuclear Science SB AS USSR) 188

[7] Zhuravkov S P, Pustovalov A V, Lobanova G L, Kvashnina O S and Yavorovsky N A 2014 Mechanical Engineering, Automation and Control Systems: Proceedings of International Conference (Tomsk: TPU Publishing House, 16-18 October 2014) 1-4

[8] Fedotev N P, Alabyishev A F, Rotinyan A P, Vyacheslavov P M, Zhivotinskiy P B and Galnbek A A 1967 Applied electrochemistry (Leningrad: Chemistry) 600

IOP Conf. Series: Materials Science and Engineering **110** (2016) 012083 doi:10.1088/1757-899X/110/1/012083

Influence of oxygen on luminescent properties of $ZnWO_4$ crystals surface

L A Lisitsyna

[1] Tomsk state university of Architecture and Building 634003, Tomsk, Solyanaya sq., 2 Building 2, Russia

E-mail: lisitsyna@mail.ru

Abstract. The results of studies of emission properties of oxygen-containing $ZnWO_4$ crystals are presented. Oxygen is incorporated into single crystals in two ways: annealing of crystals at 900 C in an oxygen atmosphere and implantation of high energy (28 MeV) oxygen ions. The observed variation of the photoluminescence intensity and excitation spectrum of emission at 2.6 eV are interpreted as the occurrence of a concentration gradient of emission centers in the crystal bulk under hearting of crystals and the dependence of the photons mean penetration depth in a crystal on their energy. In order to gain better insight into the nature of the phenomenon, time-resolved luminescence measurements were carried out.

1. Introduction

The effect of the state of the surface and surface associated physical phenomena on inherent spectral-kinetic parameters of a phosphor luminescence is one of the less studied issues. It requires further investigation and more detailed analysis. The actuality of the research is determined by application of scintillators, both large-volume and nano-sized ones, in a variety of environments and temperatures, and by the need to maintain stable parameters during long-lasting operation of scintillators.

To understand the role of the surface we investigate stationary luminescence excitation spectra of the crystals subjected to different types of treatment and with nanosecond time-resolution the luminescence excited by ionizing radiation in $ZnWO_4$ crystals.

2. The tasted samples and the experimental method

Three types of sources were used for the luminescence excitation: (i) an electron beam pulse accelerator with the electron energy of 250 keV, pulse duration of 10 ns and a radiation dose absorbed by the crystal within the action of the electron pulse equal to 5.10^1 Gy; (ii) pulsed Nd-YAG laser irradiation with photon energy of 4.66 eV, pulse duration of 10 ns, an energy per pulse of 40 mJ; (iii) xenon steady state lamp with a radiation energy of 3–6.2 eV.

The studied $ZnWO_4$ single crystals were grown by the Czochralski technique at the Institute for Scintillation Materials NAS Ukraine (Kharkov).

Under investigation were the following crystals: i) crystals freshly cleaved from a crystal block; ii) the crystals stored in oxygen atmosphere at 900 C for 50 hours; iii) crystals implanted with the oxygen ions of 28 MeV. The concentration of the implanted oxygen ions was 10^{18} ion/cm^3.
The spectral-kinetic characteristics of the luminescence were studied in the region of 4–1.5 eV after the different delay times after the end of the excitation nanosecond pulse in the time interval of 10^{-8}–10^{-1} s. The spectra were not recalculated with account of PMT sensitivity.

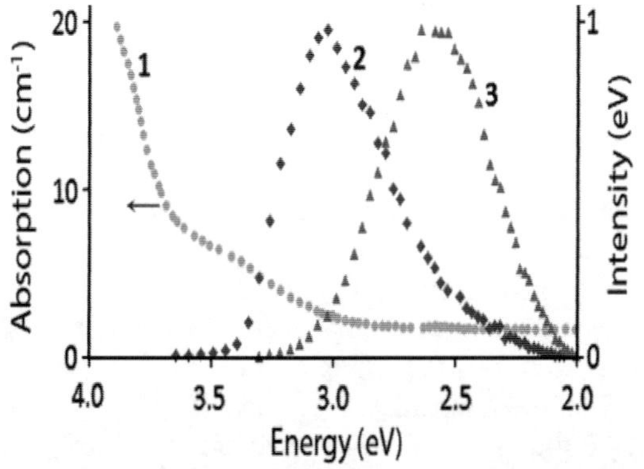

Figure 1. Absorption spectrum (1), photoluminescence spectra of ZnWO4 crystal (2,3) measured with 10 (2) and 400 ns (3) at 300 K after the action of laser pulse (spectra are normalized).

3. Results

The spectral composition of the time-resolved luminescence of $ZnWO_4$ crystals under excitation both with 4.66 eV photons and 250 keV electrons is the same. The emission spectrum of the short-lived decay component is an asymmetric band at 3.1 eV with FWHM equal to 0.5 eV and the decay time constant (τ_1) equal to 10 ns. The emission spectrum of the long-lived decay component is a band at 2.6 eV (Figure 1) with a multi-exponential character of the emission decay. The best fit to the experimental data at 300 K was obtained using three exponentials with the decay time constants of 200 ns, 3 and 25 μs. The obtained spectral-kinetic parameters of the long-lived luminescence coincide with those published for $ZnWO_4$ crystals in [1-4]. The ratio of the intensities of the emission bands at 2.6 (J_2) and 3.1 eV (J_1) in the photoluminescence (PL) spectra of freshly cleaved $ZnWO_4$ crystals is more than 1 ($J_2/J_1>>1$); the ratio of the decay time constants of the emission at these bands $\tau_2/\tau_1\approx10^3$ (where τ_1=10 ns, τ_2=25 μs (the greatest value of the time decay constants at 2.6 eV emission band at 300 K); so the inequality $J_2\tau_2>>J_1\tau_1$ is satisfied, where $J_2\tau_2$ and $J_1\tau_1$ are the light yield at 2.6 and 3.1 eV, respectively.

Figure 2 shows the excitation spectrum of the emission band at 2.6 eV for freshly cleaved $ZnWO_4$ crystals. The excitation spectrum reaches the maximum value at photon energy 3.9 eV and has a substantially constant value in the range 3.9–6.2 eV. (At 6.2 eV, this value is lower than that at 3.9 eV only by 10% in our crystals). According to [2,4–6], the emission intensity at 2.6 eV does not depend on photon energy of 4 –7 eV at all, and it is equal to its maximum value at photon energy of 4 eV.

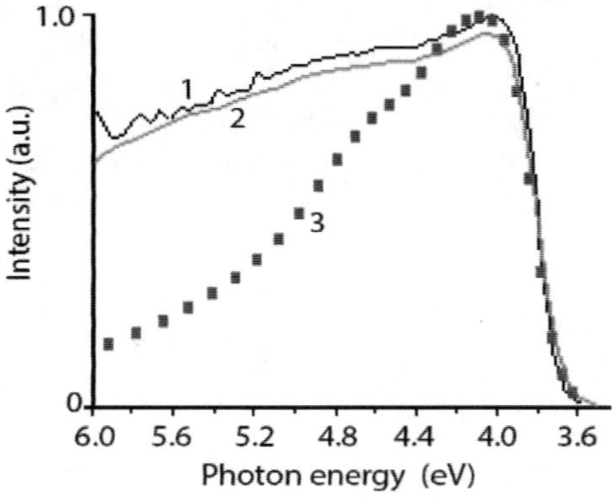

Figure 2. Excitation spectra of emission band at 2.6 eV at 300 K for different $ZnWO_4$ crystals: freshly cleaved sample (1), oxygen implanted with fluence of 10^{15} ion/cm^2 (2), heated in oxygen atmosphere at 900 C for 50 hours (3).

Implantation of oxygen ions in $ZnWO_4$ crystals with the fluence of 10^{15} ion/cm^2 changes neither the emission spectrum, nor the excitation spectrum for emission at 2.6 eV (figure 2). There is no change in relative intensities of emission bands at 3.1 and 2.6 eV as well.

Heating of crystals in oxygen atmosphere at 900 C for 50 hours causes the following changes in the spectra luminescence of these crystals.

• Intensity of the emission band at 2.6 eV under 6.2 eV photon energy excitation becomes ~ 80% lower than that under 3.9 eV (figure 2).

• The ratio of the intensities of the emission bands at 2.6 (J_2) and 3.1 eV (J_1): after heating of the crystal changes and the light yields in both bands becomes comparable, i.e., $J_1\tau_1 \approx J_2\tau_2$, whereas, before heating of the crystal $J_2\tau_2 / J_1\tau_1 > 10^3$. (The duration of the luminescence decay in both emission bands does not depend on the hearting).

The heating of crystals does not affect the ratio between the intensities of the emission bands in spectra of photoluminescence (PL) in contrast to the spectra of cathodoluminescence (CL) the hearted crystals. As an example, in the PL spectrum of the heated crystal, the ratio J_2/J_1 is 0.05, whereas, in the CL spectrum of the same crystal, the ratio J_2/J_1 is 5, and it is equal to that in the crystal before heating.

4. Discussion

Decrease of the luminescence at 2.6 eV in annealed crystals under excitation with UV energy photons could result from the following processes: (i) radiationless processes on the surface, (according to [7], it is caused by killer centers created on the crystal surface); (ii) destruction of the luminescence centers responsible for emission at 2.6 eV; (iii) reabsorption of excitation energy by the luminescence centers responsible for emission at 3.1 eV formed during heat treatment.

We propose the following model of the processes. The main statements of the model are set out below.

1) We suggest the presence of two types of emission centers (EC) in $ZnWO_4$ crystals: EC_b and EC_s responsible for the emission at 2.6 eV and at 3.1 eV, respectively. EC_b are homogeneously distributed within the bulk of the crystal. The mean value of their concentration is high and not less than ~10^{17} cm^{-3} is due to the impossibility of maintaining stoichiometry of the melt during crystal growth [8–11] EC_s are localized on the thin sample surface layer. Their concentration depends on the temperature and duration of crystal storage in air. The minimal number of EC_s is found on the surface of freshly cleaved crystals, as evidenced by the low intensity of the band at 3.1 eV in these crystals.

2) During heating of crystals in oxygen, the creation of an additional number of EC_s due to the destruction (transformation) of EC_b occurs. The concentrations of EC_s and EC_b inside the crystal at distance x from the surface are determined by~$\exp(-x^2/L^2)$ and by~$(1-\exp(-x^2/L^2))$, respectively. (L is the characteristic diffusion length). Heat treatment leads to the depletion of EC_b in the surface layer, i.e., EC_b distribution in the crystal becomes non-uniform.

3) To excite luminescence, the crystal is subjected to a photon flux. The photon flux N is equal to 10^{13} cm^{-2} c^{-1} and is the same for different values of the photon energy in range of 3–6 eV.

Inhomogeneous distribution in the crystal of the centers EC_b leads to the satisfaction of inequalities $N_b \gg N$ and $N_b \le N$ in the crystal bulk an in the heated crystal, respectively.

4) The luminescense at 2.6 eV belongs to EC_b and its intensity is determined by the number of EC_b in an emission state (N_b*). N_b* value depends on the relation between N_b and N. Obviously, if $N_b \le N$ (in the surface layer), N_b*~N_b. If $N_b \gg N$ (in the crystal bulk), N_b*~N.

5) According to the Buger-Lambert-Beer law, the number of the absorbed photons in a layer of thickness x is given by exp(-kx), where k is an absorption coefficient. The dependence k=f(hν) in the region of 4–7 eV for the $ZnWO_4$ crystal is published in [12], hence the mean penetration depth (k^{-1}) of photons in a crystal can be calculated as a function of the photon energy. According to our

estimations, if the photon energy decreases from 7 to 4.5 eV, k^{-1} varies by one order of magnitude (from 10 to 100 nm).

6) The photons with a penetration depth $k^{-1}>L$ preferentially excite the LC in the bulk of the crystal in which $N_b>>N$. The photons with a penetration depth $k^{-1}\leq L$ excite the LC in the thin surface layer of crystal in which $N_b\leq N$.

Then, taking into account paragraph (4) of this model, in crystal heated in oxygen, the value of EC_b* depends on two parameters: (i) the distance from the surface of the crystal; (ii) the energy of the photons exciting the emission (i.e., the depth of the photon penetration in the crystals).

Hence, the value of the photon energy, at which the intensity of the luminescence at 2.6 eV decreases, can be used to estimate the depth of the disturbed surface layer in which $N_b<N$. According to our estimates, the depth of the disturbed surface layer is equal to 80 nm in crystals heated in oxygen atmosphere at 900 C for 50 hours.

7) In the process of oxygen implantation, no additional EC_s is created on the crystal surface, and as a result, no concentration gradient of EC_b is observed. The ratio $N_b>>N$ is fulfilled in all regions of the crystal, and the intensity of the emission at 2.6 eV does not depend on the photon energy.

Since the implantation of oxygen ions does not result in suppression of the photoluminescence at 2.6 eV, it is obvious that this type of treatment does not cause depletion of oxygen luminescence centers.

There is no doubt that oxygen implantation has an effect on crystal surface, however, it does not lead to creation of the luminescence centers responsible for the band at 3.1 eV. We assume that implantation does cause modification of the oxygen present in the EC_b composition as it occurs during heating of crystal.

8) Photons with the energy <4 eV, for which the depth of penetration into the crystal $k^{-1}>L$, as well as the electrons with a penetration depth equal to 0.2 mm (250 keV), excite the emission in the bulk of the crystal, and the effect of the surface state is not so high. As a result, there is no effect of crystal heating on the efficiency of the luminescence excitation at 2.6 eV in the range of photon energy of 3.6–4 eV.

Within the frame of the given model, the coincidence of the excitation spectrum at 2.6 eV in $ZnWO_4$ macrocrystal heated in oxygen atmosphere at 900 C with that in $ZnWO_4$ nanocrystal, synthesized in air at 1114 K [13], indicates that in both cases the crystal surface formed at high temperature suppresses the photoluminescence at 2.6 eV.

5. Conclusions

The comparative analysis of the luminescence spectra excited by radiation with different depth of penetration into a crystal revealed the presence of spatial distribution of defects in the material bulk.

A detailed analysis of the time-resolved characteristics of the crystals allows us to suggest the existence of a spatially inhomogeneous structure of the original material imperfection. We suggest two types of LC in $ZnWO_4$ crystals. One of them, located in the crystal bulk, is responsible for the emission at 2.6 eV. The crystal surface is enriched with LC responsible for the emission at 3.1 eV.

The ratio between the two types of LC changes in favor of the surface LC during heating in oxygen atmosphere or in air, or during crystal storing in air at 300 K. The dependence of the depth of probe radiation penetration into the crystal on radiation energy make the ratio between the intensities of the emission bands of the surface and bulk centers depend on radiation energy. In excitation of the $ZnWO_4$ crystal bulk not only by high-energy electrons, as shown in this paper, and, apparently, in excitation by X-ray and γ-rays, the state of the surface does not affect the efficiency of the excitation band at 2.6 eV. An extrinsic luminescence band (3.1 eV) at the surface has an insignificant effect on the luminescence spectrum of $ZnWO_4$ crystal only if the radiation penetrates a significant distance deep in the crystal from its surface.

The observed effect can be used to estimate the energy for heavy particles of low energies.

The value of the threshold photon energy, at which the luminescence intensity at 2.6 eV decreases in the excitation spectrum, was used to estimate the depth of the disturbed surface layer. For $ZnWO_4$ heated at 900 C in oxygen, this value was found to be equal to 80 nm.

The absence of the dependence of the luminescence intensity on the energy of excitation exceeding the threshold (>4 eV) indicates that the concentration of LC_b in the surface layer does not differ from the average value in the crystal volume.

References

[1] Kraus H, Danevich F A, Henry S,. Kobychev V V, Mikhailik V B,. Mokina V M, Nagorny S S, Polischuk O G and Tretyak V I, 2009 *NIM A* **600** 594

[2] Nagirnyi V, Feldbach E, Jonsson L, Kirm M, Kotlov A, Luschik A, Nefedov V and Zadneprovski B, 2002 *NIM A* **486** 395

[3] Mikhailik V, Kraus H, 2010 *Phys. Stat. Solidi B* **247** 1583

[4] Krutyak N, Mikhailin V V, Vasil'ev A N, Spassky D A, Tupitsyna I A, Dubovik A M, Galashov E N, Shlegel V N and Belsky A N, 2013 *J.Luminesc.* **144** 105

[5] Kolobanov V N, Kamenskikh I A, Mikhailin V V,. Shpinkov I N, Spassky D A, Zadneprovsky B I, Potkin L I and Zimmerer G, 2002 *NIM A* **486** 496

[6] Wang H, Medina F, Liu D and Ya-Dong Zhous, 1994 *J. Condens. Matter* **6** 5373

[7] Groenink J.A, Blasser G., 1980 *J. Sol. State Chem.* **32** 9

[8] Atroshchanko L V, Burachas S F, Galchinetsky L P, Greenev B V, Rizhikov B D and Starzhinskiy N G, 1998 *Crystals of Scintillators and Detectors of Ionizing Radiations on their Base*, Kiev. Naukova Dumka

[9] Watterich A A, Kovacs L, Wurz R, Schon F, Hofstaetter A and Scharmann A, 2001 *J. Phys.: Condens. Matter.* **13** 595

[10] Lisitsyna. L, Lisitsyn V, 2013 *Physics of the Solid State* **55** 2297

[11] Lisitsyn V M, Valiev D T, Lisitsyna L A, Tupitsyna I A, Polisadova E F and Oleshko V I, 2013 *J. Applied Spectroscopy.* **80** 361

[12] Kolobanov V N, Mikhailin V V, Shpinkov I N, Spassky D A, Kirm M, Zimmerer G and. Makhov V N, 1999 *Proceed. of 5th Intern. Conf. of Inorganic Scintillators and their Applications, Moscow State University, Moscow,* 652

[13] Dabre K V, Dhoble S J, 2012 *Recent research in Science and technology* **4** 2

IOP Conf. Series: Materials Science and Engineering **110** (2016) 012084 doi:10.1088/1757-899X/110/1/012084

Effect of heat treatment on the structural parameters and magnetic properties of copper ferrite nanopowders obtained by the sol-gel combustion[1]

V A Zhuravlev[1], E P Naiden[1], R V Minin[2], V I Itin[2] and M R Ufimtsev[1]

[1]Tomsk State University, Lenina av., 36, Tomsk, 634050 Russia
[2]Tomsk Scientific Center SB RAS, pr. Academicheskii 10/4, Tomsk, 634021 Russia

E-mail: ptica@mail.tsu.ru

Abstract. Phase composition, structure parameters and basic magnetic characteristics obtained by the sol-gel combustion nanopowders of ferrospinel $CuFe_2O_4$ are investigated. A comparison of the properties of synthesized materials: first sample – immediately after the combustion of the gel and second sample – after annealing at 1073 K for 4 hours are performed. Annealing leads to an increase in the concentration of the phase with tetragonal crystal structure. Particle sizes and the value of anisotropy field of this phase also increased.

1. Introduction

Nanosized magnetic particles attracted attention of researchers working in different fields of solid state physic in connection with their rich and often unusual range of properties and applications. This fully applies to the copper ferrite $CuFe_2O_4$. It is known that the samples of this ferrimagnet may exist in modifications with cubic or tetragonal crystal structures. Implementation of these crystal phases depends on the temperature and conditions of synthesis [1, 2].

According to [1 – 3], the cubic crystal structure with a lattice constant a ≈ 8.42 Å exists in some temperature range below the Curie temperature. The first-order structural phase transition (SPT) into the tetragonal crystal phase with a lattice constant a = b ≈ 5.82 Å and c ≈ 8.69 Å takes place with further cooling. The SPT in the copper ferrites was studied experimentally in [2, 3] with the help of high-resolution neutron diffraction technique. The temperature dependences of the initial magnetic permeability and the parameters of ferromagnetic resonance in the vicinity of the SPT were studied in [4]. In papers [2 – 5] the magnetocrystalline anisotropy of copper ferrites was investigated. It is shown that in the tetragonal phase $CuFe_2O_4$ refers to materials with easy magnetization plane (EMP) and with the large value of anisotropy fields [4, 5].

The investigation of the influence of a size effects on the structural and magnetic properties of this ferrite is of interest. The magnetic properties of nanostructured samples of copper ferrospinels synthesized by different methods were investigated in [6 – 8]. The methods used in this works were: mechanochemical reactions [6], co-precipitation [7] and SHS [8]. It is noted that after the transition to a nanostructured state the value of saturation magnetization increases by about 20 % as compared with bulk samples.

[1] This Research is supported by Tomsk State University Competitiveness Improvement Program reg. N 8.1.23.2015

The results of studies of the properties nanostructured copper ferrite powders obtained by the sol-gel combustion are presented in this paper.

2. Preparation of samples

To synthesize the nanoparticles of copper ferrite we used the chemical reaction:

$$Cu(NO_3)_2{\cdot}3H_2O+2Fe(NO_3)_3{\cdot}9H_2O+C_6H_8O_7+O_2=CuFe_2O_4+6CO_2+25H_2O+5NO_2+3NO.$$

The aqueous solutions with concentration of 1 M: copper nitrate 3-aqueous $Cu(NO_3)_2{\cdot}3H_2O$, iron nitrate 9-aqueous $Fe(NO_3)_3{\cdot}9H_2O$, and citric acid $C_6H_8O_7$ were pre-prepared. Then they were mixed in accordance with the relationship:

$$\{[Cu(NO_3)_2{\cdot}3H_2O]:[Fe(NO_3)_3{\cdot}9H_2O]\}:C_6H_8O_7=\{1:2\}:2\,.$$

Concentrated ammonium hydroxide NH_4OH was added dropwise with constant stirring to the obtained solution until pH $\approx$ 2. The resultant sol was heated for 3 – 5 hours without boiling. At the evaporation of water it turned into a viscous brown-green gel, which then ignited and burnt. Thus the initial sample No 1 was prepared. The sample No 2 was obtained after the calcinations of initial powder at a temperature of 1073 K for 4 hours.

3. The results of x-ray researches

The x-ray analysis was performed using a SHIMADZU XRD-6000 polycrystalline diffractometer. The computer database of x-ray powder diffractometry PDF4+ was used for a qualitative analysis of the phase composition. The quantitative analysis of the phase composition and the refinement of the structural parameters of phases being detected were performed using the program for full profile analysis Powder Cell 2.4.

The results of X-ray diffraction analysis of the samples are summarized in Table 1. According to the table, the product of combustion gel (sample No 1) is a mixture of copper ferrospinels with cubic and tetragonal crystal structure at about equal proportions. This result is quite understandable. The relatively rapid cooling of the reaction product after the combustion process is equivalent to quenching of the gel. The existing at high-temperature non-equilibrium state is retained. Some amount of the hematite phase α–Fe_2O_3 is also in sample No 1.

Table 1. Phase composition and structural parameters of synthesized materials after combustion sol-gel (sample No 1) and after calcinations (sample No 2).

Sample	Phase composition	Phase content vol. (%)	Parameters of the crystal lattice a (Å)	c (Å)	Lattice distortion $\gamma_T = c/a\sqrt{2}$	CSR (nm)
No 1	$CuFe_2O_4$– tetra.	48	5.830	8.610	1.04	22
	$CuFe_2O_4$– cube.	46	8.335	-	-	80
	α–Fe_2O_3	6	5.063	13.736	-	15
No 2	$CuFe_2O_4$– tetra.	95	5.826	8.665	1.05	80
	$CuFe_2O_4$– cube.	4	8.416	-	-	16
	α–Fe_2O_3	1	5.135	13.640	-	15

Calcinations of the sample No 1 followed by slow cooling leads to a significant increase in the content of the tetragonal phase and a significant increase in its coherent scattering region (CSR). The distortion of tetragonal lattice also increased and approached to the thermodynamically equilibrium value that equals to 1.06 [1]. This suggests the diffusion of Cu^{+2} ions from tetrahedral to octahedral sublattices during annealing. The obtained values of the lattice constants of the tetragonal and cubic phases are close to the data in [1 – 3].

4. Research of the magnetization curves

A study of the magnetization curves in pulsed magnetic fields up to 14 kOe was conducted by the method described in [9]. The values of the specific residual magnetization (σ_r) and the specific saturation magnetization (σ_s) are presented in Table 2. In the last two columns of the table the calculated according to the formula $M = \sigma \cdot \rho$ values of the magnetizations of unit volume are shown. Here $\rho = 5.42$ г/см^3 is the X-ray density of ferrimagnets $CuFe_2O_4$ [10].

Table 2. The research results of the magnetization curves

Sample	σ_r Gs·(cm^3/g)	σ_s, Gs·(cm^3/g)	M_r, Gs	M_s, Gs	M_r/M_s
No 1	20.6	31.5	112	171	0.66
No 2	22.1	30.9	120	168	0.71

According to Table 2 the saturation magnetization values (σ_s, M_s) of samples No 1 and No 2 are closed. The ratio of M_r/M_s after annealing increases, which indicates an increase in the coercive force of the material. It should be noted that the value M_s is more by ~ 20% then one given in [1, 10] for a bulk sample magnitude $M_s \approx 135$ Gs. The same increase in the saturation magnetization of the copper ferrite in the nanostructured state was observed in [6 – 8].

5. The study of the magnetic anisotropy by the FMR method

The ferromagnetic resonance (FMR) spectra were measured by the standard waveguide transmission technique in the frequency range 26 – 37 GHz using an automated radio spectroscope. For FMR investigations the powders of samples No 1 and No 2 were charged into quartz tubes with inner diameter of 0.7 mm and length of 10 mm. The density of powder samples was identical and equal to ≈ 2.6 g/cm^3.

Investigations of the ferromagnetic resonance spectra of powder and polycrystalline oxide ferrimagnetic materials allowed a number of magnetic parameters of these materials important for practical applications to be determined [11, 12]. Such as: magnitude and sign of magnetocrystalline anisotropy fields (H_{ai}); saturation magnetization (M_S); value of the gyromagnetic ratio $\gamma = ge/2mc$, where g is the effective g-factor of the examined material, e is the electron charge, m is the electron mass, and c is the velocity of light.

Figure 1 shows imaginary parts of the diagonal components of the permeability tensor (lines) calculated as described in [12] and measured experimental curves of the FMR (points) samples No 1 (Fig. 1 a) and No 2 (Fig. 1 b). Measurements and calculations were performed for the frequency 36 GHz. The experimental curves are normalized to the theoretical ones. The calculated curves 3 at Figure 1 are total resonance curves, consisting of contributions from the phases with a cubic structure (curve 1) and a tetragonal structure (curves 2).

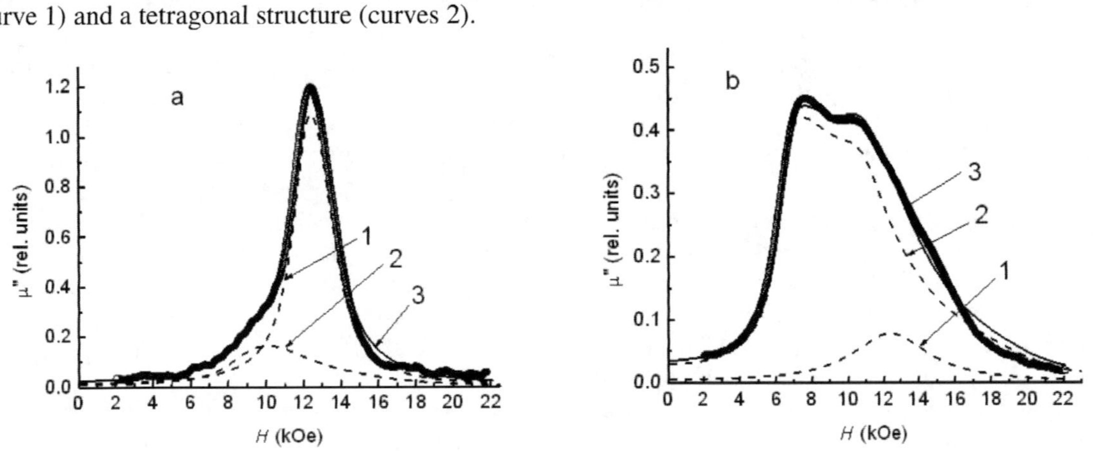

Figure 1. The curves of the FMR for sample No 1– a and No 2 – b. Points are the experimental curves. Lines 1 – calculations for the cubic phase, 2 – for the tetragonal phase, 3 – the sum of the curves 1 and 2. Parameters of calculated curves are shown in Table 3.

The shape of experimental FMR curve of sample No 1 corresponds to the materials with low values of the anisotropy fields. The value of the resonant field of this sample is close in magnitude to the resonance field for isotropic material $H_{res} = \omega/\gamma = 12.5$ kOe. The rise of the low-field slope of the resonance curve indicates the presence in this material a phase with a large value of anisotropy field. The shape of the experimental resonance curve of the sample No 2 with a predominance of the tetragonal crystal phase is typical for of a materials with magnetic ordering type of EMP (the anisotropy field $H_\theta \leq 0$).

The experimental FMR spectra of samples No 1 and No 2 were measured in the frequency range 26 - 37 GHz. The processing of spectra was carried out in two stages. In the first stage, the frequency dependences of the resonance fields were drawn. Then the estimations for magnetomechanical ratios γ and anisotropy fields for cubic (H_a) and tetragonal (H_θ, H_{a3}) crystal phases were performed by the least squares method. At the second stage, more accurate values of these parameters and volume concentration of cubic (c_K) and tetragonal (c_T) phases were determined by a detailed comparison of the shapes of experimental and calculated resonance curves at different frequencies. Note that the values of M_s for cubic and tetragonal crystal phases are unknown separately. Thus we took their magnetizations M_s from Table 2 at the calculation of the partial contributions to the total FMR absorption.

Table 3 shows the parameters used to calculate the curves of the FMR samples No 1 and No 2. Here c_K, c_T – are volume concentrations of the phases with cubic and tetragonal structures. H_a, H_θ, H_{a3} – are the values of anisotropy fields, $\gamma/2\pi$ – are the magnetomechanical ratios and α – are decay constants in the Landau-Lifshitz-Gilbert equation for a monocrystalline grain.

Table 3. Parameters of the calculated curves

Sample	The phase with low anisotropy (cubic + superparamagnetic tetragonal)				The phase with tetragonal structure				
	c_K (%)	$\gamma/2\pi$ GHz/kOe	H_a, kOe	α	c_T (%)	$\gamma/2\pi$ GHz/kOe	H_θ kOe	H_{a3} kOe	α
No 1	75 (48)	2.88	-1.0	0.08	25 (46)	3.10	-5.0	3.0	0.16
No 2	20 (4)	2.88	-1.0	0.15	80 (95)	3.10	-8.0	6.0	0.12

The tabulated for the cubic phase values of the anisotropy fields are rather rough upper estimate. However, they are correlated with the value of $H_a \approx -890$ Oe obtained from the data of [10] for room temperature. It is interesting to compare our estimation of parameters for the tetragonal crystal phase with the data obtained on the single-crystal sample in [5]. The listed in Table 4 values of $\gamma/2\pi$ are considerably more than estimation $\gamma / 2\pi \approx 2.6$ GHz / kOe from [5]. The value of the anisotropy field H_θ of sample No 2 is three times less than H_θ in [5]. As regards the value of the anisotropy field in the basal plane H_{a3}, it is conversely more then given in [5] at approximately one order of magnitude.

Table 3 shows that the volume concentrations of the phases determined from an experiment on FMR markedly differs from the X-ray diffraction data (Table 1). According to the FMR research of the sample No 1 the concentration of phase with large values of the anisotropy field is almost 25% less than one given from the X-ray diffraction. For the phase with a small quantity of H_a the situation is reversed. This behavior takes place in the sample №2, wherein the content of phase with low anisotropy field H_a also increased, and the phases with large anisotropy fields H_θ, H_{a3} reduced approximately at 15 % compared with X-ray diffraction data.

Possible reasons for these differences are following. Values of particle sizes on the CSR shown in Table 1 are raw estimations. Part of the particles with small sizes having a tetragonal crystal structure may exist in a superparamagnetic state with a small anisotropy. FMR in such particles occurs at the same values of the magnetic field as in particles with a cubic crystal structure [13].

Smaller values of the anisotropy fields for phase with tetragonal structure of sample No 1 (CSR ≈ 22 nm) as compared with the sample No 2 (CSR ≈ 80 nm) can also be explained by the influence of size effects. As noted in the paper [13], with decreasing of a particle sizes the influence of thermal fluctuations on the magnetic properties of the material are increased and the value of the anisotropy field is reduced.

6. Conclusion

In the paper the effect of heat treatment on the structural and magnetic properties of copper ferrite nanopowders synthesized by sol-gel combustion is studied. It is shown that the as-prepared powder consists of a mixture of phases with cubic and tetragonal crystal structures with approximately the same concentrations. Annealing leads to a significant increase of content of thermodynamic equilibrium at room temperature tetragonal phase. The value of the saturation magnetization does not change and the anisotropy fields of the tetragonal phase increases significantly. Measured by FMR method parameters of nanostructured copper ferrite differ significantly from the parameters obtained on bulk monocrystalline sample in [5].

References

[1] Krupicka S 1976 Physik der Ferrite und der Verwandten Magnetischen Oxide, vol. I *Mir, Moscow* 353 (in Russian).
[2] Balagurov A M, Bobrikov I A, Pomjakushin V Yu, Sheptyakov D V and Yushankhai V Yu 2015 *J. Magn. Magn. Matter.* **374** 591
[3] Balagurov A M, Bobrikov I A, Maschenko M S, Sangaa D and Simkin V G 2013 *Crystallography Reports* **58** 710
[4] Onyszkiewicz I, Malafaev N T, Murakhovskii A A and Pietrzak J 1982 *Phys. Stat. solidi (a)* **73** K243
[5] Malafaev N T, Murakhovskii A A, Popkov J A and Onyszkiewicz I 1990 *J. Magn. Magn. Matter.* **89** 8
[6] Jiang J Z, Goya G F and Rechenberg H R 1999 *J. Phys.: Condens. Matter.* **11** 4063
[7] Pajić D, Zadro K, Vanderberghe R E and Nedkov I 2004 *J. Magn. Magn. Matter.* **281** 353
[8] Kumar E R, Jayaprakash R, Devi G S and Reddy P S P 2014 *J. Magn. Magn. Matter.* **355** 87
[9] Kreslin V Yu and Naiden E P 2002 *Prib. Tekh. Eksp.* **1** 63
[10] Smit J and Wijn H P J 1959 Ferrite. Physical properties of ferrimagnetic oxides in relation to their technical applications. *Philips' Technical Library* 369
[11] Zhuravlev V A 1999 *Physics of the Solid State* **41** 956
[12] Zhuravlev V A and Meshcheryakov V A 2014 *Russian Physics Journal* **56** 1387
[13] Raikher Yu L and Stepanov V I 1994 *Phys. Rev.* **50** 6250

IOP Conf. Series: Materials Science and Engineering **110** (2016) 012085 doi:10.1088/1757-899X/110/1/012085

Study of interrelation between electromagnetic radiation and rock strength

L V Yavorovich[1], A A Bespal'ko[1], P I Fedotov[1] and E K Pomishin[1]

[1] Problem Research Laboratory of Electronics, Dielectrics and Semiconductors, Nondestructive Testing Institute, Tomsk Polytechnic University, Lenin Avenue, 30, Tomsk, 634050, Russia

E-mail: pif@tpu.ru

Abstract. The paper presents the study of the electromagnetic radiation of real rock samples under acoustic impact. The test samples are made of kerns from the iron ore mine with different petrophysical features and strength. The study has shown that the electromagnetic radiation parameters are related to the sample strength. However, this dependence is not observed for samples with magnetite that can be due to individual structural-textural features and quartz inclusions.

1. Introduction

The basic mechanism of mechanoelectric transformations (MET) in rocks is to be revealed to work out a method based on measuring electromagnetic radiation (EMR) to monitor the change in the rock stress-stained state (SST) and its strength and to forecast geodynamic events. To obtain MET patterns, we used the method of physical modeling in vitro.

When rocks crack, electrons, positive ions, photons and radio waves are emitted. In the study, we consider the EMR in the frequency range up to 1 MHz.

Stepanov was the first to observe the electric charge on the NaCl crystal surface as a result of flowage in the absence of an external electric field in 1933 [1]. He attributed this phenomenon to crystal slayers rubbing against internal inhomogeneity and microcrack originating.

Further investigations on this phenomenon were conducted by Fishbach and Novikov (1955) [2]. They studied increase in the electric charge on the NaCl crystal surface under pressure and attributed it to charged dislocation. Urusovskaya (1958), Martyshev (1965) and Kornfel'd (1971) continued the study [3–5]. A number of investigations were performed for piezo- and nonpiezoelectric, crystalline and amorphous, metallic and dielectric materials and rocks under different effects [6–11].

After that, interest in EMR was of applied relevance and focused on the earthquake forecast [12–14], forecast of rock destruction in underground mines [15–17], study of explosion [18–19], nondestructive testing of defect structure and strength.

It should be noted that a significant part of the research into this problem belongs to a research group under the guidance of Vorob'ev A.A. (1970), Tomsk Polytechnic University.

He coined the term "mechanoelectric transformation" related to EMR. Vorob'ev noted that transformation of mechanic energy into electromagnetic one depends on the properties of interacting systems which possess energy. He pointed out that all types of energy can transform into mechanic energy which in the interior of the Earth can transforms into electric energy [20]. Transformation of free charges or charged particles occurs under the influence of an electric field. Thereby, EMR is caused by increase or change in the charge on the dielectric surface.

Over recent years, great attention has been paid to studying electromagnetic signal (EMS) parameters, induced by acoustic impact [21–24] to find out the mechanism of EMR processes. The majority of the researchers consider an electric dipole occurring on the opposite sides of the crack to be the source of EMS [25–27]. The electric dipole is also formed by double electric layers (DEL) in interaction of different materials through pores in materials. The change in the dipole moment causes EMS.

Theoretical underpinning of the EME method for testing dielectrics quality is provided in the research by B.B. Lasukov [28]. It shows that stimulating mechanic vibration by a single normed stroke causes the bias current that is a function of the dipole moment and of the speed of change in the volume density of elementary sources. The quantity of charge, its volume structure and amplitude depend on physicochemical properties of the sample inner regions.

Therefore, the paper aims to study the change in the EMR parameters under acoustic impact for rock samples with different physicochemical properties, granules and micro-cracks. Charged particles with higher probability concentrate in inhomogeneities, produce an electric field and, as consequence, mechanic energy transforms into electromagnetic energy. In addition, the study of EMR emitted from the rocks is of practical interest as a key to developing the method of monitoring and forecasting geodynamic events in rock massifs.

2. Test object

The test objects are rock samples from Tashtagol ore mine (Kemerovo region). These are skarn samples with different petrographic composition without magnetite and magnetite ore samples with different content of magnetite. Cylindrically-shaped samples with a diameter of 42±1 mm and a height of 80±2 mm were cut out of the core. Before testing, the sides of the samples were polished until flatness with the discrepancy no more than 0.5±0.1 °. The angle between the sample sides and its axis was 90±1 °. The image of the samples is shown in Figure 1.

Figure 1. Samples image: left is skarn sample, right is ore sample.

3. Study technique

Acoustic impact was performed by a little steel ball launched from a spring pistol. It is a dynamic method of impact. The block diagram of the dynamic method is shown in Figure 2.

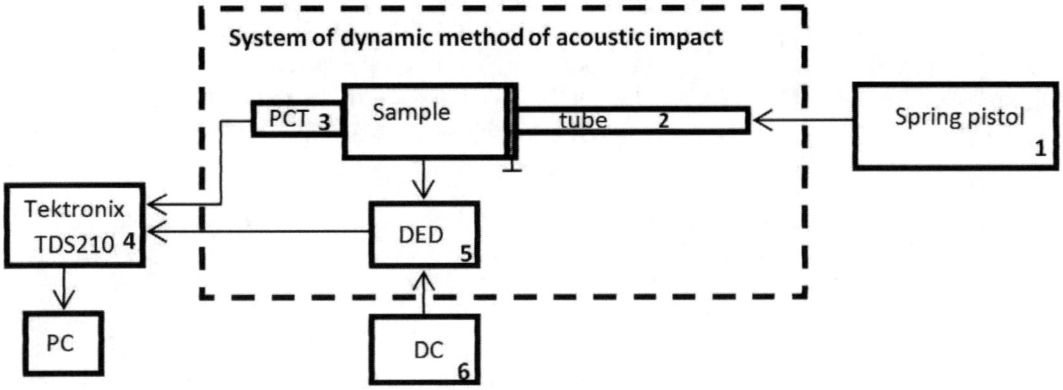

Figure 2. Block diagram of the dynamic method of acoustic impact

The ball is launched by pistol (1), passes through tube (2) and strikes a metal grounded plate that is connected to the sample by means of mineral oil. The acoustic pulse passes through the sample and is recorded by piezoceramic transducer (PCT) (3). A signal from PCT (3) initiates recording by the scope Tektronix TDS2024B (4). The EMS generated by the acoustic wave is received by differential electric detector (DED) (5) measuring only the electric part of the EMS. DED connected to the scope differentiates and amplifies the signal. The signals from the scope are transferred to a PC for the analysis. To change the duration and amplitude of the acoustic signal, we used balls with different mass of $0.25\cdot 10^{-3}$ and $0.86\cdot 10^{-3}$ kg that corresponds to the acoustic signal front of $10\cdot 10^{-6}$ and $14\cdot 10^{-6}$ s.

4. Experimental results

To determine an ultimate strength of the investigated samples, they were subjected to uniaxial compression with a press.

The task for iron ore samples was to estimate the relation of the quantity of the magnetite content in the samples to their ultimate strength and the maximum EMS amplitude. Table 1 summarizes the data on the magnetite content in the iron ore samples, their ultimate strength and the maximum EMS amplitude.

Table 1. Data on the magnetite content in the iron ore samples, their ultimate strength and the maximum EMS amplitude.

Sample	Mass (g)	density (g/cm^3)	volume (cm^3)	Magnetite in sample (g)	Magnetite in sample (%)	Ultimate strength (kN)	max EMS (mV)
1	360	3.21	112	57	16	158.6	350
2	380	3.45	110	118	31	279.8	20
3	393	3.60	109	157	40	196.5	800

The magnetite mass in the sample is calculated by the formula:

$$M_m = \rho_m \cdot V_o \cdot \left(1 - \frac{\rho_0 - \rho_m}{\rho_{vm} - \rho_m}\right),$$

where M_m is magnetite mass, g; ρ_m is magnetite density, g/cm^3; V_o is sample volume, cm^3; ρ_o is sample density, g/cm^3; ρ_{vm} is density of the enclosing rock, g/cm^3.

As can be seen in Table 1, the ultimate strength of the samples does not correlate with the magnetite content. The EMS amplitude does not correlate to the ultimate strength as well. However, this does not mean that there is no dependence between the parameters. Structurally-textural features of the samples and the quartz content may considerably affect the parameters. This is proved by the study which shows ten-fold increase in the EMS amplitude in the sample containing quartz in comparison with the sample containing no quartz under acoustic impact [29]. Analogous results were obtained by other researchers [30–31]. In particular, in [31], in the experiment on dynamic compression of rocks, it was found that the EMS amplitude for a gabbro sample with 2 % of quartz was six times less than that for a granite sample with 36 % of quartz. Quartz inclusions are randomly arranged by volume in magnetite ores samples, and this can explain significant difference in the EMS amplitude in the samples in our experiments.

The task for enclosing rock samples was to estimate the relations between the ultimate strength and the parameters of the recorded EMS. For this purpose, the analysis of analogue electromagnetic signals was carried out. Each sample was subjected to dynamic impact for 2 times. As the EMS amplitude oscillations varied from 4 to 10 %, their values were averaged. The average EMS amplitude, average duration of the signal and the mean value of the EMS amplitudes were calculated. Table 2 shows the data of the analysis.

Table 2. EMS parameters of skarn samples under acoustic impact.

Sample	Strike	EMS peak-to-peak (V)	Averaged peak-to-peak (V)	Signal duration (ms)	Averaged signal duration (ms)	Amplitude EMS (V)	Averaged amplitude EMS (V)	Ultimate strength (kN)
4	1	0.0296	0.0284	11.5	13.05	0.018	0.0164	222.8
4	2	0.0272		14.6		0.0148		
5	1	0.0428	0.045	37.5	37.15	0.0232	0.0264	127.6
5	2	0.0472		36.8		0.0296		
6	1	0.0948	0.1002	22.9	30.1	0.0432	0.0388	125.2
6	2	0.1056		37.3		0.0344		
7	1	0.0736	0.072	21.8	22.8	0.0324	0.0314	110.6
7	2	0.0704		23.8		0.0304		

The analysis of the data presented in Table 2 shows that the averaged EMS amplitude and its duration decrease as the ultimate strength increases under acoustic impact on skarn samples by the dynamic method.

5. Conclusions

Real samples of the iron ore deposit have been tested. These were skarns as an enclosing rock and ore samples with magnetite. The iron ore deposit is useful because of ore extraction is performed through blasting. All works in mine are accompanied by various acoustic effects, and acoustic waves passing through inhomogeneities produce EMS. Samples with different ultimate strength were tested to reveal the dependence of EMS parameters on rock properties. This is an integral characteristic that depends on petrographical features. The dynamic method of acoustic impact has shown that as the ultimate strength of enclosing rocks increases, the average EMS amplitude and its duration decrease. However, this dependence is not characteristic of samples containing magnetite that can be attributed to its complicated structurally-textural features and quartz inclusions present in the volume.

Acknowledgements

Work is performed with financial support of a grant of the Russian Federal Property Fund No. 14-08-00395 and within the state task of the Ministry of Education and Science of the Russian Federation.

References

[1] Stepanov F W 1933 Uber den Mechnismus der plastichen Deformation *Zs.Phys.* **81** No 5 560–3
[2] Fischbach D B, Nowick A S 1955 Deformation – induced charge flow in NaCl crystals *Phys.Rew.* **99** No 4 333–5
[3] Urusovskaya A A 1958 Research of crystals by the etching method [in Russian] *Crystallography* **3** No 5 726–31
[4] Martyshev Yu N 1965 Research of the glow and electrolizing of LiF crystals at their deformation [in Russian] *Crystallography* **10** Issue 2 224–7
[5] Kornfeld M I 1971 Electrical charges on the surface of alkali-halide crystals [in Russian] *FTT* **13** Issue 2 474–9

[6] Gol'd R M, Markov G P, Mogila P G et al. 1975 Electromagnetic radiation of minerals and rocks exposed to the mechanical loading [in Russian] *J. of USSR Academy of Sciences. Physics of the Earth* **7** 71–4
[7] Nitsan U 1977 *Geophys. Res. Lett.* **4** 333–6
[8] Warwick J W, Stoker C and Meyer T R 1982 *J. Geophys. Res.*B **87** 2851–9
[9] Ogawa T, Oike K and Miura T 1985 *J. Geophys.Res.* **90** 6245–9
[10] Yamada I, Masuda H 1989 *Phys. Earth Planet.* Int.**57** 157–68
[11] O'Keefe S G and Thiel D V 1995 *Phys. Earth Planet. Inter.* **89** 127–35
[12] Mastov Sh R, Lasukov V V 1989 Theoretical model for the generation of an electromagnetic signal in the process of brittle fracture [in Russian] *J. of USSR Academy of Sciences. Physics of the Earth* **6** 38–48
[13] King C Y 1983 *Nature* **301** 377
[14] Hayakawa M. et. el 1993 *Phys. Earth Planet. Inter.* **77** 127–35
[15] Yavorovich L V, Gol'd R M, Yevseyev V D, Khorsov N N 2000 Investigation of distributions of electromagnetic-signal parameters during uniaxial compression of rocks *J. Min. Sci.* **36** No 6 536–40
[16] Kurlenya M, Vostretsov A, Kulakov G, Yakovitskaya G 2000 Recording and Processing of Electromagnetic Emission Signals of Rock [in Russian] *Izd. Sib. Otdel.RAN* Novosibirsk
[17] Vozoff and Frid 2001 *Symp. on Geological Hazards (Rafferety's Resort Newcastle, 15–16 November 2001*
[18] Sakai H, Oda H, Nakayama T and Doi H 1992 *J. Phys. Earth* **40** 447–58
[19] Tomizawa I and Yamada I 1995 *J. Geomagn. Geoelect.* **47** 313–24
[20] Vorobyov A A, Zavodovskaya E K, Salnikov V N 1975 Changes of electrical conductivity and radio emission of rocks and minerals [in Russian] *DAN* **220** No 1 82–5
[21] Carpinteri A, Lacidogna G, Niccolini G 2006 Critical behavior in concrete structures and damage localization by Acoustic Emission," *Key Eng. Mat.* **312** 305
[22] Carpinteri A, Lacidogna G, Manuello A, Niccolini G, Schiavi G, Agosto A 2012 Mechanical and electromagnetic emissions related to stress-induced cracks," *Experimental Techniques* **36** No 3 53
[23] Bespal'ko A A, Gol'd R M, Yavorovich L V, Datsko D I 2002 Influence of textural features of siltstone samples on the parameters of an electromagnetic signal under acoustic excitation [in Russian] *Physical and technical problems of mining* **2** 27–31
[24] Bespal'ko A A, Yavorovich L V, Gol'd R M, Datsko D I 2003 Excitation of electromagnetic radiation in the laminated rock under acoustic excitation [in Russian] *FTPRPI* **2** 8–14
[25] Bespal'ko A A, Gol'd R M, Yavorovich L V 2004 The impact of electrification of calcite on the parameters of electromagnetic signals under pulse acoustic excitation [in Russian] Physical mesomechanics, **7** No 5 95–9
[26] Matsuda T, Yamanaka C, and Ikeya M 2001 Behavior of stressinduced charges in cement containing quartz crystals *Phys. Stat. Sol. A* **2** 359
[27] Niccolini1 G, Xu J, Manuello A, Lacidogna G, Carpinteri A 2012 Onset time determination of acoustic and electromagnetic emission during rock fracture *Progress In Electromagnetics Research Letters* **35** 51–62
[28] Gordeev V F, Lasukov V V 2001 Physics of the electromagnetic emission method of quality control of the materials and its prospects *Russian physics journal* **44** No 7 771-8
[29] Bespal'ko A A, Yavorovich L V, Fedotov P I 2005 Communication parameters of electromagnetic signals with electrical characteristics of rocks under acoustic and quasi-static exposures [in Russian] *Izv. Tomsk Polytech. Univ.* **308** 7 18-23
[30] Kobayashi Hidetoshi, Keitaro Horikawa, Kinya Ogawa, KeikoWatanabe 2014 Impact compressive and bending behavior of rocks accompanied by electromagnetic phenomena *Phil. Trans. R. Soc. A* **372** 20130292

[31] Wan G X, Li X B, Hong L 2008 Piezoelectric responses of brittle rock mass containing quartz to static stress and exploding stress wave respectively *J. Central South Univ. Technol.* **15** 344–9

The foam-glass material for a radio frequency echoless chambers

O Kazmina[1], V Suslyaev[2], K Dorozhkin[2], V Kuznetsov[3], E Lebedeva[1]

[1]National Research Tomsk Polytechnic University, Tomsk, Russian Federation
[2]National Research Tomsk State University, Tomsk, Russian Federation
[3]Boreskov Institute of Catalysis, Novosibirsk, Russian Federation

E-mail: kazmina@tpu.ru

Abstract. The conducted experiment of foam glass modification by carbon nanotubes shows increased radio absorbing properties in comparison with a foam glass without additives. Addition of carbon nanodimensional tubes in number of 1,5 wt.% increases a tangent of dielectric losses angle by 2,5 times. The coefficient of electromagnetic radiation absorption in the range of frequencies of 120 - 260 GHz increases for a foam glass with carbon nanotubes (1,5 wt. %) twice in comparison with a foam glass without additives. The foam glass modified by carbon nanotubes is recommended as the effective radio absorbing material for the device of anechoic cameras. This material is fire safety, nonflammable, environmentally friendly, rather light-weight.

1. Introduction

Materials which interact efficiently with electromagnetic radiation (EMR) find wide application to solve problems of electromagnetic compatibility, to protect biological objects against harmful impact, to decrease radio visibility, to manufacture various elements of high frequency radio electronics [1, 2]. Facing of the inner surfaces of echoless chambers, which began to use widely in last decades, is one of direction of application of such materials. From one side it is connected with tendency of complication and increasing of accuracy of antenna measurements and panel tests of radio complexes, from other one it is connected with widespread use of the chamber in the technique for testing new wide band radio wave absorbing materials. A radio frequency echoless chamber is a closed electro magnetic screen which inner surfaces are coated completely with radio wave absorber. To install chambers one uses a porous pyramid shaped material. Such surface form widens frequency range, increases additionally absorbing capacity and decreases reflection of electro magnetic waves from the walls. One uses the material, which contains current conducting carbonaceous particles (soot and graphite) and magnetic components (ferrites), as material of a radio wave absorber. Such composition in combination with porous structure provides efficient absorbance of electromagnetic waves in wide frequency range.

The material must have the high absorption coefficient and be fire proof, durable, strong, have relatively low density, and be mounted and be processed easily. Foam glass [3] meets these

[1] To whom any correspondence should be addressed.

requirements, but it is inferior to traditional radio wave absorbers. Therefore, the study of foam glass modifications with various additives which increase its absorption capacity [4 – 6] has practical significance. An added modifier changes the physical and mechanical properties, size and shape of the foam glass pores. It is shown in [7] that the radio wave absorbing ability of foam glass can be increased with the addition of ilmenite. It is known that carbon nanotubes (CNT) having the complex of unique mechanical and heat-physical properties interact effectively with the electromagnetic radiation that allows to use then in radio wave absorbing composites [8]. The purpose of this work is to investigate the CNT impact on foam glass properties and its radio wave absorbing ability.

2. Raw materials and methods of investigations

Feedstock for production of foam glass is of lamp glass powder with specific surface area of 6000 cm^2/g, which composition is given in Table 1. This glass belongs to the group of silicate and contains together with the main components barium oxide, which is introduced into furnace – charge as barium sulfate that increases foaming ability of fluid glass. We used soot as gasifier and its amount in all samples was 0.5 wt. %. Multilayer carbon nanotubes of relatively high purity (Table 1) have been used as modifying additive.

CNTs were synthesized by catalytic CVD technique using ethylene decomposition at 680°C [9] Bimetallic Fe-Co/Al2O3 catalyst with Fe2Co alloy as active component [10] was used to produce CNT. The study was carried out using CNTs average diameter of 20-25 nm, длину около 1 мкм. Carbon nanotube oxidation was performed by refluxing of CNTs (1–2 g) in the excess of nitric acid for ~90 min [11]. Dimensions of tubes have been chosen to reduce the percolation threshold, after which conductivity and reflectance coefficients increase. For this reason, the amount of CNT introduced into the foam glass has been changed within small range from 0.5 to 1.5 wt. %.

Table 1. Chemistry of glass and CNT

material	Oxides and dirt content (wt.%)							
	SiO_2	Al_2O3	Fe_2O_3	CaO	MgO	Na_2O	K_2O	BaO
glass	71,9	0,60	0,10	5,50	3,20	16,10	0,80	1,80
CNT	0,0169 (Si)		0,0011 (Fe)		0,005 (Ca)	следы (Ti, Co, Cu, Al, Sn)		

The process of production of foam glass includes such technological operations as preparation of glass powder, mixing it with gasifier and modifying additive, foaming of the foaming mixture with following gradual cooling and processing of the finished material up to given sizes and shapes (Figure 1).

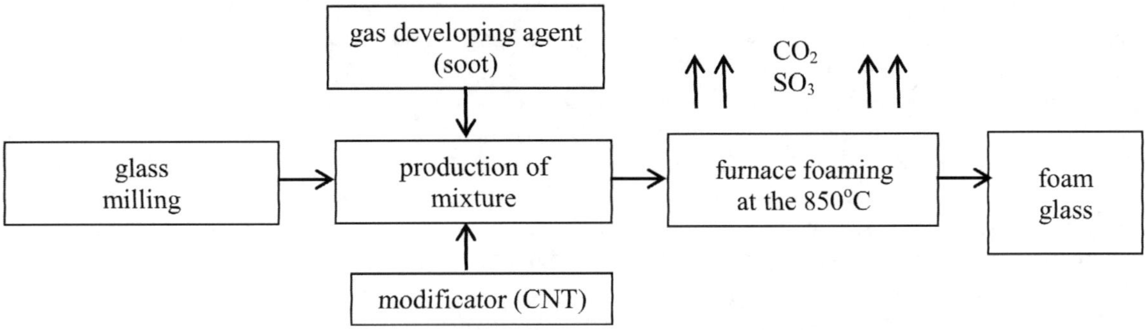

Figure 1. Technological scheme of foam glass producing

The electromagnetic characteristics were investigated using the equipment of the «Center of Radio Measurements at Tomsk State University». The reflection and transmission coefficients were measured by the method of free space using a radio spectroscope built around an E8363B Vector Network Analyzer (Agilent Technologies) in the range 26–36 GHz and an STD-21 teraherz spectrometer in the range 60–260 GHz [12, 13].

3. Experimental results

The necessary condition to obtain high quality foam glass with uniform fine pore structure is homogeneous foaming mixture. Given the difficulty of uniform distribution of CNT in the mixture the method of inserting of nanotubes in a dry form has been developed in the course of the experiment. For this purpose, previously the premix was prepared by mixing of the compo-nents in a defined sequence with addition of surfactant which improves wettability of particles. At that modifying additives which have been mixed in the rotary-vane-mixer, are sufficiently evenly distributed by powder volume. Further, the binder mixture was added in the form of a 15% solution of sodium silicate in amount of 5%. The mixture was stirred in the mixer that leads to appearance of small granule germs. This operation allows to obtain premix in the form of a granular powder, which is not dusty, is not foliated, is not prone to aggregation and is easy dosed. Prepared premix, containing the carbonaceous gasifier in full volume and CNT, is mixed with glass powder in a planetary mill for 1 minute. Foaming has been performed at maximum temperature of 800 ° C with an exposure of 15 minutes.

Radio-wave absorbing properties of the foam glass depend on the value of complex dielectric permittivity (DP) of the glass at the given frequency, system porosity and pore diameter. The absorbing material will correspond to the destination in the event if it has no reflection of electromagnetic waves from the outer surface, and the energy penetrating into the material, will be absorbed completely. Accomplishment of these conditions are achieved by appropriate choice of construction (for example: single-layer, multi-layer) and selection of electrical and magnetic properties of the material, owing to coordination with the free space and transformation of EMR into thermal energy is achieved. For better absorption of radiation it is preferable to use materials with higher values of the dissipation factor of a dielectric.

The measurements of the electromagnetic response of the samples to be studied showed that if concentration of modifying additives in foam glass is increased, losses are increased faster than real component of DP. The samples with 1.5 wt. % of CNT have the highest loss tangent values. It is established that introduction of carbon nanoscale tubes in amount of 1.5 wt. % increases dissipation factor of a dielectric by 2.5 times (Figs. 1 and 2).

Qualitative indicators of material are absorption, transmission and reflection coefficients of wave. The reflection coefficients for all samples practically do not differ and are near to zero. Table 2 shows the results of measurements of the absorption and transmission coefficients of electromagnetic radiation in the high frequency range from 120 to 260 GHz for samples of foam glass with the addition of the CNT and without it. The absorption coefficient of modified foam glass, on average, is twice as high as in comparison with foam glass without additives.

Density, mechanical strength, heat conductivity and water absorption are important exploitation characteristics of foam glass together with its electrical and physical properties. Values of these indicators are listed in Table 3 from which one can see that density and strength of material are slightly decreased at increase of inserted CNT. Physical and mechanical properties are directly connected with porous structure of the material. It is found that the pore size of modified foam glass is increased on average up to 2.5 mm in comparison with pore size of foam glass without additives (0.5-1 mm). Thus, differences in electrophysical properties of porous samples are caused by not only with changes in the material composition, but also its macrostructure.

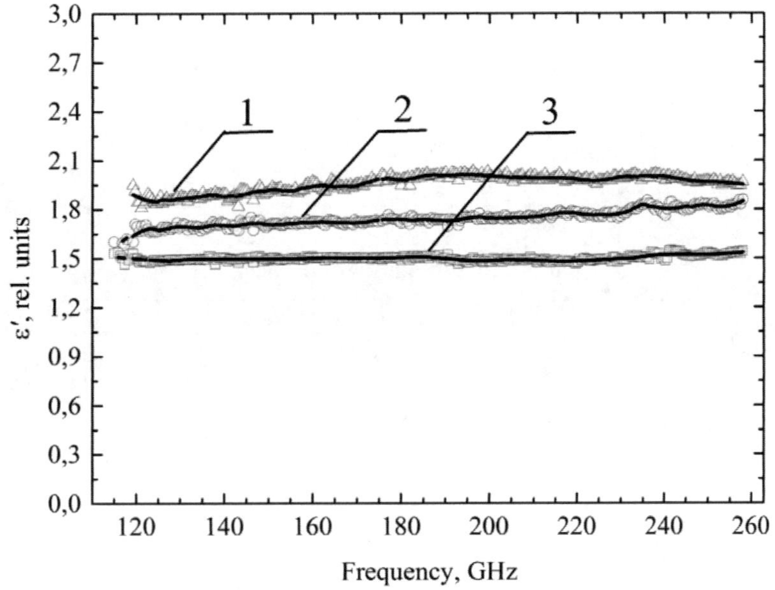

Figure 1. Output of the dielectric constant versus frequency for foam glass containing carbon nanotubes(weight %): 1 – 1,5; 2 – 1; 3 – 0,5

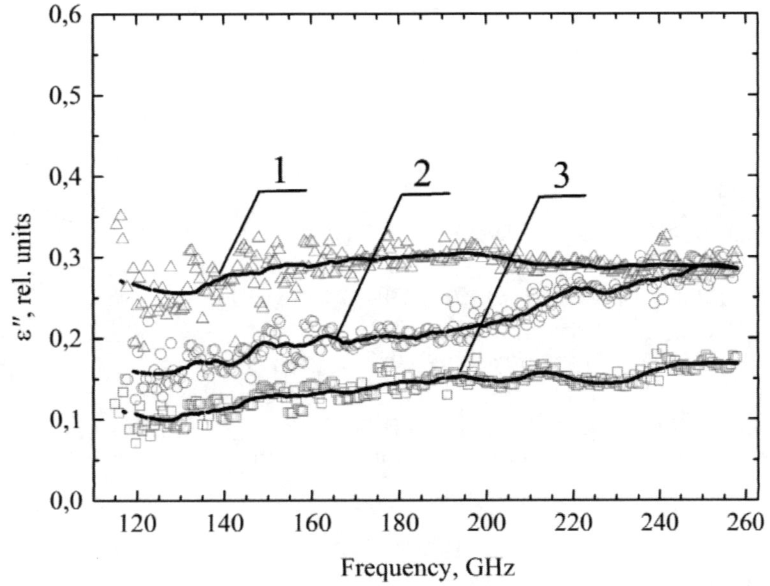

Figure 2. Output of the dielectric constant versus frequency for foam glass containing carbon nanotubes (weight %): 1 – 1,5; 2 – 1; 3 – 0,5.

Table 2. Transmission and adsorption coefficients

number of CNT, (wt. %)	coefficients at frequency, dB			
	transmission		adsorption	
	120 GHz	260 GHz	120 GHz	260 GHz
0	-4,0	-11,2	-2,2	-0,15
0,5	-7,0	-24,0	-2,2	-0,1
1,0	-6,9	-20,5	-2,3	-0,3
1,5	-10,5	-24,3	-0,9	-0,3

Table 3. Foam glass properties

number of CNT, (wt. %)	property values			
	density, kg/m^3	water adsorption, wt.%	durability, MPa	thermal conductivity, W/m·K
0	175	1,2	1,2	0,073
0,5	163	1,4	1,1	0,070
1,0	156	1,6	1,1	0,068
1,5	151	1,5	1,0	0,065

4. Conclusions

The effect of CNT on the physical and mechanical and radio wave absorbing properties of foam glass has been revealed. Introduction of CNT nanotubes in amount of 1.5 wt. % increases imaginary component of dielectric permittivity of foam glass by five times and real component by two times. The absorption coefficient of electromagnetic radiation in the frequency range of 120 - 260 GHz is increased in average for foam glass with carbon nanotube (1.5 wt. %) by two times in comparison with foam glass without additives. Foam glass, modified with carbon nano-tubes, is recommended as the effective radio absorbing material to arrange echoless chambers. At that the material meets the requirements of fire safety, is non-flammable, environmentally friendly and relatively easy (density in the range of 180 kg/m^3).

References

[1] Qin F and Brosseau C 2012 *J. Appl.Phys.* 111 061301
[2] Park Ki-Yeon et al 2006 *J. Compos. Sci. Technol.* 66 576–584
[3] Kuo Chen et al 2011 *J. Mat.Sci. Eng.* B 176 1239–1242
[4] Xiuhua Li et al 2012 *J. of Non-Cryst. Solids* 358 2917–2921
[5] Ilbeom Choi et al 2012 *J. Compos. Struct.* 94 3002–3008
[6] Yongbao Feng et al 2012 *J. Magn. Magn. Mater.* 324 3034–3039
[7] Kazmina O et al 2014 24th International Crimean Conference Microwave & Telecommunication Technology Conference Proceedings September 7–13, Crimea, Russia 920–921
[8] Limeng Chen et al 2010 *J. Polymer* 51 2368–2375
[9] Usoltseva A et al 2007 *J. Phys. Stat. Sol.* B 244 3920
[10] Kuznetsov V et al 2012 *J. Phys.Stat. Sol.* B 249 2390
[11] Mazov I et al 2012 J. Appl. Surf. Sci. 258 6272–6280
[12] Suslyaev V et al 2013 *Russian Physics Journal* 55 970–975
[13] Suslyaev V et al 2014 *Adv.Mater. Research* 1040 137–141

IOP Conf. Series: Materials Science and Engineering **110** (2016) 012087 doi:10.1088/1757-899X/110/1/012087

Development of data optimization methodology for non-destructive testing of concrete strength by the parameters of the electric response to impact excitation

T V Fursa[1], A P Surzhikov[1] and M V Petrov[1]

[1]National Research Tomsk Polytechnic University, 634050, Tomsk, 30 Lenin Avenue, Russia

E-mail: fursa@tpu.ru

Abstract. The paper presents the research results by the improvement of the non-destructive testing method of concrete strength by the parameters of the electric response to impact excitation. The electric response parameters from the set of identical concrete samples sized of 100x100x100 mm were studied. It is shown that the use of linear filtering procedure reduces the variance of diagnostic electric parameter for concrete strength determination and is in a good agreement with the elastic characteristics of the material.

1. Introduction

Concrete is widely used in the power structures, therefore the control of its strength is of practical importance. For these purposes, the existing methods for the concrete strength control are improved [1, 2] and new methods are developed [3-7]. Most of these methods has some disadvantages and doesn't find practical application.

The method based on the phenomenon of mechanoelectric transformations is developed [8, 9] to solve the problem of the concrete strength determination. This method is based on the fact that under mechanical impact excitation of the heterogeneous non-metallic materials, an electric response appears. Electric response arises when the elastic wave induced by the impact excitation deforms and displaces the signal sources in a heterogeneous material. Electrical double layer and piezoelectric inclusions on the border of components are the main sources of mechanoelectric transformations in concrete. These inclusions are contained in the concrete in the form of sand or coarse aggregate. Investigations have shown that the piezoelectric inclusions play the main role in the mechanoelectric transformations in concrete [10]. Piezoquartz is contained in sand, which is used for the manufacture of cement-sand samples. X-ray diffraction analysis showed that sand consists of 80-90% quartz. X-ray diffraction analysis showed that in the gravel, which is used for the manufacture of heavy concrete, grains containing no quartz aren't detected, while the proportion of grains, which consist only of quartz is about 20%. The charge is forming on the edge of piezoelectric quartz under the influence of mechanical stresses caused by acoustic waves due to the direct piezoelectric effect. The consequence of this is the emergence of an external electric field, which is recorded by signal receiver located near the object of investigation. Thus, the parameters of the electric response relate with characteristics of elastic waves within the material and can be used for their contactless measurement.

IOP Conf. Series: Materials Science and Engineering **110** (2016) 012087 doi:10.1088/1757-899X/110/1/012087

The algorithm for determination of the concrete samples strength sized of 50x50x100 mm, based on the use of the generalized parameter of the electric response is developed [11]. Generalized parameter (P) defined by:

$$P = \frac{S}{r_{max}} \quad (1)$$

where, r_{max} is the maximum correlation coefficient of the spectrum of the electrical signal from the test sample with the spectrum of the standard, S is a frequency offset value of the spectrum of the electrical signal from the test sample relative to a reference spectrum in which there is a maximum cross-correlation coefficient, s^{-1}.

As a mathematical standard of electrical signal from the defect-free sample of this form a spectrum of freely damped harmonic oscillations that occur under the law is used:

$$y(t) = Ae^{-\beta t}(\sin 2\pi f t) \quad (2)$$

where, A is the oscillation amplitude, β is the damping coefficient; f is the frequency of oscillations.

The concrete sample with size 50x50x100 mm has the bar shape. Therefore, when the impact excitation occurs along the length of concrete sample the longitudinal acoustic oscillations are formed along the length way of the sample.

In this paper investigated adaptation of proposed algorithm of non-destructive testing on standard cube samples, which are used in manufacturing for evaluation of strength characteristics of concrete products.

2. Methods of investigations

Investigations were performed using the hardware-software complex allowing to produce impact excitations on concrete sample and record electric responses. The figure 1 shows the image of the measuring system:

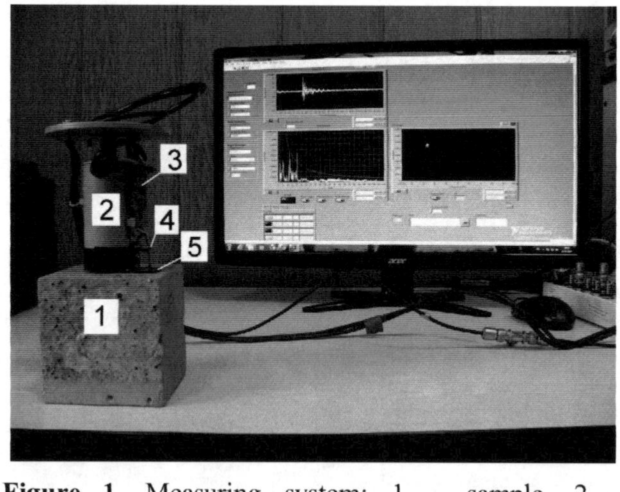

Figure 1. Measuring system: 1 – sample, 2 – electromechanical impact device, 3 – differential amplifier, 4 – compensating electrode, 5 – measuring electrode.

Pulse mechanical excitation of the samples is performed using electromechanical impactor with normalized force of the impact. Impact excitation is performed using device based on electromagnet. Differential electric sensor is used for recording electric response. The receiver of the electric sensor consists of two metal plates size 20x20 mm. One plate is measuring receiver, and the second plate is compensating receiver. The electric measuring receiver is located at a distance of 2 mm from the surface, thus creating the conditions for non-contact reception. The compensating receiver is located at a height of 30 mm from the measuring receiver. The measuring receiver treats both a useful signal from the sample and a signal of external noise. The compensating receiver is distant from the source of useful signal and practically only accept external noise. This measurement scheme allows to improve significantly the signal-to-noise ratio. Receivers are fixed on the body of the electromechanical impactor device through the acoustic isolation. As a result, the distance from the electric receiver to the point of impact is always the same (20 mm).

Electric signals from the sensor are recorded using the I/O board "NI PCI-6251", combined with PC. More detailed method of measurement is presented in the article [12].

To measure the speed of sound ultrasonic pulse generator/receiver OLYMPUS PANAMETRICS 5058PR and piezoelectric sensors OLYMPUS PANAMETRICS V1011 VIDEOSCAN were used.

3. Experimental investigations and discussion

In this work a batch of heavyweight concrete samples with size 100x100x100 mm, consisting of 16 samples was investigated. Samples were prepared in accordance with GOST 10180-2012 with the factory metal molds. The weight ratio of cement, sand and coarse aggregate was 1:2:4. Water-cement ratio equals 0.4. The maximum size of coarse aggregate was 20 mm. The river gravel was used as of coarse aggregate.

Figure 2 shows the spectra of the electric responses of the heavy concrete samples of various shapes.

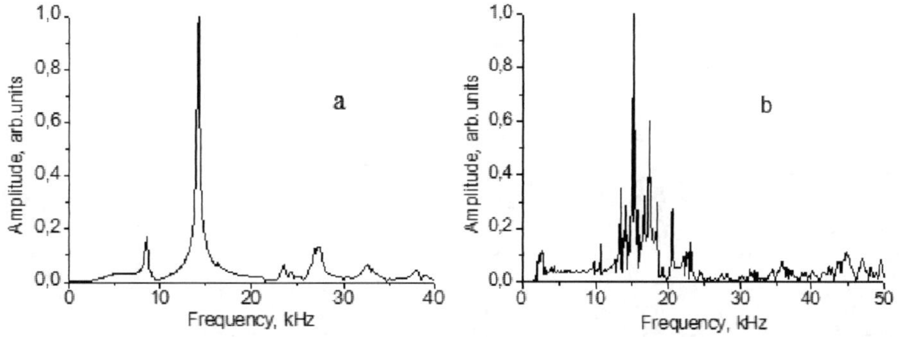

Figure 2. Spectra of electric responses of concrete samples size 50x50x100 mm (a) and 100x100x100 mm (b).

As shown in figure 2 the spectrum of the electric response from the sample in the form of a bar has almost one dominant peak and the spectrum of the signal from the sample of cubic shape are more complicated.

. Along with having a dominant peak a considerable amount of significant spectral peaks located at either side of the main peak is observed. This is due to the formation of more complex wave processes in a cubic sample.

Investigation showed that even in the samples having the same composition and made from the same materials there are significant differences in the ratio of the spectral peaks located near the dominant peak.

Using the program developed in the programming environment LabView, the generalized parameter according to the formula (1) was calculated.

Figure 3 shows the results of the calculation of the generalized parameter and averaging the data of multiple tests.

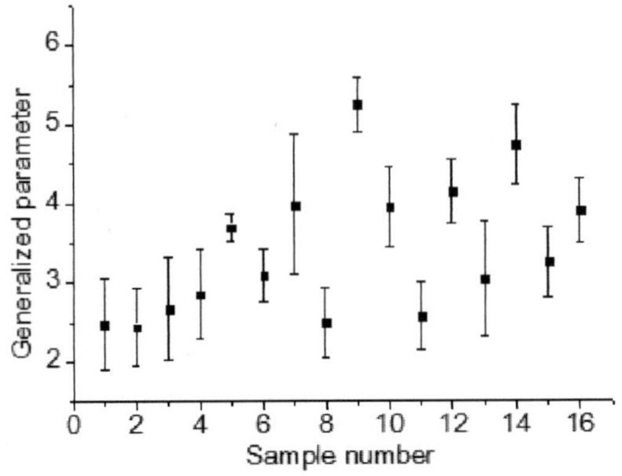

Figure 3. Variation of the generalized parameter when using the algorithm described above.

The figure 3 shows a significant spread of values in multiple tests of the same sample and between samples. To reduce the spread of values, it was proposed to produce linear filtering of spectra in the mode of moving average. The program for the calculation of the generalized parameter was supplemented this procedure. Figure 4 shows a front panel of the program.

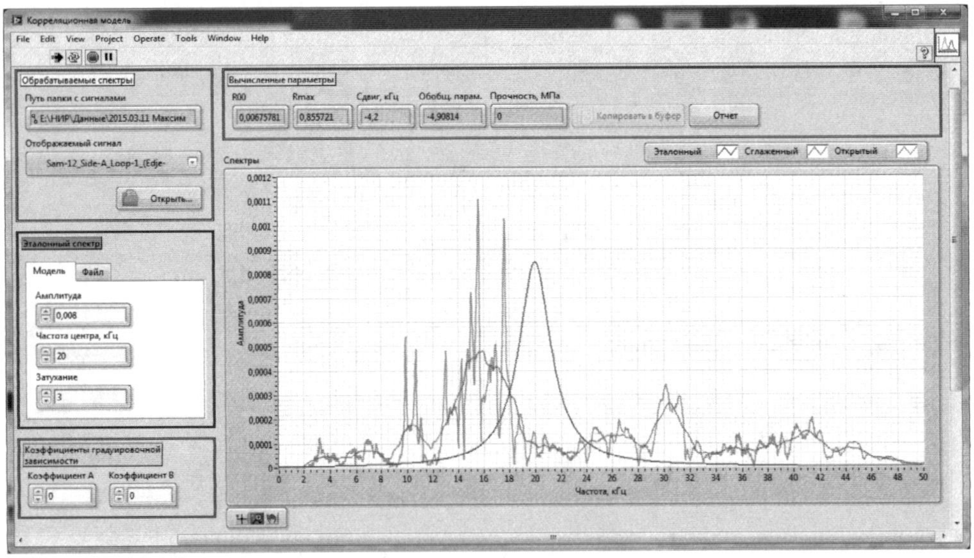

Figure 4. Front panel of program for the calculation of the generalized parameter.

The program allows specifying the necessary filtering options, depending on the specific experimental data, and changing the standard spectrum (amplitude and attenuation coefficient). The results of the calculation are written to files for further processing and analysis.

Figure 5 shows the results of the calculation of the generalized parameter when comparing the standard spectrum with a smoothed spectrum.

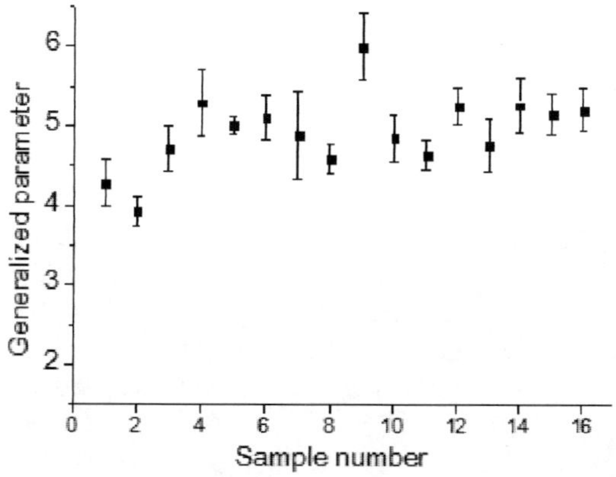

Figure 5. Variation of the generalized parameter when using the linear filtering procedure.

Investigations have shown that the use of linear filtering procedures can significantly reduce the spread of the values of the diagnostic parameter for identical samples.

If there are no defects in the samples, the strength is proportional to the Young's modulus of concrete. To evaluate the modulus of variations in the samples sound velocity and density of the samples were measured. Calculation of Young's modulus was produced by the formula:

$$E = \frac{c_p{}^2 * (1 + \vartheta) * (1 - 2\vartheta) * \rho}{1 - \vartheta} \tag{3}$$

where, E is Young's modulus; c_ρ is velocity of longitudinal wave; ϑ is Poisson's ratio.

Table 1 shows the values of sound speed, density and calculated Young's modulus.

Table 1. The calculated Young's modulus.

Sample number	Speed of sound [10^3 m/s]	Density [10^3 kg/m^2]	Young's modulus [GPa]
1	4.13	2.36	36.30
2	4.27	2.31	37.91
3	4.10	2.28	34.48
4	4.17	2.37	37.02
5	4.13	2.33	35.75
6	4.27	2.36	38.83
7	4.17	2.34	36.58
8	4.24	2.31	37.36
9	4.13	2.25	34.55
10	4.07	2.35	34.95
11	4.17	2.27	35.47
12	4.20	2.31	36.69
13	4.10	2.29	34.63
14	4.07	2.29	34.03
15	4.07	2.32	34.52
16	4.13	2.33	35.81

Analysis of the data in the table shows that the Young's modulus in the batch of samples has similar values. The maximum value of the Young's modulus is equal to 38.83 GPa and minimum is equals 34.03 GPa. The average spread of the values is order of 1%. Such close values of Young's modulus of the samples indicate the identity of samples in the batch. Spread of the values of the generalized parameter without the use of linear filtering procedures for these samples is about 6%. Linear filtering procedure leads to the fact that variations in the generalized parameter become equal to 1.7%. Consequently, the use of linear filtering procedure allows normalizing generalized parameter relatively elastic properties of concrete.

4. Conclusions

The purpose of the investigations presented in this paper was the development of methodology to improve reliability of non-destructive testing of concrete strength, based on the phenomenon of mechanoelectric transformations. The paper proposed to produce linear filtering of spectra of electrical responses in the mode of moving average. It is shown that the use of linear filtering procedures can significantly reduce the spread of the diagnostic parameter values for identical concrete samples.

5. Acknowledgements

The work was supported with the Ministry of Science and Education of the Russian Federation (state assignment "Science").

References

[1] Carino N J 1994 Nondestructive testing of concrete: history and challenges *Concrete Technology – Past, Present and Future* ed P K Mehta (Detroit, MI) ACI SP-144 pp 623–78

[2] Gudra T and Stawiski B 2000 Non-destructive strength characterization of concrete using surface waves *Ndt&E International* **33** 1–6

[3] Shah A A, Alsayed S H, Abbas H and Al-Salloum Y A 2012 Predicting residual strength of non-linear ultrasonically evaluated damaged concrete using artificial neural network *Construction and Building Materials* **29** 42–50

[4] Taillet E, Lataste J F, Rivard P and Denis A 2014 Non-destructive evaluation of cracks in massive concrete using normal dc resistivity logging *NDT & E International* **63** 11–20

[5] Chekroun M, Le Marrec L, Abraham O, Durand O and Villain G 2009 Analysis of coherent surface wave dispersion and attenuation for non-destructive testing of concrete *Ultrasonics* **49** 743–51

[6] Abraham O, Piwakowski B, Villain G and Durand O 2012 Non-contact, automated surface wave measurements for the mechanical characterisation of concrete *Construction and Building Materials* **37** 904–15

[7] Song G, Gu H and Mo Y L 2008 Smart aggregates: multifunctional sensors for concrete structures – a tutorial and a review *Smart Materials and Structures* **17** (No. 3) 1–17

[8] Fursa T V, Lyukshin B A and Utsyn G E 2013 Relation between the electric response and the characteristics of elastic waves under shock excitation of heterogeneous dielectric materials with piezoelectric inclusions *Technical Physics* **58** 263–6

[9] Fursa T V, Surzhikov A P and Osipov K Yu 2007 Development of an acoustoelectric method for determining the porosity of dielectric materials *Russian Journal of Nondestructive Testing* **43** 95–9

[10] Fursa T V and Dann D D 2011 Mechanoelectrical transformations in heterogeneous materials with piezoelectric inclusions *Technical Physics* **56** 1112–7

[11] Fursa T V, Osipov K Yu and Dann D D 2011 Development of a nondestructive method for testing the strength of concrete with a faulted structure based on the phenomenon of mechanoelectric transformations *Russian Journal of Nondestructive Testing* **47** 323–8

[12] Fursa T V, Osipov K Yu, Lyukshin B A and Utsyn G E 2014 The development of a method for crack-depth estimation in concrete by the electric response parameters to pulse mechanical excitation *Meas. Sci. Technol.* **25** (No. 5) 10

IOP Conf. Series: Materials Science and Engineering **110** (2016) 012088 doi:10.1088/1757-899X/110/1/012088

Influence of cyclic freeze-thaw on the parameters of the electric response to the pulse mechanical excitation of concrete reinforced by glass fibre reinforced polymer bars

T V Fursa[1], M V Petrov[1] and I N Korzenok[1]

[1]National Research Tomsk Polytechnic University, 634050, Tomsk, 30 Lenin Avenue, Russia

E-mail: fursa@tpu.ru

Abstract. Studies of the influence of cyclic freeze-thaw on the parameters of electric response from samples of concrete reinforced by glass fibre reinforced polymer (GFRP) bars were conducted. It is found that an increase in the number of freeze-thaw cycles increases the attenuation coefficient of energy of electric responses and moves the centre of gravity of spectrum to the low-frequency area. The results can be used to develop a method of non-destructive testing of reinforced concrete.

1. Introduction

In the operation of structures of reinforced concrete in the natural conditions under the influence of seasonal temperature and humidity a gradual destruction is occurring. Concrete damage caused by freeze-thaw is a major problem in cold climates. Concrete reinforced by GFRP bars is increasingly used as an alternative to steel reinforced concrete structures.

Currently, there are no enough simple and reliable methods for determining damages in the construction of reinforced concrete. Therefore, it's necessary to develop methods of control to ensure the safe operation of engineering structures and buildings. Currently, acoustic methods [1-4] are mostly used to control defects in concrete. To solve this problem the non-destructive testing method, based on the phenomenon of mechanoelectric transformations under the pulsed mechanical excitation of heterogeneous non-metallic materials [5] may be used.

This paper is devoted to study changes in the parameters of the electric response of concrete reinforced by GFRP bars under conditions of cyclic freeze-thaw.

2. Experimental procedures

Experimental studies were carried out using the hardware and software complex which allows producing a single normalized by force impact on the object of study and recording the electrical response resulting from the impact.

Figure 1 shows a photo demonstrating the process of measurement of electrical response from the concrete sample.

Figure 1. The measuring system photo: 1 – sample; 2 – electromechanical impact device; 3 – measuring receiver; 4 – compensating receiver; 5 – differential amplifier; 6 – NI board terminal block; 7 – power supply.

The technique of measurement of the electric response to mechanical excitation is described in more detail in [6].

For the processing of received data used the program Origin and special programs, developed in environment LabView.

3. Experimental procedures

For this research the laboratory models of concrete, reinforced by GFRP bars were made. The models were represented by the heavy concrete samples sized of 100x100x100 mm. One GFRP bar with diameter of 10 mm and length of 150 mm was placed in each sample during its forming.

Studies were conducted on the batch of models of reinforced concrete, consisting of 12 samples. Before performing the climatic test the samples were soaked in water. There were performed 8, 14 and 20 cycles of freeze-thaw. Freezing was carried out in a climatic chamber at a temperature of minus 400 °C, and thawing in a universal test chamber at a temperature of (20 ± 5) °C and 95% humidity. Three samples were not subjected to climatic test and stored in room conditions for a comparative analysis.

After the climatic tests, the electric response to impact excitation was measured. Figure 2 shows the electrical responses from the samples after different cycles of freeze-thaw.

IOP Conf. Series: Materials Science and Engineering **110** (2016) 012088 doi:10.1088/1757-899X/110/1/012088

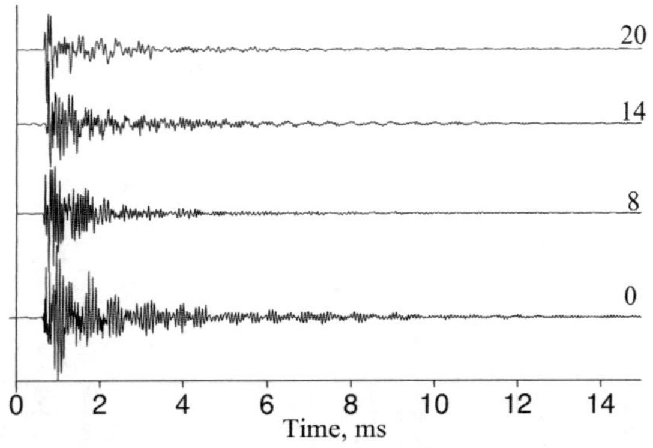

Figure 2. Typical electric responses from reinforced concrete samples that have passed different number of cycles of freeze-thaw (numerals indicate the number of cycles).

As shown in Figure 2 with increasing the number of cycles of freeze-thaw, a change of character of attenuation of electrical responses is observed. The attenuation coefficients of energy of the electric response from samples that have passed different number of cycles of freeze-thaw were calculated using the time-frequency analysis [7-8]. Figure 3 is a graph showing the variation of attenuation coefficient in the process of climatic tests.

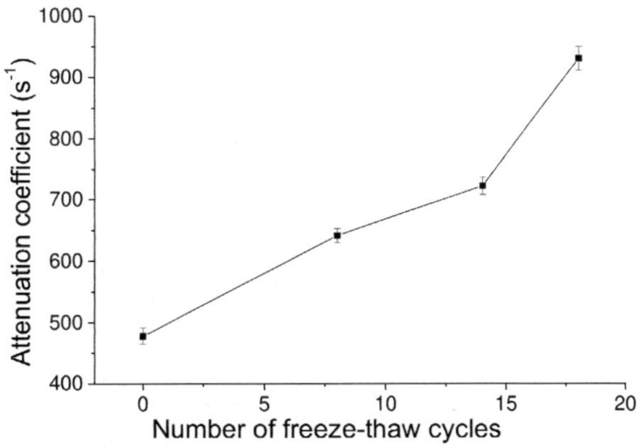

Figure 3. The dependence of the attenuation coefficient of energy of electric response on the number of cycles of freeze-thaw.

As seen in Figure 3 with an increase the number of cycles of freeze-thaw consistent increase in the attenuation coefficient of energy of electric response is observed. Figure 4 shows the spectra of the electric signals detected from the samples subjected to the climate testing.

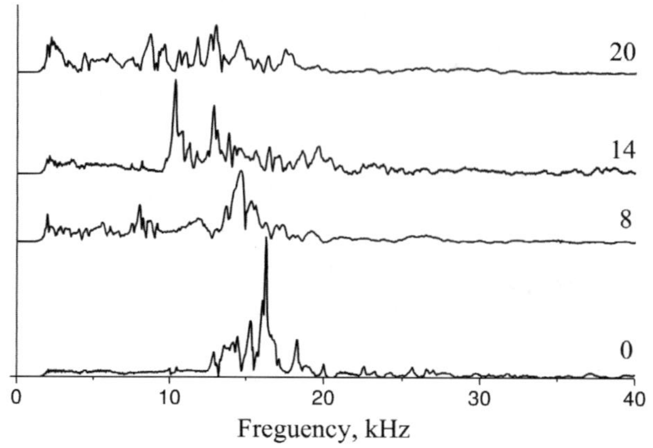

Figure 4. Typical frequency response of electrical responses from the samples of reinforced concrete, subjected to different number of cycles of freeze-thaw (numerals indicate the number of cycles).

As can be seen in figure with an increase the number of freeze-thaw cycles the spectral composition of electrical responses is changed and the spectrum is moved to the low-frequency region. Frequencies of the center of gravity of the spectrum after a different number of freeze-thaw cycles were determined using the special program developed in environment LabView.

Figure 5 is a graph showing the nature of the displacement of signal spectra from the reinforced concrete under cyclic freeze-thaw conditions.

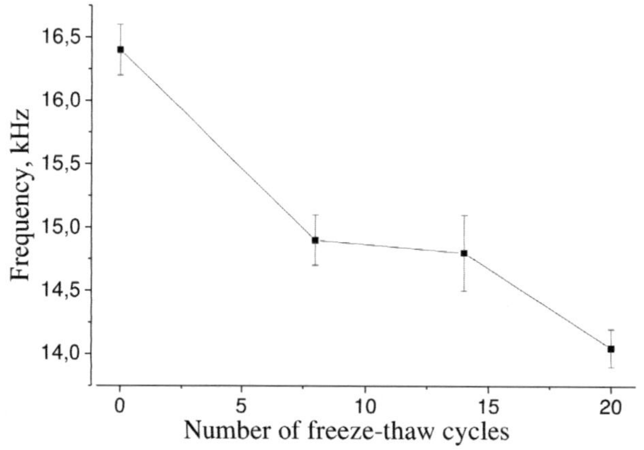

Figure 5. The dependence of frequency of center of gravity of the electric signals on the number of freeze-thaw cycles.

As shown in Figure 5 an increase the number of freeze-thaw cycles leads to moving of the center of gravity of the spectrum of electrical response to the low frequency region.

4. Conclusions

The purpose of the present work was to study the parameters of the electric response from concrete reinforced by GFRP bars under the cyclic freeze-thaw conditions. Visual observations showed that

IOP Conf. Series: Materials Science and Engineering **110** (2016) 012088 doi:10.1088/1757-899X/110/1/012088

appearance of surface cracks is not observed up to 20 cycles during the climatic test of reinforced concrete.

Absence of surface cracks indicates that the change in the characteristics of electric signals during the freeze-thaw is associated with the internal structural damages, including the contact failure of GFRP bar with concrete.

Studies have shown that in the process of increasing the number of freeze-thaw cycles the parameters of electric response are changed in time and frequency domain. It was found that with increasing number of freeze-thaw cycles an increase in the attenuation coefficient of energy of electric response and spectrum of signals moves to the low frequency region.

Therefore, the parameters of electrical response to impact excitation are sensitive to changes in the quality of the contact of the GFRP bar with concrete and it can be used for the development of nondestructive testing of reinforced materials.

The work is done under the State task “Science”.

References

[1] Aggelis D G, Mpalaskas A C and Matikas T E 2013 Investigation of different fracture modes in cement-based materials by acoustic emission *Cement and Concrete Research*. **48** 1–8

[2] Liu Pei-Ling and Yeh Po-Liang 2011 Spectral tomography of concrete structures based on impact echo depth spectra *NDT & E International* **44** (Is. 8) 692–702

[3] Antonaci P, Bruno C L E, Gliozzi A S and Scalerandi M 2010 Monitoring evolution of compressive damage in concrete with linear and nonlinear ultrasonic methods *Cem. Concr. Res.* **40** 1106–13

[4] Bui D, Kodjo S A, Rivard P and Fournier B 2013 Evaluation of concrete distributed cracks by ultrasonic travel time shift under an external mechanical perturbation: study of indirect and semi-direct transmission configurations *J. Nondestruct. Eval.* **32** 25–36

[5] Fursa T V, Surzhikov A P and Osipov K Yu 2007 Development of an acoustoelectric method for determining the porosity of dielectric materials *Russian Journal of Nondestructive Testing* **43** (No. 2) 95–9

[6] Fursa T V, Osipov K Yu, Lyukshin B A and Utsyn G E 2014 The development of a method for crack-depth estimation in concrete by the electric response parameters to pulse mechanical excitation *Meas. Sci. Technol.* **25** 10

[7] Quiviger A, Payan C, Chaix J F, Garnier V and Salin J 2012 Effect of the presence and size of a real macro-crack on diffuse ultrasound in concrete *NDT&E International* **45** 128–32

[8] Osipov K Yu and Fursa T V 2013 Evaluating the depth of open cracks in concrete from parameters of electric response to elastic impact excitation *Technical Physics Letters* **39** 481–3

Dynamic processes happening during the evaporation of films of fusible materials

L N Orlikov[1], N L Orlikov[2], S A Ghyngazov[2] and S M Shandarov[1]

[1]The Chair of Electronic Devices, Tomsk State University of Control Systems and Radioelectronics, 40 Lenin Ave, 634050, Tomsk, Russia
[2]Tomsk Polytechnic University, 30 Lenin Ave, 634050, Tomsk, Russia

E-mail: oln4@yandex.ru

Abstract. The formation of optically-transparent films of fusible materials like zinc, lead, bismuth on glass substrates to obtain single-mode optical diffusive waveguides is considered. The quality of films for waveguides depends on the amount of oxygen in residual environment, the gas exchange rate in the working chamber, the rate of oxygen desorption from the heated elements in vacuum chamber, the evaporator and the substrate. The regularities of variations of local layer thickness and the procedure phase diagram from the local gas exchange rate are determined.

1. Introduction

Optical waveguides on glass substrates are a promising area in their application in simple and cheap optoelectronic devices. As shown in [1], the highest refractive index is achieved during the formation of waveguides by oxidized film diffusion. However, realization of a number of electro-optical effects is restrained by probabilistic repeatability of wave guiding layers which holds down the development of optoelectronics [1-3]. This happens due to the fact that film formation in gas exchange mode isn't explored enough. One of the reasons of probabilistic repeatability of local thickness and film composition is dynamic processes which happen during the material evaporation. The regularities of evaporation, which were earlier found by Knudsen, Langmuir and other scientists for point sources, fail when it comes to the line where one material escape into another state. Most materials, which have three states – solid, liquid, gaseous – at ambient pressure heating, in vacuum, lose their liquid state partly or completely. Moreover, the film distribution over the substrate is quite unclear because of the poor study of molecular vapor flow and substrate interaction.

2. Methodology

One of the ways to solve the stated problem is to analyze the structure of molecular flow and to apply the regularities of vapor flow and material-substrate interaction [4-8].

3. Simulation of vapor flow formation

We know several models of gas and vapor formation.

Knudsen model (fig. 1a). Knudsen flow suggests spherical free molecule flow from the point source without taking into account the interaction with the substrate. Herewith, it is the realization of the cosine law of film distribution over the substrate with mass transfer which is inversely proportional to the squared distance to the substrate. Some works [2, 5] introduce evaporation and condensation probability functions.

Hydrodynamic model (fig. 1b, c). The 1990s witnessed the development of gas and vapor formation model in the approximation of Navier–Stokes equations [6-8]. Below are some of its statements.

We know that the evaporation temperature is the temperature at which vapor pressure tends to 1 Pa. During the formation of films, the pressure in vacuum chamber is held at the value of about 0,01 Pa and lower to provide the molecule transition to the substrate and minimal interactions of residual gas with metal vapor. As a result, the vapor flow is overlapped by pressure difference by the 2-3 orders and more. The flow shape, temperature, pressure, extension angle and flow rate are defined by the functions [6] which depend on the ratio of evaporation pressure (about 1 Pa) to ambient pressure (~0,1-0,01 Pa). The flow effusion can change from the shape of torch to the shape of sphere.

Figure 1a shows the conditional distribution of complete film coating thickness in accordance with Knudsen model and Hydrodynamic model (fig. 1b, c). Figure 2 shows the functions of gas flow temperature and pressure distribution [6].

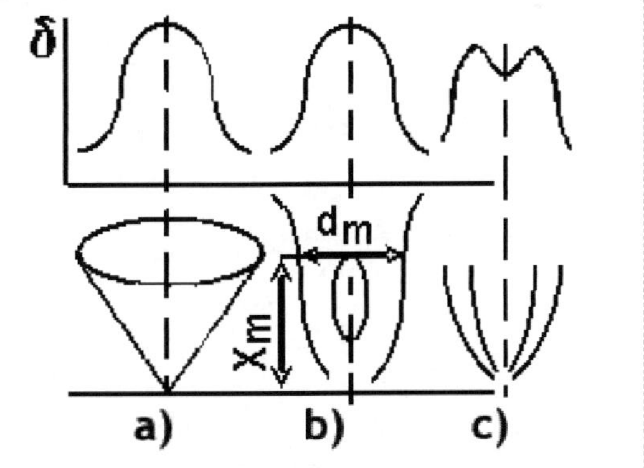

Figure 1. Film distribution over the substrate: a) according to Knudsen model; b), c) according to hydrodynamic model.

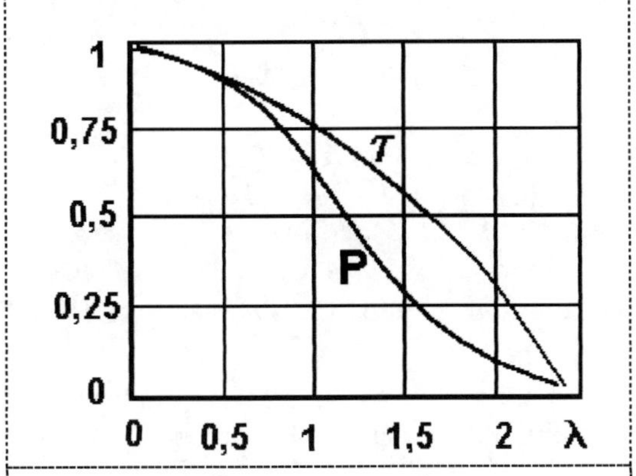

Figure 2. Flow parameters function.

The main characteristic of the flow is the rate coefficient $\lambda = V/V'$, which is equal to ratio of flow particle velocity to the effusion cutoff velocity. When $\lambda=0$, the flow is static; when $\lambda=1$, the flow achieves the sound velocity; when $\lambda>1$, the speed is supersonic. The speed of temperature change is defined by the equation:

$$\tau(\lambda) = T/T^* = [1-(k-1)/(k+1)\lambda^2] \tag{1}$$

The equation of the full pressure change is:

$$\pi(\lambda) = P/P^* = [1-(k-1)/(k+1)\lambda^2]^{k/(k-1)}, \tag{2}$$

where k=1,3 is adiabatic exponent of hot vapor.

Figure 3 shows Neller diagram [9], which demonstrates that at small densities the flow is formed in accordance with molecular effusion model.

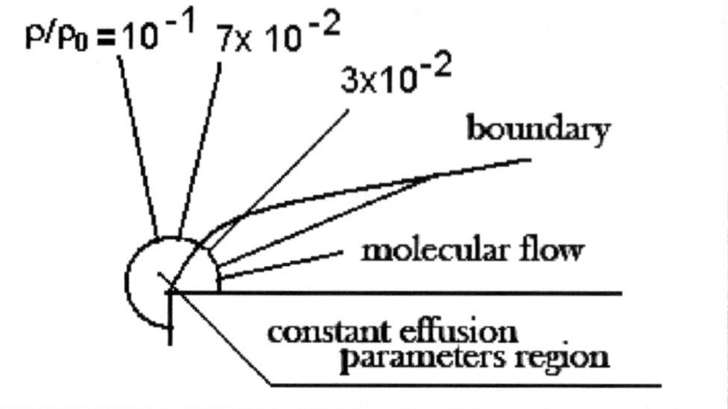

Figure 3. Neller diagram.

4. Experimental studies

Visualization of lead vapor flow in magnetron discharge plasma [10] showed that the vapor flow from the crucible to substrate gets the shape of sphere at the chamber pressures (0,1 - 0,01 Pa). When the chamber pressure is more than 0,1 Pa the flow gets the shape of several torches (cycles). It happens in accordance with Neller diagram. The transversal and longitudinal dimensions of the first expansion cycle (dm и Xm), after which the film thickness variation is less than 2%, follow the laws of gas flow dynamics [6] and depend on the ratio of evaporation pressure (PJ ~ 1 Pa) to the pressure Pk in vacuum chamber (P_j / P_k).

$$X_m = (0,7-1,34)d\left(kP_j/P_k\right)^{0,5} \tag{3}$$

$$d_m = d(kP_j/P_k)^{0,5} \tag{4}$$

where d is gas flow diameter at the crucible; P_J is flow cutoff pressure.

Figure 4 shows the diagram [11] which explains the tendency of local lead film thickness variations for different substrate lengths L, different ratios (P_j / P_k) and different distances h to the substrate according to laser measurements. Flow expansion starts at the point where the flow sooner reaches the vacuum. If the crucible is mitered, there appears the pressure gradient deflecting vapor flow. Figure 5 shows lead film thickness variation for different crucible miters.

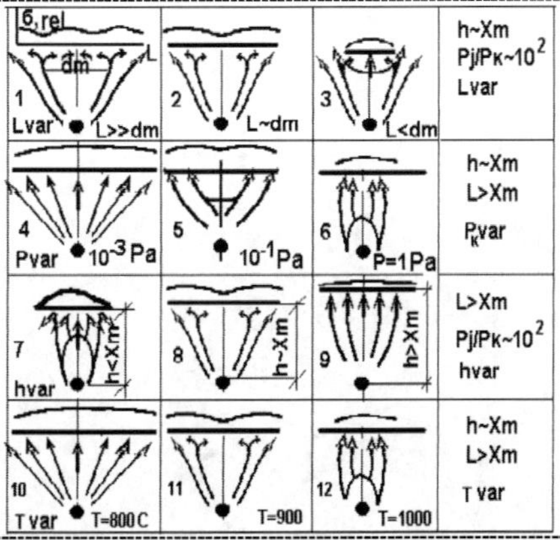

Figure 4. Vapor flow and substrate interaction diagram.

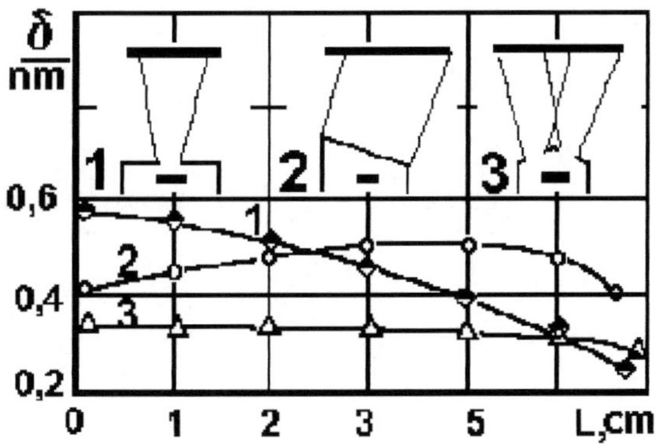

Figure 5. Film thickness distribution over the substrate for different miters.

4.1. The influence of substrate. Due to the holders, the picture of substrate L flow-around (1, 2, 3) even of small dimensions corresponds to the flow-around model of small infinite in size obstructions in transonic flow.

4.2. The influence of chamber pressure. As the chamber pressure reaches the evaporation pressure (4, 5, 6), the shape of flow effusion changes from sphere to torch, and agrees with Kneller diagram.

4.3. The influence of distance. As the substrate retreats from the evaporation cutoff (7, 8, 9), the vapor flow shape changes from the high-amplitude mode to free flow mode. The film thickness changes from cosine to more uniform one. The evaporation temperature change (10, 11, 12) is adequate to the evaporation cutoff pressure increase and ratio decrease.

4.4. Parametric thickness synthesis law control. Changing the local temperature at a certain point of crucible miter results in thermobaric effect at which temperature differential causes pressure differential, which also contribute to vapor flow deflection.

5. Results and conclusions

The experiments on formation of lead and zinc films demonstrated that the locally coldest parts of the substrate or chamber elements are preferred centers of film condensation. Zinc is a bright flash evaporator. Vapors super cooling and their aggregating happen in the region between the evaporator and the substrate. Parametric estimations showed that the ratio of local zinc vapor pressure to vacuum chamber pressure is ~100, the rate coefficient is □~2, the temperature function ( ) = 0,2 at the evaporation temperature 6000 C. It means that the flow temperature fell by a factor of 5. Microscopic study showed that at the variable condensation function the film gets the shape of cumulo nimbus.

According to mass spectroscopy, during film formation the intensity of hydro carbonic component increases by a factor of 1-2 from 10^{-6} to 10^{-5} Wt due to gas emission from heated elements. The direction of gas-phase reaction is estimated with respect to absorbed gas Q from vacuum chamber with the volume V in a time unit [12].

$$Q = V\frac{dP}{dt} \tag{5}$$

With the temperature increase, segregate makeup of near surface phase is arranged in the following order: hydride, oxide, carbide, and nitride. The flow rate change (achieved by the evaporation rate change and chamber vacuum level) results in density and temperature functions change. The amount of gas in films can be estimated from the evaporation pressure change and from the isobaric potential diagram change.

$$\Delta Z = -RT\ln P_i \tag{6}$$

where R is a universal gas constant.

Taking into account gas-dynamic functions (1) and (2) the isobaric potential is:

$$\Delta Z = f(R,T,P_i) = f(R,\tau(\lambda),\pi(\lambda)) \tag{7}$$

where $\tau(\lambda), \pi(\lambda)$ are gas-dynamic functions of temperature and full pressure distribution.

As a result,

$$\Delta Z = -R\tau(\lambda)\cdot\tau(\lambda)^{k(k-1)} \approx -R\tau^2(\lambda) \tag{8}$$

That is, the decrease of isobaric potential is proportional to temperature change function squared. Using numeric values of temperature functions $\tau(\lambda)$, one can see the following: when the flow is static $\lambda\to 0$ $\tau^2=1$ the isobaric potential (ΔZ) doesn't change. When the working flow is close to sound velocity $\lambda\to 1$, $\tau^2=0{,}64$, $\Delta Z=0{,}64$; where $\lambda\to 2{,}5$, $\tau^2\to 0$, $\Delta Z\to 0$.

One can see that the flow rate variation results in the change of density and temperature functions, which is followed by phase diagram shift by the flow rate coefficient squared. Figure 6 shows the isobaric potential.

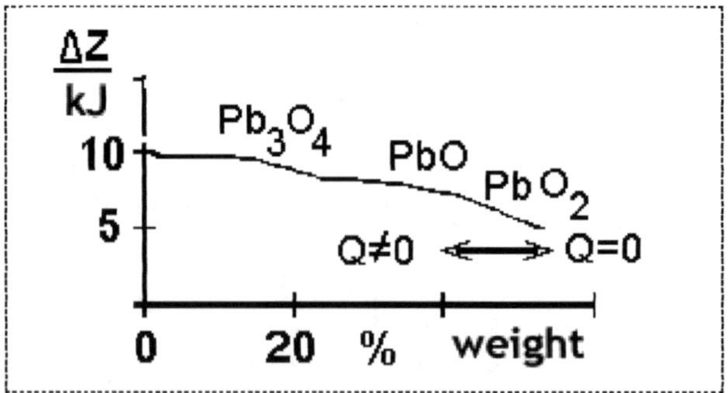

Figure 6. The diagram of isobaric potential.

The magnitude of isobaric potential and the percentage of absorbed gas at working gas inleakage decreases due to the decrease of local gas pressure.

IOP Conf. Series: Materials Science and Engineering **110** (2016) 012089 doi:10.1088/1757-899X/110/1/012089

6. Conclusion

According to gas dynamics laws we were able to find conditions which allow forecasting the rules of local optical layer thickness distribution and gas-phase reaction route at the minimal distance from the substrate to evaporator and maximal material utilization rate. The obtained results on geometric and parametric vapor flow control were used in formation of zinc oxide waveguides on glass substrate [13-14]. The dynamic approach enables to increase the uniformity of film thickness in the process of Gallium Arsenide molecular-beam epitaxy.

Acknowledgements

This work was financially supported by The Ministry of Education and Science of the Russian Federation in part of the science activity program.

References

[1] Peitmann K, Hukriede J, Buse K and Kratzig E 2001 Phus. Rev. **61** 7 4615
[2] Esseling M, Holtmann F, Woerdemann M and Denz C 2010 Optics Express. **18** 16 17404
[3] Arestov S and Orlikov L 2013 Izvestiya vuzov, ser. Fizika **56** ½ 178-181
[4] Makhov I., Makhova P., Pomazkova E 1990.– Electron. Promyshl. 1990, No 10, P.64-65.
[5] Ohashi M, Ozeki M and Cui J 1999 Rev. Sci. Instrum. **85** 7522-27
[6] Abramovichm G N 1976 Prikladnaya gazovaya dinamika Moscow Nayka 808
[7] Zarvin A, Sharafutdinov R u.a 1979 Diagnosis of rarefied flows gaza Novosibirsk ITF 179-188
[8] Vasenkov A, Sharafutdinov R, Belikov A and Kuznetsov O 1995 Microelectronika **24** 3 163-165
[9] Neller G 1976 Dynamic rarefied gas editor A. Isclinski Moskau, Mir, 195-206
[10] Kimura T 1977 AIAA Journal **5** 611-612.
[11] Orlikov L and Orlikov N 2001 Technical Physics Letters ZHTF **27** 10 881-882
[12] Rosanov L 1990 Vacuumnaja Technica Moskau 320
[13] Orlikov N, Orlikov L and Shandarov S 2001 Radiotekhnika i elektronika **11** 112-115
[14] Orlikov L N, Orlikov N L, Arestov S I, Mambetova K M and Shandarov S M 2015 *IOP Conf. Ser.: Mater. Sci. Eng.* **81** 012043

IOP Conf. Series: Materials Science and Engineering **110** (2016) 012090 doi:10.1088/1757-899X/110/1/012090

Estimation of thermomagnetometry method sensitivity for magnetic phase determination

A L Astafyev, A P Surzhikov and E N Lysenko

Tomsk Polytechnic University, 30 Lenin Ave, 634050, Tomsk, Russia

E-mail: astafeval@tpu.ru

Abstract. In this article, the sensitivity of thermomagnetometry method for magnetic phase determination in lithium substituted ferrites was estimated by thermogravimetric analysis in magnetic field of lithium-zinc ferrite and iron dioxide ($Li_{0.4}Fe_{2.4}Zn_{0.2}O_4+\alpha$-$Fe_2O_3$) components mixture with different mass proportions: 2, 4, 6, and 100 mass% $Li_{0.4}Fe_{2.4}Zn_{0.2}O_4$ phase in total mixture. Thereby, analyzed samples are mixture of magnetic and nonmagnetic phases. Results of thermomagnetometric analysis were supplemented with the X-Ray diffraction analysis data. It was shown that the thermomagnetometry method allows to determine a magnetic phases with mass content not less than 2 mass% in analyzed mixture. In this case, we can clearly estimate the position and intensity of the peak on derivative thermogravimetric curve, which connected with a magnetic phase transition in ferrite at Curie temperature.
Keywords: lithium-zinc ferrite; lithium-substituted ferrospinel; ferrospinel; thermomagnetometry method, X-Ray analysis.

1 Introduction

Ferrites are the mostly usable elements for electronic and radio devices in our days [1-9]. During production of multicomponent ferrite materials, the great attention is paid to maximization of powder homogeneity in the synthesis stages [10, 11]. One of the more simple technological methods for increase of phase powders homogeneity is intermediate grinding and mixing operation, which is additional step in synthesis technology [12-15]. To optimize the time-temperature regimes of the technological model, it is necessary to control the phase composition of the powders after each stage of annealing.

X-Ray diffraction analysis (XRD) is the mostly used method for control phase composition in ferrites. However, for multicomponent ferrites synthesis, it's needed an additional procedure of phase verification due to formation of wide variety of ferrospinel phases with close values of the lattice parameters. Thus, reflections from these phases in the registered X-Ray diffraction patterns merge into a single reflection; and inability to make a correct separation of X-Ray reflections leads to necessarily of development a new estimation method of phase composition, which would be as an addition to XRD analysis. One of those methods is thermomagnetometric analysis, which is thermogravimetric analysis (TG/DTG) with applied magnetic field to the samples [16]. Possibility of TG/DTG method for estimation of both the qualitatively and quantitatively magnetic phase analyzes for lithium substituted ferrites was shown in several works [17-20]. However, for extensive use of this technique in practice, should be carefully considered the conditions of its application.

Thereby, in present work, we tried to define a sensitivity of thermomagnetometric analysis for magnetic phase detection in $Li_{0.4}Fe_{2.4}Zn_{0.2}O_4$ lithium-zinc ferrites. Results of thermomagnetometric analysis were supplemented with the XRD data.

2 Experimental

Lithium-zinc ferrite was prepared by solid phase synthesis method in accordance with the reaction $Li_2CO_3+6Fe_2O_3+ZnO \rightarrow 5Li_{0.4}Fe_{2.4}Zn_{0.2}O_4+CO_2$ using resistance furnace at temperature of 800°C in air atmosphere during 6 hours with additional intermediate grinding and mixing operations after each 120 minutes. XRD analysis showed 100% presence of $Li_{0.4}Fe_{2.4}Zn_{0.2}O_4$ phase after synthesis. After that, lithium ferrite was mixed with α-Fe_2O_3 (nonmagnetic phase) powder in an agate mortar in different mass proportions, so that the samples had 2, 4, 6 and 100 mass% of $Li_{0.4}Fe_{2.4}Zn_{0.2}O_4$ phase. Thereby, samples were mixture of $Li_{0.4}Fe_{2.4}Zn_{0.2}O_4+Fe_2O_3$ magnetic with nonmagnetic phases with different mass content.

X-ray diffraction analysis was performed using an ARL X'TRA diffractometer with a semiconductor Peltier Si(Li) detector and CuK_α radiation. The Powder Cell 2.4 software was used for a full-profile analysis of the X-ray diffraction patterns. The phase composition was identified using PDF-4 powder database of the International Centre for Diffraction Data (ICDD).

Thermomagnetometric analysis was carried out on STA 449C Jupiter (Netzsch, Germany) thermal analyzer. The magnetic assemblage consisting of two permanent magnets creating a field of ~5 Oe was attached on the outer side of measurement cell for control of the magnetic state of samples. The heating of samples was carried out in corundum crucibles in air atmosphere with line speed of heating 50°C/min. The results were analyzed by Proteus Analysis (Netzsch, Germany).

3 Results and discussion

XRD patterns are shown in Figure 1 for different samples. It can be seen that all peaks for $Li_{0.4}Fe_{2.4}Zn_{0.2}O_4+Fe_2O_3$ mixture are peak overlapping from reflections of $Li_{0.4}Fe_{2.4}Zn_{0.2}O_4$ and Fe_2O_3 powders; and just one low intensity peak at $2\theta \approx 30{,}4°$ can be assigned only to $Li_{0.4}Fe_{2.4}Zn_{0.2}O_4$ phase. Figure 2 shows such peaks for samples with different mass content of $Li_{0.4}Fe_{2.4}Zn_{0.2}O_4$ phase. Here, the peak is more intense with increase the $Li_{0.4}Fe_{2.4}Zn_{0.2}O_4$ phase in the mixture.

It was shown in [17], the content of ferrite phase can be determined from the mass jump (Δm) on TG curve, when the magnetic phase transition at Curie temperature occurs during heating of ferrite sample in an external magnetic field (Fig. 3, TG solid line). Curie temperature can be determined by position of peak maximum on DTG curve (Fig. 3, DTG dotted line). Mass change was not observed in absence of the magnetic field (Fig. 3, TG dashed line).

Figure 4 shows TG and DTG curves for $Li_{0.4}Fe_{2.4}Zn_{0.2}O_4+Fe_2O_3$ samples with different content of lithium-zinc ferrite phase. The values of mass jump and Curie temperature for all samples are summarized in Table 1. According to TG/DTG analysis for pure lithium-zinc ferrite (Fig. 4 a), magnetic phase transition corresponds to Curie temperature for $Li_{0.4}Fe_{2.4}Zn_{0.2}O_4$ [17]. The height of mass jump on TG curve is maximum and equal to 0.324%. By reduction of magnetic phase in mixture (Fig. 4 b, c, d), a decrease in Δm are observed up to 0.003 % for mixture with 2 mass% $Li_{0.4}Fe_{2.4}Zn_{0.2}O_4$. In this case, we can clearly estimate the position, height and half-width of the peak on DTG curve and it was shown in [18], the values of DTG peak areas can be compared with a quantitative content of the ferrite phase.

Thus, the results showed that the thermomagnetometry method allows to determine a magnetic ferrite phases with mass content not less than 2 mass%. However, this work should be continued in terms of determine the sensitivity of thermomagnetometric method for analysis of ferrites with different magnetization.

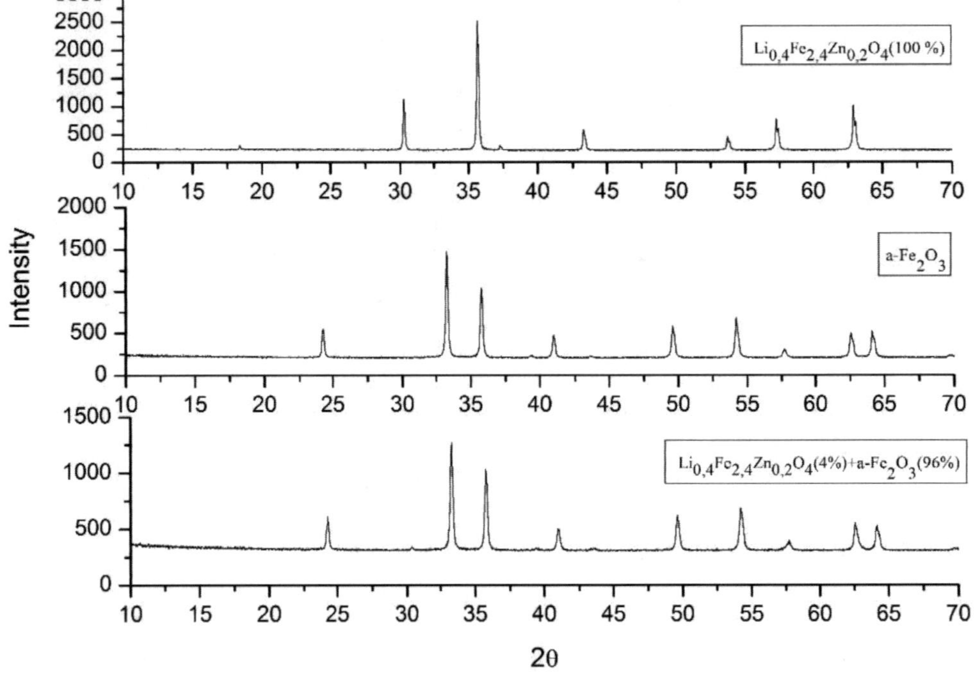

Figure 1. X-Ray diffraction patterns

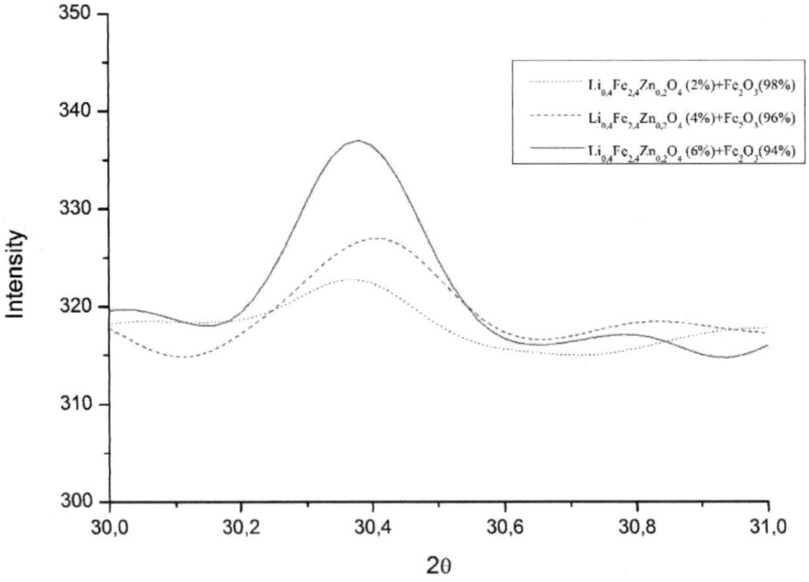

Figure 2. Part of XRD patters for samples with different mass content of $Li_{0.4}Fe_{2.4}Zn_{0.2}O_4$ phase

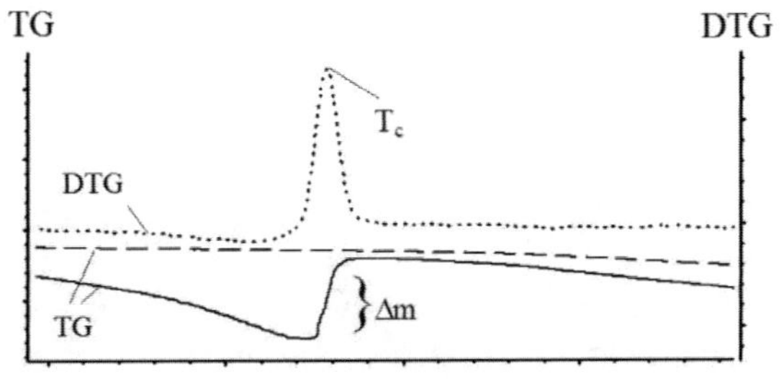

Figure 3. TG/DTG curves for ferrite materials during heating in magnetic field (TG solid line and DTG dotted line) and without magnetic field (TG dashed line)

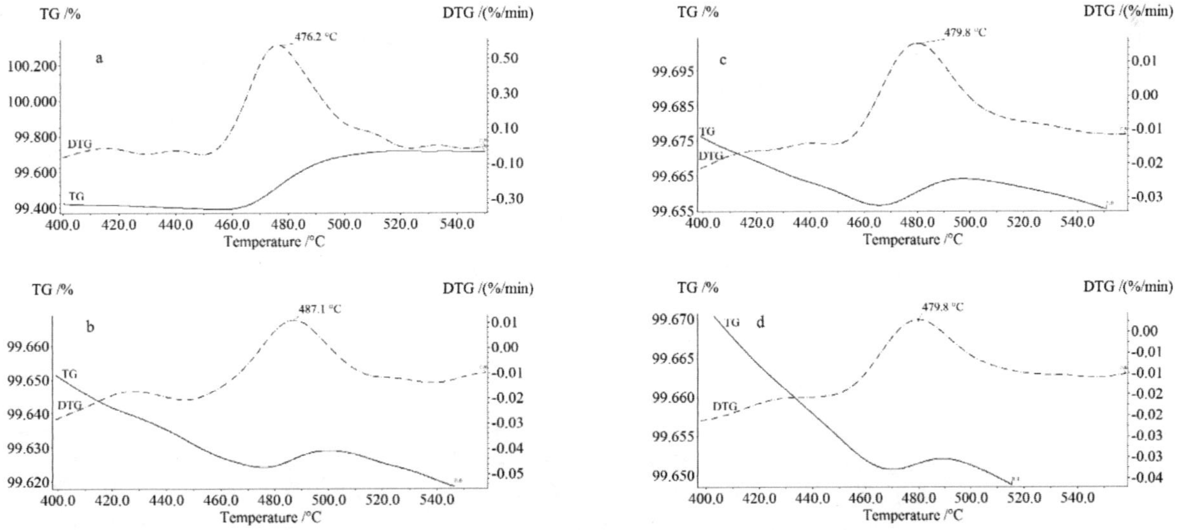

Figure 4. TG/DTG curves for $Li_{0.4}Fe_{2.4}Zn_{0.2}O_4+Fe_2O_3$ samples with different content of lithium-zinc phase in mixture: a – 100 mass %, b - 2 mass %, c- 4 mass %, d - 6 mass %.

Table 1. Parameters of magnetic phase transition for $Fe_2O_3+Li_{0.4}Fe_{2.4}Zn_{0.2}O_4$

Content of $Li_{0.4}Fe_{2.4}Zn_{0.2}O_4$ phase in a mixture (mass %)	Mass jump, Δm (%)	Curie temperature, T_c (°C)
100	0,324	476,2
6	0,01	486,4
4	0,01	478,3
2	0,003	486,4

4. Acknowledgements This work was supported by The Ministry of Education and Science of the Russian Federation in part of the Science program.

References

[1] Smit J and Wijn H P J 1959 *Ferrites: Physical properties of ferrimagnetic oxides in relation to their technical applications.* (Philips Technical Library, Eindhoven) p 369
[2] Baba P, Argentina G, Courtney W, Dionne G and Temme D 1972 Fabrication and properties of microwave lithium ferrites *IEEE Trans. Magn.* **8** 83–94
[3] Levin B E, Tret'yakov Y D and Letyuk L M 1979 *Physicochemical principles of preparation, properties, and applications of ferrites.*(Moscow: Metallurgiya, Russia) p 470
[4] Fazio E De, Bercoff P G and Jacobo S E 2011 Electromagnetic properties of manganese-zinc ferrite with lithium substitution *J. Magn. Magn. Mater.* **323** 2813–7
[5] White G O and Patton C E 1978 Magnetic properties of lithium ferrite microwave materials *Magn. Magn. Mater.* **9** 299–317
[6] Patton C E, Edmondson C A and Liu Y H 1982 Magnetic properties of lithium zinc ferrite *J. Appl. Phys.* **53** 2431–3
[7] Sláma J, Šoka M, Grusková A, Dosoudil R, Jančárik V and Degmová J 2013 Magnetic properties of selected substituted spinel ferrites *J. Magn. Magn. Mater.* **326** 251–6
[8] Grusková A, Sláma J, Dosoudil R, Ušáková M, Jančárik V and Ušák E 2008 Microwave properties of some substituted LiZn ferrites *J. Magn. Magn. Mater.* **320** 860–4
[9] Manjula R, Murthy V R K and Sobhanadri J 1986 Electrical conductivity and thermoelectric power measurements of some lithium-titanium ferrites *J. Appl. Phys.* **59** 2929–32
[10] Surzhikov A P, Lysenko E N, Vasendina E A, Sokolovskii A N, Vlasov V A and Pritulov A M 2011 Thermogravimetric investigation of the effect of annealing conditions on the soft ferrite phase homogeneity *J. Therm. Anal. Calorim.* **104** 613–7
[11] Surzhikov A P, Pritulov A M, Lysenko E N, Sokolovsky A N, Vlasov V A and Vasendina E A 2012 Influence of solid-phase ferritization method on phase composition of lithium-zinc ferrites with various concentration of zinc *J. Therm. Anal. Calorim.* **109** 63–7
[12] Berbenni V, Marini A, Matteazzi P, Ricceri R and Welham N J 2003 Solid-state formation of lithium ferrites from mechanically activated Li2CO3–Fe2O3 mixtures *J. Eur. Ceram. Soc.* **23** 527–36
[13] Surzhikov A P, Lysenko E N, Malyshev A V, Pritulov A M and Kazakovskaya O G 2012 Influence of mechanical activation of initial reagents on synthesis of lithium ferrite *Russ. Phys. J.* **6** 672–7
[14] Widatallah H M, Johnson C and Berry F J 2002 The influence of ball milling and subsequent calcination on the formation of LiFeO2 *J. Mater. Sci.* **37** 4621–5
[15] Kavanlooee M, Hashemi B, Maleki-Ghaleh H and Kavanlooee J 2012 Effect of annealing on phase evolution, microstructure, and magnetic properties of nanocrystalline ball-milled LiZnTi ferrite *J. Electronic. Mater.* **41** 3082–8
[16] Gallagher P K 1997 Thermomagnetometry *J. Therm. Anal. Calorim.* **49** 33–44
[17] Surzhikov A P, Pritulov A M, Lysenko E N, Vlasov V A, Vasendina E A and Malyshev A V 2013 Analysis of the phase composition and homogeneity of ferrite lithium-substituted powders by the thermomagnetometry method *J. Therm. Anal. Calorim.* **112** 739–45
[18] Astafyev A L, Lysenko E N, Surzhikov A P and Neudahina N A 2014 Development of control method for ferrite phase composition using thermomagnetometric analysis *IOP Conf. Ser.: Mat. Sci. Eng.* **66** 012037
[19] Astafyev A L, Surzhikov A P and Lysenko E N 2015 Investigation of the Phase Composition of Lithium-Titanium Ferrites by Thermo-Magnetometric and X-Ray Analysis *Adv. Mat. Res.* **1085** 233–6
[20] Surzhikov A P, Pritulov A M, Lysenko E N, Sokolovskii A N, Vlasov V A and Vasendina E A 2012 Dependence of lithium–zinc ferrospinel phase composition on the duration of synthesis in an accelerated electron beam *J. Therm. Anal. Calorim.* **110** 733–41

Electrical properties of lithium ferrite with addition of ZrO_2

S A Lamonova, A P Surzhikov and E N Lysenko

Tomsk Polytechnic University, 30 Lenin Ave, 634050, Tomsk, Russia

E-mail: sal17@tpu.ru

Abstract. The study of electrical properties of composite ceramics based on lithium ferrite $LiFe_5O_8$ with the addition of ZrO_2 (1 and 2 wt%) was carried out. The samples were prepared by standard ceramic technology. Synthesis of lithium ferrite was conducted at 800°C for 120 minutes. The zirconium dioxide was added to synthesized powder by mixing in planetary ball mill, and then the composite powders were sintered at temperatures of 1100°C and 1050ºC. The electrical conductivity of the samples was studied using two-probe spreading resistance method. It was found that addition of ZrO_2 and an increase of its up to 2 wt% leads to increase in the activation energy and electrical resistivity.

1. Introduction

Lithium ferrite ($LiFe_5O_8$) is widespread as the microwave ferrite material. The high Curie temperature (above 600°C) and saturation magnetization (about 3600 G) allowed to create a number of substituted lithium ferrites, which have been used in various microwave and high frequency devices [1-2]. For such ferrite systems, an important parameter is the high electrical resistivity ($\rho > 10^6$ Om·cm) providing low dielectric loss of material. Previously, the preparation of $LiFe_5O_8$ by the standard ceramic technology was studied [3-7]. Also, the dielectric and magnetic properties of these compounds were investigated [1, 2, 8-12].

It was previously demonstrated [13, 14], the small amounts (< 5 wt%) of ZrO_2 may be used as an effective additive in the manufacture of some ferrite materials, which significantly influences on the microstructure and magnetic properties of ferrites. Thus, it was found that the introduction of ZrO_2 into Li-Zn ferrite during synthesis leads to increase the value of initial magnetic permeability of ferrites and reduce the magnetic losses [15]. Moreover, the addition of ZrO2 during high temperature sintering improves the mechanical properties of nickel ferrite, such as a bending strength, microhardness and fracture toughness [16]. It was shown that ZrO2 is insoluble in the ferrites at high temperatures, and the composite materials, such as $ZrO_2/NiFe_2O_4$ in [17], are obtained after sintering.

In this paper, the influence of ZrO_2 addition (1 and 2 wt%) on the electrical conductivity of $LiFe_5O_8$ lithium ferrite was studied.

2. Experimental

Lithium ferrite samples were prepared by standard ceramic technology. Iron oxide (Fe_2O_3) and lithium carbonate (Li_2CO_3) with a ratio of 1:5 were used as the initial reagents to produce lithium ferrite. Before weighing, initial powders were dried for three hours at a temperature of 200ºC in a laboratory furnace. The mixed powder was milled in ball mill during 10 minutes to homogenize the mixture of initial reagents. Synthesis of lithium ferrite was conducted at 800°C for 120 minutes in "HimLabo" standard laboratory oven. The synthesized lithium ferrite powder was divided into several parts. The samples from the first part were sintered at a temperature of 1050°C for 2 hours (LF_1050). The samples from the second part were sintered for 2 hours at a temperature of 1100°C (LF_1100).

The zirconium dioxide with 1 and 2 wt% was added to synthesized powder from third and fourth parts by mixing in planetary ball mill. Then, the composite powders were sintered in ceramic samples at temperatures of 1100°C and 1050°C (LF_1_1050, LF_1_1100 - the lithium ferrites with 1 wt% ZrO_2 were sintered at a temperature of 1050°C and 1100°C, respectively; LF_2_1050, LF_2_1100 - the lithium ferrites with 2 wt% ZrO_2 were sintered at a temperature of 1050°C and 1100°C, respectively).

The electrical conductivity of the lithium ceramic was investigated by spreading resistance method [10, 18]. In this case, the electrical resistivity was calculated by the equations:

$$\rho = \frac{U \cdot r_0 \cdot \pi}{I}, \quad (1)$$

where U – applied voltage, I – current intensity, r_0 – contact diameter.

Since the spreading resistance is mainly determined by the contact region, the thickness of which is r_0, the method is characterized by high local measurement. The resolution of the method corresponds to the diameter of the contact, and is ~ 10 μm.

3. Results and discussion

Figures 1 and 2 shows the temperature dependence of the conductivity current for samples sintered at a temperature of 1050°C (figure 1) and 1100°C (figure 2). It can be seen that the curves for electrical conductivity current versus temperature are characterized by linear dependence in logarithmic coordinates for all lithium ferrite samples. The electrical conductivity of ferrite samples increases with the increasing temperature. A similar character of Ln(I)=f(1/kT) dependencies indicates that the addition of ZrO_2 has not significant effect on the mechanism of electric transport in lithium ferrites, which is determined by electron hopping between Fe^{2+} and Fe^{3+} heterovalent ions [8].

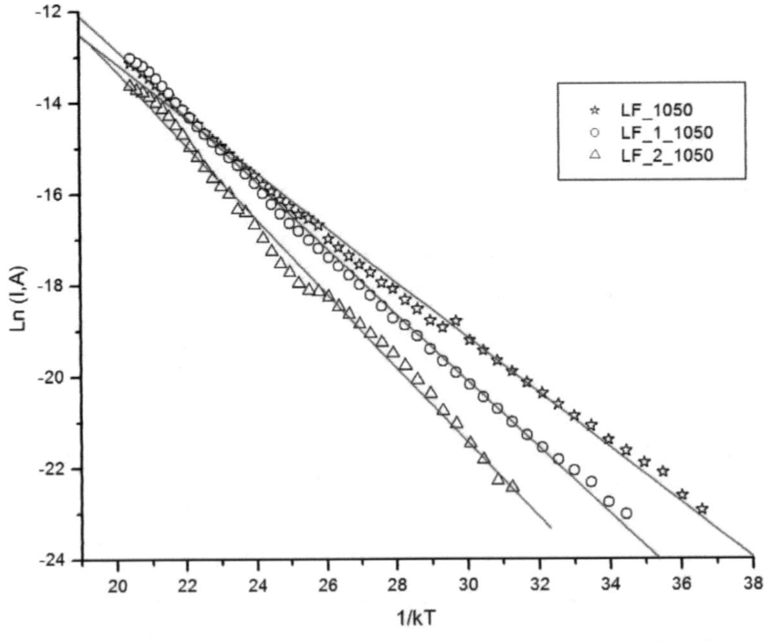

Figure 1. The temperature dependence of conductivity current for samples sintered at 1050°C.

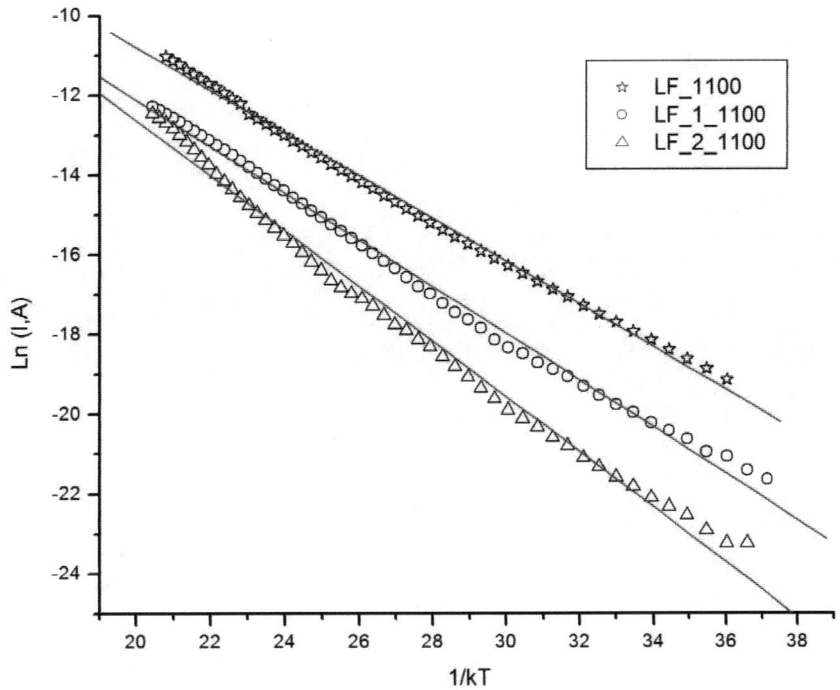

Figure 2. The temperature dependence of conductivity current for samples sintered at 1100°C.

From electrical conductivity results for all samples, the values of activation energy for electronic conduction (E_a) were found as the tangent of the angle slope of the curves in the temperature dependence of conductivity current. Moreover, the electrical resistivity for all samples was calculated using the equations (1). These electrophysical parameters are shown in table 1. The presented data show that the samples are characterized by the high values of the electrical resistivity, which is typical for lithium ferrite. However, the increase in the sintering temperature for samples, as previously were repeatedly shown [19, 20], reduces the electrical resistance of the samples, which is associated with an increase in the concentration of Fe^{2+} ions.

The results showed that the activation energy and electrical conductivity values increase with increasing the ZrO_2 addition. This can be attributed to changes in the microstructure of ferrite samples.

Table 1. Electrophysical parameters of ferrite samples.

Samples	E_a, (eV)	ρ, (MOm·cm)
LF_1050	0.602	2.61
LF_1_1050	0.726	5.45
LF_2_1050	0.816	16.354
LF_1100	0.536	0.127
LF_1_1100	0.585	0.882
LF_2_1100	0.69	4.015

4. Conclusion

Lithium ferrites with the addition of ZrO_2 have semiconducting properties. The results showed that electrical conductivity increases with increasing temperature, and characterizes by linear dependence.

The electrophysical parameters, such as the activation energy for electronic conduction and electrical resistivity, were found for all samples, and are characterized by the high values typically for lithium ferrites. It was assumed that the conductivity mechanism are unchanged in the temperature range of 20 - 300°C, and is determined by the hopping of electrons between Fe^{2+} и Fe^{3+} ions.

It was found that addition of ZrO_2 and an increase of its up to 2 wt% leads to increase in the activation energy and electrical resistivity. This can be attributed to change in the microstructure of ferrite samples.

Acknowledgements
This work was financially supported by The Ministry of Education and Science of the Russian Federation in part of the science activity program.

Reference

[1] White G O and Patton C E 1978 Magnetic properties of lithium ferrite microwave materials *J. Magn. Magn. Mater.* **9** 299
[2] Baba P D, Argentina G M, Courtney W E, Dionne G F and Temme D H 1972 Fabrication and properties of microwave lithium ferrites *IEEE Trans. Magn.* **8** 83
[3] Edelio B, Mercier T le and Quarton M 1995 Microstructure and physicochemical studies of pure and zinc-substituted lithium ferrites sintered above 1000°C *J. Am. Ceram. Soc.* **78** 365–8
[4] Surzhikov A P, Lysenko E N, Malyshev A V, Pritulov A M and Kazakovskaya O G 2012 Influence of mechanical activation of initial reagents on synthesis of lithium ferrite *Russ. Phys. J.* **6** 672–7
[5] Surzhikov A P, Frangulyan T S, Ghyngazov S A and Lysenko E N 2012 Investigation of structural states and oxidation processes in $Li_{0.5}Fe_{2.5}O_{4-\delta}$ using TG analysis. *J. Therm. Anal. Calorim.* **3** 1207–12
[6] El-Shobaky G A and Ibrahim A A 1987 Solid-solid interactions between ferric oxide and lithium carbonate and the thermal stability of the lithium ferrites produced *J. Thermoch. Acta* **118** 151–8
[7] Berbenni V, Marini A and Capsoni D 1998 Solid state reaction study of the system Li_2CO_3–Fe_2O_3 *J. Phys.* **12** 997–1003
[8] Matsui T and Wagner J B 1977 Electrical properties of $LiFe_5O_8$ *J. Elecrochem. Soc.* 1141–3
[9] Smit J and Wijn H P J 1959 *Ferrites: physical properties of ferrimagnetic oxides in relation to their technical applications* (Eindhoven, The Netherlands: Philips Technical Library) p 369
[10] Surzhikov A P, Frangulyan T S, Ghyngazov S A, Lisenko E N and Galtseva O V 2006 Investigation of electroconductivity of lithium pentaferrite *Russ. Phys. J.* **49** 506
[11] Green J J and Van Hook H J 1977 Microwave properties of lithium ferrites *IEEE Trans. Microwave Theory Tech.* **25** 155
[12] Surzhikov A P, Lysenko E N and Malyshev A V 2014 Study of the Radio-Wave Absorbing Properties of a Lithium-Zinc Ferrite Based Composite *Russ. Phys. J.* **57** 621
[13] Li L, Lan Z, Yu Z, Sun K, Xu Z and Ji H 2008 Microstructure, Electrical, and Magnetic Properties of ZrO_2 Added MnZn Ferrites *IEEE Trans. Magn.* **44** 2107–12
[14] Wang S -F, Yang H -C, Hsu Y -F and Hsieh C -K 2015 Effects of SnO_2,WO_3, and ZrO_2 addition on the magnetic and mechanical properties of NiCuZn ferrites *J. Magn. Magn. Mater.* **374** 381–7
[15] Sattar A A, El-Sayed H M, Agami W R and Ghani A A 2007 Magnetic Properties and Electrical Resistivity of Zr^{4+} Substituted Li-Zn Ferrite *Am. J. Appl. Sci.* **4** 89–93
[16] Long X, Liu Y, Yao G, Du J, Zhang X, Cheng J and Hua Z 2013 Microstructure and mechanical properties of $NiFe_2O_4$ ceramics reinforced with ZrO_2 particles with different sintering temperatures *J. Alloys Comp.* **551** 444–50

[17] Hua Z, Yao G, Ma J and Zhang M 2013 Fabrication and mechanical properties of short ZrO_2 fiber reinforced $NiFe_2O_4$ matrix composites *J. Cerc. Int.* **39** 3699–708
[18] Clarysse T and Vandervorst W 1998 Qualification of spreading resistance probe operations *J. Vac. Sci. Technology B: Microelectronics and Nanometer Structures* **16** 260–71
[19] Surzhikov A P, Ghyngazov S A, Lysenko E N and Frangylyan T S 2004 Grain boundary diffusion in polycrystalline ferrites with different intergranular potential barrier *Russ. Phys. J.* **47** 319–23.
[20] Ghyngazov S A, Lisenko E N, Petyukevich M S and Frangulyan T S 2007 Interaction of lithium-titanium ceramics with air during sintering *Russ. Phys. J.* **50** 134–9.

TG study of the $Li_{0.4}Fe_{2.4}Zn_{0.2}O_4$ ferrite synthesis

E N Lysenko, E V Nikolaev and A P Surzhikov

Tomsk Polytechnic University, 30 Lenin Ave, 634050, Tomsk, Russia

E-mail: lysenkoen@tpu.ru

Abstract. In this paper, the kinetic analysis of Li-Zn ferrite synthesis was studied using thermogravimetry (TG) method through the simultaneous application of non-linear regression to several measurements run at different heating rates (multivariate non-linear regression). Using TG-curves obtained for the four heating rates and Netzsch Thermokinetics software package, the kinetic models with minimal adjustable parameters were selected to quantitatively describe the reaction of Li-Zn ferrite synthesis. It was shown that the experimental TG-curves clearly suggest a two-step process for the ferrite synthesis and therefore a model-fitting kinetic analysis based on multivariate non-linear regressions was conducted. The complex reaction was described by a two-step reaction scheme consisting of sequential reaction steps. It is established that the best results were obtained using the Yander three-dimensional diffusion model at the first stage and Ginstling-Bronstein model at the second step. The kinetic parameters for lithium-zinc ferrite synthesis reaction were found and discussed.

1. Introduction

The solid-state synthesis is the main method of commercial production of microwave ferrites including zinc substituted lithium ferrites [1-3]. It involves two-stage high-temperature heating of reagent mixtures including synthesis of ferrites with the formation of single-phase ferrite compositions and further sintering at a higher temperature to produce high-density ferrite ceramics [4-10]. The dielectric and magnetic properties of lithium-substituted ferrites were intensively studied. However, the kinetic parameters for lithium ferrite synthesis reaction were not investigated, especially, for lithium-zinc ferrite with the chemical formula $Li_{0.4}Fe_{2.4}Zn_{0.2}O_4$, which is widely used in microwave technology.

The study of kinetic regularities of synthesis has two main objectives. The first is experimental determination of degree or rate of conversion depending on the conditions of the reactions and the mathematical description of the found dependence. The second is evaluation of kinetic parameters of the reaction and their interpretation in relation with the nature of ongoing processes.

Mathematical treatment of the kinetic data can be carried out in two ways:

1) Finding of equation that best describes experiment, and parameters which do not have a specific physical meaning;
2) Using of kinetic equations based on certain models of the interaction in solids. The parameters of these equations are related with the nature of ongoing processes.

For kinetic analysis of the synthesis reaction, equation is generally assumed:

$$de/_{dt} = -k \cdot (T(t)) \cdot f(e,p) \tag{1}$$

IOP Conf. Series: Materials Science and Engineering **110** (2016) 012092 doi:10.1088/1757-899X/110/1/012092

where t – time; T – temperature; e - concentration of the initial material; p – concentration of the product, $f(e,p)$ – type of reaction. The reaction models $f(e,p)$, most commonly used in heterogeneous kinetics, are presented somewhere else [11]. It is assumed that the conversion function, $U(t,T,e,p)$, can be described by two different functions $k(T)$ and $f(e,p)$:

$$(t,T,e,p) = k(T(t)) \cdot f(e,p) \tag{2}$$

For single-stage reactions $f(e,p)$ has a less complex form of $f(x)$, i.e. $e = 1 - \chi e$ and $x = p$ (x - degree of conversion). Thus, the complete separation of variables in the equation (2) and analytical solution of the differential equation (1) is only possible for the one-step reaction. For complex multistage of processes solution (1) leads to a system of differential equations for which it is impossible to separate the variables and there is no analytical solution.

From the technical aspect, the kinetic analysis will be examined as a tool for data reduction. The information will be extracted from a series of measurements with many data points in the form of one model with few parameters. This model will then serve in the preparation of predictions under another temperature program than used for the experiments and for process optimization.

The aim of this model-fitting kinetic approach is to find a kinetic model with minimal adjustable parameters which quantitatively describes the kinetics reaction [12]. This formal kinetic model can then be used for the kinetic analysis of thermal measurements, including the process of synthesis.

In this work, the kinetic model describing the process of lithium-zinc ferrite synthesis was obtained using the model-fitting kinetic analysis of ferrite formation employing thermogravimetric (TG) analysis. Also the kinetic parameters quantitatively describing the kinetics of synthesis process were obtained. The Netzsch Thermokinetics software (Version 2006.08, Netzsch-Geratebau GmbH) package was used for the non-linear regression model-fitting analysis.

2. Experimental

Commercial powders of zinc oxide – ZnO, iron oxide – Fe_2O_3 and lithium carbonate Li_2CO_3 were used as the initial reagents of reaction mixture for $Li_{0.4}Fe_{2.4}Zn_{0.2}O_4$ lithium-zinc ferrite synthesis. Li_2CO_3– ZnO–Fe_2O_3 mechanical mixture of initial reagents was thoroughly mixed in an agate mortar. Thermogravimetric (TG) and differential scattering calorimetric (DSC) measurements were performed using STA 449C Jupiter analyzer (Netzsch-Geratebau GmbH, Germany). Samples with a mass of ~ 10 mg were placed in the alumina crucibles. Both TG and DSC baselines were corrected by subtraction of predetermined baselines run under identical conditions except for the absence of sample. Samples were heated up to 800°C with programmed heating rate of 2.5, 5, 10, 20 °C/min. All experiments were performed under atmospheric pressure conditions. The Netzsch Proteus software packages were used for TG/DSC data analysis.

3. Results and Discussion

Figure 1 shows the TG/DSC curves for sample heated at 20 °C/min. The curves for samples heated at other heating rates were similar. The TG curve indicates a mass loss process at ca. 500-730 °C, which is connected to the reaction between the initial reagents according to the equation $Li_2CO_3 + 6Fe_2O_3 + ZnO \rightarrow 5Li_{0.4}Fe_{2.4}Zn_{0.2}O_4 + CO_2\uparrow$. A mass loss corresponds to the theoretical value of output CO_2.

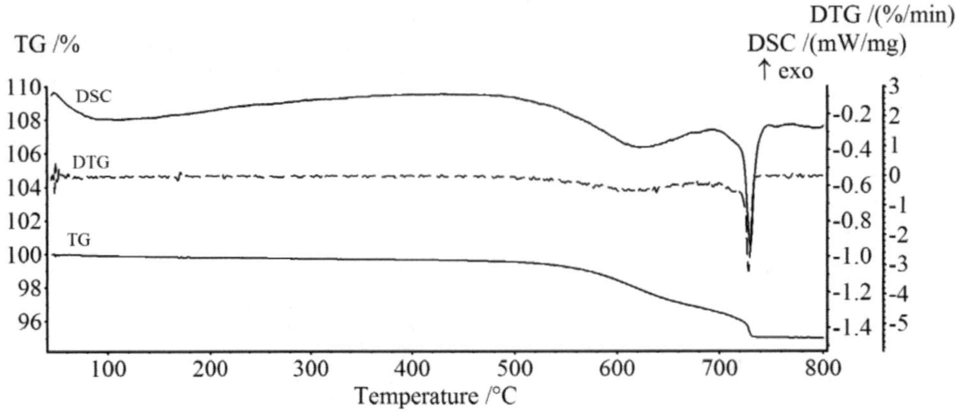

Figure1. TG/DSC curves for lithium-zinc ferrite synthesis

The TG curves can be divided on two main stages, in which the temperature ranges depend on heating rate. From the derivate of TG (DTG) peak, the first stage is located at ca. 500-680°C. In the second stage, the maximum peak of DTG curve is at ca. 723 °C, and this stage corresponds to melting of the lithium carbonate [13, 14].

It is obvious that the reaction mechanism of Li-Zn ferrite synthesis is complex and changes during the reaction [15, 16]. Thus, it is unlikely that the total reaction can be described by a simple two-step reaction. Consequently, multistage model must be used to describe the kinetics of the reaction.

For the initial assessment of the kinetic parameters, the model-independent evaluation of the kinetic parameters was performed. For this, a model-independent method by Friedman analysis was used [17]. Figure 2 and figure 3 show the results of analysis by Friedman for all heating rates. From Friedman analysis, the pre-exponential factor and the activation energy of synthesis were used as input values for the model of non-linear regression.

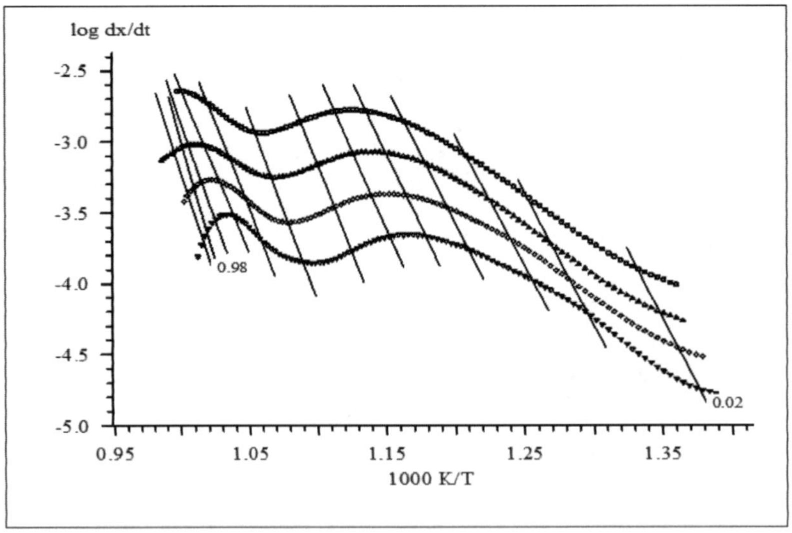

Figure 2. Friedman graphic for the reaction

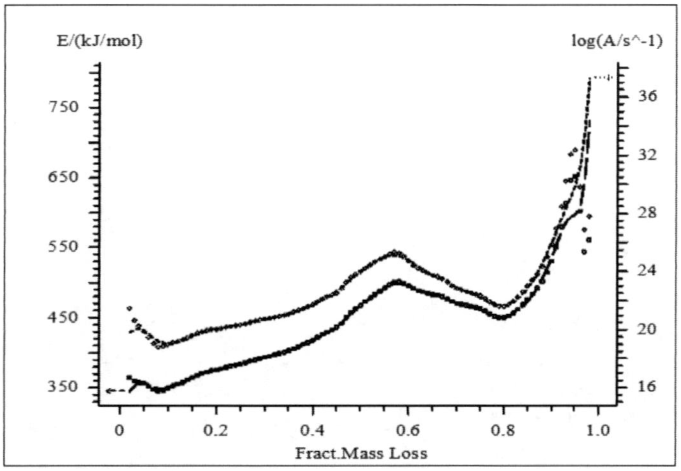

Figure 3. Arrhenius parameters derived from the Friedman analysis

The fitting of the TG curves by the model of non-linear regression was produced by the several kinetic models: model of Ginstling-Bronstein, Yander, model Abraham-Erofeev, etc. The best results were obtained using in the first stage the diffusion model of Yander and diffusion model of Ginstling-Bronstein in the second stage. Figure 4 shows the results from the non-linear regression analysis which summarized in Table 1.

According to the results, the activation energy at the first stage has a high value and is ~292 kJ/mol. Then, the activation energy increases and reaches values of ~ 397 kJ/mol. Such values is significantly higher than that of lithium ferrite ($LiFeO_2$) with a value of 111 kJ/mol [18] and ball milled zinc ferrite ($ZnFe_2O_4$) with a value of 138 kJ/mol [19]. This result may be due to the fact that the non-milled initial reagents. This result may be due to the fact that the non-milled reactants are characterized by low reactivity.

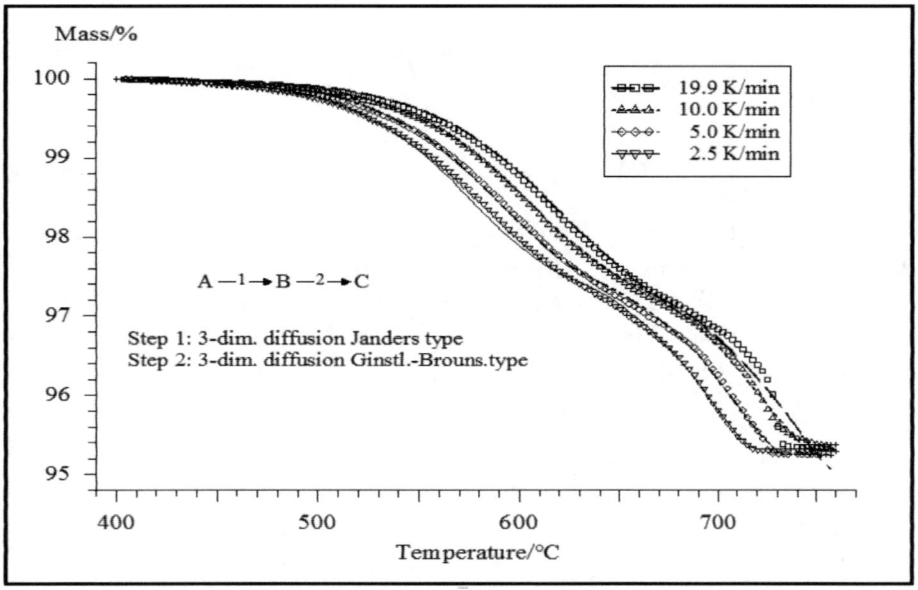

Figure 4. Fit for the non-linear regression of the TG kinetic data. Symbols - experimental points, lines - calculated curves

Table 1. Kinetic results derived from non-linear regression analysis

№	Parameter	Optimal value
0	log A1/s^-1	14.27
1	E1 (kJ/mol)	291.98
2	log A2/s^-1	17.75
3	E2 (kJ/mol)	397.34
4	FollReact. 1	0.48
5	Mass Loss1 (%)	-4.99
6	Mass Loss2 (%)	-4.69
7	Mass Loss3 (%)	-4.74
8	Mass Loss4 (%)	-4.67

Conclusions

The kinetic of Li-Zn ferrite synthesis up to temperatures 800°C was studied in air using non-isothermal thermogravimetric analysis. It was shown that the reaction mechanism of lithium-zinc ferrite synthesis is complex and changes during the reaction. A model-fitting kinetic analysis based on multivariate non-linear regressions was conducted for fitting of experimental TG curves.

It was established that the reaction can be described by a two-step reaction scheme consisting of sequential reaction steps. The best results were obtained using the diffusion models of Yander and Ginstling-Bronstein at the first and second stages, respectively. The fitting results the activation energy values for the ferrite synthesis of ~ 292 kJ/mol and ~ 397 kJ/mol, which correspond to the proceeding of the reaction in the first and second stages, respectively.

The values of the kinetic parameters can be advantageously used for further prediction of the ferrite synthesis in isothermal conditions.

Acknowledgements

This work was supported by The Ministry of Education and Science of the Russian Federation in part of the "Science" program.

Reference

[1] Baba P D, Argentina G M, Courtney W E, Dionne G F and Temme D H 1972 Fabrication and properties of microwave lithium ferrites IEEE *Trans. Magn.* **8** 83–94

[2] Smit J and Wijn H P J 1959 Ferrites. *Physical properties of ferrimagnetic oxides in relation to their technical applications* (Eindhoven: Philips Technical Library)

[3] Surzhikov A P, Pritulov A M, Lysenko E N, Sokolovskii A N, Vlasov V A and Vasendina E A 2012 Influence of solid-phase ferritization method on phase composition of lithium-zinc ferrites with various concentration of zinc *J. Therm. Anal. Calorim.* **109** 63–7

[4] Pointon A J and Saull R C 1969 Solid-state reactions in lithium ferrites *J. Amer. Ceram. Soc.* **52** 157–60

[5] Surzhikov A P, Lysenko E N, Malyshev A V, Pritulov A M and Kazakovskaya O G 2012 Influence of mechanical activation of initial reagents on synthesis of lithium ferrite *Russ. Phys. J.* **6** 672–7

[6] White G O and Patton C E 1978 Magnetic properties of lithium ferrite microwave materials *Magn. Magn. Mater.* **9** 299–317

[7] Patton C E, Edmondson C A and Liu Y H 1982 Magnetic properties of lithium zinc ferrite *J. Appl. Phys.* **53** 2431–3

[8] Sláma J, Šoka M, Grusková A, Dosoudil R, Jančárik V and Degmová J 2013 Magnetic properties of selected substituted spinel ferrites *J. Magn. Magn. Mater.* **326** 251–6
[9] Ghyngazov S A, Lisenko E N, Petyukevich M S and Frangulyan T S 2007 Interaction of lithium-titanium ceramics with air during sintering *Russ. Phys. J.* **50** 134–9
[10] Grusková A, Sláma J, Dosoudil R, Ušáková M, Jančárik V and Ušák E 2008 Microwave properties of some substituted LiZn ferrites *J. Magn. Magn. Mater.* **320** 860–4
[11] Brown M E, Dollimore D and Gallway A K 1980 *Reaction in Solid State. Comprehensive Chemical Kinetics* (Amsterdam, Elsevier Scientific Publishing Co.)
[12] Opfermann J 2000 Kinetic analysis using multivariate non-linear regression *J. Therm. Anal. Calorim.* **60** 641–58
[13] Berbenni V, Marini A and Capsoni D 1998 Solid state reaction study of the system Li_2CO_3/Fe_2O_3 *J. Phys. Sci.* **53** 997–1003.
[14] Surzhikov A P, Pritulov A M, Lysenko E N, Sokolovskii A N, Vlasov V A and Vasendina E A 2012 Dependence of lithium–zinc ferrospinel phase composition on the duration of synthesis in an accelerated electron beam *J. Therm. Anal. Calorim.* **110** 733–8.
[15] El-Shobaky G A and Ibrahim A A 1987 Solid-solid interactions between ferric oxide and lithium carbonate and the thermal stability of the lithium ferrites produced *Thermoch. Acta* **118** 151–8
[16] Surzhikov A P, Pritulov A M, Lysenko E N, Vlasov V A, Vasendina E A and Malyshev A V 2013 Analysis of the phase composition and homogeneity of ferrite lithium-substituted powders by the thermomagnetometry method *J. Therm. Anal. Calorim.* **112** 739–45
[17] Friedman H L 1964 Kinetics of thermal degradation of char-forming plastics from thermogravimetry. Application to a phenolic plastic *J. Polym. Sci. C.* **6** 83–95
[18] Povarov V G and Blyandur E P 1999 Kinetics of the formation of ferrite in the system Li_2CO_3-Fe_2O_3 *J. Kinetica i katalyz.* **40** 520–4
[19] Baranchikov A E, Ivanov V K and Murav`eva G P 2004 Kinetics of the formation of zinc ferrite in ultrasonic field *J. Dokl. Chem.* **397** 146–8

Investigation of oxidation process of mechanically activated ultrafine iron powders

E N Lysenko, E V Nikolaev, V A Vlasov and S P Zhuravkov

Tomsk Polytechnic University, 30 Lenin Ave, 634050, Tomsk, Russia

E-mail: lysenkoen@tpu.ru

Abstract. The oxidation of mechanically activated ultrafine iron powders was studied using X-ray powder diffraction and thermogravimetric analyzes. The powders with average particles size of 100 nm were made by the electric explosion of wire, and were subjected to mechanical activation in planetary ball mill for 15 and 40 minutes. It was shown that a certain amount of FeO phase is formed during mechanical activation of ultrafine iron powders. According to thermogravimetric analysis, the oxidation process of non-milled ultrafine iron powders is a complex process and occurs in three stages. The preliminary mechanical activation of powders considerably changes the nature of the iron powders oxidation, leads to increasing in the temperature of oxidation onset and shifts the reaction to higher temperatures. For the milled powders, the oxidation is more simple process and occurs in a single step.

1. Introduction

Iron powders are widely used in powder metallurgy for catalysts and other magnetic fluids producing [1]. With the development of nanotechnology, the various methods of ultrafine metal powders producing have been developed. These methods include a technique based on an electrical wire explosion [2-4]. Ultrafine iron powders obtained from this technology are differ from micropowders by their characteristics, primarily, ultrafine powders have high reactivity.

In the works [5-8], the strongly influence of the structure and particle size on the oxidation rate of ultrafine iron powders was shown, that leads to decrease in temperature and an increase in initial rate of the oxidation process. It was also shown that the oxidation of the iron powder is a complicated mechanism; the oxidation occurs through the formation of intermediate products [9], and depends on particle size, temperature, and partial pressure of oxygen in the gas phase, the rate of heating of the powder.

It is known that the mechanical activation in high-energy ball mill is one way of preparing of powder material in a highly dispersed state with disordered structure [10, 11]. It should be assumed that a change in the character of the oxidation of nanopowders, subjected to preliminary mechanical activation in a planetary mill, occurs.

Thus, using X-ray powder diffraction and thermogravimetric analyzes, the effect of mechanical activation on the oxidation of ultrafine iron powders obtained electro-explosion method are investigated. It should be noted that the thermogravimetric analysis is a popular method to study the oxidation processes in materials [12, 13].

2. Experimental

Ultrafine iron powders were made by the electric explosion of wire (EEW) in argon using the technique detailed in [14-16]. The powder was stored in hexane to prevent oxidation during shipping and storage. From Brunauer-Emmett-Teller (BET) surface area, the average particles size was calculated (Table 1) and value was equal 100 nm.

The powder was dry milled in air using an AGO-2S planetary ball mill with stainless steel balls and vials. The powder to ball mass ratio was 1:1. The powder was milled according to different regimes, and was divided into three parts: the initial non-milled powder (sample A) and the powders are mechanically activated for 15 (sample B) and 40 minutes (sample C).

The powder oxidation was studied by thermogravimetric analysis, which performed using a STA 449C Jupiter (Netzsch-Geratebau GmbH, Germany) instrument. Samples with a mass of ~10 mg were placed in the alumina crucibles and heated up to 800°C with programmed heating rate of 10 C/min. Both TG and DSC baselines were corrected by subtraction of predetermined baselines run under identical conditions except for the absence of sample. All oxidation experiments were performed under atmospheric pressure conditions using air as the oxidant. The gas flow rate was controlled at 20 mL/min. The Netzsch Proteus software packages were used for data analysis.

Phase composition of the initial powders and oxidation products were determined by XRD phase analysis using ARL X'TRA diffractometer with CuK_{α} radiation. The Powder Cell 2.4 software was used for a full-profile analysis of XRD patterns.

3. Results and Discussion

Figure 1, 2, 3 show the X-ray diffraction patterns for initial ultrafine iron powder (Figure 1) and milled powders during 15 min (Figure 2), 40 min (Figure 3). For the initial powder, all observed reflections belong to the phase of a pure α-Fe. For the mechanically activated powders, a decrease in the intensities and an increase in the half-width of superstructure reflections for α-Fe were observed with increase of the duration of mechanical activation. This clearly indicates that the crystallite size has been decreased with the increase of the milling time. The crystallite size of all samples was calculated using Debye-Scherrer formula and is shown in Table 1.

According to XRD-data shown in Table 1, a monoxide of iron (FeO) is produced as a result of oxidation of iron powder during mechanical activation; and the concentration of FeO increases with the increase of the milling time.

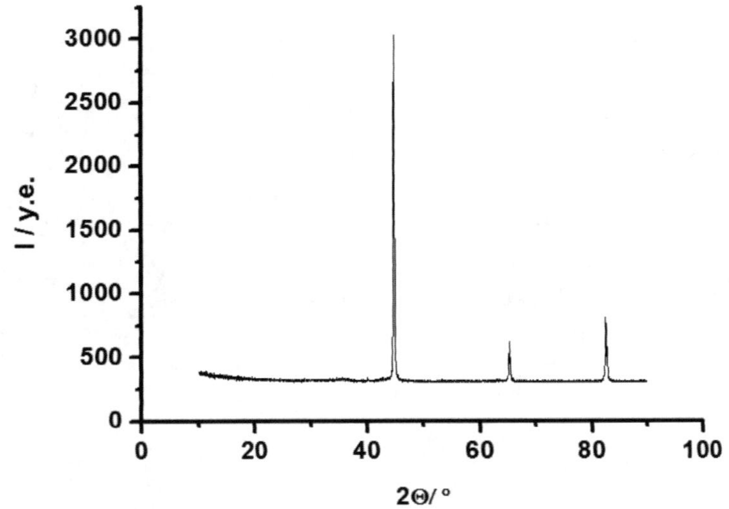

Figure 1. XRD pattern recorded from the non-milled ultrafine iron powder

IOP Conf. Series: Materials Science and Engineering **110** (2016) 012093 doi:10.1088/1757-899X/110/1/012093

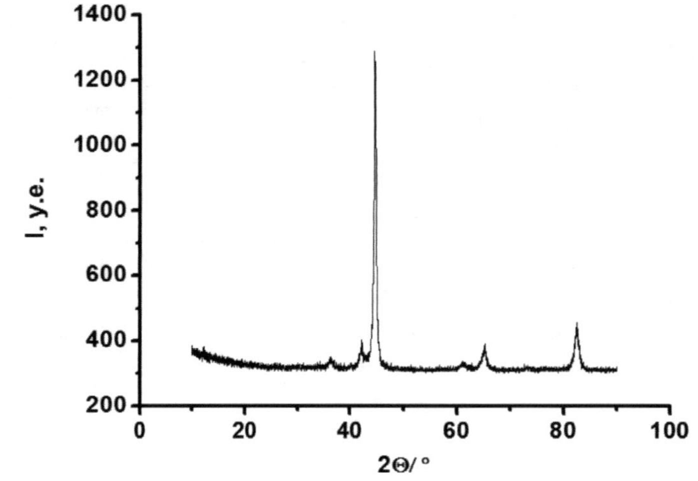

Figure 2. XRD pattern for 15 min milled ultrafine iron powder

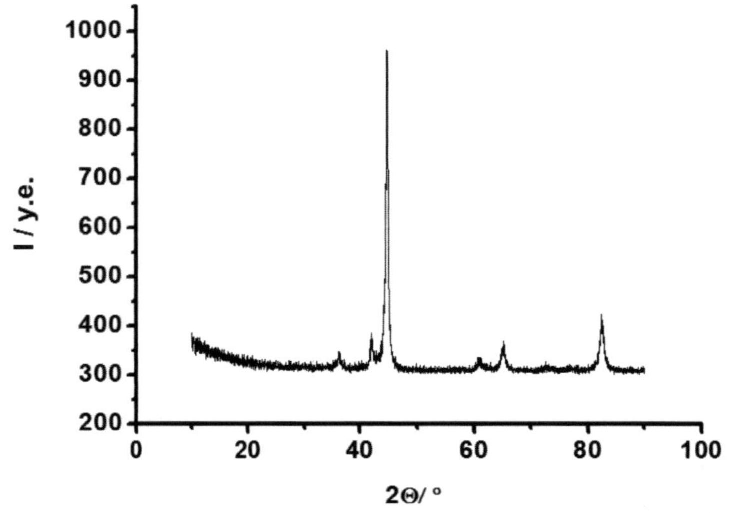

Figure 3. XRD pattern for 40 min milled ultrafine iron powder

Table 1. The results of XRD analysis

Sample	Phase composition (%)		Average particle size (nm)
	α–Fe	FeO	
A	100	-	100
B	81.5	18.5	100
C	77.3	22.7	49

Figure 4 shows the TG curves obtained during the heating of iron powders. Derivative thermogravimetric curves (DTG) are shown in Figure 5. For all samples, the mass increasing is observed up to 800°C, and the mass change values were 39.48 % for sample A, 34.33 % for sample B, and 31.75% for sample C.

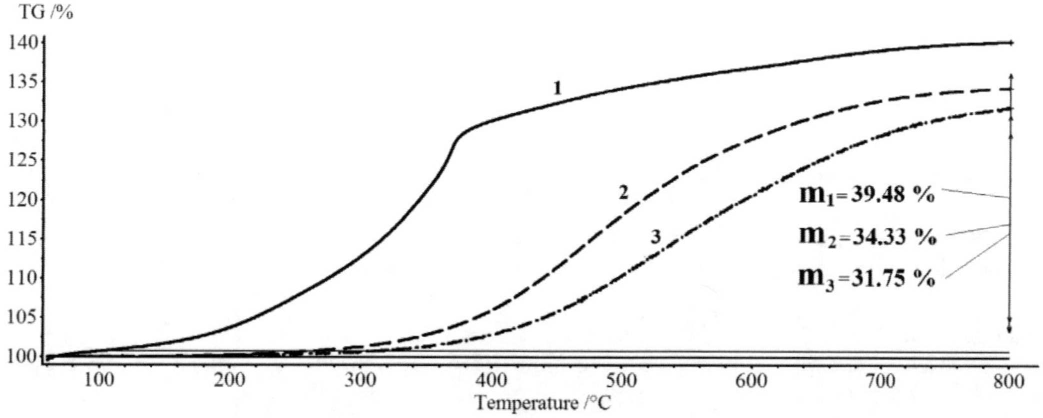

Figure 4. TG curves for sample A (curve 1), sample B (curve 2), and sample C (curve 3)

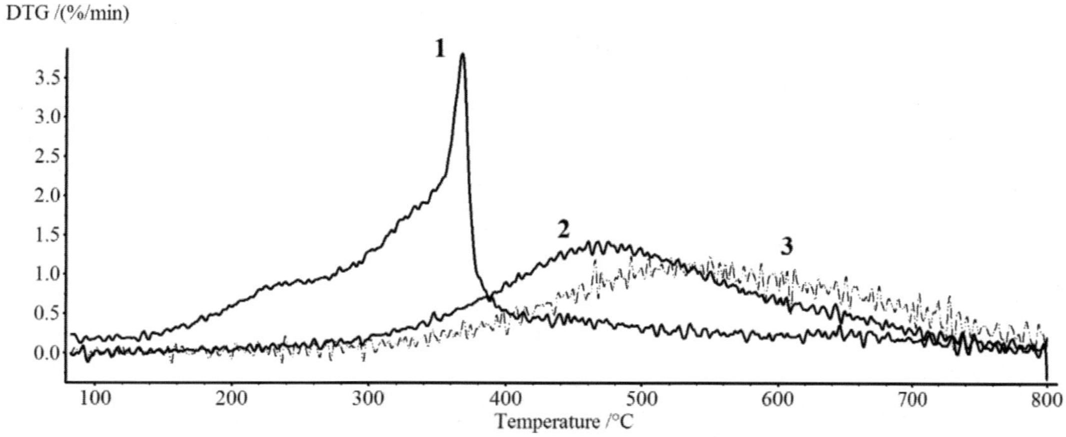

Figure 5. DTG curves for sample A (curve 1), sample B (curve 2), and sample C (curve 3)

For initial powder, the mass change in the end of reaction is close to the theoretical value, which correspond to the reaction $4Fe+3O_2 \rightarrow 2Fe_2O_3$. In this case, the DTG curve (Figure 5, curve 1) indicates a three-step oxidation process. Obviously, the mechanism of the oxidation reaction for initial α-Fe is complex and changes during the reaction. It was shown in [7], the first step is the oxidation of the iron with simultaneously production of Fe_2O_3 and Fe_3O_4 according to the following equation:

$$4Fe + 3O_3 \rightarrow 2Fe_2O_3$$

$$3Fe + 2O_2 \rightarrow Fe_3O_4$$

In the second stage, magnetite is oxidized to hematite to the following equation:

$$2Fe_3O_4 + O_2 \rightarrow 3Fe_2O_3$$

Different nature of oxidation is observed for mechanically activated samples. Start oxidation is shifted to higher temperatures, and one-step process of oxidation takes place. Thus, mass gain decreases with increasing of milling time that is consistent with the results of XRD analysis, which indicates the presence of partially oxidized iron powder.

4. Conclusion

The processes of oxidation of ultrafine iron powders mechanically activated in a planetary ball mill for 15 and 40 minutes were investigated. It was shown that the mechanism of oxidation for initial powder is complex and changes during the reaction. For the milled ultrafine powders, a monoxide of iron is produced as a result of oxidation of iron powder during mechanical activation; and the concentration of FeO increases with the increase of the milling time. In this case, the oxidation processes are simpler and occur in one stage for both milled iron oxides. However, oxidation starts at a higher temperature compared with the non-milled powder. The data, which were obtained for the oxidation processes, can be used to simulate the processes of oxidation and calculate the kinetic parameters of oxidation.

Acknowledgements

This work was financially supported by The Ministry of Education and Science of the Russian Federation in part of the science activity program.

Reference

[1] Brown M E, Tribelhorn M J and Blenkinsop M G 1993 Use of thermomagnetometry in the study of iron-containing pyrotechnic systems *J. Therm. Anal. Calorim.* **40** 1123-30

[2] Kwon Y S, Jung Y H, Yavorovskii N A, Illyn A P and Kim J S 2001 Ultra-fine powder by wire explosion method *Scripta Materialia* **44** 2247–51

[3] Nazarenko O B and Ilyin A P 2008 Nanopowders production by electrical explosion of wires: environmental applications *(In Proceedings of the 3rd Environmental Physics Conference, Aswan)* 135–40

[4] Burtsev V A, Kalinin K V and Luchinsky A V 1990 *Wires electrical explosion and its application in electrophysical plants* (Moscow, Energoatomizdat)

[5] Chen R Y and Yuen W Y D 2003 Review of the high-temperature oxidation of iron and carbon steels in air or oxygen *Oxidation of Metals* **59** 433–68

[6] Wen D, Song P, Zhang K and Qian J 2011 Thermal oxidation of iron nanoparticles and its implication for chemical-looping combustion *J. Chem. Technol. Biot.* **86** 375–80

[7] Lysenko E N, Surzhikov A P, Zhuravkov S P, Vlasov V A, Pustovalov A V and Yavorovskii N A 2014 The oxidation kinetics study of ultrafine iron powders by thermogravimetric analysis *J. Therm. Anal. Calor.* **115** 1447–52

[8] Korshunov A V 2011 Effect of the dispersion of iron powder on patterns of their oxidation at heated in air *J. Izv. TPU* **3** 5–11 (in Russian)

[9] Sanders J P and Gallagher P K 2003 Thermomagnetometric evidence of γ-Fe_2O_3 as an intermediate in the oxidation of magnetite *Thermochim. Acta* **406** 241–3.

[10] Surzhikov A P, Lysenko E N, Malyshev A V, Pritulov A M and Kazakovskaya O G 2012 Influence of mechanical activation of initial reagents on synthesis of lithium ferrite *Russ. Phys. J.* **6** 672–7

[11] Lysenko E N, Surzhikov A P, Vlasov V A, Malyshev A V and Nikolaev E V 2015 Thermal analysis study of solid-phase synthesis of zinc- and titanium-substituted lithium ferrites from mechanically activated reagents *J. Therm. Anal. Calorim.* DOI 10.1007/s10973-015-4849-9

[12] Surzhikov A P, Frangulyan T S, Ghyngazov S A and Lysenko E N 2012 Investigation of structural states and oxidation processes in $Li_{0.5}Fe_{2.5}O_{4-\delta}$ using TG analysis *J. Therm. Anal. Calorim.* **3** 1207–12

[13] Surzhikov A P, Frangulyan T S, Ghyngazov S A and Lysenko E N 2010 Investigation of oxidation processes in non-stoichiometric lithium–titanium ferrites using TG analysis *J. Therm. Anal. Calorim.* **102** 883–7

[14] Beketov I V, Safronov A P, Medvedev A I, Alonso J, Kurlyandskaya G V and Bhagat S M 2012 Iron oxide nanoparticles fabricated by electric explosion of wire: focus on magnetic nanofluids *J. AIP Adv.* **2** 022154

IOP Conf. Series: Materials Science and Engineering **110** (2016) 012093 doi:10.1088/1757-899X/110/1/012093

[15] Yavorovsky N A, Davydovich V I and Bil B A 1995 Apparatus for ultrafine powders manufacturing of inorganic materials by electrical explosion *Ru Patent* 2048278

[16] Yavorovsky N A and Sedoy V S 2009 Method of obtaining ultrafinely powders *Ru Patent* 2359784

Highly-Effective Purification of Air on the Fibrous Filtering Nozzles

O V Galtseva[1], S V Bordunov[2] and S N Torgaev[1, 3, 4]

[1]Associate Professor, National Research Tomsk Polytechnic University, Tomsk, Russia
[2]Superviser, Public Corporation Scientific - Innovation Enterprise “Echteh”, Tomsk, Russia
[3]Associate Professor, National Research Tomsk State University, Tomsk, Russia,
[4]Junior Research Fellow, V.E. Zuev Institute of Atmospheric Optics Russian Academy of Sciences, Siberian Branch, Tomsk, Russia

E-mail: bvv@academ.tsc.ru

Abstract. A series of experiments by air purification on fibrous filtering nozzles was made. It is experimentally shown that the fibrous filter can operate in a wide rate range. The degree of trapping of fine aerosols of glass was 99 % at a linear rate of 0.01 *m/s*. the degree of capture decreased to 85 % at the increasing of filtration rate up to 0.06 *m/s*. Dustiness of the air ranged from 3 to 5 g/m^3 at the course of the experiment. Hydraulic resistance changed from 5 to 25 *mm* of water column. The calculated data of resistance and falling of pressure on fibrous filters are given; these data were received on the equations from various sources in comparison with experimentally obtained data. According to the results of series of experiments the amendment of the well-known Fuchsian equation is calculated for calculation of the resistance of fibrous air filter. This amendment considers a form and defects of surface of the fibers received by centrifugal-spinneret method.

1. Introduction

Currently the polymeric fibrous materials are widely used in the industry as filters for air purification from dust and aerosols, for capture of hydrocarbons. Compared with the fibers from vegetable raw materials the polymeric fibrous materials have greater chemical resistance and can work in wide range of temperatures and pH environment [1]. Besides, these materials have a low moisture capacity, high resistance to ionizing radiation and to bacterial influence.

Possibility of gas purification in devices of the filtering type considerably extends on the basis of the fibrous polymeric filtering materials. In the USSR for thin purification of ventilating emissions of the enterprises (including the nuclear industry) the cell filters with the volume nonwoven filtering fabric (for example, the *FVNR* or *FPP* type) were used. The filtering materials of *FPP* type (Petryanov's filters) were manufactured from perchlorvinyl fibers with diameter from 1.5 to 2.5 microns by *TU* 16-16-2813-84, these materials are used in filters of thin purification and for production of means of individual protection of respiratory organs (*MIPRO*) of "Petal" type. *FVNR* fibrous material is a nonwoven material of polyethylene or polypropylene fibers with diameter of 25

40 μm. The production of these materials was left outside of Russia, and therefore it was necessary to find competitive alternative to these materials.

The following advantages are characteristics of the fibrous polymer of nozzles:

- a higher degree of gas purification from suspended particles compared to gas purification devices of other types;
- versatility (the ability to capture solid particles in dry form and liquid particles of the fog);
- ability of filters to work at any pressure gases;
- good degree of purification at low concentrations of suspended particles in the purified gas (portions of *mg* per 1 m^3 of cleaned gases);
- stability of purification and less dependence from the physical and chemical properties of entrained particles, expense of gases than, for example, at the electrical purification;
- possibility of using of chemically resistant materials;
- easy operation.

2. Experimental

Accordingly we developed a centrifugal-aerodynamic method and installation for receiving fiber on its basis. As a result the layers of fibers are parallel to surface of form-building mandrel. Also the interlacings of threads are on depth of sample of fibrous material. According to the standard opinion this structure is considered nondirectional. Thus elementary fibers are characterized by tortuosity and thickness with different values. The material has pores of complex geometry; dimensions of pores are defined by many technological, physicochemical and other factors. Material differs in high mechanical durability due to interaction between fibers.

The essence of the method is the formation of film of melt of thermoplastic inside of the rotating reactor and the formation of the fibers on the edge of this reactor. The difference of this method from known is that the heating of melt of thermoplastics until the desired viscosity is ensured by film mode of flow inside of rotating reactor with linear rate more than 10 m/s on the edge by heating of reactor walls. The molten fiber at the outlet from the reactor is cooled by air flow [2]. Control of temperature and flow rate of the melt in the installation, the rate of rotation of reactor and some other parameters allowed us to produce fibers with originally specified diameter and other physical and mechanical properties. These fibers have high sorption properties with respect to hydrocarbons and other air pollutants.

The filtration characteristics of the materials obtained from the wastes of polyethylene and polypropylene are significantly higher than those fibers obtained from commodity polymers by standard techniques. This is due to the influence of high-temperature method of producing fibers in air, which leads to structural changes of the polymer at the molecular level. In turn, it leads to an increase of specific surface area and surface energy of the resulting polymeric fibers and also to the appearance of carbonyl and hydroxyl functional groups on the surfaces, which give the properties of ion exchangers to the fibers. The obtained fibrous materials advantageously differ in their physical and chemical characteristics from the traditionally used the Petryanov's fabric and can be recommended for systems of submicron purification of air from pollutants of natural and anthropogenic origin.

The installation for testing of fibrous filters is shown in figure 1.

The generator of dustiness was used for the formation of contaminated dust-gas mixture supplied to the volume of the filter; the researched filtering partition from a mixture of polypropylene fibers was installed in this filter. Particles of glass with particle sizes from 0.1 to 3.0 μm were used as the pollutant of air. Air was supplied to the lower part of the generator of dustiness by blower through the receiver and the valve.

Due to this, the suspended particles of glass were carried away from the generator of dustiness; particles were moved into the studied filter and then into the sampler. The sampler with porous partition with pore size of 0.001 μm (porous metal ceramics) is intended for catching of the dispersed particles in dusty air, passing through the studied filtering partition in the filter. Controlling of air consumption and, therefore, the rate of filtration was carried out by indicators of rotameter, and this process regulated by the valve. The particles of glass were moved from the hopper by auger into the generator of dustiness. The dust content in air was regulated by the number of turns of the auger. The exemplary manometer was used for measuring of the aerodynamic resistance of the filter.

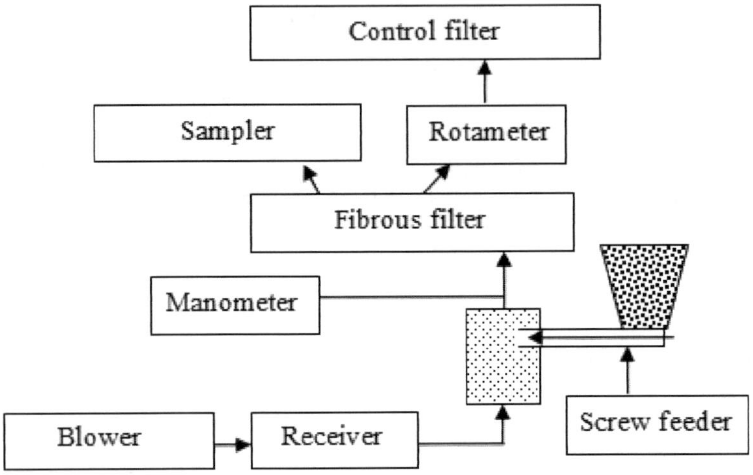

Figure 1. Scheme of experimental installation for testing of fibrous filters on efficiency of air purification from dust.

A control filter was used for additional control catching of dust particles; in the case of work with toxic substances it was used for preventing of hitting of those particles into the atmosphere of the laboratory space. Determination of the efficiency of work of the filter was carried out by the gravimetric method by weighing the filter and sampler on an analytical balance. The technique allows sufficiently accurately determining the efficacy of the filter. Particles of glass with particle sizes from 0.01 to 1.0 *μm* were used in experiment.

As a result of the conducted research it was found that the fibrous filter can operate in a wide rate range. The degree of catching of fine aerosols of glass was 99% at a linear rate of 0.01 *m/s*. At increasing rate of filtration up to 0.06 *m/s* the degree of catching decreased to 85 %. Dustiness of the air ranged from 3 to 5 g/m^3 in the course of the experiment. Hydraulic resistance changed from 5 to 25 *mm* of water column. The possibility of using of designed fibrous filter material at rates of 0.06 *m/s* is the undoubted advantage, as fibrous filters from fiberglass, cermet, ceramic and granular filters operate at the rate range from 0.01 to 0.02 *m/s*; the Petryanov's filters (*FPP*-15-1.7) can operate only at the rates smaller than 0.01 *m/s*.

The special catching device was used to carry out the experiments for catching of dust on the fibrous filters from recycled materials with the parallel use of standard filters *AFA*. It consists of the filter holder, filter from hydrophobic material of *FP* type with a working area of 20 cm^2 (*AFA-CPS*-20), aspirator, allowing the passage of air through the filter with volume rate of 20 *l/min*, and the above-described non-standard device.

Multi-stage air cleaning system used for cleaning and conditioning of atmospheric air with an initial dust content of up to 1.0 mg/m^3 (at the possible short-term content of dust up to 10 mg/m^3) moved into "clean" industrial premises.

On first stage the air filters of third class were used and intended for catching of dust particles with size from 5 to 50 *μm*; those filters were located before the electric air heaters. The filter consists of casing, the fixed zigzag lattice of rods strengthened in casing. The filter material is manually placed on a lattice and fixed by the clamping brackets on bends. Thus corrugations are formed; the filtering surface develops by four times in comparison with the entrance section of the filter. Dusty material removed from the filter and replaced on new at reaching the maximum of allowable resistance.

3. Results and considerations

In our opinion an approximate processing of the obtained results by the method of Davis the most accurately describes the process of purification of gases. A more precise definition of the geometric structure of fibers of the investigated filter has shown that mechanism of touch and diffusive mechanism are the main mechanisms of catching of aerosol particles. These can explain the insignificant hydraulic resistance and ability to regeneration of fibrous filters. In addition, according to the basic provisions of the theory of filtration the most effective filters are made from compositions of fibers from different materials and diameters. The developed production technology of fibers from secondary raw materials allows receiving the filtering material with such structure.

Table 1 shows the calculated data of the resistance and pressure drop in fibrous filters obtained from equations of various sources compared to experimental data.

Table 1. The results of calculations of the resistance coefficient and pressure drop in fibrous filter with diameter of fiber 100 μm.

H_f, *m*	α	*u*, *m/s*	K_f: 1	K_f: 2	K_f: 3	ΔP_l, *Pa*
0.16	0.19	0.0885	$K = 4\alpha$ / -1.15·$ln\alpha$ – 0.75 (1)	[3]	0.666	943
0.16	0.19	0.0885	$K = 2B\alpha$ / - 0.5·$ln\alpha + \alpha - \alpha^2/4 - 0.75$ (2)	[5]	$B = 2.0$ $K = 4.72$	6684
					$B = 1.5$ $K = 3.5$	5013
0.16	0.19	0.0885	$K = 4\alpha$ / -1.15·$ln\alpha$ – 0.5 (3)	[6]	0.539	763
0.16	0.19	0.0885	$K = 4\alpha$ / -1.15 $ln\alpha$ – 0.5 (4)	[6]	0.539	763
0.46	0.222	0.0461	$K = 4\alpha$ / -1.15·$ln\alpha$ – 0.75 (5)	[3]	0.88	1870
0.46	0.222	0.0461	$K = 2B\alpha$ / -0.5·$ln\alpha + \alpha - \alpha^2/4 - 0.75$ (6)	[5]	$B = 2.0$ $K = 8.3$	17681
					$B = 1.5$ $K = 6.2$	3262
0.46	0.222	0.0461	$K = 4\alpha$ / -1.15 $ln\alpha$ – 0.5 (7)	[6]	0.722	1531

The calculations are performed for model filter with a filtration area of $19.625 \cdot 10^{-4}$ m^2. The following symbols are shown in the columns of the table 1: 1 - the equation, 2 - source of information, 3 - result of the calculation, H_f - height of the filter (*m*), *A* - relative packing density (numerically equal to the volume of fibers in volume of the filter), *u* – the rate of filtration, *B* - the coefficient of heterogeneity of the packing fibers (at b = 2 all fibers parallel to each other and perpendicular to the flow), K_f - resistance coefficient of filter, ΔP_l - head loss on the filter, defined by the equation (8) [4]

$$K_f = \Delta P_l \cdot r^2 / u\mu H . \quad (8)$$

Calculations by Fuchsian equation (1) in [3] show the closest to the experimental results of pressure drop and resistance coefficient of the fibrous filter. The experimental data obtained on the installation according to figure 1.

Figure 2 shows the results of calculations of the resistance coefficient by equation (1) [3] and results experimentally obtained on the filters with fiber with average nominal diameter 2.2, 30, 50 and 100 *mkm* with porosity 80.56, 82.2, 82.6 and 83.6%, respectively.

As can be seen from figure 2 the experimentally obtained coefficient of resistance increases in comparison with the calculated coefficient with the increase in fiber diameter; it is due to one factor. The number of defects on the surface of the fibers and shape of section of the fibers increase with increase of diameter of the fiber. It increases the turbulence of the air flow in the filter and, as a consequence, the coefficient of resistance.

These defects of fiber form almost did not influence on the results of experiments at the filtering purification of water with higher density and viscosity, but these defects became apparent in experiment with purification of air.

However, the experimental results are described quite accurately by the linear equation (9)

$$K_e = -0.0016D + 0.6955, \tag{9}$$

where D is the diameter of the fiber, which can be used as a correction factor to equation (1) [3], and then:

$$K = (4\alpha / -1.15 \cdot ln\,\alpha + 0.75) + K_e = 4\alpha / -1.15 \cdot ln\,\alpha + 0.016D - 0.0545, \tag{10}$$

K_e is actually a measure of difference of the resistance coefficient of the fibers (obtained by centrifugal-spunbond method) from the typical fibers with section in the form of a circle.

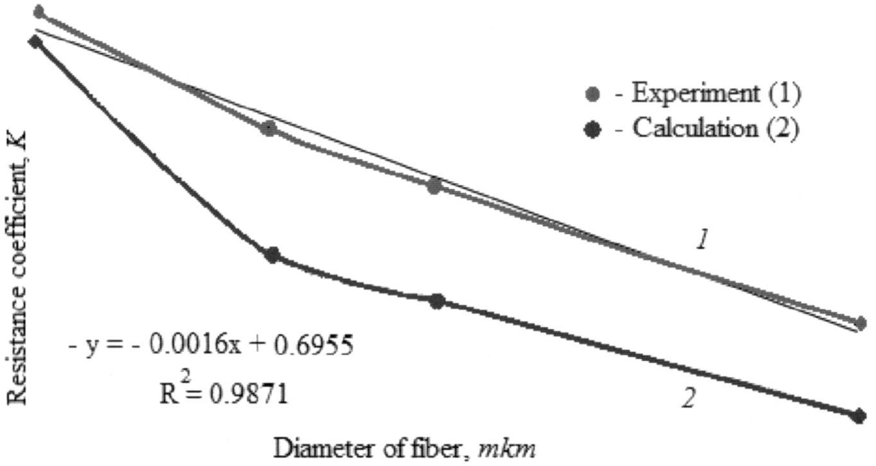

Figure 2. Dependence of the resistance coefficient on the diameter of the fiber.

4. Summary

According to the results of series of experiments the amendment of the well-known Fuchsian equation is calculated for calculation of the resistance of fibrous air filter with regard to form and surface defects of the fibers.

References

[1] Galtseva O V and Bordunov S V 2015 *IOP Conf. Series: Materials Science and Engineering* **81** 112010 DOI: 10.1088/1757-899X/81/1/012062
[2] Bordunov S V 2010 RF Patent 2388854 Byull. Izobret. **13**
[3] Fux N A and Stechkina I B 1962 *Doklady Akademii Nauk USSR* **147 (5)** 1144
[4] Whitby K T *et al* 1962 *American Society of Heating, Refrigerating and Air-conditioning Engineers J.* **79 (6)** 79
[5] Langmuir I 1962 *The Collected Works* **10** 394
[6] Happel J 1959 *AIChE J.* **5** 174

IOP Conf. Series: Materials Science and Engineering **110** (2016) 012095 doi:10.1088/1757-899X/110/1/012095

Assessment of quartz materials crystallinity by x-ray diffraction

M Korovkin[1], L AnanIeva[1], T Nebera[2], A Antsiferova[1]

[1]National research Tomsk Polytechnic University, Russia
[2]National research Tomsk state University, Russia

E-mail: mvk@tpu.ru

Abstract. The estimated degree of crystallinity of natural and synthetic grown quartz and quartzite by calculating the x-ray diffraction patterns. It is shown that the index of crystallinity of natural quartzite varies widely, reflecting the different degree of their transformation. The highest values of the index of crystallinity are characterized natural and synthetic single crystals of quartz.

1. Introduction

Despite of the fact there were numerous studies of structural and physical properties of quartz, the interest in quartz content is not waned. This is because of the wide use of products made from natural quartz, the usage of synthetic quartz crystals with the desired properties and natural polycrystalline quartzite as a source of high-pure quartz raw materials, as well as to study the conditions of their mineralization and recrystalization.

To evaluate the degree of conversion and perfection of crystalline quartz [1] it was offered to use the "crystallinity index", measured by the parameters of the x-ray diffraction spectra. In the paper [2] it was proposed to use the method of infrared spectroscopy for evaluation the degree of chalcedonies conversion.

The evaluation of the degree of crystallinity can be detected with the most pure forms of natural quartz material – quartzite [3...5], which can be used in high technologies. This paper describes the study of crystallinity index of natural, synthetic quartz crystals and natural polycrystalline silica quartzite that was held with the XRD method.

2. Research methodology

To determine the degree of quartz material conversion, samples of fine-grained quartzite of the Antonov group of deposits [6] (the Deposit "Sopka-248" and "Belokamensky"), quartzite deposits (Borel-Saridag, Buryatia) [7], granular quartz (a field is Malokutulak, Yakutia), natural smoky quartz (Polar Urals), natural Brazilian optical quartz were studied.

In addition, the samples of artificially grown quartz from different conditions were studied for comparison with obtained results. These are: highly pure quartz synthesized in the laboratory of the Institute of Mineralogy SB RAS, Novosibirsk, Russia (samples provided by L. G. Shcherbakova), samples of synthetic quartz with an admixture of aluminum and polychrome zoned crystal of quartz with an admixture of two or trivalent iron [8] grown in the all-Union research Institute of synthesis of mineral raw materials (samples provided by V. Balakirev).

Furthermore, samples of fused quartz and of technical quartz glass was also studied with XRD method.

For this research fine-pounded samples were taken, which were compressed into a "pill". The measurements were performed on the diffractometer X"Pert PRO. Radiographs were taken with increments of about 0.02 in the range 5...70 deg., with the rotation of 30 rpm and a shutter speed of 0.1 sec at a point.

After filming on the diffractometer X Pert PRO x-ray samples under Cu Kα–radiation with a minimum speed and higher sensitivity, x-ray spectra was received. They cover the whole range of possible angles.
If necessary, radiographs X-axis can be derived interplanar distances, calculated in accordance to the software in the form of labels on the Y – axis the intensity of diffracted lines. The results of the registration of x-ray diffraction spectra were processed by a special program with data output in digital form [9].
The index of crystallinity Kci was carried out according to the method of Murata and Norman [1] who proposed to use a multi-blade block parameters (quintuplets) peak in the region of 67°and 69° on the x-ray diffraction pattern. All the obtained radiographs of the estimated maximum quintuplets peak corresponds to the value of 2θ = 67,7797° ± 0, 0003, unlike proposed in [1], the values of 2θ = 67,74°.

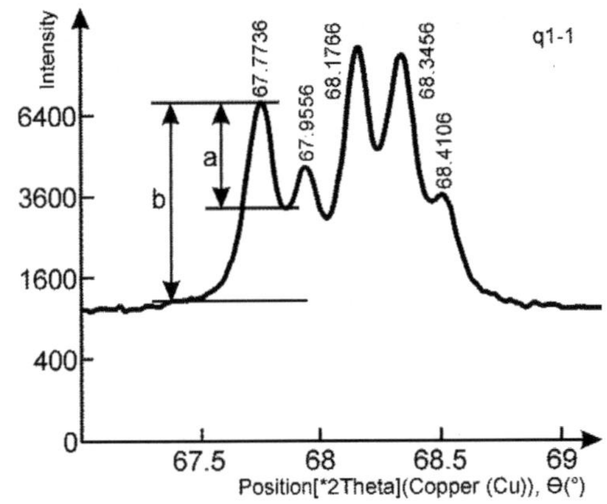

Fig. 1. An explanation of the calculation of the crystallinity index by the method of Murata and Norman [1] in the parameters produce multiple peak in the region of 67°and 69° on the x-ray diffraction pattern

To calculate the index of crystallinity according to the formula K=10 F a/b proposed in work [1] the experimental values of the intensity of the peak at 2θ = 67,77° (Fig. 1) was used: a/b – ratio of the intensity of the peak at 2θ = 67,77° to the intensity of the background; F is the scaling factor used for assessment of the crystallinity index of K = 10 reference spectra crystal optical quartz (Madagascar). In the calculations, the scaling factor was taken equal to the unit.

3. Experimental results and discussion

In diffraction patterns of fused quartz the multiple peak at 67°and 69° is not observed indicating the absence of crystalline phases (table), that is also confirmed by the results of the studies by infrared spectroscopy.

Table. The index of quartz crystallinity of different materials, calculated by the method of x-ray diffraction

Sample	Crystallinity index Kci
Fused quartzite / quartz	0
Quartzite (deposit Sopka-248)	2.88…6.78
Quartzite (deposit Belokamensky)	4.65…6.71
Superquarzities (deposit Bural-Sardag)	6.66…7.34
Granular quartz (deposit Malokutulak, Yakutia)	7.84…8.24
Natural smoky quartz	7.33…7.48

Synthetic quartz (Institute of synthesis of mineral raw materials)	7.32...7.84
Synthetic quartz (Institute of Mineralogy SB RAS)	7.20...7.32
Brazilian optical quartz	7.51…7.55
Quartz glass	8.48…8.57

Crystalline quartzite of "Sopka-248 and Belokamensky" deposits are characterized with a wide range of calculated values of crystallinity index. The cleanest fine-grained varieties of quartzite have low values of $K \approx 2.88$.

There is a change quartzite chemical composition and color with depth, as well as from the Central portions of the ore body to the periphery; the degree of their crystallinity increases up to values of 6.67…5.38. In local areas, especially in areas of high fragmentation, initially chemically pure quartzites under the influence of supergene processes decrease their quality characteristics and the degree of crystallinity increases up to 6.7.

Increasement in the degree of crystallinity of quartzite from Antonov group of deposits is connected to the irregular processes of metamorphism of siliceous biogenic sediments, resulting in the crystallization of amorphous silica and a crystalline phase α-quartz, which is also confirmed by infrared spectroscopy [10].

Quartzites with superquality from field Bural-Sardag are comprised larger grains with inclusions of porphyritic elongated, subparallel orientation, which indicates their formation (recrystallization) under prolongated dynamic stress [7]. In the thin cross sections the regenerated fragments of quartz were found. High values of the index of crystallinity (7.34…6.66) indicate a high degree of their transformation.

Granular quartz of Malokutulak deposit came under the processes of dissolution due to the passing of the intergranular boundaries of the mineral-forming solutions and the subsequent dynamic recrystallization, which led to formation of polygonal structure of quartz with sinuous grain boundaries, characterized by high K values in the range of 7.84…8.24 comparable to the estimates of the crystalline natural quartz.

Crystalline samples (natural and synthetic) quartz have the highest values of the index of crystallinity. Minimal variation in Kci shows the sustained processes of crystallization and formation of a committed and well-ordered crystals.

Technical quartz glass is characterized by the absence of the far right in its structure. It consists of microcrystal grains of quartz, and in the spectra of x-ray diffraction reflects the presence of crystallinity (see table).

4. Conclusion

Obtained by x-ray diffraction values of the index of crystallinity, reflect the degree of conversion of quartz raw materials: the highest values characterized by granular and crystalline (natural and synthetic) quartz. Despite of the fact that values calculated by the proposed method, the index of crystallinity of quartzite are very relative, it can be used for comparative evaluation of the degree of metamorphism of the quartzites (in combination with the methods of infrared spectroscopy) within the same field.

References

[1] Murata K and Norman M 1976. Am. J. Sci. **276** 1120
[2] Plusnina I. 1978 Doklady Akademii nauk USSR **240 (4)** 839
[3] Korovkin M, Ananjeva L, Nebera T and Razva O 2013 Abstr. Int. Conf. Crystallogenesis and Mineralogy (Novosibirsk, Russian) p. 176
[4] Korovkin M, Ananieva L and Antsiferova A 2012 Isvestaya Tomsk Polytechnic University **320 (1)** 47

IOP Conf. Series: Materials Science and Engineering **110** (2016) 012095 doi:10.1088/1757-899X/110/1/012095

[5] Razva O, Anufrienkova A and Korovkin M 2014 Sovremennie naukoemkie technologii. **7** 27
[6] Zakharov V 2000 Mining J **7** 7
[7] Nemchinova N and Kletz V 2004. J. of fundamental research **5** 48
[8] Balakirev V et al 1979. Mineralogy and crystallophysica jewelry varieties of silica. Moscow, Nedra Pabl. 149 p.
[9] Description for X-ray diffractometer X'Pert PRO (2013) The Analytical X-ray Company 20 p.
[10] Korovkin M, Ananieva L, Antsiferova A. 2011 Proc. 10 Int. Congress for Applied Mineralogy (ICAM) Trondheim, Norway Springer press p 403

Wear resistance of a metal surface modified with minerals

S V Kislov[1], V G Kislov[1], P V Balasch[2], A V Skazochkin[3,5], G G Bondarenko[4] and A N Tikhonov[4]

[1] Research and Production Association "Geoenergetika", Kaluga, Russia
[2] Scientific and Production Center "Mico-Tech" Limited Liability Company, Moscow, Russia
[3] Kaluga branch of the Russian Presidential Academy of National Economy and Public Administration, Kaluga, Russia
[4] National Research University "Higher School of Economics", Moscow, Russia

[1, 2] E-mail: geoen@mail.ru
[3, 5] E-mail: avskaz@rambler.ru
[4] E-mail: gbondarenko@hse.ru

Abstract. The article describes the advantages of the new technology of mineral coating of metal products for the friction pair of mechanical systems. It presents the research results of the wear rate of the samples made of 12X13 steel (X12Cr13) with mineral layers, in the experiments with a piston ring sliding inside a cylinder liner with grease. The wear rate of the samples with mineral layers is lower almost by two factors than that of the samples made of grey foundry iron and untreated samples. As the result of slip/rolling abrasion tests of parts with mineral layers under conditions of high contact pressure, a suggestion was made concerning probable mechanics of surface wear.

Introduction

The core indicators of the quality of machine's parts, assemblies and mechanisms are reliability and lifespan, which are greatly defined by the properties of the surface layers of parts and junctions: wear resistance, corrosion resistance, friction rate, contact rigidity, seating stability, connection tightness. The reason for a short lifespan of parts and other structural components may be significantly associated with tear-and-wear and corrosive deterioration of their surface layers. To eliminate or slow down the processes taking place on the border between metal and environment, which have a negative effect on the operability of materials, various types of surface treatment are used.

Besides multiple methods involving coatings, films and protective layers, there is a rapidly developing modern technology of changing physical and chemical properties of surface layers of metal parts by means of modification. In particular, industrial application of natural minerals for hardening surface layers of structural materials is finding its niche, due to the effects, combination of which leads to enhanced wear resistance, anti-friction, anti-corrosion and other specific properties of a friction pair.

The basis for a new original technology of creating mineral layers enhancing wear resistance of metal parts was created in Russia in 2009-2013 [1]. The mineral layers protect parts of all-purpose mechanical systems from wearing off, extend their lifespan and reduce energy consumption due to significant elimination of mechanical losses in machines and mechanisms [2]. The properties of mineral layers created by RPA "Geoenergetika", have been confirmed by experiments and field tests

[5] To whom any correspondence should be addressed.

conducted by partners from the Federal Institute for Materials Research and Testing, BAM, Germany [3].

The technology of creating mineral layers was introduced to the Russian and European markets in 2012-2015 and has already found application in various industries: metallurgy, machine building, power engineering, ship construction.

The mineral layers created enhance wear resistance of friction pairs from 3 to 10 times, feature unique tribotechnical parameters and can be applied to water lubricated metal parts, in environments with abrasive particles, and under conditions of unlubricated friction [2, 4, 5].

The created layers have unique properties, because parts coated by mineral layers of RPA "Geoenergetika", demonstrate great performance:

- in seawater, salt mist, hydrogen sulphide, increased humidity, abrasive dust and other aggressive environments;
- under thermal cyclic loads in a wide temperature range.

Technology of mineral layers' creation:

- does not require special basins, furnaces, vacuum chambers or specific conditions;
- has no fundamental limitations of mass and dimensional properties;
- does not change the geometry of the treated parts, does not require alteration of designs;
- is environment-friendly, has no harmful effect on human health and environment.

The present work investigates the wear rate of the samples before and after application of mineral coatings, as well as in comparison with some other samples.

Experiment details, results and their discussion

The technological process of mineral layers' creation consists in embedding nanocomposite mineral powders into the crystal lattice of the treated metal by means of ultrasound and electrospark treatment, as well as ball and roller knurling. The details of the technological process of mineral layers' creation, as well as structural changes on the surface and subsurface layers of metal samples (chromium steels, titanium alloys) are described in the work [4]. As the result of the combination of technological procedures changing the surface, microhardness of the surface layers is increased. Microstructural researches have demonstrated that a modified layer is observed, with granular structure, clearly different from the remaining mass [4]. The enhancement of hardness and strength properties is conditioned by the reduction of grain size.

This work investigates deterioration of samples with mineral layers and without special treatment, by means of an experiment including sliding of a piston ring in a cylinder liner lubricated with machine oil (the methodology is described in works [6, 7]) and an experiment of slip/rolling under conditions of high contact pressure (a thorough description of the method can be found in work [8]).

For all the experiments modified mineral layers according to NPO "Geoenergetika" technology were created on the surface of metal samples produced by lathe turning from 12X13 steel (Russian equivalent of EU X12Cr13 steel) rods in the shape of piston rings [1]. The thickness of the mineral layers created varied from 5 to 20 micron.

Wear rate was compared individually for a piston ring with untreated surface and a piston ring with modified surface. Synthetic oil based on polybutyleneglycol (GLYMOT PBG B20) and machine oil were used in the experiments. The mode involving machine oil lubrication was chosen because lube is typically present in most practical situations.

"PU", "PV", "PW" and "PT" samples in figure 1 designate various combinations of surface treatment conditions and metal from which the samples are produced.

Wear rates of the rotating disks were compared with wear rates of GGL20HCN European brand grey foundry iron disks with high carbon content (3.7 weight percent of carbon and 2.0 weight percent of silicon) which is normally used for production of cylinder liners and brake disks. The curly bracket and "X12Cr13" title in figure 1 denote the results acquired for a 12X13 (X12Cr13) steel piston ring without any treatment.

Abrasion slip-rolling experiments were conducted under conditions of high contact pressure P = 2.25 GPa. Experiments involving sliding with machine oil were conducted under much lower contact pressure, of about 100 MPa.

Figure1 presents wear rates of the surface of PT series rings (PT-502, PT-503, PT-504 and PT-201, PT-202, PT-203) made of 12X13 steel with mineral layers which were compared with wear rates of the surface of untreated rings made of 12X13 steel, 20X13 steel and grey foundry iron.

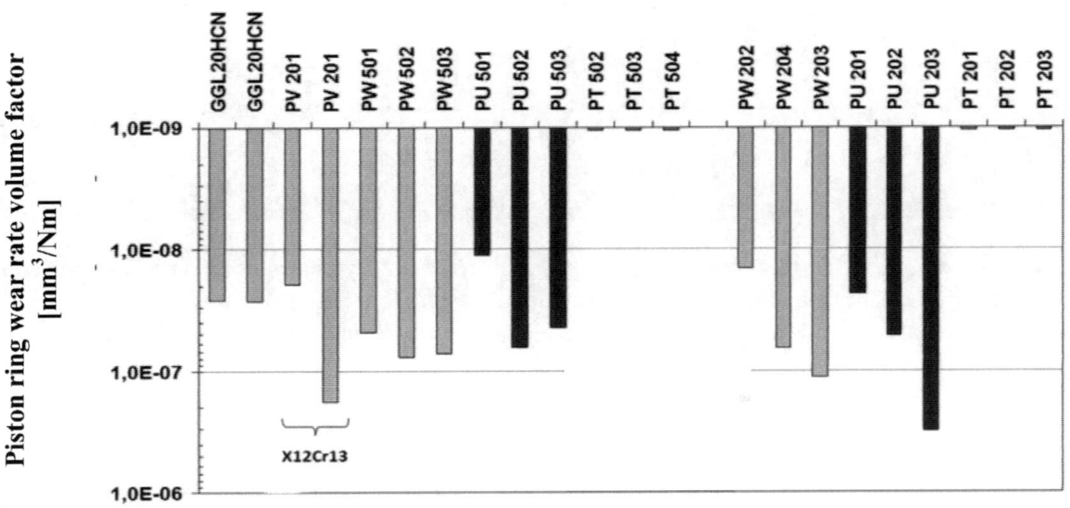

Figure 1. Wear rate volume factors for piston rings sliding under mixed lubrication conditions over cast iron and 12X13 steel treated by RPA "Geoenergetika" in machine oil (F_N=50 H; v =0,3 m/s; T_{oil} = 170°C; s = 24 km)

Evidently, creating mineral layers on the surface of PT series rings provided wear rate which is by two factors lower than that of standard samples made of grey foundry iron and untreated 12X13 and 20X13 steel rings.

Figures 2 and 3 clearly demonstrate that after 10 million cycles of slip-rolling abrasion test the surfaces with mineral coating, exposed to friction, became smoother. It is possible that the technology of creating a thin modified layer on a metal surface, leading to its local strengthening and enhanced hardness [4], preconditioned good slip-rolling friction resistance of the samples made of 12X13 and ШХ15 (Russian equivalent of 102Cr6 German steel) steels, which was probably defined by general residual compression stresses.

Figure 2. The picture of a wear track on a treated 12X13 steel surface performed by means of an optical microscope (T = 120 ^{0}C, P = 1.5 GPa, 10 million cycles, slip-rolling).

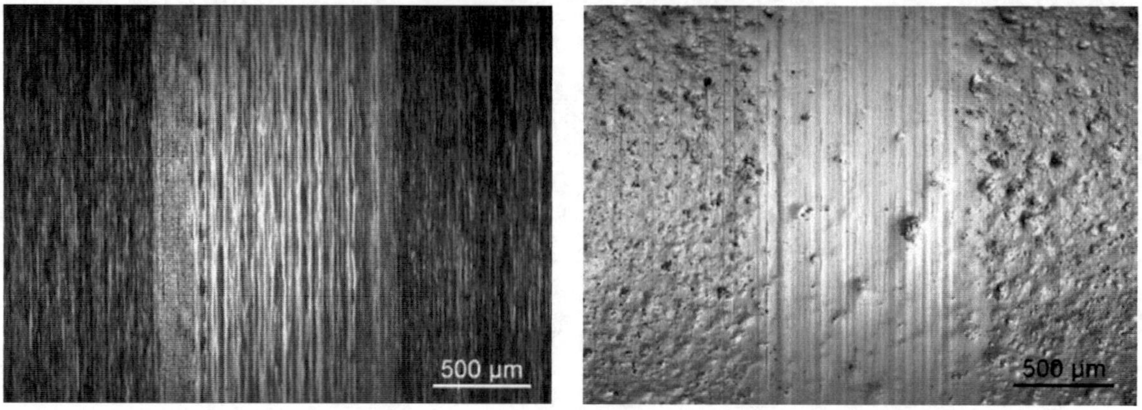

a **b**

Figure 3. The picture of a wear track after slip-rolling performed by means of an optical microscope (a - surface of the sample made of ШХ15 steel without mineral coating; b - cylindrical sample made of ШХ15 steel with mineral coating; T = 120 ^{0}C, P = 1.5 GPa, 10 million cycles, slip-rolling).

The thin layer of the hardened surface of steel samples modified by minerals has demonstrated good resistance to slip-rolling procedure [7, 8] after ten million cycles at the temperature of 120 ^{0}C and pressure P = 1.5 GPa. Under such conditions the effective thickness of the oil film h_{min} at 120 ^{0}C was 0.027 micron [8] which is the evidence of boundary lubrication mode, so the wear of the surfaces is mainly conditioned by physical and chemical interaction taking place at the spots of actual surface contact. Besides, the pictures in figure 2 show microcavities randomly scattered over the working surface of the sample. It is due to these created microcavities that the oil is detained when lubrication is suspended, which may be listed among the advantages of the technology [4] because performance capabilities and equipment safety are enhanced during start-and-stop operations, e.g. spindle bearings in case of insufficient lubrication.

Conclusions

1. The mineral layers created on the surface of the samples presented by 12X13 steel rings provided wear rate which is almost by two factors lower than that of the samples made of grey foundry iron and untreated 12X13 and 20X13 steel samples during experiments involving piston ring sliding in a cylinder liner with machine oil lubrication.
2. When testing slip-rolling abrasion of parts with mineral layers under high contact pressure conditions, boundary lubrication mode is possible. That is why the surface wear is only conditioned by physical and chemical interaction at the spots of actual surface contact.

Acknowledgements

The authors are grateful to Dr. Mathias Woydt, head of Tribology and Wear protection division of Federal Institute for Materials Research and Testing, BAM, Germany, for performing sample measuring operations and consulting.

References

[1] Kislov S V, Kislov V G and Lazarev S Yu 2011 Creation method of combined mineral surface layer for metal products protecting it from action of aggressive medium and with designated tribotechnical parameters *Patent on invention №2421548 from 20.06.2011, patent application 2009142259/02 from 18.11.2009*

[2] Elagina O Yu, Gantimirov B M, Tomskij K O, Mazurkevich AN and Kislov S V 2013 Research of tribotechnical parameters of coat by metal to metal friction without oil *Perspective materials* **7** 56-60

[3] Kislov S, Kislov V, Ostrovsky P and Woydt M 2012 Localized hardening and structuring of steel by a simple process. *53 Tribologie Fachtagung «Reibung, Schmierung und Verschleiß» der GfT Göttingen, Herstellung: Druckservice Zillekens* **2** 49/1-7

[4] Kislov S V, Kislov V G, Skazochkin A V, Bondarenko G G and Tihonov A N 2015 Effective mineral coats for surface hardening of metal products *Metals* **4** 56-63

[5] Polovinkin V N 2009 *Nanotechnology in ship construction* Academician A.N. Krylov Reserch institute St-Peterburg 132

[6] Woydt M and Kelling N 2003 Testing the tribological properties of lubricants and materials for the system "piston ring/ cylinder liner" outside of engines *Industrial Lubrication&Tribology* **55 5** 213-222

[7] Löffelbein B, Woydt M and Habig K-H 1993 Sliding friction and wear of ceramics in neutral, acid and basic aqueous solutions *WEAR* **162-164** 220-228

[8] Scholz C, Spaltmann D and Woydt M 2011 Slip-rolling resistance of thin films and high toughness steel substrates under high Hertzian contact pressures *WEAR* **270 7–8** 506–514

Study the properties of activated carbon and oxyhydroxide aluminum as sorbents for removal humic substances from natural waters

L N Shiyan[1], K I Machekhina[1*], E N Gryaznova[2]

[1]Department of General and Inorganic Chemistry, Institute of High Technology Physics, Tomsk Polytechnic University, 634050, Tomsk, Russia.
[2]Institute of Water, Tomsk Polytechnic University, 634050, Tomsk, Russia.

*E-mail: mauthksu@yandex.ru

Abstract. The present work relates to the problem of high-quality drinking water supply using processes of adsorption on activated carbon and aluminum oxyhydroxide for removal humic-type organic substances. Also the paper reports on sorbtion properties of the activeted carbon Norit SA UF and oxyhydroxide aluminum for removal humic substances. It was found out that the maximum adsorption capacity of activated carbon to organic substances is equal to 0.25 mg/mg and aluminum oxyhydroxide is equal to 0.3 mg/mg. It is shown that the maximum adsorption capacity of activated carbon Norit SA UF to iron (III) ions is equal to 0.0045 mg/mg and to silicon ions is equal to 0.024 mg/mg. Consequently, the aluminum oxyhydroxide has better adsorption characteristics in comparison with the activated carbon for removal of humic substances, iron and silicon ions. It is associated with the fact that activated carbon has a large adsorption surface, and this is due to its porous structure, but not all molecules can enter into these pores. Therefore, the fibrous structure of aluminum oxyhydroxide promotes better sorption capacity. The presented results suggest that activated carbon Norit SA UF and aluminum oxyhydroxide can be used as sorbents for removal humic substances or other organic substances from groundwater and natural waters.

1. Introduction

The present work relates to the problem of high-quality drinking water supply using processes of adsorption on activated carbon and aluminum oxyhydroxide for removal organic substances from natural water. The present of organic substances in natural water is one of the reasons of formation colloid solutions with high stability [1-3]. The stability of the colloidal system has both positive and negative sides. On the one side stable colloid systems are used to increase productivity in chemical industry and technology, agriculture and animal industry. They are also used to manufacture highly effective pharmaceutical products [4] and to obtain lubricants with high tribological properties [5]. On the other side it is necessary to reduce the stability of colloidal systems. Various types of colloid systems exist in the nature or arise spontaneously in industrial and natural processes, e.g., on extraction and purification of natural waters. This is due to high stability of colloid solutions based on iron compounds. Colloids of iron compounds are found in natural waters of Germany, Finland, Russia, USA, etc. It is shown that humic-type organic substances play an important role in the formation of iron colloid compounds [6, 7]. In Russia, where boggy areas occupy half the country and iron ions are present almost in all water sources, the problem of colloid removal is of immediate interest.

Marshes and small rivers, which contain humic-type substances, are most likely to contribute to the formation of stable colloid iron compounds. There are different methods of water treatment from humic-type substances such as aeration, ozonation, filtration and etc. [8]. However, some of them are not effective or economically impractical. Therefore, the question of finding methods of treatment remains open.

Currently, many research groups engaged in the study of the processes of absorption for the removal colloid substances [9, 10]. In addition there is work studying the adsorption processes of organic substances on activated carbon [11, 12]. Therefore, this work is relevant.

The objective of the work is to investigate the adsorption processes for removal humic substances from natural water on activated carbon and on aluminum oxyhydroxide.

2. Experimental approach

For synthesis and analytical study, reagents $FeSO_4 \cdot 7H_2O$, $Na_2SiO_3 \cdot 9H_2O$ and distilled water were used. Organic substances for experiments were extracted from peat bogs in the Tomsk region. The organic substances concentration was determined by analysis of the total organic carbon (TOC) content. The initial organic substances concentration varied in the range 0.05...4.0 mg/L. The organic substances molar mass was determined by the gel-chromatography method [13]. The samples contained fractions of 200–20000 Da. All model solutions were synthesized at room temperature. A shaker GFL 3005 (Germany) with a shaking frequency 100 rounds / min was used for mixing.
The iron and silicon concentrations in the solution were determined using an ICP-OES plasma optical emission spectrometer (Varian, USA). The pH value was measured using a WTW Multiline P4 multifunctional device (WTW GmbH, Germany).

For the adsorption experiments activated carbon Norit SA UF was used, which was produced by Norit, table 1. The specified sorbent has high kinetic characteristics due to the superthin structure of its particles. The second sorbent which was used it is aluminum oxyhydroxide. For synthesis aluminum oxyhydroxide, the method was used according technique [14].

Table 1. Specification of the sorbents Norit SA UF and aluminum oxyhydroxide.

Characteristic	Value	
Sorbent	Norit SA UF	Aluminum oxyhydroxide
Produce	Norit	Laboratory product
Name	Norit SA UF	Aluminum oxyhydroxide
Total surface area (m^2/g)	1200	215
Particles size,(μm)	7	-
Apparent density, (kg/m^3)	160	-
Moisture, mass., (%)	5	-

Adsorption experiment consisted of two basic parts. The first part is analysis of the adsorption kinetics and second one is analysis of the adsorption isotherm. Activated carbon and aluminum oxyhydroxide was used as adsorbent for the analysis of the adsorption kinetics. The mass of activated carbon was 12.5 mg in 250 ml solution (50mg/L). The activated carbon (m=12.5mg) was put into a flask (volume 250 ml) filled with model solution. After that the model solution with activated carbon was shaked. The flasks were removed after 15, 30, 60, 120, 180 and 1440 minutes after the beginning of shaking. After that solutions were filtrated through a membrane with a pore size of 0.4μm (Millpore, USA). The mass of aluminum oxyhydroxide was 100 mg in 250 ml solution (400 mg/L). The aluminum oxyhydroxide (m=100 mg) was put into a flask (volume 250 ml) filled with model solution. After that the model solution with aluminum oxyhydroxide was shaked. The flasks were removed after 15, 30, 60, 120, 180, 240 and 360 minutes after the beginning of shaking. After that solutions were filtrated through a membrane with a pore size of 0.4μm (Millpore, USA).

Three pharmaceutically active compounds were used: clofibric acid, diclofenac and carbamazepinee. All of them have different properties (table 2).

IOP Conf. Series: Materials Science and Engineering **110** (2016) 012097 doi:10.1088/1757-899X/110/1/012097

Table 2. Pharmaceutically active compounds used for the experiments.

	Clofibric acid (97%), Aldrich	
Application	Blood lipid regulator agent	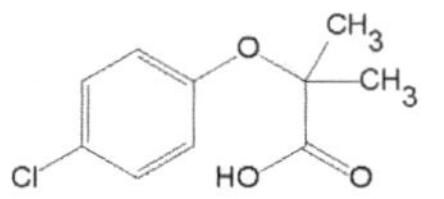
Formula	$C_{11}H_{11}ClO_3$	
M (g/mol)	216	
Charge	negative	
	Diclofenac sodium salt, Sigma	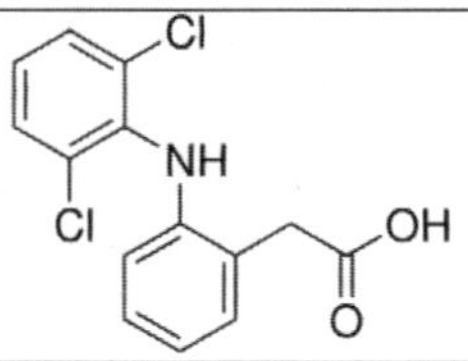
Application	Anti-rheumatic	
Formula	$C_{11}H_{11}Cl_2NO_2$	
M (g/mol)	296	
Charge	neutral	
	Carbamazepinee (99%), Aldrich	
Application	Anticonvulsant agent	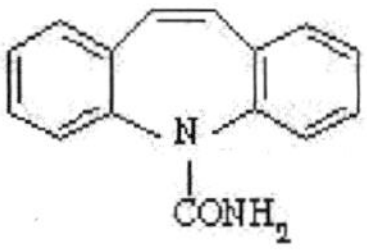
Formula	$C_{15}H_{12}N_2O$	
M (g/mol)	236	
Charge	neutral	

3. Results and discussion

It was experimentally found that silicon and iron ions do not practically adsorb on activated carbon but the concentration of humic-type organic substances reduces by 3 times. The time of adsorption is equal to 3 hours. The concentrations of iron and silicon ions did not change in the initial solutions and it was 6.73 mg/L for iron ions and 25.61 mg/L for silicon ions. The initial concentration of organic substances was varied in the range 1.2...30.1 mg/L. The Langmuir model was applied for the calculation of the adsorption isotherm. For this purpose eq. 1. was used in the form formula:

$$\frac{C}{q}=\frac{C}{q_m}+\frac{1}{q_m \cdot K_l}, \qquad (1)$$

where

q is the amount of the substance adsorbed on 1 m² of a surface (or 1g) adsorbent, g/m ² or mg/mg;
q_m is the maximum amount of the substance adsorbed on 1 m² of a surface (or 1g) adsorbent, mg/mg; C is the equilibrium concentration of the adsorbent in the gas phase or the solution; K_l is the Langmuir adsorption coefficient which depends on the structure of the adsorbed molecules.

Figure 1 shows the adsorption isotherm which was constructed according to the Langmuir model.

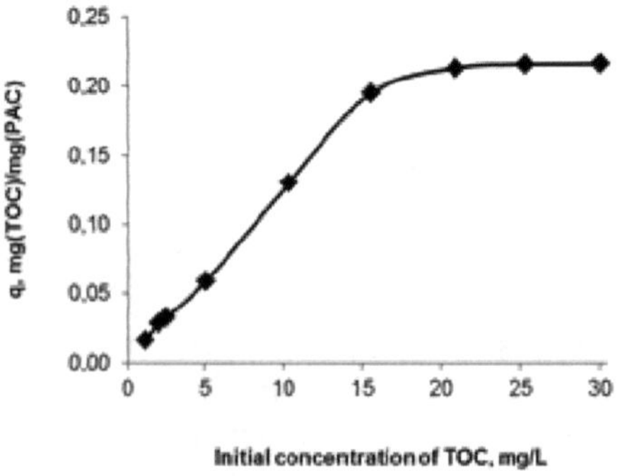

Figure 1. The adsorption isotherm of model solution on activated carbon for organic

substances

As it is seen in Fig. 1, the maximum adsorption capacity of activated carbon to organic substances is 0.25 mg/mg.

For comparison further adsorption experiments have been performed with diclofenac, clofibric acid and carbamazepine in demineralized water because these substances can be in natural waters [15]. Comparing the maximum adsorption capacity (q_{max}), which was calculated with the Langmuir model (table 3) it is possible to draw the conclusion that at the given experimental conditions, diclofenac adsorbs better on activated carbon than the humic-type organic substances of the model solution, but carbamazepine and clofibric acid did not.

Table 3. Process of adsorption on activated carbon.

Substances	Adsorptive capacity (q_{max})
Humic-type organic substances	0.25 mg TOC/mg PAC
Diclofenac	0.45mg Diclofenac /mg PAC
Clofibric acid	0.12 mg Clofibric acid/mg PAC
Carbamazepine	0.23 mg Carbamazepine/mg PAC

For the aluminum oxyhydroxide, the effects of humic-type organic substances are similar, but the time of adsorption is equal to 2 hours. Comparison of activated carbon shows that silicon and iron ions practically adsorb on aluminum oxyhydroxide. It was necessary to increase the iron and silicon concentration in the solution to obtain the adsorption isotherm. But it is impossible due to the model colloid system is stable at a certain ratio of iron - silicon - organic substances. Therefore, another method is used to construct isotherms. The mass of aluminum oxyhydroxide was varied in the range 0.1...60.0 mg. Figure 2 shows the adsorption isotherm which was constructed according to the Langmuir model for organic substances.

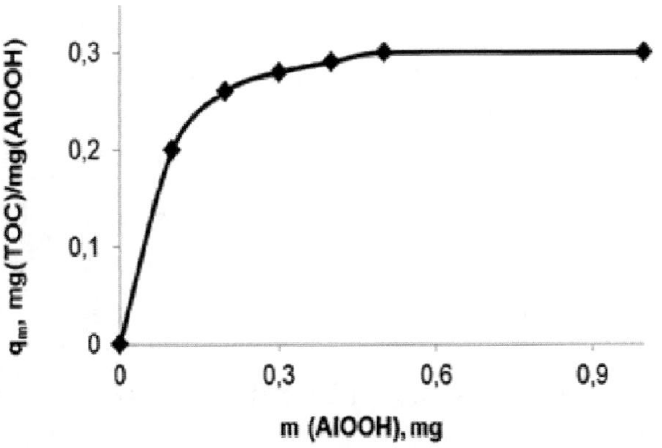

Figure 2. The adsorption isotherm of model solution on aluminum oxyhydroxide for organic substances

As it is seen in Fig. 2, the maximum adsorption capacity of aluminum oxyhydroxide to organic substances is 0.3 mg/mg. Comparison of activated carbon shows that adsorption capacity of aluminum oxyhydroxide more than it. This may be due to the fact that the aluminum oxyhydroxide has a positive charge on the surface, while the humic-type organic substances have negative charge, as shown earlier [6]. Figure 3 show the adsorption isotherm which was constructed according to the Langmuir model for iron and silicon ions.

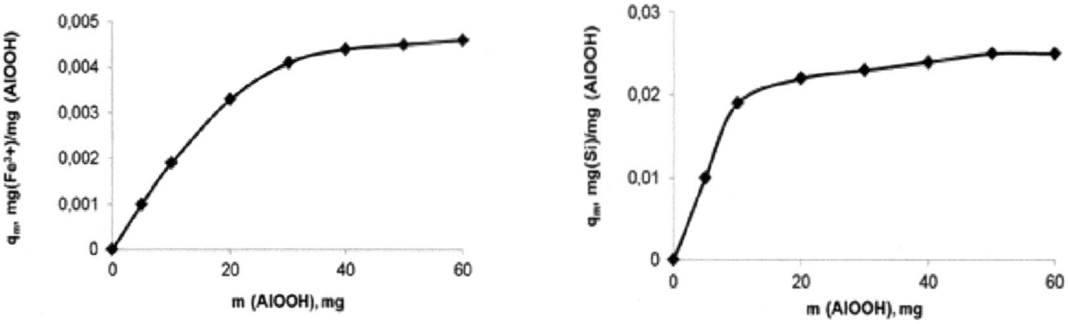

a b

Figure 3. The adsorption isotherm of model solution on aluminum oxyhydroxide for: a) iron ions; b) silicon ions.

Figure 3 demonstrates that the maximum adsorption capacity of aluminum oxyhydroxide to iron ions is 0.004 mg/mg and to silicon ions is 0.024 mg/mg. The results obtained in the work suggest that the aluminum oxyhydroxide has better sorption capacity than activated carbon is associated with the fact that activated carbon has a large adsorption surface, and this is due to its porous structure, but not all molecules can enter into these pores. Therefore, the fibrous structure of aluminum oxyhydroxide promotes better sorption capacity.

Thus, the destruction of iron colloid substances occurs due to the removal of organic substances from the model colloid solution through the adsorption process on aluminum oxyhydroxide with formation of a precipitate in the form of $Fe(OH)_3$.

4. Conclusion

The application of aluminum oxyhydroxide is most effective in relation to organic substances, iron and silicon ions. The adsorption capacity of aluminum oxyhydroxide was determined and it is equal to about 0.30 mg/mg for humic-type organic substances. Consequently, the aluminum oxyhydroxide has better adsorption characteristics in comparison with the activated carbon (0.25 mg/mg). Furthermore, aluminum oxyhydroxide can be used for treatment of colloidal solution for natural waters containing silicon and iron ions.

Acknowledgements

The study was supported by National Research Tomsk Polytechnic University, the project No. 7.1504.2015.

References

[1] Ma J, Li G B, Chen Z L, Xu G R, Cail G Q 2001 *Water science and Technology* 51-61
[2] Shiyan L N, Machekhina K I, Tropina E A, Gryaznova E N, An V V 2013 *Advanced Material Research* **872** 237-240
[3] Schafer A I, Schwicker U, Fisher M M, Fane A G, Waite T D 2000 *Journal of membrane science* 151-172
[4] Buchs B, Evangelou M, Winkel L, Lenz M 2013 *Environ. Science Technology* 2401-2407
[5] An V, Irtegov Y 2014 *Journal of Nanomaterials* ID 731073
[6] Voyno D A, Machekhina K I, Shiyan L N 2014 *Advanced Materials Research* **971-973** 266-269
[7] Aitken G R, Hsu-Kim H, Ryan J N 2011 *Environ. Science Technology* 3196-3201
[8] Machekhina K I, Shiyan L N 2014 *Advanced Materials Research* **1040** 342-346
[9] Schmitt P, Taylor H E, Aiken G R, Roth D A, Frimmel F H 2002 *Environ. Sci. and Technol.* 2932-2938.

[10] Vincent B 1974 *Adv. Colloid Interface Sci.* **4** 193
[11] Hegazy A K, Abdel-Ghani N T, El-Chaghaby G A 2014 *Applied Water Sci.* 273-281
[12] Yaws C L, Narasimhan P K, Pike R W 2003 *Oil and Gas Journal* 80-81
[13] Perminova I V, Frimmel F H, Kudryavtsev A V, Kulikova N A, Abbt-Braun G, Hesse S, Petrosyan V S 2003 *Environ. Sci. Technol.* **37** 2477-2485
[14] Shiyan L, Yavorovskii N, Pustovalov A, Gryaznova E *Advanced Materials Research* 2014 **1040** 59-64
[15] Ternes T A, Meisenheimer M, Mcdowell D, Sacher F, Brauch H J, Haist-Gulde B, Preuss G, Wilme U, Zulei-Seibert N 2002 *Environ. Sci. and Technol.* 3855-3863

Electropulse treatment of water solution of humic substances in a layer iron granules in process of water treatment

G L Lobanova[1] , T A Yurmazova[2] , L N Shiyan[2], K I Machekhina[2*]

[1]Laboratory 12, Institute of High Technology Physics, Tomsk Polytechnic University, 634050, Tomsk, Russia.
[2]Department of General and Inorganic Chemistry, Institute of High Technology Physics, Tomsk Polytechnic University, 634050, Tomsk, Russia.

*E-mail: mauthksu@yandex.ru

Abstract. The present work is a part of a continuations study of the physical and chemical processes complex in natural waters containing humic-type organic substances at the influence of pulsed electrical discharges in a layer of iron pellets. The study of humic substances processing in the iron granules layer by means of pulsed electric discharge for the purpose of water purification from organic compounds humic origin from natural water of the northern regions of Russia is relevant for the water treatment technologies. In case of molar humate sodium – iron ions (II) at the ratio 2:3, reduction of solution colour and chemical oxygen demand occur due to the humate sodium ions and iron (II) participation in oxidation-reduction reactions followed by coagulation insoluble compounds formation at a pH of 6.5. In order to achieve this molar ratio and the time of pulsed electric discharge, equal to 10 seconds is experimentally identified. The role of secondary processes that occur after disconnection of the discharge is shown. The time of contact in active erosion products with sodium humate, equal to 1 hour is established. During this time, the value of permanganate oxidation and iron concentration in solution achieves the value of maximum permissible concentrations and further contact time increase does not lead to the controlled parameters change.

1. Introduction

Currently, water purification processes from impurities by means of various electrical discharge types' application are widely explored [1, 2]. In the paper [3] the effect of glow and diaphragm discharges of aqueous solutions of organic dyes is investigated. It is shown that the effect of glow and diaphragm discharges on water solutions of organic substances leads to their destruction. Along with that [4] the effects of pulsed electric discharge (PED) in the layer of metal pellets which are placed in water solutions of organic substances are investigated.

The difference of PED in the layer of metal granules from other discharges types is a complex phenomenon. The primary effects are initiated by discharges channels in the developing system. Reducing properties of active surface of eroding metal and the tendency of substances present in the solution to redox reaction stimulate the development of secondary phenomena. These phenomena are able to determine the further way of chemical reactions which are developing in electroerosion reactors.

The pulsed electric discharge is a unique tool to influence either liquid [5] or solid phase [6] in a heterogeneous system with conducting particles. The destruction of the metallic phase in the areas of discharge channel formation leads to a mass transfer processes intensification, the water solution activation, the gaseous products formation and the conjugate oxidation - reduction reactions development.

In the papers [7-10] the mechanism of the reactions which are taking place at electric pulse impact on water solutions of methylene blue, eosin and furatsilina are studied. It is shown that redox reactions and adsorption processes occur as a result of pulsed electric discharge. Process efficiency depends on the material load. The investigated water solutions of organic substances are involved in redox reactions and adsorption processes in different ways due to the different structure and properties of methylene blue, eosin and furatsilina.

The present work is a part of a continuations study of the physical and chemical processes complex in natural waters containing humic substances at the influence of pulsed electrical discharges in a layer of iron pellets.

The objective of the work is to investigate the influence of phenomena initiated by the discharge channels and secondary redox processes on the stability of humic substances in the model solution and groundwater.

2. Experimental approach

For the experiments concerning the effect of the discharge on the stability of humic substances, the installation was used according to the paper [6]. The installation consisted of a 1.5-liter reactor fitted with electrodes and a switching power supply. The iron pallets weighing 300 g were located in the reactor. Voltage pulses were supplied to the electrodes from a pulse generator with the energy output of 0.5 J and variable pulse repetition rate. The voltage amplitude was 500 V, pulse repetition frequency was 500 imp·s^{-} , the energy and the duration of voltage pulses was 15 ms. The volume of the model solution was varied in the range 0.3…0.9 L.

For synthesis and analytical study, the model solution of humic substances was used. Humic organic substances for experiments were extracted from peat bogs in Tomsk region [11, 12].

The organic substances concentration was determined by the analysis of the chemical oxygen demand (COD), permanganate oxidation (PO) and the total organic carbon (TOC) content.

The iron and silicon concentrations in the solution were determined by means of ICP-OES plasma optical emission spectrometer (Varian, USA). The concentration of iron (II) was determined according technique State Russia Standard 4011-72 with o-phenanthroline. The pH value was measured by means of a WTW Multiline P4 multifunctional device (WTW GmbH, Germany).

Centrifugation was carried out immediately after the PED solutions processing to stop the oxidation-reduction reactions in the systems "eroded metal – investigated substances". The suspension was centrifuged by means of a centrifuge Allegra 54 R (Beckman-Coulter, USA) with a shaking frequency 20000 rounds / min for 10 min.

The influence of active particles was evaluated by exclusion. To do this, the appropriate amounts of distilled water were treated with PED. The suspensions made were injected into the humic substances solution with the contact time was 10 seconds. Thus, humic substances were not involved in the processes initiated by the discharge channels, but only came into contact with eroded metal.

3. Results and discussion

The model solutions of sodium humate were used to determine the water treatment process parameters. The sodium humate concentration in model solution 1 was 8 mg/L whereas in solution 2 it was 16 mg/L. This sodium humate concentration corresponded to that in the groundwater of the Western Siberian region [12].

It is known [8] that as a result of pulsed electric discharge, water solutions are characterised by eroded iron particles formation, redox reactions of $Fe^{o} \rightarrow Fe^{2+} \rightarrow Fe^{3+}$ transition occurance, pH changes in the range of 6.5 ... 9.5.

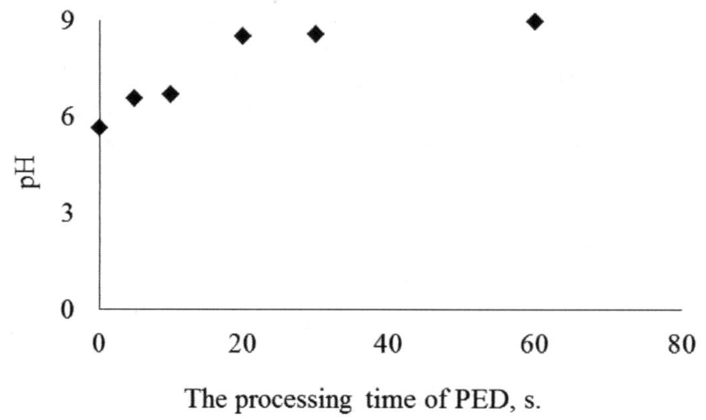

Figure 1. pH change in the model solution from processing time of PED.

The influence of processing time on the humic substances insoluble form was investigated to determine the conditions of effective of humic substances coagulation in the process of electropulse effect.

Fig. 2 demonstrates that the colour of humate sodium solution and concentration of iron (II) formed by the reaction of iron eroded with water depend on the processing time of PED.

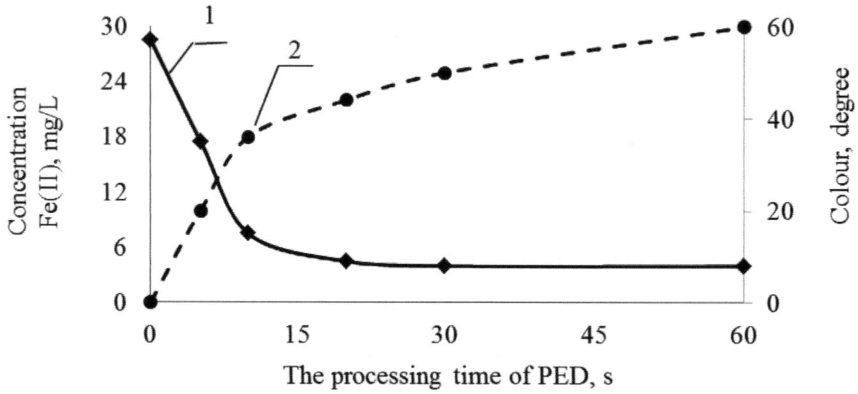

Figure 2. Dependence of the processing time of PED on colour of humate sodium solution (1) and the iron (II) concentration (2).

It is seen in Fig. 2 that the most intense color reduction to maximum permissible concentration values occurs when the processing time is not more than 10 seconds, whereas the iron ions (II) concentration in solution corresponds to 18 mg/L. Experimental treatment time is 10 seconds, which is sufficient to obtaining a humic substances to iron ratio of 2: 3. A further increase in the processing time can be considered inappropriate. The iron ions reactions with the sodium humate occur followed by coagulation to form precipitate at the determined ratio. The residual concentration of iron ions does not exceed the maximum permissible concentration for drinking water which equals 0.3 mg/L in the solution.

Reduced color solution as an indicator of the sodium humate concentration was confirmed by measuring the permanganate oxidation (PO), chemical oxygen demand (COD) and measuring the concentration of total organic carbon (TOC). The dependence of these parameters on the processing time of the solution by a pulsed electric discharge is shown in the Fig. 3. This figure demonstrates that the change of COD and PO are well correlated with the changes in solution color (Fig. 2). The maximum decrease in the concentration of humic substances is also achieved with the processing time of 10 seconds.

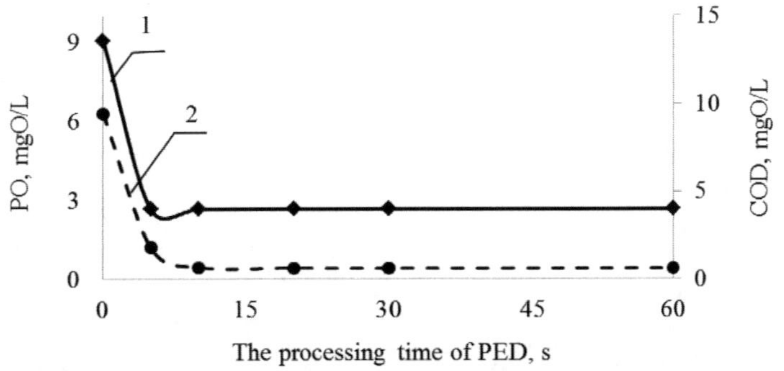

Figure 3. Dependence of the processing time of PED on value of COD (1) and PO (2).

As shown earlier [7, 8], the secondary processes occurring after discharge shutdown make a major contribution to the process of electro-processing solutions involving metal loading.

The distilled water was treated for 10 seconds and the solution added to a suspension of sodium humate after discharge disconnection to study the suspension activity. The sodium humate concentration was selected in the amount to equal the concentration of a stock solution. The suspension activity was evaluated relative to dissolved sodium humate at stated intervals (1-6 hours). The result of the interaction was observed for changes in the total iron concentration and PO values and it was compared with indicators of a stock sodium humate solution.

Fig. 4 shows the dependence of the iron total concentration and PO on the contact time with the active iron erosion products.

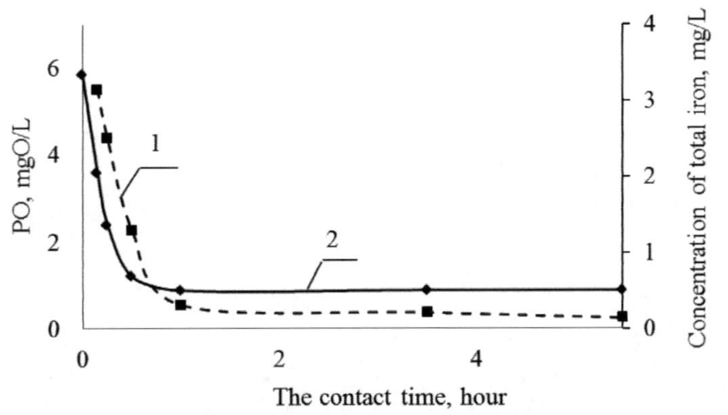

Figure 4. Dependence of the contact time on total iron concentration (1) and PO value (2).

It is seen in Fig. 4 that the optimum contact time of sodium humate with active erosion products is one hour. The value of PO and the iron concentration in solution come up to the maximum permissible concentration and further increase contact time does not lead to a change in the monitored parameters. The contact time experimentally ascertained shows that the load remains active for at least 1 hour after discharge disconnection for a sodium humate solution at TOC concentration 8 mg/L. The load activity is connected with the interaction of iron erosion products, water and sodium humate.

The groundwater of village Belyy Yar of Tomsk region was used as natural water. That water has a complex chemical composition and it contains such impurities as humic substances, iron ions and silicon compounds. Several papers [12, 13] show that the ability to form colloidal compounds in this type of water causes problems in the process of water purification. The difficulty of removing colloid impurities are related to their stability to physical and chemical effects used in water treatment technology.

Fig. 5 shows the dependence of the permanganate oxidation and the color solution at the processing time of PED. The minimum processing time is 5 seconds,the maximum is 60 seconds.

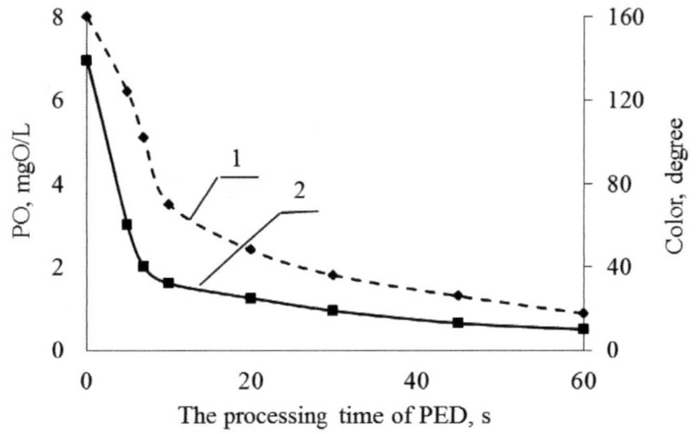

Figure 5. Dependence of the processing time on PO value (1) and color of solution (2).

Fig. 6 show, the changes of total iron concentration and silicon at the same processing time of PED.

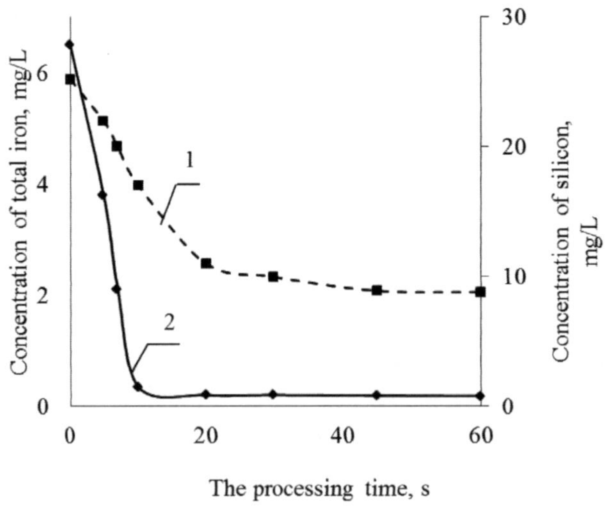

Figure 6. Dependence of the processing time of PED on silicon (1) and total iron (2) concentration.

Fig. 5 and 6 demonstrate that humic substances concentration is decreased and also the iron ions and silicon concentration are decreased at the groundwater treatment of village Belyy Yar, Tomsk region. The maximum parameters decrease is observed at the processing time close to 10 seconds. It fully corresponds to the results obtained for model humic solutions.

4. Conclusion

1. It is shown that the most intense color reduction to the maximum permissible concentration occurs at the processing time equal to 10 seconds. Experimental processing time is sufficient to obtain molar humic substances iron ion at the ratio of 2:3 and further processing time increase can be considered inappropriate.
2. It is found out that load activity is maintained for 1 hour after the discharge disconnection.
3. The role of secondary processes that occur after discharge disconnection, which are conditioned by suspension of activity is shown. The contact time of active erosion products with sodium humate, equal to 1 hour is established. The value of permanganate oxidation and iron concentration in solution achieves maximum permissible concentration during this time. Furthermore, the increase of contact time does not lead to the change in the monitored parameters.

Acknowledgements
The study was supported by National Research Tomsk Polytechnic University, the project No. 7.1504.2015.

We are indebted to Dr. Zhanna S. Anikina from National Research Tomsk Polytechnic University, Department of Foreign Languages, Institute of Power Engineering for linguistik support.

References

[1] Piskarev I M 2000 *Journal of Physical of Chemistry Journal High Energi Chemistry* **74** 546-551
[2] Piskarev I M, Spirov G M, Selemir V D, Karelin V I, Shlepkin S I 2007 *Journal High Energi Chemistry* **41** 334-336
[3] Maksimova A I, Xlustova A V, Subbotina N N 2009 *Chemistry and Chemical Technology* **52** 116-120
[4] Boev S, Yavorovsky N A 1999 *Digest of Technical Papers-IEEE International Pulsed Power Conference* **1** 181-184
[5] Zerba A A, Petrichenko S V 2004 *Technical electrodynamics* **3** 27-32
[6] Danilenko N B, Savelev G G, Yavorovsky N A Yyrmazova T A *2008 Russian Journal of Applied Chemistry* **81** 803-810.
[7] Aristova N A, Piskarev I M 2003 *Journal* of Physical of Chemistry **77** 723-726
[8] Lobanova G L, Shiyan L N, Yyrmazova T A, Galanov A I 2014 *Proceedings of the universities. Physics* **57** 141-145
[9] Parkansky N 2013 *Journal Plasma Chemistry Plasma Process* **33** 907-919
[10] Kim J R 2013 *American Journal of Analytical Chemistry* **4** 115-122
[11] Shiyan L N, Tropina E A, Machekhina K I, Gryaznova E N, An V V 2014 *SpringerPlus* DOI: 10.1186/2193-1801-3-260, URL: http://www.springerplus.com/content/3/1/260.
[12] Machekhina K I, Shiyan L N 2014 *Advanced Materials Research* **1040** 342-346
[13] Voyno D A, Machekhina K I, Shiyan L N 2014 *Advanced Materials Research* **971-973** 266-269

IOP Conf. Series: Materials Science and Engineering **110** (2016) 012099 doi:10.1088/1757-899X/110/1/012099

Influence of concentration modifier on the structure and functional properties of aluminum oxyhydroxide modified

L N Shiyan[1], E N Gryaznova[2*], S I Galanov[3] and O I Sidorova[3]

[1]Institute of High Technology Physics, Tomsk Polytechnic University, Tomsk, 634050, Russia
[2]Institute of Water, Tomsk Polytechnic University, Tomsk, 634050, Russia
[3]Faculty of Chemistry, Tomsk State University, Tomsk, 634050, Russia

*E-mail: t_elena@mail2000.ru

Abstract. Studying the properties of nanomaterials is an important task, but nanomaterials with desired properties is a promising direction. The aim of this work is to investigate the influence of the value of the concentration of the modifier (ions Mn^{2+}) on the structural and functional properties of modified aluminum oxyhydroxide. In this paper, using methods such as the X-ray diffraction studies, differential thermal analysis, electron microscopy, chromatography. The paper found that increasing the concentration of the modifier result in significant changes in the morphology, the appearance of metallic aluminum, which is well seen on X-ray data samples. The influence of thermal effects on a modified aluminum oxyhydroxide argon. Set the phase transition temperatures in the synthesized samples. It is shown that with increasing sodeozhaniya manganese in the composition of the synthesized samples decreases the value of specific surface area. Study of the functional properties showed that the synthesized material has catalytic properties in the oxidation of methane. It is shown that the effective sample is a sample with a manganese content of 2.7 wt. %. By XRD results calcined in air samples modified aluminum oxyhydroxide was shown that only in the sample with a manganese content of 2.7 wt. % $MnAl_2O_4$ phase is formed, which is catalytically active phase.

1. Introduction

Nanoparticles are increasingly being used in medicine, electronics, tribotechnology, catalysis [1, 2]. For example, multi-component nanoparticles serve as the basis for the creation of next-generation drugs, built on new principles, performing the role of the media not only drugs but also the functional part of the drug. Nanofibers aluminum oxyhydroxide (AlOOH) are used for the manufacture of dressings which have a high efficiency in the treatment of wounds and burns. On the basis of these fibers were developed adsorbents for water purification from microorganisms and viruses [3–7]. It is known that the scope of the nanoparticles is determined by their properties, so the preparation of nanoparticles with new properties is the actual direction of science and technology. Giving new properties of aluminum oxyhydroxides can be due to its modification [8, 9]. Due to the very small size of the material surface modification can not be accomplished, however, the volume in the modification process of direct synthesis of the material can be realized. In [10] described in detail the method of modifying aluminum oxyhydroxide manganese ions (II). As manganese modifier was selected as this is an active metal component in the catalysts of the oxidation of organic substances [11–14], which will bring the catalytic properties of aluminum oxyhydroxide.

The purpose of this paper is to establish the dependence of the change of the structural and functional properties of modified aluminum oxyhydroxide on the concentration of modifier in its composition.

At present, many research teams are studying the properties and field of application of aluminum oxyhydroxide [3–7]. Thus, this study is significant.

The objective of the work is to study the dependence of the phase constitution and functional properties of modified aluminum oxyhydroxide on the concentration of modifier in its composition.

2. Materials and methods

In this work electro explosive aluminum nanopowder was used, which was produced by electric wire explosion in an argon environment. The Fig. 1 is the microphotography of the electro explosive aluminum nanopowder.

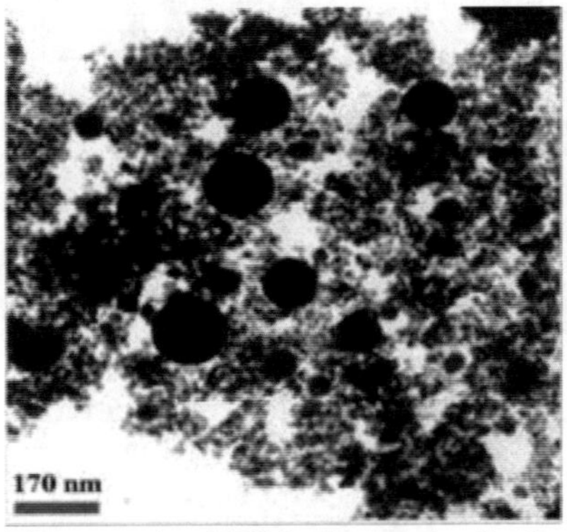

Figure 1. TEM image of the electroexplosive aluminum nanopowder.

Modifying agent was $MnSO_4 \times 5H_2O$. Manganese ions are used as catalysts for many chemical reactions such as decomposition of hydrogen peroxide, oxidation of iron (II) and others. In the experiments the different concentration of manganese ions were used, which are represented in table 1.

Table 1. Characteristics of modified nanofibers of aluminum oxyhydroxide.

The number of samples	Initial concentration of manganese ions in suspension, (mg/L)
0	-
1	1.0
2	30.0
3	60.0

The synthesis process of nanofibers in salt solutions was carried out under the same conditions as the synthesis in distilled water [5]. The volume of solution was 200 ml and the weight of aluminum nanopowder was 0.075. The synthesis of AlOOH nanofibers consists in three state. The first one was the preparation of aluminum nanopowder suspension and the salt solution, dispersing of the suspension using an ultrasonic bath for 5 min. The second stage involved the synthesis of nanofibers in a solution of $MnSO_4$ at constant temperature 60 °C during 6 hours. The third stage was the drying of the finished product at 60 °C.

The quantitative content of manganese ions in the modified nanofibers was determined by chemical analysis, according to GOST 4974-72. Morphological structure of the initial and modified nanofibers were studied on transmission electron microscope - JEM-2100F, Japan. The changes of specific surface areas were determined by the analyzer, "Air heater-M" by the BET method (low temperature adsorption of nitrogen-helium mixture). Quality of the products of synthesis was determined by X-ray diffraction on Shimadzu XRD-7000 diffractometer. The range of scanning angles was from 0 to 1000 with step 0.50 and scanning speed 1 degree per minute. The thermal stability of the modified aluminum oxyhydroxide and aluminum oxyhydroxide samples for the thermal oxidation in air and in argon were controlled with using thermal analyzer SDT Q-600.

3. Results and Discussion

In this work was prepare 4 samples of the modified aluminum oxyhydroxide. The value of specific surface area was a quantity of the effectiveness of the modification of AlOOH nanofibers. The table 2 shows the results of measurement of specific surface area of samples synthesized in solutions with different concentrations of salt.

Table 2. Characteristics of modified nanofibers of aluminum oxyhydroxide.

The number of samples	Initial concentration of manganese ions in suspension, (mg/L)	The concentration of manganese ions in samples, (mas %)	The specific surface area, (m^2/g)
0	-	-	196.16
1	1.0	0.4	219.48
2	30.0	1.4	211.76
3	60.0	2.7	110.10

The results in the table 2 are shown that the value of concentration of manganese ions is influenced on the value of specific surface area of the synthesis products. The increasing of specific surface area can be associated with an low concentrations of manganese ions in solution was adsorbed on the faces of the slow-growing nanowires, thereby blocking their growth, and fast-growing end face manganese ions can be incorporated into the crystal, without hindering its growth. However, if the manganese concentration in the solution the amount of active growth centers aluminum oxyhydroxide is reduced by blocking their manganese ions, which leads to a decrease in specific surface area. The increasing of the proportion of the amorphous AlOOH phase in the presence of Mn (II) was confirmed by X-ray diffraction of the obtained samples. Figure 2 presents the X-ray diffraction patterns of the initial (sample no. 0) and modified sample nos. 1, 2, and 3.

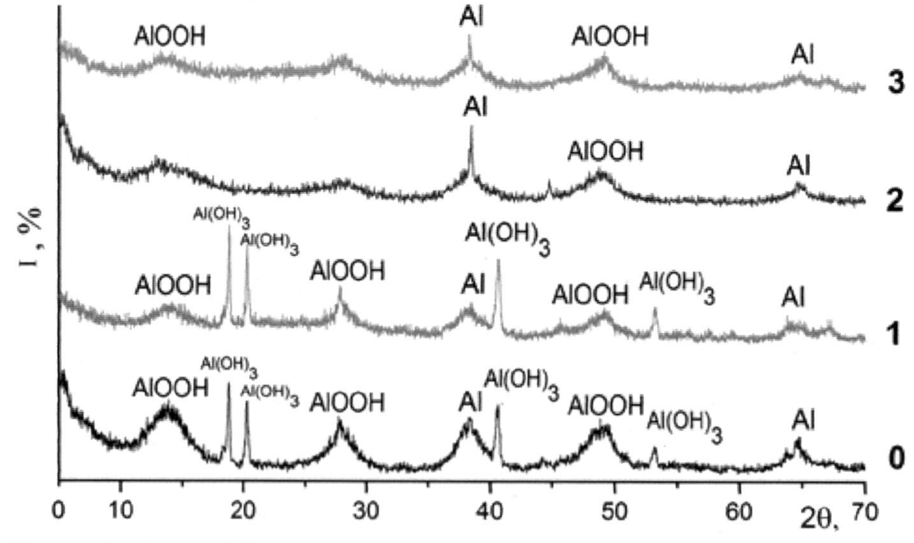

Figure 2. X-ray diffraction patterns of the nanofibers: (0) Initial and (1–3) manganese-modified samples.

Analysis of the X-ray diffraction patterns showed that sample no. 0 actually contains crystalline $Al(OH)_3$ and amorphous AlOOH phases, as well as metallic Al phase. The X-ray diffraction pattern of sample no. 1 is virtually identical to that of the initial sample, except for the occurrence of significantly broadened peaks of AlOOH and metallic Al phases. Sample nos. 2 and 3 exhibit virtually identical patterns which differ from those of sample nos. 0 and 1 by the lack of the $Al(OH)_3$ phase and occurrence of a high background level. Although manganese was not detected by the X-ray diffraction technique, its participation in synthesis is indicated by the change in the phase composition of the products with increasing manganese ions concentration in solution. The difference in the phase compositions of the products for the reactions of aluminum with liquid water and salt solutions was confirmed by the DTA data. Figure 3 shows the thermograms of the initial and modifyed samples.

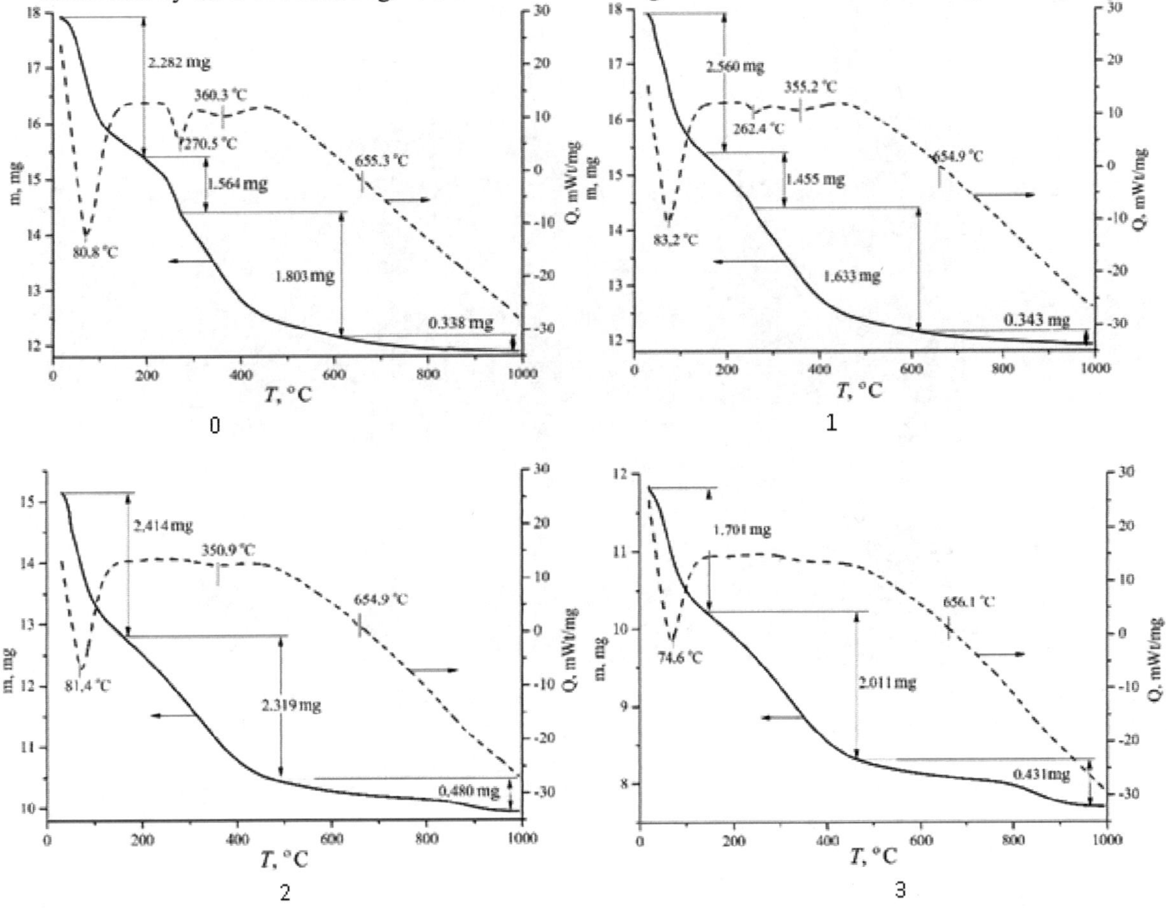

Figure 3. Thermograms of the nanofibers: (0) Initial and (1–3) manganese-modified samples.

It is seen that the mass loss (TGA) and heat effect (DTA) curves for sample nos. 0–2 are similar though different from those of sample nos. 3. The first endothermic peak (80.8 °C) is to the removal of physisorbed water. The weight reduction at a high temperature (270.5 °C) associated with transformation of Al(OH)3 to AlOOH (boehmite) and the third dehydration endothermic peak associated with dehydroxylation of boehmite (removal of OH groups). Thermograms of the sample 3 are shown that the third endothermic peak is absent because of a decrease in the yield of fibrous boehmite.

Figure 4 shows the electron micrographs of the initial and modified samples. It is seen that the sample no. 0 has a fibrous structure which is similar in principle to that of sample nos. 1 and 2 with the lowest manganese content. Sample no. 3 does not exhibit a prominent fibrous structure and contains unreacted spherical metallic aluminum particles. This fact confirms the presence of metallic aluminum phase on the X-ray sample 3 diffraction patterns (Fig. 2).

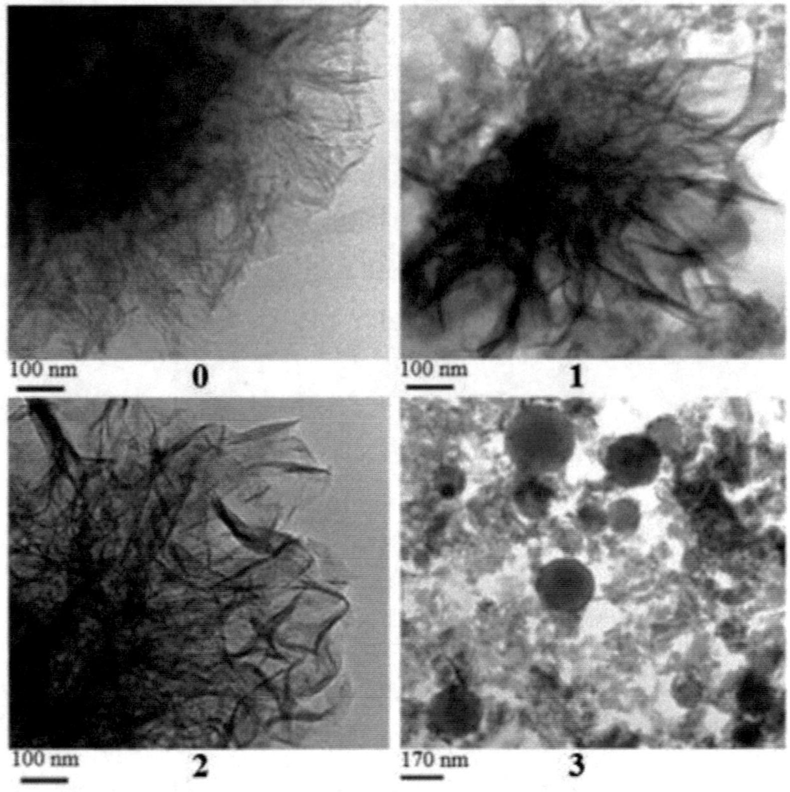

Figure 4. TEM image of the (0) initial and (1–3) modified samples.

The new functional properties of modified aluminum oxyhydroxide was studied on the reaction of the oxidation of methane, because the manganese are widely used as catalyst for oxidation reactions of organic substances. It was founded that heat treatment at 850 °C of modified aluminum oxyhydroxide leads to the formation of the combined structures of aluminum oxide with manganese and stable phases The results of investigation present on Figure 5 and table 3.

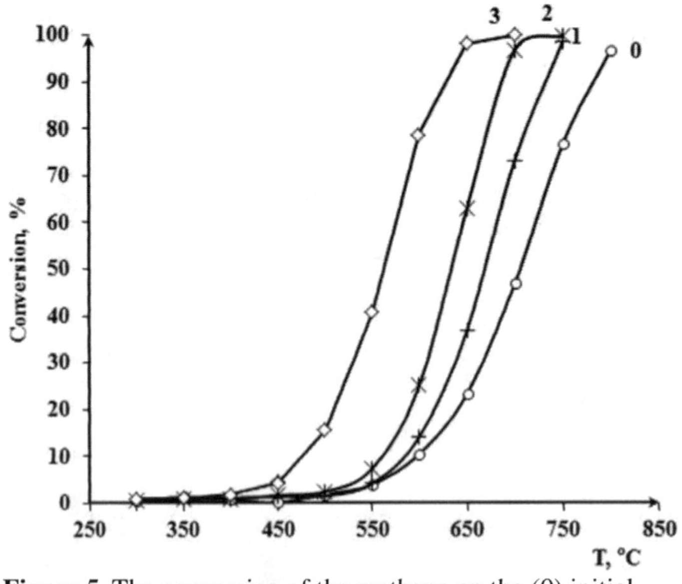

Figure 5. The conversion of the methane on the (0) initial and (1–3) modified samples.

Table 3. The phase constitution of samples after heat treatment at 850 °C

The number of samples	The concentration of manganese ions in samples, (mas %)	The phase constitution of samples after heat treatment at 850 °C
0	-	γ-Al_2O_3
1	0.4	γ-Al_2O_3
2	1.4	γ-Al_2O_3
3	2.7	δ-Al_2O_3, Al_2O_3, $MnAl_2O_4$

Figure 5 shows that the modification leads to the catalytic properties of the modified samples. It is shown that the amount of manganese in the modified samples affects the temperature methane. Figure 5 shows that the complete conversion of methane for sample 3 reached at 600 °C.

4. Conclusion

Thus it was shown that the modification process of manganese ions (II) of aluminum oxyhydroxide leads to a change in the phase composition, morphology and surface area. With the concentration of manganese increasing the contents of the hydroxide phases decreases and as a consequence of the specific surface area decreases.

Found that the most effective in the oxidation of methane is an industrial sample containing manganese ions (II) 2,7 wt. % Due to the formation of a spinel structure by calcination phase composition and unchanging during operation.

Acknowledgement
The work was supported by State task of «Science» of 7.1326.2015.

References

[1] Irtegov Y, An V *Advanced Materials Research* 2014 **1040** 171-175
[2] Shiyan L, Yavorovskii N, Pustovalov A, Gryaznova E *Advanced Materials Research* 2014 **1040** 59-64
[3] Jaworowskii N, Shiyan L, Saveliev G, Galanov F *Nanotechnics* 2008 40-45
[4] Shutilov A, Zenkovets G, Tsybulya S, Gavrilov V *Kinetics and Catalysis* 2012 **53** 128–140
[5] Streletskii A, Kolbanev I, Borunova A, Butyagin P *Colloid Journal* 2005 631-637
[6] Alwitt R *Journal of The Electrochemical Society* 1974 1322-1328
[7] Fernando J, Chung D *Journal Por. Mater* 2002 211-219
[8] Paul B, Martens N, Frost R *Journal of Colloid and Interface Science* 2011 **360** 132-138
[9] Lojkowski W, Gedanken A, Grzanka E, Opalinska A, Strachowski T, Pielaszek R, Tomaszewska-Grzeda A, Yatsunenko S, Godlewski M, Matysiak H, Kurzydłowski K *J. Nanopart Res* 2009 **11** 1991–2002
[10] Gryaznova E, Shiyan L, Yavorovskii N, Korobochkin V *Rus. J. of Appl. Chem.* 2013 **86** 360-365
[11] Bulavchenko O, Tsybulya S, Afonasenko T, Tsyrulnikov P *Applied Catalysis A: General*. 2013 **459** 73–80
[12] Craciun R, Nentwick B, Hadjiivanov K, Knözinger H *Applied Catalysis A: General* 2003 **243** 67–79
[13] Frey K, Iablokov V, Safran G, Osan J, Sajo I, Szukiewicz R, Chenakin S, Kruse N *Journal of Catalysis* 2012 **278** 30–36
[14] Iablokov V, Frey K, Geszti O, Kruse N *Catalysis Letters* 2010 **134** 210–216

Use of continuous solid-phase synthesis to obtain phosphors based on strontium aluminate

B Kerbel[1], L Katsnelson[2], Yu Falkovich[3], D Prokopyev[4]

1 Seversk Technological Institute NRNI [National Research Nuclear University] MEPhI [Moscow Engineering Physics Institute] 636036, Russia, 636036, Seversk, Tomsk region, pr. Kommunisticheskiy, 65
2 RPE [Research and Production Enterprise] OJSC [Open Joint-Stock Company] Tekhnologika, Russia, 863033, Rostov-on-Don, Portovaya St., 303
3 National Research Tomsk Polytechnic University, Russia, 634030 Tomsk, pr. Lenina, 30
4. National Research Tomsk Polytechnic University, Russia, 634030 Tomsk, pr. Lenina, 30

E-mail: BMKerbel@mephi.ru

Abstract. The effect of conditions of continuous solid-phase synthesis on particle size distribution of nanostructured powders of strontium aluminate was studied. It was shown that continuous solid-phase synthesis allows for: synthesis of strontium aluminate in the form of nanostructured powders with controlled particle size distribution directly during its synthesis; in the presence of a liquid phase strontium aluminate is synthesized with a high level of monophasity. It was shown that in order to optimize the illuminating parameters of phosphors based on strontium aluminate, it is advisable to use continuous solid-phase synthesis.

1. State-of-The-Art

One of the important areas of modern production actively using rare earth elements (REE) to solve their problems is the design and synthesis of phosphors. In this regard the materials that are of particular interest those that have not only the required illuminating elements, but also low-hazard effects on the human body, which in turn makes it possible to expand the scope of their application, and significantly reduces the cost of production as a whole. Such phosphors include phosphors based on strontium aluminates.

However, the optimization of the conditions of the synthesis of phosphors based on them is a rather complex technological problem [1-5], which is associated with:

- the criticality of chemical reactions in $SrO - Al_2O_3$ due to the rather large number of polymorphic transformations of starting reagents in the process of their heat treatment;
- the necessity of the introduction of activators and co-activators (Eu. Dy, etc.) in the crystal lattice of the phosphor base with the formation of a specific configuration of their order.

In this regard, the optimization of the probability of synthesis of phosphors based on strontium aluminates [6-10] is determined by the technological capabilities of the method of synthesis and depends, usually, on:

- the level of homogeneity achieved of the starting mechanical mixture of the synthesized strontium aluminate;
- the dispersion and the single-phase reagents used;
- the rate of synthesis of the desired material, as well as the controllability of technological factors governing the optimization of the processes.

Among the most common methods for the synthesis of phosphors based on strontium aluminates are classical solid-phase and sol-gel techniques. Less commonly used methods are combustion and induction heating. Solid-phase synthesis is usually accomplished in the range of sufficiently high temperatures of 1200 – 1400 °C for 4-6 hours using the inert or redox environment and the subsequent grinding of the desired product.

Virtually no method of synthesis provides a high level of monophasity of the desired material with a uniform distribution of REE in its entirety and, consequently, the maximum possible illuminating parameters. In this regard, the process procedure is a rather complex and multi-step regimen which, nevertheless, does not guarantee a sufficiently high level of repeatability of achievable parameters.

Recently, particular attention has been paid to solid-phase synthesis using strontium aluminate fluxing, i.e., in the presence of a liquid phase. This is due to the fact that the synthesis in the presence of a liquid phase usually occurs at a higher quality level, allowing in this regard for the significant reduction of the temperature of the synthesis. So, it should be noted that not every liquid phase provides the optimal solution to the problem, since in the process of heat treatment the liquid phase, coming into contact with the reagents of the starting mixture of the synthesized material, forms the intermediate phase and does not always promote the formation of the monophasity of the desired product.

It is known that solid-phase synthesis of inorganic photoluminophors is carried out with step calcination, in the final stage of which high temperatures and prolonged exposure cycles are conventionally used with the use of an inert or redox environment and the subsequent grinding of the desired product.

For example, yttrium aluminum garnet $(Y_3GdCe)3Al_5O_{12}$ is synthesized in three stages at a temperature at the finishing stage of 1520-1600 °C and a holding time of 8-10 hours, synthesis of strontium aluminate ($SrAl_2O_4$:Eu, Dy and $Sr_4Al_{14}O_{25}$:Eu, Dy) occurs in two stages to withstand temperatures at the finish of 1320 °C for 4-6 hours. In this regard, the achievement of homogeneous single-phase structures is not guaranteed, which ultimately has a negative effect on the brightness of stationary and spontaneous emission. Nevertheless, among the most common methods of synthesis of phosphors based on strontium aluminates are classical solid-phase and sol-gel techniques. Less commonly, the methods of combustion and induction heating are used. By virtue of the above reasons, almost no synthesis method used guarantees achievement of a high level of monophasity of the desired material with uniform distribution of REE in its entirety and, consequently, the maximum possible illuminating parameters. In this regard, the process procedure is a rather complex and multi-step regimen which, nevertheless, does not guarantee a sufficiently high level of repeatability of the achievable parameters.

Use of a liquid phase at the stage of synthesis allows the improvement of the quality of both phase composition and the powders of the synthesized materials. In this regard, the use of special purpose glasses (of the optimal composition with respect to the problem being solved) as a liquid phase allows, usually, achievement of a constructive compromise in solving the problems of general and special purpose, including in the case of synthesis in order to achieve monophasic strontium aluminate compounds and to enhance the illuminating parameters of the phosphors based on them.

In general, the use of such special purpose glasses allows for solid-phase synthesis in the presence of a minimum quantity of the liquid phase. The efficiency achieved in this regard is determined by the probability of optimization of the balance of activity of the special purpose glass of the desired purpose with respect to the synthesized material, which ultimately allows ease of solid-phase synthesis to supplement the advantage of liquid phase synthesis, namely the acceleration of the transfer processes and increase the homogeneity of the desired product [12].

The complexity and ambiguity of the processes characterizing the luminescence properties of specific phosphors does not allow us today to talk about the priority of certain technological factors having a dominant influence on the formation of their illuminating parameters and forming, in particular, the nanostructured composition of the synthesized powders and their particle size distribution.

Generally, the effectiveness of the results achieved is associated with the development of the materials, including REE additives of various concentrations. The result of the research accomplished is the emergence of new materials and technologies possessing the maximum values of their functionally important parameters. In this regard, it should be understood that the involvement of the results achieved in the new research and development continually expands the use of REE and their role in the development of scientific and technological progress as a whole.

It is clear that the development of new materials is associated with existing technologies and especially with those whose capabilities most optimally meet the requirements of the problems being solved. In this sense, the technology of production of phosphors has a lot to do with the production technology of functional ceramics, particularly at the stage of solid-phase synthesis. Today, in the sphere of development of materials of different functionality, sufficiently large scientific, experimental and technological material has been accumulated to allow for the development of the most optimal algorithm for the new research. Obtaining of ultra- and nano-powders for the production of functional oxide ceramics with improved electrophysical, chemical and mechanical parameters is still one of the most urgent tasks of ceramic technology relating to the number of critical technologies of the federal level.

Nevertheless, the role of modern nanotechnologies in this matter clearly occupies a leading position. However, the fact that most of these technologies are of a laboratory character speaks to their constructive use at the stage of fundamental research of the tasks at hand, but does not provide rapid industrial implementation of the results obtained.

The practical solution of the problems of these critical technologies in the synthesis of high quality ultra- and nano-dispersed powders for the production of oxide ceramics with a stable chemical, phase and granulometric composition was found in the development of the technology of continuous solid-phase synthesis [13]. Conceptually, the continuity of solid-phase synthesis is associated with the transfer of the flow of solid phase reactions at the level of the individual granule of the starting homogeneous mechanical mixture introduced into the reaction chamber, preheated to the particular synthesis temperature of the oxide material. The thermal shock and temperature gradient emerging (at the initial time it reaches 8,000-100,000 deg/Mm.) so intensify the diffusion processes that the completion of the solid-phase synthesis is accomplished over a period of time measured in minutes rather than hours, as in classical solid-phase synthesis [13,14].

Research has shown that powders of oxide materials synthesized by solid-phase synthesis are a combination of nanostructured macro-objects formed from the nanophase by self-assembly (bottom-up) directly during synthesis. Thus, the nanostructured macro-object is not a single crystal formation, but is a plurality of discrete crystalline particles held within the scope of the macro-object by more significant forces than gravitational.

The purpose of this work was to study the effect of conditions of continuous solid-phase synthesis on the particle size distribution of nanostructured powders of strontium aluminate, because the issues of the effect the rate of synthesis, the composition of the gaseous medium, the particle size of the synthesized phase and minimizing of the defects of the precursor particles are being raised today by researchers of the synthesis of phosphors in general, and phosphors based on strontium aluminate in particular.

2. Experimental part and discussion of the results

Used for synthesis of strontium aluminate ($SrAl_2O_4$) were strontium carbonate ($SrCO_3$) and aluminum oxide (Al_2O_3) of chemically pure grade. Homogenization of the starting mechanical mixture was carried out with conventional ceramic technology by mixing the starting reagents using distilled water. Synthesis of strontium aluminates was carried out with the technology of continuous solid-phase synthesis at temperatures up to 1250°C and a synthesis time of 30 min. The particle size distribution of the synthesized powders was determined by an Analysette 22 laser particle analyzer, X-Ray analysis on an ARL X'TRA diffractometer ($Cu_{k\alpha}$ radiation)

Figure 1 shows comparative data of particle size distribution of strontium aluminate synthesized by classical (CT) and continuous technologies. of solid-phase synthesis (SPS) using a batch process of the starting mechanical mixture (SMM) whose particle size distribution is as shown in Figure 1.

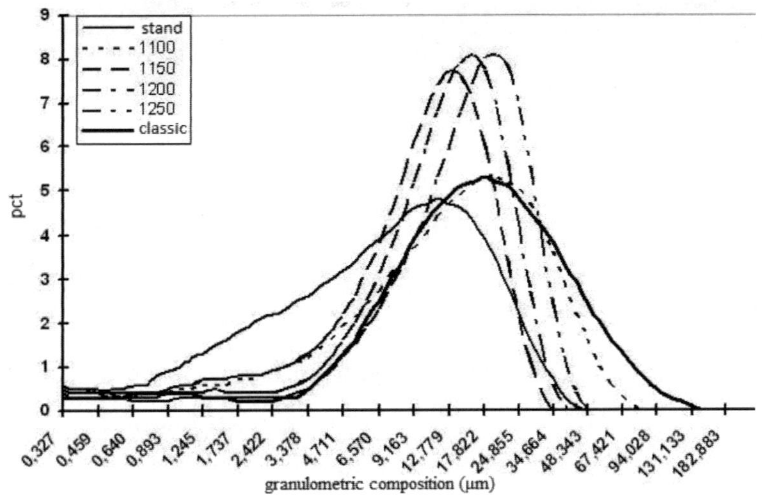

Figure 1. Particle size distribution $SrAl_2O_4$ (t=15minutes)

As can be seen, the particle size distribution of powders synthesized by the classical technology practically coincides with the particle size distribution of powders synthesized by solid-phase synthesis at a synthesis temperature of 1100°C for 15 minutes.

According to a scanning microscope (ZEIZZ SUPRA – 25), the formation of particle size distribution of the synthesized oxide material is accomplished directly during the synthesis process by continuous self- optimization of its composition depending on the conditions of the synthesis. Thus, the prevalent influence on the optimization of the sizing of the nanostructured macro-objects formed is the synthesis temperature and the time duration of synthesis for the optimal ratio between them.

Analysis of the research showed that the process of formation of a growing nanostructured macro-object is accomplished by incorporating smaller macro-objects of varying complexity formed in turn from the initial nanoparticles sized in the order of 20-50 nm while obeying the conditions of self-optimization of the overall system.

This model is in good agreement with the results of the synthesis of strontium aluminate which follows from the analysis of particle size distribution of powders having a bias tending toward their increase at a synthesis temperature from 1150 to 1250 °C and maintaining a period of time of synthesis equal to 15 minutes.

Figure 2 shows the diffraction patterns of strontium aluminate synthesized in continuous and classical solid-phase techniques.

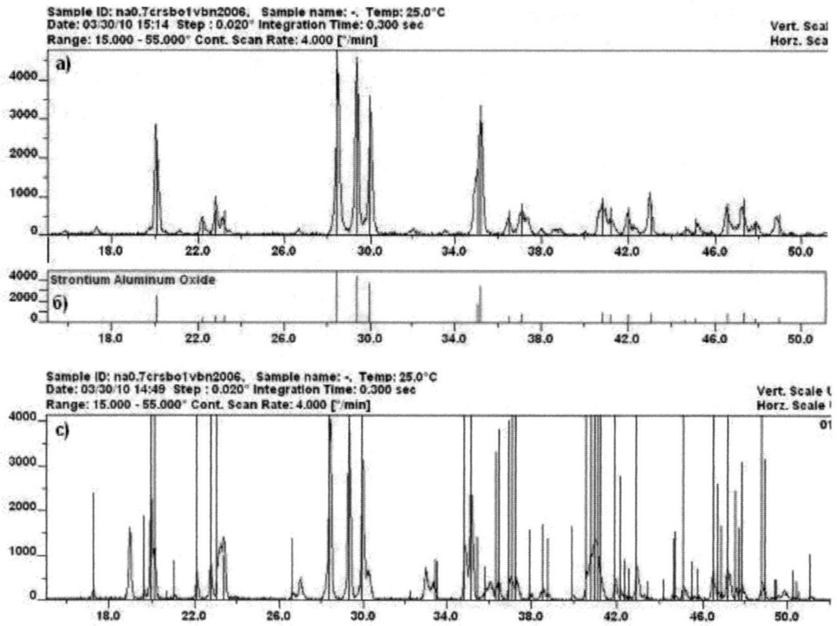

Figure 2. X-ray phase analysis of products of the synthesis of strontium aluminate by continuous (2a) and classical (2c) techniques, 2b – international standard bar chart

The results of research of the phase composition synthesized in conditions of continuous solid-phase synthesis in the presence of special purpose glass in an amount of 0.5-0.75% wt. (typically administered 2-8% wt.) showed that in the temperature interval of 1175-1225 °C and synthesis time 25-30 minutes, the monophasity of the synthesized strontium aluminate reaches 99±0.5 %.

Comparative analysis of these diffraction patterns from an international standard bar chart (Fig.2b) favors the monophasity of strontium aluminate synthesized by solid-phase technology in the presence of a liquid phase.

As can be seen from the data, the use of effective science-intensive technologies can solve not only general and special purpose problems at a higher level of quality, but also more actively promote the rapid filling of the internal market with high technology products.

The advantages of the proposed solution for technical, consumer and economic performance in comparison with known domestic analogs are shown in the table.

Technical Specifications	Classical Technology	Solid-Phase Synthesis Technology
Material	$(SrAl_2O_4)$:Eu,Dy,Y.	$(SrAl_2O_4)$:Eu,Dy,Y.
Temperature of synthesis, °C	1320	1225
Total time of synthesis, min.	≥ 1440	10-15
Particle fineness of synthesized material, μm.	Synthesized material, average grain size after grinding 15-20 μm	Sizes less than 25-50 nm, no further grinding required

3. Conclusions

1) Continuous solid-phase synthesis allows for the synthesis of strontium aluminate in the form of nanostructured powders with controlled particle size distribution directly during synthesis.

2) Continuous solid-phase synthesis in the presence of a liquid phase purpose allows for the synthesis of strontium aluminate with a high level of monophasity.

3) Continuous solid-phase synthesis is advisable to use to optimize illumination parameters of phosphors based on strontium aluminate.

References

[1] Wang D and Wang M 1999 Synthesis, crystal structure and X-ray powder data of the phosphor matrix 4SrO - 7Al2O3 *J.of Materials science* **Vol 34** pp 4959-64.

[2] Smets B, Rutten J, Hocks G Verlijsdonk J and Smets B 1989 2SrO - 3Al2O3:Eu and 1,29(Ba,Ca)O - 6Al2O3 – Two New Blue Phosphors *J. of the Electrochem. Soc* **Vol 136** pp 2119-23

[3] Capron M and Dony A 2002 Strontium dialuminate SrAl2O4: synthesis and stability *J. Am.Ceram.Soc* **Vol 85** № 12 pp 3036-40

[4] Von A. R, Schulze H and Muller-Buschbaum 1981 Zur Struktur von monoklinem SrAl2O4 *J. Anorg.all.chem* **Vo 475** pp 205 – 210.

[5] Fukuda K and Fukushima K 2005 Crystal structure of hexagonal SrAl2O4 at 1073K *J. of Solid State Chemistry* **Vol 178** pp 2709-14.

[6] Ravichandran D,.Johnson S T, Erdei S, Roy R and White WB 1999 Crystal chemistry and luminescence of the Eu2+ - activated alkaline earth aluminates phosphor *Displays* **Vol 19** pp 197 – 203.

[7] Peng T, Huajun L, Yang H and Yan C 2004 Synthesis of SrAl2O4:Eu, Dy phosphor powders by sol-gel processes and its optical properties *Materials Chemistry and Physics* Vol 85 pp 68–72

[8] Wang D, Yin Q, Li Y and Xiong Y 2005 SrAl2O4:Eu2+, Dy3+ phosphor thin films derived from the sol-gel process *J. of the Electrochem. Soc* **Vol 152** , iss 1 pp. H12–H14

[9] Sharma P, Haranath D, Chander H and Singh S 2008 Green chemistry-mediated synthesis of nanostructures of afterglow phosphor *Applied Surface Science* **Vol 254** pp 4052-55

[10] Yu X et al 2004 The influence of some processing conditions on luminescence of SrAl2O4:Eu2+ nanoparticles produced by combustion method *Materials Letters* **Vol 58** pp 1087–91

[11] Hovhannisyan R M et al 2003 Processes of borate formation taking place in batches of alkaline earth aluminium borate glasses *Glass Technology* **Vol 44** pp 101–111

[12] Katznelson LM 1996 The nature of the effect of "memory" of dispersed crystalline state to the piezoceramic. /Diss.kand.fiz.-mat.nauk. Rostov n / d -192s.]

[13] Katsnelson L M, Kerbel Б M 2012 Continuous solid-phase synthesis of ultra- and nano powders of oxide materials for producting high-functional ceramics *Non-ferrous metal.* **№1** p 34-37

[14] Katznelson L M and B M 2013 Kerbel technologically equilibrium oxide production of functional materials with high electrophysical parameters of the South Siberian scientific bulletin №1 (3) C 34-46] URL: http://s-sibsb.ru/issues-of-the-journal.html?catid=21.

[15] S.V. Bogdanova, S.A. Chornya, B.M. Kerbelb, A.N. Zhiganovb, Yu.V. Falkovichc, Applying Intelligent Management Comprehensive System in Rental Estimation of Russian Rare-earth Deposits of Global Resource Cluster Procedia Chemistry1st International Symposium on Inorganic Fluorides: Chemistry and Technology, ISIF 2014 Volume 11, 2014, Pages 63–72

IOP Conf. Series: Materials Science and Engineering **110** (2016) 012101 doi:10.1088/1757-899X/110/1/012101

Modeling of temperature fields in the working chamber of the process furnace for REE synthesis

Kerbel B M [1], Ageev A Yu[1], Payusov A Yu[1], Katsnelson L M,[2] Tereshchenko E V[1] , Verkhoturova V V[3]

1. Seversk Technological Institute, MEPhI, 636036, Russia, Seversk, Tomsk region, Kommunistichesky avenue 65.
2 JSC SPE "Technologic", Russia, 863033, Rostov-on-Don, Portovaya street, 303
3. National Research Tomsk Polytechnic University, 634050, Russia, Tomsk, Lenin avenue street 30.

Email: BMKerbel@mephi.ru

Abstract. The results of mathematical modeling of temperature fields in the working chamber of the process furnace for special purposes are shown. Studied laboratory furnace is test equipment, which is used for practicing the stages of the technological process of continuous solid-phase synthesis of nanopowders of various purpose, such as obtaining of luminophore powders with rare earth elements oxides in its composition. Mathematical model adequacy is tested empirically

Studied laboratory furnace is test equipment, which is used for practicing the stages of the technological process of continuous solid-phase synthesis (CSPhS) of nanopowders of various purpose [1], for example, obtaining luminophore powders [2], with rare earth elements oxides in its composition.

Mathematical modeling of temperature fields is carried out to create an automated control system of the technical system, which appears as shaft-type laboratory furnace in this paper. Solid-phase synthesis of REE oxide powders is carried out in a high-temperature furnace chamber, it is necessary to keep determinate distribution of thermal field in closed space of the working chamber.

This distribution is caused by special features of technological process of continuous solid-phase synthesis. Design layout view of studied furnace is shown in Figure 1.

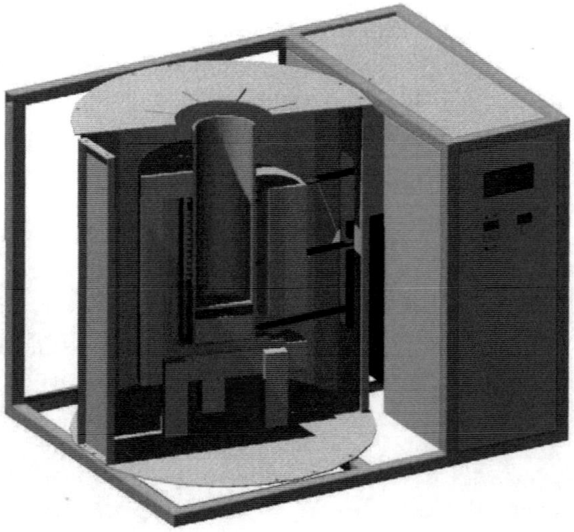

Figure 1 - High-temperature shaft-type furnace

Distinctive features of the furnace design are: relatively small dimensions, 1200°C as the upper limit of the temperatures operating range, the ability of automatic program control of the process when the furnace reaches fully operating conditions and automatically maintain this mode. In addition, distinctive features of the furnace design include dual-zone heating by the high level of the working chamber and the absence of heating on the sides (at the top (cover) and lower part (bottom)), as well as the presence of thermal capacitance "cushion" at the bottom of the chamber. Main technical characteristics of the studied furnace are shown in Table 1.

Table 1 – Main technical characteristics of the shaft-type furnace

№	Furnace parameters, items	values
1	The maximum allowable temperature, °C	1350
2	The number of thermocouples	2
3	Accuracy of temperature maintenance, not worse than, °C	±2
4	Working dimensions, not more than, mm	
	Diameter	100
	The furnace depth	300

While optimizing the furnace work, the conditions of the solid-phase reactions behavior in continuous synthesis mode should be taken into account, as well as high-speed sintering conditions. In this regard, focused research [3, 4] has been conducted to determine the factors, influencing the formation of the chemical continuity process conditions.
Analyzing obtained data, the following requirements appear that dictate the optimum temperature conditions, which are necessary to provide within the considered experimental furnace:

- The higher the specific surface area of the synthesized material, the lower temperature and the shorter time period are needed to complete the continuous synthesis of a particular material;
- Synthesis temperature has the most significant impact on minimizing its duration;

- Boundary conditions of the continuous synthesis are determined by prehistory of mechanical mixture, including its specific surface area;
- ceteris paribus, the intensity of the solid-phase reactions behavior in continuous synthesis is determined by the temperature-time characteristics of the synthesis;
- maximum quality of the synthesized oxide material is achieved at maximum intensity coefficient value of continuous solid-phase synthesis.

It has been found that the synthesis required quality could be achieved under certain interdependency of synthesis temperature and time. The tendency is that that the reduction of the synthesis time leads to synthesis temperature increase and, accordingly, has a direct impact on the installation capacity, in general. It is clear that the time-temperature interdependence has the restrictions, determined by criticality of particular oxide material to the conditions of synthesis.

During the oxide materials synthesis using the technology of continuous solid-phase synthesis, nanomodule formation occurs with following self-assembly of nanostructured macro-objects having constant particle size distribution in one technological process:

- control of the oxide material dispersion directly during the process of its synthesis allows avoiding grinding of the synthesized material from the classical ceramic technology specifications, which naturally brings to the end all the problems associated with the technological specifications, that are particularly significant for the functional materials with sensitivity to contamination;
- regulation of the powders particle size composition during CSPhS process eliminates the problem of molding ultra- and nano-level powders, which is rather complicated technological problem for real ceramic manufacturing;
-
- ceteris paribus, ceramic materials get the most preferred parameters values in comparison with traditional ceramic technology of their producing.

Initial oxides grinding, occurring in the first few seconds, as if precedes a new phase synthesis, although, it is not possible to separate these two processes, because they occur simultaneously at the initial moment (seconds), and, then, the quality and dispersion of synthesis are formed.

The necessity to carry out the experiments on baking the powders batches of various volumes using CSPhS technology determined the necessity to assess the parameters of the temperature of non-gradient zone (NGZ) in the furnace working chamber.

Traditionally, quantitative description of the heat transfer process is based on the theory of mathematical physics, using a system of differential equations in partial derivatives, with imposing the necessary boundary (Neumann coefficients) and initial (Dirichlet coefficients) conditions [5] on this system. However, due to the bulkiness of the mathematical apparatus, such approach is not always convenient.

Alternative method, which is successfully applied in everyday engineering practice, appears as the calculation of the temperature in a given three-dimensional object space using analytical expressions, obtained with processing of experimental data.

Let us consider the experimental data as a result of some process realization disturbed with the interference, which is certainly determined by the unknown function of the $y = f(x)$ form. In most cases, the experimental data of this kind are represented as tables or charts, which display the tables

Obviously, the charts are the simplest means, widely used for a long time in various fields of human activity, for cognitive representation of the experimental data that allows visualizing

the qualitative features of the process and evaluate them, despite the interference and measurement inaccuracy.
The charts that show the same process and the characteristics of a certain object may differ significantly from one another by scale, by the number of measurements used, by the level of interference, and so on. At the same time, the distinctive features of the chart type describe the parameters of the imaged object or the process. Automatic or automated processing of such charts includes the comparison of their types to determine whether different charts describe the same or different processes or objects.

Wherein, the situation, when the experimental data set with sufficient accuracy for engineering practice could be approximated by a function, is typical enough.

The most common method of the experimental data approximation is the method of the least squares [6, 7]. This method allows using approximating functions of any kind, and belongs to the global methods group. Polynomial approximation of various degree is one of the least squares method variation.
The requirement for the minimum sum of squared deviations from the approximating function to the data points is the criteria of the least squares method accuracy:

$$\mathrm{F} = \sum_{i=1}^{n}(y_i - f(x_i))^2 \rightarrow min,$$

where y – some known values, x - a set of unknown (desired) variables.

Thus, this method does not require from approximating function to go through all given points. This is significant at approximation of experimental data with some inaccuracy.

Applying to the studied shaft-type furnace, series of measurements were carried out on the temperature profile of the furnace - temperature change at the vertical geometric axis of the working chamber depending on the depth of immersion therein. These measurements were conducted for different sections of the working temperature range of the furnace - for the initial zone, the central zone and high zone. The numerical values of the set the temperatures in this case were: 200, 500, 800, 1000 and 1200 ° C.

The results of measurements were taken into account for building appropriate interdependences - furnace temperature profiles for taking the temperature set points. For example, Figure 2 shows a series of temperature profiles obtained (the points on the charts show the temperature values for each of its profiles).

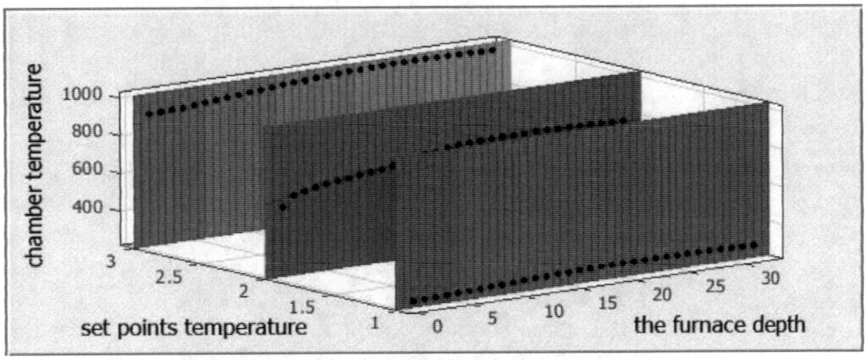

Figure 2 – Experimental temperature profiles of the studied furnace

For the mathematical processing of the experimental data polynomial approximation with polynom of the 5th order was used. Thus, each of the temperature profile was approximated by this function. The results of the temperature profiles approximation for set points of 200 and 800 ° C are shown in Figure 3. As it is seen from the Figure, the deviation of the calculated and experimental values of the temperature does not exceed X°C or Y%, which is sufficient accuracy for engineering calculations.

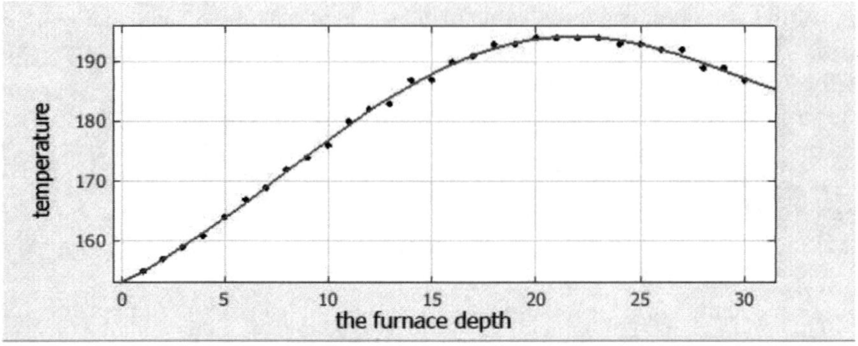

a) the experimental and calculated furnace profiles at 200 °C set point

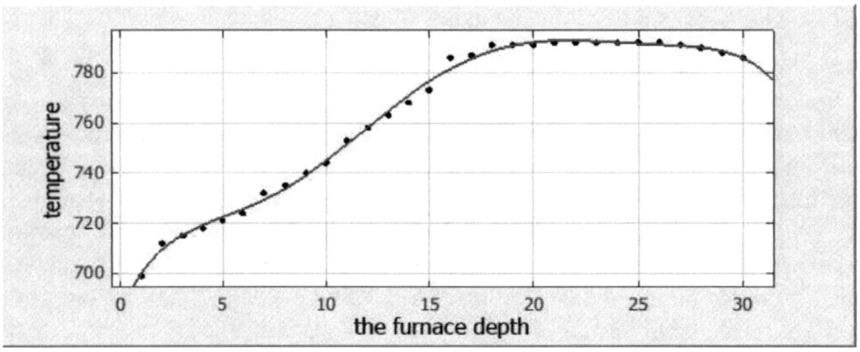

b) the experimental and calculated profiles at 800 °C set point

Figure 3 – Comparison of the calculated and experimental furnace profiles

Surface approximation was performed to obtain more universal tool, which allows analytically evaluating the temperature profile parameters of the studied furnace at arbitrary set point (Figure 4).

Figure 4 shows the approximated surface of the temperature profiles of the studied furnace, which is described by expression:

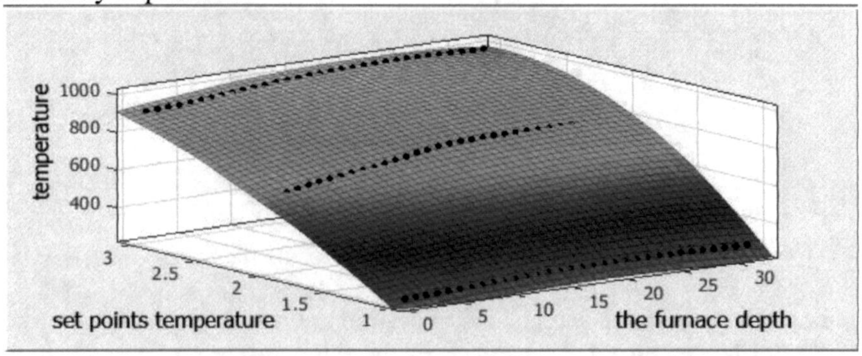

Figure 4 – Temperature profiles approximation with surface

$$f(x,y) = p00 + p10x + p01y + p20x^2 + p11xy + p02y^2 + p30x^3\ p21x^2y + p12xy^2,$$

Where the values of the respective coefficients are:

$$p00 = -276{,}9\,(-307{,}8;\ -246);$$
$$p10 = -6{,}441\,(-9{,}65;\ -3{,}233);$$
$$p01 = 577{,}4\,(545{,}1;\ 609{,}6);$$
$$p20 = -0{,}05262\,(-0{,}2112;\ 0{,}1059);$$
$$p11 = 14{,}9\,(12{,}97;\ 16{,}84);$$
$$p02 = -62{,}3\,(-70{,}16;\ -54{,}44);$$
$$p30 = -0{,}001724\,(-0{,}004664;\ 0{,}001217);$$
$$p21 = -0{,}01679\,(-0{,}04589;\ 0{,}0123);$$
$$p12 = -3{,}396\,(-3{,}811;\ -2{,}98).$$

The resulting approximated surface allows calculating the furnace temperature profile for any arbitrary set point at the temperature.

To assess the adequacy and accuracy of the values, obtained using the approximated surface, a comparison of the calculated and experimental parameters was carried out for a number of set points, which were not used in the original mathematical processing. For example, Figures 5 and 6 show the calculated and experimental temperature profiles of the furnace for 300 и 1000 °C set points, respectively.

As it is seen from the Figures, the discrepancy in the parameters, determined with the approximated surface and measured temperature profiles, does not exceed ±3 °C, which is in accordance with the requirements for practical engineering calculations.

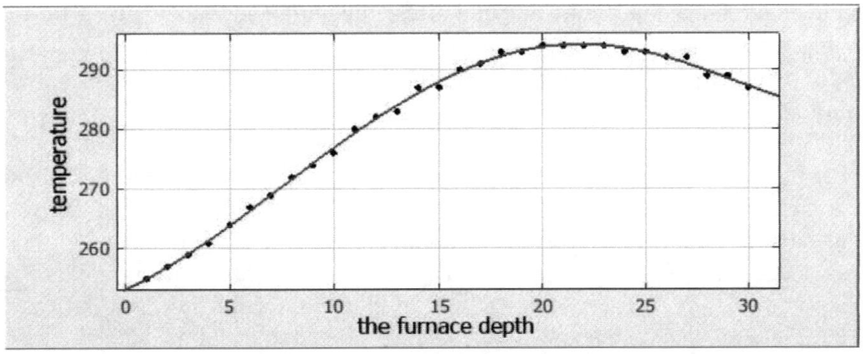

Figure 5 – Calculated and measured profiles for 300 °C set point

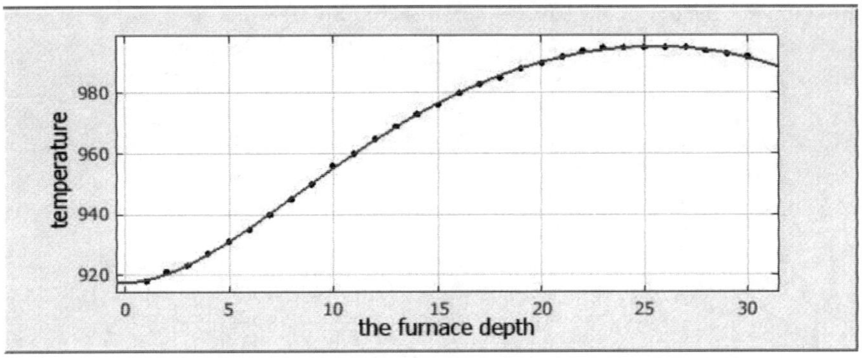

Figure 6 – Calculated and measured profiles for 1000 °C set point

The resulting analytical expression, which describes approximated surface of the temperature profiles of the studied furnace, allows evaluating only the vertical size of the non-gradient zone (NGZ) of the working chamber . The studies are the first step in working towards the application of approximating expressions for assessing the full set of NGZ parameters of the shaft-type furnace. At subsequent stages the evaluation of the radial temperature change in the furnace working chamber will be conducted.

Thus, this work shows the permissibility of using relatively simple approximating analytical expressions to assess the parameters of non-gradient zone (NGZ) of the shaft-type furnace working chamber. The furnace temperature profile may be analytically determined and graphically constructed for any arbitrary set point from the operating temperature range. The accuracy of the temperature profile design and an assessment of the NGZ axial dimension with its use is acceptable for practical engineering calculations.

Additional positive feature of this approach is the convenience and ease of programming the considered approximated surface and its use in the software of the furnace controller. This gives the opportunity of almost instantaneous construction of the furnace calculated temperature profile on the computer screen of the operator when setting a new set point by the temperature.

Developed furnace design and the levels of temperature fields in its working chamber allows carrying out the experiments on a continuous solid-phase synthesis of almost any oxide materials: ferrites, luminophores, semiconductors, posistors, piezoelectric materials, high-temperature superconductors, structural, condensed and other materials.

References

1. Katsnelson L M, Kerbel B.M. Continuous solid-phase synthesis of ultra- and nano powders of oxide materials for producting high-functional ceramics *Non-ferrous metal* 2012, **1,** 34-37
2. Kerbel B.M., Katsnelson L.M., Buynovskiy A.S., Tereschenko E.V., Daneikina N.V. Synthesis of photoluminophors using a continuous solid-phase technology *Procedia Chemistry,* 2014, **1**, 152–157
3. Katznelson L.M., Kerbel B.M. Production of technologically equilibrium oxide functional materials with increased electrophysical parameters. South Siberian Scientific Bulletin; 2013, **1(3),** 34-46; Online access: http://s-sibsb.ru/issues-of-the-journal.html?catid=21.
4. Katznelson L.M., Kerbel B.M. Installation of a continuous solid-phase synthesis of nanostructured powders and ultra-oxide materials in a real ceramic production. *Non-ferrous metals.,* 2012, **1**, 75 – 78.
5. Laptev G.I., Laptev G.G. Mathematical physics equations. M.: *Nauka*, 2003.
6. Numerical methods: using MATLAB: translated from English. / D.G. Matthews, K.D. Fink; Ed. Y.V. Kozachenko. – **3-rd Ed**. - Moscow: Williams, 2001. - 720 p.
7. Heat and mass transfer. /A.S. Telegin, V.S. Shvydkoi, Yu.G.Yaroshenko, Moscow, Metallurgy, 1995

Effect of powerful pulsed and continuous ion beams on the Al-Cu-Mg alloy structure

N V Gushchina[1], F F Makhin'ko[1], V V Ovchinnikov[1, 2], V I Gusel'nikov[3] and G E Remnev[3]

[1] Institute of Electrophysics, Ural Branch of Russian Academy of Sciences, Amundsena Street 106, Yekaterinburg, 620016, Russia
[2] Ural Federal Technical University named after the First President of Russia B.E. Yeltsin, Mira Street 19, Yekaterinburg, 620002, Russia
[3] Institute of Physics and Technology, Tomsk Polytechnic University, pr. Lenina 30, Tomsk, 634050 Russia

e-mail: viae05@rambler.ru

Abstract. The paper considers the results of an electron microscopy study of the VD1 alloy of the Al-Cu-Mg system after cold working and subsequent irradiation with a powerful pulsed ion beam (70% C^+ + 30% H^+, E = 180 keV) in the pulsed-periodic mode (τ = 80 ns, f = 0.1 Hz, j = 200 A/cm^2, F = $1\cdot10^{14}$ cm^{-2}) and under the conditions of the generation of only one pulse (τ = 180 ns, j = 100 A/cm^2, F = $2\cdot10^{15}$ cm^{-2}). It is established that this irradiation noticeably affects the microstructure of the cold-worked 3 mm thick sheets of VD1 alloy. The initial cellular dislocation structure transforms into a subgrain one. The intensity of structural transformations in the alloy increases with ion current density of a pulse. A similar transformation of a dislocation structure over the entire thickness of the sample is observed under irradiation with continuous Ar^+ ion beams (E = 20-40 keV) with not high fluences (10^{15}-10^{16} cm^{-2}).

1. Introduction

The nature of *long-range effect* under both continuous and pulsed ion implantation is still under investigation. The exposure of powerful ion beams (with the ion energy is 10^5-$3\cdot10^5$ eV, the energy density in the pulse is 1-3 J/cm^2, and the pulse duration is ~100 ns) on metal materials leads to changes in the structural state of its surface layers, the thickness of which exceeds the projective ion range by 2-3 orders of magnitude and the layer thickness heated during pulse action exceeds it by 1-2 orders of magnitude [1-3]. The changes at large depths are associated with quenching effects due to the rapid (>10^9 K/s) cooling of the heated layer and waves of mechanical stresses (thermoelastic waves) formed due to thermoelastic processes and recoil momentum during the pulsed evaporation of the surface layer of the target. Such waves can strongly affect the structure and properties of materials and lead to the formation of various radiation and deformation defects.

It is established that structural-phase transformations under irradiation with not only a *powerful pulsed* but *continuous* beams of midenergy (from a few tens to several hundred keV) ions take place not only in the ion penetration zone, but also in the region that exceeds the projected ranges of ions by more than a factor of 10^3-10^4 [4-13]. To explain the nature of these long range effects under continuous ion irradiation, a hypothesis has been proposed, which is associated with the propagation of intense post-cascade solitary waves resulted from the evolution of dense cascades of atomic collisions [14, 15] and are able to restructure the metastable media [10, 16].

In connection with this, it is of interest to compare the influence of the pulse-repetitive and continuous ion beams on the structure and properties of the alloy at similar fluences, as well as to compare these results with the effect of a single powerful ion pulse when the fluence of implanted ions is minimal and cannot play a noticeable role in changing the composition of the surface layer.

An electron-microscopy examination of the exposure of nanosecond pulses (with a high current density in the pulse) on the structure and phase composition of the VD1 alloy of the Al-Cu-Mg system were performed with the goal to compare the results with previously obtained data on the influence of continuous Ar^+ ion beams with an energy of 20-40 keV on structural-phase state of the VD1 alloy [11-13].

2. Experimental

The samples to be investigated were cut from cold-worked VD1 alloy (Al - 1.84 Cu, 0.79 Mg, 0.29 Mn, wt %) sheets ~3 mm thick (ε = 60%) fabricated at the Kamensk-Ural'skii metallurgical works. The sheets were cladded with technically pure aluminum from each side. The thickness of the clad layer was 60 μm.

The VD1 alloy samples were irradiated from one side by a powerful ion beam (70% C^+ + 30% H^+) under conditions of single-pulse generation and in a pulse-repetitive mode (at a pulse repetition rate of f = 0.1 Hz) using a TEMP accelerator (Tomsk Polytechnical University) at a pulse duration τ = 80 ns. The ion energy was 180 keV. In the case of a single pulse, the ion current density was 200 A/cm^2 that corresponds to a fluence of $1 \cdot 10^{14}$ cm^{-2}. The pulse-repetitive mode irradiation was performed at an ion current density of 100 A/cm^2 to a fluence of $2 \cdot 10^{15}$ cm^{-2}. The average temperature of the samples during irradiation did not exceed 30°C.

The electron microscopy analysis was carried out by the method of thin foils using a JEM-200 CX transmission electron microscope. TEM examination was carried out in a cross section and in the section parallel to the irradiated surface at a distance of ~150 μm from it.

3. Results

The VD1 alloy in the initial cold-worked state is characterized by a dislocation cellular structure with narrow boundaries between the cells (Figure 1*a*). The cell boundaries are dense dislocation tangles. The diameter of the cells is 0.5-2 μm. Inside the cells the dislocation networks are observed (Figure 1*b*). Some regions in the sample exhibit subgrain structure (Figure 1*c*), but the fraction of such regions is relatively small. Slip bands 0.3-0.5 μm wide (Figure 1*d*) are observed in the cross section of the sample. The bands contain randomly distributed dislocation tangles.

The study of the phase composition of the alloy has shown the clusters of equiaxed Al_8Fe_2Si particles ~50-100 nm in diameter (Figures 1*c*, 1*d*) inside the cells along both longitudinal and transverse sections of the alloy sample. Figure 1*e* illustrates an electron diffraction pattern with reflections from that phase.

Irradiation with single pulses. The investigation of the longitudinal section of the VD1 alloy sample at a distance of 150 μm from the irradiated surface after one-sided irradiation with *single pulses* of 70% C^+ + 30% H^+ ions at the ion current density j = 200 A/cm^2 (fluence of 10^{14} cm^{-2} per pulse) revealed almost complete transformation of the cell structure into a subgrain one over the examined section (Figure 2). Subgrains formed have an equiaxed shape and an average diameter of 1-3 μm. The subgrains are almost free of dislocations. The signs of the cellular structure, namely, imperfect subgrain boundaries, have been retained only in several regions of the sample, which occupy the insignificant volume of the sample.

Some subgrains, similar to the initial state, include a small number of equiaxed Al_8Fe_2Si particles ~100 nm in diameter.

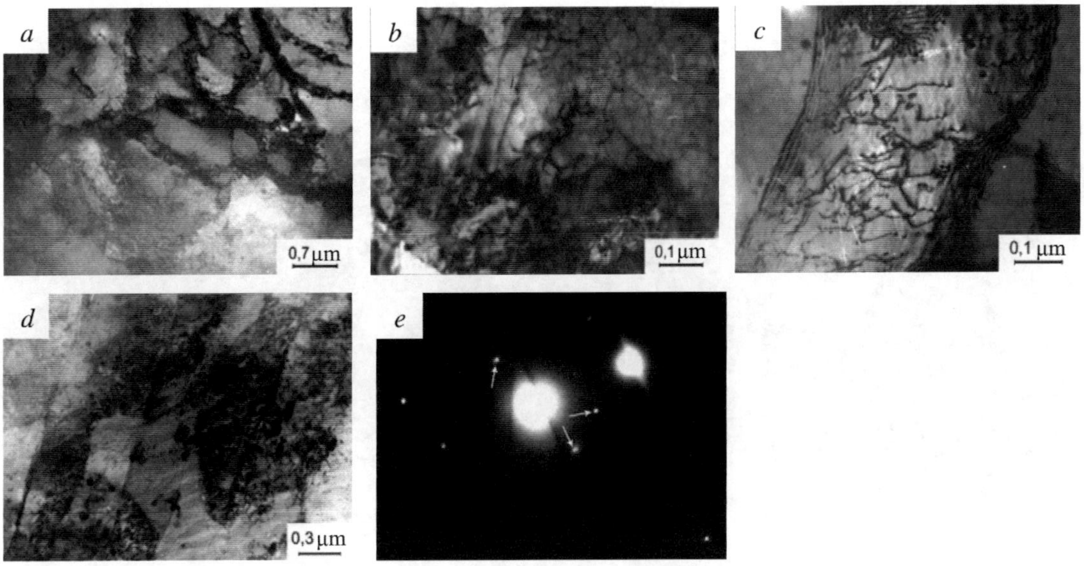

Figure 1. Microstructure of the VD1 alloy in the initial cold-worked state: *a-c* – longitudinal section of the sample: *a, b* – cellular structure; *c* – subgrain structure; *d* – slip bands in the cross section; *e* – electron diffraction pattern with reflections from the Al_8Fe_2Si phase.

The investigation of the cross section of the sample confirmed that irradiation in the *single pulse* mode caused the transition of the structure from cellular to the subgrain one over the entire sample volume (Figure 2*a*). For example, elongated or equiaxed-shaped subgrains are observed inside slip bands near the irradiated surface. They are particularly free of dislocations (Figure 2*c*). The opposite surface of the sample 3 mm thick also have slip bands (which contain individual subgrains) free of randomly distributed dislocations, but they are narrower then those observed on irradiated surfase. The subgrains exhibit either the finest subgrains or they are separated into individual fragments by cell boundaries (Figure 2*d*). Thus, irradiation affected the structure of the VD1 alloy over the entire depth of the sample, but this effect decreased with distance from the irradiated surface.

Pulse-repetitive mode. Irradiation of the sample in the pulse-repetitive mode at a lower ion current density $j = 100$ A/cm^2 and a higher fluence $2\cdot10^{15}$ cm^{-2} resulted in the formation of a mixed structure in the alloy: in some regions, a cellular structure dominates (Figure 3*a*), in other, subgrain one (Figure 3*b*), whereas the signs of the both structures are observed in the third regions (Figure 3*c*). The average diameter of subgrains varies from 0.5 to 1.5 µm and is comparable with the cell diameter.

The figures demonstrate a small number of equiaxed Al_8Fe_2Si particles ~100 nm in diameter in several subgrains.

Slip bands are clearly observed in the cross section of the sample. The bands become narrower at a distance from the irradiated surface (Figures 3*d*, 3*e*). Similar to that in longitudinal section, the mixed structure is observed in the bands located near the irradiated surface (Figure 3*d*): some of them contain fine subgrains, whereas others are fragmented into separate blocks by dense dislocation tangles. There are almost no subgrains within slip bands near the opposite surface, whereas dislocation tangles are wider. However, this structure is more perfect as compared with that of the deformed state. Our results confirmed the effect of irradiation over the entire depth of the sample 3 mm thick.

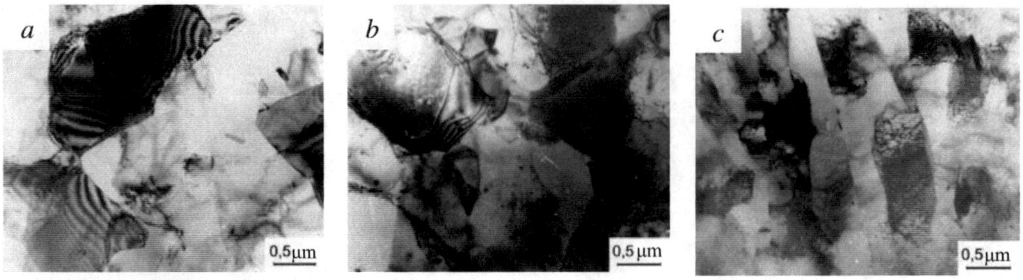

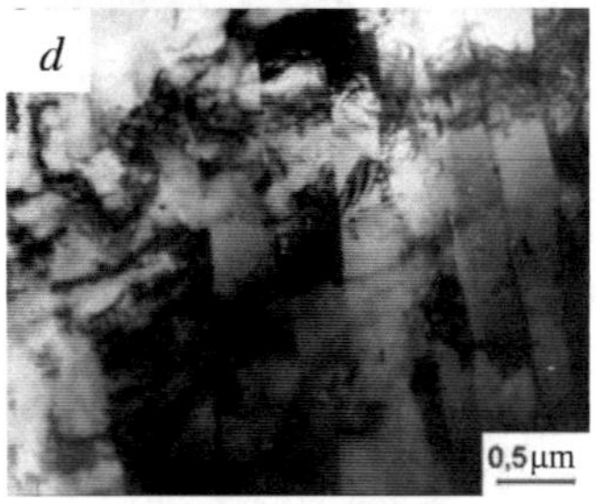

Figure 2. Subgrain structure in the VD1 alloy after pulsed irradiation with 70% C^+ + 30% H^+ ions at E = 180 keV, j = 200 A/cm^2, F = 10^{14} cm^{-2}: a, b – longitudinal section of the sample; c, d – cross section of the sample: c – near the irradiated surface, d – near the unirradiated surface.

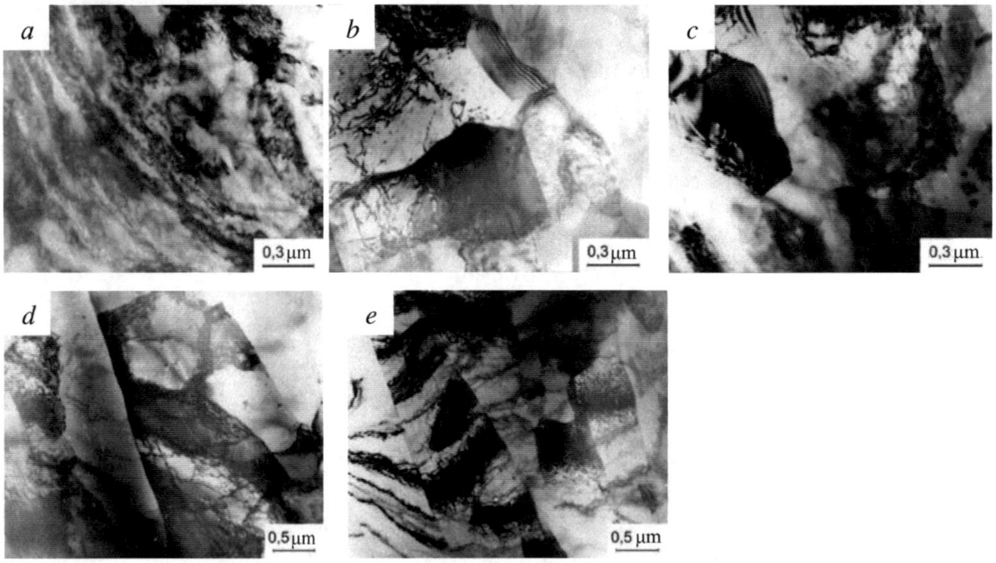

Figure 3. Microstructure of the VD1 alloy after pulsed irradiation with 70% C^+ + 30% H^+ ions at E = 180 keV, j = 200 A/cm^2, F = 10^{15} cm^{-2}: *a-c* – longitudinal section of the sample, *d*, *e* – cross section of the sample: *d* – near the irradiated surface, *e* – near the unirradiated surface.

Continuous mode. It was established previously in [11-13] that the effect of continuous Ar^+ ion beams with an energy of 20-40 keV on the cold-worked VD1 alloy also led to the transformation of the structure from cellular dislocation to subgrain at relativity low fluences (10^{15}-10^{16} cm^{-2}) and to the formation of a coarse-grained structure with a grain size of more than 10 μm at elevated fluences (~ 10^{17} cm^{-2}). Furthermore, at elevated fluences, supersaturated solid solution decomposition was observed in the alloy with the precipitation of hardening θ'(θ") phase ($CuAl_2$). The above structural changes occur with a high rate (during a few seconds or *tens of seconds* of irradiation) over the entire sample volume 3 mm thick, which exceeds the projective Ar^+ ion range by a factor of more than tens of thousands. The average projected *range* of Ar^+ ions with an energy of 40 keV in the studied alloys is ~40 nm, according to calculations. There is no a prolonged holding of samples at elevated temperatures, as in the case of conventional furnace annealing [11-13].

Thus, under the effect of powerful nanosecond pulsed ion beams (70% C^+ + 30% H^+, E = 180 keV, j = 200 A/cm^{-2}, F = 10^{14} cm^{-2}), the transformation of the dislocation structure is observed over the entire sample volume of the cold-worked VD1 alloy of the Al-Cu-Mg system, which is similar to that occurring under irradiation of this alloy with continuous ion beams (Ar^+, E = 20-40 keV, j = 150-400 μA/cm^2) at comparable fluences (10^{15}-10^{16} cm^{-2}) [11-13]. Irradiation in the pulse-repetitive mode at a lower ion current density has a slightly less effect on the state of the alloy.

4. Summary

Thus, irradiation with single and repetitive pulses of 70% C^+ + 30% H ions substantially affects the microstructure of the cold-worked VD1 alloy. The initial cellular dislocation structure of the alloy transforms into the subgrain structure. The influence that the ion current density has on the degree of transformation of the initial structure of the alloy was found. More intense structural changes took place under the effect of a single pulse with a higher ion current density (j = 200 A/cm^2), despite a lower value of fluence of – 10^{14} см$^{-2}$ (at j = 100 A/cm^2, samples were irradiated to a fluence of $2\cdot 10^{15}$ cm^{-2}).

The electron microscopy study of the VD1 alloy samples in both the section parallel to the irradiated surface at a distance of ~150 μm from it and the transverse section indicates that the depth of the effect of pulsed powerful ion beams exceeds the projective ion paths in this alloy by a factor of more than 1000. The initial cellular dislocation structure transforms into a subgrain one. A similar transformation of a dislocation structure is observed under irradiation of this alloy with continuous Ar^+ ion beams with small fluences (10^{15}-10^{16} cm^{-2}) [11-13].

Acknowledgments

The research was carried out within the state assignment of FASO of Russia (No. 0389-2014-0002), supported in part by RFBR (project No. 15-08-06744-A).

References

[1] Ligachev A E, Pogrebnyak A D, Remnev G E and Chistyakov S A 1987 *Fizika* **52**
[2] Shulov V A and Remnev G E 1993 *Laser and particle Beams* **11** 14 707
[3] Davis H A, Remnev G E, Stinnett R V and Yatsui K 1996 *MRS Bulletin* **XXI** 8 58
[4] Borodin S N, Kreindel Yu E, Mesyats G A and Ovchinnikov V V 1989 *Pis'ma Zh. Tekh. Fiz.* **15** 51
[5] Kreindel Yu E and Ovchinnikov V V 1990 *Vacuum* **42** 1/2 81
[6] Ovchinnikov V V, Kogan Yu D, Gavrilov N V and Shtoltz A K 1994 *Surface and Coating Technology* **64** 1
[7] Ovchinnikov V V, Chernoborodov V I and Ignatenko Yu G 1995 *Nucl. Instrum. and Meth. In Phys. Res.* B **103** 313
[8] Goloborodsky B Yu, Ovchinnikov V V and Semionkin V A 2001 *Fusion Sci. Technol.* **39** 1217
[9] Ovchinnikov V V, Goloborodsky B Yu, Gushchina N V, Semionkin V A and Wieser E 2006 *Appl. Phys.* A **83** 3
[10] Ovchinnikov V V 2008 *Phys. Usp.* **51** 955
[11] Ovchinnikov V V, Gushchina N V, Makhin'ko F F, et al 2008 *The Physics of Metals and Metallography* **105** 4 375
[12] Ovchinnikov V V, Gavrilov N V, Gushchina N V, Kameneckich A C, Emlin D R, Mozharovskii S M and Kaigorodova L I 2010 *Russian metallurgy (Metally)* **3** 207
[13] Gushchina N V, Ovchinnikov V V, Klepikova A A and Kaigorodova L I 2014 *Izv. Vyssh. Uchebn. Zaved., Fiz.* **57** 3/3 288
[14] Zhukov V P and Boldin A A 1987 *At. Energ.* **63** 6 375
[15] Zhukov V P and Demidov A V 1985 *At. Energ.* **59** 1 29
[16] Ovchinnikov V V 1994 in: Proc. XVI International Symposium on Discharges and Electrical Insulation in Vacuum **2259** ed. Mesyats G A (Moscow-St. Petersburg. SPIE) pp. 605-608

Ion irradiation-induced decomposition of Al + 4 wt. % Cu supersaturated solid solution

N V Gushchina[1], V V Ovchinnikov[1, 2] and A Mücklich[3]
[1] Institute of Electrophysics, Ural Branch, Russian Academy of Sciences, ul. Amundsena 106, Ekaterinburg , 620016, Russia
[2] Yeltsin Ural Federal University, Yekaterinburg, 620002 Russia
[3] Institute of Ion Beam Physics and Materials Research, Helmholtz-Zentrum Dresden-Rossendorf (HZDR), 01314, Dresden, PF 510119, Germany

e-mail: guscha@rambler.ru

Abstract. The decomposition process of the model precipitation-hardening Al + 4 wt. % Cu supersaturated solid solution induced by Ar^+ ions irradiation. Using X-ray diffraction, high-resolution electron microscopy methods and microhardness measurements, it was established, that, already at low temperatures ($T < 60$ °C), ion irradiation causes accelerated decomposition of solid solution, with precipitation of θ′ and θ-phase particles at a depth greatly exceeding the Ar^+ ions projected range.

1. Introduction

Many experimental facts [1-12] gathered in the course of studying different properties and microstructure of ion-implanted materials testify to the fact that the effect of accelerated ion beams on materials is not limited to the ions penetration zone, but extends to much greater distances.

The terms "long-range action" and "long-range effects" as applied to the bombardment of solids with accelerated ions relate to the experimental facts of changes to the structure and properties of materials at depth L, which is much in excess of the mean projected range R_p of ions. It was found [6, 7, 9-12] that the maximum response to the ion irradiation as well as the most extended impact zone is observed for metastable media with a high stored energy.

It would interesting to find out the extent of these phenomena (i.e., whether such processes are initiated under irradiation with accelerated ions in all or at least in the majority of metastable materials). Investigation of alloys displaying various types of structural phase [6, 7] and intra-phase transformations caused by ion irradiation deserves attention. This relates in particular to the formation of new phases by diffusion [7, 9] and diffusion-free mechanisms [4, 6], including details of the initial stages of transformation, e.g., irradiation-induced disperse phase nucleation and precipitation.

The present paper describes experimental investigations of the effect of irradiation with Ar^+ ions on the process of decomposition of a supersaturated solid solution on example of the Al + 4 wt. % Cu precipitation hardening alloy.

As is known, decomposition of this supersaturated solid solution proceeds in several stages, with formation (1) of coherent (with the initial matrix) Guinier-Preston zones (*ZGP*); (2) of coherent precipitates of the metastable θ''-phase; (3) of partially coherent particles of the θ'-phase; and (4) of incoherent particles of the stable $CuAl_2$ θ-phase [14]. The described sequence of phase transformations in the course of decomposition is responsible for primary strengthening (and following softening) of this alloy. Significant strengthening at natural and artificial aging results from the formation of *ZGPs* and intermediate metastable θ" and θ'-phases impeding the movement of dislocations. The subsequent stages, i.e., loss of coherence between the matrix and the metastable phase precipitates and formation and coagulation of stable phase precipitates, cause loss of strength.

2 Experimental

Samples of a Al + 4 wt. % Cu alloy in the form of 100 μm-thick plates were irradiated with Ar^+ ions (E = 20 keV, j = 160 μA/cm^2) after their quenching in water from 520 °C in order to obtain the supersaturated solid solution state. Irradiation with a continuous argon ion beams was carried out using a reflex-glow-discharge based ion source [14]. The sample was fixed to a copper collector with a heat-conducting compound for better heat removal. The sample temperature during irradiation was kept under control and never exceeded 60 °C, even under the longest (30 min) exposure to the ion beam. The ion implantation parameters are given in the Table 1.

Table 1. Irradiation parameters of Al + 4 wt. % Cu alloy samples.

Ord. No.	Ion beam parameters				Time, min	Fluence, cm^{-2}
	Ar^+ mass, a.m.u.	j, μA/cm^2	E, keV	R_p, nm		
1	39.962	160	20	22.5	0.017 (1 s)	10^{15}
2					0.33 (20 s)	$2 \cdot 10^{16}$
3					1	$6 \cdot 10^{16}$
4					5	$3 \cdot 10^{17}$
5					15	$9 \cdot 10^{17}$
6					30	$1.8 \cdot 10^{18}$

To investigate the structural-phase state and the microplastic resistance of the samples before and after Ar^+ ion irradiation, the following methods were used: X-ray diffraction analysis, high-resolution electron microscopy (HREM), Vickers microhardness and kinetic microhardness tests.

The X-ray diffraction analysis in order to determine the lattice spacing was carried out using the DRON-4-07 diffractometer and $Cu_{K\alpha}$ radiation. The electron microscopy study was performed with the help of a Philips CM-300 transmission electron microscope at an acceleration voltage of 300 kV.

The *microhardness test* at 40 g load was carried out using the Neophot microscope (the indenter penetration depth was between 6-8 μm). Besides, measurements of *kinetic microhardness* of irradiated samples were carried out by continuous indentation under a 5 g load in the Shimadzu DUH-2002 instrument. Continuous automatic recording of parameters in the course of indentation (such as load on indenter, depth of its penetration into material and duration of loading) allows correct measurements to be made of microplastic properties of thin surface layers of materials as well as of films and coatings at small and super small loads (0.1-5.0 g). The indenter penetration depth under a 5 g load was approximately 2-3 μm.

3 Results and discussion

At the lowest fluence $D = 1 \cdot 10^{15}$ cm^{-2} (the irradiation time is 1 s), the greatest *increase* in microhardness is observed, as compared with the initial quenched state (Figure 1 a, b). With further increase of the irradiation dose, microhardness gradually *decreases*.

IOP Conf. Series: Materials Science and Engineering **110** (2016) 012103 doi:10.1088/1757-899X/110/1/012103

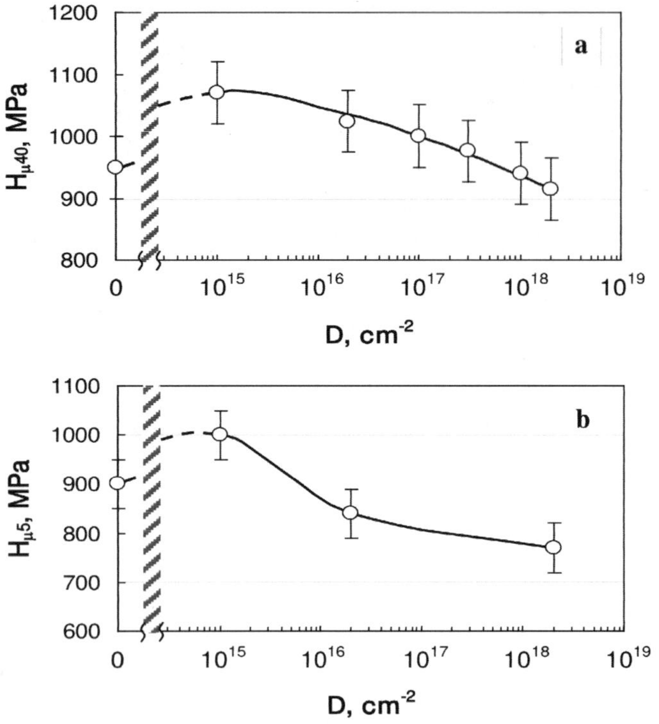

Figure 1. Changes in microhardness $H_{\mu,40}$ (a) and $H_{\mu,5}$ (b) of the Al + 4 wt. % Cu alloy in dependence on Ar^+ ion irradiation dose.

Comparison of the results obtained under different loads shows that the *relative* changes in microhardness caused by irradiation are more pronounced under smaller load. This points to the fact that the degree of radiation-induced decomposition of the supersaturated solid solution (proved later with the X-ray lattice parameter measurements and XTEM examinations) at any *given* fluence is higher near to the irradiated surface. However, changes in microhardness under 40 g load (indenter penetration 6-8 μm) are also quite significant and testify that the supersaturated solid solution decomposes not near the irradiated surface only, but also at a depth many times exceeding the ions projected range.

The X-ray analysis data show an increase in the solid solution lattice parameter (Figure 2), which testifies to the fact that already at small irradiation doses (~10^{15}-10^{16} cm^{-2}) and low temperatures (< 60 °C) solid solution decomposition goes on with precipitation of second-phase particles which are observed after thermal annealing at a temperatures exceeding 150-200 °C. It should be noted that the X-ray diffraction method reflect the structural state over the ~50 μm thick surface layer [15]. Heating of a quenched sample in oil up to 60 °C and exposure during 30 min (reproducing the target heating mode under high-dose irradiation) causes no changes in the microhardness and the lattice parameter.

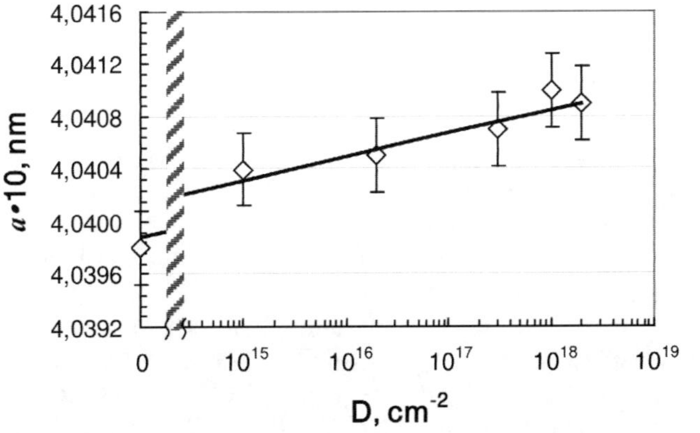

Figure 2. Changes in the lattice parameter a of the Al + 4 wt. % Cu alloy in dependence on Ar^+ ion irradiation dose.

Cross-sectional transmission electron microscopy investigations (XTEM) confirmed that ion irradiation causes decomposition of the solid solution with formation of θ"-, θ'- and θ-phase particles. The microstructure of a sample after Ar^+ irradiation to a fluence of $2\cdot10^{16}$ cm^{-2} is shown in Figure 3. The electron microscope images display a contrast from fine plate-shaped particles parallel to $<001>_{Al}$ (Figure 3 a). To identify the type of the formed particles, the electron diffraction patterns of individual regions and high-resolution images (Figures 3 b-d) were analyzed. The values of interplanar spacings for planes $\{001\}_{\theta''}$, $\{101\}_{\theta'}$ and $\{200\}_{\theta}$ determined by Fourier transformation (0.75, 0.3 and 0.34 nm, respectively) are in satisfactory agreement with the known data (0.77, 0.3, and 0.33 nm) [13].

The investigated layer depth (in the cross-sectional image) was ~1000 nm, and the whole volume of this layer contained the θ-, θ'- and θ"-phase precipitates whereas the mean *projected range* R_p for Ar^+ ions, as calculated with the use of the TRIM method, being 22.5 nm only (see Table 1).

Precipitates of the stable θ-phase prevail near the irradiated surface, while the dominating θ'-phase is observed mainly at a larger depth. The extent of the zone of θ-phase (and particularly of θ'-phase) precipitation is, according to data of electron microscopy, microhardness tests and X-ray diffraction, at least two-tree orders of magnitude greater than the indicated above ions projected range of 20 keV Ar^+ ions in Al + 4 wt. % Cu alloy.

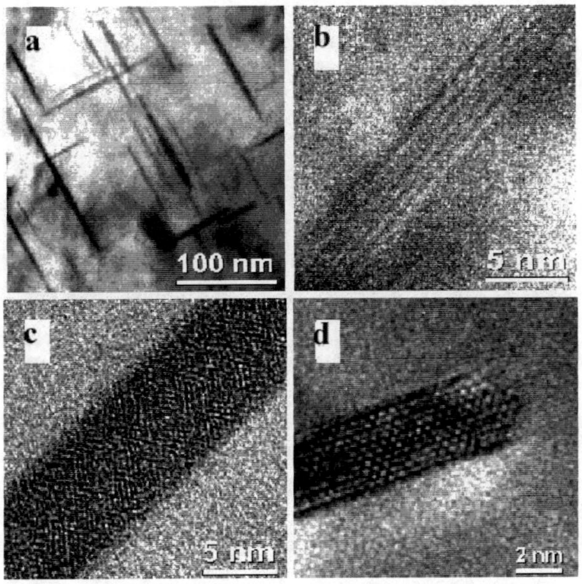

Figure 3. Results of electron-microscopic examination of cross-section samples irradiated with Ar^+ ions at $2\cdot10^{16}$ cm^{-2} (exposure time 20 s, j = 160 $\mu A/cm^2$): precipitates of θ"- θ'- and θ-phase (a) and their HREM images (b, c, d).

The process of phase ageing manifests itself by the data of microhardness and lattice parameter measurements already at the fluence of $1\cdot10^{15}cm^{-2}$, which corresponds to just 1 s of irradiation. This allows us to doubt the diffusion nature of the observed changes.

According to [12], radiation-stimulated (radiation-enhanced) migration processes are capable of just a several times increase of the depth of ions penetration into the material. So, thermostimulated and radiation-stimulated diffusion at T =500 and 700 °C during 2 h increases the depth of Sb^{+} (50 keV) ions penetration in silicon by approximately 50 to 100 nm. In our case, irradiation during only 20 s to a fluence of $2\cdot10^{16}cm^{-2}$ at T < 60 °C, causes, according to direct XTEM data, decomposition of Al + 4 wt.% Cu solid solution in a zone of the order at least 1000 nm deep. In view of microhardness and lattice parameter measurements this rating may be increased to $(10^{2}\text{-}10^{3})R_{p}$.

A multiple increase in the exposure depth and the rate of solid solution decomposition can be explained by the radiation-dynamic action of accelerated ions [10, 11, 16]. The essence of this effect is that the nanosized regions of atomic displacement cascades (Thermal Spikes [17, 18]) are zones of explosive energy release. The temperature of the cascade regions, which are thermalized within a time of the order of 10^{-12} s, may reach, according to Monte Carlo and molecular dynamics calculations, 5000-6000 K and even higher values.

Experimental proof of the formation of such extremely heated regions [18] has recently been obtained from the analysis of the emission spectra of metal targets (Fe, Zr and W) in the course of ion irradiation. Estimated temperatures are close to the simulated data.

The rapid heating of these regions and pressure boosting give rise to post-cascade shock waves. The pressure at the front of shock waves may exceed not just the real, but also the theoretical yield point of condensed matter [11, 18]. For comparison, the characteristic time of explosive energy release in a nuclear and in a chemical explosion is ~10^{-8} and 10^{-5} s, respectively.

As it was shown in [7, 9, 11], the structural and phase transformations in metastable media with increased stored energy cover macroscopic volumes of the material and occur with high rates (often within several seconds of exposure) at rather small fluences (D = 10^{15}-10^{16} cm^{-2}) and at reduced (by 150-200 K) temperatures compared to similar thermoactivated processes.

Radiation jolting of material with post-cascade waves may bring down the effective energy of diffusion activation [11] due to propagation in the volume of post-cascade shock waves. On the base of analysis of theoretical models [10, 11] it was shown that, in *metastable media*, such waves may become *unattenuating* (or slowly attenuating) at the expense of the energy liberated by structural-phase transformations at their front.

Metastable media with high stored energy, including supersaturated solid solutions as it has been shown by the present investigation, cold-worked materials [11] and alloys disordered by quenching or plastic deformation [7, 11], undergo maximum influence. Ion irradiation initiates fast-going (self-propagating into the depth of the material) processes transforming such media into the states with lower free energy. As a rough analogy, we may refer to the effects of explosive crystallization of supercooled liquid and decomposition of a supersaturated water solution as a result of impact or shaking.

The processes of phase formation/dissolution [11], atomic ordering and delamination [6, 7, 10], recrystallization of cold deformed materials [9, 11] (accompanied with explosive transformation of dislocation structure, with formation of point defects and activation of grain boundaries) and some others effects take place at anomalously low temperatures (100-200 °C below of similar thermostimulated processes).

The results of the investigation performed concerning *acceleration* of the processes of metastable medium transformation under irradiation, *bringing down* the temperature of phases precipitation, significant *increase* of the process depth (as compared with the ions penetration depth) are in agreement with the trends found in the available studies and supplement them with concrete actual data relating to the Al + 4 wt. % Cu system. The obtained data may be explained within the framework of the above mentioned approach.

4 Conclusion

Thus, it was established that irradiation of an Al + 4 wt. % Cu supersaturated solid solution with Ar^+ ions strongly (at least 10^3 times) accelerates its decomposition, ensuring precipitation of intermediate θ', θ'' and equilibrium θ phases, taking place already at low temperatures (at which, in the course of usual heating, only the zonal stage of aging occurs, and even that requires long-term exposure). At that, the depth, at which phase precipitation proceeds, exceeds the ions penetration depth by at least two-tree orders of magnitude (which follows from analysis of the microhardness, electron microscopy, and X-ray diffraction data).

In view of the real difficulties of explaining the fast processes of ageing of alloy Al + 4 wt. % Cu in the course of ion bombardment at *anomalously low* temperatures and at an *anomalously large* depth of $(10^2\text{-}10^3)R_p$ within the scope of traditional migration models, the authors, with account for the earlier performed research [7, 10, 11], come out with a suggestion about a significant role of alternative mechanisms of radiation-dynamic nature.

Acknowledgements This work was performed in the framework of the State Task, no. 0389-2414-0002, supported in part by RFBR (project No. 15-08-06744-A). We thank E. Wieser for the idea and participation in discussing the results.

References

[1] Guseva M I and Gordeeva G V 1986 *Phys. Status Solidi* A **95** 385
[2] Braun M 1988 *Vacuum* **38** No. 11 973
[3] Roy S and Dev B N 2012 *Phys. Status Solidi* A **209** 1511
[4] Johnson E, Grabak L, Johansen A, Sarholt-Kristensen L, Borgesen P, Scherzer B M U, Hayashi N and Sakamoto I 1989 *Nucl. Instrum. Methods Phys. Res.* B **39** 567
[5] Didenko A N, Rjabchikov A J, Isaev G P, Arzubov N M, Sharkeev Yu P, Kozlov E V, Pushkareva G V, Nikonova I V and Ligachev A E 1989 Mater. Sci. E*ng*. A **115** 337
[6] Kreindel Yu E and Ovchinnikov V V 1990 *Vacuum* **42** No. 1/2 81
[7] Ovchinnikov V V, Chernoborodov V I and Ignatenko Yu G 1995 *Nucl. Instrum. Methods Phys. Res.* B **103** 313
[8] Perry A J, Treglio J R, Schaffer J P, Brunner J, Valdova V and Rafaja D 1994 *Surf. Coat. Technol.* **66** 377
[9] Ovchinnikov V V, Goloborodsky B Yu, Gushchina N V, Semionkin V A and Wieser E 2006 *Appl. Phys.* A **83** 83
[10] Ovchinnikov V V 1994 in: Proc. XVI International Symposium on Discharges and Electrical Insulation in Vacuum 2259 ed. Mesyats G A (Moscow-St.Petersburg. SPIE) pp. 605-608
[11] Ovchinnikov V V 2008 *Phys.-Usp.* **51** No. 9 955
[12] Goloborodsky B Yu, Ovchinnikov V V and Semionkin V A 2001 *Fusion Sci. Technol.* **39** 1217
[13] Phillips V A 1975 *Acta Metall.* **23** 6
[14] Gavrilov N V and Emlin D R 2003 *Technical Physics. Russ. J. Appl. Phys.* **48** No. 9 1186
[15] Mirkin L I Handbook of X-ray Analysis of Polycrystalline Materials (Gosudarstvennoe Izdatelstvo Fiziko-Matematicheskoi Literatury, Moscow, 1961).
[16] Zhukov V and Ryabenko A 1984 *Radiation Effects Defects Solids* **82** 85
[17] Thompson D A 1981 *Radiation Effects Defects Solids* **56** 105
[18] Ovchinnikov V V, Makhin'ko F F, Solomonov V I, Gushchina N V and Kaigorodova O A 2012 *Tech. Phys. Lett.* **38** 42

Determining the effect of the duration of alternating acoustic excitation on electromagnetic response parameters of the composite

V P Surzhikov, L V Yavorovich and B D Zhanchipov

Tomsk, 634050, Russia, Tomsk Polytechnic University

E-mail: iva1954@ mail.ru

Abstract. The paper presents the results of the experimental investigation dealing with the effect of the alternating excitation acoustic pulse duration on the parameters of the electromagnetic emission caused by interaction of acoustic vibrations with inhomogeneities in the sample structure, piezoelectric inclusions and media interfaces of different dielectric properties. The model sample was epoxy resin with quartz sand as filler. The electrical component of the response was recorded with a capacitive transducer. The acoustic pulse was generated by piezoelectric transducer at frequencies of 57 kHz, 65 kHz, 74 kHz, 87.5 kHz and 94.5 kHz with the pulse duration varying from 10 μs to 100 μs. It is shown that the reduced duration of the acoustic action leads to dominance of the intrinsic frequency in the spectrum. A contribution of tensile pulses to acoustic electromagnetic transformation response generation is revealed.

1. Introduction

The phenomenon of electromagnetic emission was first used in Tomsk Polytechnic University to develop methods of geodynamic phenomena forecasting (earthquakes, rock bursts and landslides) [1–5]. Later studies were aimed at using this phenomenon in nondestructive detection of defects and stress-strain state testing [6–8]. In [9–11], the authors showed the efficiency of the characteristics of electromagnetic emission under pulsed acoustic excitation to detect changes in the stress-strain and the degree of sample imperfection.

To further advance the method and develop the proposed non-destructive testing criteria by the parameters of the electromagnetic response (EMR) of solid nonmetallic materials to acoustic impact, including composite materials, the optimal conditions of EMR excitation under acoustic electromagnetic transformations (AET) are to be studied and determined. The impact of the geometry of acoustic excitation and EMR measurement, as well as excitation pulse parameters, frequency and duration in particular, is to be found out.

This paper presents the results of the experimental studies of electromagnetic responses under excitation with pulses of different duration and frequency.

2. Materials and experimental technique

The model sample was epoxy resin with a size of 60.5×78×87.5 mm with quartz sand filler. The volume fraction of the filler was 35%. The electrical component of the response to the acoustic excitation was recorded with a capacitive transducer. A PZT-19 ceramic based piezoelectric transducer located on the edge of the sample with a size of 60.5×87.5 mm was used to generate ultrasonic vibrations. The acoustic pulse was initiated by an electric harmonic pulse generated by the master generator at fixed frequencies of

57 kHz, 65 kHz, 74 kHz, 87.5 kHz and 94.5 kHz. The pulse duration varied from 10 μs to 100 μs. The capacitive transducers were installed parallel to the edges of the sample 87.5×78 mm in size at a distance of 1 mm from the surface. The signals from the capacitive transducer were amplified by the differential amplifier. The data acquisition board NI PCI-6133 recorded the electromagnetic responses from the AET sources excited by 140 acoustic pulses spaced with an interval of 7 ms. For further analysis, the responses were averaged over the entire time period. The experimental error did not exceed 5%. A detailed description of the measurement technique is provided in [6,8].

3. Results and discussion

The results obtained in measuring the amplitude-frequency parameters of the electromagnetic responses to acoustic excitation tend to be identical for all the pulse frequencies used in the experiments. Fig. 1 shows the EMR amplitude changes at a frequency of 57 kHz, which feature this tendency under discretely changing duration of the ultrasound packet: 10; 30; 70 and 100 μs.

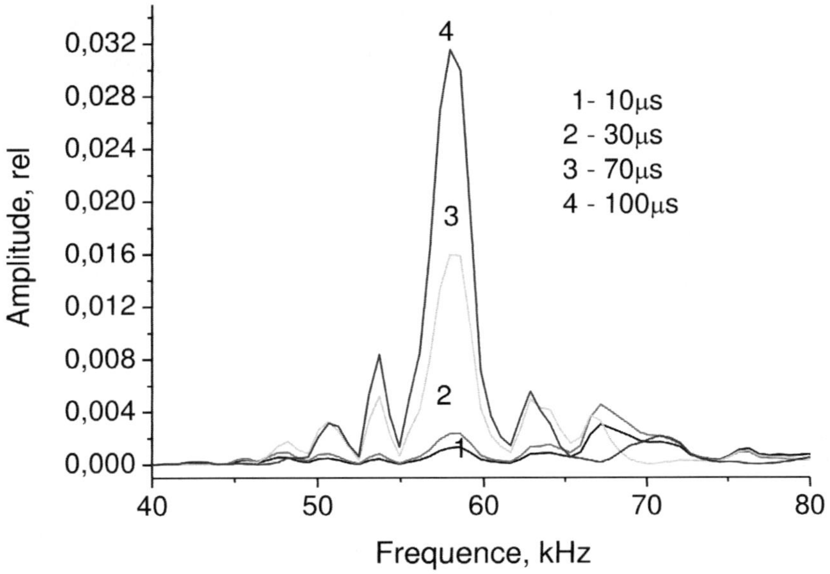

Figure 1. Spectra of the response power at a pulse frequency of 57 kHz for different pulse durations

The acoustic pulses were excited by piezoelectric transducer. The diameter of the contact surface of the transducer with the sample was 20 mm. As a result, the vibrations at the frequency of the master generator and forced vibrations at the intrinsic frequencies of the test object were initiated in the model sample. It is known that the electromagnetic signal parameters are determined by the characteristics of the acoustic impact [5,8]. Under excitation pulse duration of 10 μs and 30 μs, the vibrations at intrinsic frequencies are dominant in the spectrum. As pulse duration increases, the maximum at the frequency of the excitation pulse becomes dominant.

Fig. 2 shows the matched dependence of the integral of the response vibrational energy on the excitation pulse duration (Curve 1) and the sinusoidal voltage of a definite duration of the piezoelectric radiator from the master generator (Curve 2). To determine the energy of the integrated signal, the amplitude for each point was squared and the resulting values were summed over the entire response duration. The compression maximum in the generated acoustic wave corresponds to the positive voltage pulse maximum, and the tension maximum corresponds to the negative voltage pulse minimum. Curve 1 exhibits a number of "steps" with the duration equal to the period of the vibrations generated by the master generator with a sharp jump in the amplitude values. The amplitude jump coincides with the leading edge of the negative half-wave of the generator responsible for generation of the tensile stress in the acoustic wave.

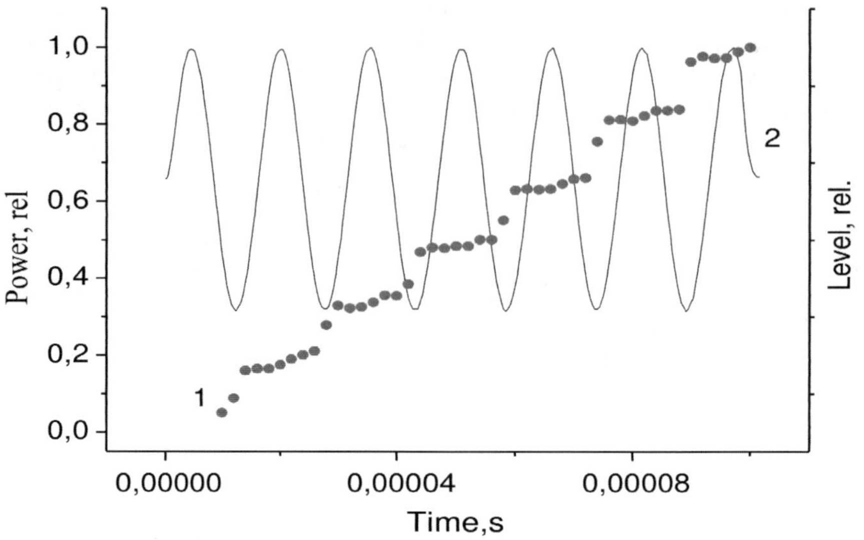

Figure 2. The integral of the vibrational energy in the electromagnetic response of the model sample versus excitation pulse duration (1); excitation pulse at 65 kHz generated by piezoelectric radiator (2).

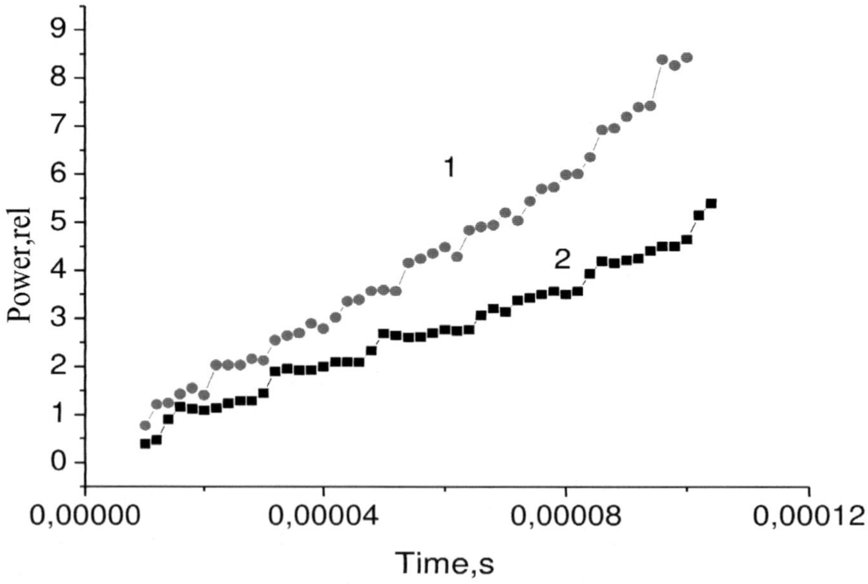

Figure 3. The integral of the vibrational energy in the electromagnetic response of the model sample versus pulse duration of the acoustic excitation at 94.5 kHz (1) and 57 kHz (2)

Fig. 3 shows the dependences of the integral of the electromagnetic response vibrational energy on the excitation pulse duration for frequencies of 94.5 kHz and 57 kHz. The dependencies indicate alternating sections with a smooth rise within the time equal to the period of the vibrations from the master generator with a sharp jump in the amplitude values coinciding with the emergence of the tensile pulse in the sample.

In order to eliminate the impact of the superposition of vibrations on the shape of the dependencies (Figures 2 and 3), the control measurements were performed with non-differential circuit when all the measurements were performed with one capacitive transducer (Fig. 4).

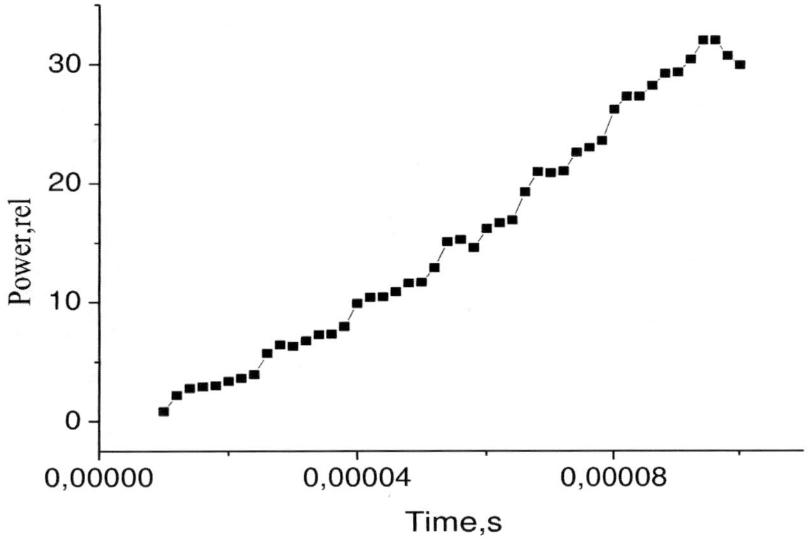

Figure 4 The integral of the vibrational energy in the electromagnetic response of the model sample versus the acoustic excitation pulse duration at 74 kHz (for non-differential amplifier input)

It can be seen that in this case, the feature of the integral dependence of the vibrational energy in the electromagnetic response of the model sample on the acoustic excitation duration is similar to that obtained via the differential measurement circuit as well. The steps in the curve under short excitation pulse durations become more distinct. This may be due to the fact that under short acoustic pulse durations, the electromagnetic response at intrinsic frequencies becomes dominant in the spectrum.

4.Conclusions

Thus, simulation of the impact of the alternating acoustic excitation duration on the parameters of the electromagnetic response of the model composite sample made of epoxy resin with quartz sand filler indicated non-linear dependence of the electromagnetic signal vibrational energy integral on the action time. The change in pulse duration causes redistribution in the AET response spectrum. Reduced duration of the acoustic excitation makes the intrinsic frequencies of the electromagnetic signal dominant in the spectrum. It is found that under alternating acoustic excitation, tensile pulses greatly contribute to the formation of the AET response. The obtained results can be used for development and application of the techniques to detect defects in solid dielectric materials by the parameters of the electromagnetic responses to alternating acoustic impact.

Acknowledgments

This work was financially supported by The Ministry of Education and Science of the Russian Federation in part of the science activity program.

References

[1] Yavorovich L V, Bespal'ko A A, Fedotov P I and Popelyaev A I 2015 Interconnection between parameters of rock samples electromagnetic signals and content of magnetite in the samples *IOP Conf. Series: Materials Science and Engineering* **81** No 1 012055

[2] Bespal'ko A A, Surzhikov A P, Yavorovich L V, Fedotov P I 2012 *Russian Journal of Nondestructive Testing* **48** (4) 34

[3] Bespal'ko A A, Yavorovich L V, Fedotov P I 2011 *Russian Journal of Nondestructive Testing* **47** (10) 41

[4] Bespal'ko A A, Yavorovich L V, Viitman E E, Fedotov P I 2010 V. A. *Journal of Mining Science* **46** (2) 136

[5] Bespal'ko A A, Yavorovich LV, Fedotov P I 2012 *Journal of Mining Science* **43** (5) 472

[6] Fursa T V, Asanov V A, Saveljev A V 2003 *Russian Journal of Nondestructive Testing* **8** 84
[7] Fursa T V, Demihova A A, Vlasov V A 2014 *Russian Journal of Nondestructive Testing* **5** 14
[8] Fursa T V 2012 *Russian Journal of Nondestructive Testing* **8** 55
[9] Surzhikov V P, Khorsov N N , Khorsov P N 2012 *Russian Journal of Nondestructive Testing* **48** 85
[10] Surzhikov V P, Khorsov N N , Khorsov P N 2015 *Technical Physics in Russian* **85** 153
[11] Surzhikov V P, Fedotov P I , Khorsov N N 2015 *Technical Physics* **60** 385

Sintering of zirconia ceramics using microwave and spark heating techniques

A S Ivashutenko, T S Frangulyan, S A Ghyngazov and A B Petrova

Tomsk Polytechnic University, 30 Lenin Ave, 634050, Tomsk, Russia

E-mail: ghyngazov@tpu.ru

Abstract. The paper presents the results of an complex study of structural and mechanical properties of zirconia ceramics sintered using different techniques. The samples were sintered via the conventional method of heating, in the field of microwave radiation and spark plasma. The experimental data indicates that a microwave field and spark plasma have a stimulating effect on zirconia ceramics sintering. In contrast to the microwave sintering, spark plasma sintering provides ceramics with improved properties at similar time-temperature annealing modes. Moreover, the properties of the ceramics under spark plasma sintering at T=1300 °C are similar to the properties of the ceramics sintered in a microwave field at T=1400 °C.

1. Introduction

Zirconia ceramics is a promising material that has been widely used in various fields of science and technology. Ultrafine powders (UFP) used in its manufacturing provide high-density ceramics of fine structure with unique physical and chemical properties. This expands the scope of its application, for construction and instrumental purposes in particular.

It is well known that both structural characteristics and physical properties of ceramics are primarily determined by the characteristics of the ultrafine powder medium such as particle size distribution, morphology of particles, their size and degree of agglomeration. Depending on the technique used to synthesize zirconium oxide powder, these characteristics can vary greatly.

Investigations conducted over the years [1–6] aimed at producing zirconia ceramics from UFP synthesized by the plasma chemistry method. This method is based on thermal decomposition of aqueous solutions of yttrium and zirconium nitrates in high-frequency discharge plasma. It features high performance [7, 8].

However, the studies showed that the properties of the ceramics sintered from these powders via conventional technological techniques are not as good as those of the ceramics produced from commercial UFP, the well-proven powder by Tosoh Corporation (Japan) in particular [9,10]. This is due to the reduced efficiency of plasma chemical powders (PCP) sintering which is caused by the presence of a large number of hard particle agglomerates in PHP.

This paper aims to study activation of sintering compacts from PCP through nonconventional techniques such as microwave radiation and spark plasma.

2. Experimental procedure

Zirconia ceramics was sintered from PCP of stabilized zirconium dioxide (in mol%) ZrO_2– 3 Y_2O_3. The compacts were prepared through static compression. Thermal sintering of the compacts was carried in a resistance furnace SNOL. The rate of sample heating and cooling was 10 °C/min.

IOP Conf. Series: Materials Science and Engineering **110** (2016) 012105 doi:10.1088/1757-899X/110/1/012105

Microwave (MW) sintering and spark plasma (SP) sintering of ceramics was performed at the Research Center in Karlsruhe (Germany). MW sintering was carried out in a microwave oven made by Gycom (Russia). The setting parameters were as follows: frequency of 30 GHz; power of 15 kW. The sintering mode: the rate of the temperature rise was 50 °C/min; the exposure time was 10 min. For SP sintering, a sintering furnace (HPD 10, FST Systeme Gmbh, Germany) was used. The heating rate was 200 °C/min and the pressure applied was 32 MPa. The bulk density of the ceramic samples was measured through hydrostatic weighing in distilled water using Shimadzu analytical balance supplied with special attachment for this purpose. An X-ray diffraction analysis of the powders and sintered ceramics was performed with ARL X'tra diffractometer using SuK□ monochromatic X-rays. The obtaioned X-ray diffraction regularities were processed by full-profile analysis with the software package Powder Cell 2.4. The microstructure of the ceramic samples was investigated using scanning electron microscopy (SEM) with the Hitachi TM-3000 electron microscope. The microhardness was determined with the ZHV1 micro Vickers hardness tester (Germany). The indentation load was 200 g, and the time of exposure under load was 10 s. The sample surface was subjected to 10 indentations.

3. Experimental results

The X-ray analysis showed that the phase composition of the zirconia ceramics obtained in different sintering techniques was similar. In all the cases, the diffraction pattern of the samples indicated the lines corresponding to tetragonal zirconium dioxide only.

Studies have been conducted aimed at establishing the optimal regime of thermal firing of ceramics, which results are presented in Table 1.

Table 1. Characteristics of zirconia ceramics sintered at different firing regimes

T (°C)	t (min)	ρ_{rel}	H_v (Гпа)	**T (°C)**	t (min)	ρ_{rel}	H_v (Гпа)
	0	0.79	5.9		0	0.91	9.7
1300	30	0.84	7.6	**1500**	30	0.92	10.4
	60	086	8.1		60	0.92	10.4
	180	0.88	8.4		180	0.91	10.2
	0	0.86	8.1		0	0.92	10.4
1400	30	0.91	9.2	**1600**	30	0.89	10.0
	60	0.92	9.8		60	0.88	9.2
	180	0.93	10.4		180	0.85	9.1

Note: T is the sintering temperature; t is duration of the isothermal stage; ρ_{rel} is relative density; H_v is microhardness.

The regularities of sintering zirconia ceramics from UFP PCP in thermal heating were studied in [1]. The dilatometric analysis of the densification kinetics of the samples revealed that during isothermal exposure at high temperatures T>1400 °C, they are observed to expand and their density decreases.

This effect can be attributed to the formation of a large number of gas-filled closed pores, which are very difficult to remove even at high temperatures and long isothermal exposure. At higher sintering temperatures, gas pressure in the closed pores increases. Since zirconium dioxide is characterized by high-temperature superplasticity, this factor may contribute to sample swelling during isothermal exposure, which negatively affects its properties. An optimum mode of thermal sintering of zirconia ceramics (T=1400 C, 3 hour annealing) was found. The characteristics of the ceramics obtained under these conditions are presented in Table 1.

As seen from Table 1, zirconia ceramics is characterized by high porosity. As is known [11], increased porosity typically reduces the ceramic hardness. Therefore, the sintered ceramic samples were of insufficiently high microhardness. Thus, thermal sintering zirconia ceramics from PCP does

IOP Conf. Series: Materials Science and Engineering **110** (2016) 012105 doi:10.1088/1757-899X/110/1/012105

not allow production of the material that satisfies the requirements of the product for constructional and instrumental purposes.

The results of the complex study of structural and mechanical properties of the zirconia ceramics sintered by different techniques are summarized in Table 2.

Table 2. Structural and mechanical properties of zirconia ceramics produced by thermal (T), microwave (MW) and spark plasma (SP) sintering

Sintering technique	Sintering mode	L (nm)	$\Delta d/d \cdot 10^3$	ρ (g/cm^3)	Θ (%)	$R_з$ (nm)	H_V (GPa)
T	T=1400°C, t= 180 min	101	0.2	5.6	7–8	340	9.5
MW	T=1200°C, t=10 min	-	-	5.26	13–14	148	8.3
MW	T=1300°C, t=10 min	15.5	0.7	5.65	6–7	206	10–11
MW	T=1400°C, t=10 min	44	0.7	5.86	3–4	220	12.6
MW	T=1500°C, t=10 min	108	0.8	5.3	12–13	455	8.4
SP	T=1100°C, t=10 min	64	0.8	4.3	30	-	4.0
SP	T=1200°C, t=10 min	38	1.0	5.52	9–10	-	10.5
SP	T=1300°C, t=10 min	50	0.6	5.8	4–5	200	11.8
T (Tosoh)	T=1400°C, t= 180 min	95	0.3	6.03	1.0	270	13

Note: L is the size of the coherent X-ray scattering region; $\Delta d/d$ is the value of the crystal lattice microstrain; ρ is pycnometric density; Θ is porosity; H_V is microhardness; $R_з$ is the mean value of the grain size.

A comparative analysis of the results presented in Table 2 indicates the following. MW heating of the samples does not change the optimum temperature of sintering; however, it essentially reduces the duration of the annealing process. This forms the ceramic structure of increased density, hardness and smaller grain size.

In contrast to MW sintering, SP sintering allows production of ceramics with improved properties at the same temperature-time annealing modes (Table 2). Moreover, the properties of the ceramics in SP sintering at T=1300 °C are similar to the properties of the ceramics sintered in the MW field at T=1400 °C.

The obtained experimental data shows that the microwave field and spark plasma have a stimulating effect on zirconia ceramics sintering. A theoretical basis of the activation of the oxide powder sintering under above heating conditions is reported in [12,13]. According to the concepts developed in these studies, high-energy action activates the diffusion mass transfer in heterogeneous structures. This is due to the formation of temperature gradients in the region of particle-particle contacts, which in turn causes thermal diffusion flows of the substance.

The data in Table 2 show that non-conventional sources of material heating such as microwave radiation and plasma spark provide the properties of the zirconia ceramics sintered from PCP close to those of the ceramics thermally sintered from the commercial powder by Tosoh Corporation.

4. Conclusion

The results obtained indicate the use perspectiveness of the considered techniques of activated PCP sintering to produce ceramics with improved properties compared to those obtained in thermal sintering.

Acknowledgements

This work was financially supported by The Ministry of Education and Science of the Russian Federation in part of the science activity program.

References

[1] Surzhikov A P, Ghyngazov S A and Frangulyan T S 2012 Rus. Phys. J. **55** (4) 345–52
[2] Surzhikov A P, Frangulyan T S and Ghyngazov S A 2014 J. Therm. Anal. Calor. **115** (2) 1439–45
[3] Khasanov O, Reichel U, Dvilis E, and Khasanov A 2011 IOP Conference Series: Materials Science and Engineering **18** article number 082004 Kulkov S N 2008 Physical Mesomechanics **11** 1–2 29–41
[4] Ivashutenko A S, Martyushev N V, Vidayev I G and Kostikov K S 2014 Advanced Materials Research **1040** 845–849
[5] Kulkov S N 2012 Journal of Surface Investigation **6** (3) 413-415
[6] Patent 2076069, C01G25 / 02 (Russia) Dorda F A, Dedov N V, Korobtsev V P, Stepanov I A, Utkin S V, Publ. 27.03.1997
[7] Larin V K, Kondakov V M, Malui E N, Matykha V A, Dedov N V et al 2003 Izvestiya Vyschikh Uchebn. Zaved. Tsvetn. Metallurg. **5** 59–64 (in Russian)
[8] Mazaheri M, Simchi A and Golestani-Fard F 2008 Journal of the European Ceramic Society **28** 15 2933–39
[9] Theunissen G S, Bouma J S, Winnubst A J A and Burggraaf A J 1992 Journal of Materials Science **27** 16 4429–38
[10] Buyakova S P and Kulkov S N 2005 Ogneupory i Tekhnicheskaya Keramika (in Russian) **11** 6–11
[11] Annenkov Yu M and Ivashutenko A S 2005 Bulletin of Tomsk Polytechnic University (in Russian) **308** 7 30–34
[12] Akarachkin S A, Annenkov Yu M and Ivashutenko A S 2012 *Butlerov Communications* **31** 9 130–137

Electromagnetic properties of $LaCa_3Fe_5O_{12}$ in the microwave range

V V Golenkina[1], S A Ghyngazov[2], V I Suslyaev[3], E Yu Korovin[3], G E Kuleshov[3], D A Kaykenov[4], E S Mustafin[4], T S Mylnikova[2]

[1]Tomsk State University of Control Systems and Radioelectronics, Tomsk, Russia
[2] National Research Tomsk Polytechnic University, Tomsk, Russia
[3]Tomsk State University, Tomsk, Russia E.A.
[4]Buketov Karaganda State University, Karaganda, Kazakhstan

E-mail: ghyngazov@tpu.ru

Abstract. The X-ray diffraction analysis of the $LaCa_3Fe_5O_{12}$ ferrite (lanthanum ferrite) prepared through high-temperature synthesis via ceramic technology was performed. It was found that ferrites belong to tetragonal system. The electromagnetic response from a flat layer of the composite based on this material under electromagnetic radiation in the frequency range of 0.01–18 GHz was investigated. It is shown that the developed material effectively interacts with electromagnetic radiation. The interaction effectiveness is directly proportional to ferrite concentration. Increased concentration of ferrite leads to growth of the reflection coefficient due to high conductivity of the material and visible decrease in the transmission coefficient in the frequency range of 4–14 GHz.

1. Introduction

A new class of ferrites, lanthanum ferrites, has been intensively studied in the last years. Ferrites of the type $LaM_3Fe_5O_{12}$ (M_3=Mg, Ca, Sr), $TmM_3Fe_5O_{12}$ (M_3=Ca, Sr, Ba), $Tm_2M\ _3Fe_5O_{12}$ (M_3=Li, Na, K), $ErMFe_2O_{5,5}$ (M–Ca, Sr, Ba) and $YbSrFe_2O_{5.5}$ are distinguished from other ferrites in this class. Their peculiarity lies in the fact that they contain rare earth metals which determine their new unique properties. They are promising for practical application due to a large number of studies of their properties [3–7].

The calorimetric study of the specific heat of the $LaM_3Fe_5O_{12}$ (M_3 = Mg, Ca, Sr) ferrite was carried out in [3]. Synthesis and X-ray analysis of $TmM_3Fe_5O_{12}$ (M_3 = Ca, Sr, Ba) and $Tm_2M\ _3Fe_5O_{12}$ (M_3 = Li, Na, K) compounds are described in [4,5]. The X-ray study of the double ferrite $ErMFe_2O_{5,5}$ (M–Ca, Sr, Ba) was carried out in [6]. In addition to the X-ray analysis, thermodynamic and electrical properties of $YbSrFe_2O_{5.5}$ and $YbBaFe_2O_{5.5}$ ferrites were studied in [7].

The $LaCa_3Fe_5O_{12}$ ferrite (lanthanum ferrite) is of particular interest. Due to high electrical conductivity and oxygen permeability of the ferrite, solid solutions are promising materials for producing gas electrodes and sensors, and SOFC. They are also used as catalysts for complete oxidation of gas cleaning reactions.

Since the requirements for multifunctionality of the developed materials are growing due to advances in modern science and technology, it is important to assess the result of interaction of the lanthanum ferrite-based composite with electromagnetic radiation (EMR) in a microwave range. The effectiveness of the interaction between this composite with EMR has not been studied, although it is known that ferrites are widely used as radar absorbing materials for environmental protection of biological objects against harmful effects, for prevention of information leak through radio-frequency channels, noise elimination by means of communication, stealth-technology, etc. [8–12].

The aim of this research is to carry out the X-ray analysis of $LaCa_3Fe_5O_{12}$, to prepare samples and to study the electromagnetic response from a flat layer of the composite based on this material under electromagnetic radiation in the frequency range of 0.01–18 GHz.

2. Experimental procedure

The $LaCa_3Fe_5O_{12}$ ferrite was obtained by high-temperature synthesis via ceramic technology at different temperatures in three stages [13]. The raw material (lanthanum oxide, ferric oxide (III), calcium carbonate) were weighed to the fourth decimal point and stirred. The mixtures were thoroughly grinded and then fired in a furnace, first at 800 ^{0}C for 10 hours, then at 1300 ^{0}C for 10 hours, and finally at 400 ^{0}C for 20 hours. The X-ray analysis of the $LaCa_3Fe_5O_{12}$ ferrite powder was performed using the ARL X'TRA diffractometer (Thermo Fisher Scintific).

Radiophysical measurements of the samples were carried out in the frequency range from 0.01 to 4.00 GHz with the analog meter of the modulus of the transmission transmission coefficient and the voltage standing wave ratio (VSWR) P2M-04 (home made) by "Micran", and the vector network analyzer E8363V (Agilent technologies) was used to perform measurements in the frequency range from 4.00 to 18.00 GHz.

The samples for radiophysical measurements were made of the ferrite powder under study and epoxy resin in the form of disks with an outer diameter of 7 mm and an internal diameter of 3 mm. The thickness of the disk ranged from 0.5 to 2 mm. The disk mold was filled with 3 g of ferrite and 3 g of the binding agent per 30 wt%, 50 wt% and 65 wt% concentration of the substance.

3. Experimental Results

3.1. X-ray diffraction analysis (XRD)

A typical diffraction pattern for the material under study is presented in Figure 1. These diffraction patterns were used for XRD.

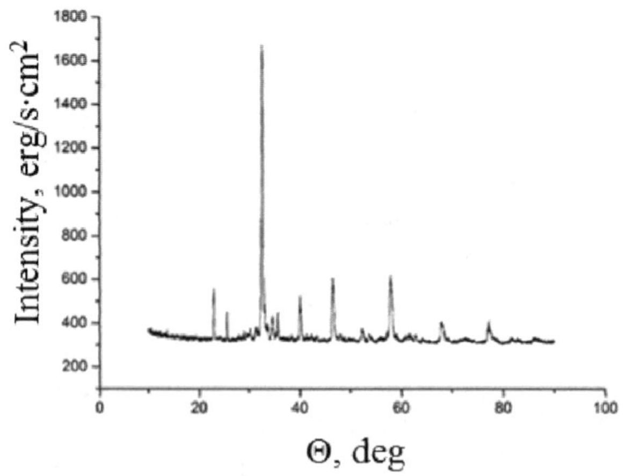

Figure 1. Dependence of the intensity on the scattering angle.

The results of XRD analysis of the $LaCa_3Fe_5O_{12}$ powder are shown in Table 1.

Table 1. Lattice parameters of the $LaCa_3Fe_5O_{12}$ ferrite.

Type of the system	Lattice parameters (Å)		V^o ($Å^3$)	$V^o_{el. cell}$ ($Å^3$)	Z	Density (g/cm^3)	
	a	c				X-ray	pycnometer
tetra gonal	10.84	16.58	1948.20	243.50	8	5.80	5.82 ± 0.04

where V^o is the volume of the elementary cell;
$V^o_{el. cell}$ is the number of formula units per cell;
Z is the number of formula units in a cell.

Homology method [14] was used to carry out indexing of the X-ray patterns (Table 2) of the studied compound powder. A distorted structural type of peroxide was taken as a homologue.

The relationship between the lattice parameters and the interplanar distance in the case of the tetragonal system (a=b) in indexing the X-ray pattern by this method is as follows:

$$\frac{1}{d_{hkl}} = \frac{h^2+k^2}{a^2} + \frac{l^2}{c^2} \quad (1)$$

where a and c are crystal lattice parameters, d is interplanar distance.

In addition to good agreement between the calculated and determined values of $1/d^2$ (maximum deviation of $1/d^2_{calc.}$ from $1/d^2_{exper.}$ should not exceed a probable measurement error), two more criteria can testify to indexing correctness:

a) the ratio of the number of theoretically possible lines in the X-ray pattern to the number of lines found experimentally is to be close to unity, when calculating the number of possible lines, a systematic extinction is considered;

b) good agreement must be observed between the experimental and calculated values of density ($\rho_{calc.}$), i.e. the number of formula units per elementary cell should be close to an integer, which is typically low. The minimal multiplicity of this space group (if it has been determined) can be considered. The number of formula units (Z) is calculated by the equation:

$$Z = \frac{V\rho_{exp.}}{1.66M} \quad (2)$$

where V the cell volume, M is the formula weight.

Otherwise, either $\rho_{exp.}$ or $\rho_{calc.}$ or V are incorrect, i.e. indexing is performed incorrectly. Systematic differences between $\rho_{exp.}$ and $\rho_{calc.}$ can be caused by the presence of a large number of lattice imperfections; however, these cases are extremely rare. In many cases, a simple chemical formula of the compound does not correspond to its stoichiometric formula. The deviation of Z from the integer can be explained by statistical filling of one regular point system with atoms of different types (i.e. random filling of similar positions with different atoms).

The volume of the elementary cells (V^0) for the studied phases was determined by the formula:

$$V^0 = a^2 \cdot c \quad (3)$$

The X-ray (calculated) density ($\rho_{roent/x\text{-}ray}$) of the studied compound was determined by the formula:

$$\rho_{roent} = \frac{1.66 \cdot Mr \cdot Z}{V^0} \quad (4)$$

where Mr is the molecular weight of the compound, Z is the number of formula units in the cell.

The ferrite density was measured by the technique described in [15] in 1 ml glass pycnometer. To measure the sample density, toluene, well wetting the studied substances and chemically inert, was used as indifferent liquid. It should be noted that the dependence of toluene density on temperature is not significant (ρ^{20}=0.8659 g/cm^3; ρ^{25}=0.8634 g/cm^3).

Table 2. Indexing scheme for the $LaCa_3Fe_5O_{12}$ powder.

I/I_0, %	d, Å	$10^4/d^2_{exp.}$	hkl	$10^4/d^2_{calc.}$
13	3.8739	666.3	220	660.5
8	3.6523	749.6	300	743.0625
4	3.0448	1079	320	1073.313
100	2.7531	1319	400	1321
36	2.6786	1393	410	1403.563
21	2.6500	1424	116	1424.405
7	2.6027	1476	330	1486.125
26	2.5069	1591	206	1589.53
15	2.3928	1747	413	1718.383
15	2.2500	1975	423	1966.07
7	2.2256	2019	306	2002.343
11	2.1947	2076	316	2084.905
5	2.1048	2257	415	2278.063
25	1.9521	2624	434	2623.743
4	1.8780	2835	009	2833.38
18	1.8330	2976	600	2972.25
7	1.7458	3281	603	3287.07
20	1.6909	3498	229	3493.88
4	1.6600	3629	418	3642.283
4	1.6170	3825	2010	3828.25
31	1.5970	3921	615	3929.313
3	1.5392	4221	606	4231.53
13	1.4839	4541	626	4561.78
11	1.4519	4744	617	4768.833
1	1.4448	4791	730	4788.625
13	1.3857	5208	1112	5202.245
5	1.3757	5284	800	5284
7	1.3117	5812	609	5805.63
9	1.2394	6510	664	6504.18

The densities of the compounds were calculated by the formula:

$$\rho = \frac{M_3 - M_0}{\dfrac{M_1 - M_0}{\rho_1} - \dfrac{M_4 - M_3}{\rho_2}} \tag{5}$$

where M_0 is the mass of the empty pycnometer, g; M_1 is the mass of the pycnometer filled with water, g; M_2 is the mass of the pycnometer filled with toluene, g; M_3 is the mass of the pycnometer filled with the studied substance, g; M_4 is the mass of the pycnometer filled with toluene and the substance, g; ρ_1 is water density at a specified temperature, g/cm^3 (reference value); ρ_2 is toluene density, g/cm^3.

The densities of the samples were measured 4–5 times, and the results were averaged.

IOP Conf. Series: Materials Science and Engineering **110** (2016) 012106 doi:10.1088/1757-899X/110/1/012106

Indices *hkl* were determined analytically (selection, search for sequences of interplanar distances assuming that the inverse square of the distance depends on Miller indices according to formula 1.

3.2. Investigation of the interaction of $LaCa_3Fe_5O_{12}$ with electromagnetic radiation

The dependences in the frequency range of 0.01–4 GHz obtained using different devices (the analog meter of the modulus of the transmission coefficient and VSWR P2M-04 and the vector network analyzer E8363V) coincided. Therefore, we used the data obtained with the vector network analyzer E8363V. The results of radiophysical measurements are shown in Figures 2–4.

Figure 2 shows that the reflection coefficient of $LaCa_3Fe_5O_{12}$ in the studied range depends on the frequency, and it is described by the curve with a maximum which depends on the concentration of ferrite in the composite.

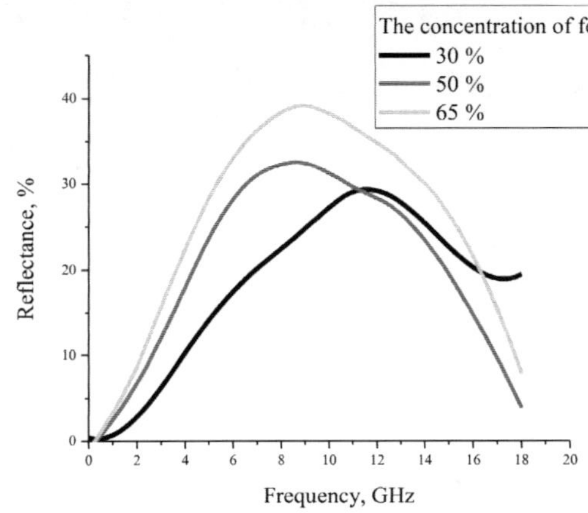

Figure 2. Dependence of the reflection coefficient on the electromagnetic radiation frequency.

The curves of the transmission coefficient relative to frequency are shown in Figure 3, which shows that the transmission coefficient also depends on the concentration of ferrite in the composite, and it is described by the curve with a minimum. The minimum occurs within the frequency range of 6–16 GHz.

The absorption coefficient of the studied material grows continuously as the frequency increases, and as the ferrite concentration in the composite increases, it grows up at a fixed frequency (Figure 4).

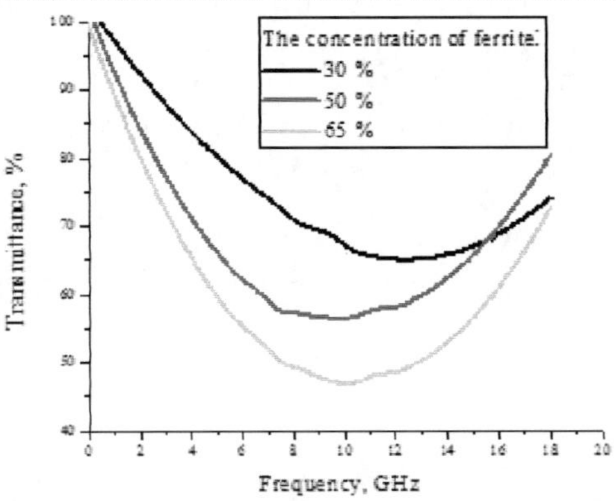

Figure 3. Dependence of the transmission coefficient on the electromagnetic radiation frequency.

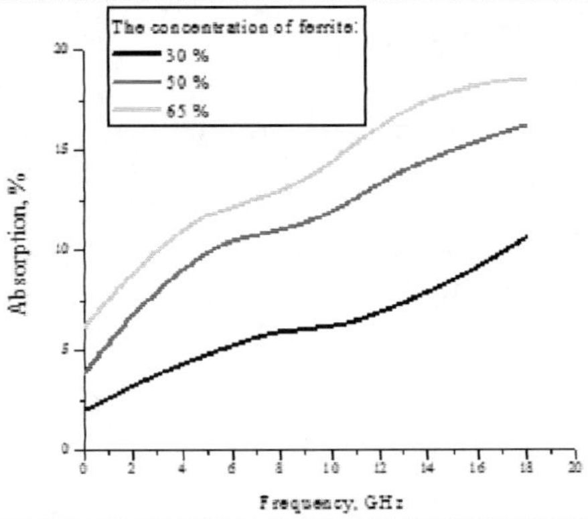

Figure 4. Dependence of the absorption coefficient on the electromagnetic radiation frequency.

4. Conclusions

The obtained results show that LaCa3Fe5O12 ferrites belong to the tetragonal system. The lattice parameters a and c, the elementary cell volume (Vo), the number of formula units per 1 cell (Voel.cell), the number of formula units in the cell (Z) have been determined.

The developed material effectively interacts with electromagnetic radiation. The effectiveness of this interaction is directly proportional to the concentration of ferrite. Increased concentration of ferrite provides growth of the reflection coefficient, which is caused by high conductivity of the material and distinct reduction of the transmission coefficient in the frequency range of 4–14 GHz, which can be used to make screening devices for solving problems of electromagnetic compatibility of modern high-frequency radio equipment. The operating frequency range and effectiveness of interaction with electromagnetic radiation can be changed through selecting the composite thickness.

Acknowledgements

This work was financially supported by The Ministry of Education and Science of the Russian Federation in part of the science activity program.

References

1. Tolokonnikov E G, Mustafin E S, Kasenov B K, et al. 2005 Russ. J. Inorg. Chem. **50** 148
2. Kasenov B K, Mustafin E S, Kasenova Sh B, et al. 2004 Russ. J. Inorg. Chem. **49** 102
3. Mustafin E S, Kasenov R Z, Pudov A M, Blyalev S A, Kaikenov D A, Muratbekova A A 2014 Vestn. KGU **3** 21–25
4. Mustafin E S, Omarov Kh B, Kasenov R Z, Pudov A M, Kaikenov D A, Satymbaeva A S 2012 Russ. J. Inorg. Chem. **57** 1259–1261
5. Mustafin E S, Omarov Kh B, Kasenov R Z, Pudov A M, Kaikenov D A, Satymbaeva A S 2012 Inorg. Mat. **48** 6 622 - 624
6. Kassenov B K, Mustafin E S, Zhumadilov E K 2010 Inorganic Chem. RAS **3** 489–491
7. Kassenov B K, Tolokonnikov E G, Mustafin E S, Kasenova Sh B, Davrenbekov S Zh, Sagintaeva Zh I, Zhumadilov EK 2006 Russ. J. Inorg. Chem. **51** 1 368 - 373
8. Zhuravlev V A, Suslyaev V I 2006 Russian Physics Journal **49** 8 840-846
9. Suslyaev V I, Naiden E P, Korovin E Y, Itin V I, Zhuravlev V A, Terekhov O G Patent R F 2382804 February 27 2010
10. Minin R V, Itin, V I, Korovin E Y Russian Physics Journal 2011 **53** 9 974-982
11. Zhuravlev V A, Suslyaev V I, Dotsenko O A, Babinovich A N Russian Physics Journal 2011 **53** 8 874-876
12. Naiden E P, Zhuravlev V A, Minin R V, Suslyaev V I, Itin V I, Korovin E Yu 2015 International Journal of Self-Propagating High-Temperature Synthesis **24** 3 148-151
13. Mustafin E S, Omarov Kh B, Kasenov R Z, Pudov A M, Kaikenov D A, Satymbaeva A S 2012 Inorg. Mat. **48** 6 622 - 624
14. Kovba L M, Trunov V K 1976 X-ray analysis M.: MSU 232
15. Kivilis S S 1959 Technique of measuring the density of liquids and solids M.: Standard 191

Evolution of mechanical properties of boron/manganese 22MnB5 steel under magnetic pulse influences

A P Falaleev, V V Meshkov, A A Vetrogon and A V Shymchenko

Department of Automobile transport, Sevastopol State University, 33 Universitetskaya str., 299053, Sevastopol, Russian Federation

E-mail: a_falaleev@mail.ru, vv_meshkov@ukr.net, windgone1980@gmail.com avsh_91@mail.ru

Abstract. The boron/manganese 22MnB5 steel can be noted as the widely used material for creation of details, which must withstand high amount of load and impact influences. The complexity and high labor input of restoration of boron steel parts leads to growing interest in the new forming technologies such as magnetic pulse forming. There is the investigation of the evolution of mechanical properties of 22MnB5 steel during the restoration by means of magnetic pulse influence and induction heating. The heating of 22MnB5 blanks to the temperature above 900^{0}C was examined. The forming processes at various temperatures (800, 900 and 950^{0}C) were performed during the experiments. The test measurements allowed to obtain the relationships between the strain and the operation parameters such as induced current, pulse discharge time and the operation temperature. Based on these results the assumption about usage of these parameters for control of deformation process was made. Taking into account the load distribution and the plasticity evolution during the heating process, the computer simulation was performed in order to obtain more clear strain distribution through the processed area. The measurement of hardness and the comparison with the properties evolution during hot stamping processes confirmed the obtained results.

1. Introduction

Improvements of manufacturing and production process allow non ferrous alloys such as aluminum, magnesium and titanium alloys, to occupy a significant niche among the construction materials. The high strength, good wear and corrosion resistances, as well as lower density comparing with the steel made these alloys a good option for aeronautical, nautical, and automotive purposes.

In order to keep their part of market, steel manufacturers had to improve their products and provide better relation between costs, weight, and properties. The advanced high strength steels (AHSS) came as the material that capable to provide these requirements. The cost issue is solved by saving of available manufacturing powers for casting, forming, and machining, since the creation of parts from AHSS use the existing production and manufacturing technologies. Although AHSS has density commensurate to the traditional mild steels, the weight reduction can be achieved by decrease of part thickness due to significant values of material strength. Despite the fact that AHSS have different structures, hardening and strengthening mechanisms, they were grouped together due to their considerable strength parameters. Values of yield and ultimate strengths for these materials are varied in the range from 550 MPa and 650 MPa for High-Strength Low-Alloy (HSLA) steels to 1250 MPa and 1500 MPa for Martensitic (MS) steels respectively.

Among the AHSS the boron steel should be highlighted, as the steel that has the highest strength-to-weight ratio that makes this steel is cost effective choice for strong and critical constructions. The hardening and strengthening mechanisms for boron steel are strongly associated with heat treatment and bake hardening that allows to obtain an almost full martensite structure, which provide the highest strength properties. [1]

Parts made from boron steels as well as from other types of advanced high-strength steels have some limitations during the production and restoration operations, since the applying of the colossal strength is required for forming parts made from this steel or restoration these parts after their deformation. However, this problem is solved by use of the correct sequence, and precision control of forming and heat treatment. The well-known and wide used manufacturing technology is hot stamping process also known as press hardening. The traditional hot stamping process includes:

- Heating of the blank in a furnace to temperatures above AC_3-temperature point in order to transform ferrite- bainite structure into austenite;
- Commonly the heating rate is higher than 10 ^{0}C/s and after reaching the temperature, material is kept at this temperature at least 3-5 min;
- After preprocessing the heated blank is transferred to high speed hydraulic press. The cooling and forming process are going simultaneously with the cooling rate higher than 29^{0}C/s in order to achieve the complete transformation of ductile austenite into hard and brittle martensite. [2]

The production and manufacturing processes are well studied and developed. However, the restoration of elements made from boron steels is time-consuming process and demanding to the compliance of heat and cooling rate requirements. One of the possible ways is the usage of magnetic technologies such as induction heating and usage of magnetic field for material forming. According to [3] the induction heating allows to obtain high heating rates that necessary to avoid grain coarsening during heating process. This method was successfully approved in works of Bariani et. al. [4] [2]

The usage of magnetic field allows to improve the material plasticity and as result increase the productivity operation process. [5] [6] The some articles demonstrate the increase of material plasticity and reduce of residual stresses. In the work [7] it was shown that the elongation of HS 37/23 steel increase by 4.95% and 7.50% under influence of magnetic field with magnetic induction 0.4 T and 1.07 T respectively.

In this particular study, the main object is to analyze the properties of boron/manganese 22MnB5 steel during the forming operation performed by means of induction heating and magnetic forming. The method used for pulsed magnetic forming process is described in the articles [8] [9].

2. Experimental method

2.1. Materials parameters

As it is mentioned above, the material investigated in this study is the boron/manganese 22MnB5 steel as the most widely used and well-known grade of boron steels. The experimental measurements are carried out few times with metal blanks that have thickness about 0.001 m. The chemical composition and mechanical properties of studied boron steel are shown in table 1 and table 2.

Table 1. Chemical composition of the boron/manganese 22MnB5 steel (wt.%).

Steel grade	C	Mn	P	S	Si	Cr	N	Ti	B	Al
22MnB5	0.22	1.25	0.013	0.004	0.25	0.14	0.004	0.03	0.003	0.040

Table 2. Mechanical properties of boron/manganese 22MnB5 steel.

Steel grade	Ultimate Tensile strength (MPa)	Yield strength (MPa)	Elongation(%)
22MnB5	1500	1000	8

2.2. Experimental parameters

The schematic representation of the experimental setup for this research is presented on the figure 1. It is well known that the discharge of electromagnetic impulse produced by the coil inductor leads to induction of eddy currents in both the metal blank and accessory screen. Since two parts with codirectional current flow are located in a variable electromagnetic field, then it is possible to assume the appearance of attracting and repelling forces between them. However, the researches performed in works [9] and [8] demonstrates that the relation between these force is strongly depended on the electromagnetic field frequency. The obtained results show that at the frequency ≤ 2 kHz, the repelling force is much more lower and can be neglected comparing with attractive action.

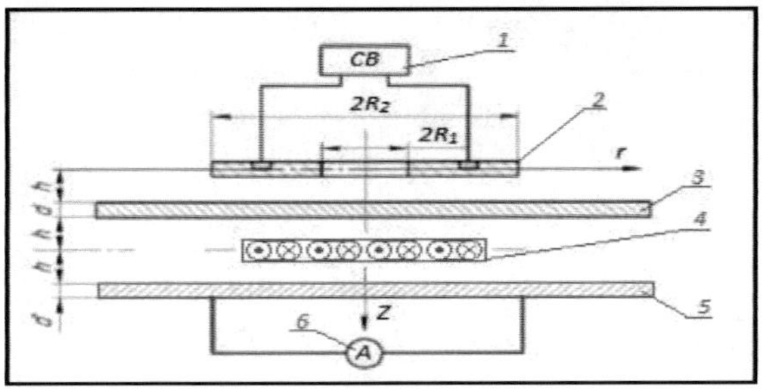

Figure 1. Schematic of experimental setup: 1 is the control block; 2 is the multi-turn induction coil; 3 is the accessory attracting screen; 4 is face inductor heater; 5 is the sheet metal blank; 6 is included into a chain Ampermeter.

For purpose of investigation of boron steel behavior during forming process by means of pulse electromagnetic attraction, the test frequency is 2 kHz. In order to generate magnetic pulses, the experimental setup includes induction coil with the inner and outer radii equal R_1 = 0.07 m and R_2 = 0.075 m, respectively. The used experimental setup allows to generate pulses with times equaled 1, 2.5 and 5 µs and to induce the currents I_{ind} in the blank at the range 40...75 kA.

However, to provide the material deformation and simplify the samples forming, the blanks must be heated above the AC_3 – temperature in order to obtain complete transformation of already existing microstructure into austenite phase. In most cases, the boron/manganese 22MnB5 steel is heated above 900 ^{0}C and kept at this temperature at least 3 min. [3] [10] [1] The test measurements are performed at operation temperatures 950, 900 and 800 ^{0}C due to the low strength and high plasticity of material at these temperatures.

It is important to control the temperature during forming operation as well as during heat/cooling processes. For this purpose, the laser pyrometer Pyrometer IMPAC IGA 15 plus is used that can provide the accurate and digital noncontact measurements of sample temperature. The range of its measurement is 250...1800 ^{0}C and the measurement error is less than 1.5%.

The presence of some difference in mechanical properties of tested samples can be established by measurements of material hardness. This measurement is performed by using Vickers hardness test as consistent with ISO 6507-1 standard.

3. Experimental method

During the experimental measurements, the relationship between current induced in the blank and relation deformation of material was obtained. In some cases, the induced current was not sufficient for plastic deformation of tested blanks. These results are represented on the figure 2 as the graphical interpretations of obtained results.

According to the obtained results, the relative strain obtained after pulse magnetic impact strongly depends on the operation parameters such as time of pulse discharge, operation temperature and induced in blank current. While the current and its distribution determines the value of attracting interaction, the operation temperature influences on the plasticity evolution of processed material. According to obtained results, with the temperature increase the significant reduction of flow stress as well as the decrease of work hardening is observed. The main reason of this is the tempering of microstructure, softening of material and the removing of its residual stresses.

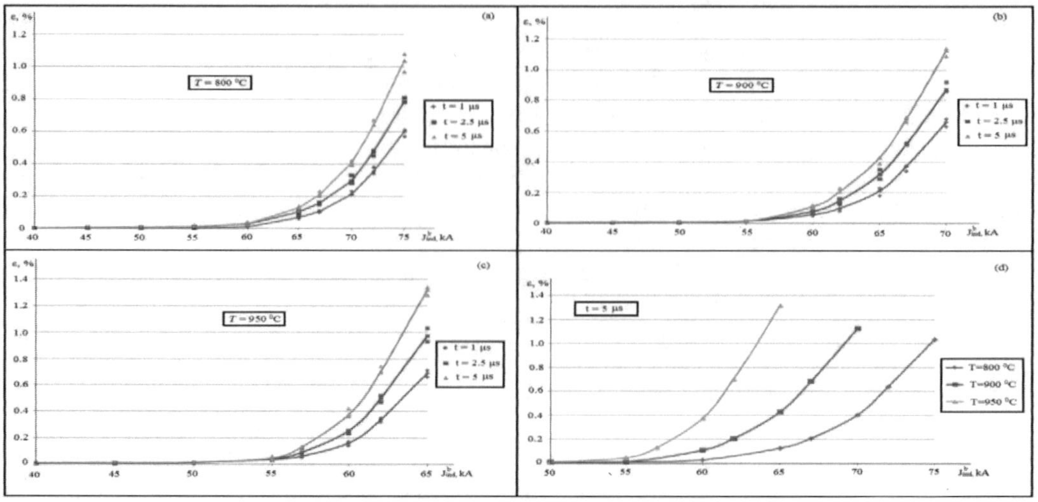

Figure 2. Results of experimental measurements: (a),(b), (c) – the dependence of relative strain from the current induced in the blank at 800, 900 and 950 ^{0}C respectively; (d) – influence of temperature on the forming process.

The discharge time is also has some influence on the plastic deformation of processed material. During operation process, the duration of pulse is responsible for the duration of magnetic field treatment as well as for the duration of forming operation that leads to variation of strain rate. From the obtained results it can be seen that during the operation by pulses with duration of 5 μs, the boron steel demonstrates less strength and better formability. One of the possible reason is more prolonged influence of magnetic field on the material microstructure that leads to changing of amount and distribution of various structural defects such as vacancy, itinerant electron etc. These defects are the reason of the material crystal lattice deformation that interrupts the movement of dislocations during the loading process. The influence of magnetic field leads to self elimination of these defects that allows to improve material plasticity.

In the works [9] [8] the authors present describe the attractive force as the function of currents induced in the processed blank and attractive screen. This relation can be expressed by equation (1):

$$F_{attr} = \mu_0 \cdot J_{ind}^{b} \cdot J_{ind}^{s} \cdot \frac{r}{(2h)}, \text{ N} \qquad (1)$$

where J_{ind}^{b} , J_{ind}^{s}- current inducted in the sheet metal blank and the accessory attracting screen, respectively, A;
r – the radius of the operation area (in the range from 0 to $R_1 = 0.07$ m);
h – the distance between the coil inductor, the accessory attracting screen and the sheet metal blank. In this particular study, the distance h is 0.00015 m;
μ_0 - magnetic constant, which equals $1.26 \cdot 10^{-6}$ N·A^{-2}.

The hardening mechanism and mechanical properties of boron/manganese 22MnB5 steel after heat treatment is extensively studied due to frequent use of hot stamping operations for this particular material. According to works [11] the plasticity evolution during the heating process can be expressed by equation (2).

$$\sigma_{22MnB5}(\varepsilon, T_H) = K_1 \cdot \exp\left(\frac{\beta_4}{T_H}\right)(\varepsilon_0 + \varepsilon)^{n(T_H)}, \qquad (2)$$

where T_H is the heat temperature, ^{0}C;

$n(T_H)$ is the strain hardening exponent, that can be calculate as:

$$n(T_H) = n_0 \cdot \exp(-c_n(T_H - T_0)), \quad (3)$$

The other coefficients are empirical constants. According to the [11] the values of these coefficients for 22MnB5 steel are: $K_1 = 34.48$; $\beta_4 = 2186.04$; $\varepsilon_0 = 0.0025$; $n_0 = 0.2034$; $c_n = 0.0024$.

This equation allows to make computer simulation in order to establish the strain distribution during the forming process. The simulation was performed by means of ABAQUS 6.11-3 Standard. The material properties are calculated in accordance with the UMAT procedure. The results of the performed simulation are presented on the figure 3 as the distribution of strain across the deformed area of tested sample.

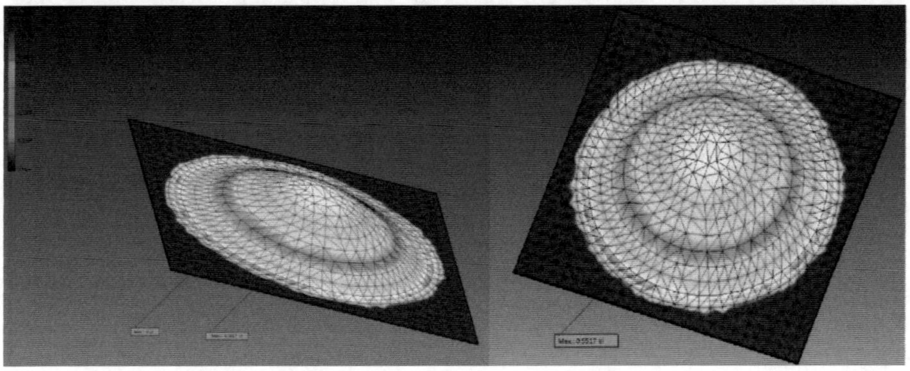

Figure 3. The strain distribution during the magnetic pulse influences.

It can be seen that the strain obtained during forming operation is mostly concentrated at the edge of operation area. However, the irregular strain distribution as well as the temperature heating can lead to the partial martensite transformation and decrease of strength properties of tested samples.

The measurements of the material hardness allow to analyze the changes in material microstructure after finishing of forming processes. According to obtained results the hardness of heated to 800, 900 and 950^{0}C samples are 450, 400 and 350 HV0.3 respectively. At the same time the hardness of 22MnB5 steel before heating is higher and equals to 500 HV0.3. It allows to assume that the material microstructure is non-uniform and includes other impurity phases in addition to the martensite.

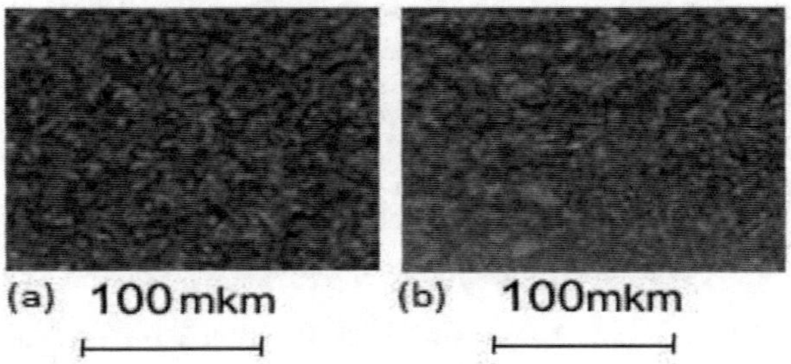

Figure 4. Optical microscopic images of the 22MnB5 steel before (a) and after (b) heating to 950 ^{0}C

The figure 4 demonstrates the martensite microstructure with some amount of ferrite phase. However, after heating the microstructure contains higher amount of ferrite and austenite that demonstrates degradation of martensite transformation. Moreover, the figure 4 (b) illustrate growth and enlargement of grains that leads to softening of 22MnB5 steel. These results well confirm the obtained hardness decrease mentioned earlier.

4. Conclusion

Within this paper, the new forming method based on the induction heating and action of the magnetic pulses is examined. This method is considered as the alternative way to restore and repair damaged details made from boron/manganese 22MnB5 steel. The experiment measurements demonstrate the influence of operation temperature, discharge time, and induced current on the degree of obtained plastic deformation. According to this results, the increase of temperature as well as the increase of discharge time leads to improvement of material plasticity. However, the usage of magnetic field for generation of processing load leads to non-uniform distribution of strain through the operation area. The presence of plastic strain localized in localized areas as well as heating process prevents full transformation of austenite into martensite phase. As the result repaired parts demonstrate the decrease of mechanical strength as well as the decrease of hardness. These results demonstrate that proposed method may be used for restoration of some parts. However, it is undesirable for the restoration of critical and vital components.

5. Acknowledgments

The research was funded by the Ministry of Education and Science of the Russian Federation according to base part of state tasks of the Sevastopol State University №2014/702 (Project 3867).

References

[1] Merklein M and Lechler J 2006 Investigation of the thermo-mechanical properties of hot stamping steels. *J. Mater. Process. Technol.* **177** 452 - 455

[2] Turetta A, Ghiotti A and Bruschi S 2006 Investigation of 22MnB5 formability in hot stamping operations. *J. Mater. Process. Technol.* **177** 396 - 400

[3] Kolleck R, Veit R, Merklein M, Lechler J and Geiger M 2009 Investigation on induction heating for hot stamping of boron alloyed steels. *CIRP Annals - Manufacturing Technology* **58** 275 - 278

[4] Bariani P F, Bruschi S, Ghiotti A and Turetta A 2008 Testing formability in the hot stamping of HSS. *Annals of the CIRP* **57** 265 - 268

[5] Malyushevsky P P 1983 *Basics of the pulse technology [in Russian]*. Naukova Dumka (Kiev, USSR)

[6] Delyusto L G 2005 *Fundamentals of metal rolling in a constant magnetic fields [in Russian]*. Mashinostroyeniye (Moscow, Russia)

[7] Grinkevich V A, Shevchenko T N, Krayev M V, Krayeva V S and Bondarev S V 2013 Experimental study of plastic deformation of St3 steel in a magnetic field [in Russian]. *Obrabotka metallov davleniyem* **37(4)**, 79 - 82

[8] Batygin Y V, Golovashchenko S F and Gnatov A V 2014 Pulsed electromagnetic attraction of nonmagnetic sheet metals. *J. Mater. Process. Technol.* **214** 390 - 401

[9] Batygin Y V, Golovashchenko S F and Gnatov A V 2013 Pulsed electromagnetic attraction of sheet metals - Fundamentals and perspective applications. *J. Mater. Process. Technol.* **213** 444 - 452

[10] Maikranz-Valentin M, Weidig U, Schoof U, Becker H H and Steinhoff K 2008 Components with optimised properties due to advanced thermo-mechanical process strategies in hot sheet metal forming. *Steel Research Int.* **79** 92 - 97

[11] So H, Fabmann D, Hoffmann H, Golle R and Schaper M 2012 An investigation of the blanking process of the quenchable boron alloyed steel 22MnB5 before and after stamping process. *J. Mater. Process. Technol.* **212** 437 - 449

IOP Conf. Series: Materials Science and Engineering **110** (2016) 012108 doi:10.1088/1757-899X/110/1/012108

Radiation treatment of the ceramic and polymer implants

M V Korobeynikov, A A Bryazgin, V V Bezuglov, E A Shtarklev, A Yu Vlasov, L A Voronin and V O Tkachenko

Budker institute of nuclear physics SB RAS,11 Lavrentiev avenue, Novosibirsk 630090, Russia

E-mail: mkor@inp.nsk.su

Abstract. Implants are used in medical practice during decades. The ceramic implants are the new trend in medicine. The polymer implants are used for many years, and they are mainly sterilized by the radiation treatment. The article describes the new ceramic and polymer implants that were treated in the Budker Institute of Nuclear Physics.

1. Introduction

Implants are used in medical practice during decades, and the search of the new ones and development of new ideas are going on.

The main requirements for the implants – they ought to be accepted by body (not to be rejected), can be incorporated into the tissues and perform the functions of the replaced tissue.

The ideal implant ought to replace the tissue, be soaked with the interstitial fluids, permit the body cells to settle on and inside it and to form the corresponding tissue and then after passing some time the implant ought to gradually dissolve leaving the tissue on the place of the implant.

We observe the increase in number of illnesses requiring the transplantation, so the need in the implants replacing the organs is increasing.

2. Ceramic implants

Joint stock company "NEVZ-KERAMIKS" situated in Novosibirsk is producing the ceramic medical implants based on Al_2O_3 and ZrO_2 ceramics – implants and endoprostheses [1]. The goods have the bioactive surface from nanostructured hydroxyapatite and can be successfully used in men surgical treatment as well as in stomatology and in oral surgery.

Porous structure firmness is provided by combination of the submicron and nano-sized components of the ceramics.

The hydroxyapatite implants' surface with nanoparticles facilitate the new bone tissue formation on the border and long and steady implant fixation inside the body.

The ceramic implants can be sterilized only by ionizing radiation – all other sterilization processes are inadmissible due to regulation.

The implants were hermetically packed into the individual packs and then the groups of 8 packed implants were hermetically packed into the big packs. The 2-sided irradiation was carried out by electron beam generated by ILU-6 pulse linear accelerator in the Budker Institute of Nuclear Physics, the electron energy was 2.45 MeV.

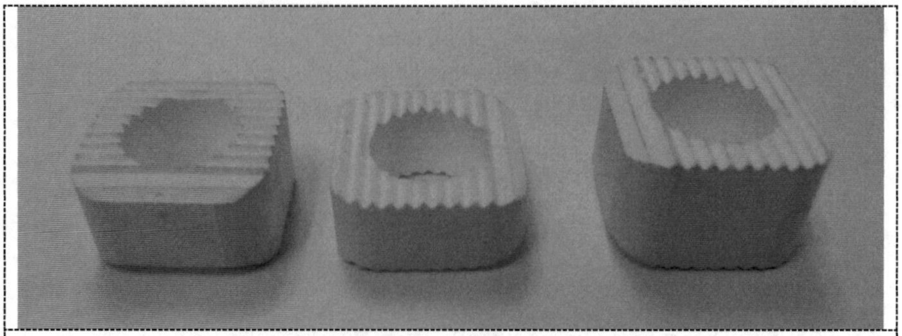

Figure 1. Vertebral disk implants.

The vertebral disk implants of various sizes are shown in Figure 1, Figure 2 presents the implant's drawing, Figure 3 shows the packed implants prepared for sterilization.

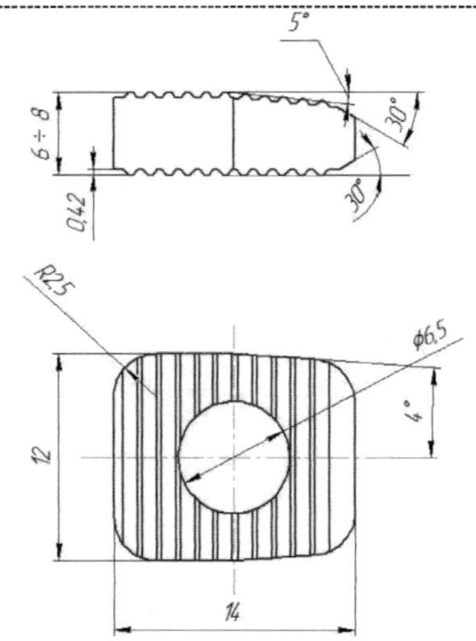

Figure 2. Vertebral disk implant drawing.

Figure 3. Packed implants prepared for sterilization.

First time in Russia (and in the whole world) the ceramic vertebral disk implants were implanted in men in Novosibirsk in 2013.

The ceramic implants permit to use the magnetic resonance tomography to monitor the results of the operation – the metal implants do not allow to use this method because of their strong response.
And such implants will not cause the problems for the people during passing the metal detectors (inspections in the airports, etc.). Figure 4 shows computer tomography images of the ceramic vertebral disk implant in patient.

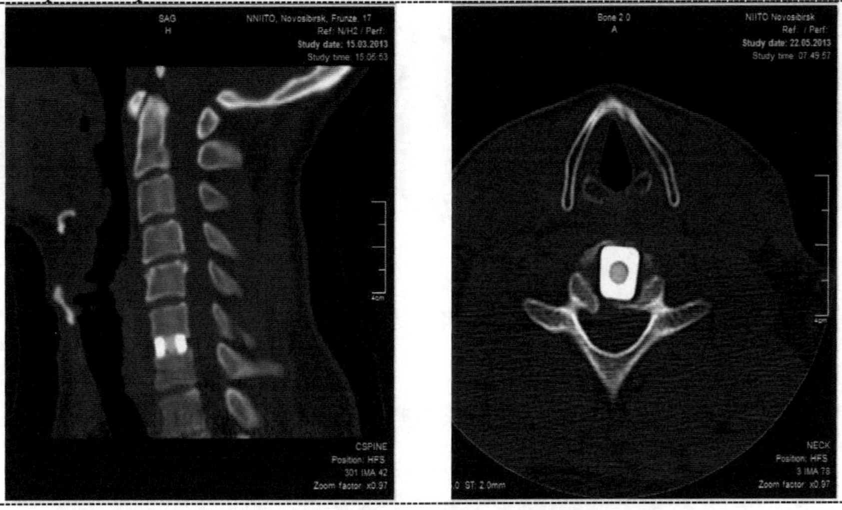

Figure 4. Ceramic vertebral disk implant in patient - computer tomography images.

3. Vascular Implants Produced by Electron Beam Treatment

Vascular implants are now in demand due to increase in operations number. The Institute of Chemical Biology and Fundamental Medicine (Siberian Branch of Russian Academy of Science) in collaboration with Institute of Blood Circulation Pathology named by E.N. Meshalkin is developing the brand new blood vessels implants based on the polymer electrospun 3D matrices.
Electrospinning application areas:

- Cartilage
- Bones
- Muscles
- Nervous tissue
- Skin
- **Vessels**

The requirements to the vascular implants:

- Biocompatibility, including the compatibility with the blood, good cells' adhesion, the possibility for the cells to migrate into the structure and to germinate inside.
- Stable biodegradation, biodegradation rate ought to correspond the replacement tissue grow rate, absence of the biodegradation toxic products.
- The mechanical firmness enough to endure the mechanical loads during the new tissue formation time (including the fatigue resistance).

The basic sheets were produced by electrospinning from poloycaprolactome, nylon, polylactic-co-glicolide and their mixtures dissolved in hexafluoro-2-propanol. The fibers have the diameter of about

IOP Conf. Series: Materials Science and Engineering **110** (2016) 012108 doi:10.1088/1757-899X/110/1/012108

1 micron. The sheets were soaked in the aqueous gelatine solutions containing factors stimulating the vascular cells growing and then were treated by electron beam for grafting.

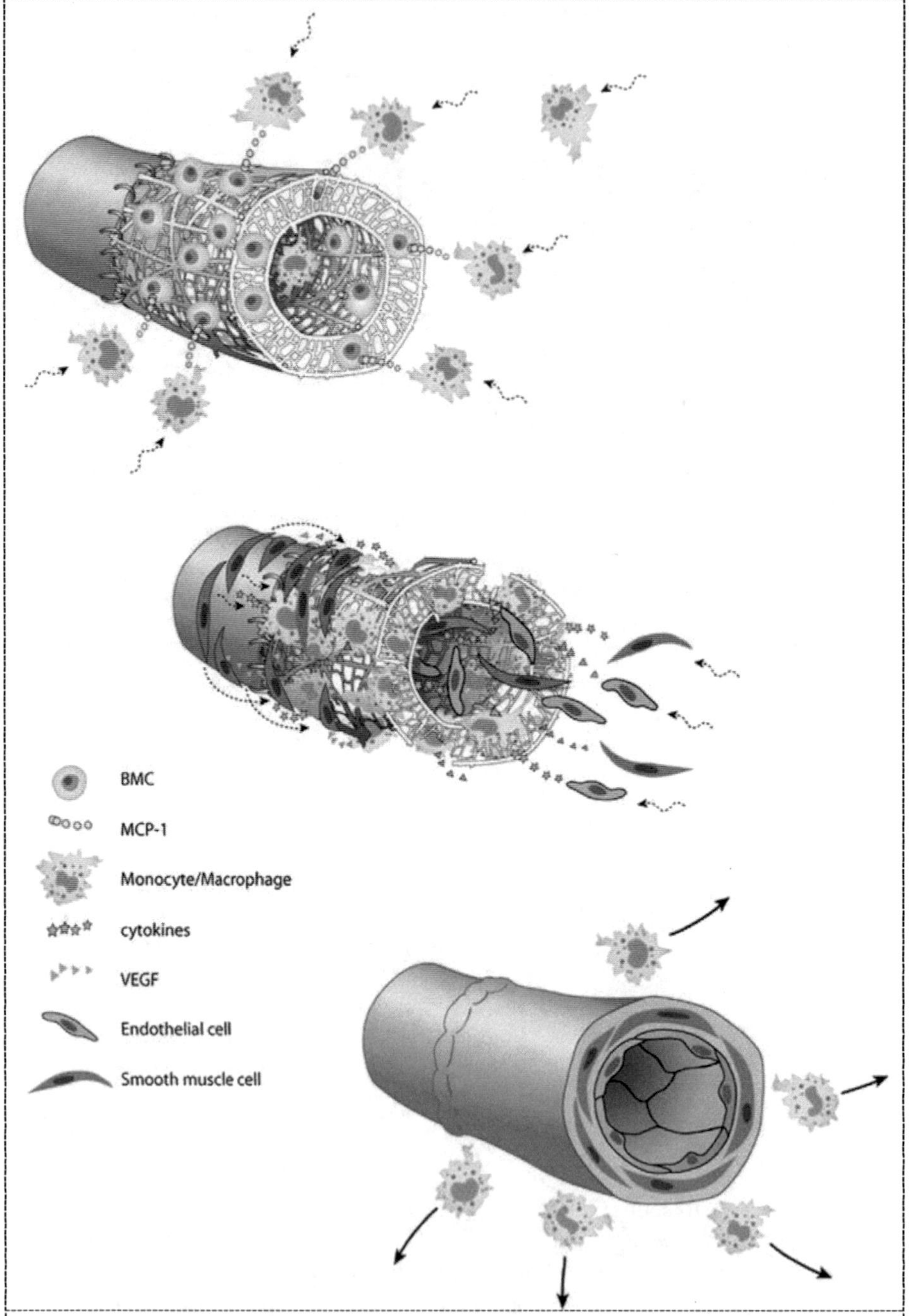

Figure 5. Blood vessel formation from the vascular implant.

Figure 5 shows the blood vessel formation process – the various cells are filling the space inside the implant, proliferating and gradually forming the new blood vessel tissue, meanwhile the polymer matrix is subjecting to biodegradation, and so the new blood vessel is forming.

The irradiation of the nylon electrospun 3D matrixes with doses of 50-100 kGy increases elastic properties of the grafts and is a convenient approach for increasing their mechanic properties [2]. It was shown that the dosage up to 25 kGy does not influence on the mechanical properties, while doses in the range from 50 kGy to 100 kGy increase rigidity of the material with subsequent destruction of the materials in dosages higher than 150-200 kGy.

Polymers used in electrospinning technique in the described experiments:

- Nylon 6
- Polycaprolactone (PCL)
- Polycaprolactone (PCL) + gelatin
- Poly lactic co glycolic acid (PLGA)

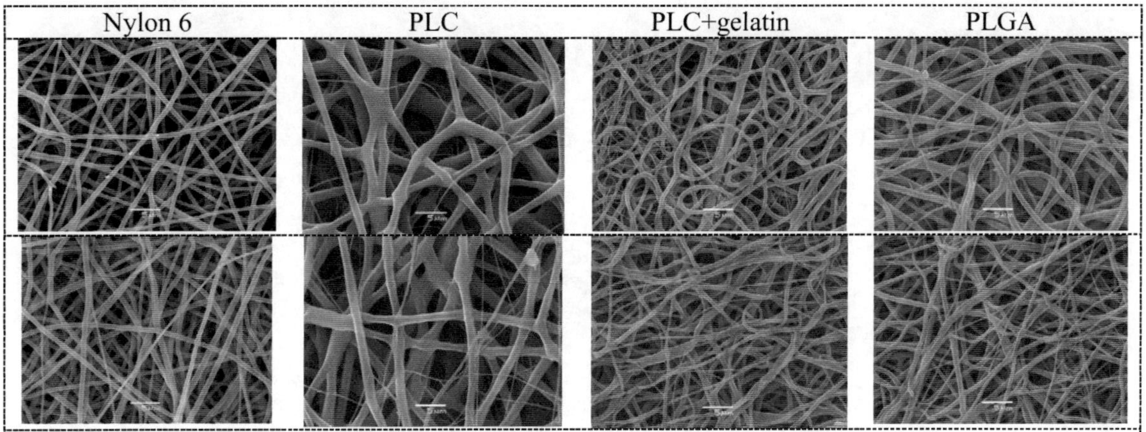

Figure 6. Scaffolds morphology before and after irradiation with 100 kGy dose.

Figure 6 shows the scaffolds morphology before and after irradiation with dose of 100 kGy. Figure 7 shows the endoteliocytes (upper line) and fibroblasts (lower line) growing on different materials. Scanning electron microscopy was performed in the Department of Physical and Chemical Methods of the Institute of Catalysis, Siberian Branch of Russian Academy of Science (LEO-1430 Microscope, Jeol, Japan).

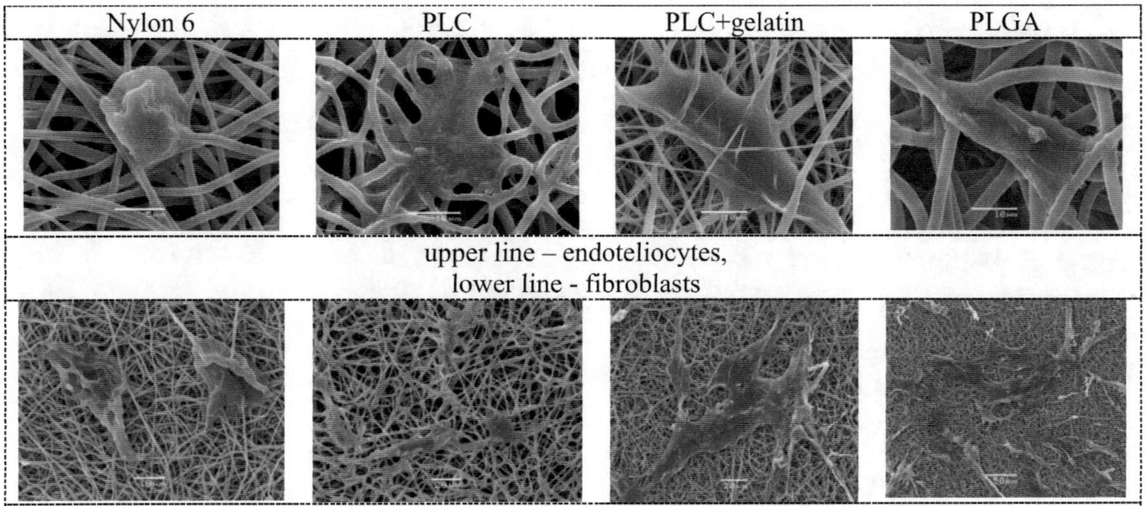

Figure 7. Different electrospun scaffolds as cell growth substrates.

IOP Conf. Series: Materials Science and Engineering **110** (2016) 012108 doi:10.1088/1757-899X/110/1/012108

4. Conclusions

The ceramic vertebral disk implants were introduced into clinical practice and implanted in men.
The modern chemical technologies and physical methods permit to produce the 3D polymer matrices having the properties approaching to the parameters of living tissues. The brand new polymer vascular implants were subjected to various tests and then tested in mice.

The trials show that fibroblasts and endotheliocytes can fasten on the implants and are successfully proliferating. The endotheliocytes are much more sensible to the materials than fibroblasts, and are proliferating only on certain surfaces.
The artificial vascular implants are on the way to clinical practice.

References

[1] NEVZ-KERAMIKS Products Bioceramics http://www.nevz-ceramics.com/ru/produktyi-i-materialyi/biokeramika.html

[2] Laktionov P P, Lebedeva A O, Korobeinikov M V, Yunoshev A S, Karpenko A A, Popova I V, Sergeevichev D S, Pokushalov E A 2014 European Journal of Vascular and Endovascular Surgery **47** 6 692

Type the title of your paper here Effect of the focused light from the xenon arc lamp on the surface tension of the molten enamel

A D Aleutdinov, S A Ghyngazov, T S Mylnikova and K A Aleutdinov

National Research Tomsk Polytechnic University, 30 Lenin Ave., Tomsk, Russia

E-mail: a_aleut@mail.ru

Abstract. The effect of exposure to the focused light from the xenon arc lamp on the surface tension of molten enamels was studied with a designed light beam setup as compared to that observed in conventional heating in a resistance furnace. The objects under investigation were enamels No. 261, UES-200 and UES-300. The power density of the light beam was varied in the range of (30–80) W/cm^2. When exposed to light, the surface tension is shown to be an order of magnitude lower than that obtained in conventional furnace heating.

1. Introduction
Glass-enamel coatings of metal products are the most effective protection against corrosion and, in addition, they provide the surface of enameled metal products with a number of useful properties inherent in oxide glasses, namely, hardness, smoothness, acid and alkali resistance, attractive appearance and durability [1]. Along with viscosity and linear temperature expansion coefficients, the surface tension of enamels is an important technological parameter. For successful enameling, the surface tension of enamels at temperatures close to the firing point (750–1050 °C) should be no more than 230–340 mN/m [1]. It is known that decreased surface tension improves the wettability of the surface of the metal oxidized with molten enamel which facilitates release of gas bubbles from the melt and increases coating adhesion. That is, the decreased surface tension of the melt provides high-quality defect-free coating [1]. Conventional technology for producing enamel coating involves heating in resistance furnaces. A setup for treating the material with the focused light from the xenon short-arc lamp (XSAL) DKsShRB-10000 is described in [2–3]. This setup provides local heating to temperatures of about 1000 °C. This type of heating can be used for enameling small metal parts, and it is promising for repairing imperfections and re-enameling cast iron products [3]. To optimize the enameling modes, the effect of the XSAL focused light power density on the surface tension of the molten enamel should be studied.

This paper aims to study the effect of the XSAL focused light power density on the surface tension of the molten enamel of different composition.

2. Experimental procedure
We investigated enamels No. 261, UES-200 and UES-300. These enamels are acid and alkali resistant that determines their wide practical use. They can be used in liquid media with pH values from 1 to 14. The composition of the studied enamels is presented in Table 1.

Table 1. Chemical composition of acid and alkali resistant enamel frits (mass%).

Oxide	No. 261	UES-200	UES-300
Li_2O	5	4	4
Na_20	14	13.34	13.34
K_2O	1	4.38	4.38
CaO	4	2.28	2.28
MgO			
CoO	0.30	0.30	2.5
TiO_2	1		
ZrO_2	3	8	9.5
Cr_2O_3	4.7	2.2	
$Si0_2$	64	62	62
B_20_3	2	2	2
Al_2O_3			
P_2O_5			
Na_2SiF_6	1		
Na_3AlF_6			

2.1. The value of the surface tension coefficients of enamels studied under conventional furnace heating

For conventional furnace heating, the surface tension (σ) of the molten enamel of specific composition can be calculated by known methods based on the principle of additivity. Each component of the enamel contributes to the total property of this complex substance. This contribution is proportional to the content of each of the component. The proportionality factor takes into account the interrelation of the components, and it is called a partial coefficient [4]. The most precise calculation for the studied enamels can be obtained by the method of A. Dietzel [1]. The formula for calculation of the enamel surface tension according to Dietzel is as follows:

$$\sigma = \sum c_i * \sigma_i \tag{1}$$

where c_i is the mass fraction of the component (%) and σ_i is the partial coefficient according to Dietzel. Data for calculation of the enamel surface tension according to Dietzel (900 °C) are shown in Table 2.

Table 2. Data for calculation of the enamel surface tension according to Dietzel (900 °C) [1]

Oxide	σ_i, (mN/m)	No. 261		UES-200		UES-300	
		c_i	$c_i*\sigma_i$	c_i	$c_i*\sigma_i$	c_i	$c_i*\sigma_i$
$Si0_2$	340	64.00	217	62	211	62	211
CaO	480	4.00	19	2.28	11	2.28	11
B_20_3	80	2.0	1	2.0	1	2.0	1
K_2O	10	1.00	0.1	4.38	0.43	4.38	0.43
Na_20	150	14.00	21	13.34	19.5	13.34	19.5
Li_2O	460	5.0	23	4.0	18.5	4.0	18.5
TiO_2	250	1.0	2.5	-		-	
ZrO_2	410	3.0	10.5	9.5	39	8.0	33
CrO_3	-590	5.0	-29.5	2.2	-13	-	-
CoO	450	-		0.30	1.5	4.0	18
Na_2SiF_6	130	1.00	1	-	-	-	-
		$\sum c_i*\sigma_i$	265.5	$\sum c_i*\sigma_i$	289	$\sum c_i*\sigma_i$	312

IOP Conf. Series: Materials Science and Engineering **110** (2016) 012109 doi:10.1088/1757-899X/110/1/012109

For the temperature of the furnace roasting of 900 °C, the calculated value of the surface tension of molten enamel No. 261 was 265.5 mN/m, that of UES-200 enamel was 289 mN/m, and for UES-300 enamel, it was 312 mN/m.

2.2. Experimental determination of the surface tension of the studied enamels in light beam heating

The falling-drop method was used to experimentally determine the coefficient of the surface tension of the enamels under light beam heating.

The experimental setup [2–3] providing continuous exposure was used as a source of light beam heating. An optical schematic of the setup is shown in Fig. 1.

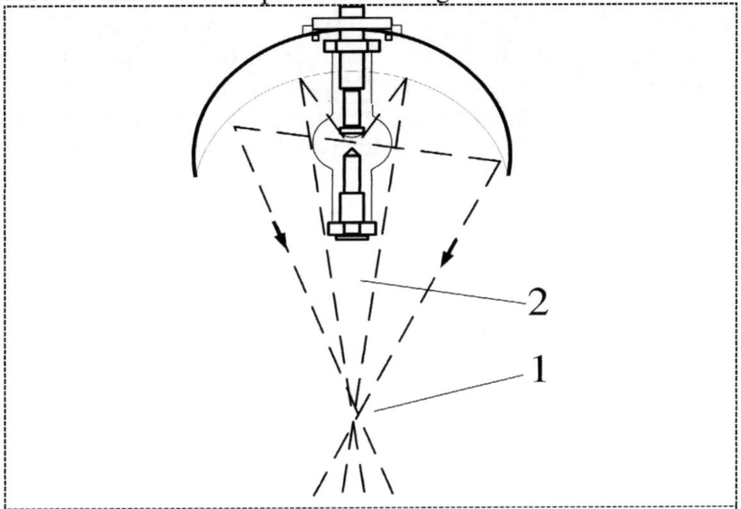

Figure 1. Optical schematic for focusing the light from the xenon arc lamp: area of highest light concentration (1), shadow cone (2).

This optical schematic allows placement of the main part of the enamel rod in the shadow cone under the lower lamp electrode, and that of the bottom end of the rod in the area of the highest light concentration. When arranged this way, the molten drop is formed at the bottom end of the rod under heating by XSAL focused light. It provides an opportunity to study the effect of the XSAL focused light power density on the surface tension of the molten enamel. The technique for determining this dependence is as follows.

The initial length of the rod with the diameter d_l= (1.5–1.8) mm is chosen equal to 10 cm. The shadow cone length is 9 cm. The initial position of the enamel rod is set largely in the shadow cone. The bottom end part of the order of (3–5) d_l is placed in the area of the highest light concentration. When exposed to XSAL focused light, as the molten drop grows, the rod is moved down to keep the molten drop in the peak-heated region. When the weight of the drop exceeds the surface tension at the point of its detachment from the rod, the drop falls. This process can be described by the expression:

$$\sigma \pi d = mg , \tag{2}$$

where σ is the surface tension of the molten enamel, d is the diameter of the drop neck at the instant of its detachment, m is the drop mass and g is the acceleration of gravity. Hence, the surface tension of the molten enamel can be calculated by the expression:

$$\sigma = \frac{mg}{\pi d} \tag{3}$$

Since the diameter of the drop neck at the instant of its detachment is smaller than the diameter of the enamel rod (d_1), the surface tension of the molten enamel can be experimentally determined by the formula:

$$\sigma = \left(\frac{mg}{\pi d_1}\right)\left(1 + \frac{d_1}{d_2}\right), \tag{4}$$

where m is the average mass of the drop, d_1 is the diameter of the enamel rod and d_2 is the average diameter of the enamel drop.

After drop detachment, the rod is kept moving down until the formation and subsequent detachment of the next drop. Falling drops are collected in a container filled with water located at a distance of about 1 m under the installation. After that, the average drop diameter for a given XSAL focused light power is measured. The dimension of the drop diameter is taken to determine the surface tension of the molten enamel at a given XSAL focused light power density by formula (4). This procedure is repeated for heating similar enamel rods at different XSAL focused light power densities. After that, the value of the surface tension of the molten enamel is plotted against the XSAL focused light power density.

3. Experimental results and discussion

For each of the enamels, the measurements were performed at lamp currents equal to 10, 15 and 20 A. In the experiment, the diameter of the light spot (1 in Fig.1) was maintained equal to 1 cm. The setup efficiency of the conversion of electrical energy into light is 10%. Therefore, at a current density of the lamp equal to 10 A, the XSAL focused light power density (q) in the area of the highest light emission concentration was about 38 W/cm2, at a lamp current of 15 A, it was 57 W/cm2, and at 20 A, it amounted to77 W/cm2. For enamel No. 261, at 77 W/cm2, the experiment failed as the enamel was melting too fast.

Fig. 2 shows the graph of the experimental results.

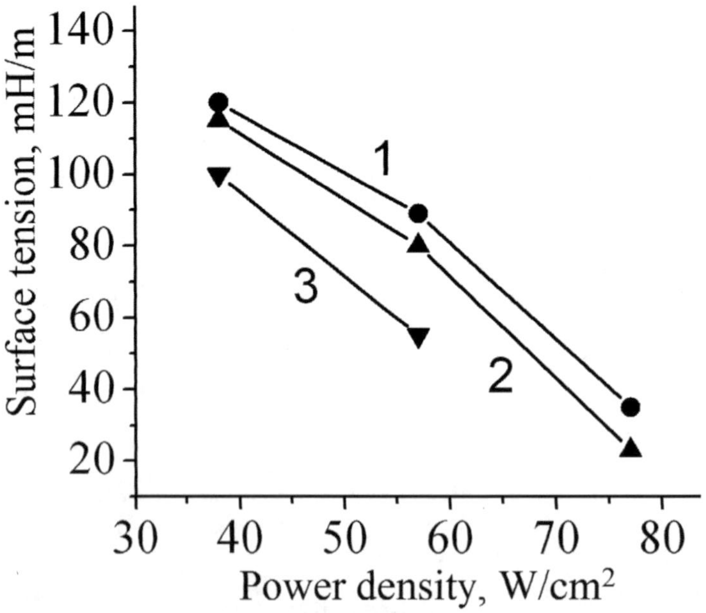

Figure 2. The coefficient of the enamel surface tension versus the light flux power density. UES-300 (1), UES-200 (2) and enamel No. 261(3).

The calculated and experimental values of the enamel surface tension are shown in Table 3.

Table 3. Correlation of the calculated and experimental coefficients of the enamel surface tension

Enamel	σ_1 (mN/m)	$q \approx 38$ (W/cm^2)		$q \approx 57$ (W/cm^2)		$q \approx 77$ (W/cm^2)	
		$\sigma_{2.}$	$\sigma_{1,}/\sigma_2$	σ_2	σ_1/σ_2	σ_2	σ_1/σ_2
No. 261	265.5	96	2.7	55	4.8	-	-
UES-200	289	115	2.5	80	3.6	28	10.3
UES-300	312	120	2.6	89	3.5	35	8.9

Note σ_1 is calculated coefficients and σ_2 is experimental coefficients of the enamel surface tension.

Surface tension is a thermodynamic characteristic of the surface of the interface between two phases determined as a portion of the internal energy of the substance. [5] In this case, it is the characteristic of the interface between the molten enamel and gas mixture of vapors of enamel and air. It is known that in conventional furnace heating, the surface tension of the molten enamel depends on the composition of the gaseous mixture above the molten enamel surface. The presence of the water vapor reduces the surface tension of the melt, and high content of carbon dioxide increases the surface tension of the molten enamel [1].

Therefore, the surface tension of the molten enamel depends on the state of the enamel and gaseous mixture of vapors of the enamel and the air above the melt, primarily, on the temperature of the melt and gaseous mixture [6]. The emission spectrum of the xenon short-arc lamp consists of 10% of ultraviolet radiation, 35% of visible radiation and 55% of infrared radiation. This spectral range is effective for heating both metals and metal oxides.

The features of the high-energy effect of exposure to the focused light from the arc source are determined, primarily, by a significant portion of ultraviolet radiation in the polychromatic light flux generated by XSAL. Enamel melting in exposure to the light-beam is accompanied by both strong ionization of the air above the melt and the molten enamel. As the power density increases, the degree of excitation of the melt and the air grows. This explains an anomalous decrease in the surface tension of the molten enamel in heating by polychromatic light.

4. Conclusions

It is found that in light beam heating, the surface tension of enamels is strongly dependent on the power density of exposure, and it can be more than 10 times less than that obtained in furnace heating. When the power density of the light beam is 38 W/cm2, the value of the enamel surface tension is 2.5–2.7 times less than that obtained in furnace heating at 900 °C, and at a power density of 77 W/cm2, the ratio increases by 8–10 times.

Acknowledgements

This work was financially supported by The Ministry of Education and Science of the Russian Federation in part of the science activity program.

References

1. Petsold A, Peshman G 1990 Emal i emalirovanie: Spravochnik M. Metallurgija 576
2. Aleutdinov A D 2013 Systems. Methods. Technologies **2** (18) 124–127
3. Ghyngazov S A, Aleutdinov A D, Mylnikova T S, Luchnikov P A 2015 IOP Conf. Ser.: Mater. Sci. Eng. **81** 012069
4. Appen A A 1974 Chemistry of glass L: Chemistry 352
5. Weisskopf V F 1985 American Journal of Physics **53** 618—619
6. Khaidarov G G, Khaidarov A G, Mashek A Ch, Mayorov E E 2012 Vestnik St. Petersburg University Series 4 (Physics, Chemistry) **1** 24–28

Crystal phase structure investigation in the process of radiation-thermal transformations in systems SrO-Fe_2O_3 , $SrCO_3$-Fe_2O_3 (perovskite) and garnet Y_2O_3-Fe_2O_3

U V Ancharova[1,4], M A Mikhailenko[1], B P Tolochko[1], N Z Lyakhov[1], M V Korobeinikov[2], A A Bryazgin[2], Z S Vinokurov[3] and A G Selyutin[3]

[1]Institute of solid state chemistry and mechanochemistry SB RAS, Kutateladze, 18, Novosibirsk, Russia
[2]Budker institute of nuclear physics SB RAS, Lavrentieva, 11, Novosibirsk, Russia
[3]Boreskov institute of catalysis SB RAS, Lavrentieva, 5, Novosibirsk, Russia
[4]Novosibirsk State University, Pirogova, 2, Novosibirsk, Russia

E-mail: ancharova@gmail.com; mkor@inp.nsk.su

Abstract. Chemical reactions products structural investigations during thermal and radiation-thermal processes are presented. The reaction mixtures were prepared by mechanical activation of stoichiometric compounds corresponding to synthesis reaction of strontium ferrite with perovskite structure ($SrFeO_{3-\delta}$) or ferro-garnet ($Y_3Fe_5O_{12}$). The phase formation staging changes dramatically depending on the temperature and the way of the high-temperature treatment.

1. Introduction

The effect of the radiation-stimulated diffusion was discovered in the 1980-s when the powerfull electron beam with energies of more than 1 MeV became available for experiments. The Russian pioneers in this field were the researchers from Novosibirsk and Tomsk as the powerful electron accelerators were produced and accommodated for the research activities in Budker Institute of Nuclear Physics in Novosibirsk.

The electrons of the beam generate lots of secondary electrons that have the energies in the range of 10 to 200 eV. The impact of such electron can knock out atom away from the crystal lattice site. In case of the intensive electron beam treatment the quantity of the knocked-on atoms can greatly exceed the level corresponding the normal Maxwellian distribution, and the diffusion is enhanced.

Initially we used the radiation-thermal treatment to reduce the process temperature and to speed up the process. But the detailed studies of the radiation-thermal synthesis process in some mixtures revealed that even the chemical reactions staging in the ferrite synthesis differ in the radiation-thermal processes comparing with the thermal one.

2. Experimental

Synthesis and treatment experimental methods are well described earlier in our publication [1]. The stoichiometric mixtures of iron oxide and strontium carbonate or yttrium oxide, corresponding to the

IOP Conf. Series: Materials Science and Engineering **110** (2016) 012110 doi:10.1088/1757-899X/110/1/012110

reaction of synthesis for $SrFeO_3$ or $Y_3Fe_5O_{12}$ were mechanically activated in a ball mill to produce homogeneous mixtures for further treatment. The synthesis was carried out at high temperature chamber in case of *in situ* experiments for investigation of the staging and the rate of the phase transformation. In case of radiation-thermal technology for complex oxides synthesis the accelerator ILU-6 was used (located in BINP SB RAS), the technology is very well described in details in [1] as well. The products of radiation-thermal reaction were then investigated *ex situ*.

Both *ex situ* and *in situ* structural investigations of reaction products were carried out using X-Ray diffraction methods. The synchrotron radiation was generated by VEPP-3 storage ring (SSTRC [2] located in BINP SB RAS), the measurements were done by the diffractometer installed on the 6th beamline "Precision difractometry". The same settings were used as described in [1].

3. Results

The high-temperature strontium ferrites thermal synthesis experiments were carried out *in situ* to investigate the staging of the phase transformation at different temperatures: 700, 800 and 900°C. The results are presented in Figures 1 and 2, the high-temperature treatment pattern included two stages: heating up with the rate of 100°C/minute, then shelf with the certain temperature during several tens of minutes. In the stage of heating up the gradual increase in intensity and the sharpening of the diffraction reflections of reactants was fixed. That points to the annealing of the defects in the reaction mixture caused by the mechanical activation. At the shelf stage at a certain temperature the strontium ferrite $SrCO_3+FeO_{3/2} \rightarrow SrFeO_3$ synthesis reaction starts.

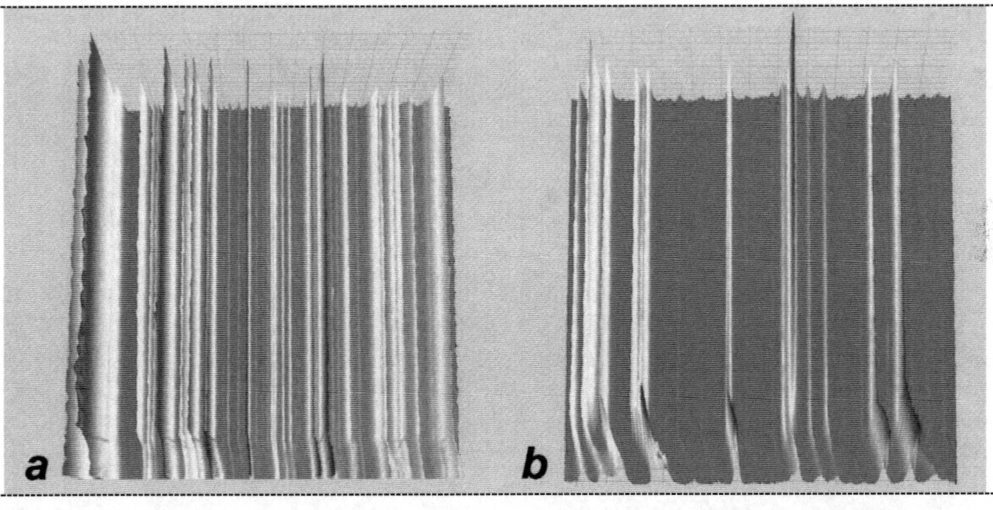

Figure 1. *In situ* X-ray diffraction experiments for the synthesis of strontium ferrites during the high-temperature treatment at the temperatures of 700°C (*a*) and 800 °C (*b*), here the diffraction angle 2θ is in the range of 20-50 degrees (wavelength λ=1.01Å) on the horizontal axis, time is up to 70 minutes on the vertical axis, the intensity of the X-ray diffraction is perpendicular.

At the temperatures of 700-800°C the reaction goes without the intermediate phases formation, directly to the product (Figure 1). However, at the temperature of 900°C the reaction staging dramatically changes: strontium carbonate decomposes to strontium oxide and carbon dioxide: $SrCO_3 \rightarrow SrO + CO_2$, with a very high rate at the first stage of the reaction (Figure 2*a*) over the time scales smaller then strontium ferrite formation rate. Just for ~ 10 minutes the strontium carbonate concentration decreases in ~ 20 times, while the strontium oxide concentration sharply increases from zero, reaching its maximum value. That involves into reaction of synthesis mostly strontium oxide (which is the product of strontium carbonate decomposition) instead of strontium carbonate.

Therefore, the staging of the reaction at 900°C without irradiation differs dramatically compared to the reactions at lower temperatures. It occurs by presence of such intermediate phases like strontium hexaferrite $SrFe_{12}O_{19}$ and Ruddlesden-Popper (RP2) phase $Sr_3Fe_2O_7$ in the reaction mixture during the synthesis along with reactants and product of strontium ferrite $SrFeO_3$ (Figure 2*b*).

The similar experiments as above (for full-profile phase analysis of the reaction, held *in situ* during thermal treatment), were made using intense beams of high-energy electrons as a heater of the reaction mixture to consider radiation-thermal processes and to find the differences between staging and rate of the phase formation in thermal (T) and radiation-thermal (RT) reactions. The reactions were investigated *ex situ* using synchrotron radiation X-Ray diffraction and further full-profile phase analysis as well.

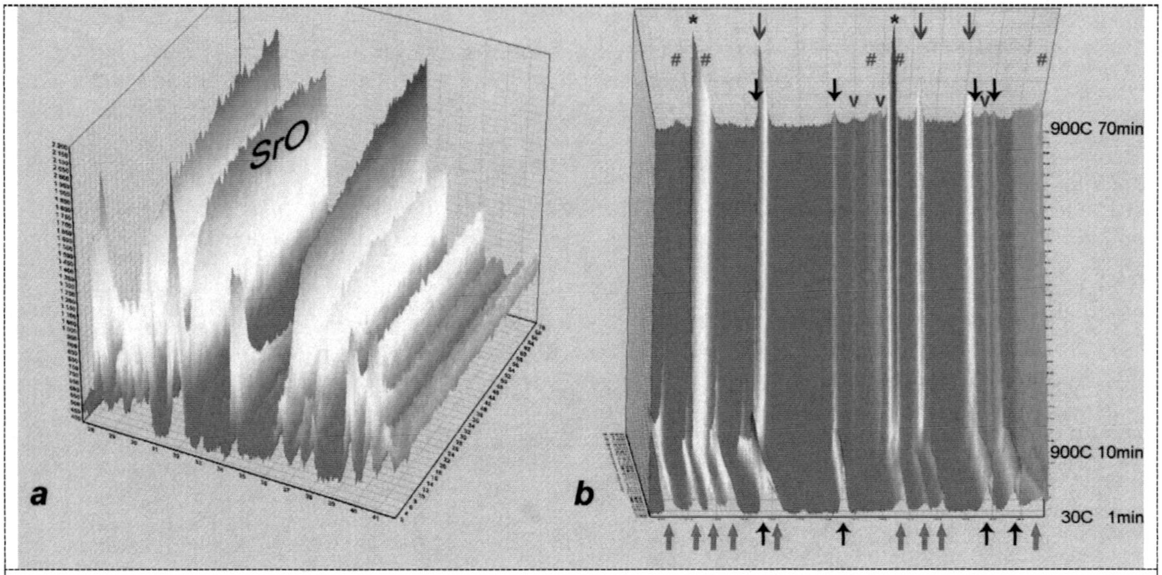

Figure 2. *In situ* X-ray diffraction experiments for the synthesis of strontium ferrites during the high-temperature treatment to the temperatures 900°C, here the diffraction angle 2θ in the range 20-50 degrees (wavelength λ=1.01Å) is on the horizontal axis, time up to 70 minutes is on the vertical, the intensity of the X-ray diffraction is perpendicularly; Reactants are marked by arrows (Fe_2O_3 by dark arrows, $SrCO_3$ by lighter arrows on the bottom, SrO by lighter arrows on the top), intermediate phases and product are marked by (v) $SrFe_{12}O_{19}$; (#) $Sr_3Fe_2O_7$; (*)$SrFeO_3$.

Both for thermal and radiation-thermal treatment the different exposure times from 10 to 70 minutes were considered. As a marker for characterization of the reactions the value of the conversion degree to the final product (strontium ferrite) over a characteristic time of 30 minutes was selected (the table of results is already presented in our previous work at [1]).

The comparison between thermal and radiation-thermal reactions shows that without decomposition of strontium carbonate at temperatures below 850°C the rate and staging of the reactions doesn't differ significantly: both T and RT reactions interaction of reactants takes place directly without the formation of intermediate phases. Thus the irradiation with intense electron beam does not accelerate the reaction of interaction between strontium carbonate phase with iron oxide, despite the well-known effect [3-6] of diffuse acceleration of the reactants by irradiating with intense beam of high-energy electrons. It can be concluded that, in this case, the reaction rate is not limited by mutual diffusion of the reactants. Apparently, the limiting factor at the temperatures below the temperature of decomposition of strontium carbonate is the carbon dioxide CO_2 elimination from the system.

However, at temperatures above 850°C there is a number of differences in chemical reactions for the synthesis of strontium ferrite depending on the method of heating.

Ex situ phase analysis of products (Figure 3, table in [1]) in high-temperature synthesis reaction with irradiation with intense electron beam showed that with radiation-thermal treatment of the reaction mixture the carbonate decomposition does not occur over the temperature range up to 1000°C, that is confirmed in experiments on radiation-thermal treatment on the pure strontium carbonate. Since the strontium carbonate enters directly to the chemical reaction without stage of decomposition to the oxide, a set of intermediate phases of the reaction differs in case of radiation-thermal synthesis: instead of the phase RP2 ($Sr_3Fe_2O_7$), the phase RP3 ($Sr_4Fe_3O_{10}$) was formed. Despite the fact that the reaction occurs by a different mechanism then in the thermal process, the concentration of the reaction product phase (strontium ferrite), grows faster in the radiation-thermal process.

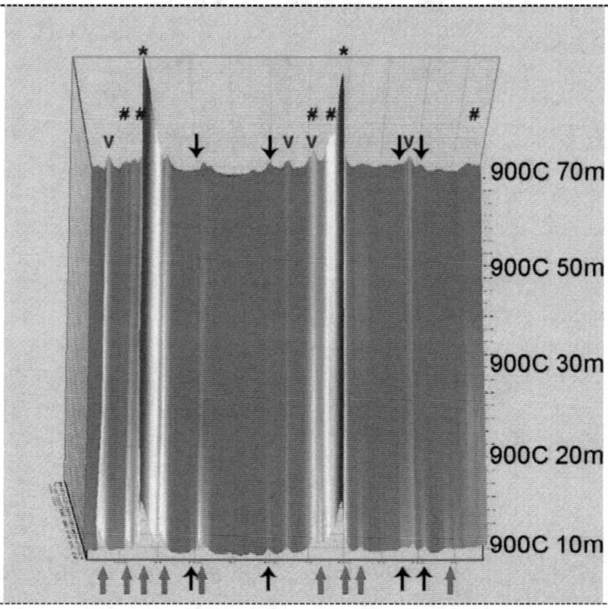

Figure 3. *Ex situ* X-ray diffraction experiments for the synthesis of strontium ferrites during the radiation-thermal treatment to the temperatures 900°C, here the diffraction angle 2θ in the range 20-50 degrees (wavelength λ=1.01Å) is on the horizontal axis, time up to 70 minutes is on the vertical, the intensity of the X-ray diffraction is perpendicularly; Reactants are marked by arrows (Fe_2O_3 by dark arrows, $SrCO_3$ by lighter arrows), intermediate phases and product are marked by (v) $SrFe_{12}O_{19}$; (#) $Sr_4Fe_3O_7$; (*)$SrFeO_3$.

Radiation-thermal treatment of the mixture for synthesis of ferrogarnet ($Y_3Fe_5O_{12}$), were held at the different temperatures from 700°C to 1050°C by irradiation with intense beam of high-energy electrons. *Ex situ* X-Ray diffraction patterns are presented at the Figure 4. The full-profile phase analysis showed that the synthesis takes place with intermediate phase of the yttrium ferrite with the ortho-perovskite structure ($YFeO_3$), from iron and yttrium oxides directly at the temperatures up to 1000°C without the formation of ferrogarnet, which appear and grow very fast only at temperature of 1050°C. At the temperature of 700°C no diffraction reflections from the intermediate phase were found, beginning from 750°C, at 800°C and 850°C broadened peaks of the yttrium orthoferrite were observed, only a small amount of the phase $YFeO_3$ (2, 5 and 9% correspondingly, for the treatment time of 30 minutes) appears. At the higher temperatures between 900°C and 1000°C, the yttrium orthoferrite phase grows faster and appears already as strong diffraction reflections (18, 25 and 40% correspondingly). The most intense reaction takes place at the temperature of 1050°C, where along with the yttrium orthoferrite phase the yttrium ferrogarnet phase appers (26 and 55% correspondingly). That means that at the temperature of 1050°C the mechanism of the reaction switches dramatically, involving the reactions between yttrium orthoferrite and iron oxide.

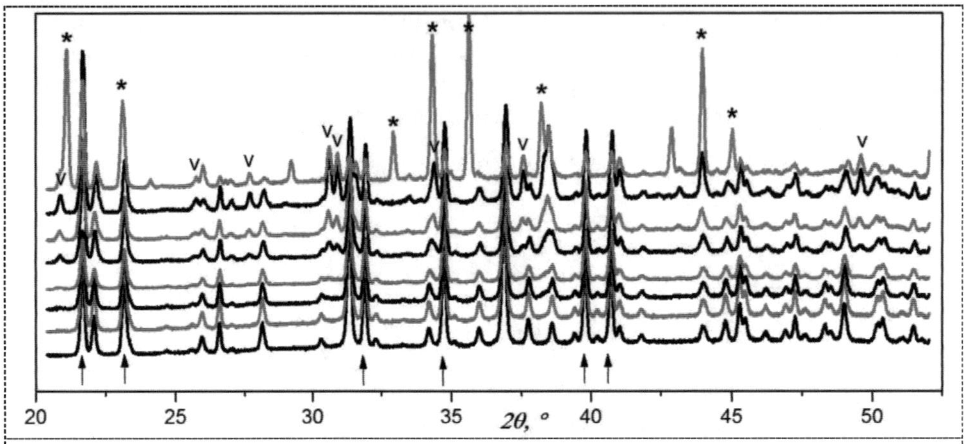

Figure 4. *Ex situ* X-ray diffraction experiments for the synthesis of ferrogarnet during the radiation-thermal treatment (30 minutes) to the temperatures: (from bottom to top) 700°C, 750°C, 800°C, 850°C 900°C, 950°C, 1000°C, 1050°C; (wavelength λ=1.01Å); Fe_2O_3 marked by dark arrows, Y_2O_3 unmarked, the product is marked by (*) $Y_3Fe_5O_{12}$, the intermediate phase by (v) $YFeO_3$.

4. Conclusions

The kinetics and the chemical reactions staging in the ferrite synthesis differ in the radiation-thermal processes comparing with the thermal one. This effect can be caused by intensive defects formation in the crystal accelerating the mass transport processes inside the reaction mixture. The second reason can be related to the different atmosphere composition during the irradiation: the processes of mass exchange involves the oxygen from the atmosphere, which present in an ionized and atomized states during the irradiation. That also can cause the strontium carbonate decomposition suppression: air ionizing processes inhibit the mass exchange between the carbonate and environment.

Acknowledgments
This work was carried out on the equipment belonging to the shared research center "SSTRC" that is supported by the Ministry of Education and Science of the Russian Federation.

References

[1] Ancharova U V, Mikhailenko M A, Tolochko B P, Lyakhov N Z, Korobeinikov M V, Bryazgin A A, Bezuglov V V, Shtarklev E A, Vlasov A Yu and Vinokurov Z S 2015 IOP Conf. Ser.: Mater. Sci. Eng. **81** 012122
[2] Kuzin M V Shared research center "Siberian Synchrotron and Terahertz Radiation Centre" http://ssrc.inp.nsk.su/CKP/
[3] Lyakhov N Z, Boldyrev V V, Voronin A P, Gribkov O S, Bochkarev L G, Rusakov S V and Auslender V L 1995 Journal of Thermal Analysis **43** 21
[4] Boldyrev V V, Voronin A P, Gribkov O S, Tkachenko E V, Karagedov G R, Yakobson B I and Auslender V L 1989 Solid State Ionics **36** 1

[5] Voronin A P, Neronov V A, Boldyrev V V, Melekhova T F, Auslender V L, Aleksandrov V V, Lyakhov N Z, Polyakov V A and Savinkina M A 1982 *Soviet Powder Metallurgy and Metal Ceramics* **21** 632

[6] Auslender V L, Bochkarev I G, Boldyrev V V, Lyakhov N Z and Voronin A P 1997 *Solid State Ionics* **101-103** 489

Institute of Physics
Dirac House, Temple Back
Bristol BS1 6BE UK

ISSN: 1757-8981
ISBN 978-1-5108-2086-9

9 781510 820869

CURRAN ASSOCIATES INC.
proceedings
.com